D0845762

ASTRONOMY AND
ASTROPHYSICS LIBRARY

Martin Harwit

Astrophysical Concepts

Fourth Edition

 Springer

Martin Harwit
511 H St.
SW Washington, DC 20024
USA and
Cornell University,
Ithaca, NY 14853-6801
USA

Library of Congress Control Number: 2006922469

Cover picture: The galaxy Messier 51 observed at X-ray wavelengths by Andrew Wilson and Yuichi Terashima, NASA/Chandra X-ray Observatory Center (CXC), University of Maryland, USA, and Institute of Space and Astronautical Science, Japan. This image, obtained with the Advanced Charge-Coupled-Device Imaging Spectrometer (ACIS-S) on the Chandra X-ray Observatory was processed by Andrea Prestwich and Holly Jessop. Courtesy of Harvey Tananbaum. For details see Chapter 1, page 8.

ISSN 0941-7834
ISBN 10: 0-387-32943-9
ISBN 13: 978-0387-32943-7
eISBN 0-387-33228-6

Printed on acid-free paper.

Printed in the United States of America.

9 8 7 6 5 4 3 2 1

springer.com

Preface to the Fourth Edition

Thirty-three years have passed since the first edition of *Astrophysical Concepts* appeared. During this time astrophysics has undergone major revolutions. We have gained new perspectives on the Universe with the aid of powerful gamma-ray, X-ray, and infrared telescopes, whose sensitivities could not have been imagined three decades earlier. We have become expert at snaring neutrinos to gain insight on nuclear processes at work in the Sun and supernovae. We have direct evidence for the existence of neutron stars and gravitational waves, and persuasive arguments for the detection of black holes on scales of individual stars as well as galactic nuclei. Primordial fluctuations, remnants from the first moments in the expansion of the Universe have revealed themselves in the faint ripples marking the microwave sky. These ripples also document the first appearance of dark matter now known to have dominated the formation and evolution of all cosmic structure. And dark energy has gradually had to be acknowledged to be the dominant factor driving the expansion of the Universe today.

With so much that is new, and so many new problems revealed by knowledge already gained, much of the book had to be fully revised. My principal aim in this fourth edition, however, has continued to be the presentation of a wide range of astrophysical topics in sufficient depth to give the reader a general quantitative understanding. The book outlines cosmic events but does not portray them in detail — it provides a series of astrophysical sketches. I think this approach still befits the prevailing uncertainties and rapidly evolving views in astrophysics.

The first edition of *Astrophysical Concepts* was based on notes I prepared for a course aimed at seniors and beginning graduate students in physics and astronomy at Cornell. This course defined the level at which the book is written.

For readers who are versed in physics but are unfamiliar with astronomical terminology, Appendix A is included. It gives a brief background of astronomical concepts and should be read before starting the main text.

The first few chapters outline the scope of modern astrophysics and deal with elementary problems concerning the size and mass of cosmic objects. However, it soon becomes apparent that a broad foundation in physics is needed to proceed. This base is developed in Chapters 4 to 7 by using, as examples, specific astronomical processes. Chapters 8 to 14 enlarge on the topics first outlined in Chapter 1 and show how we can obtain quantitative insight into the structure and evolution of stars, the dynamics of cosmic gases, the large-scale behavior of the Universe, and the origins of life.

Throughout the book I emphasize astrophysical concepts. This means that objects such as asteroids, stars, supernovae, or quasars are not described in individual chapters or sections. Instead, they are mentioned wherever relevant physical principles are discussed. Thus, features common to various astronomical phenomena are emphasized, but information about specific astronomical objects remains distributed. For example, different aspects of neutron stars and pulsars are discussed in Chapters 5, 6, 8, Appendix A, and elsewhere. To compensate for this treatment, a comprehensive index is included.

I have sketched no more than the outlines of several traditional astronomical topics, such as the theories of radiative transfer, stellar atmospheres, and polytropic gas spheres, because a complete presentation would have required extensive mathematical development to be genuinely useful. However, the main physical concepts of these subjects are worked into the text, often as remarks without specific mention. In addition, where appropriate, I refer to other sources that treat these topics in greater detail.

The greatly expanded list of references is designed for readers who wish to cover any given area in greater depth. Beginning students should not feel compelled to look these up. They are included for those who would like to research any given topic to greater depth or read about a subject in the discoverer's own words. Occasionally I also refer to informative popular articles designed to keep the larger scientific community abreast of developments.

A book that covers a major portion of astrophysics must be guided by the many excellent monographs and review articles that exist today. It is impossible to properly acknowledge all of them and to give credit to the astrophysicists whose viewpoints strongly influenced my writing. I am grateful for the many suggestions offered by colleagues and by several generations of Cornell students who saw earlier editions of this book evolve from a series of informal lecture notes.

I thank Harry Blom and Chris Coughlin, my editors at Springer, for their advice, Frank Ganz for his unfailing courtesy in sharing his expertise on working with LaTeX, Valerie Greco and Natacha Menar for their meticulous copy editing, and Natacha also for seeing the book through production. They all made working with Springer easy and enjoyable.

Finally, I acknowledge with pleasure my especial indebtedness to Andrew Wilson and Yuichi Terashima, NASA/Chandra X-ray Observatory Center (CXC), University of Maryland, USA and Institute of Space and Astronautical Science, Japan for the image of the galaxy Messier 51 that appears on the cover of the book. This image, obtained with the Advanced Charge-Coupled-Device Imaging Spectrometer (ACIS-S) on the Chandra X-ray Observatory was processed by Andrea Prestwich and Holly Jessop. I thank Harvey Tananbaum, Director of CXC, for the friendly reception that my request for this image received.

Colleagues from all over the astronomical community were gracious in permitting me to incorporate figures and tables they had produced, sometimes updating original figures to make them current. For help with these I thank Joao Alves, Jennifer Barnett, Charles L. Bennett, Michael Blanton, Brian Boyle, Alain Coc, John Cowan, Scott Croom, Tamara Davis, Frank Eisenhauer, Xiaohui Fan, Masataka Fukugita, Zoltan Haiman, Jonathan Hargis, Günther Hasinger, Alan Heavens, Alexander Heger, W. Raphael Hix, Stanley D. Hunter, D. Heath Jones, Maciej Konacki, Charlie Lada, Elizabeth Lada, Charley Lineweaver, Bradley Meyer, Philip Myers, Ken'ichi Nomoto, Jim Peebles, Saul Perlmutter, Michael Perryman, Juri Poutanen, Clement Pryke, Adam Riess, Sara Seager, David Spergel, Volker Springel, John Stauffer, Max Tegmark, Lih-Sin The, Ethan Vishniac, and J. Craig Wheeler.

My greatest debt, however, is to my wife Marianne. In the thirty-three-year history of the book, she has at various times taken on the roles of mail clerk, proofreader, editorial assistant, sales manager, and publisher. Throughout, she has also remained my most loyal critic.

Martin Harwit

Contents

1 An Approach to Astrophysics

In a sense each of us has been inside a star; in a sense each of us has been in the vast empty spaces between the stars; and — if the Universe ever had a beginning — each of us was there!

Every molecule in our bodies contains matter that once was subjected to the tremendous temperatures and pressures at the center of a star. This is where the iron in our red blood cells originated. The oxygen we breathe, the carbon and nitrogen in our tissues, and the calcium in our bones, also were formed through the fusion of smaller atoms at the center of a star.

Terrestrial ores containing uranium, plutonium, lead, and many other massive atoms were formed in a supernova explosion — the self-destruction of a star in which a sun's mass is hurled into space at huge velocity. Most of the matter on Earth and in our bodies went through such a cataclysmic event!

To account for a fraction of the light elements, lithium, beryllium, and boron, which we find in traces on Earth, we have to go back to a cosmic explosion signifying the birth of the entire Universe. A separate portion of these same elements originated through cosmic-ray bombardment in interstellar space. These two constituents became admixed long before the Earth we now walk on was formed from a cloud of gas and dust spread so tenuously that a gram of soil would have occupied a volume the size of the entire planet.

How do we know all this? And how sure are we of this knowledge?

This book was written to answer such questions and to provide a means for making astrophysical judgments.

We are just beginning a long and exciting journey into the Universe. There is much to be learned, much to be discarded, and much to be revised. We have excellent theories, but theories are guides for understanding the truth. They are not truth itself; we must continually revise them if they are to keep leading us in the right direction.

In going through the book, just as in devising new theories, we will find ourselves baffled by choices between the real and the apparent. We will have to learn that it may still be too early to make such choices, that reality in astrophysics has often been short-lived, and that — disturbing though it would be — we may some day have to reconcile ourselves to the realization that our theories had recognized only superficial effects — not the deeper, truly motivating, factors. We may therefore do well to avoid a preoccupation with astrophysical "reality," and rather take a longer view, looking more closely at those physical concepts likely to play a role in the future evolution of our understanding.

The development of astrophysics in the last few decades has been revolutionary. We have discarded what had appeared to be our most reliable theories, replaced them, and frequently found even the replacements lacking. The only constant in this revolution has been the pool of astrophysical concepts. It has provided a continuing source of material for our evolving theories.

This pool contained the neutron stars 35 years before their discovery, and it contained black holes three decades before astronomers started searching for them. The best investment of our efforts may lie in a deeper exploration of these concepts.

In astrophysics we often worry whether we should organize our thinking around individual objects — planets, stars, pulsars, and galaxies — or whether we should divide the subject according to physical principles common to the various astrophysical processes.

The book's emphasis on concepts makes the second approach more appropriate, but also raises problems. Much of the information about individual types of objects has had to be distributed throughout the book, and can be gathered only through use of the Index. This leads to a certain unevenness in the presentation.

The unevenness is made even more severe by the varied mathematical treatment. No astrophysical picture is complete if we cannot assign a numerical value to its scale. In this book, we will consistently aim at obtaining rough orders of magnitude characteristics of the different phenomena. In some cases, this aim leads to no mathematical difficulties. In other problems we will have to go through rather complex calculations before even the crudest answers emerge.

Given these difficulties, which appear to be partly dictated by the nature of modern astrophysics, let us examine the most effective ways to use this book: For those who have no previous background in astronomy, Appendix A may provide a good starting point. It briefly describes the astronomical objects we will study and introduces astronomical notation. This notation will be used throughout the book and is generally not defined in other chapters. Those who have previously studied astronomy will be able to start directly with the present chapter, which presents the current searches going on in astrophysics — the problems we will be pursuing, the questions that we will seek to answer as we progress through the book. Chapters 2 and 3 show that, while some of the rough dimensions of the Universe can be measured by conceptually simple means, a deeper familiarity with physics is required to understand the cosmic sources of energy and the nature of cosmic evolution. The physical tools we need are therefore presented in the intermediate Chapters 4 to 7. We then gather these tools to work our way through the formation and evolution of stars, the processes that take place in interstellar space, the evolution of the Universe, the synthesis of chemical elements mentioned right at the start of this section, the formation of galaxies and clusters of galaxies, and the astrophysical setting for the origins of life.

This is an exciting, challenging venture; we have a long way to go.

Let us start.

1:1 Channels for Astronomical Information

Imagine a planet inhabited by a blind civilization. One day an inventor discovers an instrument sensitive to visible light and this device is found to be useful for many purposes, particularly for astronomy.

Human beings can see light and we would expect to have a big head start in astronomy compared to any civilization that was just discovering methods for detecting visible radiation. Think then of an even more advanced culture that could detect not only visible light but also all other electromagnetic radiation, *cosmic rays, neutrinos*, and *gravitational waves*. Clearly, that civilization's knowledge of astronomy could be far greater than ours.

Four entirely independent channels are known to exist by means of which information can reach us from distant parts of the Universe.

(a) Electromagnetic radiation: gamma rays, X-rays, ultraviolet, visible, infrared, and radio waves.

(b) Cosmic-ray particles: These comprise high-energy electrons, protons, and heavier nuclei as well as the (unstable) neutrons and mesons. Some cosmic-ray particles consist of antimatter.

(c) Neutrinos and antineutrinos: There are three known types of neutrinos and antineutrinos, each associated with electrons, μ-mesons, and τ-mesons.

(d) Gravitational waves.

Most of us are familiar with channel (a), currently the channel through which we obtain the bulk of astronomical information. However, let us briefly describe channels (b), (c), and (d).

(b) There are fundamental differences between cosmic-ray particles and electromagnetic or gravitational waves: (i) cosmic rays move at very nearly the speed of light, whereas electromagnetic and gravitational waves move at precisely the speed of light; (ii) cosmic-ray particles can be electrons, neutrons, or nuclei of atoms, all with positive rest-mass; iii) when electrically charged, these particles are deflected by cosmic magnetic fields. The direction from which a charged cosmic-ray particle arrives at the Earth is usually unrelated to the actual direction of the source.

Cosmic-ray astronomy is far more advanced than either neutrino or gravitational wave work. Through cosmic-ray studies we hope to learn about the chemistry of the Universe on a large scale, eventually to single out regions in which, as yet unknown, grandiose accelerators produce these highly energetic particles (Bi97).

(c) Neutrinos, have extremely low rest-mass. They have one great advantage in that they can traverse great depths of matter without being absorbed. Neutrino astronomy could give us a direct look at the interior of stars, much as X-rays can be used to examine a metal block for internal flaws or a medical patient for lung ailments. Neutrinos could also convey information about past ages of the Universe because, except for a systematic energy loss due to the expansion of the Universe, the neutrinos are preserved in almost unmodified form over many æons.[1] Much of

---

[1] One æon $\equiv 10^9$ yr $\equiv 1$ Gyr.

the history of the Universe must be recorded in the ambient neutrino flux, but so far we do not know how to tap this information.

A search for solar electron neutrinos at first seemed to show that the flux received at Earth is lower than had been predicted, based on expected nuclear reactions in the Sun. However, we now know that neutrinos oscillate between electron-, μ- and τ-neutrino states. The originally predicted rate at which electron neutrinos are radiated by the Sun appears to have been correct, but a large fraction of these neutrinos are converted into μ- and τ-neutrinos before they reach Earth (Ba96).

The 1987 explosion of a supernova in the nearby Large Magellanic Cloud provided the first direct evidence for copious generation of neutrinos in these huge eruptions and gave neutrino astronomy a new boost (Hi87), (Bi87).

(d) Gravitational waves, when reliably detected, will yield information on the motion of very massive bodies. Gravitational waves have not yet been directly detected, though their existence is indirectly inferred from observations on changes in the orbital motions of closely spaced pairs of compact stars. We seem to be on the threshold of important discoveries that are sure to have a significant influence on astronomy.

In addition to use of channels (a)–(d) information on the solid constituents of the ambient interstellar and interplanetary medium can also be gained by collecting and chemically analyzing interstellar dust grains that penetrate into the Solar System and meteorites that orbit the Sun.

It is clear that astronomy cannot be complete until techniques are developed to detect all of the principal means by which information can reach us. Until then astrophysical theories must remain provisory.

Not only must we be able to detect these information carriers, but we will also have to develop detectors sensitive to the entire spectral range for each type of carrier. The importance of this is shown by the great contribution made by radio-astronomy. Eight decades ago, all our astronomical data were obtained in the visible, near infrared, or near ultraviolet regions; no one at that time suspected that a wealth of information was available in the radio, infrared, X-ray, or gamma-ray spectrum. Yet today, the only complete maps we have of our own Galaxy lie in these spectral ranges. They show, respectively, the distributions of pulsars and molecular, atomic or ionized gas; clouds of dust; bright, hot X-ray emitting stars and X-ray binaries; and giant gamma-ray flares from soft gamma-ray repeaters, believed to be *magnetars*, neutron stars with magnetic fields ranging up to $10^{14} - 10^{15}$ gauss (Sc05).

Just as we have made our first astrophysically significant neutrino observations and are reaching for gravitational wave detection, a variety of new carriers of information have been proposed. We now speak of axions, photinos, magnetic monopoles, tachyons, and other carriers of information which — should they exist — could serve as further channels of communication through which we could gather astrophysical information. All these hypothesized entities arise from an extension of known theory into domains where we still lack experimental data. Theoretically, they are plausible, but there is no evidence that they exist in Nature. *Photinos, ax-*

ions, and other *weakly interacting massive particles, WIMPS,* could, however, be making themselves felt through their gravitational attraction, even though otherwise unobserved. The hot, massive gaseous haloes around giant elliptical galaxies suggest that these galaxies contain far more mass than is observed in stars and interstellar gases. The surprisingly high speeds at which stars and clouds of hydrogen orbit the centers of spiral galaxies even when located at the extreme periphery of their galaxies' disks lead to the same conclusion. These inferences, however, assume that the force of gravity declines with the square of the distance, and we have no direct observational proof that this is so on the scale of galaxies. But if it is, galaxies quite generally must contain an abundance of some form of dark matter. Could this consist of exotic particles?

1:2 X-Ray Astronomy: Development of a New Field

The development of a new branch of astronomy often follows a general pattern: vague theoretical thinking tells us that no new development is to be expected. Not until some chance observation focuses attention onto a new area are serious preliminary measurements undertaken. Many of these initial findings later have to be discarded as techniques improve.

These awkward developmental stages are always exciting. Let us outline the evolution of X-ray astronomy, as an example, to convey the sense of advances that should take place in astronomy and astrophysics in the next decades, as we venture further into neutrino observations and search for ways to detect gravitational waves.

Until 1962 only solar X-ray emission had been observed. This flux can solely be detected with instruments taken above the Earth's atmosphere and is so weak that no one expected a large X-ray flux from sources outside the Solar System. Then, in June 1962, R. Giacconi, H. Gursky, and F. Paolini of the American Science and Engineering Corporation (ASE) and B. Rossi of MIT flew a set of large area Geiger counters aboard an Aerobee rocket (Gi62). The increased area of these counters was designed to permit detection of X-rays scattered by the Moon but originating from the Sun. The counters were sensitive in the wavelength region from 2 to 8 Å.

No lunar X-ray flux could be detected. However, a source of X-rays was discovered in a part of the sky not far from the center of the Galaxy and a diffuse background flux of X-ray counts was evident from all portions of the sky. Various arguments showed that this flux probably was not emitted in the outer layers of the Earth's atmosphere and therefore should be cosmic in origin. Later flights by the same group verified their first results.

At this point, a team of researchers at the U.S. Naval Research Laboratory became interested. They had experience with solar X-ray observations and were able to construct an X-ray counter some ten times more sensitive than that flown by Giacconi's group. Instead of the very wide field of view used by that group, the NRL team limited their field to 10 degrees of arc so that their map of the sky could show somewhat finer detail (Bo64a).

An extremely powerful source was located in the constellation Scorpius about 20 degrees of arc from the Galactic center. At first this source, catalogued as Sco X-1, remained unidentified. Photographic plates showed no unusual objects in that part of the sky. The NRL group also discovered a second source, some eight times weaker than the Scorpio source. This was identified as the Crab Nebula, a remnant of a supernova explosion observed by Chinese astronomers in 1054 A.D. The NRL team, Bowyer, Byram, Chubb, and Friedman, believed that these two sources accounted for most of the emission observed by Giaconni's group.

Many explanations were advanced about the possible nature of these sources. Arguments were given in favor of emission by a new breed of highly dense stars whose cores consisted of neutrons. Other theories suggested that the emission might come from extremely hot interstellar gas clouds. No decision could be made on the basis of observations because none of the apparatus flown had sufficient angular resolving power.

Then, early in 1964, Herbert Friedman at NRL heard that the Moon would occult the Crab Nebula only seven weeks later. Here was a great opportunity to test whether at least one cosmic X-ray source was extended or stellar. For, as the edge of the Moon passes over a well-defined point source, all the radiation is suddenly cut off. In contrast, the flux from a diffuse source diminishes gradually.

No other lunar occultation of either the Scorpio source or the Crab Nebula was expected for many years; so the NRL group went into frenzied preparations and managed to prepare a payload in time. The flight had to be timed to within seconds, because the Aerobee rocket to be used only gave five minutes of useful observing time at altitude. Two possible flight times were available: one at the beginning of the eclipse, the other at the end. Because of limited flight duration it was not possible to observe both the initial immersion and subsequent egress from behind the Moon.

The first flight time was set for 22:42:30 Universal Time on July 7, 1964. That time would allow the group to observe immersion of the central 2 minutes of arc of the nebula. Launch took place within half a second of the prescribed time. At altitude, an attitude control system oriented the Geiger counters. At 160 seconds after launch, the control system locked on the Crab. By 200 seconds a noticeable decrease in flux could be seen and by 330 seconds the X-ray count was down to normal background level. The slow eclipse had shown that the Crab Nebula is an extended source. We could definitely state that at least one of the cosmic X-ray sources was diffuse. Others might be due to stars. But this one was not (Bo64b).

A few weeks after this NRL flight, the ASE–MIT group was also ready to test angular sizes of X-ray sources. Their experiment made use of a collimator designed by the Japanese physicist, Minoru Oda (Od65). This device consisted of two wire grids separated by a distance D that was large compared to the open space between wires, which was slightly less than the wire diameter d. The principle on which this collimator works is illustrated in Fig. 1.1.

When the angular diameter of the source is small compared to d/D, alternating strong and weak signals are detected as the collimator aperture is swept across the

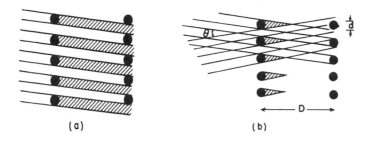

Fig. 1.1. Principle of operation of an X-ray astronomical wire-grid collimator. (*a*) For parallel light the front grid casts a sharp shadow on the rear grid. As the collimator is rotated about an axis parallel to the grid lines, light is alternately transmitted and stopped depending on whether the shadow is cast on the wires of the rear grid or between them. (*b*) For light from a source whose angular dimension $\theta \gg d/D$, the shadow cast by the front grid is washed out and rotation of the collimator assembly does not give rise to a strong variation of the transmitted X-ray flux.

source. If $\theta \gg d/D$ virtually no change in signal strength is detected as a function of orientation.

In their first flight the MIT–ASE group found Sco X-1 to have an angular diameter small compared to $1/2°$. Two months later a second flight confirmed that the source diameter was smaller still — less than $1/8°$. A year and a half later this group found that the source must be far smaller yet, less than $20''$ in diameter. On this flight two collimators with different wire spacings were used. This meant that the transmission peaks for the two collimators coincided only for normal incidence and, in this way, yielded an accurate position of Sco X-1 (Gu66). An optical identification was then obtained at the Tokyo Observatory and subsequently confirmed at Mount Palomar (Sa66). It showed an intense ultraviolet object that flickered on a time scale less than one minute.

The brightness and color of neighboring stars in the vicinity of Sco X-1 showed that these stars were at a distance of a few hundred light years from the Sun, and this gave a good first estimate of the total energy output of the source. A search on old plates showed that the mean photographic brightness of the object had not changed much since 1896.

Interestingly, the 1969 discovery that the Crab Nebula contains a pulsar emitting radio waves sent X-ray astronomers back to previously collected data. Some of these records showed up the pulsar's characteristic millisecond pulsations also at X-ray energies, and demonstrated that an appreciable fraction of the flux — 10 to 15% — comes from a point source now believed to be a neutron star formed in the supernova explosion. Our views of the Crab Nebula as a predominantly diffuse X-ray source had to be revised.

Myriad other Galactic X-ray sources have by now been located and identified; and frequently they have a violet, stellar (pointlike) appearance similar to Sco X-

1. These objects sometimes suddenly increase in brightness by many magnitudes within hours. Others pulsate regularly, somewhat like the Crab Nebula pulsar.

The first galaxy to be observed at X-ray frequencies was M87, a galaxy known to be a bright radio source (By67). It is a peculiar galaxy consisting of a spherical distribution of stars from which a jet of gas protrudes. The jet is bluish in visible light; it shines by virtue of highly relativistic electrons spiraling in magnetic fields and emitting *synchrotron radiation* (see Chapter 6).

Four decades after the first feeble sighting of X-rays from beyond the Solar System, we know that many varieties of stars emit X-rays at some level, as do galaxies and hot clouds of gas surrounding massive galaxies at the centers of galaxy clusters. With the aid of the Röntgen satellite, ROSAT, roughly a million X-ray sources, many of them quasars or galaxies exhibiting violently active nuclei, have been observed. X-ray maps of clusters of galaxies show that most atomic matter in the Universe is not contained in stars and cool interstellar clouds of gas as previously thought. Hot, ionized, X-ray emitting gas surrounding the central galaxies in a cluster appears to comprise several times more mass than stars and interstellar gases combined (Mu97). In addition to these discrete sources, a ubiquitous, diffuse, isotropic background flux is observed, a significant fraction of which appears to be emitted by *Active Galactic Nuclei, AGNs*.

The cover of the book exhibits an X-ray image of the spiral galaxy M51, showing both the primary spiral and its smaller companion. The X-ray luminosity of M51 in the 0.3 to 8 keV band is 4×10^{40} erg s^{-1}, typical of many spiral galaxies. The colors in the image are red for the energy range from 300 eV to 1 keV; green for the range $1 - 2$ keV and blue for the $2 - 8$ keV range. Roughly 65% of the X-rays are emitted by compact unresolved sources; the remainder is a diffuse component at energies mostly below 1 keV. In many spirals the diffuse component arises largely where massive stars have recently formed. Of the 117 compact sources in the image as many as 27 may be background quasars or AGNs. Compact sources intrinsic to the galaxy most probably are low-mass X-ray binary stars. Also seen, however, are high-mass X-ray binaries in regions of star formation and a few X-ray bright supernova remnants. Of particular interest in M51 are several ultraluminous X-ray sources with luminosities in excess of the *Eddington luminosity* of a neutron star. At the Eddington luminosity, also called the *Eddington limit*, the radiation emitted by the star exerts such a high pressure on infalling material that attracting gravitational forces are overcome; material ceases to fall onto the star's surface, shutting off the supply of energy it requires to continue radiating. The high observed luminosity suggests that the compact source might be the remnant of a star so massive that it collapsed under its own gravitational attraction and became a *black hole* (Pr04).

The range of X-ray and gamma-ray energies at which observations have by now been carried out covers many orders of magnitude. Radio, infrared, visual, ultraviolet, X-ray, and gamma-ray spectra are now available for many sources, and provide complementing information. Four decades after its discovery, Sco X-1 is now known to lie at a distance of $\sim$3 kpc. It consists of a neutron star with mass $\sim$1.4 $M_\odot$ orbited every 19 hours by a star of mass $\sim$0.4 $M_\odot$. Relativistic jets of gas stream out in opposite directions from this system at speeds of ~45% the speed of light (Fo01, St02). This *microquasar* continues to be a tantalizing system calling for further study. Other interesting X-ray sources are black holes surrounded by accretion disks onto which matter tidally stripped from a nearby star rains down, liberating vast amounts of kinetic energy radiated away as X-rays.

The fundamental nature of astrophysical discoveries being made — or remaining to be made — leaves little room for doubt that our knowledge remains fragmentary and a large part of current theory will be drastically revised as we learn more over the next decades. In parts of astrophysics — notably in cosmology — our lack of observations, our limited knowledge, influences the very way in which we think and may hinder our approach to scientific problems. It is therefore useful to examine the starting point from which our reasoning always embarks.

1:3 The Appropriate Set of Physical Laws

Today *astrophysics* and *astronomy* have almost become synonymous. In earlier times it was not at all clear that the study of stars had anything in common with physics. But physical explanations for the observations not only of stars, but of interstellar matter and of phenomena on the scale of galaxies, have been so successful that we confidently assume all astronomical processes to be subject to physical reasoning.

Several points, however, must be kept in mind. First, the laws of physics that we apply to astrophysical processes are largely based on experiments that we can carry out with equipment in a very confined range of sizes. We measure the speed of light over regions that maximally have dimensions of the order of 10^{14} cm, the size of the inner Solar System. Our knowledge of large-scale dynamics is also based on detailed studies of the Solar System. We then extrapolate the dynamical laws gained on such a small scale to processes on a cosmic scale of ~10^{18} to 10^{28} cm. We have no guarantee that this extrapolation is warranted.

It may well be true that these local laws do span the entire range of cosmic mass and distance scales; but we only have to recall that the laws of quantum mechanics, which hold on a scale of 10^{-8} cm, are quite different from the laws we would have expected on the basis of classical measurements carried out with objects 1 cm in size.

A second point, similar in vein, is the question of the constancy of the laws of Nature. We now postulate that empty space — vacuum — once carried vast reservoirs of energy that controlled the evolution of the early Universe. The Universe we currently observe greatly differs from such an original state, and the laws of physics that earlier were in effect may have been quite different from those observed today.

A third question concerns the observational basis of science. Current theories suggest that the Universe stretches well beyond a cosmic horizon where galaxies recede at the speed of light, and thus well beyond the most distant regions from which light could ever reach us, domains forever beyond observational reach. We are thus confronted with a scientific assertion about the size of the Universe that may in principle be unverifiable. Should we accept it on faith? Where will this lead us?

How far will this approach work for us? How soon will the philosophical difficulties connected with the uniqueness of the Universe arise? In observing the cosmic microwave background, first predicted by Ralph Alpher and Robert Herman (Aℓ48) and independently discovered by Arno Penzias and Robert W. Wilson (Pe65), we have already encountered one such limit. The low-frequency undulations in surface brightness observed across the celestial sphere are substantially weaker than expected; but we have no other universe to which we could compare this finding to see whether it is significant or merely a statistical fluctuation restricted to our particular locale in the Universe. The inability to explain such apparent cosmic anomalies for lack of observations beyond the cosmic horizon is called the *cosmic variance* problem.

Until we encounter many more uniqueness limits of this kind, we address ourselves to concrete problems which, although still unsolved, nevertheless are expected to have solutions that can be reached using the laws of physics as we know them. Among these are questions concerning the origin and evolution of galaxies, stars, and planetary systems. We also think we will be able to fully explain the origins and abundances of the various chemical elements. Perhaps the origin of life itself will become clearly established as astrophysical and biochemical processes become better understood.

This then is the current situation. We know a great deal about some as yet apparently unrelated astronomical events. We feel that a connection must exist, but we are not sure. Not knowing, we divide our knowledge into a number of different "areas": cosmology, galactic structure, stellar evolution, cosmic rays, and so on. We do this with misgivings, but the strategy is to seek insight by solving individual small problems. All the time we expect to widen the domains of understanding until some day contact is made between the diverse areas and a firm path of reasoning is established. The next few sections sketch some of the more important problems we are currently investigating.

1:4 The Formation of Stars

We believe that no star has existed forever — because sooner or later its energy supply must run out — and so we must account for the birth of stars. Inasmuch as those stars that we believe to be young are always found close to clouds of interstellar dust and gas, we argue that such clouds of cosmic matter must be contracting slowly, giving rise to increasingly compact condensations, some of which eventually collapse down to stellar size.

This picture makes a good deal of sense. Dust grains in interstellar space are very effective at radiating away heat. When a hydrogen atom in a cloud of dust and gas collides with a cold dust grain, the grain becomes slightly heated and radiates away this energy in the infrared part of the electromagnetic spectrum. This results in a net loss of kinetic energy of the gas, which gravitates toward the center of the cloud, gains some kinetic energy in falling, and again transfers a part of this to ambient dust grains to repeat the cooling cycle. The gas also transfers some of its centrally directed momentum to grains, thus also causing the grains to drift in toward the center of the contracting cloud. The cloud as a whole contracts.

Grain radiation is not the only radiative process that rids a *protostellar cloud* of energy. As it collapses, the protostar becomes progressively hotter, and various molecular and atomic states are excited through collisions. The excited particles emit radiation to return to their ground states. As radiation escapes, the net loss of energy cools the cloud (Fig. 1.2).

Attractive though it is, there are difficulties with this picture. First, the protostar cannot just lose energy in forming a star. It must also lose angular momentum.

The amount of matter needed to form a star from an interstellar cloud with a density 10^3 atoms cm^{-3} requires the collapse of gas from a volume whose initial radius r would be of order 10^{18} cm. Over such distances, the observed rotational velocity v, about the cloud's center might be $\sim 10^4$ cm s^{-1} for a cold Galactic molecular cloud, so that the angular momentum per unit mass $rv \sim 10^{22}$ cm^2 s^{-1}. In contrast, the observed surface velocities of typical stars indicate an angular momentum per

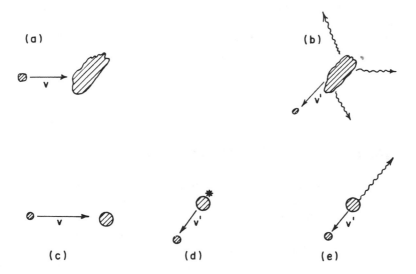

Fig. 1.2. Cooling processes in *protostellar clouds*. (*a*) An atom with velocity v hits a grain. Its kinetic energy is $v^2/2$ multiplied by the atomic mass m. (*b*) The grain radiates away the absorbed energy and the atom leaves with reduced velocity v' and reduced kinetic energy $mv'^2/2$. In (*c*), (*d*), and (*e*) an atom collides with another atom or with a molecule. This second particle goes into an excited (higher) energy state, denoted by an asterisk (*), and then emits radiation to return to its initial state. The first atom loses kinetic energy also in this process. If the emitted radiation escapes from the cloud, the entire *protostellar mass* loses energy and slowly contracts to form a star.

unit mass many orders of magnitude lower: 10^{16} to $10^{18.3}$ cm^2 s^{-1}. For the Sun, it is only $\sim 10^{15}$ cm^2 s^{-1}, but the angular momentum of the Solar System taken as a whole corresponds to 10^{17} cm^2 s^{-1}. Entries in Table 1.4 allow us to conclude that most of the angular momentum resides in the motion of Jupiter orbiting the Sun. A comparison of the rotational velocities of stars given in Table A.4 of Appendix A, further indicates that the angular momentum of the entire Solar System equals that of more massive stars of spectral types F and A. It is therefore tempting to associate the observed low angular momentum of less massive stars with the formation of planetary systems. The initially contracting cloud of interstellar matter somehow contrives to redistribute almost all of its angular momentum to a gaseous disk that eventually gives rise to orbiting planets. Only a small fraction of the angular momentum is retained by the star.

A similar problem concerns the *magnetic field* initially present in the interstellar medium. If this field is predominantly oriented along some given direction, then the final field after contraction of the cloud to form a star would also have that direction. The flux density B of the magnetic field permeating a cloud is inversely proportional to the cross-section of the area of the cloud as it contracts, as long as the magnetic lines of force act as if frozen to the partially ionized gas (Section 6:2). Thus, the number of these lines of force threading through the cross-sectional area stays constant. A field, B, initially as weak as 10^{-6} gauss would become some 10^{14} times stronger as the protostellar radius decreased from 10^{18} down to 10^{11} cm. Actual fields found on the surfaces of stars like the Sun are of the order of one gauss, and the highest fields observed in a few peculiar stars only range up to tens of thousands of gauss. Protostellar contraction must therefore be accompanied by destruction or loss of magnetic field lines permeating the interstellar material. How this loss occurs is still under active investigation.[2]

1:5 The Hertzsprung–Russell and Color-Magnitude Diagrams

Granted that we do not know very much about how stars are born, can we say anything about how they evolve after birth? The answer to this is a convincing "Yes," although many questions persist.

When the absolute brightness of a set of stars is plotted against surface temperature, as measured either from an analysis of the stellar spectrum or by the star's color, we find that only certain portions of such *Hertzsprung–Russell* and *color-magnitude diagrams* are populated appreciably. The concentration of stars in select parts of these diagrams, Figs. 1.3 to 1.7, help us to understand how stars of different masses evolve as they age.

Color-magnitude diagrams and Hertzsprung–Russell plots share a number of common features. Stars on the left of these diagrams are hot, having extreme surface

---

[2] White dwarfs and the even denser neutron stars, respectively, have surface fields of order 10^5 and 10^{12} gauss — just the field strengths we would expect if the Sun were to shrink to the size of these stars without shedding field lines.

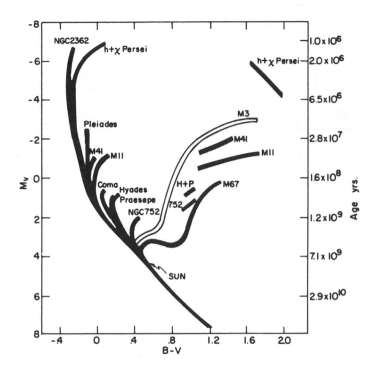

Fig. 1.3. Color-magnitude diagram for galactic clusters within the Galaxy and for the globular cluster M3. These clusters show different turn-off points from the main sequence, the diagonal line running from upper left to lower right. From the theory of nuclear evolution we can determine the ages of stars at their turn-off points. Those ages are shown on the right of the figure (after A. Sandage (Sa57)).

temperatures ranging to nearly 10^5 K. Stars on the right are cool. Luminous stars are found at the top of the diagram and faint stars at the bottom. A large majority of the stars falls on the *main sequence*, a track that runs diagonally from top left to bottom right. Stars on the *subgiant* and *red-giant branch* (Fig. 1.4) are comparatively rare and belong to a population that is more or less spherically distributed about the Galactic center in a *halo*. These halo stars, are also referred to as *Population II stars* to distinguish them from the *Population I stars* that lie in the Milky Way plane and make this portion of the Galaxy appear particularly bright. Halo stars — globular cluster stars among them — tend to be stars with masses $M \leq 1 M_\odot$.

Stellar evolution theory attempts to explain the distribution of stars within the H–R diagram, showing not only why certain regions are populated and others not, but also why some regions — particularly the main sequence — are heavily populated, whereas stars are sparse elsewhere. The basic assumptions of the theory are that nuclear reactions in a star's interior provide the energy the star emits as starlight.

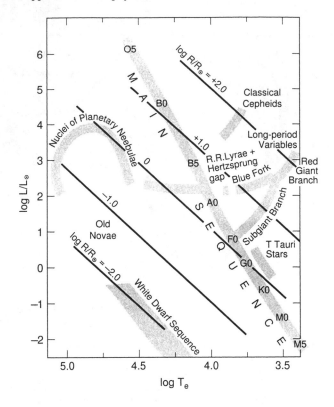

Fig. 1.4. Schematic Hertzsprung–Russell diagram. The lines of constant slope represent stars having identical radii. The effective temperature of a star, T_e is defined in Section 4:13.

As the star evolves, a progression of different nuclear processes set in. While on the main sequence, the star sustains itself by converting hydrogen to helium.

Figure 1.3 plots the range of colors and magnitudes for stars in several of the loosely agglomerated *galactic clusters* found in the Milky Way plane — clusters that must be very young because the stars they contain are too bright to have existed a long time on their limited fund of nuclear energy.

Figure 1.4 shows the characteristics of a wide variety of different stars in a Hertzsprung–Russell diagram that plots a star's luminosity as a function of its effective surface temperature. The location of a star in this plot often is one of the first clues to its identity. Figures 1.5 and 1.6 exhibit the evolution of protostellar objects, to which we will return below, while Fig. 1.7 provides a plot of the visual magnitude versus color of some of the very oldest stars in the Galaxy. These are faint stars that have slowly used up their nuclear fuel over some 10 Gyr. They are members of the *globular cluster* Messier 3. Such clusters are spherically symmetric aggregates consisting of hundreds of thousands of stars (Fig. A.1(c)). Many may have formed at the time the Galaxy itself was born, and are found primarily in the nearly spherical Galactic halo extending well above and below the *Milky Way plane*.

A star's position on the main sequence in the H–R diagram is determined by its surface temperature and luminosity when its central temperature has risen sufficiently for nuclear reactions to set in. The first nuclei to be consumed liberating energy are deuterium, lithium, beryllium, and boron admixed in trace quantities with hydrogen, the star's main constituent. But these minor energy sources are quickly exhausted. The central temperature then has to rise somewhat higher to initiate conversion of hydrogen into helium with the release of sufficient energy to maintain a steady state. For each gram of hydrogen converted into helium $0.007\,c^2 = 6 \times 10^{18}$ erg of energy can leave the star's surface to travel out into space. A slight motion through the Hertzsprung–Russell diagram persists even during this phase. In the course of several billion years, a star moves from the zero-age main sequence, expanding somewhat and becoming more luminous. The Sun may have had a zero-age luminosity $L = 2.78 \times 10^{33}$ erg s^{-1} and a radius $R = 6.608 \times 10^{10}$ cm, compared to its current $L_\odot = 3.84 \times 10^{33}$ erg s^{-1} and $R_\odot = 6.96 \times 10^{10}$ cm, 4.5 Gyr later (Bℓ99)). It will keep shining at these approximate rates for another 5 Gyr.

O and B stars, Population I objects found solely in the Milky Way plane, have far shorter life spans than the Sun. They are the bluest, most luminous main sequence stars. Their high luminosity tells us that they must be young. Figure 1.5 shows a $15M_\odot$ main sequence B star to be $\sim 3 \times 10^4$ times more luminous than the Sun. Though its nuclear fuel supply is 15 times that of the Sun, its projected lifespan is thousands of times shorter, lasting just a few million years. O and B stars cannot be older than this when their supply of fuel runs out. In contrast, Population II objects are probably $\sim 10^{10}$ yr old, judging by their low luminosities and by the fact that the brighter members of this population are just turning into *red giants* — stars that have consumed all the hydrogen in a central core and are beginning to release other nuclear resources.

Though a star spends most of its life on the main sequence, it is not born there. Its life begins in the dense core of a cloud consisting primarily of molecular hydrogen and traces of carbon monoxide and other small molecules. Dust grains permeate these clouds making them impenetrable to starlight.

How do we know all this? What is the observational evidence?

1:6 The Birth of Low-Mass Stars

Observations at infrared wavelengths can penetrate dense dusty molecular clouds to detect nascent stars. Most of these stars have low masses, since low-mass stars, like the Sun, are far more prevalent than stars with masses at or above 10 $M_\odot$, about whose formative stages we still know little. Current observations indicate that the birth of a low-mass star proceeds from the formation of a dense pre-stellar molecular cloud through roughly three protostellar stages, respectively labeled Class 0, Class I, and Class II. As Fig. 1.6 indicates the entire evolutionary process may require of order thirty million years.

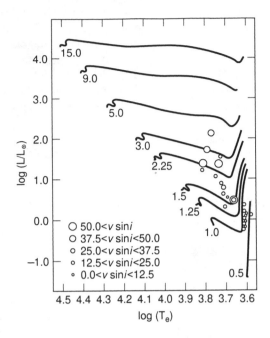

Fig. 1.5. Contraction of stars toward the main sequence. The path of the stars across the Hertzsprung–Russell diagram proceeds toward the left. The left end of the curves roughly coincides with the main sequence. The star with mass $15M_\odot$ completes the transit shown here in $\sim 6 \times 10^4$ yr; the $0.5M_\odot$ star in 1.5×10^8 yr. The steep portion on the right is called the *Hayashi track* (Wh04). The curves denoting very low mass stars correspond to pre-main-sequence evolutionary tracks of T Tauri stars. Circular symbols indicate the projected surface rotational velocities, where i is the usually unknown inclination of a star's rotational axis to the line of sight. Massive stars generally rotate rapidly (Ha66), (Bo86) (after Iben (Ib65)).

    i) In a *pre-stellar phase* a dusty cloud of molecular hydrogen builds up a dense, rotating core. Dust grains radiate away heat, and as the cloud cools to 10 K, it begins to contract.

    ii) In the ensuing protostellar phase the magnetic field appears to leak out of the core, and turbulent motions are damped. If the cloud has excess angular momentum as it contracts, it may fragment into smaller cores or flatten into a disk. Strong bipolar, highly collimated outflows appear. Their origins are uncertain; they may represent the escape of gas compressed to excessive pressures by high-angular-momentum material falling onto an accretion disk orbiting the central protostar. The temperature of the central condensation gradually rises to $\sim$70K, radiates at submillimeter and far-infrared wavelengths, but remains optically invisible. This is a *Class 0 object*; its lifespan is $\lesssim 10^5$ yr.

    iii) As the protostar accretes more mass and an increasingly massive disk is formed, infalling material from the remnant cloud falls on the disk at supersonic

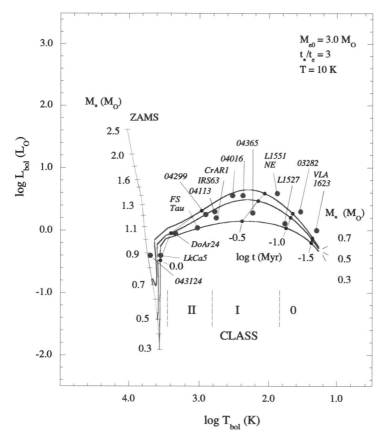

Fig. 1.6. Temperature and bolometric luminosity of protostars. The evolutionary tracks of protostars of 0.3, 0.5, and $0.7 M_\odot$ are shown proceeding from right to left where the stars reach the zero-age main sequence (ZAMS) and begin consuming their nuclear fuel. A protostar begins life as a Class 0 object, a dense cloud at temperatures in the 8 to 12 K range, glowing at submillimeter wavelengths. Strong bipolar outflows of gas accompany contraction of the central core. Shown here, along with the positions in the H–R diagram of a number of young stellar objects (YSOs), is a set of modeled evolutionary tracks represented by solid lines. The time to fall inward to form a star t_* is taken to be three times the duration of the outflow t_e that dissipates an initially enveloping cloud — $t_*/t_e = 3$. The bipolar outflow weakens by the time the protostar evolves to Class 1, and abates totally thereafter. A Class 0 object gradually evolves, its temperature rising as it passes through the Class I and II stages before arriving on the ZAMS ∼30 million years after the onset of contraction. *Isochrones* — lines of equal age — indicating equal times before arrival on the ZAMS are drawn as straight line segments connecting small dots on the evolutionary tracks. For the tracks shown, the initial mass M_{e0} of the gaseous envelope in which the protostar originally formed has been assumed to be six times the final mass $M_{*\infty}$ of the star. Note the sharp drop in luminosity of these low-mass stars on final approach to the main sequence. In Fig. 1.5, the Hayashi track, shown at the extreme right, corresponds to this drop and shows the same temperature, ∼4000 K. Courtesy of Philip C. Myers (My98).

velocities, generating heat that constitutes the main source of energy radiated away. A weakened and less collimated outflow persists. The protostar, whose temperature gradually rises to ∼600 K, and its surrounding disk, become observable at near- and mid-infrared wavelengths. The protostar is now a *Class 1 object*. It transits this phase in ∼1 to 6×10^5 yr.

iv) In the following phase, the outflows have stopped. *Class II objects* exhibit low accretion rates and strong infrared emission from the enveloping disk. Some of these objects are *young stellar objects, YSO*, long known as *T Tauri stars* after the archetypical star for this family.

v) Finally, the infall of the dusty ambient cloud material clears and visible light escapes to reveal a T Tauri star with an optically thin disk and far less infrared emission. In this phase the evolving star is called a *Class III object*. The fate of the disk is still uncertain. Some of it may break up to form a system of planets or a single major orbiting companion. A fraction may eventually fall onto the star, making the disk an *accretion disk*, while the remainder may be ejected back into interstellar space. Observations indicate that circumstellar disks are rare around stars older than 300 – 400 Myr, suggesting that when planets do form, they form rather quickly — within a few hundred million years after the star reaches the main sequence (Ha99). A low-mass star may then continue to shine for another 10 Gyr.

Occasionally, young stars exhibit strong polar outflows, now emanating directly from the star with velocities as high as several hundred kilometers per second. Named after George Herbig and Guillermo Haro who first noted them in the 1950s, these outflows are referred to as *Herbig–Haro objects*.

1:7 Massive Stars

Because massive stars are rare, we know much less about conditions favoring their formation. They are often found in associations containing numerous massive stars. These circumstances suggest that the formation of such stars may be triggered by supersonic shocks. Massive stars exhibit strong outflows, and intense radiation. Both can induce shock compression in ambient molecular clouds, possibly favoring the successive formation of massive stars in a *stellar association*. Shocks must also pervade regions where molecular clouds in merging galaxies collide. Here also massive stars invariably appear. These signs are suggestive, but more incisive observations will be needed before we feel we reliably understand how massive stars are formed, and why the masses of observed Galactic stars never exceed $\sim 150 M_\odot$ (Fi05a).

1:8 The Late Stages of Stellar Evolution

Structural changes that result when all the hydrogen at a star's center has been used up give rise to a change in the star's surface temperature and brightness. The star moves off the main sequence in the Hertzsprung–Russell diagram. In the Galactic cluster $h + \chi$ Persei (Fig. 1.3), evidence for such a move is a curling from the main sequence toward the right — toward lower temperatures — and a new grouping in the right-hand, upper corner of the plot where bright red stars are to be found. Detailed calculations based on model stars, and on the rates at which the nuclear reactions proceed in them, indicate that stars just turning off the main sequence in $h + \chi$ Persei cannot be more than two million years old.

In contrast, the Galactic cluster M67 shows no main sequence stars bluer than spectral type F. These stars therefore all are relatively small — not much more massive than the Sun. Stars of this luminosity complete the hydrogen burning in their central regions in roughly $3 - 4 \times 10^9$ yr. We think that this is the present age of M67.

This cluster also has a well-developed giant branch. Stars that leave the main sequence travel out into this branch. Because the actual number of stars lying along the branch is small compared to the number populating the main sequence, we conclude that the stars do not spend much time in the subgiant or red-giant stages before going on to some other stage. Otherwise, we would expect this portion of the diagram to be more densely populated — perhaps as densely as the main sequence near the *turn-off point*, because the subgiants and giants all are former main sequence stars that resided on the portion of the main sequence just above the present turn-off point.

If we assume that all members of a cluster were formed at one and the same epoch, the density of stars observed in different portions of the H–R diagram tells us not only where various stars reside, but also the initial distribution of stellar masses: the *initial mass function* when the cluster first formed, and how old the cluster may be. If we know the luminosities and masses of stars just turning off the main sequence, having just used up their central supply of hydrogen, we can calculate how old they and the clusters are.

We note one other feature shown in Fig. 1.3. As a star moves off the main sequence it becomes redder and, if anything, more luminous than it was on the main sequence. At lower temperatures, however, an object always emits less radiation per unit area. Correspondingly, the only way that this particular course of evolution can proceed is for the star to grow as it moves off the main sequence. The stars become giants. Their radii increase by factors of 10 to 100.

Let us see how such stages evolve.

For this purpose the study of globular clusters is most instructive. They contain some of the oldest stars in the Galaxy, whose faint, red, turn-off points indicate their age. Figure 1.7 shows the color-magnitude diagram for the globular cluster M3. Once a star leaves the main sequence (MS) at its turn-off (TO), it continues to burn hydrogen in a shell around the hydrogen-depleted core and moves up the subgiant branch (SGB), becoming brighter and redder and eventually reaching the red-giant branch (RGB), in a first red-giant phase in which *hydrogen shell burning* is the main source of energy. The *helium core* slowly increases in mass while the burning shell moves outward converting more hydrogen into helium as it proceeds (see Figs. 8.8 and 8.9).

At the tip of the red giant branch the massive core has heated up through contraction to a temperature at which three helium nuclei can fuse to form carbon in

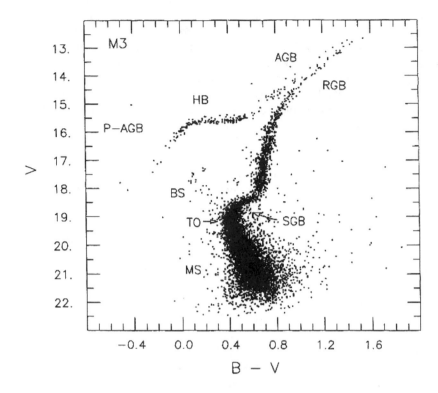

Fig. 1.7. Color-magnitude diagram for the globular cluster M3. The acronyms referring to different parts of the diagram, counterclockwise from the bottom stand for main sequence (MS); turn-off point from the main sequence (TO); subgiant branch (SGB); red-giant branch (RGB): asymptotic giant branch (AGB); horizontal branch (HB); post-AGB(P-AGB), where planetary nebulae can be found; and blue stragglers (BS). Reprinted with permission from the *Annual Review of Astronomy and Astrophysics* (Re88).

the so-called *triple-alpha process*. An added helium nucleus can also convert some of the carbon to oxygen. The triple-alpha process occurs very rapidly, burning helium in what is called a *helium flash*. Roughly 3% of the core, whose total mass is $\sim 0.5 M_\odot$, is burned in the flash. This is enough to heat the core to the point where it expands against gravitational forces and then keeps burning helium in a convectively churning core mass.

Interestingly, during the flash, the energy conversion rate in the core may equal the total energy being released by all the stars in a cluster. But this energy cannot immediately reach the surface of the star. Instead, it goes into effecting an expansion of the highly gravitationally bound core.

The horizontal branch (HB) phase in the life of the star is believed to represent a subsequent stage in which the main source of energy comes from helium burning in the core — with hydrogen burning continuing farther out in a shell. The star moves along the HB to the left, but then reverses its path and roughly retraces its steps in a loop along the HB that lasts some 10^8 yr. In comparison, the evolution from the turn-off point to the red-giant branch must have lasted rather longer, judging from the significantly larger population along that track.

When the star reaches the right-hand limit of the (HB), the core has become exhausted of helium and a stage of helium shell burning sets in. The segment leading up the asymptotic giant branch (AGB) represents a phase in which the star consists of an inert *carbon–oxygen core*, surrounded by a *helium-burning shell*, a further helium zone in which no nuclear conversion is going on, a hydrogen-burning shell, and finally an envelope made up of the original unconverted matter from which the star formed. The chemical makeup of the outermost portions of the star, the only portions we actually see, seldom show clear evidence of the complex changes that have occurred in the star's interior, even at this advanced stage of evolution.

During this *asymptotic giant branch (AGB) stage*, the star follows almost the same path through the H–R diagram that it earlier traced in going up — and later down — the red-giant branch. The AGB and RGB have an outwardly similar appearance.

In each flash of helium shell burning, the peak luminosity may be of order $10^5 L_\odot$. There are indications that, at least in certain model stars, the helium shell flash causes the star to move off the red-giant branch in a loop that leads to the left and back again in a matter of about 10^3 yr. This motion takes the star into a portion of the diagram occupied by Population II *Cepheid variables* — luminous pulsating stars whose pulsation periods change significantly in the course of a million years. Eventually, the pulsations become unstable and the star's surface layers are ejected in violent outbursts that separate off an outer shell of the star lying above a point somewhere between the hydrogen- and helium-burning shells. A planetary nebula is formed.

Planetary nebulae are hot *central stars* surrounded by a shell of ejected material.

Abundances of the elements He, C, N, O, Ne, S, Ar, and in some cases Mg and Si have been measured for a variety of planetary nebulae. These indicate that these stars can follow two quite distinct evolutionary paths, depending on whether the star leaves the highly reddened asymptotic giant branch (AGB) as a hydrogen- or helium-burning object. This affects both the rate at which the star moves across the H–R diagram and where it moves across it. The rapid mass loss, through material blown off the star's surface into space at this stage of its evolution, channels stars from a wide range of initial masses into a narrow range of core mass. Material appears to be dredged up from the interior of these stars in three different stages to appear on the surface. The first dredge-up occurs as the star becomes a red giant for the first time and involves convection from regions that have undergone partial burning of carbon, nitrogen, and oxygen. A second dredge-up occurs in stars more massive than 3 to 5 $M_\odot$, when the hydrogen-burning shell is extinguished and convection again brings up processed material. The third dredge-up occurs during the thermally pulsing AGB phase in which, after each helium-burning pulse, convection brings up material rich in ^{4}He, ^{12}C, and heavier elements. A further stage of *hot-bottom burning* occurs in AGB stars more massive than $M > 3M_\odot$ when convection in the star's envelope cycles matter through the hydrogen-burning shell during the interpulse phase (Do97). Each of these phases brings up to the stellar surface mixtures of isotopes that permit us to reconstruct details of the nuclear processes that have taken place at successive stages deep in the interior of these evolving stars.

The central star of the planetary nebula first appears very hot and bright, as indicated by the loop drawn on the left in Fig. 1.4, but then cools down, slowly contracting to the *white dwarf* stage in the lower left of the H–R diagram occupied by white dwarfs. The central portion of the star contains a relatively low-mass core, rich in carbon and oxygen. The mass of the white dwarf at this stage might be no more than $\sim 0.7M_\odot$, the rest of the mass having been ejected during the outbursts that produced the gaseous envelope characterizing planetary nebulae.

A star more massive than $\sim 8M_\odot$ evolves much more rapidly than low-mass stars, and its thermonuclear reactions proceed further as the star becomes a red supergiant. Helium burning at the star's core proceeds beyond carbon and oxygen, continuing on to form the most stable of nuclei, iron ^{56}Fe. Surrounding the iron core, nuclear burning continues in shells of elements whose nuclei contain even numbers of both protons and neutrons — calcium ^{40}Ca, silicon ^{28}Si, magnesium ^{24}Mg, oxygen ^{16}O, and others. Although the star at this stage has a radius comparable to or even exceeding the Earth's orbital radius about the Sun, the radius of its core may be smaller than that of the Earth (Gℓ97*).

The size of the highly ionized iron core continues to grow. Its self-gravity and tendency to collapse is resisted only by the pressure of the embedded free electrons at relativistic energies. As the hydrostatic pressure due to gravity rises, the pressure on the electrons forces them into the iron nuclei, in an *inverse beta decay* that enriches the nuclei with energetically favored neutrons. This diminishes the electron pressure and permits further collapse. The core has reached its limiting mass and suddenly implodes — in less than a second! The inverse beta decay *neutronizes* the core; a flood of neutrinos is released. The neutrinos collide with the imploding nuclei, whose high density prevents the neutrinos from escaping. Eventually, the imploding core rebounds, sending shock waves outward. In the ensuing turmoil, which is not at all well understood, the outer shells of the star are hurled into interstellar space at speeds of tens of thousands of kilometers per second. A *supernova* has just exploded!

Supernovae that do not exhibit emission lines of hydrogen are designated supernovae of type I, SNe I. Those that do are designated SNe II. SN 1987A, which exploded in the Large Magellanic Cloud in 1987, was a supernova of this second type. The total energy released in such an explosion is of the order of 10^{53} erg. Only one percent of this energy resides in the massive outflow of matter, which persists for thousands of years as a readily identified, expanding, radiating, spherical shell — a *supernova remnant*. The remaining energy is released largely in an outpouring of neutrinos lasting no more than about ten seconds. The core cools and forms a *neutron star*. Its density is about 10^{14} g cm^{-3}, and its mass is roughly $1.4M_\odot$. As

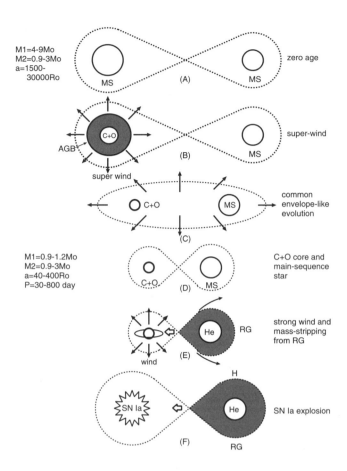

Fig. 1.8. One potential way to form a supernova of type Ia from stars in a close binary. The two stars forming the binary arrive on the main sequence, one having a mass in the range $4 - 9 \; M_\odot$, while its less massive companion might have a mass of $1 - 3 \; M_\odot$. Their orbital separation — twice the orbital semi-major axis a — at this stage may range from 3000 to 60,000 $R_\odot$ or $\sim 15 - 300$ AU. The more massive star is first to evolve to the AGB phase characterized by a carbon and oxygen core. It sheds its outer envelope and becomes a carbon–oxygen white dwarf. The wind emitted by the giant star preferentially reaches escape velocity along the direction of orbital motion, so that the system loses angular momentum. In addition, as the giant's outer layers expand, the companion loses momentum as it becomes enveloped in this denser medium. Both effects cause the two stars to spiral in toward each other. The less massive companion next reaches the red-giant phase, in which its extended outer envelope can be gravitationally captured by the more compact white dwarf. Eventually the white dwarf becomes sufficiently massive to collapse and explode as a Type Ia supernova. Courtesy of K. Nomoto (Ha99a).

the core collapses, its angular momentum has to be conserved, and so the neutron star is born rapidly spinning. Its magnetic field, highly compressed during collapse, may be as high as $\sim 10^{13}$ G. The spinning neutron star is a *pulsar*.

Pulsars are found only in some supernova remnants. In many instances they appear to be missing, most probably due to the high-velocity kicks that pulsars receive in the course of a supernova explosion. Many pulsars are observed to travel through the Galaxy at speeds of order 1000 km s^{-1}, leaving behind the supernova remnant that resulted from the same explosion but whose expansion may have been slowed in sweeping up ambient interstellar gases.

Like supernovae of type II, some supernovae of type I, those designated SNe Ib and SNe Ic appear to originate in the collapse of massive isolated stars. SNe Ia, in contrast, are formed when a white dwarf rich in carbon and oxygen accretes sufficient mass from a less compact binary companion to trigger the central collapse followed by a massive explosion, as sketched in Fig. 1.8.

White dwarfs are sufficiently compact to tidally strip material from a close binary companion star. As the amount of matter accreted onto the white dwarf increases, the star may grow so massive that matter in its core can no longer support the hydrostatic pressure. The dwarf star implodes to form a neutron star while ejecting matter in a giant supernova explosion. Supernovae of this type are designated SNe Ia. Type Ia supernovae may also be formed through the merger of two white dwarfs. Neither of these two processes is well understood.

SNe Ia are remarkably homogeneous in their luminosities, rise times of fifteen to twenty days and a subsequent decline of several weeks. The post-explosion luminosity is powered by the radioactive decay of ^{56}Ni into ^{56}Co and then ^{56}Fe. The radioactive decay rate determines the rate at which the luminosity declines. At maximum luminosity, SNe Ia tend to be ~ 5 times more luminous than SNe II. The homogeneity of SNe Ia permits their use as distance indicators. Other factors being equal, the fainter the supernova, the farther away the explosion must have occurred. This property, together with their high luminosities, when combined with red shifts determined for their parent galaxies, makes SNe Ia useful markers for charting the expansion rate of the Universe.

For some massive collapsing stars, the implosion is believed to form not a neutron star, but rather a *black hole* so dense that its internal pressure cannot support the hydrostatic pressure of overlying material. The collapse proceeds until the star swallows itself and shrinks to a singularity.

1:9 Abundance of the Chemical Elements in Stars and the Solar System

The abundance of chemical elements found in the Sun, in meteorites that fall onto Earth from interplanetary space, and in terrestrial matter, are remarkably similar provided we discount the very high abundances of hydrogen and helium in the Sun, which the Earth is unable to gravitationally bind. This uniformity stretches beyond the Solar System to other stars and nearby galaxies. Where significant differences arise, the deviant abundances most often provide insights on stellar evolution or chemical fractionation. Table 1.1 shows chemical abundances in the Solar System derived mainly from *carbonaceous chondrites*, meteorites considered to be most representative of the *primeval Solar Nebula* and hence probably also of the solar sur-

Table 1.1. Abundance, by Number of Atoms, of the Chemical Elements in the Solar System. Compilations of Abundance Normalized to $Si = 10^6$. (Reprinted from E. Anders and N. Grevesse (An89) © with kind permission from Elsevier Science Ltd, The Boulevard, Langford Lane, Kidlington OX5 1GB, UK.)[a]

Element	Abundance	Element	Abundance	Element	Abundance
1 H	2.79×10^{10}	29 Cu	522	58 Ce	1.136
2 He	2.72×10^9	30 Zn	1260	59 Pr	0.1669
3 Li	57.1	31 Ga	37.8	60 Nd	0.8279
4 Be	0.73	32 Ge	119	62 Sm	0.2582
5 B	21.2	33 As	6.65	63 Eu	0.0973
6 C	1.01×10^7	34 Se	62.1	64 Gd	0.3300
7 N	3.13×10^6	35 Br	11.8	65 Tb	0.0603
8 O	2.38×10^7	36 Kr	45	66 Dy	0.3942
9 F	843	37 Rb	7.09	67 Ho	0.0889
10 Ne	3.44×10^6	38 Sr	23.5	68 Er	0.2508
11 Na	5.74×10^4	39 Y	4.64	69 Tm	0.0378
12 Mg	1.074×10^6	40 Zr	11.4	70 Yb	0.2479
13 Al	8.49×10^4	41 Nb	0.698	71 Lu	0.0367
14 Si	1.00×10^6	42 Mo	2.55	72 Hf	0.154
15 P	1.04×10^4	44 Ru	1.86	73 Ta	0.0207
16 S	5.15×10^5	45 Rh	0.344	74 W	0.133
17 Cl	5240	46 Pd	1.39	75 Re	0.0517
18 Ar	1.01×10^5	47 Ag	0.486	76 Os	0.675
19 K	3770	48 Cd	1.61	77 Ir	0.661
20 Ca	6.11×10^4	49 In	0.184	78 Pt	1.34
21 Sc	34.2	50 Sn	3.82	79 Au	0.187
22 Ti	2400	51 Sb	0.309	80 Hg	0.34
23 V	293	52 Te	4.81	81 Tl	0.184
24 Cr	1.35×10^4	53 I	0.90	82 Pb	3.15
25 Mn	9550	54 Xe	4.7	83 Bi	0.144
26 Fe	9.00×10^5	55 Cs	0.372	90 Th	0.0335
27 Co	2250	56 Ba	4.49	92 U	0.0090
28 Ni	4.93×10^4	57 La	0.4460		

[a] This table represents a best estimate for primitive solar matter, and is based as much as possible on abundances of Type 1 carbonaceous chondrites — stony meteorites containing millimeter-sized silicate spherules — because volatile substances probably escape least from this type of meteorite (Ca68). For H, C, N, O, and noble gases, solar and other astronomical data were used. Abundances of Mg, S, and Fe are based on an average of mean values for individual meteorites. For the remaining elements, a straight average of all acceptable analyses was used.

face composition some 4.5 Gyr ago (Ca68). Because not all elemental abundances can be reliably determined in this way — some volatile elements, for example, may have escaped from the meteorites through diffusion — the table has been augmented using information obtained from solar spectra and from the cosmic rays emitted by the Sun. We note that the heaviest elements, which can be readily determined in meteorites, are not easily obtained in spectra of stellar atmospheres. The two types of information therefore complement each other and serve also to point out agreement or differences for elements for which direct comparisons are available.

Spectroscopic determinations of the abundances of chemical elements in the atmospheres of stars other than the Sun provide us with information on the chemical composition of the medium from which those stars were formed. The theory of stellar structure shows that for most types of stars, the outer layers remain unaffected by the nuclear processes that liberate energy at the stars' centers. Deuterium, lithium, beryllium, and boron, however, do not remain representative of the protostellar material, because their nuclei are readily destroyed during an early convective contraction that mixes material from the protostar's surface into the hot central portions of the star.

Table 1.2 shows that the relative abundances of some of the more abundant el-

Table 1.2. Elemental Abundances $\log n$ for "Normal" Stars Relative to $\log n = 12$ for Hydrogen.

		Abundance: $\log n$					
		Sun					
Atomic Number	Element	Goldberg, Müller, Aller (1960, 1967)	Various Other Sources	τ-Scorpii B0 V	ζ-Persei B1 Ib	Planetary Nebulae	$\mathrm{Log}(n/n_\odot)$ ε-Virginis G8 III
1	H	12.00	12.00	12.00	12.00	12.00	0.00
2	He		11.2	11.12	11.31	11.25	
6	C	8.51	8.51	8.21	8.26	8.7	−0.12
			8.55				
7	N	8.06	7.93	8.47	8.31	8.5	
8	O	8.83	8.77	8.81	9.03	9.0	
10	Ne			8.98	8.61	8.6	
11	Na	6.30	6.18				+0.30
12	Mg	7.36	7.48	7.7	7.77		+0.04
13	Al	6.20	6.40	6.4	6.78		+0.14
14	Si	7.24	7.55	7.66	7.97		+0.13
15	P	5.34	5.43				
16	S	7.30	7.21	7.3	7.48	8.0	+0.09

^{a} Based on a more extensive compilation from various sources by A. Unsöld (Un69)*.

ements in stars with different ages: (a) the Sun, which is 5×10^9 yr old; (b) a very young B0 star, Tau Scorpii; (c) planetary nebulae; (d) a red giant ε Virginis; and many other "normal" stars, all have the same chemical composition, within the limits of observational error. This is significant because the ages of these objects cover much of the lifetime of the Galaxy since the earliest stage of star formation during which the globular cluster red-giant precursors are thought to have formed.

These analyses show that throughout most of the life of the Galaxy interstellar matter has had an almost unchanged composition. This appears surprising, at first, because the continual formation of heavy elements in stars, and the ejection of matter from these stars at the end of their lives, should gradually have enriched the interstellar medium and the composition of stars most recently formed.

The reason why we do not observe a gradual enrichment may be due to two factors. In massive supernova explosions, ejecta rich in heavy elements may be hurled out of the Galaxy to enrich the intergalactic, rather than interstellar, medium. Although this will not happen to material ejected at lower velocities as planetary nebulae form, fresh material diluting these enriched gases continues to be tidally swept up by the Galaxy from intergalactic clouds and small galaxies rich in gas that has never been processed in stars (Se04). The infall rate appears to amount to two solar masses each year, which is about equal to the star formation rate in the Galaxy (La72).

Evidence for an early growth in the abundance of heavy elements, when the Galaxy was very young, nevertheless is persuasive. We suspect that low-mass stars belonging to the Galaxy's halo population are among the stars that formed earliest in the birth of the Milky Way, roughly ten billion years ago. Some of these show abundances of the elements from carbon to barium that are up to a couple of orders of magnitude lower than in younger stars like the Sun. Table 1.3 shows the ratios of abundances of selected atoms from sodium through lanthanum, relative to the same ratios found in the Sun. This low but nevertheless significant *metal abundance* — in this context the word "metal" denotes any atom heavier than helium — is striking. Although these stars exhibit a metal deficiency relative to hydrogen, helium is not significantly deficient. Its origin dates back to an epoch when the Universe was only a few minutes old and temperatures were of the order of 10^9 K.

Primordial Star Formation

We are still attempting to determine how stars may have formed at early epochs in the history of the Universe when no heavy elements existed and stars had to form from dust-free, helium–hydrogen mixtures. Though the details are still lacking, indications are that the first stars were very massive and erupted in exceptionally powerful supernova explosions, giving rise to the first chemical elements more massive than 7 atomic mass units (Ab02, He02).

The very oldest stars in the Galaxy have metal abundances systematically low, compared to their hydrogen content, by as much as a couple of orders of magnitude. Two of the most metal-deficient stars discovered to date have iron deficiencies

Table 1.3. Abundances in Globular Cluster Stars.[a]

Abundance	Metal Rich[b]	Intermediate[c]	Metal Poor[d]
[Fe/H]	−0.80	−1.35	−2.25
[Na/Fe]	0.25	0.05	0.25
[Mg/Fe]	0.20	0.30	−0.10
[Si/Fe]	0.30	0.25	
[Sc/Fe]	0.10	0.05	0.05
[Ti/Fe]	0.30	0.35	0.30
[V/Fe]	0.25	0.25	0.35
[Cr/Fe]	0.10	0.05	−0.10
[Mn/Fe]	−0.10	−0.30	−0.15
[Co/Fe]	−0.00	0.00	
[Ni/Fe]	−0.20	−0.15	0.05
[Cu/Fe]	−0.20	−0.30	−0.40
[Y/Fe]	0.10	−0.05	−0.20
[Zr/Fe]	−0.15	0.05	
[Ba/Fe]	−0.35	−0.20	−0.15
[La/Fe]	0.00	0.05	

[a] Ratios shown in square brackets are comparisons to solar ratios and given in logarithms to the base ten.
[b] These abundances are straight means of the results for the clusters NGC 104 (47Tuc), NGC 6352, and NGC 6838 (M71).
[c] These abundances are means of the results for 17 clusters with $-0.8 \geq [Fe/H] \geq -1.9$.
[d] These abundances are averages of the abundances of M15 and M92.
From (Wh89), with permission from the *Annual Reviews of Astronomy and Astrophysics* ©1989, Annual Reviews, Inc.

[Fe/H] = −5.3 and −5.45, meaning that their iron abundances relative to hydrogen are, respectively, 2×10^5 and 3×10^5 times lower than in the Sun (Su04), (Ao06).

Was there an even earlier stage of star formation in which the very first heavy elements were formed — a stage that left no apparent survivors? All available evidence indicates this, but no genuinely primordial star devoid of all elements heavier than helium has yet been identified.

The most distant galaxies and quasars observed date back to a time when the Universe was less than a billion years old. Yet they have spectra that reveal metal abundances similar to those of the oldest Galactic stars. Our theories tell us that these elements could only have formed inside stars. The ratios of abundances observed for iron, oxygen, calcium, and neon are reminiscent of the ratios observed in the ejecta of supernovae of type II. This drives us to conjecture that the Universe had already gone through at least one star-forming epoch before even these earliest observed quasars and galaxies formed. Stars of some sort apparently formed early within the first billion years in the life of the Universe.

Extremely massive stars could have formed in the very earliest condensations the Universe produced in a collapse of a primordial mixture of nearly pure hydrogen and helium. Such putative stars, called *Population III* stars, are thought to have been very massive and to have formed when the Universe was $\lesssim 10^8$ yr old. Because of their high mass, the stars rapidly evolved to the supernova stage in $\sim 10^6$ yr and ejected a first dose of heavy elements that appears to have been incorporated into every generation of stars subsequently formed.

Population III stars may have existed at such early epochs that the original wavelength λ of radiation reaching us today would have suffered a red shift $z \equiv \Delta\lambda/\lambda \sim 20$. Such early luminous objects would then be apparent to us only through observations at infrared wavelengths around 5 to $10\,\mu$m. As we build instruments to look ever farther into the Universe and back to earlier times, a set of first hints on the formation of stars and galaxies should emerge. What it will teach us remains uncertain.

The relative abundance of different chemical elements observed in the Sun, on Earth, and in meteorites has been studied for many decades. A triumph of the theory of stellar evolution is that the chemical elements produced in the stars agree so well with the predictions of the theory. By virtue of the theory, the central temperatures of stars of different masses and at different stages of evolution can be calculated, experiments conducted at accelerators can be used to calculate nuclear reaction rates at these temperatures, and complex chains of successive nuclear reactions can then be concatenated to determine the final abundances of the chemical elements that should be ejected from different types of stars. The concord between the observed abundances of the chemical elements and the abundances predicted by theory is, by now, found to be quite striking, even if some anomalies persist.

1:10 Origin of the Solar System

Sometime around the era when the Sun formed, the system of planets became established too. Was the *Solar System* formed after the Sun already was several hundred million years old, or were the Sun and the *planets* (Table 1.4) born in one and the same process? Did the Solar System form out of a single cloud of matter surrounding the Sun, or was there another star involved in the birth of the planets?

Both questions seem to have been answered, at least in part, by observations cited in Section 1:6. Most probably the Sun and a surrounding dusty disk formed from a single contracting cloud of gas and dust. After a few hundred million years, the disk gave birth to the planetary system now observed.

Planets are fairly common around stars in the Sun's vicinity. They probably are common everywhere but, at the moment, our telescopes do not reach out sufficiently far to check this. Many of the planetary systems discovered to date appear very different from our own. They seem to contain a single massive planet orbiting the parent star in an eccentric orbit that brings the planet close to the star once each orbit. Most techniques for discovering planets are indirect; they measure the gravitational tug of the planet on the parent star. Because a single massive planet passing close by the star exerts the greatest tug, this type of system is most readily detected. Nevertheless, a few systems have been discovered that contain several planets. We still do not know how frequently each type of system actually occurs. For this to

Table 1.4. Characteristics of the Planets

Planet	Mercury	Venus	Earth	Mars	Jupiter	Saturn	Uranus	Neptune	Pluto
Orbital semimajor axis (AU)	0.387	0.723	1.000	1.524	5.203	9.55	19.2	30.1	39.4
Sidereal period (yr)	0.241	0.615	1.000	1.881	11.86	29.46	84.02	164.8	248
Eccentricity	0.206	0.007	0.017	0.093	0.048	0.056	0.046	0.009	0.246
Inclination	7°0'	3°24'	0°00'	1°51'	1°19'	2°30'	0°46'	1°47'	17°10'
Equatorial radius (km)	2439	6052	6378	3397	71,400	60,330	25,400	24,300	1100
Mass (g)	3×10^{26}	4.9×10^{27}	5.97×10^{27}	6.4×10^{26}	1.90×10^{30}	5.7×10^{29}	8.7×10^{28}	1.0×10^{29}	1.2×10^{25}
Density (g cm^{-3})	5.44	5.25	5.52	3.94	1.31	0.69	1.21	1.67	1 ?
Number of known moons	0	0	1	2	> 16	> 17	> 15	2	1
Ring system					Yes	Yes	Yes	Yes	
Visual magnitude m_v at maximum angle from the Sun	−0.2	−4.2		−2.0	−2.5	+0.70	+5.5	+7.9	+14.9
Horizontal magnetic equatorial field (G)	0.0035	$< 3 \times 10^{-4}$	0.31	$< 10^{-3}$	4.3	0.2	0.2	0.1	(?)
Surface gravity (cm sec^{-2})	370	887	980	371	2312	896	~777	1100	72
Main atmospheric constituents		CO_2	N_2 O_2	CO_2 O	H_2, He NH_3, CH_4	H_2, He NH_3, CH_4	H_2, He NH_3, CH_4	H_2, He NH_3, CH_4	
Rotation period	58.6 d	243 d	23.93 hr	24.6 h	9.925 h	10.66 hr	17.2 hr	16 hr	6.4 d
Oblateness	0.0	0.0	0.0034	0.007	0.06	0.1	0.02	0.02	(?)
Subsolar surface temperature (K)	620	250 (cloud top)	295	270	140 ± 10	138 ± 6	125 ± 15	134 ± 18	(?)
					(at ~ 1 cm radio wavelengths, whole disk)				
Albedo: fraction of light that is reflected	0.06	0.85	0.4	0.15	0.58	0.57	0.8	0.71	0.61
Excess luminosity (Watt)					$\sim 3.3 \times 10^{17}$	$\sim 8.5 \times 10^{16}$	$< 1.5 \times 10^{15}$	$(1-3) \times 10^{15}$	

become known, we will need to develop techniques that are capable of detecting a wider range of planetary systems. In addition, we can learn a great deal more by studying our own Solar System in considerably greater depth. Here, too, a variety of tools are needed to answer different classes of questions.

Let us look at some of these.

(a) Dynamical Questions

Newton's laws of motion describe the orbits of the planets about the Sun; they also describe the changes in these orbits as the planets interact with each other. This is the subject matter of *celestial mechanics*. From the motion of the moons around individual planets, and from the short-term interactions of neighboring planets, we can judge the masses of the major bodies that make up the Solar System. Knowing these masses, and knowing the instantaneous orbits, we might think that we should be able to look back in time to calculate how the Solar System evolved in the past, what its appearance was a few hundred years ago, a few million years ago, and possibly billions of years ago!

A revival of celestial mechanics, based on the prodigious capability of computers to conduct large numbers of repeated steps might be thought to give shape to the dream of reconstructing a dynamic history of the Solar System by computing the evolution of planetary orbits backward in time. However, even small unknown disturbances, possibly due to a relatively close approach of a passing star, or a collision among planets far back in time, lead to extrapolations that diverge, becoming meaningless as we calculate ever farther back in time.

An example of this difficulty is this: we know that the planets all orbit close to the Earth's orbital plane, the *ecliptic*. Only Mercury, the smallest and nearest planet to the Sun has an *inclination* as high as $7°$. Pluto has a higher inclination yet, greater than $17°$; but its inclination, in contrast to that of the other planets, is believed to vary rapidly under the perturbing influence of its far more massive neighboring planets. Generally, then, the *orbital angular momentum* axis for all the planets lies along the same direction. The mean angular momentum has a direction nearly normal to the orbital plane of Jupiter, the most massive planet.

Surprisingly this angular momentum has a direction $7°$ away from the Sun's spin axis. The Sun's equatorial plane is inclined that strongly relative to the ecliptic.

How can this be? Does it mean that the Sun and planets could not have been formed from one and the same rotating mass? Does it mean that some other massive body was present and instrumental in the birth of the planets? On a more detailed basis, could it be that Mercury, whose orbit does have about the same inclination to the ecliptic as the Sun's equator was formed later than the more distant planets, or else — because of its proximity to the Sun — interacted more strongly with our parent star?

The number of questions raised by this single consideration is large. It may therefore not be a very productive line of pursuit. Perhaps some future theory involving much more complex arguments will automatically also produce the proper relationships among orbital inclinations as a natural side product; but the side product alone may not be a sufficient clue to the overall structure of that theory, and may not help us very much right now. Many such questions remain puzzling (De04).

A second example involves Bode's law. The planets occupy orbits that are regularly spaced according to a pattern first noted two centuries ago and then popularized by the German astronomer Bode (see Fig. 1.9).

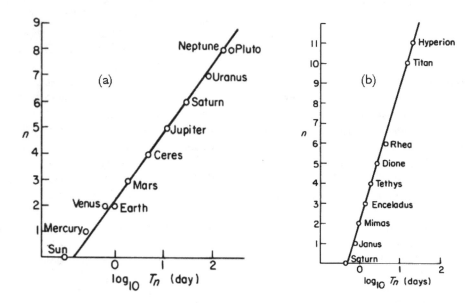

Fig. 1.9. The orbital relationship $T_n = T_0 A^n$ applied (a) to the Solar System and (b) to the moons of Saturn. T_n is the orbital period of the nth planet or moon. T_0 is chosen, respectively, close to the rotational period of the Sun or the parent planet (after Dermott (De68). With the permission of the Officers and Council of the Royal Astronomical Society.) In addition to the large regular satellites of Saturn found close to the parent and listed in (b), smaller irregular satellites, much darker in appearance, are found at greater distances. These two groups may have had distinct origins.

Bode discovered that the distance of the planets from the Sun follows a regular progression. More recently Dermott (De68) was able to show that a slightly new phrasing of this law permits us to include not only planetary orbits around the Sun, but also the orbits of *moons* around their parent planets. In either case, the *orbital period T_n* of the nth body of the orbital system can be written in terms of a basic period T_0, close to the *spin period T_p* of the parent body

$$T_n = AT_0^n .$$

Is *Bode's law* just a coincidence of numbers, or does it describe some deeper interrelation among the planets' orbits? In particular, does that relationship provide any insights into the early history of the Solar System, or is it an arrangement that would hold for any system of bodies orbiting about a central mass, given only enough time for these bodies to reach some state of dynamic equilibrium? Perhaps Bode's law is just telling us that a rich system of planets or satellites requires their orbits to be sufficiently far apart to prevent collisions (De73).

(b) Radioactive Dating

Radioactive dating allows us to conclude that at least some rocks on Earth must have solidified some 3.8 Gyr ago, and that meteorites falling onto Earth from inter-planetary space have ages of 4.5 to 4.6 Gyr. The Earth as a whole seems to be about 4.5 Gyr old. Lunar surface samples brought back to Earth indicate ages in excess of 4.4 Gyr so that the Moon and Earth may have comparable ages. This would be in agreement with the belief that the Moon was torn out of the Earth by a giant asteroidal impact that could have occurred between 4.5 and 4.6 Gyr ago (De94).

The age of solidified rocks can be determined from the ratio of radioactive *parent* and *decay products* found in a sample. Specifically, the uranium isotope ^{238}U decays into lead ^{206}Pb, by sequentially emitting eight alpha particles. The half-life for this decay is 4.5 Gyr. If the rock is not porous, the alpha particles become trapped as helium atoms after combining with some of the electrons that are released as the nuclear charge diminishes in the alpha decay. By measuring the ratio of ^{238}U to ^{206}Pb and helium present in the rock, some estimate of the age can be obtained. This estimate must take into account that other radioactive decay may be going on simultaneously. The uranium isotope ^{235}U, for example, decays into lead ^{207}Pb with the release of seven alpha particles in a half-life of 0.7 Gyr; thorium ^{232}Th decays into lead ^{208}Pb and six helium atoms in 13.9 Gyr; rubidium ^{87}Rb turns into strontium ^{87}Sr in 46 Gyr; and potassium ^{40}K turns into argon ^{40}Ar in 1.25 Gyr (Wh64). A complete age determination usually involves several of these decays. Only if all the dates obtained agree can we feel safe in setting an age for a studied sample.

Studies of this kind indicate that the Earth and the meteorites solidified about 4.5 Gyr ago, within a time of $\sim 10^8$ yr. Other theories, involving nuclear processes going on in stars, predict the abundance ratios for the various isotopes of a given element at the time that matter was ejected from a star. The current abundance ratios for some of the radioactive isotopes found in the Solar System, therefore, also can be used to fix the time of formation of these elements in the interior of an earlier star. Somewhat surprisingly that time is only of the order 6 Gyr, so that the Sun must have formed within ~ 1 Gyr after the heavier elements in the Solar System were formed, perhaps in the explosion of an earlier generation star. The entire process of star and planet formation probably took place within a time span of only 1 Gyr. Figure 1.10 provides a rough early history of our planet.

(c) Chemical and Mineralogical Evidence

Remarkably detailed information on the early history of the Solar System comes from the study of meteorites. Many *stony meteorites* show an abundance of milli-

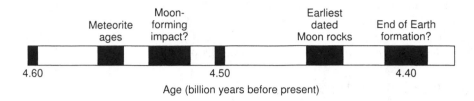

Fig. 1.10. Time line for the formation of meteorites, the earliest dated Moon rocks, and the formation of the Earth (De94). (Reprinted with permission from *Nature* ©1994 Macmillan Magazines Limited.)

meter-sized spheres held together by a matrix of silicates. Were these spherical *chondrules* already present in the early Solar Nebula and do they therefore contain information that could be used to infer primitive conditions? Studies on the X-ray flaring of T Tauri stars suggest that chondrules may form in T Tauri disks through flash-heating. Magnetic fields connecting the rotating central star and the surrounding disk can suddenly reconnect to release magnetic pressure and energy, and accelerate charged particles to high energies. These might heat and melt dust aggregates, which could then cool and condense as chondrules (Gr97), (Sh97).

Iron meteorites show crystalline structure that can only form at very high pressures. Does this mean that these meteorites originated in the interior of a larger planetary body that once broke up? Can the high-pressure conditions needed for the crystalline structures be provided by shocks that would naturally occur when asteroids from time to time collide? The few close-up pictures we have of asteroids show them to be thoroughly cratered, battered by many successive high velocity collisions.

(d) Comets, Asteroids, and Kuiper Belt Objects

The early Solar System may have comprised mainly comet-sized objects. Comets contain frozen gases such as water, ammonia, and methane. Large amounts of hydrogen are trapped in these molecules or larger *mother molecules* that can break up into NH_3, OH, CO_2, and CH on exposure to solar radiation. Embedded in these ices are silicate grains and other solids. Some comets approach the Sun from distances as remote as 10^{18} cm. They appear to be bound members of the Solar System that have been at great distances from the Sun most of their life and are approaching now after an absence of a hundred million years, or perhaps considerably more. In these comets we may be seeing the primordial matter from which the planets formed. The comets apparently were pushed out to large distances from the Sun early in the formation of the Solar System, and have been orbiting there ever since. They may represent deep-frozen samples of matter preserved from the early Solar System and, therefore, are extremely interesting objects to study if the history of the Solar System is to be reconstructed. To date we have been able to study comets only as they fall apart on approaching the Sun, where solar heating evaporates some of the frozen gases and liberates solid materials that were held together by the ices. Some of these solid particles later strike Earth's atmosphere and heat or burn up because of their high approach velocity to Earth. This heating and burning gives rise to emission of light whose spectrum can be analyzed for the presence of various elements. From this and from the spectra of gases released by the comet on approaching the Sun, we can make crude chemical analyses of the comets' contents. By and large, we find that comets comprise chemical elements that are also abundant in the outer planets. A precise analysis of material constituting the nucleus of a comet may have to await a spacecraft probe that lands on a nucleus and examines the local matter before evaporation, dissociation, or ionization by sunlight. In the meantime, space probes that approached comet Halley in 1986 were at least able to analyze the gases in close vicinity of the nucleus and establish the dust-grain size-distribution there.

The gravitational influence of Jupiter is so great that it can significantly alter the orbits of at least some comets and bring them appreciably closer to the Sun. These comets are captured from the highly elliptical orbits that have taken them to the most distant parts of the Solar System and placed them into relatively small, short-period orbits with *aphelion points* near Jupiter's orbit.

The continual heating by solar irradiation can then evaporate most of the frozen gases of a short-period comet. The comet nucleus itself is too small to hold on to these gases through gravitation. Over $\sim 10^2$ orbits, the comet disintegrates. If it has a solid core, only that core remains intact after a few thousand years. It is possible that at least some asteroids — bodies whose sizes range largely from a few kilometers down to fractions of kilometers — are the remnants of earlier comets. They certainly have orbits very similar to the short-period comets and might, therefore, have a common origin. The largest asteroids, however, are more than 100 km in diameter, much larger than observed comets seen in the past, and it is likely that these larger asteroids do not represent cometary remnants.

In the early 1990s, a whole new family of asteroidal and cometary bodies was discovered. They orbit the Sun at distances of 50 to 100 AU, well beyond the orbits of even the outermost planets, Neptune and Pluto. The number of comets in this *Kuiper belt* may run into the hundreds of millions or billions. We are only just beginning to investigate the significance of these bodies in the history of the Solar System.

(d) The Chemical Makeup of Planets

The differences in *density* and *chemical composition* of the planets may provide evidence about how they were formed. The inner, *terrestrial planets* are much more dense than the outer *giant planets*. They contain silicates and iron, which solidify at relatively high temperatures and hence could have solidified close to the Sun. They contain lesser amounts of hydrogen, because hydrogen is readily evaporated from a small planet close to the Sun where temperatures are high. Because of this evaporation, the atmospheres of the inner planets as seen today may be quite different from their enveloping atmospheres during early times in the Solar System. The Earth's atmosphere in particular is thought to have been reducing — meaning that hydrogen was prevalent and that oxygen was tied up in molecules and unavailable for combining with other elements. Today, of course, the atmosphere is definitely *oxidizing* with its 20% abundance of free oxygen gas.

We can see from Table 1.4 that the major planets are less dense but more massive than the inner planets. They contain a large fraction of their mass in the form of hydrogen and are able to retain it because of the low temperatures determined by their relatively large distances from the Sun and because of their stronger gravitational pull.

The distribution by volatility suggests that elements with low vapor pressures were able to solidify at small distances from the Sun in the early life of a gaseous *pre-planetary nebula* surrounding the Sun. Initially the size of condensations may have been no bigger than dust grains, but these grains could have aggregated by successive collisions, some of which would have vaporized colliding grains, while others would have caused the grains to stick. Both vaporization and sticking would act to narrow the velocity ranges of successively condensing dust grains until they were able to clump gravitationally. As such clumps grew to 1000 km proportions, they could start sweeping up a wider swath of matter through their gravitational attraction, and grow at the expense of multitudes of smaller bodies. A more recently suggested alternative to this process is that the protoplanetary disk orbiting the star becomes unstable and fragments into rings that collapse to form planets. Both ideas are currently being pursued; both may contribute to a more complete understanding of planet formation.

The natural abundance of heavy elements found in the Sun appears to be mirrored in the composition of the Solar System as a whole. We know this for the terrestrial planets; and though the atmospheres of the giant planets appear primarily to contain hydrogen and helium — also in the abundances observed in the Sun — they no doubt harbor heavier elements deep in their interior.

1:11 The Galaxy and the Local Group

Some of the phenomenological evidence for the evolution of galaxies comes from the *Local Group* of galaxies. The Galaxy and the Andromeda Nebula appear to be a pair of gravitationally bound galaxies with a substantial number of smaller companions. By 1994, a careful search had displayed a Local Group numbering some 27 identified members (Table 1.5), with another 18 suspected. There are certain to be several others obscured by the absorbing matter within the Galaxy, and probably many more that have been too faint to detect by present-day techniques.

The group contains a number of *dwarf spheroidal systems*. These are very small galaxies, devoid of gas and dust, looking very much like extremely large globular clusters but with very low surface brightness. One of these systems, Fornax, contains five apparently normal globular clusters and, therefore, must be considered to be more like a galaxy than like a cluster of stars.

As will be discussed in Chapter 3, a loosely bound group of stars, such as any one of these spheroidal systems, cannot come too close to a massive, gravitation-

Table 1.5. Members of the Local Group of Galaxies Known in 1994 (after van den Bergh (vd68), (vd72), (vd94)).

Name	R.A. h m	Dec. °	Type	$M/M_\odot$	M_v	Distance kpc	(b/a)*	Radius pc
M31 = NGC 224	00 42.7	+41 16	Sb I-II	3.1×10^{11}	−21.1	690		
Galaxy	17 45.7	−29 00	Sb or Sc	1.3×10^{11}	−20.6?			
M33 = NGC 598	01 33.9	+30 39	Sc II-III	3.9×10^{10}	−18.9			
LMC	05 23.6	−69 47	Ir III-IV	6×10^9	−18.1	50		
M32 = NGC 221	00 42.7	+40 52	E2		−16.4			
NGC 6822	19 44.9	−14 46	Ir IV-V	1.4×10^9	−16.4			
NGC 205	00 40.3	+41 41	S0/E5p		−16.3			
SMC	00 52.8	−72 54	Ir or Ir IV-V	1.5×10^9	−16.2	60		
NGC 185	00 38.9	+48 20	DSph/DE3p		−15.3			
NGC 147	00 33.1	+48 31	DSph/DE5		−15.1			
IC 1613	01 04.9	+02 07	IrV	3.9×10^8	−14.9			
WLM = DDO 221	00 02.0	−15 28	Ir IV-V		−14.1			
Fornax	02 39.6	−34 31	DSph		−13.7	∼ 180	0.65	900
And I	00 45.7	+38 00	DSph		−11.8	40**		
And II	01 16.3	+33 25	DSph		−11.8	125**		
Leo I = Regulus Sy.	10 08.4	+12 18	DSph		−11.7	∼ 220	0.69	200
Aqr = DDO 210	20 46.9	−12 51	DIr		−11.5			
Sculptor	00 59.9	−33 42	DSph		−10.7	∼ 84	0.65	300
And III	00 35.3	+36 31	DSph		−10.3	60**		
Psc = LGS 3	01 03.9	+21 53	DIr		−10.2			
Sextans	10 13.0	−01 37	DSph		−10.0			
Phoenix	01 51.1	−44 27	DIr/DSph		−9.9			
Tucana	22 41.9	−64 25	Dsph		−9.5			
Leo II	11 13.5	+22 10	Dsph		−9.4	∼ 220	0.99	200
Ursa Minor	15 08.8	+67 07	Dsph		−8.9	∼ 67	0.45	200
Carina	06 41.6	−50 58	Dsph		−8.9			
Draco	17 20.2	+57 55	Dsph		−8.6	∼ 67	0.71	130

* (b/a) is the ratio of minor to major axes. Among the Galaxy's companions only Leo II appears round.
** distance from M31.

ally attracting center, before being pulled apart by the difference in the gravitational forces acting on its near and far sides. This is evidenced by the *Magellanic Stream*, an extended trail of gas falling into the Galaxy, apparently stripped from the Magellanic Clouds at some recent close approach to the Galaxy. Most of the other minor galaxies in the Local Group have apparently escaped this fate and are unlikely to ever have come very close either to the Galaxy or to the *Andromeda nebula, M31*.

Dwarf systems not tied to the Galaxy could be more or less uniformly distributed throughout the Local Group, and there could be some 200 of them. We would only see the nearest members because they are too faint to be seen far away. It may, however, be that all these objects are bound to either the Galaxy or to M31 and, in that case, the total number would be smaller. We would then suspect that such systems are formed at the edge of a galaxy, in some protogalactic stage, and that no close approach to the center ever occurred. Interestingly, the colors of the stars in the dwarf systems are rather different — and their H–R diagrams differ strongly — from H–R diagrams for components of the Galaxy. This indicates a different helium or metal abundance in the dwarf systems. That view is also supported by studies of individual variable stars in these objects.

In these systems that apparently have always been well isolated from the Galaxy itself, we therefore seem to have the interesting possibility of studying the evolution of stars having a different initial chemical composition from that found in most of the stars in the Galaxy. Not all these small galaxies are alike. While the dwarf spheroidal systems have no apparent gas content, the Magellanic Clouds are rich in gas with respective mass fractions of $\sim 10\%$ (LMC) and $\sim 30\%$ (SMC), considerably higher than the gas contents of the Galaxy and M31, respectively 3% and 1% by mass.

Related to the question of Local Group membership is the possible existence of globular clusters whose velocities are so great that even though the clusters are near to the Galaxy, they might not be physically bound unless the mass of the Galaxy is as high as $10^{12}\,M_\odot$ (Pe85). Tidal considerations also show that some globular clusters could never have been close to the Galactic center, suggesting a quite separate chemical evolution, independent of the evolution of the Galaxy.

Abundant evidence exists that the stars, at least in our Galaxy and in M31, have an increasing metal abundance near the center. The nuclear region appears to be particularly metal rich, and this seems to indicate that the evolution of chemical elements is somehow speeded up in these regions and is not uniform throughout the galaxy. The various chemically distinct stellar populations in the Galaxy may provide useful tests of the theory of buildup of chemical elements and perhaps offer insight into events that led to chemical differentiation, both within the Galaxy, and between the Galaxy and its chemically isolated companions.

1:12 The Formation of Large-Scale Structures

A major area of contemporary studies concerns the formation and evolution of large-scale cosmic structures. Matter is inhomogeneously distributed throughout the Universe. Much of the Universe is empty; but embedded in these empty spaces we find large aggregates of galaxies, intracluster and intercluster gas, and large voids nearly devoid of galaxies. At large distances we observe all of these in varying proportions, as in Fig. 1.11. A number of different physical processes compete in determining the

Fig. 1.11. A deep sky exposure obtained of the Galaxy Cluster 0024+1654 in Pisces. North is up and east is left. This massive cluster acts as a gravitational lens, magnifying a far more distant background galaxy whose image appears broken up by the lens into five elongated, ring-shaped arcs, three of which are highly magnified — two in the southeast and one in the northwest. The resolution is greater along the long axis of the arcs. The image also reveals the many small galaxies that apparently existed at early epochs. These evidently merged at a later epoch to form the sizeable galaxies we see today. These distant galaxies are largely blue, suggesting a population of luminous young stars. Hubble Space Telescope image, courtesy of NASA (Co96).

characteristics and makeup of these aggregates. The physical processes that lead to the substantial differences that distinguish globular clusters from galaxies, and individual galaxies from clusters of galaxies, are largely unknown. Though we speak of *hierarchical structure formation*, we do not know why there should be abrupt distinctions as we switch from a scale of 10^{20} cm to 10^{23} cm, to 10^{25} cm, and lastly to 10^{26} cm. These are the respective dimensions of *globular clusters, galaxies, clusters of galaxies, sheets or walls of galaxies*, and the *voids* they enclose (Fig. 1.12). Larger yet, by a factor of $\sim 10^2$, is the scale of the entire Universe whose horizon is roughly 10^{28} cm distant.

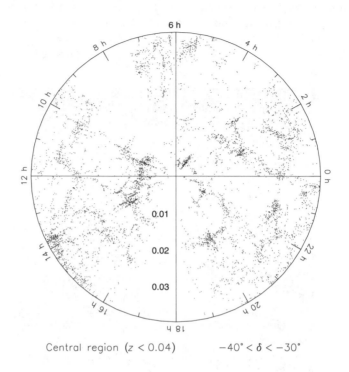

Fig. 1.12. The nearby structure of the Universe at red shifts $z < 0.04$, in a cone-shaped swath of the sky in the declination range $-40° < \delta < -30°$. This distribution of galaxies in right ascension and red shift was obtained at near-infrared wavelengths in the 2.2μm K-band, as part of the 6dF Galaxy Survey. This survey, carried out with the UK Schmidt Telescope in Australia, was obtained through a tiling of successive $6°$ fields (6dF) on the sky. A galaxy with a red shift $z = 0.01$ is expanding away from us at a speed of 1% the speed of light. At an expansion velocity of 70 km s^{-1} Mpc^{-1}, a galaxy at $z = 0.01$ lies at a distance of $\sim$43 Mpc. Clearly shown in this map are the walls of galaxies, and the voids they enclose. These are the largest structures observed in the Universe. (Note the two blank wedges, respectively around 8^h and 17^h, reflecting a lack of data.) Compare the structures seen here to those on scales a factor of 6 larger, in Fig. 13.6, and note both the similarities and differences. Courtesy of D. Heath Jones (Jo04).

Let us ask next how galaxy formation could have taken place in the Universe when it was rapidly expanding at all epochs. What gave rise to these condensations? How did they form? How did the many small galaxies revealed in Fig. 1.11 evolve into the large, fully formed galaxies that are abundant today, though rare at early epochs?

The formation of *quasars* also needs to be understood. Quasars are the most distant sources readily detected in the Universe. They are extremely luminous and rapidly consume their energy supply. We think they cannot shine for longer than a few million years at the highest observed luminosities $\sim 10^{48}$ erg s^{-1}. Whether isolated quasars can form directly from the intergalactic medium through contraction, or exist solely within galaxies is likely to be observationally answered within the next few years.

The puzzle of concentrating mass into quasars and galaxies in a universe that is rapidly expanding, appears to be resolved by postulating quantum fluctuations dating back to primordial times. The *inflationary origin* of the Universe proposes that the galaxies we observe today formed around such higher density regions after the Universe had sufficiently cooled, some 10^8 yr after its birth. The growth of these fluctuations which gravitationally attract ambient matter was helped by the prevalence of *dark matter*, so-called because it makes itself apparent only through its gravitational attraction for normal baryonic matter — stars, dust, and gas — but emits no apparently detectable electromagnetic radiation.

Rotation curves in spiral galaxies (Fig. 1.13). provide strong evidence for the ex-

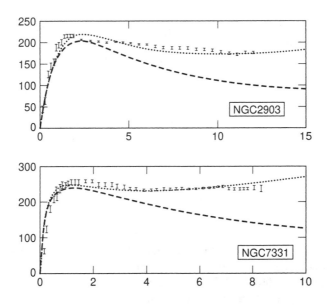

Fig. 1.13. Rotation curves for the spiral galaxies NGC2903 and NGC7331. Units on the abscissa are multiples of the radius of each galaxy's optical disk. Data points refer to radio observations of hydrogen gas velocities (in km s^{-1}). While the data show a flattened curve out to large distances from the galaxy's center, we would expect a slow drop in rotational velocity, as indicated by the dashed curves if mass in these galaxies were concentrated as centrally as the light-emitting stars. Although the stellar mass appears to determine the shape of the rotation curve in the inner parts of the galaxies, an additional factor, usually considered to be an as yet unspecified *dark matter* is required to account for the high rotational velocities at larger distances from the center. The dotted curves represent a suggested theoretical fit (Ma97).

istence of dark matter. For most spirals, the observed rotational velocities far from the center would be excessive if the only gravitationally attracting mass present were that of stars and gas clouds. Similarly, some central elliptical galaxies in large clusters would be unable to gravitationally bind their observed haloes of hot, X-ray emitting plasma, if the entire aggregate were not far more massive than its stellar component suggests. This has led to the postulate that an unknown form of matter, *dark matter*, pervades these galaxies out to large distances. Other alternatives, however, have also been proposed, including modifications to Newtonian dynamics and, in particular, the inverse square law of gravitational attraction over distances comparable to the dimensions of galaxies (Mi95a).

Accounting for the dark matter is currently one of the prime unsolved problems of astrophysics. If, as is widely assumed, dark matter rather than an entirely new theory of gravitation is involved, its prevalence is readily estimated. An examination of condensations on all scales suggests that the overall density of dark matter in the Universe is roughly five times that of *baryonic matter* — another name for atomic matter.

Equally demanding of attention as dark matter is the apparent ubiquity of an even larger contributor to the mass density of the Universe — *dark energy*. Dark energy makes itself apparent only through an acceleration it induces in the expansion of the Universe. *General relativity* can account for an energy term somewhat resembling dark energy. Einstein called it a *cosmological constant*, implying that its density should not vary as the Universe expands. We do not know whether dark energy remains constant or changes during cosmic expansion.

1:13 Black Holes

The observational evidence for stellar black holes has been accumulating for many years. Several X-ray binary stars are known in which an invisible star, whose mass is several times greater than the Sun's, is orbited by a visible companion. The invisible star is embedded in an accretion disk that emits X-rays as matter falls onto it from the companion. The emission tends to be sporadic and some such stars flare up as recurrent novae (Ca96), (Ka97), (Na97a). The X-Ray Nova XTE J11118+480 has a dwarf secondary that orbits the primary star at a velocity of $\sim$700 km s^{-1}, with a period of 4.08 hours, indicating that its primary has a mass of at least $6M_\odot$, which greatly exceeds the maximum mass a neutron star could have (Mc01). This and perhaps a dozen other such X-ray novae all appear to be stellar black holes.

We do not know how individual stars convert their cores into black holes. One uncertainty is the *equation of state*, the relation between pressure and density, for material at densities in excess of $\sim$10^{15} g cm^{-3}. At such densities and pressures the stiffness of the material, its ability to resist pressure before yielding, is not in hand. Knowledge of this property is critical to understanding whether, how, and when a massive star may collapse to form a black hole.

The evidence for massive black holes at the center of galaxies is also strong. The Galaxy, whose center is located at the position of the radio source Sgr A*, appears to house a black hole of relatively low mass $\sim 4 \times 10^6 M_\odot$ (Gh03). High velocities of stars orbiting the nuclei of other galaxies indicate the existence of compact central concentrations with masses of order $10^9 M_\odot$ or even higher, confined to regions perhaps less than a parsec in radius (Ta95). Such regions are not small enough to guarantee the existence of a black hole, but our observing techniques are not yet sufficiently refined to discern the actual size of the nucleus, which might be orders of magnitude smaller. Still, lower limits to the density of matter determined for such nuclear regions already exceed normal densities elsewhere in galaxies by many orders of magnitude. The mass density within the radio source Sgr A* at the center of the Galaxy exceeds $6 \times 10^{21} M_\odot$ pc^{-3} (See Figure 3.5c). In contrast, the mass density in the Galactic plane in the Sun's neighborhood is $\sim 0.15 M_\odot$ pc^{-3} (Sc03).

A black hole is formed whenever an aggregate of mass M is confined within a sphere of radius $R = 2MG/c^2$, where G is the gravitational constant and c is the speed of light. For a star five times as massive as the Sun, this radius is only 15 km. Before its collapse, the star's central density would have been $\sim 10^{15}$ g cm^{-3}. For a black hole of mass $10^9 M_\odot$, the radius is $\sim 3 \times 10^{14}$ cm, and its density is only 0.02 g cm^{-3} — comparable to the density of air at a pressure of a few atmospheres. If sufficient mass can gradually be added to the nucleus of a galaxy, while heat is radiated away, a massive black hole will ultimately form. Whether this happens in Nature is a different question. But the physics of such processes is more readily understood than the physics governing the formation of stellar black holes.

1:14 Magnetohydrodynamics and Turbulence

One vital theoretical tool that we still lack is an adequate theory of turbulence, particularly in partially or fully ionized gases permeated by magnetic fields. The absence of an adequate theory frustrates us on every level in astrophysics, denying us insight on the origins of cosmic magnetic fields, the formation of galaxies and galaxy clusters, stars and star clusters, and planets. Turbulent processes often dominate the transport of energy and matter and the shedding of angular momentum. Where magnetic fields are known to play a dynamic role, the mathematical problems of magnetohydrodynamic turbulence, and the generation and collapse of magnetic fields, are even more daunting.

Convection on all scales involves turbulence. In stars, heat and chemical elements are frequently transported to the surface by convection. In accretion disks around protostars, neutron stars, or black holes, turbulence may dominate the rate at which orbiting matter can lose angular momentum to spiral in and fall onto the central mass. The number of protostellar fragments into which cold, dense, contracting interstellar clouds divide, and the masses of stars that then form, hinge on turbulent processes. In protostellar or protoplanetary clouds the size distribution of turbulent eddies is likely to determine whether a binary stellar system forms or a group of planets is born orbiting a single star. Astronomical observations alone will not solve these problems. A most urgent need of theoretical astrophysics today is greater insight into turbulent processes.

1:15 Problems of Life

A fascinating problem of astrophysical science concerns the origin of life. Because physical and chemical methods have consistently shown themselves able to clarify biological problems, there now is great confidence that the origins of life, and the conditions under which life can originate, will some day be understood.

We do not know just how to define life and all it entails. Is a virus alive? Or is virus formation just a matter of the reproduction of rather complex forms, just as crystal formation is a reproduction of a complex form? To what extent are natural *mutation* and eventual death requisite features of living matter? Somewhere a line between animate and inanimate matter must be drawn, and we do not yet know just how to do that.

Even when we understand how to define life and living matter, we still will have to investigate whether entirely different physical or chemical bases of life might not be possible, and whether life on quite different scales might some day be found in the Universe.

Even in the more restricted problem of life as we know it on Earth, we are faced with formidable difficulties. We know of millions of different forms of life on our planet. We also know that species die out and that new, quite different, species are born. Why? Do conditions on Earth change sufficiently, so that the habitat becomes too unfriendly for one kind of life and more hospitable to another? Apparently!

The primitive Earth formed from a nebula surrounding the Sun, and had an atmosphere whose hydrogen and CO_2 content was far greater than today's. The form that life took at that time must have been entirely *anaerobic*. As the atmosphere slowly became rich in oxygen, and life changed to take advantage of oxygen as a source of energy, some anaerobes remained and sought refuge where oxygen could not penetrate and where competition from the *aerobes*, or oxygen-metabolizing organisms, was not severe (Op61a,b), (Sh66).

One of the interesting problems of astrophysics, then, is to try to understand the chemistry of the primitive Earth. By noting the overall composition of solar surface material and the chemical composition of the atmospheres of other planets where conditions may have always remained stable, we may come to understand what changes have taken place on Earth. The chemistry of comets may also help to produce an understanding of the initial conditions that existed on the young Earth.

Is life, even as we know it, abundant in the Universe? The answer to that question is still thoroughly speculative. If we estimated conservatively, we might suggest that life exists only on planets around stars having the same general characteristics as the Sun; but even then, we might need to postulate the existence of a planet just at the distance where water neither freezes nor boils. This is referred to as a star's *habitable zone*. Even when observations tell us the number of habitable planets around Sun-like stars, we will still need to estimate the likelihood of life spontaneously arising on such a planet.

Increasingly sophisticated laboratory tests are now possible. They seek to establish the kind of lifelike molecule that could occur under conditions assumed to have held on the primitive Earth. These experiments are beginning to synthesize lifelike primitive organisms out of component parts. But whether this will permit us to estimate the probability of spontaneous formation of life in Nature is not at all clear.

We do not yet know the full range of habitats that host life on Earth. Some bacteria thrive in hot springs where temperatures exceed $100\,°C$; they exist in the depth of the oceans and in volcanic openings; and they might exist deep inside the Earth where we have not had the means to find them.

Entirely different possibilities should also be considered. Perhaps life is more in the nature of an infection that, having started in a given planetary system, is then able to spread from one system to another, either through natural causes, or through the intervention of intelligent beings who would like to see life propagate over wide regions. If this were true, then life would have had to be formed only once, and from then on no further spontaneous formation would have been necessary. The search for a spontaneous origin of life on a primitive Earth would then be misguided.

The assumption of intelligent life existing in other regions of the Galaxy or the Universe is of course fascinating. Can we contact such life? How would we communicate? If an intelligent civilization far more advanced than ours exists, is it trying to communicate with us? Is there some unique best way of communicating, which a better understanding of physics and astrophysics will some day provide? Do we have to communicate by means of electromagnetic signals, or are there perhaps faster than light particles — tachyons — that we will discover later on and that would almost certainly be used by an intelligent civilization bent on saving time? These are highly speculative conjectures, but they should not be rejected out of hand.

If other civilizations exist, should we visit them, or is that even possible outside our Solar System? After all, the purpose of a visit is to see, talk, and touch; all of that could be done by improved communication techniques provided only that the distant civilization is able, and also willing, to communicate. There are relatively few things that cannot be settled that way, although without some actual exchange of mass, we probably could not decide whether a given civilization was made of matter or antimatter.

There are many fundamental questions of life on which astrophysics can throw new light and the interest of astrophysicists in biological problems is bound to increase in coming years.

1:16 Unobserved Astronomical Objects

In Appendix A we list a wide variety of astronomical objects, and we might think that we know enough to reasonably describe the world around us.

To avoid this trap of complacency, we should complete our list of astronomical objects by citing those that may not yet have been observed. We might think that this would be difficult; but it is not. To illustrate this, we first restrict ourselves to optical observations of diffuse objects. An extension to other techniques then becomes obvious.

We produce a plot comparing the absolute photographic magnitude and the logarithm of the diameter of different objects (Fig. 1.14). This was first conceived by

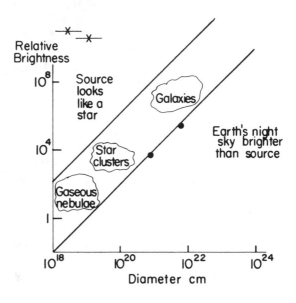

Fig. 1.14. Diagram showing the diameter–brightness strip onto which extended objects observed at visual wavelengths through the atmosphere tend to fall. Objects in the upper left-hand corner are compact and are not readily distinguished from ordinary stars. In the lower right-hand corner atmospheric night sky emission interferes with observations. The upper and lower crosses, respectively, represent the highly compact quasars 3C 273 and 3C 48. The upper and lower filled circles represent the Fornax and Draco galaxies — minor members of the Local Group of galaxies. (Based on a drawing by Arp (Ar65).)

Halton Arp (Ar65).

We note that all objects normally discovered in optical observations with ground-based telescopes have to lie on a strip between the two slanted lines on the Arp plot. Objects lying to the left or above this strip appear stellar. But because there are about 10^{11} stars that can appear on optical images obtained from within the Galaxy, abnormal or highly compact objects with a stellar appearance cannot readily be separated from bona fide stars without an inordinate amount of labor.

To detect something unusual about objects falling into this upper region on the chart, some other peculiar earmark must be found. Quasars, which lie above the strip, were first discovered by virtue of their radio emission, and were only later identified by their spectra as distant, red-shifted objects.

To the right and below the strip, the surface brightness of a diffuse object is so low that the foreground glow emitted by the night sky outshines the object, making it undetectable. Exceptions to this are Local Group minor galaxies such as Fornax and

Draco in which individual stars can be counted. If these objects were more distant, these stars would not be detected and the galaxies could not be discerned.

We note that the strip of observable objects covers only a small portion of the available area on the plot. This means that we have not yet had the opportunity to see many different varieties of objects that probably occur in Nature. It would be too much of a coincidence to expect all classes of objects in the Universe to fit neatly into a pattern defined by our own instrumental capabilities — and to fall onto the strip of observable sources in the Arp diagram. This point has recently been brought home through the use of novel, high-contrast techniques to search for faint diffuse galaxies. In the parts of the sky where such observations have been carried out, a substantial number of new galaxies have been discovered, suggesting that current catalogues may be missing at least one-third of all galaxies in the nearby universe (Sp97).

By taking instruments above the atmosphere in rockets and satellites, we are able to get above the atmosphere's night sky emission. The demarcation line on the right can therefore be moved downward and further to the right — though not very far. Sunlight-scattering interplanetary dust imposes a limit which is almost as severe as atmospheric emission. On the other hand, telescopes taken above the atmosphere avoid distortions produced by atmospheric scintillation. This permits them to obtain sharper images, and moves the line on the left of the strip upward and to the left. The combined effect is to widen the strip and to allow us to identify a larger variety of objects than are accessible from the ground. This is one reason for launching an observatory into an orbit high above the atmosphere. It accounts for some of the successes of the relatively small, 2.4 m Hubble Space Telescope, whose light-gathering power is more than an order of magnitude lower than that of the largest ground-based telescopes.

Of course, not all objects emit visible radiation, and so we cannot expect to find out all there is to know in astronomy simply by making visual observations. The ubiquitous background radiation reaching us from a time when the Universe was less than a million years old is only detected at microwave frequencies. Short bursts of gamma rays, lasting no longer than a few tens of seconds, flashing spasmodically at a rate of a few bursts per week, and showing no preference for any particular part of the sky, for thirty years could be detected only at gamma-ray frequencies. Not until early in 1997 were the first few *gamma-ray bursts* identified also at X-ray and optical wavelengths, and then only with great difficulty. Interstellar and circumstellar masers, with few exceptions, are detected only at radio wavelengths. They could not have been discovered at optical or X-ray frequencies. Nor would they have been identified without the very high spectral resolution that radioastronomical techniques permit. Quasars similarly might not have been identified had radioastronomers not evolved techniques that provided sub-arc-second images. Such techniques, however, would have been useless for detecting the microwave background radiation, which required high sensitivity to diffuse radiation reaching us isotropically — with equal intensity — from all parts of the sky.

These examples show how important it is to observe not only at all wavelengths transmitted across the Universe, but also at high and at low spatial resolving power, spectral resolution, and even time resolution.

We note from the highly subjective Fig. 1.15 that most of our knowledge about

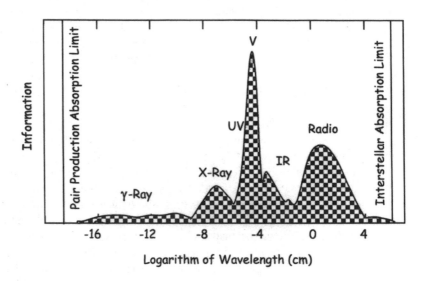

Fig. 1.15. Subjective drawing indicating the amount of information that has been gained through observations in the various portions of the electromagnetic spectrum. The ordinate has a quite arbitrary scale, probably more nearly logarithmic than linear. The peak *V* represents observations in the visual part of the spectrum.

the Universe still comes from visual and radio observations, mainly because more observations have been carried out in these two wavelength domains than in other parts of the spectrum, though a wide variety of infrared, extreme-ultraviolet, X-ray, and gamma-ray sensing instruments, have by now been placed in Earth orbit where they obtain a clear view of the Universe without atmospheric obstruction.

Our natural aim is to perfect observational techniques throughout (Ha81)*:

(a) The entire electromagnetic spectrum, going all the way from the lowest frequencies in the hundred kilocycle radio band, where interstellar plasma absorbs and blocks passage of radiation, up to the highest energy gamma rays. In some of these wavelength regimes the very structure and contents of the universe may limit the distances across which we are able to survey. Figure 1.16 illustrates this difficulty.

(b) The entire modulation frequency spectrum, going up to megacycle frequencies. Pulsars would never have been discovered had it not been for electronic innovations that permitted observations of intensity changes over millisecond time intervals. Using photographic plates, where exposure times of the order of an hour were representative, astronomers could not have discovered objects with periodic

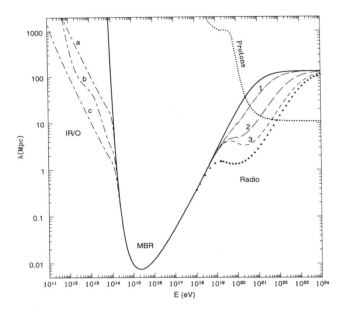

Fig. 1.16. Distances across the Universe from which highly energetic photons and protons are able to reach us. Collisions with the cosmic microwave background radiation (CMBR) break up energetic photons to produce electron–positron pairs and prevent penetration of the intergalactic medium from beyond the distances shown. Uncertainties in distance-limiting effects through collisions with infrared and optical (IR/O) or radio photons whose extragalactic fluxes are still quite uncertain, are shown, respectively, by the spread between curves a, b, and c, and curves 1, 2, 3 and the chain of triangles. The distances from which energetic protons can reach us are also limited by collisions with extragalactic photons. The highest-energy particles observed, to date, have energies of $\sim 3 \times 10^{20}$ eV, and could not have arrived from distances beyond ~ 30 Mpc (O'H98). (See also Sections 5:10 and 5:11.) (After Coppi and Aharonian (Co97)).

brightness undulations much shorter than an hour. At the other extreme, analysis of old photographic plates cannot yield discernible variations for phenomena whose period is far longer than a century, because photographic records only stretch that far back.

(c) The entire spatial frequency domain. As already indicated, many observing techniques are good for stellar or at least highly compact objects, but are not capable of detecting a uniform background. Other techniques permit background measurements but not the observation of faint compact objects. Until we have observed the entire range of possible angular sizes, from the highest angular resolution for detecting the most compact sources, down to the lowest resolving powers for detecting a uniform background, we may miss potentially interesting astronomical sources.

Table 1.6. The Mass–energy Density of Different Cosmic Constituents Averaged over the Entire Universe.[a]

Constituent	Fractional Density Parameter[b]	Equivalent Mass Density[b] g cm^{-3}
Dark Energy	0.72 ± 0.03	$(7 \pm 0.3) \times 10^{-30}$
Dark Matter	0.23 ± 0.03	$(2.2 \pm 0.3) \times 10^{-30}$
Primeval Electromagnetic Radiation	5×10^{-5}	4.9×10^{-34}
Primeval neutrinos	$(1.3 \pm 0.3) \times 10^{-3}$	$(1.3 \pm 0.3) \times 10^{-32}$
Binding Energy from Primeval Nucleosynthesis	-8×10^{-5}	-8×10^{-34}
Gravitational Binding from Primeval Structure	$(-8 \pm 1) \times 10^{-7}$	$(-8 \pm 1) \times 10^{-36}$
Total Baryonic Rest–Mass	0.045 ± 0.003	$(4.4 \pm 0.3) \times 10^{-31}$
Warm Intercluster Plasma	0.40 ± 0.003	$(3.9 \pm 0.3) \times 10^{-30}$
Intracluster Plasma	$(1.8 \pm 0.7) \times 10^{-3}$	$(1.8 \pm 0.7) \times 10^{-32}$
Main Sequence Stars in Spheroids and Bulges	$(1.5 \pm 0.4) \times 10^{-3}$	$(1.5 \pm 0.4) \times 10^{-32}$
Main Sequence Stars in Disks and Irregulars	$(5.5 \pm 0.4) \times 10^{-4}$	$(5.4 \pm 0.4) \times 10^{-33}$
White Dwarfs	$(3.6 \pm 0.8) \times 10^{-4}$	$(3.5 \pm 0.7) \times 10^{-33}$
Neutron Stars	$(5 \pm 2) \times 10^{-5}$	$(5 \pm 2) \times 10^{-34}$
Black holes	$(7 \pm 2) \times 10^{-5}$	$(7 \pm 2) \times 10^{-34}$
Substellar Objects	$(1.4 \pm 0.7) \times 10^{-4}$	$(1.4 \pm 0.7) \times 10^{-33}$
Planets	10^{-6}	10^{-35}
H I + He I	$(6.2 \pm 1) \times 10^{-4}$	$(6 \pm 1) \times 10^{-33}$
Molecular Gas	$(1.6 \pm 0.6) \times 10^{-4}$	$(1.6 \pm 0.6) \times 10^{-33}$
Dust	$(2.5 \pm 1.2) \times 10^{-6}$	$(2.4 \pm 1) \times 10^{-35}$
Baryons Sequestered in Massive Black Holes[c]	$4 \times 10^{-6}/(1 - \epsilon)$	$4 \times 10^{-35}/(1 - \epsilon)$
Binding Energy from Gravitational Settling	-10^{-5}	-10^{34}
Binding Energy from Stellar Nucleosynthesis	-5×10^{-6}	-5×10^{-35}
Radiant Energy Originating in Stars	2×10^{-6}	2×10^{-35}
Neutrinos from Stellar Core Collapse	3×10^{-6}	3×10^{-35}
Cosmic Rays and Magnetic Fields	$\sim 5 \times 10^{-9}$	$\sim 5 \times 10^{-38}$
Kinetic Energy in the Intergalactic Medium	$(1 \pm 0.5) \times 10^{-8}$	$(1 \pm 0.5) \times 10^{-37}$

[a] Based on a more comprehensive tabulation by Fukugita and Peebles (Fu04).

[b] Based on a Hubble constant $H_0 = 70$ km s^{-1} Mpc^{-1} and a Euclidean model of the Universe, density parameter $\Omega_0 = 1$.

[c] ϵ is the mass–energy radiated away in the formation of the black hole.

(d) The entire set of communication channels: electromagnetic and gravitational radiation, cosmic rays, neutrinos, and, if they exist, any others. These channels again can be expected to exhibit the existence of new phenomena in a Universe rich far beyond our most adventurous speculation.

(e) Some astronomical objects, however, may be hard to find without exploratory voyages. If 10% of the mass of our Galaxy consisted of snowballs (fist-sized chunks of frozen water freely floating through interstellar space) we would never know it. The amount of light scattered from these objects would be too low to make them detectable. They would not be able to penetrate the Solar System without evaporating in sunlight. Only when spaceships began travel beyond the Solar System would they be detected, and then as a major nuisance. A spaceship moving at nearly the speed of light could be completely destroyed on colliding with one of these miniature icebergs.

(f) Another set of uncertainties greets us when we analyze those aspects of the Universe that our observations do provide. As Table 1.5 shows, more than 95% of the mass–energy in the Universe appears to consist of two components, neither of which fits into current theories of matter and radiation. A predominant fraction, ~72% of the mass–energy density governing the expansion of the Universe is in dark energy. Another ~23% consists of dark matter. Less than 5% of the mass–energy density of the Universe is in baryonic matter. The mass density due to electromagnetic radiation, as seen from the table is minute even when compared to the small fraction found in baryonic matter. If we understand the physics of less than 5% of the contents of the Universe, does this mean that our understanding of the Universe is also less than 5%?

From this perspective, it almost seems premature to construct sophisticated cosmological theories and cosmic models. On the other hand, these theories and models often suggest novel observational tests that produce new results. We should therefore think of astrophysical theory not so much as a structure that summarizes all we know about the Universe. Rather, it is a continually changing pattern of thought that permits us to find our way forward. The compilation of Table 1.6 can help us in this effort by at least exhibiting what we do and what we do not know.

Now that we have worked our way through the landscape of the Universe as it appears to us today, and have found where much of the uncharted territory lies, it is time to start learning the use of the tools that have brought us this far. These will help us to push on farther to see how much more we may learn about the grand structure of the Cosmos, its origins, and its evolution.

The chapters ahead of us will provide a guide.

2 The Cosmic Distance Scale

2:1 Size of the Solar System

A first requirement for the establishment of a cosmic distance scale is the correct measurement of distances within the Solar System. The basic step in this procedure is the measurement of the distance to Venus. A precise way of obtaining this distance is through the use of radar techniques. Another method is described in Problems 2–2 and 2–3 at the end of this chapter.

A radar pulse is sent out in the direction of Venus, and the time between its transmission and reception is measured. Since time measurements can be made with great accuracy, the distance to Venus and the dimensions of its orbit can be established to within a kilometer — a precision of one part in a hundred million.

Once the distance to Venus is known at closest approach a, and most distant separation b, and these measurements are repeated over a number of years, the diameters and eccentricities of both the orbits of Earth and Venus can be computed. The mean distance from the Earth to the Sun is then directly available as the mean value of $(a + b)/2$ (Fig. 2.1). This distance is called the *astronomical unit, AU*

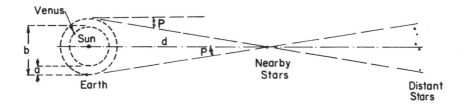

Fig. 2.1. Measurement of the astronomical unit and trigonometric parallax.

(1 AU $\sim 1.5 \times 10^{13}$ cm). A check on the Earth–Venus distance is obtained from trajectories of space vehicles sent to Venus.

2:2 Trigonometric Parallax

When observations are made from opposite extremes in the Earth's orbit about the Sun, a nearby star appears displaced relative to more distant stars in the same part of the sky. The *parallax p* is defined as half the apparent angular displacement measured in this way. The distance d to the star is then

$$d = \frac{1\,\text{AU}}{\tan p} \tag{2-1}$$

or

$$d = 1.5 \times 10^{13}(\tan p)^{-1}\,\text{cm}.$$

A star whose parallax is one second of arc is at a distance of 3×10^{18} cm, since $\tan 1'' = 5 \times 10^{-6}$. This distance forms a convenient astrophysical unit of length, and is called the *parsec*, pc. 1 pc $= 3 \times 10^{18}$ cm.

With observations from the *Hipparcos* satellite, which has obtained the most accurate positional observations on more than 120,000 stars, the trigonometric parallax can be reliably determined with an overall accuracy of ~ 0.001 arcsec (Pe95). At distances of about 100 pc this yields an accuracy of $\sim 10\%$.

2:3 Spectroscopic Parallax

Once the distance to nearby stars has been determined, we can correlate absolute brightness with spectral type. Bright stars of recognizable spectral type then become distance indicators across large distances where only the brightest stars are individually recognized and a trigonometric parallax is too small to be measured.

2:4 Superposition of Main Sequences

This method is based on the assumption that main sequence stars have identical properties in all galactic clusters. This means that the slope of the main sequence is the same, and that main sequence stars of a given spectral type or color have the same absolute magnitude in all clusters (see Fig. 1.3). On this assumption we can compare the brightness of the main sequence of the Hyades cluster of stars (Figure 2.2) and any other galactic cluster. The vertical shift necessary to bring the two main sequences into superposition gives the relative distances of the clusters.

PROBLEM 2-1. (a) If the shift in apparent magnitudes is $\Delta m = m_{GC} - m_{Hya}$ show that the relative distances are

$$\Delta m = 5 \log \frac{r_{GC}}{r_{Hya}} + A', \tag{2-2}$$

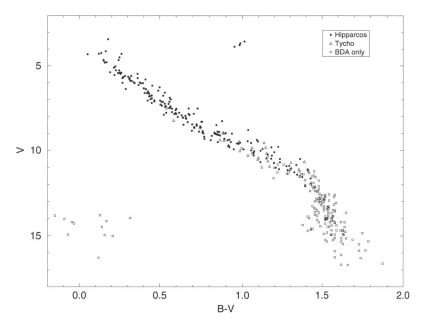

Fig. 2.2. Color-magnitude diagram of the Hyades Cluster. Most of the data shown were obtained with the Hipparcos satellite, as indicated by the filled circles. Data from two other data bases, the Tycho catalogue, and the Bas des Amas, BDA, are included and show the fainter end of the main sequence. Stars at the lower left are dwarfs. Stars at upper right lie on the red-giant branch. (Courtesy of Michael Perryman (Pe98a))

where A' is a correction for the difference in interstellar reddening of the galactic cluster, GC, and the Hyades. The derivation is analogous to the work leading to equation (A–2). The factor A' can be determined through use of stellar line spectra as explained in Section A:8.

(b) In Chapter 3 we will see how spectroscopic measurements on visual binaries — binary stars whose orbits can be clearly resolved — can provide complete information on orbital velocities and separations of the two stars, and thereby also their distances from Earth (see Fig. 3.5 (b)). The distance to such a Hyades cluster binary has been determined in this way and found to be 47.8 ± 1.6 pc (To97). Assuming that interstellar reddening can be neglected for nearby clusters, compare the main sequences for the Hyades and Pleiades clusters (Figs. 2.2 and A.2). Convince yourself that it makes sense for the Pleiades to be about 2.5 times more distant. The distance to the Pleiades has been estimated to be 116 pc.

To measure the distance to globular clusters, which generally lie much farther away, we can proceed on one of three different assumptions:

(a) The Hertzsprung–Russell diagram of the globular cluster has a segment that runs essentially parallel to the galactic cluster main sequence. We can assume that this segment coincides with the main sequence of the Hyades cluster. The distance of the globular cluster can then be calculated in terms of equation (2–2).

(b) Alternatively we can assume that the segment coincides with the main sequence defined by a group of dwarf stars in the Sun's immediate neighborhood. The distance to these dwarf stars is determined by trigonometric parallax.

(c) Finally we can assume that the mean absolute magnitude for short-period variables (RR Lyrae variables) is the same in globular clusters and in the solar neighborhood (see Section 2:5, below).

None of these three choices is safe in itself. However, when applied to the globular cluster M3 (Fig. 1.7), all three methods give distance values in fair agreement with each other. This verifies that the main sequences of different groupings of stars coincide reasonably well and can be used as distance indicators.

2:5 RR Lyrae Variables

We find that the apparent magnitudes of all RR Lyrae variables in a given globular cluster are the same regardless of the variables' periods, though the metallicity of the stars in any one cluster — the abundance, for example, of iron observed in the atmospheres of such stars — does affect their periodicity. Because these stars are intrinsically luminous, and because their short pulsation periods make them stand out, they serve as ideal distance indicators. We assume that the absolute magnitudes of these stars is the same not only within a given cluster, but also elsewhere. The relative distances of two clusters can then be determined by the inverse square law corrected for interstellar extinction (equation (2–2)).

2:6 Cepheid Variables

At the end of the nineteenth century, Cepheid variables in the Magellanic Clouds were found to have periods that are a function of luminosity. The Magellanic Clouds are dwarf companions to the Galaxy. They are small galaxies in their own right and are sufficiently compact so that all their stars can be taken to be at essentially the same distance from the Sun. By comparing the magnitudes of Cepheids in the Magellanic Clouds to those in globular clusters, one was able to obtain relative distances to these objects.

However, there was a pitfall. The Cepheids in the Magellanic Clouds are *Population I* stars — stars normally found in the disk of a galaxy. Globular clusters, on the other hand, belong to the halo *Population II* component that is more or less spherically distributed about the center of a galaxy.

In 1952, Walter Baade analyzed the magnitudes of Cepheids in the Andromeda nebula, comparing Population I with Population II regions. He found that Population

I Cepheids were about 1.5 magnitudes brighter than Population II Cepheids. The *distance modulus* of M31 had previously been derived by comparing these brighter Cepheids with type II Cepheids in clusters within our own Galaxy. The distance to M31 had therefore been erroneously underestimated by a factor of two. Baade's measurements showed that this distance and the distance to all other galaxies had to be doubled.

2:7 Novae and HII Regions

Novae have an absolute magnitude that is related to the decay rate of the luminosity after an outburst. The great intrinsic brightness of a nova makes it a useful distance indicator for nearby galaxies.

The diameters of bright HII (ionized hydrogen) regions also form good yardsticks by which to judge the distances of such galaxies.

2:8 Supernovae

Supernova outbursts have, by now, been observed in distant galaxies red-shifted by as much as $z = 1.75$ (Ri01). They exploded at an epoch when the Universe was only half its present age. While there are several different types of these highly luminous explosions, supernovae of type SN Ia have a predictable luminosity if the color and rate of decline of luminosity are taken into account. By making observations at different wavelengths it is also possible to calibrate out any absorption by dust in the distant galaxy that might dim the apparent luminosity of the supernova and lead to a false distance measure (see Fig. 11.6) (Pe98b, Ri95).

2:9 The Tully–Fisher and Faber–Jackson Relations

The Tully–Fisher relation is an empirical relation between the apparent magnitude of a galaxy and its rotational velocity as measured by the Doppler width of observed spectral lines. Figure 1.13 shows that rotational velocities are fairly uniform out to quite large distances from a galaxy's center. They also seem to be well correlated with the luminosity of a galaxy, even though this luminosity is not directly related to the galaxy's total mass that determines rotational speeds. Figure 2.3 shows the Tully–Fisher distance calibrated against the distance judged from observations on Cepheid variables. The agreement is quite good.

A similar empirical relationship holds for gas-poor elliptical galaxies. There, the widths of stellar absorption lines, Doppler broadened by the stars' random velocities in the galaxy's gravitational field, are a measure of the galaxy's absolute magnitude in blue light, M_B (Fa76). Neither relationship has a satisfactory explanation, particularly since the velocities depend on mass, determined by the dominance of dark matter, whereas the light output of a galaxy is determined by the starlight it emits.

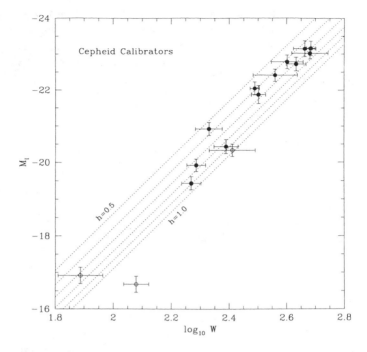

Fig. 2.3. Comparison of the Tully–Fisher relation and Cepheid distance indicators. M_I is the galaxies' I-band absolute magnitude (see Table A.1) based on Cepheid calibration of the galaxies' distances. W is the rotational velocity width of atomic hydrogen spectral lines measured in km s^{-1} (see Fig. 1.13). The dotted lines $h = 0.5$ and $h = 1.0$, respectively, indicate a Hubble constant of 50 and 100 km s^{-1} Mpc^{-1}. The interspersed dotted lines correspond to values of the Hubble constant spaced 10 km s^{-1} Mpc^{-1} apart. Open circles denote galaxies for which the Tully–Fisher relation should not be applied (Gi97).

2:10 Distance–Red-Shift Relation

The distances to galaxies can be gauged through a comparison of bright sources that populate them. Suitable candidates are O stars, novae, Cepheid variables, and HII regions. These individual objects can be detected to distances about as far out as the nearer Virgo cluster galaxies. By comparing the distances estimated from the apparent magnitudes of such stars and the sizes of HII regions, it is possible to show that the spectral red shift of light from these galaxies is linearly related to distance: $\Delta\lambda/\lambda \equiv z \propto r$.

We can also compare the magnitudes of individual galaxies to estimate relative distances. Here we must be careful to compare galaxies of the same general type. To minimize errors due to statistical variation in brightness, we sometimes compare not the brightest, but rather, say, the tenth brightest galaxies in two different clusters. By this device we hope to avoid selecting galaxies that are unusually luminous.

The data show a linear distance–red-shift relation (see Fig. 11.6). It is not clear how far this linearity persists, but for many cosmological purposes we use the red shift as a reliable indicator of a galaxy's distance. This procedure may not be sufficiently accurate, even in our own part of the Universe, wherever massive clusters attract galaxies and accelerate them to high velocities.

We should still note that distance measurements are not easy, and that errors cannot always be avoided. In 1958, Sandage (Sa58) discovered that previous observers had mistaken ionized hydrogen regions for bright stars. This had led them to underestimate the distance to galaxies by a factor of ~ 3 beyond the error previously unearthed by Baade. Within a space of five years the dimensions of the Universe therefore had to be revised upward by a total factor of ~ 6. It is not unlikely that, from time to time, other corrections may lead to further revisions of the cosmic distance scale. However, Fig. 2.4 shows that we can frequently check astro-

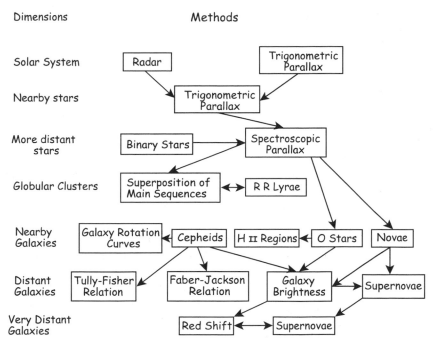

Fig. 2.4. Flow chart of distance indicators.

nomical distances by several different methods, and eventually we should be able to derive a reliable distance scale. At present, a red-shift velocity of $7000 \, \text{km s}^{-1}$ is estimated to indicate that a galaxy is at a distance of $\sim 100 \, \text{Mpc}$. The velocity–distance proportionality constant — the *Hubble constant, H,* named after Edwin Hubble who, in 1929, convincingly showed that the Universe is expanding — has a value $H \sim 70 \,\mathrm{km}\,\mathrm{s}^{-1}\,\mathrm{Mpc}^{-1}$. We will adopt this value for purposes of estimates throughout the book, but uncertainties in the value are of order $\pm 15\%$.[1]

2:11 Distances and Velocities

It is important to have several different ways of measuring extragalactic distances. Galaxies are known to have random velocities of many hundreds of kilometers per second, meaning that their red shifts can give distance measures in error by as much as $\gtrsim 10 \,\mathrm{Mpc}$. By making use of Cepheid variable and supernova distance indicators, we are able to place much better constraints on the distances to individual galaxies than red shifts can provide. Use of these more reliable distance indicators, together with red shift data, has further permitted the charting of large-scale flows of galaxies. We find that members of our Local Group and other nearby galaxies are systematically streaming toward a region of the sky, now named the *Great Attractor* on the assumption that the galaxies are gravitationally attracted toward this region. The approximate direction of this stream is shown in Fig. 12.1.

2:12 Seeliger's Theorem and Number Counts in Cosmology

Once we know the distances to the various galaxies, we can estimate typical intergalactic distances and typical number densities of galaxies. The variation of number density with distance can, in principle, be used to determine the geometric properties of the Universe. By such means we may hope to determine whether the Universe is open or closed, and whether it is finite or infinitely large. We will return to such questions in Chapters 11 to 13, but a simple argument based on Euclidean geometry is worth keeping in mind.

If a set of emitting objects is homogeneously distributed in space, then the ratio of the number of objects whose apparent magnitude is less than m to those whose apparent magnitude is less than $(m-1)$ is $N_m/N_{m-1} = 3.98$. This is called *Seeliger's theorem*. Let us see how this result is obtained.

Let $m-1$ be the apparent magnitude of a star at distance r_1 (see Fig. 2.5). Then the distance r_0 at which its apparent magnitude would be m is $r_0 = (2.512)^{1/2} r_1$. At that distance its apparent magnitude is reduced by $(r_0/r_1)^2 = 2.512$, as follows directly from our definition of the magnitude scale in Section A:7.

If stars are uniformly distributed in space and have a fixed brightness, they will appear brighter than apparent magnitude m out to a distance r_0, but brighter than

---

[1] Because of such uncertainties, astronomers often specify the assumed Hubble constant on which their calculations have been based, and write, for example, the abbreviation $h = 0.7$ or perhaps $h = 0.75$ to denote that they have adopted a Hubble constant of, respectively, 70 or 75 km s^{-1} Mpc^{-1}. If the assumed value of h is specified in this way, a computed result may later be recalculated when a better value of h becomes established. This notation can be found in Figs. A.7 and 13.9, among others.

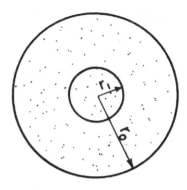

Fig. 2.5. Diagram to illustrate Seeliger's theorem.

$m - 1$ only out to a distance r_1. The ratio of the number of stars brighter than a certain magnitude, N_m/N_{m-1}, is proportional to the volume occupied.

$$\frac{N_m}{N_{m-1}} = \frac{r_0^3}{r_1^3} = (2.512)^{3/2} = 3.98. \tag{2-3}$$

Because this is true for stars of any given magnitude, it will also be true for any homogeneous distribution of stars, regardless of their luminosities. Equation (2–3) states that the flux obtained from a source is proportional to r^{-2}, and the number of sources observed down to a limiting flux density is proportional to r^3. Here, we define the *flux density* $S(\nu)$ at spectral frequency ν, as the energy received from a source in unit time, per unit telescope collecting area, and in unit spectral frequency band at frequency ν. Hence, the number of sources that have a flux density greater than $S(\nu)$ is

$$N \propto S(\nu)^{-3/2} \quad \text{because} \quad N \propto r^3 \quad \text{and} \quad S(\nu) \propto r^{-2}. \tag{2-4}$$

This proportionality, which already was of interest in classical stellar astronomy, has become even more important in modern cosmology, where it is usually found in a somewhat different form. If we take the logarithm of both sides of equation (2–4) we find

$$\log N \propto -\frac{3}{2} \log S(\nu). \tag{2-5}$$

A comparison of $\log N$ and $\log S$, often called the $\log N - \log S$ plot, shown in Figure 2.6, means this: if the logarithm of the number of sources brighter than a given magnitude is plotted against the logarithm of the flux density at the spectral frequency at which the instrument operates, then the slope of the plot should be constant, with a value of $-\frac{3}{2}$, provided: (a) the sources are homogeneously distributed in space; (b) space is Euclidean; and (c) we compensate for any cosmic red shift in apparent brightness. This latter requirement comes about because observations are made at one given frequency ν. If the source is intrinsically very bright at high frequencies, then red shift to lower frequencies would make it look deceptively bright.

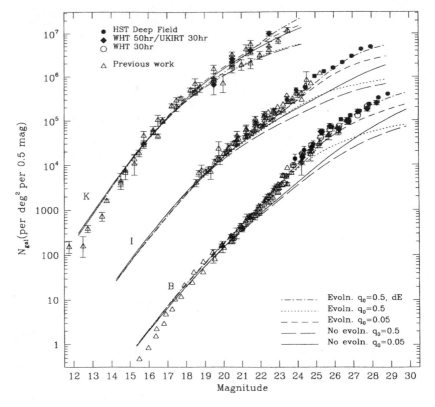

Fig. 2.6. Plot of $\log N$ against $\log S$, where N is the number of sources per unit solid angle at or below a given magnitude and $\log S$ is the magnitude of the source, which is already logarithmic. The plot shows data in the K and I infrared bands (see Table A.1) and in the blue band B. For clarity, the I-band counts have been multiplied by a factor of 10 and the K-band counts by a factor of 100. At large distances and correspondingly faint magnitudes, the deviation from a slope of $\frac{3}{2}$ is appreciable. The dashed and dotted lines refer to different cosmological models. The parameter q_0 is a measure of the acceleration or deceleration of the cosmic expansion; we will encounter it again later in Section 11:6. Some of the curves take into account that galaxies evolve — even though we do not yet know precisely how; others assume galaxies always had their present-day appearance (Me96). Reprinted with permission from *Nature* © 1996 Macmillan Magazines Limited.

The correction for such expected spectral features of a source is called a *K correction*. A further correction is also required because a cosmically red-shifted source already appears weaker just from the time dilation effect, that is, from the increased apparent spacing between the emission times of photons. How this correction is to be made is deduced in Sections 11:6 to 11:9.

In Section 11:7 we will show how number counts are to be calculated in cosmological models that are non-Euclidean. Actual observations, however, indicate that care must be taken in the interpretation of data, mainly because galaxies now are known to substantially evolve during the time it takes for their radiation to reach us. Small galaxies merge to form larger, often more luminous galaxies. At different distances, then, one is counting quite different types of galaxies.

Problems Dealing with the Size of Astronomical Objects

The methods described in the following problems are not those normally used by astronomers. However, they allow us to obtain insight into the dimensions of planetary and stellar systems without recourse to the more sophisticated methods covered in this chapter. The first seven steps were already known to Isaac Newton (Ne–).

2–2. The distance, R, to Venus can be obtained by triangulation when Venus is at its point of closest approach. Two terrestrial observers separated by 10^4 km along a line perpendicular to the direction of Venus find the position of Venus on the star background to differ by $49''$ of arc. Calculate the distance of Venus at closest approach.

2–3. At this distance, the angular diameter of Venus is $64''$, whereas at greatest separation its angular diameter is $10''$. Assume that the orbits of both the Earth and Venus are circular and concentric, and compute the two orbital radii.

2–4. The mean angular diameter of Saturn at smallest separation is about 1.24 times as great as at largest separation. (These mean angular diameters have to be averaged over many orbital revolutions, because Saturn's orbit about the Sun is appreciably eccentric). Calculate the semimajor axis, a, of Saturn's orbit about the Sun.

2–5. Both the Sun and the Moon subtend an angular diameter of $\sim\frac{1}{2}^{\circ}$ at the Earth. The lunar disk at full moon is only about 2×10^{-6} as bright as the Sun's disk. Knowing that the Moon is much nearer the Earth than the Sun, compute the reflection coefficient K of the lunar surface, assuming that the light is reflected isotropically, into 2π sterad. Show that this reflection coefficient is appreciably lower than that of terrestrial surface material (which is estimated to have a mean reflection coefficient of order 0.3). Actually the Moon scatters light mainly in the backward direction, so the result obtained here gives an artificially elevated value for K.

2–6. Assume that Saturn subtends an angular diameter of $\sim 17''$ at the Sun. Let its distance from both the Earth and the Sun be considered to be 9.5 AU. If the light received from Saturn is 0.86×10^{-11} that received from the Sun, compute the reflection coefficient of Saturn's surface. Note that Saturn is known to shine primarily by reflection, since its moons cast a shadow on the surface when they pass between the planet and the Sun.

2–7. Saturn appears to emit 0.86×10^{-11} as much light as the Sun. How far would the Sun have to be removed from the Earth to appear to have a magnitude identical to that of Saturn, that is, to appear as a first magnitude star?

2–8. Assuming the Sun to be a typical star, we conclude that the nearest stars are of the order of 5×10^{18} cm distant. We further assume that this is the characteristic distance between stars in the disk of the Andromeda spiral galaxy M31. We note that

M31 appears to be a system viewed more or less perpendicular to the disk containing the spiral arms. Other spiral galaxies viewed in profile indicate that the thickness of the disk is about $0.003L$, where L is the diameter of the galaxy. In terms of the distance D of M31, show that the flux received would be

$$\sim \left(\frac{0.003SL^3}{D^2} \right) \left(\frac{\pi}{4} \right) \left(\frac{1}{5 \times 10^{18}} \right),$$

where S is the flux we would expect to receive from the Sun if it were 5×10^{18} cm from Earth.

2–9. If the bright central region of M31 subtends an angular diameter of $3°$ at Earth, and if the galaxy is a fifth magnitude object, calculate the galaxy's distance. Show that this region's diameter is ~ 5 kpc. (Note that the full diameter of M31 is actually an order of magnitude larger.)

2–10. Find the distance of the smallest resolved galaxies, on the assumption that all spiral galaxies are of the size of M31 and that space is Euclidean. The smallest extragalactic sources resolved with currently available telescopes are of the order of $0.05''$ of arc in diameter.

2–11. Olbers's paradox: Let there be n stars per unit volume throughout the Universe.

(a) What is the number of stars seen at distances r to $r + dr$ within a solid angle Ω?

(b) How much light from these stars is incident on unit area at the observer's position, assuming each star to be as bright as the Sun?

(c) Integrating out to $r = \infty$ how much light is incident on unit detector area at the observer?

This problem will be discussed at length in Chapter 11.

Answers to Selected Problems

2–2. $R \sim 4.2 \times 10^7$ km.

2–3. $R_e \sim 1.5 \times 10^8$ km, $R_v \sim 1.1 \times 10^8$ km.

2–4. $(a + 1)/(a - 1) = 1.24$. Hence $a \sim 9.5$ AU.

2–5. If $L_\odot$ is the solar luminosity, r is the radius of the Moon, and R is the distance of the Moon — and Earth — from the Sun, then $S = (\pi r^2/4\pi R^2)L_\odot$ is the radiation accepted by the Moon. This light is spread into a 2π solid angle so that, at the distance D from the Moon, the flux per unit area is $(K \cdot S)/2\pi D^2$, which has to be compared with $L_\odot/4\pi R^2$ coming directly from the Sun.

$$\therefore \quad \frac{Kr^2}{2D^2} = 2 \times 10^{-6} \quad \text{and} \quad K \sim 0.2.$$

2–6. Saturn's diameter is $2r \sim 7.8 \times 10^{-4}$ AU.

$$\therefore \quad \frac{\dfrac{\pi r^2 L_\odot}{4\pi(9.5)^2} \cdot \dfrac{K}{2\pi(9.5)^2}}{\dfrac{L_\odot}{4\pi(1)^2}} = 0.86 \times 10^{-11}.$$

Hence $K \sim 0.90$.

2–7. The distance at which the Sun would appear to be a first magnitude star is $r \sim 5 \times 10^{18}$ cm.

2–8. If $L_\odot$ is the Sun's luminosity, and D is the distance, the flux from the galaxy is

$$\frac{(\text{volume of galaxy})(\text{number density})L_\odot}{4\pi D^2}$$

$$\sim \frac{(\pi/4)(L^2)(0.003L) \cdot \left(\dfrac{1}{5.2 \times 10^{18}}\right)^3 \cdot L_\odot}{4\pi D^2}$$

$$= \frac{(\pi/4)(L^2)(0.003L)}{D^2} \cdot \frac{S}{(5.2 \times 10^{18})}.$$

2–9. Comparing the magnitude of M31 to a first magnitude star, and taking $\theta = 3/57 = L/D$ we see from Problem 2–8 that $L \sim 5$ kpc, $D \sim 0.1$ Mpc. The actual distance to M31 is given in Table 1.5.

2–10. Distance $\sim 6 \times 10^{28}$ cm. This is at a distance at which galaxies would be receding at a speed appreciably exceeding that of light. With 0.05 arc second resolution, galaxies of the size of M31 can be resolved at all distances at which they are not excessively red-shifted.

2–11. (a) $\Omega n r^2 \, dr$.

(b) $\Omega n r^2 \, dr \dfrac{L_\odot}{4\pi r^2} = \dfrac{\Omega n}{4\pi} L_\odot \, dr$.

(c) The integral of (b) out to infinity would diverge if distant stars were not eclipsed by nearer stars. When eclipses are taken into account, the flux at the observer is finite and is equal to the flux at the surface of the Sun.

3 Dynamics and Masses of Astronomical Bodies

The motion of astronomical bodies was first analyzed correctly by Isaac Newton (1642-1727). He saw that a variety of apparently unrelated observations all had common features and should form part of a single theory of gravitational interaction. To formulate the theory, he had to invent mathematical techniques that described the observations and showed their interrelationship. His struggles with the mathematical problems are recorded in his book *Principia Mathematica* (Ne–).

The intervening three centuries since Newton's discoveries have allowed his mathematical formulation to be streamlined, so that it can now be presented in brief form; but the underlying astrophysics remains unchanged.

The aim of this chapter will be to show how astronomical observations lead to the conclusions reached by Newton. We will then show the importance of Newtonian dynamics in determining the masses of all astronomical objects. It is interesting that a correct evaluation of these masses was not obtained until more than a century after Newton's work. We will discuss the gravitational interaction of matter with antimatter and finally mention some of the limitations of Newton's work.

3:1 Universal Gravitational Attraction

A number of astronomical observations and experimental results were known to Newton when he first tried to understand the dynamics of bodies. Many of the experimental results dealing with the motion of falling bodies had been found by Galileo (1564–1642) (Ga–). The astronomical observations, which treated the motions of planets, had been gathered over many years by Tycho Brahe (1546–1601). Johannes Kepler (1571–1630) had then analyzed these data and summarized them in three empirical laws. Newton postulated that the work of Kepler and of Galileo was related. We will not retrace his reasoning here, but rather will outline the evidence with some of the advantages of three centuries of hindsight.

We know from experiments with sets of identical springs and sets of identical masses that a single mass accelerated through the release of, say, two stretched springs mounted side by side, is accelerated at twice the rate experienced by the same mass when impelled by one spring alone (Fig. 3.1). Of course, the springs have to be stretched to the same length. Measurements of this kind lead us to assert that an acceleration is always produced by a directly proportional force.

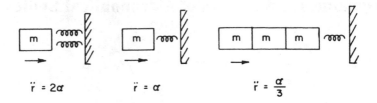

Fig. 3.1. Definition of inertial mass (See also Section 3:7).

$$F \propto \ddot{r} \,. \tag{3-1}$$

In a related experiment three connected masses m accelerated by releasing a single spring would be accelerated at only one-third the rate experienced by one mass acted on by the same spring. This second type of measurement shows that the acceleration produced is inversely proportional to the mass of the impelled body:

$$\ddot{r} \propto \frac{1}{m} \,. \tag{3-2}$$

Combining relations (3–1) and (3–2) we obtain the proportionality

$$\ddot{r} \propto F/m \,. \tag{3-3}$$

This is a brief way of stating Newton's first and second laws. The acceleration of a body is proportional to the force acting on it and inversely proportional to its mass. When the impelling force is zero, the body remains unaccelerated; its velocity stays constant and may be zero.

We can go one step further and say that the force is equal to the mass times the acceleration. This *defines* the unit of force in terms of the other two quantities:

$$F \equiv m\ddot{r} \,. \tag{3-4}$$

With these ideas in mind we can draw a significant conclusion from Galileo's experiments, which showed that two bodies placed at identical points near the Earth fall (are accelerated) at equal rates, even though their masses may be quite different. This independence of mass, interpreted in terms of (3–3), shows that the accelerating force is proportional to the mass of the falling body. We will need to make use of this point in the arguments that follow.

Galileo's work on ballistics showed that a projectile launched at a given angle falls to Earth at a greater distance if its initial velocity is increased. We can ask what would happen if the initial velocity were increased indefinitely. The projectile would keep falling to Earth at progressively greater distances and, neglecting atmospheric effects, it could presumably circle the Earth if given enough initial velocity. If the projectile still retained its original velocity on returning to its initial position, the circling motion would continue. The projectile would orbit the Earth much as the Moon.

Newton already knew a number of facts about the motion of the Moon and he performed calculations to show that the Moon behaves in every way just as a projectile placed into an orbit around Earth.

In addition to the experiments of Galileo, Newton was also aware of Kepler's observational deductions. *Kepler's laws* summarize three principal observations:

(i) The orbits along which planets move about the Sun are ellipses.

(ii) The area swept out by the radius vector joining the Sun and a planet is the same in equal time intervals. The angular velocity about the Sun is small when the planet is distant and is large when the planet is close to the Sun. The Moon shows the same behavior as it orbits the Earth.

(iii) The period a planet requires to describe a complete elliptical orbit about the Sun is related to the length of the semimajor axis of the ellipse. The square of the period P is proportional to the cube of the semimajor axis a (Fig. 3.2). This law

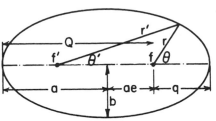

f, f' The two foci of the ellipse.
a Semimajor axis.
b Semiminor axis.
e Eccentricity: the focus of the ellipse is displaced from the center by a distance ae.
q Distance of the pericenter; we can see that $q = a(1 - e)$.
Q Distance of the apocenter; $Q = a(1 + e)$.
θ True anomaly, the angle between the radius vector r, and the major axis as seen from the focus f.
r The radius vector from the focus f.
r' The radius vector from the focus f'.

also describes the motion of satellites (moons) about their parent planets.

Newton therefore had three pieces of information:

(i) He knew that projectiles fall because they are gravitationally attracted toward the Earth.

(ii) He knew that there are certain similarities between the motions of projectiles and the motion of the Moon about the Earth.

(iii) He knew that the motion of the Moon is similar to that of Jupiter's and Saturn's satellites and that those motions appeared to be governed by the same laws that described the motions of planets about the Sun.

These ideas led him to attempt an explanation of all these phenomena in terms of accelerations produced by gravitational attraction.

Newton already suspected that in the interaction of two bodies, equal but oppositely directed forces act on both bodies (Newton's third law). The fact that a planet is attracted by the Sun, but can also attract a satellite by gravitational means, indicates that there is no real difference between the *attracting* and the *falling* body. If the force acting on one of Galileo's falling bodies was proportional to its own mass — as stated above — then the force must also be proportional to the mass of the Earth. The gravitational force of attraction between two bodies must then be proportional to the product of their masses m_a and m_b:

$$F \propto m_a m_b \ .$$ (3-5)

Because the acceleration of distant planets is smaller than that of planets lying close to the Sun, this force must also be inversely dependent on the distance between the bodies. Similarly, the distance and orbital period of the Moon show it to have an acceleration toward the Earth much smaller than that of objects at the Earth's surface. F must drop faster than r^{-1}, because otherwise the effects of distant stars would influence a planet's orbital motion more strongly than the Sun.[1] Using arguments similar to those of Problems 2–6 and 2–7, Newton surmised the distances to other stars and knew that there were a large number of stars surrounding the Sun. As a reasonable choice of distance dependence, Newton tried an inverse square relationship $F \propto r^{-2}$. We will show in Section 3:5 that a force law of the form

$$F \propto m_a m_b r^{-2}$$ (3-6)

allows us to derive Kepler's laws of motions. To turn this proportionality relation into the form of an equation, we write

$$F = \frac{m_a m_b}{r^2} G \ ,$$ (3-7)

where the proportionality constant G is the *gravitational constant*. Sometimes called the *Newtonian gravitational constant*, G is a fundamental constant of Nature whose value must be experimentally determined as discussed in Section 3:6 below.

3:2 Ellipses and Conic Sections

Since the planets are known to describe elliptical orbits about the Sun, it is convenient to start the discussion of their motions by defining a set of parameters in terms of which the elliptical paths can be described.

We can define an ellipse as the set of all points the sum of whose distances, $r + r'$, from two foci is constant (see Figure 3.2). Because the ellipse is symmetrical about the two foci, we can see that this constant must have the value $2a$:

---

[1] If *differential* acceleration of the Sun and Earth is considered, the effect of the distant stars is not so striking. However, with an r^{-1} force, the Sun would still rob the Earth of its Moon.

$$r + r' = 2a. \tag{3-8}$$

Hence $b = \sqrt{a^2 - a^2 e^2}$ by the theorem of Pythagoras. The figure also shows that

$$r \sin \theta = r' \sin \theta' \tag{3-9}$$

and that

$$r \cos \theta - r' \cos \theta' = -2ae . \tag{3-10}$$

These two equations, respectively, represent the laws of sines and of cosines for plane triangles. Squaring (3–9) and (3–10) and adding these expressions gives

$$r^2 + 4aer \cos \theta + 4a^2 e^2 = r'^2 . \tag{3-11}$$

Substituting from (3–8) then yields

$$r = \frac{a(1 - e^2)}{1 + e \cos \theta} , \tag{3-12}$$

an equation that we will need below. Actually, equation (3–12) is more general than shown here; it describes any conic section. When the eccentricity is $0 < e < 1$, the figure described is an ellipse. If $e = 0$, we retrieve the expression for a circle of radius a. If $e = 1$, a becomes infinite, the product $a(1 - e^2)$ can remain finite, and the equation describes a parabola. When $e > 1$, equation (3–12) describes a hyperbola.

3:3 Central Force

From Kepler's second law, a simple but important deduction can at once be drawn. In vector form the law states

$$\mathbf{r} \wedge \dot{\mathbf{r}} = 2A\mathbf{n} . \tag{3-13}$$

Here $\mathbf{r}$ is the radius vector from the Sun to the planet, $\dot{\mathbf{r}}$ is the planet's velocity with respect to the Sun, A is a constant, and the symbol $\wedge$ stands for the *vector product*, or *cross product*. The product of r, $\dot{r}$, and the sine of the angle between these two vectors is twice the area swept out by the radius vector in unit time. $\mathbf{n}$ is a unit vector whose direction is normal to the plane in which the planet moves.

We see that the time derivative of equation (3–13) is

$$\frac{d}{dt}(\mathbf{r} \wedge \dot{\mathbf{r}}) = \mathbf{r} \wedge \ddot{\mathbf{r}} = 0 \tag{3-14}$$

because both A and $\mathbf{n}$ are constant. Multiplying this expression by the mass of the planet m and using equation (3–4) we find that

$$\mathbf{F} \wedge \mathbf{r} = 0 . \tag{3-15}$$

Since neither the force nor the radius vector vanishes in elliptical motion, it is clear that the force and radius vectors must be collinear. The force acts along the radius vector. Such a force is called a *central force*. A planet is pulled toward the Sun at all times; and the components of a binary star are always mutually attracted.

3:4 Two-Body Problem with Attractive Force

Let us now define a coordinate system whose origin lies at the *center of mass* of bodies a and b. The positions and masses of the bodies are related (Fig. 3.3) by

$$\mathbf{r}_a = -\frac{m_b}{m_a}\mathbf{r}_b .$$

(3-16)

Because planetary motion deals with a central attractive force, and the force decreases more rapidly than the inverse first power of the distance between attracting bodies, we postulate that the attractive force is an inverse square law force. If this postulate is correct, we should obtain the motion given by Kepler's laws. We will show below that this is true.

For a central force decreasing as the square of the distance between two attracting bodies, we write the force F_a on body a due to body b as

$$\mathbf{F}_a = m_a\ddot{\mathbf{r}}_a = -\frac{m_a m_b G}{r^3}\mathbf{r} ,$$

(3-17)

where m_a and m_b are the masses of the two bodies. From the definition of $\mathbf{r}$ and the center of mass, we have

$$\mathbf{r} = \mathbf{r}_a - \mathbf{r}_b = \left(1 + \frac{m_a}{m_b}\right)\mathbf{r}_a .$$

(3-18)

Combining (3–17) and (3–18) we obtain

$$\ddot{\mathbf{r}}_a = -\frac{GM}{r^3}\mathbf{r}_a, \qquad M \equiv m_a + m_b ,$$

(3-19)

where M is the total mass of the two bodies. Subtracting a similar expression for r_b we derive

$$\ddot{\mathbf{r}} = -\frac{GM}{r^3}\mathbf{r} .$$

(3-20)

We see that the acceleration of each body relative to the other is influenced only by the total mass of the system and the separation of the bodies. If equation (3–20) is multiplied by a mass term μ, we obtain a force term that is a function only of r, M, μ, and the gravitational constant:

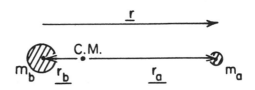

Fig. 3.3. Center of mass (CM) of two bodies a and b. The center of mass of two or more orbiting masses is also referred to as the *barycenter*.

$$\mathbf{F}(\mu, M, r) = -\frac{GM\mu}{r^3}\mathbf{r} = \frac{-Gm_a m_b \mathbf{r}}{r^3} \ . \tag{3-21}$$

If this force is to be equal to the force acting between the two masses, we must satisfy equation (3–7) which means that

$$\mu = \frac{m_a m_b}{m_a + m_b} \ ; \tag{3-22}$$

μ is called the *reduced mass*.

The equation of motion (3–20) taken together with equation (3–21), shows that the orbit of each mass about the other is equivalent to the orbit of a mass μ about a mass M that is fixed — or moves in unaccelerated motion. There is a great advantage to this reformulation. Newton's laws of motion only hold when referred to certain reference frames — stationary coordinate systems, or those in uniform unaccelerated motion (see also Sections 3:8 and 5:1). Such unaccelerated reference frames are called *inertial frames of reference* It was for this reason that we initially referred the motion of the masses a and b to the center of mass. This procedure, however, required us to keep separate accounts of the time evolution of $\mathbf{r}_a$ and $\mathbf{r}_b$. The separation $\mathbf{r}$ was only determined subsequently by adding r_a and r_b. This two-step procedure is avoided if equations (3–20) to (3–22) are used, because $\mathbf{r}$ can then be determined directly.

3:5 Kepler's Laws

Consider a polar coordinate system with unit vectors $\varepsilon_{\mathbf{r}}$ and ε_θ (Fig. 3.4). A particle is placed at position $\mathbf{r} = r\varepsilon_{\mathbf{r}}$. Since the rate of change of the unit vectors can be expressed as

$$\dot{\varepsilon}_{\mathbf{r}} = \dot{\theta}\varepsilon_\theta \ , \tag{3-23}$$

defining the rate of change (rotation) of the radial direction, and

$$\dot{\varepsilon}_\theta = -\dot{\theta}\varepsilon_{\mathbf{r}} \ , \tag{3-24}$$

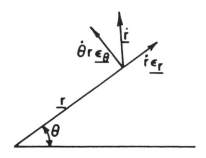

Fig. 3.4. Vector components of the velocity $\dot{\mathbf{r}}$.

giving the rate of change for the tangential direction, we can write the first and second time derivatives of r as

$$\dot{\mathbf{r}} = \dot{r}\varepsilon_{\mathbf{r}} + r\dot{\theta}\varepsilon_\theta , \tag{3-25}$$

$$\ddot{\mathbf{r}} = (\ddot{r} - r\dot{\theta}^2)\varepsilon_{\mathbf{r}} + (2\dot{r}\dot{\theta} + r\ddot{\theta})\varepsilon_\theta . \tag{3-26}$$

From expressions (3–20) and (3–26) we obtain two separate equations, respectively, for the components along and perpendicular to the radius vector

$$\ddot{r} = -\frac{GM}{r^2} + r\dot{\theta}^2 \tag{3-27}$$

and

$$2\dot{r}\dot{\theta} + r\ddot{\theta} = 0, \quad \text{so that} \quad 2\dot{r}r\dot{\theta} + r^2\ddot{\theta} = 0 . \tag{3-28}$$

The second of equations (3–28) integrates to

$$r^2\dot{\theta} = h, \tag{3-29}$$

where h is a constant that is twice the area swept out by the radius vector per unit time. This relationship has a superficial resemblance to the law of conservation of angular momentum (per unit mass). But that law would involve the distances r_a and r_b, instead of r. Equation (3–29) does state Kepler's second law, however, and that is satisfactory.

Combining equations (3–27) and (3–29) we have

$$\ddot{r} - \frac{h^2}{r^3} + \frac{MG}{r^2} = 0 . \tag{3-30}$$

PROBLEM 3–1. Choose a substitution of variables

$$y = r^{-1}, \qquad \dot{\theta}\frac{d}{d\theta} = \frac{d}{dt} , \tag{3-31}$$

to rewrite equation (3–30) in the form

$$\frac{d^2y}{d\theta^2} + y = \frac{MG}{h^2} . \tag{3-32}$$

Show that this has the solution

$$y = B\cos(\theta - \theta_0) + \frac{MG}{h^2} . \tag{3-33}$$

This leads to

$$r = \frac{1}{B\cos(\theta - \theta_0) + (MG/h^2)} . \tag{3-34}$$

This is the expression for a conic section (see equation (3–12)). It therefore represents a generalization of Kepler's first law. Gravitationally attracted bodies move along conic sections which, in the case of planets, are ellipses. We see this if we set

$$a(1 - e^2) = \frac{h^2}{MG} \tag{3-35}$$

and

$$e = \frac{Bh^2}{MG} . \tag{3-36}$$

The minimum value of r occurs for $\theta = \theta_0$.

Let r_m be a relative maximum or minimum distance between the two bodies. Then the entire velocity at separation r_m must be transverse to the radius vector, and by equation (3–29),

$$\frac{(r_m \dot\theta)^2}{2} = \frac{h^2}{2r_m^2} \tag{3-37}$$

is the kinetic energy per unit mass. The *total energy* per unit mass is the sum of *kinetic* and *potential energy* per unit mass

$$\mathcal{E} = \frac{h^2}{2r_m^2} - \frac{MG}{r_m} . \tag{3-38}$$

Solving for r_m we have

$$r_m = \left(\frac{MG}{h^2} \pm \sqrt{\frac{M^2 G^2}{h^4} + \frac{2\mathcal{E}}{h^2}} \right)^{-1} . \tag{3-39}$$

Hence the quantity B in equation (3–34) has the value

$$B = +\sqrt{\frac{M^2 G^2}{h^4} + \frac{2\mathcal{E}}{h^2}} , \tag{3-40}$$

the sign being determined by the condition that the minimum r-value occur at $\theta - \theta_0 = 0$.

Equations (3–12) and (3–35) show that the minimum value of r is

$$q = \frac{h^2}{MG(1 + e)} . \tag{3-41}$$

Substituting this into equation (3–38) we then have an expression for the energy in terms of the semimajor axis a,

$$\mathcal{E} = (e^2 - 1)\frac{M^2 G^2}{2h^2} = -\frac{MG}{2a} , \tag{3-42}$$

where we have made use of expression (3–35). To obtain the total energy of the system we can multiply $\mathcal{E}$ by μ. The total energy per unit mass is the sum of kinetic and potential energy, also per unit mass. From this we see that

$$\mathcal{E} = \frac{v^2}{2} - \frac{MG}{r} , \qquad (3\text{-}43)$$

and from (3–42) we obtain the orbital speed as

$$v^2 = MG \left(\frac{2}{r} - \frac{1}{a} \right) . \qquad (3\text{-}44)$$

We can now make a number of useful statements:

(i) If S is the area swept out by the radius vector

$$\frac{dS}{dt} = \frac{1}{2}h, \qquad S - S_0 = \frac{1}{2}ht . \qquad (3\text{-}45)$$

For an ellipse, the total area is

$$S - S_0 = \pi a b = \pi a^2 (1 - e^2)^{1/2} \qquad (3\text{-}46)$$

so that from equation (3–35) the period of the orbit is

$$P = \frac{2}{h}\pi a^2 (1 - e^2)^{1/2} = \frac{2\pi a^{3/2}}{\sqrt{MG}} . \qquad (3\text{-}47)$$

Equation (3–47) is a statement of Kepler's third law.

(ii) If the eccentricity is $e = 1$, the total energy is zero by equation (3–42) and the motion is parabolic. Astronomical observations have shown that some comets approaching the Sun from very large distances have orbits that are practically parabolic, although they may be slightly elliptical or slightly hyperbolic. At best, these comets therefore are only loosely bound to the Sun. A small gravitational perturbation by a passing star evidently can make the total energy of some of these comets slightly positive, and they escape from the Solar System to wander about in interstellar space.

We should still note that one of the big advances brought about by Newton's theory was the realization that both cometary and planetary orbits could be understood in terms of one and the same theory of gravitation. Prior to that no such connection was known.

(iii) If the eccentricity $e > 1$, the total energy is positive, and the motion of the two masses is unbound. After one near approach the bodies recede from each other indefinitely.

(iv) If the eccentricity is zero, the motion is circular with some radius R and the energy obtained from equation (3–42) is $-MG/2R$ per unit mass. Equation (3–44) then states that v^2 equals MG/R or that the gravitational attractive force per unit mass MG/R^2 must equal v^2/R, which sometimes is called the *centrifugal force* — a fictitious force that is supposed to "keep the orbiting mass at constant radius R despite the attractive pull of M."

Thus far we have shown that the motion of one mass about another describes a conic section. In addition, we can show that the orbit of each mass about the common center of mass is a conic section as well. Equation (3–19) can be rewritten as

$$\ddot{\mathbf{r}}_a = -\frac{GM}{(1 + m_a/m_b)^3}\frac{\mathbf{r}_a}{r_a^3}.$$ (3-48)

This is of the same form as equation (3–20) and we can, therefore, readily obtain equations similar in form to expressions (3–27) to (3–29), and finally (3–34). This argument also holds true if we were to talk about the vector $\mathbf{r}_b$ instead of $\mathbf{r}_a$. Hence both masses m_a and m_b are orbiting about the center of mass along paths that describe conic sections.

Let us still see how we can determine the masses of the components of a spectroscopic binary. This is the most important means we have for determining stellar masses. For such binaries we can measure the radial velocities of both stars throughout their orbits (Fig. 3.5).

It is relatively easy to determine the period of such a binary by looking at the repeating shifts of the superposed spectral lines. Equation (3–47) then gives the ratio (a^3/M) of the semimajor orbital axis cubed and the sum of the masses. If the binary, in addition, is an eclipsing binary, so that the line of sight is known to lie close to the orbital plane, then the semimajor axes of the orbits of the two components about the common center of mass can be found; and this gives the individual component masses if use is made of component equations derived from (3–18) and (3–19).

For a few visual binaries that are close enough to permit accurate observations, the motion of the individual components relative to distant background stars again permits computation of the individual semimajor axes, provided the trigonometric parallax is also known. The orbital period then allows us to compute the individual masses through Kepler's third law and equations (3–18) and (3–19).

We note that expressions such as (3–35), (3–36), (3–44), and (3–47), which connect measurable orbital characteristics to M and G always depend on the product MG and, hence, permit a determination of neither the system's total mass, nor of the gravitational constant. For a long time this presented a serious difficulty. However:

PROBLEM 3–2. Show how a rough measure of G can be obtained from falling mass experiments when the known size of the Earth and some estimate of its density are used to determine the Earth's mass. In Section 3:6, below, we show how G was eventually measured by Cavendish. Note that for an accurate determination of the Earth's density, G has to be accurately known.

3:6 Determination of the Gravitational Constant

Henry Cavendish (1731–1810), an English chemist, discovered a means of measuring the gravitational constant G, late in the eighteenth century, more than one hundred years after Newton had first shown how the motion of the planets depends on the mass of the Sun. Until Cavendish performed his experiment, the absolute

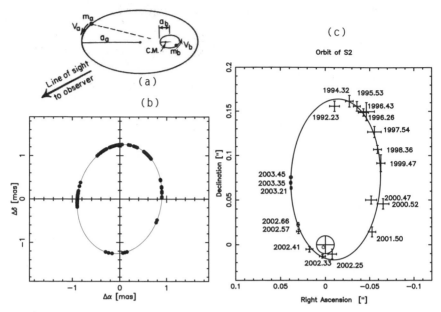

Fig. 3.5. (a) Binary star orbits and the individual semimajor axes for two stars orbiting their common center of mass. (b) The apparent orbit of the secondary star with respect to the primary in the close binary listed as HD 27483 in the Henry Draper (HD) catalogue. The two stars have almost identical masses, 1.38 ± 0.13 and $1.39 \pm 0.13 M_\odot$, and respective semimajor axes $0.02915 \pm 1.4 \times 10^{-4}$ and $0.02878 \pm 2.4 \times 10^{-4}$ AU. The distance to the binary has been determined from precision parallax measurements obtained with the aid of the *Hipparcos satellite* and is 45.9 ± 1.8 pc. The observations were carried out at optical wavelengths with a Michelson stellar interferometer (See Section 4:12). Courtesy Maciej Konacki (Ko04). (c) Orbit of the star S2 about the compact radio source Sgr A*, commonly taken to define the Galactic center at Galactic longitude $\ell = 0$ and latitude $b = 0$. The position of the star observed at infrared wavelengths is shown in Galactic coordinates for the period from 1994 to 2003, the dates being expressed in decimal form. The continuous curve is the best-fit Keplerian ellipse whose focus is shown by the small error circle, lying within a few milliarcseconds from the radio source. The size of the cross indicates a current ± 10 milliarcseconds positional uncertainty of the infrared relative to the radio astrometric reference frames. By making use both of the spectroscopic line shift of S2, and its proper motion, the deduced elliptic orbit provides a highly accurate distance to the Galactic center, 7.94 ± 0.42 kpc, as well as the mass of the purported Galactic center black hole, $3.59 \pm 0.59 \times 10^6 M_\odot$. This mass is contained within a sphere of projected radius $\lesssim 10$ mas or 10^{15} cm ~ 100 AU from the central mass. Courtesy of Frank Eisenhauer (Ei03). Sgr A* is expected to move about the barycenter of the system of stars passing close by. Radio observations show that the component of this motion perpendicular to the Galactic plane is at most 0.4 ± 0.9 km s^{-1}, as gauged by relative positions of Sgr A* and two distant quasars. This indicates that the mass of the radio emitting source must exceed $4 \times 10^5 M_\odot$ or 10% of the total. Because the size of the radio source and hence this mass concentration is <1 AU, this implies a mass density in excess of $6 \times 10^{21} M_\odot$ pc^{-3} and provides strong indication that a *supermassive black hole* of mass $\sim 10^6 M_\odot$ occupies the Galactic Center (Re04).

masses of celestial objects could not be accurately determined; there were only relative values of, say, planetary masses as judged by orbits of their moons.

In the Cavendish experiment a torsion balance is used. Typically such a device may consist of a fine quartz fiber to which a rod bearing masses m_1 and m_2 is attached, as shown in Fig. 3.6 (a). Each mass is at some distance L from the fiber.

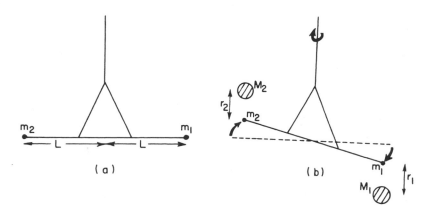

Fig. 3.6. The Cavendish experiment to determine the gravitational constant G.

We can calibrate the balance by noting the torsion that can be induced in the fiber when a small measurable torque is applied to the system by a spring with known force constant exerting a horizontal force at the position of m_2.

If masses M_1 and M_2, respectively, are placed at a small horizontal distance from masses m_1 and m_2, we may observe a twist of the fiber, in the sense shown in Fig. 3.6 (b). We can determine the distances r_1 and r_2, respectively, between m_1 and M_1, and m_2 and M_2, to find the horizontal forces acting on the ends of the bar and, hence, establish the torque acting on the quartz fiber. That torque N is

$$N = L\left(\frac{m_1 M_1 G}{r_1^2} + \frac{m_2 M_2 G}{r_2^2}\right). \qquad (3\text{-}49)$$

From the measured deflection of the masses we can determine the value of N, in terms of the calibration previously obtained on the twisted quartz wire. Because we can measure N, L, r_1, r_2, and the masses of all the different bodies, we now have the values of all quantities in (3–49) except for G, which can then be directly determined from the equation. That value is $G = 6.674 \times 10^{-8}$ dyn cm^2 g^{-2} (Ba97).

Once the value of G is known, the mass of the Sun is readily determined using Kepler's third law, equation (3–47). That law actually involves the total mass M of the Sun and planet, but by performing the calculations for a number of different planets we can verify that the mass of the Sun is very nearly equal to M and that the summed masses of the planets $\sim 0.0013\, M_\odot$ can be neglected to a good approximation. The approximate mass of the Earth, $M_\oplus = 5.974 \times 10^{27}$g, can be derived in a similar way, making use of the known orbit of the Moon.

3:7 The Concept of Mass

If we examine what we have said about the measurement of masses, we find that there are really two quite distinct ways of determining the mass of a body: (i) we can measure its acceleration in response to a measured force (equation (3–4)); or (ii) we can measure the force acting on the body when a given mass is placed at a specified distance — this is what we do when we weigh the body with a spring balance.

The first of these is a dynamic measurement; the second can be static. The mass of a body measured in the first way is called its *inertial mass*; the mass measured by means of the second method is called the *gravitational mass*.

Suppose now that we take a steel ball whose gravitational mass is m_1. We take a wooden ball that is slightly too heavy, and slowly file away excess material until its gravitational mass is also equal to m_1. If the two balls are placed on a pan balance, they should leave the balance arm in a horizontal position, because the Earth attracts both masses equally.

The question now is whether the inertial mass of these two bodies is always the same. Will the wooden ball be accelerated at precisely the same rate as the steel ball in response to a given force? We need to refine Galileo's work on falling bodies to answer this question.

The problem intrigued the Hungarian baron Roland von Eötvös, around the end of the nineteenth century. He suspended two weights of different composition but identical weight on a torsion balance, with the horizontal bar along the East–West direction (Fig. 3.7(a)). As the Earth rotated, two forces acted on each mass: (i) a gravitational attraction that is equal, because the weights of the masses are equal; and (ii) a *centrifugal force* due to the Earth's rotation. If the centrifugal force on

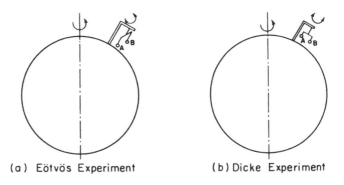

(a) Eötvös Experiment   (b) Dicke Experiment

Fig. 3.7. Eötvös and Dicke experiments.

mass A was greater or less than on mass B, this would indicate that their inertial masses differed. The bar would rotate until the torsional force in the suspending wire compensated for the inequality in the centrifugal forces.[2] Eötvös never observed such a rotation of the bar and concluded that the inertial and gravitational masses of the bodies were identical to within one part in about 10^8.

R. H. Dicke and his co-workers later refined this experiment. They suspended their weights with the bar in a North–South direction (Fig. 3.7(b)). As the Earth turns about its axis, mass A might be attracted more or less strongly toward the Sun than mass B. We should then observe a diurnal effect with the balance arm first swinging in one direction and then in the other. The absence of such an effect showed that gravitational and inertial masses are identical to within about one part in 10^{11} (Ro64). Precise observations of the lunar orbit, showing the rate at which both fall toward the Sun, and knowledge about the differences in the composition of Earth and Moon, by now have lowered this limit to one part in $\sim\!10^{13}$ (Wi04).

We may now ask whether the gravitational mass of matter is the same as that of antimatter. If there existed galaxies composed of antimatter, would they attract or repel a galaxy consisting of matter? L. I. Schiff gave a tentative answer to such questions (Sc58a). He pointed out that many atomic nuclei emit virtual positron–electron pairs. This means that part of the time a fraction of the total nuclear energy is to be found in the form of an electron and a matched positron. Such a pair of particles is continually formed and reassimilated and is never actually emitted.

Schiff pointed out that if the positrons had a negative gravitational mass, then the ratio of inertial to gravitational mass would be affected for a number of substances for which the electron–positron virtual pair formation is a major effect. The ratio of the two kinds of masses would then be different from unity by about 1, 2, and 4 parts in 10^7 for aluminum, copper, and platinum, respectively. Experiments with such substances have been performed, and inequalities of this size are ruled out by the Eötvös and Dicke experiments. It follows that matter and antimatter ought to have gravitational masses of the same sign and that galaxies interacting gravitationally with antigalaxies could not be distinguished on dynamical grounds.

Note that we have really only shown that the inertial and gravitational mass have the same sign for positrons as for electrons. But we actually know from dynamical experiments in magnetic fields that the inertial mass of the positron equals that of the electron, so that our previous conclusion should follow at once. Matter and antimatter both have positive mass.

Schiff recognized one difficulty with this argument. We are not certain that virtual electron–positron pairs behave exactly like real pairs. Could it be that the gravitational mass of a real positron differs from that of a virtual positron? Unfortunately, we will not be absolutely sure until we make a direct measurement of the positron's motion in a gravitational field. Antihydrogen atoms have by now been created at accelerators, and plans have been made to trap them and measure their rate of free-fall. This should settle the issue in the next few years (St04), (Ni95).

---

[2] This experiment works best roughly halfway between the equator and poles, say, Budapest or Princeton.

3:8 Inertial Frames of Reference — The Equivalence Principle

We noted earlier that Newton's laws of motion hold only when the motion is described in coordinates that refer to a frame of reference that is either fixed or is moving at constant velocity with respect to the distant galaxies. Such a reference frame is known as an *inertial coordinate system*.

Several perplexing questions arise when we try to understand the significance of these frames of reference. They can be described by some simple experiments.

(1) Suppose that a man were blindfolded and placed on a merry-go-round. He could determine quite accurately whether he was being spun around; he would be able to feel the centrifugal force acting on him when the merry-go-round was moving. If he adjusted the mechanism until he felt no centrifugal force he would find, on taking off his blindfold, that the merry-go-round was stationary with respect to the distant galaxies.

(2) A blindfolded man placed in a rocket ship in interstellar space could adjust his controls until he felt no forces on himself. On taking a closer look, he would find that he had adjusted the engine to give zero thrust. He might find that he was moving at constant velocity with respect to the distant galaxies. Alternatively, however, he might find that he had strayed into the vicinity of a star and was freely falling toward it! Albert Einstein first postulated that freely falling, nonrotating coordinate frames are fully equivalent to Newton's inertial frames that move at constant velocity with respect to the distant galaxies. All laws of physics have precisely the same form in both types of frames. This *equivalence principle* will prove very useful in our discussions of relativity in Chapter 5.

When we talk about a motion with respect to the distant galaxies, we really mean a motion with respect to the mean velocity of all galaxies at very large distances. Galaxies are receding in all directions but, as far as we can tell, there always exists a local frame of reference in which the motions of distant galaxies statistically appear symmetrical, no matter which direction we look.

This suggests that perhaps the local frame of zero acceleration is determined by the distribution of the galaxies in the Universe. Just how this determination comes about is a basic unanswered question of the theory of gravitation. The thought that the overall distribution of mass within the Universe should determine a local inertial framework is due to Ernst Mach and is sometimes called *Mach's principle*. Many related questions involved the same basic thought. "Is the inertial mass of a body determined by the distribution of matter in the Universe? Is the gravitational constant determined by the distribution of the distant galaxies? As a result, would the value of the gravitational constant change with time as the galaxies recede from each other? Are the atomic constants of physics related to the large-scale structure of the Universe?" We will consider these questions in Section 11:17.

3:9 Gravitational Red Shift and Time Dilation

Einstein's *principle of relativity* (Section 5:1) states that mass and energy are related in such a way that any stationary mass m has an equivalent energy mc^2 associated with it (Ei–b). Einstein showed that the separate laws of conservation of mass and of energy merge into a more general *conservation of mass–energy*. This predicts a gravitational red shift for radiation emitted at the surface of a star (Ei11). Consider two particles, an electron and a positron at rest at a very great distance from a star. The rest–mass of each particle is m_0. If the particles fall in toward the star's surface — and both have positive gravitational mass — each will acquire a total mass–energy

$$\mathcal{E} \equiv m_r c^2 = \left(m_0 c^2 + \frac{m_0 MG}{r} \right) = m_0 c^2 \left(1 + \frac{MG}{rc^2} \right) \tag{3-50}$$

at distance r from the star. The second term in the parentheses represents the conversion of potential into kinetic energy. Now let the two particles be deflected without loss of energy or momentum, so that they collide head-on, and annihilate. Two photons, each with frequency

$$\nu_r = \frac{m_r c^2}{h} \tag{3-51}$$

will be formed in this process. These photons are now permitted to escape from r but, through reflections from stationary mirrors that produce no frequency shifts, we can make them collide again at a large distance from the star.

In this collision they can form an electron–positron pair. If energy is conserved, then the photons' frequency ν_0 at a large distance from the star must be

$$\nu_0 = \frac{m_0 c^2}{h} . \tag{3-52}$$

Otherwise there would be either too much or too little energy to recreate a positron–electron pair at rest. Hence

$$\nu_0 = \frac{\nu_r}{1 + MG/rc^2} . \tag{3-53}$$

The frequency at a large distance from the star is less than the emitted frequency. For the Sun $M \sim 2 \times 10^{33}$ g, the radius $R \sim 7 \times 10^{10}$ cm, and $MG/Rc^2 \sim 2 \times 10^{-6}$ at the solar surface. For a neutron star whose mass would be about the same, but whose radius is $\sim 10^5$ times less, the fractional frequency shift

$$\frac{\Delta\nu}{\nu_r} = \frac{\nu_0 - \nu_r}{\nu_r} = -\frac{MG}{rc^2} \left[1 + \frac{MG}{rc^2} \right]^{-1} \tag{3-54}$$

becomes comparable to unity, and the frequency shift $\Delta\nu$ becomes comparable to the frequency itself.

We will see in Section 3:10 that the frequency of electromagnetic waves can give a very accurate measure of time and can therefore be used as a clock. Such a clock placed in a strong gravitational field would therefore run more slowly. Quite generally, the rate at which a clock runs is determined by the potential $\mathbb{V}(r)$ at the position r of the clock. The period P of this clock measured by an observer outside the potential field, that is, by an observer located at $\mathbb{V} = 0$, appears to be

$$P_0 = P_r \left(1 - \frac{\mathbb{V}}{c^2} \right) . \tag{3-55}$$

In Sections 3:11 and 5:14 we outline an experiment that measures this *time dilatation* as expressed through a delay in the arrival at the Earth of pulsar pulses that have passed close to the Sun.

3:10 Measures of Time

In describing the orbital motions of planets about the Sun, we have obtained expressions for position as a function of time. But how is this parameter *time* actually measured?

There are a number of ways (Sa68) of measuring time and it is interesting to see how these methods interrelate. Some rather basic questions of physics are involved. Let us first describe a set of imaginary clocks. They may not be practical but they should work in principle.

First Clock

Take an amount of tritium ^{3}H that beta decays into the helium isotope ^{3}He. If the tritium is kept frozen at a temperature around 10 K, the helium will diffuse out as it is formed. We weigh the tritium. When the mass has dropped to half its initial value, we say that a time of one unit, NT, has elapsed. We could set up a clock that struck each time the remaining mass was reduced by a factor of 2.

Second Clock

Take a quantity of the cesium isotope ^{133}Cs. It has a transition between two hyperfine levels of the ground state. We measure the frequency of the radiation (radio wave) emitted in this transition. The period of this electromagnetic wave can serve as a unit of time, AT.

Third Clock

We set up a telescope that is always pointed at the local zenith. Each time a given distant galaxy reappears exactly in the center of the telescope's field of view, we say that one unit of time, UT, has elapsed.

Fourth Clock

We note the plane described by Jupiter as it orbits the Sun. We mark the instant that the Earth crosses this plane. It does this twice per orbit — once from North to South and then from South to North. If we define the interval between successive N to S crossings as one unit of time, ET, we have still another means of measuring time.

We call these measures — NT, AT, UT, and ET — *nuclear time, atomic time, universal time,* and *ephemeris time*. As we have chosen to define them these units of time, respectively, correspond to ~ 12 years, $(9,192,631,770)^{-1}$ seconds, 1 day, and 1 year.

The basic differences between the clocks are these: the first clock uses beta decay, a weak interaction, as its basic mechanism. The second clock uses an electromagnetic process to measure time. The third clock uses the Earth's rotation to measure time; this is an inertial process. Finally, the fourth clock makes use of a gravitational force to measure time.

Because each of these clocks depends on quite different physical processes, we worry that they might not measure the same "kinds" of time. There is no reason, for example, why atomic time and ephemeris time as defined above should describe intervals having a constant ratio. At present the ratio of these time units is about $3 \times 10^{17} : 1$. Will this ratio be the same some 10^9 yr from now? Or does the strength of the gravitational field or of the weak interactions change after years in such a way that one of these clocks becomes accelerated relative to the other?

We can test such questions experimentally. They are of great importance to cosmology. For, in order to understand the nuclear history of the Universe and the formation of chemical elements, we have to know how nuclear reaction rates in stars may have been affected by the overall evolution of the Universe over past æons. This will appear more clearly after the synthesis of nuclei in stars has been discussed in Chapter 8.

The important point to realize is that we have enumerated four quite different ways of defining time.[3] UT and ET are related if the general theory of relativity holds true; and their relationship becomes a test for theories of gravitation.

In practice, comparing the rates of these clocks is difficult. Planetary perturbations make the Earth's orbit about the Sun irregular. The orbiting Moon, tides, earthquakes, and other disturbances affect the rotation rate of the Earth. An incomplete understanding of these effects makes it hard to compare the UT and ET rates with time measured by atomic clocks. Eventually, however, such practical difficulties should be overcome and a comparison of time scales may become possible.

---

[3] Actually, there are five ways. Nuclear β and α decay rates are based on weak and strong nuclear forces, respectively. We will show in Section 11:17 that these two kinds of clock apparently have run at identical rates over the past few æons.

3:11 Uses of Pulsar Time

Some binary pulsars — pulsars orbiting compact companions — emit signals with a periodicity that varies by less than one part in 10^{10} over an interval of a year (Sa97). These signals can therefore be used to define a time scale. The mechanism of the clock is not yet understood but reflects the rotational period of neutron stars. In any case, its regularity allows us to put it to scientific use. For many purposes we do not need to know how a clock works as long as its accuracy can be verified.

Many years ago, C. C. Counselman and I. I. Shapiro listed a number of interesting gravitational effects to be studied using pulsars (Co68).

(a) The orbit of the Earth can be determined with great precision. Pulsar emission acts as a "one-way" radar. Counting pulse rates from different pulsars allows us to measure the instantaneous velocity of the Earth relative to some arbitrarily defined inertial frame. Integrating these velocities over a series of time intervals yields the Earth's position as a function of time, that is, the shape of the orbit and its orientation.

Such measurements can also yield data on the positions and masses of the outermost planets. Their motions affect the position of the Solar System's *barycenter* and hence also the orbit of the Earth. The periodicity of these effects is determined by the orbital periods of the planets, and we should find corresponding periodic variations in the pulse counts (As71).

(b) A pulsar located near the ecliptic plane appears close to the Sun once each year. When the light pulses pass very close to the limb of the Sun, they are slowed down because all clocks are slowed by the presence of a strong gravitational field and because the speed of light measured locally at the Sun would still appear to be c. The arrival time of pulses at the Earth are delayed by an amount of order $100\,\mu s$, depending only on how close to the Sun the radiation passes. By keeping track of the arrival times we can compute the delay and see whether the measured delay agrees with the predictions of relativity theory. To do this, we have to first correct for the time delay due to the relatively high index of refraction of the solar corona. This is possible because the delay due to refraction is proportional to ν^{-2}, whereas the gravitational delay is independent of frequency ν. Several pulsars pass within $1°$ of the Sun and hence are suitable objects for such tests. We shall discuss this further in Section 5:14.

(c) Because pulsars are located within the Galaxy, the shear motion of stars in the Galaxy yields an acceleration relative to the Sun that can be detected by keeping track of pulse arrival times. The differential rotation of the Galaxy can therefore be mapped very accurately.

3:12 Galactic Rotation

The mass of the Galaxy is unevenly distributed. The density of matter enclosed in a sphere of radius r from the Galactic center is greatest near the nucleus. Stars near the Galactic center tend to have angular velocities $\dot{\theta}(r)$ appreciably larger than stars at greater distances; that is, $d\dot\theta/dr < 0$. Suppose for simplicity that all stars have idealized circular orbits about the Galactic center. Let the Sun be at distance r_s from the center. Relative to the Sun S, moving with velocity v_s (see Fig. 3.8), matter at Galactic longitude l, and at distance r from the center C has an approach velocity $v(r,l)$ along the line of sight.

$$v(r,l) = \left[-r_s\dot\theta(r_s)\sin l + r\dot\theta(r)\cos\theta\right] = \left[\dot\theta(r) - \dot\theta(r_s)\right]r_s\sin l\,, \qquad (3\text{-}56)$$

where the simple form of the expression on the extreme right is due to the relation $r\cos\theta = r_s\sin l$, as is evident from Fig. 3.8. We note from (3–56), and from the fact that $d\dot\theta/dr < 0$, that $v(r,l)$ is positive in the quadrants I and III so that stars and gas along these directions appear to approach and their spectra should be blue-shifted. In quadrants II and IV stellar spectra should appear red-shifted.

This is actually observed. In 1927 the Dutch astronomer Jan Oort was able to use this evidence to prove that stars in our Galaxy are in *differential rotation* about the Galactic center (Oo27a,b).

At any given Galactic longitude l, the highest velocity should be observed at point P, where the line of sight is tangential. By noting the maximum velocity at any given elongation l, we can construct a model of the Galaxy giving both its mass distribution and distance of the Sun from the Galactic center. The distance to the center obtained in this way is $r_s \sim 8.5\,\mathrm{kpc}$ (Fr96). This may be compared to the more precise distance of 7.94 ± 0.42 kpc obtained from the motion of stars about the radio source Sgr A* (See Figure 3.5(c)).

Differential rotation tends to shear aggregates of gas and dust as they orbit the Galactic center. For some time this effect was considered responsible for the appearance of spiral arms in some galaxies. However, C. C. Lin (Li67) then suggested that the spiral structure represents a local increase in density and that this enhanced

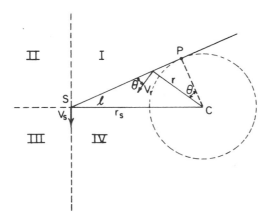

Fig. 3.8. Notation used for discussion of Galactic differential rotation.

density travels around a galaxy as a *spiral density wave*, at a *pattern velocity* different from that of the speed of the stars involved. For the Galaxy this speed is about 13.5 km s^{-1} times the distance from the center measured in kiloparsecs, kpc. At our distance from the center, this would be 110 km s^{-1}, whereas the Galactic rotation (velocity of the stars) is $\sim$220 km s^{-1}.

3:13 Scattering in an Inverse Square Law Field

When a meteorite approaches the Earth, its orbit can become appreciably changed. Similarly, a comet passing close to Jupiter can be given enough energy to escape the Solar System. In both cases the smaller object is scattered or deflected by the larger body. For a particle initially approaching from direction $\theta_\infty - \theta_0$ (Fig. 3.9), the orbital trajectory is given by equations (3–34) and (3–40).

$$\frac{1}{r} = \frac{MG}{h^2}\left[1 + \left(1 + \frac{2\mathcal{E}h^2}{M^2G^2}\right)^{1/2}\cos(\theta - \theta_0)\right]. \tag{3-57}$$

At large distances from the scatterer, the asymptotic motion is along directions (see equations (3–33), (3–42))

$$\cos(\theta_\infty - \theta_0) = -\left(1 + \frac{2\mathcal{E}h^2}{M^2G^2}\right)^{-1/2} = -\frac{1}{e}, \qquad r \to \infty. \tag{3-58}$$

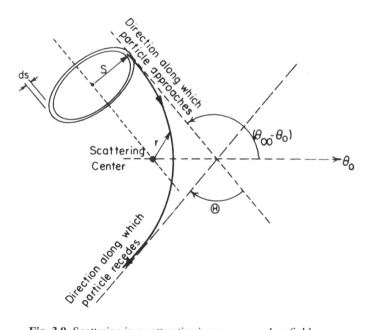

Fig. 3.9. Scattering in an attractive inverse square law field.

This has solutions for two values of $|\theta_\infty - \theta_0|$, one corresponding to the incoming, the other to the scattered asymptotic direction. The angle through which the object is deflected is $\Theta = 2(\theta_\infty - \theta_0) - \pi$. We see that

$$\sin\frac{\Theta}{2} = -\cos(\theta_\infty - \theta_0) = \frac{1}{e}\,. \tag{3-59}$$

Let s be the *impact parameter* (Figs. 3.9, 3.11) and v_0 the *approach velocity* of the scattered particle at a large distance, $r \to \infty$. Then, from (3–29), $h = sv_0$ is twice the area swept out per unit time. Since

$$\mathcal{E} = \frac{v_0^2}{2}, \qquad h^2 = 2\mathcal{E}s^2\,, \tag{3-60}$$

and

$$\sin\frac{\Theta}{2} = \left[1 + \left(\frac{2s\mathcal{E}}{MG}\right)^2\right]^{-\frac{1}{2}}\,, \tag{3-61}$$

this leads to

$$\cot\frac{\Theta}{2} = \frac{2\mathcal{E}s}{MG}\,. \tag{3-62}$$

If any object having an impact parameter between s and $s + ds$ is scattered into an angle between Θ and $\Theta + d\Theta$, we say that the *differential cross-section* $\sigma(\Theta)$ for scattering is given by

$$2\pi s\,ds \equiv -\sigma(\Theta)\,d\Omega = -2\pi\sigma(\Theta)\sin\Theta\,d\Theta\,. \tag{3-63}$$

In this equation the expression on the left represents the area of a ring through which all particles approaching from a given direction have to flow if they are to be scattered into the solid angle $d\Omega$ enclosed between two cones having half-angles Θ and $\Theta + d\Theta$, respectively. The expression on the right gives the solid angle between these two cones multiplied by the differential cross-section. The differential cross-section is therefore just a parameter that assures conservation of scattered particles. The negative sign appears because an increase in the impact parameter s results in a decreasing scattering angle Θ. The differential cross-section is proportional to the probability for scattering into an angle between Θ and $\Theta + d\Theta$, because $2\pi s\,ds$ (see equation (3–63)) is the probability for encounter at impact parameter values between s and $s + ds$. We now can rewrite (3–63) in the form

$$\sigma(\Theta) = \frac{s\,ds}{\sin\Theta\,d\Theta} \tag{3-64}$$

which, together with expression (3–62) for s yields

$$\sigma(\Theta) = \frac{1}{4}\left(\frac{MG}{2\mathcal{E}}\right)^2 \frac{1}{\sin^4(\Theta/2)}\,. \tag{3-65}$$

3:14 Stellar Drag

If a high-velocity star moves through a surrounding field of low-velocity stars, it experiences a drag because it is slightly deflected in each *distant encounter*. We can compute this drag in an elementary way through the use of the scattering theory derived above.

First, we note that the star's velocity loss along the initial direction of approach to a scattering mass — which we will take here to be another star — is

$$\Delta v = v_0(1 - \cos \Theta) \, , \tag{3-66}$$

where v_0 is the approach velocity relative to the scattering center at large distances. This is not an overall velocity loss; just a decrease in the component along the direction of approach. The change in momentum is $\mu \Delta v$ where, again, μ is the reduced mass. The force on the high-velocity star, opposite to its initial direction of motion, therefore, is

$$F = \sum_i \frac{\mu_i \Delta v_i}{\Delta t} \, . \tag{3-67}$$

Here Δt is the time during which a change Δv_i takes place; the summation is taken over all stars i encountered during this time interval. This summation can be replaced by an integration over a large aggregate of stars with number density n. In terms of the probability or cross-section for scattering into angle Θ, at any given encounter, the force becomes

$$F = 2\pi \mu v_0^2 n \int_{\Theta_{\max}}^{\Theta_{\min}} (1 - \cos \Theta)\sigma(\Theta) \sin \Theta \, d\Theta \, . \tag{3-68}$$

This assumes that all the deflections due to interactions with individual stars are small, and that the forces along the direction of motion add linearly. Using equation (3–63) we have

$$F = 2\pi \mu v_0^2 n \int_{s_{\min}}^{s_{\max}} (1 - \cos \Theta)s \, ds \, . \tag{3-69}$$

Instead of integrating over all stars, we integrate the impact parameter s over all possible values for a single star and then multiply by the stellar density n. This is an equivalent procedure because the probability of encountering a star at impact parameter s is proportional to s and to n. The extra factor v_0 that appears in the expression takes account of the increasing number of encounters, per unit time, at large velocities. If we set $\theta_0 \equiv 0$ and $\theta_\infty \equiv 0$, then

$$- \cos \Theta = \cos 2\theta \equiv \frac{1 - \tan^2 \theta}{1 + \tan^2 \theta} \, , \tag{3-70}$$

but (see Fig. 3.9)

$$\cot \frac{\Theta}{2} \equiv \tan \theta = \frac{2\mathcal{E}s}{MG} = \frac{sv_0^2}{MG} \equiv \alpha s, \tag{3-71}$$

so that, in terms of this newly defined parameter α,

$$F = 2\pi\mu v_0^2 n \int s \frac{2}{1 + \tan^2 \theta} \, ds \tag{3-72}$$

and

$$F = 4\pi\mu v_0^2 n \int \frac{s \, ds}{1 + \alpha^2 s^2}, \tag{3-73}$$

$$F = \frac{4\pi\mu v_0^2 n}{2\alpha^2} \ln(1 + \alpha^2 s^2) \Big]_{s_{\min}}^{s_{\max}}. \tag{3-74}$$

We define a slowing down time, or *relaxation time*, τ,

$$\tau \equiv \frac{\mu v_0}{F}. \tag{3-75}$$

In this calculation we have assumed that the star is moving through an assembly of stationary *field stars*. As long as the random motion of these stars is low compared to v_0, equation (3-74) holds quite well. However, when the random stellar velocities approach v_0, the particle can alternately be accelerated or slowed down by collisions and the above derivation no longer holds. For the Sun, moving with a velocity of $v_0 = 20 \, \mathrm{km \, s^{-1}}$ through the ambient star field, $\alpha = 3 \times 10^{-14} \, \mathrm{cm^{-1}}$, $n \sim 10^{-56} \, \mathrm{cm^{-3}}$, and $\mu \sim 10^{33}$ g. If $s_{\max} \sim 10^{19}$ cm, roughly the mean separation of stars, and $s_{\min}$ is much smaller, then $F \sim 10^{18}$ dyn.[4] However, even with this large force, the time is $\tau \sim 10^{21}$ s — far greater than the estimated age of the Universe. This large value of τ is disconcerting because it is symptomatic of a general problem in stellar dynamics. We find such aggregates as globular clusters to be in configurations close to those we would expect in thermodynamic equilibrium. This would mean that the stars must interact quite strongly to transfer energy to each other; and yet the above mechanism will not accomplish this at anywhere near a satisfactory rate, and neither will other mechanisms of the same general class. The interaction of these stars must be dominated by some other process. We will discuss this again in Section 3:16. However, we might note that interaction of stars with gas clouds or clouds of stars produces a larger effect than that of individual stellar encounters (Sp51a). If the mass of the cloud is $M \sim 10^6 M_\odot$ and $n \sim 10^{-65} \, \mathrm{cm^{-3}}$, F increases by 10^3, and τ decreases by 10^3. Here $s_{\max}$ might be chosen as $\sim 10^{22}$ cm.

Collisions need not always act to slow down particles. When stars in the plane of the Galaxy interact with the much more massive clouds of gas, they can actually become accelerated to high velocities. In Table A.6 we show that, relative to the Sun, older stars have higher root mean squared random velocities than younger stars. This

---

[4] When $s_{\max}$ is much larger than the mean separation, encounters begin to overlap in time, and (3-74) becomes an overestimate of F because the effects of individual stars will tend to cancel through symmetry in their distribution (Ch43)

$$\therefore \quad s_{\max} \sim n^{-1/3}. \tag{3-76}$$

may be due to collisions with such clouds. As we will see in Chapter 4 an assembly of bodies tends to arrange itself in such a way that translational energies are equal (*equipartition of energy*). The massive clouds, therefore, tend to pass some of their energy on to the less massive stars and, in so doing, accelerate them to velocities higher than the cloud velocity v_c.

A quite different class of problems in which the above calculations are useful deals with charged particles. The inverse square law electrostatic forces allow us to derive equations quite similar to (3–74) and (3–75), and we can compute the electrostatic drag on fast electrons traveling through the interstellar medium and on charged interstellar or interplanetary dust grains moving through a partially ionized medium. In Section 6:18 we will also see that the distant collisions of electrons and ions are described by equations such as (3–69) and that the opacity or emissivity of an ionized plasma can be computed making use of these equations. The radio emission from hot ionized interstellar gas can then be directly related to the plasma density, or rather to the collision frequency in a line-of-sight column through the cloud.

3:15 Virial Theorem

The theorem we will prove here again is statistical. It describes the overall dynamic behavior of a large assembly of bodies, rather than the precise behavior of any individual body belonging to the assembly.

Consider a system of masses m_j at positions $\mathbf{r}_j$. Let the force on m_j be $\mathbf{F}_j$. We now write the identity

$$\frac{d}{dt} \sum_j \mathbf{p}_j \cdot \mathbf{r}_j = \sum_j \mathbf{p}_j \cdot \dot{\mathbf{r}}_j + \sum_j \dot{\mathbf{p}}_j \cdot \mathbf{r}_j \qquad (3\text{-}77)$$

$$= 2\mathbb{T} + \sum_j \mathbf{F}_j \cdot \mathbf{r}_j , \qquad (3\text{-}78)$$

where $\mathbb{T}$ is the kinetic energy of the entire system and the time derivative of the momentum $\dot{\mathbf{p}}_j$ is equal to the force $\mathbf{F}_j$. For the moment we do not identify the left side of the equation with any physically interesting quantity. Taking the time average of both sides, we obtain

$$\frac{1}{\tau} \int_0^\tau \frac{d}{dt} \sum_j \mathbf{p}_j \cdot \mathbf{r}_j \, dt = \left\langle 2\mathbb{T} + \sum_j \mathbf{F}_j \cdot \mathbf{r}_j \right\rangle , \qquad (3\text{-}79)$$

where the brackets denote a time average. A particularly interesting problem concerns a bound system in which each member of the assembly remains a member for all time. Then all the $\mathbf{r}_j$ values must remain finite because no particle escapes from the system, and all $\mathbf{p}_j$ values must remain finite because the total energy of the system is finite.

Since $\sum_j \mathbf{p}_j \cdot \mathbf{r}_j$ remains finite, the integral of its derivative must also remain finite for all time. Now, τ, the time over which we average, can be made arbitrarily large. So, the left side of equation (3–79) approaches zero and we can set

$$\langle 2\mathbb{T} \rangle + \left\langle \sum_j \mathbf{F}_j \cdot \mathbf{r}_j \right\rangle = 0 \,. \tag{3-80}$$

If the force is derivable from a potential, this equation becomes

$$\langle 2\mathbb{T} \rangle - \left\langle \sum_j m_j \nabla \mathbb{V}(r_j) \cdot \mathbf{r}_j \right\rangle = 0 \,, \tag{3-81}$$

where $m_j \mathbb{V}(r_j)$ is the potential energy of mass m_j at position r_j. The force then is a function of position only and can be written in terms of the negative gradient ∇ of the potential:

$$\mathbf{F}_j = -m_j \nabla \mathbb{V}(r_j) \,. \tag{3-82}$$

If the potential is proportional to r^n, the gradient lies along the radial direction and

$$\sum_j m_j \nabla \mathbb{V}(r_j) \cdot \mathbf{r}_j = \sum_j m_j \frac{\partial \mathbb{V}(r_j)}{\partial r_j} r_j \,. \tag{3-83}$$

Calling the total potential energy of the entire assembly $\mathbb{V}$ we obtain

$$\langle \mathbb{T} \rangle = \frac{n}{2} \langle \mathbb{V} \rangle, \qquad \mathbb{V} \equiv \sum_j m_j \mathbb{V}(r_j), \qquad -2 < n \,. \tag{3-84}$$

This relation runs into difficulty for $n < -2$, because the total energy $\langle \mathbb{T} \rangle + \langle \mathbb{V} \rangle$ would then be positive, indicating that the system would no longer be bound. For an inverse square law force, as in gravitation or electrostatics, the potential goes as the inverse first power, $n = -1$, and

$$\langle \mathbb{T} \rangle = -\frac{1}{2} \langle \mathbb{V} \rangle \,. \tag{3-85}$$

This theorem is of great importance and finds many applications in astrophysics. It provides an estimate for the mass of clusters of galaxies, obtained by observing the spread in radial *Doppler velocities* among different galaxies in the cluster. The assumption of an isotropic distribution of velocities then yields a mean kinetic energy, and equation (3–85) gives the mean potential energy. If a typical cluster diameter is known from the cluster's distance and from the angle it subtends in the sky, we can obtain a rough estimate of the total cluster mass on the assumption that

$$-\frac{\mathbb{V}}{M} \sim \frac{MG}{R} \,. \tag{3-86}$$

Here M is the cluster mass and R is some weighted cluster radius, somewhat smaller than the observed radius of the cluster. The estimated cluster mass would normally be in error (i.e., too high) by a factor less than ~2 if the actually observed cluster radius is used in equation (3–86).

An interesting problem arises when we measure the masses of clusters of galaxies making use of the virial theorem. The masses of individual galaxies within the cluster can be determined from their rotations (Problem 3–12, below). From these we can compute the potential energy of the entire cluster when the cluster dimensions are derived from the apparent diameter and red-shift distance. An independent estimate of the potential energy, however, is obtained from (3–85) if the random velocities of the individual galaxies are taken to compute $\mathbb{T}$. To do this, we note the variations in red shifts from galaxy to galaxy and estimate the actual random velocities. Strangely the results of using (3–85) always give values of $\langle \mathbb{T} \rangle$ and, hence, $\langle \mathbb{V} \rangle$ that are about an order of magnitude higher than the total potential energy computed on the basis of individual galactic masses. We conclude that either: (i) there is a lot of undetected *dark matter* in clusters; or (ii) the clusters are breaking up; or (iii) we do not understand dynamics on such a large scale. The problem of dark matter, as it relates to the mass distribution within galaxies and clusters of galaxies, will be discussed further in Chapters 9 and 13. *Dark matter* is a generic term used for any type of matter making itself felt gravitationally, but not detectable, at least to date, by direct observational means. Problem 4–5 and the discussion following it treat the cluster problem from an observational viewpoint.

3:16 Stability Against Tidal Disruption

When a swarm of gravitationally bound particles having a total mass m approaches too close to a massive object M, the swarm tends to be torn apart. The same fate can confront a solid body held together by gravitational forces when it approaches a much more massive object.

The reason is simple. If we consider that the center of mass of the swarm is at a distance r from the mass M, and is falling straight toward it, then its acceleration toward M is $-MG/r^2$. Let r' be the swarm radius. A particle P_0 (Fig. 3.10), at the surface of the swarm nearest to M, would be accelerated at a rate $-MG/(r - r')^2$ toward M, were it not for a gravitational attraction from the center of the swarm that tends to accelerate it away from M at a rate mG/r'^2. In order for the particle to be pulled steadily away from the swarm, we must have the condition

$$MG \left(-\frac{1}{r^2} + \frac{1}{(r - r')^2} \right) > \frac{mG}{r'^2} .$$

(3-87)

Expanding the expression on the left for $r' \ll r$ and keeping only terms down to first order in r', we obtain

$$\frac{2M}{r^3} > \frac{m}{r'^3} .$$

(3-88)

Similarly for a swarm in a perfectly circular orbit about M, disruption occurs when

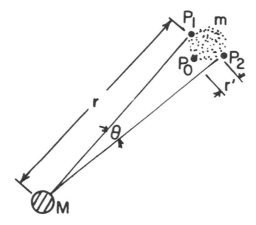

Fig. 3.10. A swarm of gravitationally bound particles — stars, atoms, molecules — can be tidally disrupted through a close encounter with a massive object M.

$$\frac{3M}{r^3} > \frac{m}{r'^3}.$$ (3-89)

PROBLEM 3–3. Derive the result (3–89). In doing this, it is helpful to think of the swarm as moving without rotation about its center, and to consider its center of mass as having a *centrifugal repulsion* per unit mass

$$F_c = r\dot\theta^2$$ (3-90)

away from M. This is different from the "repulsion" $(r - r')\dot\theta^2$ at P_0. As noted earlier, Newton's laws of motion hold only in inertial frames of reference. In rotating systems, where they do not apply, the concept of centrifugal forces is sometimes useful but needs handling with caution.

The precise ratios of the masses M and m will therefore vary with different orbits and the rotation of m will also play a role in determining its stability. What is important to note, however, is that the density of the swarm is a more important consideration than its actual mass or size taken individually.

There is a second effect that also plays an important role. Again, consider a direct infall. Here points P_1 and P_2 (Fig. 3.10) would be accelerated radially toward M and would tend to converge. The effective acceleration of P_1 and P_2 relative to each other would be roughly

$$2\frac{MG}{r^2}\frac{r'}{r} = \frac{2MGr'}{r^3}$$ (3-91)

due to this effect taken by itself. This is important whenever it is larger than the acceleration mG/r'^2 due to the mass of the swarm itself, that is, when (3–88) holds.

A lateral compression accompanies the tidal disruption and tends to concentrate the swarm, while the tidal forces attempt to tear it apart. What actually happens under these combined effects will be better understood in terms of the Liouville theorem, which we will discuss in Section 4:14.

Tidal disruption seems to play a leading role in many phenomena. Comets that approach too close to the Sun or even too close to the massive planet Jupiter have been observed to break up into fragments, and the general nature of the tidal theory seems to be borne out. Compact stars, like white dwarfs or neutron stars, are able to accrete material tidally stripped off the outer layers of more massive, but also more distended giant companions. As we saw in Section 1:11, the Magellanic Stream attests to tidal action even on the scale of galaxies. Massive galaxies are able to capture material gravitationally stripped from smaller companions.

The orbits of globular clusters cross the Galactic plane and central regions where, calculations indicate, the clusters are tidally destroyed, typically with a half-life of one Hubble time (Gn97). The Galaxy might at earlier epochs have contained substantially more globular clusters than today. We can now see why the interaction of stars within a globular cluster may only play a limited role in determining the ultimate velocity distribution of stars in the cluster. The treatment of Section 3:14, and the very long star encounter relaxation time τ predicted by equation (3–75) may not give a true picture of the actual evolution of clusters into the well-defined, compact, spherical aggregates we observe. Interaction with the Galactic nucleus may have an appreciable, perhaps dominant, influence on the distribution of stars in a globular cluster, by tidally stripping away the more loosely bound, higher-velocity stars from the cluster, leaving the residual cluster containing just its more tightly bound stars. We will touch on the globular cluster problem again in Section 4:23.

3:17 Lagrangian Equations

A physical system can be most readily understood in terms of a coordinate system that most closely mirrors its symmetries and peculiarities. A Cartesian coordinate system is not particularly convenient for treating a system that has spherical symmetry, and for more complex systems a choice of correspondingly more complex coordinates can greatly facilitate calculations. A scheme for the use of arbitrary coordinate systems involves working with *generalized coordinates*. These are variables that do not need to solely involve position and time. The evolution of a system may be more readily described in terms of positions, momenta, and time, or some other choice of variables.

We can relate some standard set of coordinates $r_j = r_1, r_2, \ldots, r_M$ through a set of *transformation equations*

$$\mathbf{r}_1 = \mathbf{r}_1(q_1, q_2, \ldots, q_N, t)$$

$$\mathbf{r}_2 = \mathbf{r}_2(q_1, q_2, \ldots, q_N, t)$$

$$\cdots \qquad (3\text{-}92)$$

$$\mathbf{r}_M = \mathbf{r}_M(q_1, q_2. \ldots, q_N, t)\,,$$

where M and N generally are unequal. N maximally equals $3M$; this maximum is attained if the M particles are totally free to move about and interact. We then require $3M$ independent variables to fully define the positions of all M particles. In contrast, if there are ℓ constraints on the particles, the total number of independent variables q_i diminishes to $3M - \ell$: The Sun is orbited by its set of planets with their respective moons. The planets are constrained to orbit the Sun, and the moons are constrained to circle their planets. These constraints lower the total number of independent variables required to describe the motions of all these bodies within the Solar System.

The time rate of the variable $\mathbf{r}_j$ denoted by $\mathbf{v}_j \equiv d\mathbf{r}_j/dt \equiv \dot{\mathbf{r}}_j$ is given by the rules of partial differentiation

$$\mathbf{v}_j = \sum_i \frac{\partial \mathbf{r}_j}{\partial q_i} \dot{q}_i + \frac{\partial \mathbf{r}_j}{\partial t}\,. \qquad (3\text{-}93)$$

We now set up the mathematical identity

$$\sum_j m_j \ddot{\mathbf{r}}_j \cdot \frac{\partial \mathbf{r}_j}{\partial q_i} = \sum_j \left\{ \frac{d}{dt}\left(m_j \dot{\mathbf{r}}_j \cdot \frac{\partial \mathbf{r}_j}{\partial q_i}\right) - m_j \dot{\mathbf{r}}_j \cdot \frac{d}{dt}\left(\frac{\partial \mathbf{r}_j}{\partial q_i}\right)\right\}\,. \qquad (3\text{-}94)$$

We can change the order of differentiation with respect to t and q_i in the last term, and from equation (3–93) obtain

$$\frac{\partial \mathbf{v}_j}{\partial q_j} = \frac{d}{dt}\left(\frac{\partial \mathbf{r}_j}{\partial q_i}\right) = \sum_k \frac{\partial^2 \mathbf{r}_j}{\partial q_i \partial q_k}\dot{q}_k + \frac{\partial^2 \mathbf{r}_j}{\partial q_i \partial t}\,. \qquad (3\text{-}95)$$

Equation (3–93) also implies that

$$\frac{\partial \mathbf{v}_j}{\partial \dot{q}_i} = \frac{\partial \mathbf{r}_j}{\partial q_i}\,. \qquad (3\text{-}96)$$

With this equation (3–94) can be written as

$$\sum_j m_j \ddot{\mathbf{r}}_j \cdot \frac{\partial \mathbf{r}_j}{\partial q_i} = \sum_j \left\{ \frac{d}{dt}\left(m_j \mathbf{v}_j \cdot \frac{\partial \mathbf{v}_j}{\partial \dot{q}_i}\right) - m_j \mathbf{v}_j \cdot \frac{\partial \mathbf{v}_j}{\partial q_i}\right\} \qquad (3\text{-}97)$$

$$= \left\{ \frac{d}{dt}\left(\frac{\partial}{\partial \dot{q}_i}\sum_j \frac{1}{2}m_j v_j^2\right) - \frac{\partial}{\partial q_i}\left(\sum_j \frac{1}{2}m_j v_j^2\right)\right\}\,.$$

Now, equation (3–4) summed over all the components of a force can be rewritten as

$$\sum_j (\mathbf{F}_j - m_j \ddot{\mathbf{r}}_j) = 0 = \sum_j (\mathbf{F}_j - m_j \dot{\mathbf{p}}_j)\,. \qquad (3\text{-}98)$$

Defining a generalized force whose components are

$$Q_i \equiv \sum_j \mathbf{F}_j \cdot \frac{\partial \mathbf{r}_j}{\partial q_i} \,, \tag{3-99}$$

and identifying $\sum_j (1/2) m_j v_j^2$ in (3–97) with the system kinetic energy $\mathbb{T}$, we can finally write

$$\sum_i \left[\frac{d}{dt} \left(\frac{\partial \mathbb{T}}{\partial \dot{q}_i} \right) - \frac{\partial \mathbb{T}}{\partial q_i} - Q_i \right] \,. \tag{3-100}$$

For N independent variables q_i we then have N independent equations

$$\frac{d}{dt} \left(\frac{\partial \mathbb{T}}{\partial \dot{q}_i} \right) - \frac{\partial \mathbb{T}}{\partial q_i} = Q_i \,. \tag{3-101}$$

If the force F can be derived from a scalar potential $\mathbb{V}(q_i)$ as $F = m_i \nabla_i \mathbb{V}(q_i)$, then $Q_i = -m_i \partial \mathbb{V}(q_i)/\partial q_i$. Such systems are called *conservative*. From (3–101) we then have

$$\frac{d}{dt} \left(\frac{\partial \mathbb{T}}{\partial \dot{q}_i} \right) - \frac{\partial (\mathbb{T} - \mathbb{V})}{\partial q_i} = 0 \,, \tag{3-102}$$

and because $\mathbb{V} \equiv \sum_i m_i \mathbb{V}(q_i)$ is a function of position alone, independent of time,

$$\frac{d}{dt} \left(\frac{\partial (\mathbb{T} - \mathbb{V})}{\partial \dot{q}_i} \right) - \frac{\partial (\mathbb{T} - \mathbb{V})}{\partial q_i} = 0, \qquad i = 1, 2, \ldots, N \,. \tag{3-103}$$

Equations (3–103) are called the *Lagrange equations*. We now define a new function called the *Lagrangian*, $\mathcal{L}$,

$$\mathcal{L} \equiv \mathbb{T} - \mathbb{V} \,, \tag{3-104}$$

in terms of which we can write the Lagrange equations as

$$\frac{d}{dt} \left(\frac{\partial \mathcal{L}}{\partial \dot{q}_i} \right) - \frac{\partial \mathcal{L}}{\partial q_i} = 0, \qquad i = 1, 2, \ldots, N \,. \tag{3-105}$$

Although much of what we have done in this section revolved around purely mathematical transformations, we should pay particular attention to the physical assumptions we made. We assumed that there exists a set of independent generalized coordinates q_i to which our set of standard coordinates is related by equations (3–92), and that the system of particles interacts through forces derivable from a scalar potential function $\mathbb{V}$ dependent on position alone.

Let us now look at a system of pointlike masses that interact through forces derived from potentials solely dependent on position. For each individual mass m_i instantaneously located at some point (x_i, y_i, z_i) we can then write

$$\frac{\partial \mathcal{L}}{\partial \dot{x}_i} = \frac{\partial \mathbb{T}}{\partial \dot{x}_i} = \frac{\partial}{\partial \dot{x}_i} \sum_i \frac{m_i}{2} (\dot{x}_i^2 + \dot{y}_i^2 + \dot{z}_i^2) = m_i \dot{x}_i \,. \tag{3-106}$$

Because $\mathbb{V}$ depends on position only, it does not appear in this expression. But we see that $\partial\mathcal{L}/\partial x_i$ is just the momentum of particle i along the x-direction.

This leads to the concept of a generalized momentum corresponding to generalized coordinates q_i and defined as

$$p_k \equiv \frac{\partial\mathcal{L}}{\partial\dot{q}_k} \, . \tag{3-107}$$

This is generally referred to as the *canonical momentum* which, we should note, does not generally have the dimensions of a linear momentum.

PROBLEM 3–4. Set up the Lagrangian for the same system of particles as in equation (3–106) but expressed in spherical polar coordinates, where the components of the velocity squared are $\dot{r}_i^2$, $r_i^2\dot{\theta}_i^2$, $r_i^2\sin^2\theta_i\dot{\phi}_i^2$, and show that the canonical momenta associated with the coordinates $(r_i, \theta_i,$ and $\phi_i)$ are, respectively $m_i\dot{r}_i$, $m_i r_i^2\dot{\theta}_i$, and $m_i r_i^2\sin^2\theta_i\dot{\phi}_i$. Only the first of these is a linear momentum. The last two are angular momenta. This shows that even for a given system of particles the choice of coordinates can decide whether a generalized momentum corresponds to a linear momentum.

If the Lagrangian of a system is not a function of some coordinate q_k then equation (3–105) reduces to

$$\frac{d}{dt}\left(\frac{\partial\mathcal{L}}{\partial\dot{q}_k}\right) = \frac{dp_k}{dt} = 0 \, , \tag{3-108}$$

meaning that p_k is constant. The coordinate q_k is then said to be *cyclic*, and the generalized momentum p_k conjugate to q_k is said to be *conserved*.

PROBLEM 3–5. Consider a planet of mass m orbiting a star of mass M that exerts a potential $\mathbb{V}(r) = -MG/r$ at the planet's position r. Show that ϕ is a cyclic coordinate.

The motion of a mass m in the vicinity of another mass M is given by the Lagrangian

$$\mathcal{L} = \mathbb{T} - \mathbb{V} = \frac{m}{2}(\dot{r}^2 + r^2\dot{\theta}^2 + r^2\sin^2\theta\dot{\phi}^2) + \frac{MmG}{r} \, . \tag{3-109}$$

From this we obtain the angular momentum component

$$p_\theta = \frac{\partial\mathcal{L}}{\partial\dot{\theta}} = mr^2\dot{\theta} \, . \tag{3-110}$$

Differentiating with respect to time and using (3–105), we have

$$\frac{dp_\theta}{dt} = \frac{d}{dt}\frac{\partial\mathcal{L}}{\partial\dot{\theta}} = \frac{d}{dt}mr^2\dot{\theta} = \frac{\partial\mathcal{L}}{\partial\theta} = m(r^2\sin\theta\cos\theta)\dot{\phi}^2 \, . \tag{3-111}$$

The spherical symmetry of the space surrounding mass M allows us to assign any value to the coordinate θ at some arbitrary point. Let us pick a point along the orbiting particle's trajectory where $\dot{\theta} = 0$ and assign the value $\pi/2$ to θ there.

PROBLEM 3–6. Convince yourself, by carrying out the actual differentiation, that $\ddot{\theta}$ must then also equal zero at this point, which means that θ must remain constant, $\theta = \pi/2$. The motion proceeds in an equatorial plane.

In deriving Kepler's Laws in Section 3:5 we had assumed motion in a plane. The spherical symmetry of the problem made this an obvious choice, because there were no forces to impel a planet to abandon the initial plane of its motion about the Sun. But now we have explicitly shown that the motion of a solitary planet orbiting a solitary star must be confined to a plane. The symmetry of the motion corresponds to the conservation of the angular momentum component p_θ. Indeed, one can show that all conservation laws correspond to symmetry properties, a relationship first demonstrated by Emmy Noether and known as Noether's theorem.

In the Solar System, with its many orbiting bodies, the most massive planet, Jupiter, tends to force all the other planets to move in orbits close to its own plane. But the mutual interactions of the planets lead to deviations from motions strictly confined to planes. The spherical symmetry of the system is broken once there are more than two gravitationally attracting bodies.

We will encounter the Lagrangian equations again in Chapter 5 where we will consider particles orbiting black holes. There, the gravitational fields are strong, and it is convenient to work with coordinates other than those of a stationary observer viewing the orbital motions from afar.

Additional Problems

3–7 The orbital period for the Earth moving about the Sun is given by equation (3–47). Averaged over the Earth's eccentric orbit, the distance of the Sun, obtained by the radar method described in Section 2:1, has a mean value of 1.5×10^{13} cm. Assuming the Earth's mass, $M_\oplus \ll M_\odot$, show that the Sun's mass is $M_\odot = 2.0 \times 10^{33}$ g.

3–8 A radar signal reflected from the Moon returns 2.56 s after transmission. The speed of light is 3.00×10^{10} cm s^{-1}. Assume the period of the Moon to be roughly 27.3 days. Find the mass of the Earth assuming the Moon's mass to be small compared to that of the Earth.

Note: In this way we can determine the mass of any planet with a moon. When a planet has no moon, its mass is determined by the perturbations it produces on the orbits of nearby planets. Such a calculation is quite time-consuming, but introduces no essentially new physical concepts. The calculations proceed within the framework of Newtonian dynamics.

3–9 Because the Moon and the Earth revolve about a common center of mass, the apparent motion of Mars has a periodicity of one month superposed on its normal orbit. The distance of the Moon is $D \sim 3.8 \times 10^5$ km. The distance of Mars at closest approach is $L \sim 5.6 \times 10^7$ km. The apparent displacement of Mars over a period of half a month then is $\sim$34 sec of arc. What is the mass of the Moon?

3–10 A meteor approaches the Earth with a speed v_0 when it is at a very large distance from the Earth. Show that the meteor will strike the Earth, at least at grazing incidence, if its impact parameter s (Fig. 3.11) is given by

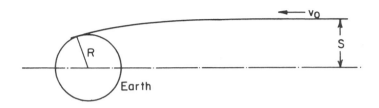

Fig. 3.11. Impact of a meteorite or a cloud of meteors on the Earth's atmosphere.

$$s \leq [R^2 + 2MGRv_0^{-2}]^{1/2} .$$

3–11 If a cloud of meteors approaches the Earth at relative speed v_0, show that the rate of mass capture is $\pi(R^2 v_0 + 2MGR/v_0)\rho$, where ρ is the mass density of the cloud. Both here and in Problem 3–10 we neglect the Sun's influence on the meteors.

3–12 A disk-shaped rotating galaxy is seen edge on. By Doppler-shift spectroscopic measurements we can determine the speed V with which the stars near the edge of the galaxy rotate about its center. Show that the mass of the galaxy in terms of the observed velocity is $\sim V^2 R/G$. State the assumptions made. R is the radius of the galaxy.

3–13 In the vicinity of young star clusters we occasionally see *runaway stars*, O or B stars that evidently were part of the cluster until recently but are receding rapidly. Blaauw (Bℓ61) suggested that the runaways initially may have been part of binaries in which the companion exploded as a supernova, leaving only part of its mass behind. Suppose that the initial motion was circular, with initial orbital velocity v for the surviving star, whose mass is m. If the initial mass of the companion was M, and its final mass after the explosion is only $M/10$, what will be the final velocity V of the runaway star at large distance from the explosion? Refer v and V to the system's center of mass.

3–14 A gravitationally bound body spins rapidly (but not at relativistic velocities). At what rotational velocity will it break up if its mass is m and its radius is r? Assume the body remains spherical until breakup — even though this assumption normally will not hold.

3–15 Observations on the compact radio source 3C 279, which is occulted by the Sun once a year, show that radio waves are bent as they pass very close to the Sun (Hi71). Show that this bending is a consequence of the equivalence principle. We will discuss the effect more thoroughly in Chapter 5.

Answers to Selected Problems

3–1.

$$\dot{r} = -\frac{1}{y^2}\dot{\theta}\frac{dy}{d\theta} = -h\frac{dy}{d\theta} \, ,$$

$$\ddot{r} = -h\dot{\theta}\frac{d^2y}{d\theta^2} = \frac{-h^2}{r^2}\frac{d^2y}{d\theta^2} \, .$$

Substituting in (3–30), we see that

$$\frac{d^2y}{d\theta^2} + y = \frac{MG}{h^2} \, .$$

Substitution of $y = B\cos(\theta - \theta_0) + MG/h^2$ satisfies the equation.

3–2. $m\ddot{r} = GmM_{\oplus}/(R_{\oplus} + H)^2$ at a height $H \ll R_{\oplus}$. If we take $M_{\oplus} = \rho_{\oplus}(4/3)\pi R_{\oplus}^3$, where the symbols represent the Earth's mass, density, and radius, we can estimate G from the measured acceleration, $G \sim g[\rho_{\oplus}(4\pi/3)R_{\oplus}]^{-1}$.

3–3. At the center of mass of the swarm, the centrifugal and gravitational forces are equal: $(r\dot{\theta})^2 = GM/r$. A particle p at the swarm's near surface, will experience a centrifugal acceleration away from M, smaller than that of the swarm's center by MGr'/r^3. It will also experience a stronger gravitational acceleration toward M, by

$$\frac{MG}{r^2}\left[-1 + \frac{r^2}{(r-r')^2}\right] \, .$$

For disruption to occur these accelerations must be stronger than mG/r'^2. Expanding this inequality for $r \ll r'$ gives (3–89)

$$\frac{3M}{r^3} > \frac{m}{r'^3} \, .$$

This solution assumes no rotation of the swarm.

3–4. This follows from the Lagrangian

$$\mathcal{L} = \frac{m}{2}(\dot{r}_i^2 + r_i^2\dot{\theta}_i^2 + r_i^2\sin^2\theta_i\dot{\phi}_i^2) - \mathbb{V}(r_i, \theta_i, \phi_i) \, . \tag{3-112}$$

3–5. Expression (3–112) inserted into (3–105) demonstrates that p_ϕ is constant, meaning that ϕ is cyclic.

3–8. As in the suggested approach to Problem (3–7), make use of equation (3–47) to arrive at an answer.

3–9. Let m be the lunar mass and $M_\oplus$ the terrestrial mass. The distance R of the Earth from the center of mass is then given by

$$RM_\oplus = (D - R)m.$$

The apparent displacement of Mars is $2R/L$, where L is the distance to Mars. Hence $2R = 1.7 \times 10^{-4} L$. $R = 4.8 \times 10^3$ km and with $M_\oplus = 6.0 \times 10^{27}$ g we can evaluate m as $\sim 7.4 \times 10^{25}$ g.

3–10. Call V the velocity the meteor has at grazing incidence, that is, when it hits the Earth tangentially. Then this velocity is perpendicular to the radius vector R. We can therefore write conservation of angular momentum as

$$s v_0 = RV \ .$$

Conservation of energy per unit meteor mass gives

$$\frac{V^2}{2} = \frac{MG}{R} + \frac{v_0^2}{2} \ .$$

Eliminating V from these equations we obtain the expression

$$s = \left(R^2 + \frac{2MGR}{v_0^2} \right)^{1/2} . \tag{3-113}$$

All meteors with impact parameter less than s can also hit the Earth. This leads to the desired expression.

3–11. The number of meteors hitting Earth per second is given by the density of meteors in space, times the volume of the cylinder of radius s swept up in unit time:

$$\pi s^2 \cdot v_0 \cdot \rho \ .$$

The impact parameter s is given in Problem 3–10.

3–12. Assume circular motion. The mass of the galaxy M acting on a star at its periphery is then given by the relation between kinetic and potential energy per unit mass of the star, as in (3-44),

$$V^2 = \frac{MG}{R} \ .$$

3–13. This problem is somewhat complex. Before the explosion the surviving star's kinetic energy $mv_0^2/2$ equals half its potential energy $mMG/2r$, where r is the separation between the stars. If the explosion is so rapid that this separation does not appreciably change before the ejecta of the exploding star expand beyond r, the binding energy on m is reduced to $mMG/10r$, so that its kinetic energy $mV^2/2$ now equals $9mMG/10r + mv_0^2/2$, which can be solved for V. In addition, however, the exploding star initially had momentum relative to the binary system's center of mass. If the explosion is spherically symmetric the ejecta escape with 90% of this momentum, so that the surviving two stars suffer a recoil. In addition, some supernovae appear to eject mass asymmetrically, leading to a further recoil of the remnant $M/10$, some of which will be transferred gravitationally to m.

3–14. The centrifugal force per unit mass exceeds the gravitational attraction $r\omega^2 > mG/r^2$; $\omega > \sqrt{mG/r^3}$.

3–15. Imagine an observer falling toward the Sun in a spaceship. Light rays passing by the Sun enter the window of his cabin. The equivalence principle states that he should see the light moving in a straight line. But because he is accelerating toward the Sun, this means that the rays must also be following a path curving toward the Sun.

4 Random Processes

4:1 Random Events

If a bottle of ether is opened at one end of a room, we can soon smell the vapors at the other end. But the ether molecules have not traversed the room in a straight line, nor in a single bound. They have undergone myriad collisions with air molecules, bouncing first one way, then another in a random walk that takes some molecules back into the bottle from which they came, others through a crack in the door, and yet others into the vicinity of an observer's nose where they can be inhaled to give the sensation of smell.

In general, molecules diffuse through their surroundings by means of two processes: (i) individual collision with other atoms and molecules; and (ii) turbulent and convective bulk motions that involve the transport of entire pockets of gas. These, too, are the mechanisms that act to mix the constituents of stellar and planetary atmospheres. Both processes give rise to random motions that can best be statistically described.

In an entirely different context, think of a broadband amplifier whose input terminals are not connected to any signal source. On displaying the output on an oscilloscope, we would find that the trace contains nothing but spikes, some large, others smaller, looking much like blades of grass on a dense lawn. An exact description of this pattern would be laborious; but a statistical summary in terms of mean height and mean spacing of spikes can be provided with ease and may in many situations present all the information actually needed.

The spikes are the noise inherent in any electrical measurement. If we are to detect, say, a radio-astronomical signal fed into the amplifier, we must be able to distinguish the signal from the noise. That can only be done if the statistics of the noise are properly understood.

Again, consider a third situation, a star embedded in a dense cloud of gas. Light emitted at the surface of the star has to penetrate through the cloud if it is to reach clear surroundings and travel on through space. An individual photon may be absorbed, re-emitted, absorbed again, and re-emitted many times in succession. The direction in which the photon is emitted may bear no relation at all to the direction in which it was traveling just before absorption. The photon may then travel about the cloud in short, randomly directed steps, until it eventually reaches the edge of the cloud and escapes. This *random walk* can be described statistically. We can estimate the total distance covered by the photon before final escape and, at any given time in its travel, we can predict the approximate distance of the photon from the star.

These three physically distinct situations can all be treated from a single mathematical point of view. In its simplest form each problem can be reduced to a random walk. We picture a man taking a sequence of steps. He may choose to take a step forward, or a step backward; but, for simplicity, we will assume that his step size remains constant. If the direction of each step is randomly determined, say by the toss of a coin, the man will execute a random walk. The toss of the coin might tell him that his first step should be backward, the next forward, the next forward again, backward, backward, forward, and so on. After 10 steps, how far will the man have moved from his initial position? How far will he be after 312 steps or after 10,000,000? We cannot give an exact answer, but we can readily evaluate the probability of his ending up at any given distance from the starting point.

4:2 Random Walk

Consider a starting position at some zero point. We toss a coin that tells the man to move forward or backward. He ends up at either the +1 or the −1 position (Fig. 4.1). If he ends up in the +1 position, the next toss of the coin will take him to the

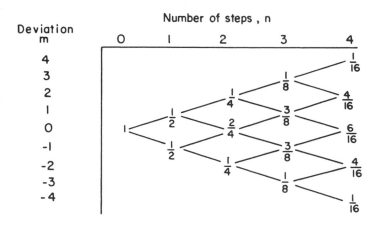

Fig. 4.1. Probability $P(m, n)$ of terminating at position m after n steps.

+2 or the 0 position, depending on whether the toss tells him to move forward or back. Similarly from the −1 position he could move to 0 or −2.

There exist two possible ways of arriving back at the zero position, and only one possible way of getting to the −2 or to the +2 position. Because all of these sequences are equally probable, there is a probability of $\frac{1}{4}$ that the man ends up in the +2 position, a probability of $\frac{1}{4}$ that he ends up at −2, and a probability of $\frac{1}{2}$ that he ends up in the zero position after two steps. The zero position is more probable because there are two distinct ways of reaching this position, whereas there is only one way to get to the $+2$ or -2 positions when only two steps are allowed.

Let us denote by $p(m, n)$ the number of ways of ending up at a distance of m steps from the starting point, if the man executes a total of n steps. We will call m the deviation from the starting position. We will call $p(m, n)$ the *relative probability* of terminating at distance m after n steps. The *absolute probability* $P(m, n)$ of terminating at position m, after n steps is displayed in Fig. 4.1 and is

$$P(m, n) = \frac{p(m, n)}{\sum_k p(k, n)}$$

$$= \frac{\text{number of paths leading to position } m}{\text{sum of all distinct paths leading to any position, } k}. \qquad (4\text{-}1)$$

The numerators $p(m, n)$ of the fractions in Fig. 4.1 have a binomial distribution; they are the same numbers that appear as coefficients in the expansion

$$\left[\frac{1}{x} + x\right]^n = x^n + nx^{n-2} + \frac{n(n-1)}{2!}x^{n-4} + \frac{n!x^{n-2r}}{(n-r)!r!} + \cdots + \frac{1}{x^n}. \qquad (4\text{-}2)$$

Knowing this, we can easily evaluate the sum of coefficients in the series $\sum_k p(k, n)$. It is the sum of the coefficients in the binomial expansion and can be obtained by setting $x = 1$ on the right side of equation (4–2).

Substituting $x = 1$ on the left side of (4–2) shows that the sum of terms must have the value 2^n:

$$\sum_{k=-n}^{n} p(k, n) = 2^n \qquad (4\text{-}3)$$

and

$$P(m, n) = \frac{p(m, n)}{2^n}. \qquad (4\text{-}4)$$

We note also that if the exponent of a given term in equation (4–2) represents the deviation m, in Fig. 4.1, then the coefficient of that term represents the relative probability $p(m, n)$. In that sense we can rewrite (4–2) as

$$\left(\frac{1}{x} + x\right)^n = \sum_{k=-n}^{n} p(k, n)x^k. \qquad (4\text{-}5)$$

Every second term of this series has a coefficient zero. If n is odd, there is no possibility of m being even, and vice versa. We now wish to determine the mean deviation from the zero position after a random walk of n steps. By this we mean the sum of distances reached in any of the 2^n possible paths that we could take, all divided by 2^n. Since there are $p(k, n)$ ways of reaching the distance k, the numerator of this expression is $\sum_k kp(k, n)$ and we see that the *mean deviation* $\langle k \rangle$ is

$$\langle k \rangle \equiv 2^{-n} \sum_{k=-n}^{n} kp(k, n)$$

$$= \frac{\text{sum of all possible terminal distances after } n \text{ steps}}{\text{number of all possible paths using } n \text{ steps}} . \tag{4-6}$$

We notice from Fig. 4.1 and from the binomial distribution (4–2) that the relative probability $p(k, n)$ of having a deviation k equals the relative probability of having deviation $-k$: $p(-k, n) = p(k, n)$. Because the summation in (4–6) is carried out over values from $-n$ to n, there will be an exact cancellation of pairs involving $k = m$ and $-m$, and the only uncancelled term is the one having $k = 0$. This shows that the value of $\langle k \rangle$ must be zero also. The mean deviation from the starting position is zero, no matter how many steps we take.

This does not mean that the absolute value of the deviation is zero. Far from it. But there are equally many ways of ending up at a positive as at a negative distance and the average position is right at the starting point itself.

This much is evident from symmetry. However, we usually need to know something about the absolute distance reached after n steps. For example, we want to know the actual distance from a star that a photon has traveled after n absorptions and re-emissions in a surrounding cloud. A useful measure of such distances is the *root mean square deviation*, known also as the *standard deviation σ*

$$\sigma \equiv \langle k^2 \rangle^{1/2} = \left[\frac{\sum_{k=-n}^{n} k^2 p(k, n)}{\sum_{k=-n}^{n} p(k, n)} \right]^{1/2} = \left[\frac{\text{sum of (distances)}^2}{\text{sum of all possible paths}} \right]^{1/2} . \tag{4-7}$$

This is obtained by first taking the mean of the deviation squared $\langle k^2 \rangle$, and then taking the root of this mean value to obtain a deviation in terms of a number of steps of unit length. If we did not take the square root, the quantity obtained would have to be measured in units of $(\text{step})^2$; this is an area, rather than a length or distance. To evaluate the sum

$$\sum_{k=-n}^{n} k^2 p(k, n) \tag{4-8}$$

we can employ a simple technique. We substitute the quantity $x = e^y$ in equation (4–5) and differentiate twice in succession with respect to y. In the limit of small y-values, we then obtain

$$\sum_{k=-n}^{n} k^2 p(k, n) = \frac{d^2}{dy^2} \sum_{k=-n}^{n} p(k, n) e^{ky} = \lim_{y \to 0} \frac{d^2}{dy^2} (e^{-y} + e^y)^n$$

$$= [n(n-1)(e^{-y} + e^y)^{n-2}(e^y - e^{-y})^2 + n(e^{-y} + e^y)^n]_{y=0}$$

$$= n2^n . \tag{4-9}$$

In summary, we can write

$$\sum_{k=-n}^{n} k^2 p(k, n) = n2^n .$$

(4-10)

Equations (4–3) and (4–10) can now be substituted into (4–7) to obtain a standard deviation

$$\sigma = n^{1/2} .$$

(4-11)

After n steps of unit length the absolute value of the distance from the starting position is approximately $n^{1/2}$ units.

The following four problems widen the applications of the random walk concept.

PROBLEM 4–1. For a one-dimensional random walk involving steps of unequal lengths, prove that the mean position after a given number of steps is the starting position.

Note that for a finite number of different step lengths, this walk can be reduced to a succession of random walks, each walk having only one step length.

PROBLEM 4–2. Prove that the root mean square deviation for a walk involving the sum of different numbers n_i of steps of length λ_i is

$$\sigma = N^{1/2} \lambda_{\rm rms} ,$$

(4-12)

where $N = \sum_i n_i$ and $\lambda_{\rm rms}$ is the root mean square value of the step length

$$\lambda_{\rm rms} = \left[\frac{\sum_i n_i \lambda_i^2}{N} \right]^{1/2} .$$

(4-13)

Such random deviations are said to add in *quadrature*.

PROBLEM 4–3. Show that the root mean square deviation in a three-dimensional walk with step length L_0 is $s^{1/2} L_0$ after s steps. To show this, take the three Cartesian components of the $i^{\rm th}$ step (see Fig. 4.2) as

$$L_0 \cos \theta_i, \qquad L_0 \sin \theta_i \cos \phi_i, \qquad L_0 \sin \theta_i \sin \phi_i .$$

(4-14)

The mean square deviations along the three coordinates are, respectively,

$$\sigma_z^2 = \sum_{i=1}^{s} L_0^2 \cos^2 \theta_i , \qquad \sigma_x^2 = \sum_{i=1}^{s} L_0^2 \sin^2 \theta_i \cos^2 \phi_i ,$$

(4-15)

$$\sigma_y^2 = \sum_{i=1}^{s} L_0^2 \sin^2 \theta_i \sin^2 \phi_i .$$

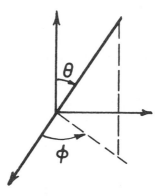

Fig. 4.2. Polar coordinate system used in describing the three-dimensional random walk.

These components can be added by the Pythagorean theorem to give the overall *mean square deviation*, also called the *variance* or *second moment*, as

$$\sigma^2 = sL_0^2 .\tag{4-16}$$

A similar situation arises if we have a volume V subdivided into equal compartments v. A man begins to randomly put either one marble or else no marble at all into successive compartments. When he has done this for all the compartments, he starts all over again. He follows this routine $2n$ times. At the end of this exercise, the mean number of marbles in each compartment is n and the standard deviation for the marbles in any one compartment, by (4–11), is $\sigma = n^{1/2}$. Suppose, next, that we combine the marbles from m successive compartments v into larger bins of volume mv. Now, the mean number of marbles per bin is $N = mn$, and the standard deviation is $N^{1/2}$. However, for an absolute comparison, it is often useful to divide the standard deviation by the mean. This ratio is called the *coefficient of variation*, $\mathcal{V}$, defined as $\mathcal{V} \equiv \Delta/N = (N)^{-1/2} \propto (mv)^{-1/2}$. Thus we see that the larger the compartment we select within the volume V, the smaller is the coefficient of variation of particles that we will find in the compartment. A set of larger aggregates appears to be more homogeneous, in the fractional differences found among them, than individual smaller aggregates. We will encounter this situation in Chapter 13, where we will be concerned with the distribution of structures condensing out of a primordial medium that started out very hot when the Universe was young, and later fragmented into galaxies and clusters of galaxies as it cooled.

PROBLEM 4–4. A hot star is surrounded by a cloud of hydrogen that is partly ionized, partly neutral. Radiation emitted by the star at the wavelength of the Lyman-α spectral line can be absorbed and re-emitted by the neutral atoms. Let the mean path traveled by a photon between emission and absorption have length L. Let the radius of the cloud be R. About how many absorption and re-emission processes are needed before the photon finally escapes from the cloud? We will return to this problem in Section 9:12, where we consider a random walk when the atoms move with high random velocities.

The random walk concept provides an essential basis for all radiative transfer computations. We will tackle such problems later in discussing the means by which energy can be transported from the center of a star, where it is initially released, to the surface layers and then through the star's atmosphere out into space. In the general theory of radiative transfer the *opacity* of the material is inversely proportional to the step length we assumed for the random walk above. The added complication that arises in most practical problems is that the mean energy per photon progressively drops as energy is transported outward from the center of a star. Energy initially released in the form of hard gamma rays eventually leaves the stellar surface as visible and infrared radiation. One gamma photon released in a nuclear reaction at the center of the star provides enough energy for about a million photons emitted at the stellar surface. The walk from the center of a star, therefore, involves not a single photon alone but also all its many descendants.

4:3 Distribution Functions, Probabilities, and Mean Values

In Section 4:2 we calculated the mean deviation and root mean square deviation after a number of steps in a random walk. Often we are interested in computing mean values for functions of the deviation and for distributions other than binomial distributions. There is a general procedure for obtaining such values.

Suppose a variable x can take on a set of discrete values x_i. Let the absolute probability of finding the value x_i in any given measurement be $P(x_i)$. If we pick a function $F(x)$ that depends only on the variable x, we can then compute the mean value that we would obtain for $F(x)$ if we were to make a large number of measurements. This mean is obtained by multiplying $F(x_i)$ by the probability $P(x_i)$ that the value x_i will be encountered in any given measurement. Summing over all i values then yields the mean value $\langle F(x) \rangle$,

$$\langle F(x) \rangle = \sum_i P(x_i) F(x_i) . \qquad (4\text{-}17)$$

Sometimes the absolute probability is not immediately available but the relative probability $p(x_i)$ is known. We then have the choice of computing $P(x_i)$ as in equation (4–1), or else we can proceed directly to write

$$\langle F(x) \rangle = \frac{\sum_i p(x_i) F(x_i)}{\sum_i p(x_i)} , \qquad (4\text{-}18)$$

where the denominator gives the normalization that is always needed when relative probabilities are used.

If x can take on a continuum of values within a certain range, the integral expressions corresponding to equations (4–17) and (4–18) are

$$\langle F(x) \rangle = \int P(x)F(x)\, dx = \frac{\int p(x)F(x)\, dx}{\int p(x)\, dx}, \tag{4-19}$$

where the integrals are taken over the range of the variable for which a mean value $\langle F(x) \rangle$ is of interest. Sometimes this range is $-\infty < x < \infty$.

We note that the expressions (4–6) and (4–7) already have the general form required by equations (4–17) to (4–19). In equation (4–6) the function $F(x)$ is just x itself, whereas in (4–7) it is x^2. We have merely substituted a new symbol x, for the values previously denoted by the position symbol k.

4:4 Projected Length of Randomly Oriented Rods

Let a system be viewed along a direction defining the axis of polar coordinates (θ, ϕ) (Fig. 4.3). A rod of length L has some arbitrary orientation θ with respect to the axis, and its projected length transverse to the line of sight is $L \sin \theta$, independent of ϕ, $0 \leq \phi < 2\pi$.

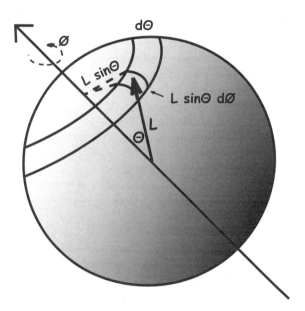

Fig. 4.3. Polar coordinate system for discussion of projected lengths.

We wish to determine the mean value of the observed length, the average being taken over all possible orientations of the rod. The probability of finding the rod with an orientation that lies within an increment $d\theta$ at angle θ is proportional to the area that the strip $d\theta$ defines on the surface of a sphere of unit radius. The normalized probability $P(\theta)$ is

$$P(\theta)\, d\theta = \frac{1}{2\pi} \int p(\theta, \phi)\, d\theta\, d\phi = \sin\theta\, d\theta . \tag{4-20}$$

We see that this is a properly normalized probability because

$$\int_0^{\pi/2} P(\theta)\, d\theta = \left. -\cos\theta \right|_0^{\pi/2} = 1 ; \tag{4-21}$$

that is, the probability of finding the rod with *some* orientation between 0 and $\pi/2$ is unity.[1] The probability of finding the rod with projected length $L\sin\theta$ is therefore $\sin\theta$, and the mean value of the projected length averaged over all position angles is

$$\frac{\int_0^{\pi/2} P(\theta) L\sin\theta\, d\theta}{\int_0^{\pi/2} P(\theta)\, d\theta} = \int_0^{\pi/2} L\sin^2\theta\, d\theta = \frac{\pi}{4}L . \tag{4-22}$$

Here, the integral in the numerator is a summation over the lengths obtained over all orientations, and the integral in the denominator assures an average value by dividing the numerator by the whole range of probabilities. This division is not strictly necessary because we already have normalized correctly. However, had we, for example, wished to find the mean projected lengths only for those rods having inclinations to the polar axis in the range $0 < \theta \le \pi/4$, the limits of integration both in the numerator and denominator would be 0 and $\pi/4$, and the integral in the denominator would no longer be trivial. Reversing the problem, we can ask for the actual value of a length S when only the random projected lengths can be observed to have mean value D. Then

$$S = \frac{4\langle D\rangle}{\pi} \tag{4-23}$$

by simple inversion of the argument developed in (4–22). We can ask a slightly different question, "Given a particular observed value of D, what is the mean of all the values S could have?" To answer this, we average $D/\sin\theta$ over the interval $0 \le \theta \le \pi/2$ for a fixed value of ϕ because the orientation ϕ is implicitly the direction along which D has been measured. This average has an infinite value because $(\sin\theta)^{-1}$ becomes large as θ approaches zero. The value of $\langle 1/S\rangle$ however is finite.

Similarly we can use our approach to decide whether elliptical galaxies are prolate (cigar-shaped), or oblate (disk-shaped). To make such an analysis, we do have to assume that all elliptical galaxies have roughly the same shape. According to this view, the globular galaxies would just be ordinary ellipticals viewed along a symmetry axis.

---

[1] The limits of integration are $0 \le \phi < 2\pi$, $0 \le \theta \le \pi/2$, since a rod with orientation (θ, ϕ) is equivalent to one with orientation $(-\theta, \phi + \pi)$.

PROBLEM 4–5. When a series of binary galaxies is observed, the total mass of each pair can be estimated roughly by measuring the projected separation between the galaxies and the projected radial component of their motions about each other. If (see Fig. 4.4) R is the distance to a pair as determined by its mean red shift,

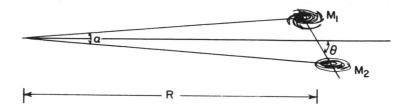

Fig. 4.4. Diagram to illustrate estimation of the total mass found in binary galaxies.

and α is the angular separation, then we can obtain the projected separation d_p. The difference between the red shifts of the two galaxies gives the projected orbital velocity component v_p. Assuming that the galaxies move in circular orbits about each other, show that the mass of the pair is statistically given by

$$M_{\text{pair}} = \frac{\langle v^2 \rangle}{G \langle 1/r \rangle} \sim \frac{3\pi}{2} \frac{\langle v_p^2 \rangle}{G \langle 1/d_p \rangle} \,, \tag{4-24}$$

where the approximation in the expression assumes that the projection of the velocity vectors is independent of the projection of the separation d_p — which is actually incorrect for circular motion. With these assumptions, however, show that $\langle v_p^2 \rangle \sim \langle v^2 \rangle / 3$ and that $\langle 1/r \rangle = (2/\pi)\langle 1/d_p \rangle$. Note that $\langle r \rangle \neq \langle 1/r \rangle^{-1}$. Because the projection angle in this case is not independent for r and v, alternative forms of (4–24) should actually be employed to take this correlation into account.

When we talk about clusters of galaxies, the same considerations apply, because the virial theorem (3–85) again sets the mean potential energy equal to twice the (negative of the) kinetic energy. The mass of the entire cluster is then substituted on the left side of equation (4–24). The right side gives the mean squared velocities of the cluster galaxies and their mean reciprocal distances from the cluster center. For sizeable clusters of galaxies, dispersion velocities are found to be $\sim 10^3$ km s^{-1} in the central parts, leveling off at distances of a few megaparsec, Mpc (Fa96). As discussed in Section 3:15, when the cluster mass is estimated in this way, it always turns out to be some 10 times higher than the sum of the masses of the individual galaxies determined as in Problem 3–12. We will return to this puzzle in Chapter 9, where we will find a need to postulate either the existence of *dark matter* or a deviation from an inverse square law of gravitational attraction over megaparsec distances.

4:5 The Motion of Molecules

An assembly of molecules surrounding an interstellar dust grain exerts pressure on the grain's surface. This pressure arises because the molecules are moving randomly and sometimes collide with the dust. A molecule initially moving toward the grain is deflected at the grain's surface and recedes following the collision. Because the particle's velocity is changed, its momentum **p** also is altered. For a brief interval the surface, therefore, exerts a force on the molecule because, by definition, a force is required to produce the change of momentum. This follows from Newton's equation (3–4), which can be rewritten as

$$\mathbf{F} = m\ddot{\mathbf{r}} = \dot{\mathbf{p}} \ . \tag{3–4}$$

If the grain exerts a force on a molecule during a given time interval τ, the molecule too must be reacting on the grain in that time. The sum of all the forces exerted by all the individual molecules impinging on unit grain area at any given time then constitutes the pressure — or force per unit area — acting on the dust.

To calculate the pressure we must first decide how many molecules hit a grain per unit time. Figure 4.5 shows a spherical polar coordinate system by means of which we can label the direction from which the particles initially approach. That direction is given by angles (θ, ϕ). If there are $n(\theta, \phi, v)$ molecules per unit volume coming from an increment of solid angle $d\Omega = \sin\theta\, d\theta\, d\phi$ about the direction (θ, ϕ) with a speed v to $v + dv$, then the number of particles incident on unit surface area in unit time is

$$\int\int\int v\cos\theta\, n(\theta, \phi, v)\sin\theta\, d\theta\, d\phi\, dv \ . \tag{4-25}$$

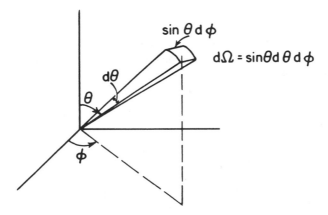

Fig. 4.5. Spherical polar coordinates for computing pressure.

The factor $\cos\theta$ has to be included because the volume of an inclined cylinder that contains all the incident particles is the product of the base area and the height (Fig. 4.6).

Expression (4–25) is proportional to v because particles with larger speeds can reach the impact area from greater distances in any given time interval.

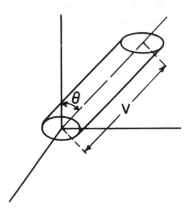

Fig. 4.6. Inclined cylindrical element containing all molecules striking the surface from direction θ, with speed v, in unit time interval.

If we assume that each molecule is reflected specularly — as from a mirror — then the angle of incidence is equal to the angle of reflection from the surface, and the total change in momentum for a reflected particle is

$$\Delta p = -2p\cos\theta . \tag{4-26}$$

Only the momentum component normal to the surface changes in such a reflection and this gives rise to the factor $\cos\theta$. We can now compute the pressure that is just (the negative of) the total change of momentum suffered by all molecules incident on unit area in unit time.

$$P = \int_0^{2\pi} d\phi \int_0^{\infty} dv \int_0^{\pi/2} d\theta \, (2p\cos\theta)\, v\cos\theta\, n(\theta, \phi, v) \sin\theta . \tag{4-27}$$

In an *isotropic gas* the number of molecules arriving from unit solid angle is independent of θ and ϕ and we can write

$$n(\theta, \phi, v)\, dv = \frac{n(v)}{4\pi}\, dv . \tag{4-28}$$

Here $n(v)$ is the number density of molecules with speeds in the range v to $v + dv$ and the factor $1/4\pi$ is a normalization constant that arises because 4π steradians are needed to describe all possible approach directions.

Expression (4–28) allows us to separate out a velocity-dependent part of the integrand in (4–27). It is independent of the direction coordinates θ and ϕ. If $v \ll c$ — where c is the speed of light — then $p = mv$, where m is the mass of a molecule. We can then write

$$\int_0^\infty n(v)v^2 \, dv \equiv n\langle v^2 \rangle , \qquad (4\text{-}29)$$

where n is the number density of particles per unit volume regardless of speed and direction, and $\langle v^2 \rangle$ is the mean squared value of the velocity. Equation (4–29) is simply a definition of the mean squared velocity.

The other part of the integral in (4–27) can now be written as

$$\frac{1}{2\pi} \int_0^{2\pi} \int_0^{\pi/2} \cos^2 \theta \sin \theta \, d\theta \, d\phi = \frac{1}{3} = \langle \cos^2 \theta \rangle . \qquad (4\text{-}30)$$

This integral defines the mean value of $\cos^2 \theta$ averaged over a hemisphere $0 \le \theta \le \pi/2$. This is the hemisphere from which all particles striking the wall must approach. Because of symmetry about $\theta = \pi/2$, the mean squared value of the cosine function actually is $\frac{1}{3}$ even if we integrate over all possible directions, rather than just one hemisphere.

Substituting equations (4–29) and (4–30) into (4–27) we can rewrite the expression for pressure as

$$P = \frac{nm\langle v^2 \rangle}{3} . \qquad (4\text{-}31)$$

Writing the product of the pressure P with the volume V that encloses N particles of the assembly, we then have the expression

$$PV = \frac{Nm\langle v^2 \rangle}{3} = N\Theta , \qquad (4\text{-}32)$$

where $N = nV$ and we define $\Theta \equiv m\langle v^2 \rangle/3$ which, as Section 4:6 will show, is proportional to the temperature.

PROBLEM 4–6. The random velocity of galaxies is thought to amount to $v \sim 100 \, \mathrm{km \ s^{-1}}$. Their number density is $n \sim 10^{-1} \, \mathrm{Mpc^{-3}}$. If typical galaxies have a mass of 3×10^{44} g, what is the cosmic pressure due to galaxies? This pressure contribution has an effect on the dynamics of the Universe. Chapters 11 to 13 discuss the role of pressure on cosmic expansion or contraction.

PROBLEM 4–7. The number density of stars close to the Sun is $n \sim 10^{-57} \, \mathrm{cm^{-3}}$. The Sun's velocity relative to these stars is $v \sim 2 \times 10^6 \, \mathrm{cm \ s^{-1}}$ and we can take the cross-section for collision with another star to be $\sigma \sim 5 \times 10^{22} \, \mathrm{cm^2}$. In the Jeans theory of the birth of the Solar System, such an encounter was considered responsible for the formation of the planets. How probable is it that the Sun would have formed planets in $P = 5 \times 10^9$ yr? How many planetary systems would we expect altogether in the Galaxy if there are 10^{11} stars and if the Sun is representative?

4:6 Ideal Gas Law

Tenuous gases obey a simple law at temperatures far above the temperature of condensation. This law relates the temperature of a gas to its pressure and density. Since it becomes exact only at high temperatures and low densities, it represents an idealization that a real gas can only approach. We speak of the *ideal gas law*. In practice, deviations from ideal behavior are small for a large variety of gases in many different situations, and the law is very useful.

To understand this law, we must first know what is to be meant by *temperature*. We can easily "feel" whether a body is hot or cold; but it is not simple to describe this feeling in terms of a measurable physical quantity. One way of measuring temperatures is in terms of a device — for example, an ordinary mercury bulb thermometer. When the thermometer is dipped into a bowl of water that feels hot, the mercury expands out of the bulb and rises in the capillary tube. When the thermometer is placed into a cold bowl of water the mercury contracts. We can attach an arbitrary scale to the capillary portion of the thermometer and take readings to obtain the temperature in terms of the location of the mercury meniscus in the capillary. To show just how arbitrary such a scale may be, we need only recall that there have been at least five different temperature scales in common use in the Western world.

Choosing a given mercury thermometer as a standard, we can make observations of the behavior of gases and eventually arrive at a relation between the density, pressure, and temperature of a given gas. This relation is called an *equation of state*. It has the functional form

$$F(T, P, \rho) = 0 . \tag{4-33}$$

The density is sometimes expressed in terms of its reciprocal, the volume per unit mass, or more often in terms of the *molar volume*, or volume per mole of gas. The *mole* is a quantity of matter represented by $\mathcal{N} = 6.02 \times 10^{23}$ molecules. $\mathcal{N}$ is called *Avogadro's number*. Avogadro's number is the number of atoms of the carbon isotope ^{12}C weighing exactly 12 grams — one *gram-atomic-weight* of ^{12}C.

Writing the molar volume as $\mathcal{V}$, we obtain the ideal gas law as

$$P\mathcal{V} = RT , \tag{4-34}$$

where R is a constant called the *gas constant*. At constant pressure the volume of a given amount of gas increases linearly with temperature. At fixed volume the pressure rises linearly with temperature. Some gases, notably helium, behave very nearly like an ideal gas and can, therefore, be used to define a *gas thermometer* temperature scale. The important point to realize is that temperature has to be defined operationally in terms of a convenient device.

We note the similarity between equations (4–32) and (4–34). When N in equation (4–32) is chosen to be Avogadro's number $\mathcal{N}$, we find that

$$\frac{RT}{\mathcal{N}} = \Theta = \frac{m\langle v^2 \rangle}{3} . \tag{4-35}$$

We can define a new constant $k = R/\mathcal{N}$, called Boltzmann's constant. Equation (4–35) then becomes

$$\frac{3}{2}kT = \frac{m\langle v^2\rangle}{2}.$$ (4-36)

The right side of equation (4–36) is the mean kinetic energy per particle in the assembly, and the temperature is therefore nothing other than an index of the mean kinetic energy. In a hot gas the molecules move at high velocity; in a cooler gas they move more slowly. The Boltzmann constant k has to be experimentally determined by direct or indirect measurement of the kinetic energy of molecules in a gas at a given temperature: $k = 1.381 \times 10^{-16}$ erg K^{-1}.

Equation (4–32) can now be rewritten as

$$PV = \mathcal{N}kT \quad \text{or} \quad P = nkT.$$ (4-37)

This is straightforward as long as we deal with one particular kind of gas or one given type of molecule. But what happens if the gas consists of a mixture of different atoms or molecules? The kinetic theory developed thus far predicts that the total pressure should still be determined by the total number density of atoms and molecules as given by equation (4–37). If there are j different kinds of particles present in thermal equilibrium, each with number density n_i, the complete relation would read

$$P = \sum_{i=1}^{j} P_i = \sum_{i=1}^{j} n_i kT = nkT,$$ (4-38)

where P_i is the *partial pressure* exerted by atoms or molecules of type i alone. Equation (4–38) expresses *Dalton's law* of partial pressures, named after the English chemist John Dalton, who first noted the effect in 1801: The total pressure of an ideal gas is the sum of the partial pressures of the various constituents.

PROBLEM 4-8. Interstellar atomic hydrogen is often found in neutral, HI clouds whose temperature is 100 K. What is the root mean squared velocity at which the hydrogen atoms travel? If the number density $n = 1$ cm^{-3}, what is the pressure in interstellar space?

PROBLEM 4–9. These clouds also contain dust grains that might characteristically have diameters 5×10^{-5} cm and unit density. Treating the dust as though it were an ideal gas, what would be the random velocity of dust grains in equilibrium at temperature T?

PROBLEM 4–10. If the gas had systematic velocity v relative to the dust grains, how much momentum would be transferred to each dust grain per unit time, and what is the acceleration? Assume that the gas density $n = 1$ cm^{-3}, $v = 10^6$ cm s^{-1}, and that the gas atoms stick to the grain in each collision.

PROBLEM 4–11. What would be the rate of mass gain for this grain? How soon would its mass increase by 1%?

PROBLEM 4–12. In an ionized hydrogen (H II) region, protons and electrons move randomly. If the temperature of this interstellar gas is 10^4 K, calculate electron and proton velocities.

4:7 Radiation Kinetics

Electromagnetic radiation is transmitted in the form of photons — discrete quanta having momentum p and energy $\mathcal{E}$. The experimentally determined relationship between the spectral frequency ν — color of the radiation — and the energy and momentum is

$$p = \frac{h\nu}{c} , \tag{4-39}$$

$$\mathcal{E} = h\nu , \tag{4-40}$$

where h is Planck's constant and c is the speed of light.

We can substitute expression (4–39) into the pressure equation (4–27), replacing v by c, and neglecting the integration over velocity because all photons have the same speed c. Expression (4–27) then reads

$$P(\nu)\,d\nu = \int_0^{2\pi} d\phi \int_0^{\pi/2} d\theta \frac{2h\nu}{c} \cos\theta\, c\, \cos\theta\, n(\theta, \phi, \nu) \sin\theta\, d\nu . \tag{4-41}$$

The two factors c cancel, and $h\nu$ can be replaced by $\mathcal{E}$. For an isotropic radiation field, $n(\theta, \phi, \nu) = n(\nu)/4\pi$, and use of equation (4–30) leads to

$$P(\nu) = \frac{n(\nu)\mathcal{E}}{3} = \frac{h\nu n(\nu)}{3} . \tag{4-42}$$

If quanta of j different spectral frequencies are present, expression (4–42) becomes

$$P = \frac{U}{3} , \tag{4-43}$$

where U is the total energy density summed over all spectral frequencies:

$$U = \sum_{i=1}^{j} n_i h\nu_i . \tag{4-44}$$

PROBLEM 4–13. In Section 4:13 we will see that the energy density of electromagnetic radiation at a temperature T is $7.57 \times 10^{-15} T^4$ erg cm^{-3}. The Universe is permeated by a microwave background radiation field at $T = 2.73$ K. What is the pressure due to this radiation and how does it compare to the pressure exerted by galaxies calculated in Problem 4–6?

PROBLEM 4–14. At Earth the radiation energy incident from the Sun on unit area per unit time is 1.37×10^6 erg cm^{-2} s^{-1}. This quantity is called the *solar constant*. Find the radiative repulsive force on a 10^{-2} cm diameter black (totally absorbing) grain, at the distance of Earth from the Sun.

PROBLEM 4–15. A spherical grain of radius $s = 10^{-4}$ cm absorbs $\frac{1}{3}$ of the solar radiation incident on its surface and scatters the remainder isotropically. Calculate the ratio of gravitational attraction to radiative repulsion from the Sun, assuming that the grain has density 6 g cm^{-3}. Show that this ratio is constant as a function of distance from the Sun.

PROBLEM 4–16. If the repulsive force of radiation on a grain is $\frac{1}{3}$ of the attraction to the Sun due to gravitation, we can define an "effective" gravitational constant $G_{\text{eff}} = \frac{2}{3}G$ where G is the gravitational constant. This will characterize the motion of the grain. What is the orbital period of such a grain moving along Earth's orbit? How does its orbital velocity compare to that of Earth?

4:8 Isothermal Distributions

We say that a gas is *isothermal* if its temperature is the same throughout the volume it occupies. Consider an isothermal, gravitationally bound, spherically symmetric gas configuration in space. The hydrostatic pressure change dP between positions r (Fig. 4.7) and $r + dr$ is given by the gravitational force acting on matter between r and $r + dr$:

$$dP = -dr\,\rho(r)\nabla\mathbb{V}(r) \,. \tag{4-45}$$

Here $\rho(r) = n(r)m$ and $\mathbb{V}(r)$ is the gravitational potential due to the mass enclosed by the sphere r. For an ideal gas (see equation (4–38)) $P/\rho = kT/m$. Dividing this expression into equation (4–45) we have

$$\frac{dP}{P} = -\frac{m}{kT}\nabla\mathbb{V}(r)\,dr$$

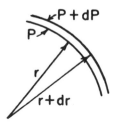

Fig. 4.7. Pressure–distance relation for a spherically symmetric configuration.

which integrates to

$$P = P_0 e^{-m\mathbb{V}(r)/kT} \ . \tag{4-46}$$

Reapplying the ideal gas law, we can also obtain the densities

$$n = n_0 e^{-m\mathbb{V}(r)/kT} \qquad \text{or} \qquad \rho = \rho_0 e^{-m\mathbb{V}(r)/kT} \ . \tag{4-47}$$

The exponential term appearing in equations (4–46) and (4–47) is called the *Boltzmann factor*. It plays an important role throughout the theory of statistical thermodynamics and, as we will see in Section 4:23, gives a useful starting point for describing the distributions of molecules in protostars, and stars in globular clusters.

4:9 Atmospheric Density

Using equation (4–47), we can readily find the density distribution in the atmosphere of a star, planet, or satellite. In what follows we will keep referring to the parent body as a planet, but the theory holds equally well for a star, moon, or any other massive body.

The gravitational potential at any location in the atmosphere is given by

$$\mathbb{V}(r) = -\frac{MG}{r} \ , \tag{4-48}$$

where r is the distance measured from the center of the planet and M is its mass. Expression (4–48) also assumes that the atmosphere is tenuous so that M can be assumed to be constant and independent of r. Let R be the planet's radius, and consider a point at height x above the surface. The difference between the potential at height x and at the surface is

$$\mathbb{V}(R + x) - \mathbb{V}(R) = -\frac{MG}{R + x} + \frac{MG}{R} = \frac{MGx}{R^2}, \qquad x \ll R \ . \tag{4-49}$$

Equation (4–47) then becomes

$$n = n_0 e^{-(mMG/kTR^2)x} = n_0 e^{-mgx/kT} \ , \tag{4-50}$$

where n_0 now represents the density at the surface and $MG/R^2 \equiv g$ is the *surface gravity* of the planet. It is clear that the atmospheric density decreases exponentially with height. We can define a *scale height*

$$\Delta \equiv \frac{kTR^2}{mMG} = \frac{kT}{mg} \ . \tag{4-51}$$

The density at height $x + \Delta$ is reduced by a factor e below the value at height x. The scale height is small for low-temperature gases composed of heavy molecules — m large — and for dense parent bodies — large M, small R.

PROBLEM 4–17. Show that an atmosphere consisting of a combination of gases has a variety of scale heights, one for each gas component. Show that the total pressure is

$$P = \sum_i P_i = \sum_i P_{i0} e^{-(m_i g x/kT)}, \qquad (4\text{-}52)$$

consistent with Dalton's law, and that the total density is

$$\rho = \sum_i n_i m_i = \sum_i n_{i0} m_i e^{-(m_i g x/kT)}, \qquad (4\text{-}53)$$

where the subscript 0 denotes a value at the base of the atmosphere. Assume no atmospheric convection. (Convection normally requires bulk motion of entire volumes of gas and gives rise to winds that do not allow complete separation of different gaseous constituents. The concept of scale height then needs to take this into account.)

At the low densities found in Earth's upper atmosphere, there is some separation of gases with different scale heights. Helium, for example, appears in appreciable concentrations only at high altitudes. In the lower atmosphere three features complicate any analysis. There are winds, temperature gradients, and atmospheric water vapor. The vapor is near the condensing point and a local atmospheric temperature drop can give rise to condensation and a decrease in pressure. This gives rise to winds. More important, the lower atmosphere is not isothermal and is subject to a variety of thermal gradients that can either induce or suppress convection.

PROBLEM 4–18. The mass of the atmosphere is negligible compared to the mass of our planet, $m_\oplus$. If the gravitational attraction at the surface of Earth is 980 dyn g^{-1}, calculate the scale height of the atmosphere's main constituent, molecular nitrogen N_2, at a temperature of 300 K.

4:10 Particle Energy Distribution in an Atmosphere

The exponential decline of particle density with height is an important clue to the velocity distribution of particles. We note that molecules at a height x_1, having an upward-directed velocity component $v_x = (2gh)^{1/2}$, have enough energy to reach a height $x_1 + h$. Whether a given molecule with this instantaneous velocity actually reaches height $x_1 + h$ cannot be predicted. The molecule might collide with another one, and lose most of its energy. However, as long as thermal equilibrium exists, and the gas temperature remains stable, we can be sure that, for every molecule that loses energy through a collision, there will be a *restituting collision* at some nearby point in which some other molecule gains a similar amount of energy. This concept, sometimes referred to as *detailed balancing*, allows us to neglect the effect of collisions in the remainder of our argument.

Because the temperature is the same at all levels of an isothermal atmosphere, the velocity distribution must also be the same everywhere, and only the number of particles changes with altitude. The ratio of the particle densities at heights $x_1 + h$ and x_1 (see equation (4–53)) is $\exp(-mgh/kT)$. Since the particles encountered at height $x_1 + h$ have all come up from the lower height x_1, to which they will eventually return — fall back — we can be certain that the fraction of particles passing through a plane at height x_1 and having speeds greater than $v_h = (2gh)^{1/2}$ is going to be precisely that fraction of particles having enough energy to reach altitudes above $x_1 + h$. We can therefore express the two-way flux of particles with vertical velocity v_x greater than v_h as

$$\frac{N(v_x > v_h)}{N(v_x > 0)} = \left[\int_h^\infty e^{-mgx/kT} \, dx \middle/ \int_0^\infty e^{-mgx/kT} \, dx \right]$$
$$= e^{-mgh/kT} = e^{-mv_h^2/2kT} . \tag{4-54}$$

Note that N is not a number density; it is *flux*, a number of particles crossing unit area in unit time.

Collisions make the velocity distribution isotropic. Hence, we consider a velocity distribution $f(v)$ that is normalized by the integral

$$\int \int \int_{-\infty}^\infty f(v_x, v_y, v_z) \, dv_x \, dv_y \, dv_z = 1 . \tag{4-55}$$

As a trial solution for the function f, we can use an exponential v_x dependence, like that given by equation (4–54). The isotropy requirement then demands a similar dependence on v_y and v_z, and equation (4–55) gives the full function as

$$f(v_x, v_y, v_z) = \left(\frac{m}{2\pi kT} \right)^{3/2} e^{-(m/2kT)(v_x^2 + v_y^2 + v_z^2)} , \tag{4-56}$$

where the coefficient is a normalization factor required by (4–55). This function is separable in the variables v_x, v_y, and v_z. To test whether it also obeys equation (4–54) we note that

$$\frac{N(v_x > v_h)}{N(v_x > 0)} = \frac{\int_{v_h}^\infty v_x e^{-(m/2kT)v_x^2} \, dv_x}{\int_0^\infty v_x e^{-(m/2kT)v_x^2} \, dv_x} = e^{-mv_h^2/2kT} . \tag{4-57}$$

The quantity v_x in the integrand plays the same role here as in equation (4–27). It takes into account that, in unit time, the higher velocity particles can reach a given surface from a larger distance and from a larger volume. We can write the distribution (4–56) in terms of the speed

$$v = (v_x^2 + v_y^2 + v_z^2)^{1/2} . \tag{4-58}$$

We then obtain

$$f(v) = \left(\frac{m}{2\pi kT}\right)^{3/2} e^{-mv^2/2kT} . \tag{4-59}$$

PROBLEM 4–19. Satisfy yourself that the normalization condition for $f(v)$ is

$$4\pi \int_0^\infty f(v)v^2 \, dv = 1 . \tag{4-60}$$

Show also that, in terms of momentum, the distribution function is

$$f(p) = \frac{1}{(2\pi mkT)^{3/2}} e^{-p^2/2mkT} \tag{4-61}$$

and

$$\int_0^\infty 4\pi f(p)p^2 \, dp = 1 . \tag{4-62}$$

Note that equations (4–56), (4–59), and (4–61) all are independent of the gravitational potential initially postulated. The equations derived here therefore have much wider applicability than just to the gravitational problem. We will discuss this further in Section 4:15.

The velocity and momentum distribution functions (4–59) and (4–61) are called Maxwell–Boltzmann distributions, after James Clerk Maxwell and Ludwig Boltzmann, two of the nineteenth century founders of classical kinetic theory. The momentum distribution is plotted in Fig. 4.8. These distribution functions have extremely wide applications.

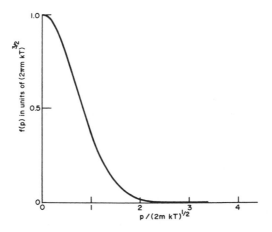

Fig. 4.8. Maxwell–Boltzmann momentum distribution.

PROBLEM 4–20. If the Moon had an atmosphere consisting of gases at 300 K calculate the mass of the lightest gas molecules for which $3kT/2 < MmG/R$. m is the mass of the molecule; M and R are the mass and radius of the Moon, respectively, 7.3×10^{25} g and 1.7×10^8 cm. Note that the quantity on the left of the inequality is related to the escape velocity at the Moon. What is this velocity? Actually, heavier molecules than those with mass m, calculated above, can escape from the Moon, because: (a) in a Maxwell–Boltzmann distribution, gases have many molecules with speeds larger than the mean speed; and (b) because the side of the Moon facing the Sun reaches temperatures of $\sim$400 K.

Despite their great usefulness, Maxwell–Boltzmann statistics cannot be applied under certain conditions, such as those encountered at high densities in the centers of stars. Neither do they apply to radiation emitted by stars. There, we need to consider quantum effects that have no classical basis. The next few sections describe these effects.

4:11 Phase Space

The quantum effects that lead to deviations from classical statistical behavior always involve particles that are identical to each other. We might deal with electrons that have almost identical positions, momenta, and spin; or we might have photons with identical frequency, position, direction of propagation, and polarization.

For electrons an important restriction comes into play. The *Pauli exclusion principle* forbids any two electrons from having identical properties. *Neutrons, protons, neutrinos*, and all other particles with odd half-integral *spin* ($\frac{1}{2}, \frac{3}{2}, \ldots$) also obey this principle. *Photons* and *pions*, on the other hand, have integral or zero spin, and any number of these particles can have identical momenta, positions, and spins. The first group of particles — those that obey the injunction of the Pauli principle — are called *Fermi–Dirac particles* or *fermions*; the others are called *bosons* and their behavior is governed by *Bose–Einstein statistics*.

Thus far we have not stated what we mean by "identical." Clearly we could always imagine an infinitesimal difference in the momenta of two particles, or in their positions. Should such particles still be termed identical, or should they not? The question is essentially answered by *Heisenberg's uncertainty principle*, which denies the possibility of physically distinguishing two particles if the difference in the momentum δp, multiplied by the difference in position δr, is less than *Planck's constant* h. This restriction derives from the uncertainty in the simultaneous measurement of momentum and position components for any given particle

$$\Delta p_x \Delta x \sim \hbar, \quad \langle (\Delta x)^2 \rangle \equiv \langle (x - \langle x \rangle)^2 \rangle = \langle x^2 - \langle x \rangle^2 \rangle, \quad \langle (\Delta p_x)^2 \rangle = \langle p_x^2 - \langle p_x \rangle^2 \rangle,$$
(4-63)

where $\hbar \equiv h/2\pi$ and h is Planck's constant, $h = 6.626 \times 10^{-27}$ erg s. The same constraints hold for $\Delta p_y \Delta y$ and $\Delta p_z \Delta z$.

We can show, quantum mechanically, that two particles are to be considered identical if their momenta and positions are identical within values

$$\delta p_x\,\delta x = h, \quad \delta p_y\,\delta y = h, \quad \delta p_z\,\delta z = h, \quad \delta x\,\delta y\,\delta z\,\delta p_x\,\delta p_y\,\delta p_z = h^3, \quad (4\text{-}64)$$

provided their spins are also identical.

In this description each particle is characterized by a position (x, y, z, p_x, p_y, p_z) in a six-dimensional *phase space*. It occupies a six-dimensional *phase cell* (Figs. 4.9 and 4.10) whose volume is $\delta x\,\delta y\,\delta z\,\delta p_x\,\delta p_y\,\delta p_z = h^3$. Particles within one phase cell are identical — physically indistinguishable — whereas those outside can be distinguished. Because δx is the dimension of the phase cell, it must be at least twice as large as Δx, the root mean square deviation from the central position. The same relation holds between δp_x and Δp_x. That is why the right side of equation (4–63) involves $\hbar$ while equation (4–64) contains the larger value, h. Figure 4.9 illustrates these differences.

We can now ask how many electrons could fit into a box with volume V? The answer depends on how high a particle momentum we wish to consider. If momenta up to a maximum value p_m are permitted, the available volume in phase space is $2(4\pi/3)p_m^3V$. The factor 2 accounts for two distinct spin polarizations, since electrons whose spins differ can always be distinguished and therefore must belong to different phase cells. This makes the number of available phase cells $[(8\pi/3)p_m^3V]/h^3$, which also is the maximum number of electrons that could occupy the box. Sometimes we may prefer to talk about *frequency space* instead of *momentum space*. Defining the particle frequency ν, by $\nu \equiv pc/h$, we obtain the number of phase cells with frequencies between ν and $\nu + d\nu$ as $[(8\pi/3)\nu_m^3V]/c^3$.

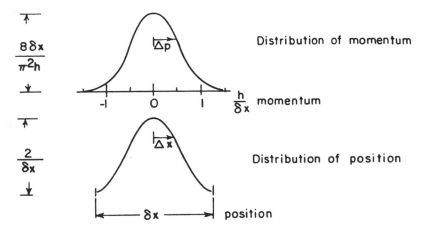

Fig. 4.9. Relation among phase cell dimensions, distribution of positions and momenta, and uncertainties in these variables. Only the simplest of a large family of distribution functions corresponding to different energies are shown.

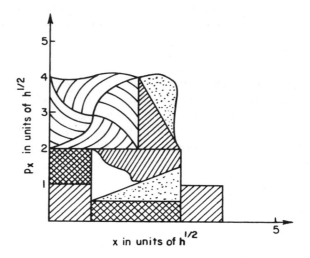

Fig. 4.10. Phase space is a six-dimensional hypothetical space having three momentum and three spatial dimensions. Projected onto the $p_x - x$ plane, individual cells always present an area h. Although their shapes may be quite arbitrary, as shown, it is often useful to think of them as square or rectangular since this makes computations simpler. In Section 4:14 (Fig. 4.13) we will see how an initially rectangular cell becomes distorted.

In general, the number of phase cells with momenta in a range p to $p + dp$, or equivalently ν to $\nu + d\nu$ is

$$Z(p)\,dp = 2V\frac{4\pi p^2\,dp}{h^3} \quad \text{of} \quad Z(\nu)\,d\nu = 2\left[\frac{4\pi\nu^2\,d\nu}{c^3}\right]V\,. \tag{4-65}$$

$Z(p)$ — and equivalently $Z(\nu)$ — are referred to as the *partition function*.

At the center of a star, ionized matter is sometimes packed so closely that all the lowest electron states are filled. Further contraction of the star can then force the electrons to assume much higher momenta than the value $(3kTm)^{1/2}$ normally found in tenuous gases. Such a closely packed gas of fermions is said to be *degenerate*. We will study this form of matter in Section 4:15 and in Chapter 8, where very dense cores of stars are discussed.

4:12 Angular Diameters of Stars

The fact that two photons sometimes occupy the same phase cell allows us to measure the angular diameter of stars. The idea is this: two photon counters are placed a distance D apart, transverse to the direction of the star. If D is small enough, we have the possibility that one photon from a cell will hit one detector, while the other photon hits the other detector, the simultaneous arrival being detected by a coincidence

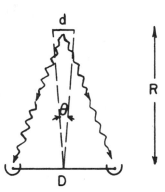

Fig. 4.11. The Hanbury Brown–Twiss interferometer.

counter. Let the diameter of the star be d and its distance R (Fig. 4.11). The angle it subtends is $\theta = d/R$. The photon pair impinging on either detector has a distribution in momentum, along the direction of D, amounting to $\Delta p_D = p\theta = (h\nu/c)\theta$ where ν is the frequency of radiation to which the detector is sensitive. But the nonzero value of Δp_D makes it necessary that D itself be small so that photons reaching either detector may be in the same phase cell. That is, it is necessary that

$$D\Delta p_D \lesssim h, \qquad \text{or} \qquad \frac{Dh\nu}{c}\theta \lesssim h, \qquad \text{or} \qquad D\theta \lesssim \lambda, \qquad (4\text{-}66)$$

where $\lambda = c/\nu$ is the *wavelength* of the radiation. By increasing D a decreasing coincidence rate is observed, and for values of D at which coincidences no longer occur the angular diameter is $\theta \lesssim \lambda/D$. The stellar *angular* diameter is

$$\theta \sim d/R \sim \lambda/D \qquad (4\text{-}67)$$

in such observations. This technique was first discovered by R. Hanbury Brown and R. Q. Twiss (Ha54). A second, related method makes use of the stellar interferometer constructed by Albert A. Michelson in 1920 to measure the angular diameter of Betelgeuse. In this interferometer only photons occupying the same phase cell coherently interfere.

4:13 The Spectrum of Light Inside and Outside a Hot Body

Any warm opaque body is permeated by a radiation bath. Atoms, molecules, or ions are continually absorbing and re-emitting quanta of light. From time to time a photon approaches the edge of the body and escapes. This diffusion of photons from the interior of the body out to its boundary, and the subsequent escape into empty space, is an important process in stars. Energy generated at the center of the star slowly diffuses outward and escapes. The escaping radiation gives the star its luminous appearance.

To understand this phenomenon in some detail we need to deduce the spectrum of the radiation as a function of temperature. Consider a photon gas embedded in material at temperature T. The radiation is in thermal equilibrium with the material if there is ample opportunity for the photons to interact with the atoms through scattering or absorption and re-emission. Two factors have to be considered:

(a) Photons are Bose–Einstein particles and can aggregate in single phase cells.

(b) If the spectral frequency of the photons aggregating in a phase cell is ν, and if there are n photons in the cell, we can consider the assembly of photons in this phase cell to be in a state with energy $(n + \frac{1}{2})h\nu$. We sometimes speak of a *quantum oscillator* in the nth state. Even when a phase cell is completely empty, in the ground state, a residual *vacuum energy* $h\nu/2$ is present.

We can compute the probability of finding a quantum oscillator in the nth excited state. The relative probability of that state is given by the Boltzmann factor $e^{-(n+1/2)h\nu/kT}$. The absolute probability is given by dividing the relative probability by the sum of all the relative probabilities:

$$P(\nu, T) = \frac{e^{-(n+1/2)h\nu/kT}}{\sum_n e^{-(n+1/2)h\nu/kT}} = \frac{e^{-(nh\nu/kT)}}{\sum_n e^{-nh\nu/kT}}, \qquad (4\text{-}68)$$

where the vacuum energy drops out. In these terms we can give the average energy $\langle \mathcal{E} \rangle$ per phase cell corresponding to frequency ν. We sum the energies of all the oscillators and divide by the total number of oscillators. Writing $x \equiv h\nu/kT$, we obtain

$$\langle \mathcal{E} \rangle = \sum_n \left(n + \frac{1}{2} \right) h\nu e^{-nh\nu/kT} \left[\sum_n e^{-nh\nu/kT} \right]^{-1}$$

$$= \frac{kT(xe^{-x} + 2xe^{-2x} + 3xe^{-3x} + \cdots)}{1 + e^{-x} + e^{-2x} + e^{-3x} + \cdots} + \frac{h\nu}{2}. \qquad (4\text{-}69)$$

The denominator in the first term of the second expression in equation (4–69) is $(1 - e^{-x})^{-1}$, as can be seen by noting that the denominator multiplied by $(1 - e^{-x})$ is unity. To evaluate the numerator, we use the same binomial expansion formula twice in succession.

$$kT\{x(e^{-x} + e^{-2x} + e^{-3x} + \cdots) + x(e^{-2x} + e^{-3x} + \cdots) + x(e^{-3x} + \cdots) + (\cdots)\}$$

$$= kT \left\{ \frac{xe^{-x}}{1 - e^{-x}} + \frac{xe^{-2x}}{1 - e^{-x}} + \frac{xe^{-3x}}{1 - e^{-x}} + \cdots \right\} = kT \frac{xe^{-x}}{(1 - e^{-x})^2}. \qquad (4\text{-}70)$$

In these terms

$$\langle \mathcal{E} \rangle = \frac{kTxe^{-x}}{1 - e^{-x}} + \frac{h\nu}{2} = \frac{kTx}{(e^x - 1)} + \frac{h\nu}{2} = \frac{h\nu}{(e^{h\nu/kT} - 1)} + \frac{h\nu}{2}. \qquad (4\text{-}71)$$

Knowing the number of phase cells per unit volume, $8\pi\nu^2\,d\nu/c^3$, and the mean energy per phase cell, we can write the energy density of photons as a function of frequency and temperature. This is the *blackbody radiation* spectrum:

$$\rho(\nu, T)\,d\nu = \frac{8\pi\nu^2\,d\nu}{c^3}\left(\frac{h\nu}{e^{h\nu/kT}-1} + \frac{h\nu}{2}\right),$$

$$(4\text{-}72)$$

$$n(\nu, T)\,d\nu = \frac{8\pi\nu^2}{c^3}\left(\frac{1}{e^{h\nu/kT}-1}\right)d\nu\,.$$

We will neglect the $h\nu/2$ term for now, and concentrate on the remainder of the expression, which can give rise to observable astronomical signals.[2] Integrated over all frequencies from 0 to ∞ the second term in parentheses in equation (4–72) would give rise to an infinite vacuum energy. Disregarding this term and integrating equation (4–72) over all frequencies from zero to infinity, we obtain the total energy density and number density of photons in terms of the generic formula (Gr80):

$$\int_0^\infty \frac{x^{\ell-1}}{e^{\mu x}-1}\,dx = \frac{1}{\mu^\ell}\Gamma(\ell)\zeta(\ell) \quad [\mu > 0, \ell > 0]\,, \qquad (4\text{-}73)$$

where $\Gamma(\ell) = (\ell - 1)!$ when ℓ is an integer > 0, and $\zeta(\ell)$ is the Riemann zeta function, $\sum_{m=1}^\infty m^{-\ell}$,

$$\rho(T) = \frac{8}{15}\frac{\pi^5}{c^3}\frac{k^4}{h^3}T^4 = aT^4 = U = 7.57 \times 10^{-15}T^4 \text{ erg cm}^{-3},$$

$$(4\text{-}74)$$

$$n(T) = \frac{8\pi}{c^3}\int_0^\infty \frac{\nu^2\,d\nu}{e^{h\nu/kT}-1} = 16\pi\left(\frac{kT}{hc}\right)^3\zeta(3) \approx 20.29T^3 \text{ photons cm}^{-3},$$

where $\zeta(3) = 1.20206$.

The coefficient of the T^4 term in equation (4–74) is a well-known definite integral. It is often denoted, as in equation (4–74), by the symbol a, the radiation density constant. We can also define another useful constant $\sigma \equiv ac/4 = 5.670 \times 10^{-5}$ erg cm^{-2} K^{-4} s^{-1}, *the Stefan–Boltzmann constant*. This constant allows us to write the energy emitted per unit area of a hot blackbody in unit time, as

---

[2] The term $h\nu/2$ cannot be observed in photon absorption or emission; but it is real nevertheless. Pulling apart two plane metallic surfaces separated by a small gap ℓ to increase this to $\ell + d\ell$ requires the application of a force. The work done results in the creation of ground state photons with wavelengths $\lambda = \ell/2$ to $(\lambda + d\lambda) = (\ell + d\ell)/2$, whose wavelengths would have been excessively long to fit into the original gap ℓ. This is the *van der Waals* force between plates and the effect is called the *Casimir* effect, which is small but measurable (La97). From (4–72) we would expect the ground state energy change to be proportional to λ^{-4} or ℓ^{-4}, and the force given by its gradient to be proportional to ℓ^{-4}. The value of the force is $F = \pi^2 c\hbar/240\ell^4$ (Ca48). The numerical factor in the denominator is so large because, as the gap increases, the remaining volume of the Universe outside the gap correspondingly decreases, and these two effects nearly cancel.

$$W = \sigma T^4 . \tag{4-75}$$

To see this, we can think of photons that escape from the surface as representative of the density of photons immediately within the surface of the body. Only those photons with velocities directed outward through the surface can be considered. So, only one-half of the photons come into consideration. These photons have an average velocity component normal to the surface equal to $c\langle \cos \theta \rangle$ where θ is the angle of emission with respect to the direction normal to the surface. We therefore have to evaluate $\langle \cos \theta \rangle$ averaged over all possible angles. This is

$$\langle \cos \theta \rangle = \frac{1}{2\pi} \int_0^{2\pi} \int_0^{\pi/2} \cos \theta \sin \theta \, d\theta \, d\phi = \left. \frac{\sin^2 \theta}{2} \right|_0^{\pi/2} = \frac{1}{2}, \tag{4-76}$$

$$\therefore \quad c\langle \cos \theta \rangle = \frac{c}{2} .$$

But because only half the photons are outward directed, the total flux is $(1/2)(c/2)(aT^4) = acT^4/4 = \sigma T^4$, as previously stated.

PROBLEM 4–21. Note that all this is strictly correct only if the index of refraction, n, in the medium is $n = 1$. For arbitrary values of n, show that

$$\rho(T) = n^3 aT^4 .$$

This is more generally the case inside a star. Show also what happens if the index of refraction is frequency dependent — which it always is.

The spectrum of most stars is closely approximated by a blackbody spectrum with individual spectral emission and absorption lines superposed. To the extent that the blackbody approximation holds, it is possible to ascertain the temperature of the star's photosphere where most of the light is emitted. Using two different wide-band filters, say the B and V filters often used in observations, we can determine the ratio of intensities in these spectral ranges. This ratio is uniquely related to the temperature. The temperature derived in this way is called the *color temperature*, T_c. A useful formula is (Aℓ63):

$$T_c = \frac{7300}{(B - V) + 0.73} . \tag{4-77}$$

PROBLEM 4–22. Using the effective wavelengths given in Table A.1 of Appendix A, compare the ratio of blue and visual radiation densities and magnitudes predicted, respectively, by equations (4–72) and (4–77) for a star at temperature 6000 K (spectral class G) and one at 10,000 K (spectral class A). Check the values given by (4–77) against Figure A.5.

Another means of defining temperature involves the luminosity of the star. Because the total power emitted per unit area is a function of temperature alone, we can calculate an *effective temperature* T_e of the star if both its luminosity and surface area can be determined:

$$L = \sigma T_e^4 4\pi R^2 . \tag{4-78}$$

If the distance of the star is known from observations of the kind described in Chapter 2, the stellar radius can be obtained using the Michelson or Hanbury Brown–Twiss interferometers discussed in Section 4:12. From (4–78) it is readily seen that

$$\log \frac{L}{L_\odot} = 4 \log \frac{T_e}{T_{e\odot}} + 2 \log \frac{R}{R_\odot} , \tag{4-79}$$

where $T_{e\odot} \sim 5,780 \, \text{K}$ and $R_\odot = 6.96 \times 10^{10} \, \text{cm}$ are the solar values. When the Hertzsprung–Russell diagram is plotted in terms of the logarithm of luminosity and effective temperature, as in Fig. 1.4, stars with identical radii lie on lines of constant slope, as required by equation (4–79).

It is worth mentioning two typical astrophysical situations in which temperature is a useful concept.

(a) Temperatures in the Solar System

The temperature of a black interplanetary object is determined by the energy equilibrium equation

$$\frac{L_\odot}{4\pi R^2} \pi r^2 = \sigma T^4 4\pi r^2 , \tag{4-80}$$

where $L_\odot$ is the solar luminosity, R is the distance from the Sun, and r is the radius of the object. If the mean efficiency for absorption (in the visible) is ε_a and the mean efficiency of reradiation (at infrared wavelengths) is ε_r, we have

$$T = \left(\frac{\varepsilon_a}{\varepsilon_r} \frac{L_\odot}{16\pi \sigma R^2} \right)^{1/4} . \tag{4-81}$$

We note that:

(i) At the Earth's distance

$$T \sim \left(\frac{\varepsilon_a}{\varepsilon_r} \right)^{1/4} \left(\frac{4 \times 10^{33}}{16\pi (5.7 \times 10^{-5}) 2.3 \times 10^{26}} \right)^{1/4} \sim 282 \left(\frac{\varepsilon_a}{\varepsilon_r} \right)^{1/4} \text{K} . \tag{4-82}$$

(ii) A gray body ($\varepsilon_a = \varepsilon_r$) has the same temperature as a black one.

(iii) For increasing distance from the Sun $T \propto R^{-1/2}$.

(iv) If the thermal conductivity of the body is small and its rotation slow, as for the Moon, the subsolar point assumes a temperature

$$T \sim \left(\frac{\varepsilon_a}{\varepsilon_r} \frac{L_\odot}{R^2 4\pi\sigma} \right)^{1/4} ,$$

which is $(4)^{1/4} \sim 1.4$ higher than the temperature of an equivalent, rapidly rotating or highly conducting body.

(b) Radio-Astronomical Temperatures

Some characteristics of radio-astronomical measurements can be understood in terms of temperatures. At very low frequencies, $\nu \ll kT/h$ — often called the *Rayleigh–Jeans limit* — the energy density in a source can be written (equation (4–72)) as

$$\rho(\nu) = \frac{8\pi kT\nu^2}{c^3} = \frac{8\pi kT}{c\lambda^2} , \tag{4-83}$$

where $\lambda \equiv c/\nu$ is the wavelength. The energy emanating from a surface in unit time, unit solid angle and unit area normal to the surface, and in unit frequency interval $\Delta\nu = 1$ at frequency ν, is called the *specific intensity* $I(\nu)$. This is the *surface brightness* of the source,

$$I(\nu) = \frac{c\rho(\nu)}{4\pi} = \frac{2\nu^2 kT}{c^2} = \frac{2kT}{\lambda^2} . \tag{4-84}$$

We shall usually express $I(\nu)$ in units of ergs s^{-1} sterad^{-1} cm^{-2} Hz^{-1}. The amount of radiant energy passing through unit area per unit time, integrated over all radiant frequencies is called the *energy flux* and has units erg cm^{-2} s^{-1}. If a specific intensity $I(\nu)$ is measured in an observation then, regardless of whether the source is thermal, we can pretend that a temperature parameter can be assigned to the observation. This is called the *brightness temperature* T_b and is defined at frequency ν as

$$T_b(\nu) \equiv \frac{I(\nu)c^2}{2k\nu^2} = \frac{I(\nu)}{2k}\lambda^2 . \tag{4-85}$$

T_b then is the temperature of an ideal blackbody whose radiant energy in the particular energy range ν to $\nu + d\nu$ is the same as that of the observed source (Ry71)*. A related concept is that of *antenna temperature* — which has nothing to do with the temperature that the antenna actually assumes under ambient climatic conditions. To examine it we must first consider some practical properties of antennas. In general, an antenna absorbs different amounts of power depending on the direction of the source. If we draw a *directional diagram* of an antenna, it usually has the shape of Fig. 4.12. The response $A(\theta, \phi)$ of the antenna is called its *effective area*. The power absorbed is

$$P \equiv \frac{1}{2} \int A(\theta, \phi)I(\nu, \theta, \phi)\, d\nu\, d\Omega \tag{4-86}$$

and for a small source,

$$P(\nu, \theta, \phi)\, d\nu = \frac{1}{2}F(\nu)A(\theta, \phi)\, d\nu, \qquad F(\nu) = \int I(\nu)\, d\Omega . \tag{4-87}$$

Here $F(\nu)$ is the flux density at the antenna, and the factor $\frac{1}{2}$ comes about because the antenna accepts only one component of polarization. If A is independent of the angle ϕ, and a diagram like Fig. 4.12 is drawn, $A(\theta)$ normally has a very large value in one particular direction, $\theta = 0$, and the large lobe around this direction is called the *main lobe*. The smaller lobes in the diagram are called side lobes. *Back lobes*

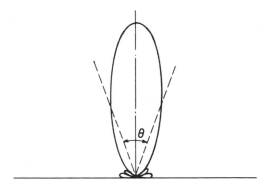

Fig. 4.12. Directional diagram of an antenna, showing a main lobe and a set of sidelobes. The angle θ is the beam width (see text).

can also occur. A well-designed radio telescope has a narrow main lobe for greatest positional accuracy and minimized sidelobes to minimize the confusion produced by sources outside the desired field of view.

We can define a mean value of the effective area of the antenna taken over all directions as

$$\langle A \rangle \equiv \frac{1}{4\pi} \int A(\theta, \phi) \, d\Omega \; ; \qquad (4\text{-}88)$$

then the *gain* of the antenna is the dimensionless quantity

$$G(\theta, \phi) \equiv \frac{A(\theta, \phi)}{\langle A \rangle} \; , \qquad (4\text{-}89)$$

which gives the ratio of the effective area in a given direction to the mean effective area. The function G has a maximum value in the direction $\theta = \phi = 0$ in a properly designed instrument. The *beamwidth* is the angle θ between points in the directional diagram at which $A(\theta, \phi) = A(0, 0)/2$.

In these terms, we can now return to the concept of *antenna temperature* T_a. If a source has directional and *specific intensity* $I(\nu, \theta, \phi)$, then a radio telescope with effective area $A(\theta, \phi)$ receives an amount of power given by (4–86). If we now disconnect the antenna and, instead, connect a resistor at temperature T to the receiver, the resistor can be shown experimentally and theoretically to produce *thermal noise* power in an amount

$$P = kT\Delta\nu \; , \qquad (4\text{-}90)$$

where $\Delta\nu$ is the receiver bandwidth. We can therefore define an antenna temperature T_a, so that

$$T_a = \frac{1}{k\Delta\nu} \cdot \frac{1}{2} \int A(\theta, \phi) I(\nu, \theta, \phi) \, d\nu \, d\Omega \; . \qquad (4\text{-}91)$$

This equation is useful for practical reasons. It is relatively easy to compare the power received from a celestial source to that received from a resistor switched to the receiver input in place of the antenna. The noise in (4–90) is sometimes called *Johnson noise* or *Nyquist noise*. J. B. Johnson (Jo28) and H. Nyquist (Ny28), respectively, supplied the experimental data and theoretical explanation leading to (4–90).

4:14 Boltzmann Equation and Liouville's Theorem

Let us define a function $f(\mathbf{r}, \mathbf{p}, t)$ as the density of particles in phase space. The number of particles in volume element $d\mathbf{r}$ at position $\mathbf{r}$, having momenta that lie in some momentum-space volume $d\mathbf{p}$ around momentum $\mathbf{p}$, is $f(\mathbf{r}, \mathbf{p}, t) \, d\mathbf{r} \, d\mathbf{p}$. We ask how the function f evolves with time. Since each particle in the assembly can be described in terms of three momentum and three spatial coordinates, the general form of the equation reads

$$\frac{\partial f}{\partial t} + \sum_i \frac{\partial f}{\partial \mathbf{r}_i} \frac{d\mathbf{r}_i}{dt} + \sum_i \frac{\partial f}{\partial \mathbf{p}_i} \frac{d\mathbf{p}_i}{dt} = \frac{df}{dt}\bigg|_{\text{collisions}}. \tag{4-92}$$

The left side of this equation gives the time rate of change of particles in the volume element $d\mathbf{r} \, d\mathbf{p}$ as a function of the coordinates $\mathbf{r}_i$, $\mathbf{p}_i$, $i = 1, 2, 3, \ldots, n$, for an n particle assembly. As the particles move, the surface enclosing them in phase space becomes distorted and the expression gives the rate of change of density through this distortion and through any other effects. The right side gives the loss or gain of particles through collisions. Equation (4–92) is called the *Boltzmann equation*.

To see how the evolution proceeds for a collisionless process in which the right side of equation (4–92) is zero, we draw a simple two-dimensional picture. In Fig. 4.13 we have an assembly of particles initially confined between positions $\mathbf{r}_1$ and $\mathbf{r}_2$ and between momentum values $\mathbf{p}_a$ and $\mathbf{p}_b$. Some time later, the momentum values are unchanged, but the particles have moved so that the higher momentum particles are now at positions between $\mathbf{r}_1'$ and $\mathbf{r}_2'$ while the lower momentum particles are at positions between $\mathbf{r}_1''$ and $\mathbf{r}_2''$. However, because the base and height of the enclosing

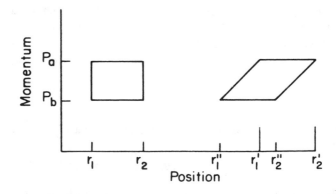

Fig. 4.13. Evolution of a collisionless assembly of particles in two-dimensional phase space.

area has not changed, the number density of particles per unit area has remained constant.

A similar argument holds when forces are applied to the particles. In that case the momenta of particles are not constant and the parallelepiped in Fig. 4.13 will also be displaced in a vertical direction. However, a similar argument can then be applied to show that the area covered by the particles still remains constant and the density of particles in this two-dimensional representation is unchanged. This is particularly easy to see if the force is the same on all particles. In that case, $d\mathbf{p}/dt$ is uniform and the difference in values $\mathbf{p}_a - \mathbf{p}_b$ is maintained constant. When different forces are exerted on different constituents of the gas, the area occupied by each constituent remains constant. These results also hold when there are gradients in the force fields.

A further extension of the argument can be applied to the full six-dimensional distribution. Unless there are some means for creating or destroying particles in the assembly — through collisions or particle–antiparticle pair formation — the density of particles will be constant along the trajectory in the six-dimensional space.

This is the sense of *Liouville's theorem*: The six-dimensional space density of particles in an assembly remains unchanged unless collisions occur:

$$df/dt = 0 \ . \tag{4-93}$$

Liouville's theorem has interesting applications to cosmic-ray particles, which move through the Galaxy, guided by magnetic field lines (see Section 6:6 for further discussion of this topic). Many of these particles are so energetic that they must be able to escape from the magnetic fields permeating the Galaxy. Their density in space outside the Galaxy could, therefore, be the same as the density that we measure in the vicinity of the Earth, provided the particles had enough time since their creation to traverse distances comparable to those of remote clusters of galaxies. Under these conditions, the spatial density of cosmic rays in extragalactic space would be the same as that measured at the Earth. This argument need not be true for low-energy particles if these particles can remain bottled up in local magnetic fields within our Galaxy.

The highest energy cosmic rays, whose energies range up to 10^{20} eV cannot be magnetically confined to the Galaxy. An extragalactic component could originate in gamma-ray bursts, active galactic nuclei, clusters of galaxies, or in as yet unidentified sources. The Liouville theorem tells us that the local density of at least these particles should be an indication of their extragalactic density. A slight caveat to this must, however, be kept in mind. Figure 1.16 and a discussion we will postpone to Section 5:10 show that particles at such high energies cannot travel large distances across the Universe before they are destroyed in collisions with microwave background radiation photons. Liouville's theorem may then permit us to say only that the observed flux of these high-energy cosmic rays represents their prevalence just in our local part of the Universe, rather than everywhere in the cosmos.

Finally we should still mention the problem discussed in Section 3:16, where a swarm of particles moves through a gravitational field. There we were concerned with tidal disruption of globular clusters, but noted that while the clusters became extended along a direction pointing toward the Galactic center, the gravitational forces also tended to produce a compression lateral to that direction. This compression produces additional transverse velocities making the evolutionary pattern quite complex. Liouville's theorem, however, gives us at least one solid guide toward understanding the overall development. It tells us that whatever detailed dynamical arguments we apply — such as those of Section 3:16 — the results must always agree at least with Liouville's requirement of a constant phase space density.

4:15 Fermi–Dirac Statistics

In a *Fermi–Dirac assembly* a phase cell can contain only one particle or none. For any given assembly there exists a *Fermi energy* $\mathcal{E}_F$ up to which all states are filled at zero temperature. At $T > 0$, excitation from a filled level at energy αkT to a higher state of energy $\mathcal{E}$ can take place. α is called the *degeneracy parameter*. The relative probabilities of being at energy $\mathcal{E}$ and αkT are, respectively,

$$e^{-(\mathcal{E}-\alpha kT)/kT} \qquad \text{and} \quad 1 \ . \tag{4-94}$$

The relative probability of occupancy of a state of energy $\mathcal{E}$, in an assembly at temperature T, therefore, is

$$\frac{e^{\alpha-\mathcal{E}/kT}}{1+e^{\alpha-(\mathcal{E}/kT)}} = \frac{1}{1+e^{(\mathcal{E}/kT)-\alpha}} \ . \tag{4-95}$$

Here we have not specified the energy αkT, but we can see that at very low temperatures, $T \sim 0$, αkT must approach $\mathcal{E}_F$ because the Fermi function

$$F(\mathcal{E}) = [1 + e^{(\mathcal{E}-\mathcal{E}_F)/kT}]^{-1} \tag{4-96}$$

has the form shown in Fig. 4.14.

We define the Fermi energy $\mathcal{E}_F$ as that energy for which $F(\mathcal{E}) = \frac{1}{2}$. Note that for $T = 0$, the exponent in (4–96) has a large absolute value, whenever $\mathcal{E} - \mathcal{E}_F \neq 0$, so that

$$F(\mathcal{E}) = 1 \quad \text{for} \quad \mathcal{E} < \mathcal{E}_F \ ,$$

$$\tag{4-97}$$

$$F(\mathcal{E}) = 0 \quad \text{for} \quad \mathcal{E} > \mathcal{E}_F \ .$$

This gives rise to the step function in Fig. 4.14. Whenever all the available energy levels are filled — which means whenever $T = 0$ — we say that the gas of *fermions* is completely *degenerate*. When $T > 0$, the step is seen to roll off more gently. The product of the probability $F(\mathcal{E})$ and $\mathcal{E}$ gives a mean value for the energy contained in all phase cells corresponding to an energy $\mathcal{E}$. Filled, as well as empty, cells have to be considered to obtain this value. From (4–94), the mean energy for cells at energy $\mathcal{E}$ is

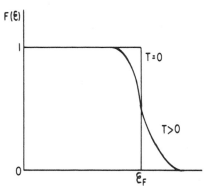

Fig. 4.14. The Fermi function $F(\mathcal{E})$.

$$\text{mean value} = \frac{\mathcal{E}}{1 + e^{(\mathcal{E}/kT)-\alpha}} . \tag{4-98}$$

We know that the number of states in the momentum range p to $p + dp$ is

$$Z(p)\,dp = \frac{8\pi p^2 V\,dp}{h^3} . \tag{4–65}$$

But

$$\mathcal{E} = \frac{p^2}{2m}, \qquad d\mathcal{E} = \frac{p}{m}\,dp ,$$

$$\therefore \quad Z(\mathcal{E})\,d\mathcal{E} = \frac{8\pi}{h^3}V\sqrt{2m\mathcal{E}}\,m\,d\mathcal{E} = \frac{4\pi V}{h^3}(2m)^{3/2}\mathcal{E}^{1/2}\,d\mathcal{E} . \tag{4-99}$$

The overall mean energy of the particles integrated over all values $\mathcal{E}$ is therefore

$$\langle \mathcal{E} \rangle = \frac{\int_0^\infty Z(\mathcal{E})F(\mathcal{E})\mathcal{E}\,d\mathcal{E}}{\int_0^\infty Z(\mathcal{E})F(\mathcal{E})\,d\mathcal{E}} . \tag{4-100}$$

Again, setting αkT equal to the Fermi energy $\mathcal{E}_F$ for an assembly of particles at temperature $T = 0$, we obtain

$$\langle \mathcal{E} \rangle_{T=0} = \left[\int_0^{\mathcal{E}_F} \mathcal{E}^{3/2}\,d\mathcal{E} \right]\left[\int_0^{\mathcal{E}_F} \mathcal{E}^{1/2}\,d\mathcal{E} \right]^{-1} = \frac{3}{5}\mathcal{E}_F . \tag{4-101}$$

One can show that for $T > 0$, $\mathcal{E}_F < \mathcal{E}_{F_0}$, the Fermi energy drops slightly.

For $\mathcal{E} - \mathcal{E}_F \gg kT$, that is, in the limit of large particle energies, we have

$$F(\mathcal{E}) \sim e^{-(\mathcal{E}-\mathcal{E}_F)/kT}, \tag{4-102}$$

which approaches a Boltzmann distribution for very energetic fermions.

At the center of stars degenerate conditions often exist. This is true mainly for electrons because at a given energy $\mathcal{E}$, $p = \sqrt{2\mathcal{E}m}$ is less, by a factor $\sqrt{m_p/m_e}$, for electrons than for protons. The lower energy electron states therefore become fully occupied — degenerate — much more readily than proton states.

PROBLEM 4–23. Suppose the Universe is filled with completely degenerate neutrinos up to an energy Φ_ν at a neutrino temperature $T_\nu = 0$. Show that, for negligibly low neutrino rest-mass, the mass density of neutrinos ρ_ν (the energy density divided by c^2), is

$$\rho_\nu = \frac{3\pi\Phi_\nu^4}{h^3 c^5} . \tag{4-103}$$

Note that neutrinos exist in only one spin state (Wa67) but that there are three neutrino species. If the corresponding antineutrinos are present as well, the mass density doubles.

To see why electrons and protons, which are actually fermions, appear to have the characteristics of Maxwell–Boltzmann particles in many astrophysical situations we note that we can derive the velocity distribution for classical particles in a way similar to the derivation of the Fermi–Dirac distribution. Assume that particles can occupy arbitrary positions in momentum and configuration space. This is equivalent to saying that the phase cells are infinitesimally small. We can obtain such a system by pretending that Planck's constant goes to zero as a limit: $h \rightarrow 0$. This makes $\mathcal{E}_F = 0$ since arbitrarily many particles can have zero and near-zero energies, and the probabilities in (4–94) become $e^{-\mathcal{E}/kT}$ and 1. We now write the number of particles in the assembly, having momenta near p:

$$n(p)\, dp \propto \frac{8\pi p^2 V}{h^3}\, dp\, e^{-p^2/2mkT} . \tag{4-104}$$

Integrating over all p values,

$$n = C \int_0^\infty \frac{8\pi V}{h^3} p^2 e^{-p^2/2mkT}\, dp , \tag{4-105}$$

where C is a dimensionless proportionality constant. This is an error function integral whose value is the total number of particles in the volume

$$n = C \frac{8\pi V}{h^3} \left(\frac{1}{4}\sqrt{\pi (2mkT)^3} \right) .$$

Hence

$$C = \frac{nh^3}{2V(2\pi mkT)^{3/2}} , \tag{4-106}$$

so that

$$n(p) = \frac{4\pi n p^2 e^{-p^2/2mkT}}{(2\pi mkT)^{3/2}} ,$$

a result already obtained in (4–61).

The Maxwell–Boltzmann statistics apply in all problems dealing with the motion of particles in the atmospheres of stars and planets, with nondegenerate matter in the interior of stars, and with gas and the random motion of dust grains in interplanetary and interstellar space. These statistics also apply in some problems of stellar dynamics in which the stars can be thought of as members of an interacting assembly. Galaxies moving within a cluster are also believed to obey the M–B statistics. The formulas developed in the next sections therefore have wide applications in astrophysics.

4:16 The Saha Equation

At high temperatures atoms in thermal equilibrium are often multiply ionized. Consider two populations of particles labeled r and $r + 1$, respectively, representing an atomic species A in states of ionization r and $r + 1$. We can think of a reaction $A_r \rightleftharpoons A_{r+1} + e^-$ that might be driven to the right by ionizing photons and collisions, and to the left by the recombination of electrons with ions. Quite generally, the number densities of populations are then related by an expression similar in form to equation (4–50), but now written as $n_{r+1}/n_r = [g'_{r+1}/g_r] \exp -[\mathcal{E}/kT]$, where g'_{r+1} and g_r are the *degeneracies* — the *statistical weights* — in the upper and lower ionization states and $\mathcal{E}$ is the total energy difference between upper and lower states. This energy includes the kinetic energy $p^2/2m_e$ given to the electron if an energy $\mathcal{E}$ were imparted to the particle in ionization state r. So far, however, we have not considered that each state of ionization also comprises several states of excitation. Calling the excitation of the particle in the rth state i, and that in the $[r+1]^{\text{st}}$ state j, we write the *Saha equation*

$$\frac{n_{r+1,j}}{n_{r,i}} = \frac{g'_{r+1,j}}{g_{r,j}} \exp -[(\chi_r + \mathcal{E}_{r+1,j} - \mathcal{E}_{r,i} + p^2/2m_e)/kT], \qquad (4\text{-}107)$$

where χ_r is the energy required to ionize the atom from the lowest excitation level in the rth state of ionization to the corresponding level in the $(r+1)^{\text{st}}$ state, m_e is the electron mass, and $\mathcal{E}_{r+1,j}$ and $\mathcal{E}_{r,i}$, respectively, are the excitation energies within the corresponding ionization states.

Now, $g'_{r+1,j}$ consists of the product of two degeneracies, $g_{r+1,j}$ and g'_e, where $g_{r+1,j}$ is the degeneracy of the ionized particle in state $[r + 1, j]$, and g_e is the electron degeneracy given by the number of states available for the liberated electron to enter upon ionization with kinetic energy $p^2/2m_e$. This electron degeneracy is (see (4–65)) just $g'_e = Z(p)\, dp/n_e V = 4\pi g_e p^2\, dp/n_e h^3$, corresponding to the number of phase space states available in volume V and momentum range p to $p + dp$ to a single electron when the electron density is n_e. Here $g_e = 2$ is the electron spin degeneracy.

Recalling equations (4–105) and (4–106) and integrating equation (4–107) over all values of p, we obtain

$$\frac{n_{r+1,j} n_e}{n_{r,i}} = \frac{g_{r+1,j} g_e}{g_{r,i}} \frac{[2\pi mkT]^{3/2}}{h^3} \exp - \left[\frac{\chi_r + \mathcal{E}_{r+1,j} - \mathcal{E}_{r,i}}{kT} \right]. \qquad (4\text{-}108)$$

We will need this expression in Chapter 8 to calculate the opacity inside a star and the time required for radiation to traverse the distance from the star's center to its surface.

PROBLEM 4–24. In the solar corona, collisional excitation of atoms predominates over other processes of excitation. Among the identified spectral lines are those of CaXIII (12 times ionized calcium) and CaXV (14 times ionized). The ionization potentials of these ions are 655 and 814 eV, respectively. The lines from CaXIII are considerably stronger than those of CaXV. This fact alone can tell us very roughly what the temperature of the corona is. What is it?

4:17 Mean Values

Once the energy, frequency, or momentum distribution of particles in an assembly are known, mean values of various functions of these parameters can be computed. For particles obeying Maxwell–Boltzmann statistics, the mean value of a function $F(p)$ is

$$\langle F(p) \rangle = \frac{\int_0^\infty Z(p) F(p) e^{-p^2/2mkT} \, dp}{\int_0^\infty Z(p) e^{-p^2/2mkT} \, dp} \, . \tag{4-109}$$

This equation has precisely the form of equation (4–19). The integrand in the denominator is the probability of finding a particle with any momentum p.

These integrals of the error function type all have the form

$$\int_0^\infty x^{n-1} e^{-ax^2} \, dx = \frac{\Gamma(n/2)}{2a^{n/2}} \, . \tag{4-110}$$

where the *Gamma function* is $\Gamma(n) = (n-1)!$ and $\Gamma(\tfrac{1}{2}) = \sqrt{\pi}$.

PROBLEM 4–25. Two frequently encountered quantities are $\langle |p| \rangle$ and $\langle p^2 \rangle$. The first of these is the mean magnitude of the momentum. The mean momentum $\langle \mathbf{p} \rangle$ is zero because momenta along different directions cancel. Show that

$$\langle |p| \rangle = \frac{\int_0^\infty |p| p^2 e^{-p^2/2mkT} \, dp}{\int_0^\infty p^2 e^{-p^2/2mkT} \, dp} = \frac{\tfrac{1}{2}(2mkT)^2 \Gamma(2)}{\tfrac{1}{2}(2mkT)^{3/2} \Gamma(\tfrac{3}{2})} = \sqrt{\frac{8mkT}{\pi}} \, . \tag{4-111}$$

Show also that

$$\langle p^2 \rangle = \frac{\int_0^\infty p^4 e^{-p^2/2mkT} \, dp}{\int_0^\infty p^2 e^{-p^2/2mkT} \, dp} = 3mkT \, . \tag{4-112}$$

PROBLEM 4–26. In Section 6:18 we will make use of the quantity $\langle 1/v \rangle$. Show that

$$\left\langle \frac{1}{v} \right\rangle = \sqrt{\frac{2m}{\pi kT}} \, . \tag{4-113}$$

PROBLEM 4–27. By observing the shape of a spectral line in astronomical spectroscopy we can only determine velocities of atoms along a line of sight. To derive the temperature of a gas whose mean squared random velocity $\langle v_r^2 \rangle$ along the line of sight is known, we therefore have to know how $\langle v_r^2 \rangle$ and T are related. For a Maxwell–Boltzmann distribution show that

$$\langle v_r^2 \rangle = kT/m. \tag{4-114}$$

The analogous integrals required for computing mean values for energies or momenta for fermions or bosons involve the Fermi–Dirac or Bose–Einstein distribution functions.

4:18 Fluctuations

Random processes invariably exhibit deviations from a mean. These *fluctuations* can be expressed in terms of the mean square deviation, which is always positive. It is obtained by taking the deviation of each value from the mean, squaring that deviation, and then averaging over all deviations. For energy $\mathcal{E}$

$$\langle (\Delta \mathcal{E})^2 \rangle = \langle (\mathcal{E} - \langle \mathcal{E} \rangle)^2 \rangle = \langle \mathcal{E}^2 \rangle - \langle \mathcal{E} \rangle^2 \, . \tag{4-115}$$

For a Maxwell–Boltzmann distribution, we can write

$$\langle \mathcal{E}^2 \rangle - \langle \mathcal{E} \rangle^2 = \frac{\int_0^\infty Z(\mathcal{E}) \mathcal{E}^2 e^{-\mathcal{E}/kT} d\mathcal{E}}{\int_0^\infty Z(\mathcal{E}) e^{-\mathcal{E}/kT} d\mathcal{E}} - \left(\frac{\int_0^\infty Z(\mathcal{E}) \mathcal{E} e^{-\mathcal{E}/kT} d\mathcal{E}}{\int_0^\infty Z(\mathcal{E}) e^{-\mathcal{E}/kT} d\mathcal{E}} \right)^2 . \tag{4-116}$$

By carrying out a partial differentiation with respect to temperature T, the right side of this equation is seen to equal

$$kT^2 \frac{\partial}{\partial T} \left(\frac{\int_0^\infty Z(\mathcal{E}) \mathcal{E} e^{-\mathcal{E}/kT} d\mathcal{E}}{\int_0^\infty Z(\mathcal{E}) e^{-\mathcal{E}/kT} d\mathcal{E}} \right) = kT^2 \frac{\partial}{\partial T} \langle \mathcal{E} \rangle \, . \tag{4-117}$$

We, therefore, obtain the energy fluctuation in a Maxwell–Boltzmann distribution as

$$\langle (\Delta \mathcal{E})^2 \rangle = kT^2 \frac{\partial \langle \mathcal{E} \rangle}{\partial T} \, , \tag{4-118}$$

an expression known as the *Einstein–Fowler equation*.

PROBLEM 4–28. Show that the relation (4–118) also holds for blackbody radiation.

4:19 The First Law of Thermodynamics

The first law of thermodynamics expresses the conservation of energy. If a gas is heated the supplied energy can act in one of two ways. It can raise the gas temperature, or it can perform work by expanding the gas against an externally applied pressure. Symbolically we write[3]

$$đQ = dU + PdV, \tag{4-119}$$

where all quantities are normalized to one mole of matter and where the left-hand side gives the amount of heat $đQ$ supplied to the system; dU is the change in internal energy and $P\,dV$ is the work performed. The nature of this last term is easily understood if we recall that work is involved in any displacement D against a force F. If the change in volume dV involves, say, the displacement of a piston of area A, then the force applied is $F = PA$ and the distance the piston moves is $D = dV/A$.

The *internal energy* U of the gas is the sum of the kinetic energy of translation as the molecules shoot around; the kinetic and potential energy involved in the vibrations of atoms within a molecule; the energy of excited electronic states; and the kinetic energy of molecular rotation.

Q is the *heat content* of the system. The heat Q that must be supplied to give rise to a one degree change in temperature is called the *heat capacity* of the system. The heat capacity depends on the amount of work that is done. If no work is involved — which means that the system is kept at constant volume — all the heat goes into increasing the internal energy and

$$c_v = \left[\frac{đQ}{dT}\right]_V = \frac{dU}{dT} . \tag{4-120}$$

The subscript V denotes constant volume.

Sometimes we need to know the heat capacity under constant pressure conditions. For an ideal gas, this relation is quite simple. In differential form, the ideal gas law (4–34) reads

$$P\,dV + V\,dP = R\,dT \tag{4-121}$$

so that the first law becomes

$$đQ = \left(\frac{dU}{dT} + R\right) dT - V\,dP . \tag{4-122}$$

For constant pressure

$$c_p = \frac{dU}{dT} + R . \tag{4-123}$$

For an ideal gas we therefore have the important relation

---

[3] $đQ$ is not an *exact differential*. This means that the change of heat $đQ$ depends on how the change is attained. For example, it can depend on whether we first raise the internal energy by dU, and then do work $P\,dV$, or vice versa.

$$c_p - c_v = R = \mathcal{N}k \,. \tag{4-124}$$

This follows from (4–120) and (4–123). $\mathcal{N}$ is Avogadro's number and the heat capacities are figured for one mole of gas.

We have already stated that the internal energy involves the translation, vibration, electronic excitation, and rotational energy of the molecules. We can ask ourselves how these energies are distributed in a typical molecule. We know that the probability of exciting any classical particle to an energy $\mathcal{E}$ is proportional to the Boltzmann factor $e^{-\mathcal{E}/kT}$. This is true whether $\mathcal{E}$ is a vibrational, electronic, rotational, or translational energy. For an assembly of classical particles, then, the mean internal energy per molecule depends only on the number of ways that energy can be excited, that is, the number of *degrees of freedom* multiplied by $kT/2$. This factor $kT/2$ is consistent with our previous finding that the total translational energy, which has three degrees of freedom, is $(\frac{3}{2})\mathcal{N}kT$ per mole. Each translational degree of freedom, therefore, has energy $\frac{1}{2}kT$ and each other available degree of freedom in thermal equilibrium will also be excited to this mean energy. This is called the *equipartition principle*.

PROBLEM 4–29. Show that an interstellar grain in thermal equilibrium with gas at $T \sim 100$ K rotates rapidly. If its radius is $a \sim 10^{-5}$ cm and its density is $\rho \sim 1$, show that the angular velocity is about $\omega \sim 10^{5.5}$ rad s^{-1}.

The equipartition principle is a part of classical physics. It does not quite agree with observations; the actual values can be explained more easily by quantum mechanical arguments. The difference between classical and quantum theory hinges to a large extent on what is meant by "available" degrees of freedom. The electronically excited states of atoms and molecules normally are not populated at low temperatures. Hence, at temperatures of the order of several hundred degrees Kelvin, electronic states do not affect the heat capacity. Even the vibrational states then make a relatively small contribution to the heat capacity because vibrational energies usually are large compared to rotational energies. Aside from the translational contribution, it is therefore the low-energy rotational states which, at low temperatures, make a major contribution to the internal energy and the specific heat at constant volume.

The rotational position of a diatomic molecule can be given in terms of two coordinates θ and ϕ. It therefore has two degrees of rotational freedom. A polyatomic molecule having three or more atoms in any configuration except a linear one requires three coordinates for a complete description and therefore has three degrees of rotational freedom. A diatomic or linear molecule makes a rotational contribution of kT to the heat capacity and a nonlinear molecule contributes $3kT/2$. Even these relatively simple rules hold only at low temperatures. At higher temperatures rotational states with progressively higher quantum numbers are excited and a quantum mechanical weighting function has to be introduced to take into account the number of degenerate (identical) states that can be excited.

We will be interested in the heat capacity of interstellar gases where temperatures are low and many of the above-mentioned difficulties do not arise. Let us define the ratio of heat capacities, respectively, at constant pressure and volume as $c_p/c_v \equiv \gamma$. Then by (4–124) we have

$$\gamma \equiv \frac{c_p}{c_v} = \frac{c_v + \mathcal{N}k}{c_v} . \tag{4-125}$$

For monatomic gases we deal with the translational internal energy and $c_v = 3k\mathcal{N}/2$, so that $\gamma = \frac{5}{3}$. For diatomic molecules two rotational degrees of freedom are available in addition to the three translational degrees, so that $c_v = 5k\mathcal{N}/2$ and $\gamma = \frac{7}{5}$.

4:20 Isothermal and Adiabatic Processes

The contraction of a cool interstellar gas cloud or, equally well, the expansion of a hot ionized gas cloud can proceed in a variety of ways. Some cosmic processes involving the dynamics of gases can occur quite slowly at constant temperature. These are called *isothermal processes*. The internal energy does not change and the heat put into the system equals work done by it. Another type of process that describes many rapidly evolving systems is the *adiabatic process* in which there is neither heat flow into the gas nor heat flowing out, $dQ = 0$.

$$dQ = c_v \, dT + P \, dV = 0 . \tag{4-126}$$

For an ideal gas

$$c_v \, dT + \frac{RT}{V} \, dV = 0 \quad \text{and} \quad c_v \frac{dT}{T} + (c_p - c_v)\frac{dV}{V} = 0 . \tag{4-127}$$

Integrating, we have

$$\log T + (\gamma - 1) \log V = \text{constant} , \tag{4-128}$$

or

$$TV^{\gamma-1} = \text{constant} ,$$
$$PV^{\gamma} = \text{constant, and} \tag{4-129}$$
$$P^{(\gamma-1)}T^{\gamma} = \text{constant} .$$

These are the adiabatic relations for an ideal gas. They govern the behavior, for example, of interstellar gases suddenly compressed by a shock front heading out from a newly formed O star or from an exploding supernova. We will study these phenomena in Chapter 9.

For thermalized electromagnetic radiation the internal energy in volume V is

$$U = aT^4 V. \tag{4-130}$$

This is just the energy density. The pressure has one-third this value, by (4–43), and for volume V we can describe an adiabatic process by

$$\text{đ}Q = dU + P\,dV = 4aT^3V\,dT + \frac{4}{3}aT^4\,dV$$

$$= 3V\,dP + 4P\,dV = 0\,, \tag{4-131}$$

$$P \propto V^{-4/3}, \qquad \gamma = \frac{4}{3}\,, \tag{4-132}$$

Because of its role in adiabatic processes, the ratio of heat capacities of a gas γ is referred to as its *adiabatic constant*. In the hot interstellar medium where, to a good approximation, we deal only with radiation, monatomic, diatomic, or ionized particles, γ ranges from $\frac{4}{3}$ to $\frac{7}{5}$ or $\frac{5}{3}$ depending on whether radiation or gas particles dominate the pressure. In very cold gas clouds dominated by molecular hydrogen, γ may have a value as high as $\frac{5}{3}$ because the temperature is insufficiently high to excite the molecules to rotate or vibrate. However, in shocked molecular regions, with which we shall deal in Chapter 9, the temperature of a molecular cloud can rise sufficiently to excite both rotational and vibrational states and γ can approach $\frac{8}{6}$ because then $c_v = 6k\mathcal{N}/2$. A reduction of γ below $\frac{4}{3}$ can also be the trigger for pre-supernova collapse, as we will see in Section 8:12.

4:21 Entropy and the Second Law of Thermodynamics

Equation (4–43) tells us that the pressure of thermal radiation is just one-third of the radiation density given in (4–130). We can therefore write the first law of thermodynamics as

$$\text{đ}Q \equiv T\,dS = dU + P\,dV = 4aT^3V\,dT + \frac{4aT^4}{3}\,dV\,, \tag{4-133}$$

where the first equality is a definition of the *entropy* S. Although $\text{đ}Q$ is not an exact differential, dS is.

PROBLEM 4–30. Show that the radiation entropy can be obtained from (4–133) as

$$S = \frac{4aT^3V}{3}\,. \tag{4-134}$$

The second law of thermodynamics asserts that the entropy of a closed system can at best remain constant, but will normally increase during any physical process. In this context it is interesting to see what Liouville's theorem tells us about the use of telescopes in concentrating light beams onto small detectors. In many applications we could obtain very high instrumental sensitivity if light from some cosmic source could be concentrated onto the smallest possible detector. Let the solid angle subtended by the astronomical object be Ω and let the telescope area be A. Then Liouville's theorem states that the smallest detector area onto which the light could be focused is

$$a = \frac{A\Omega}{4\pi} , \tag{4-135}$$

and that is only possible if light can be made to impinge on the detector from all sides. Usually we are able to make light fall onto the detector only from some smaller solid angle $\Omega' < 4\pi$ so that the minimum area of the detector becomes

$$a = \frac{A\Omega}{\Omega'} . \tag{4-136}$$

A violation of this restriction would imply that the radiation temperature at the source was lower than at the detector and that radiation was actually flowing from a cooler to a hotter object. This would violate the second law of thermodynamics which states that heat cannot flow freely from a cold to a hot object, because the combined entropy of the two objects would then be lowered.

4:22 Formation of Condensations and the Stability of the Interstellar Medium

We think that the stars were formed from gases that originally permeated the whole Galaxy, and that galaxies were formed from a medium that initially was more or less uniformly distributed throughout the Universe.

There is strong evidence that star formation is going on at the present time. Many stars are in a stage that can only persist for a few million years because the stellar luminosity — energy output — is so great that these stars soon would deplete their available energy and evolve into objects with entirely different appearance. These bright stars are generally found in the vicinity of cool, dusty, molecular clouds, and appear to have formed from this dense gas.

We now ask how a molecular cloud could collapse to form a star. To answer this question we can study the stability of the cloud and its dependence on the ratio of heat capacities γ.

Consider an assembly of molecules. Their kinetic energy $\mathbb{T}$ per mole is

$$\mathbb{T} = \frac{3}{2}(c_p - c_v)T , \tag{4-137}$$

or

$$\mathbb{T} = \frac{3}{2}(\gamma - 1)c_v T . \tag{4-138}$$

The internal energy is

$$U = c_v T. \tag{4-139}$$

Hence

$$\mathbb{T} = \frac{3}{2}(\gamma - 1)U . \tag{4-140}$$

By the virial theorem (3–85) we then have

$$3(\gamma - 1)U + \mathbb{V} = 0 \qquad (4\text{-}141)$$

as long as inverse square law forces predominate among particles. This means that the equation holds true both when gravitational forces are important and where charged particle interactions dominate the behavior on a small scale (see Section 4:23 below). Sometimes it can even hold when light pressure from surrounding stars acts to drive dust grains toward each other.

If the total energy per mole is

$$\mathcal{E} = U + \mathbb{V} \qquad (4\text{-}142)$$

we have from equation (4–141) that

$$\mathcal{E} = -(3\gamma - 4)U = - \left(\frac{3\gamma - 4}{3(\gamma - 1)} \mathbb{V} \right) . \qquad (4\text{-}143)$$

Three results are apparent (Ch39)*:

(a) If $\gamma = \frac{4}{3}$, $\mathcal{E}$ is always zero independent of the configuration of the cloud. Expansion and contraction are possible and the configuration is unstable. This case corresponds to a photon gas (4–130) and to molecular hydrogen at sufficiently high temperatures to excite both rotational and vibrational excited states. In its early stages, a planetary nebula has radiation-dominated pressure acting to produce its expansion (Ka68). It should therefore be only marginally stable. Similarly, a shocked interstellar molecular cloud should exhibit signs of instability, and this appears to be supported by the observation that protostars and young stars are often observed in clouds that also exhibit prevailing shocks.

(b) For $\gamma = 1$, $\mathbb{V}$ is always zero for any $\mathcal{E}$ value and again no stable configuration exists.

(c) For $\gamma > \frac{4}{3}$, equation (4–143) shows that $\mathcal{E}$ is always negative and the system is bound. If the system contracts and the potential energy changes by $\Delta \mathbb{V}$ then

$$\Delta \mathcal{E} = +\frac{(3\gamma - 4)}{3(\gamma - 1)} \Delta \mathbb{V} = -(3\gamma - 4) \Delta U . \qquad (4\text{-}144)$$

An amount of energy $-\Delta \mathcal{E}$ is lost by radiation

$$-\Delta \mathcal{E} = -\frac{3\gamma - 4}{3(\gamma - 1)} \Delta \mathbb{V} , \qquad (4\text{-}145)$$

while the internal energy increases by

$$\Delta U = -\frac{1}{3(\gamma - 1)} \Delta \mathbb{V} \qquad (4\text{-}146)$$

through a rise in temperature.

As a protostar contracts to form a star it therefore becomes progressively hotter. Two factors are worth mention:

(a) When theories of the kind developed here are applied on a cosmic scale, say, to formation of galaxies or clusters of galaxies, we run into difficulties in defining the potential $\mathbb{V}$. The zero level of the potential can no longer be defined using Newtonian theory alone, and some more comprehensive approach such as that of general relativity should be used. This considerably complicates the treatment of the problem.

(b) In practice, star formation may be preferentially induced by external events. This is indicated by the observed formation of massive stars in regions where other stars have just formed and where pressures from the surrounding medium are setting in. Formation, especially of massive stars, may result from the compression of cool gas clouds in collisions of galaxies, or where supersonic stellar winds, supernova explosions, or intense ultraviolet radiation from nearby hot stars shock these clouds. We will discuss such processes in Chapters 9 and 10. The stability of an isolated medium, as treated above, may therefore not be strictly relevant to the discussion. The angular momentum of a cloud or the magnetic fields threading it, also plays a role. Nevertheless, conditions stable as judged by their ratio of heat capacities γ tend to resist compression, whereas intrinsically unstable configurations more readily collapse under pressure.

4:23 Ionized Gases and Clusters of Stars and Galaxies

The behavior of large clusters of stars or galaxies can be described statistically much as we describe the behavior of gases. There are many striking similarities between the physics of ionized gases (plasma) and aggregates of stars or galaxies. These similarities come about because Newton's gravitational attraction can be written in a form similar to Coulomb's electrostatic force:

$$\text{Newton's force} \qquad \text{Coulomb's force}$$

$$\frac{(iG^{1/2}m_1)(iG^{1/2}m_2)}{r^2} \qquad \frac{Q_1Q_2}{r^2} \, . \tag{4-147}$$

Here the gravitational analogue to electrostatic charge is the product of mass, the square root of the gravitational constant, and the imaginary number, i. The correspondence can be extended to include fields, potentials, potential energies, and other physical parameters. The primary difference between gravitational processes and electrostatic interactions is that electric charges can be both positive and negative whereas the sign of the gravitational analogue to charge is always the same — mass is always positive. Let us first derive some properties of assemblies of gravitationally interacting particles and then make a comparison to plasma behavior.

Clusters of Stars and Galaxies

If we take a spherical distribution of particles — a set of stars or galaxies the attractive force acting on unit mass at distance r from the center is

$$F(r) = -\frac{1}{r^2} \int_0^r 4\pi G r^2 \rho(r) \, dr \ . \tag{4-148}$$

This means that

$$\frac{d}{dr} r^2 F(r) = -4\pi G r^2 \rho(r) \tag{4-149}$$

and setting the force per unit mass equal to a potential gradient

$$F(r) = -\nabla\mathbb{V}(r) = -\frac{d}{dr}\mathbb{V}(r) \tag{4-150}$$

we have

$$\frac{1}{r^2}\frac{d}{dr}r^2\frac{d}{dr}\mathbb{V}(r) = +4\pi G \rho(r) \ . \tag{4-151}$$

This is *Poisson's equation*. Substituting from equation (4–47) we have the *Poisson–Boltzmann equation* for a gas at temperature T:

$$\frac{1}{r^2}\frac{d}{dr}r^2\frac{d\mathbb{V}}{dr} = +4\pi\rho_0 G e^{-[m\mathbb{V}(r)/kT]} \ . \tag{4-152}$$

The potential appearing in the exponent on the right of this equation was obtained through an integration of $\nabla\mathbb{V}(r)$ that led to equation (4–46). The behavior of an assembly of stars or galaxies would, therefore, be no different if some constant potential present throughout the Universe were added to $\mathbb{V}(r)$. This is essentially the point that was already raised in Section 4:22 in connection with the stability of uniform distributions of gas.

PROBLEM 4–31. Show that the substitutions

$$\frac{m\mathbb{V}(r)}{kT} \equiv \psi \quad \text{and} \quad r \equiv \left(\frac{kT}{4\pi\rho_o mG}\right)^{1/2} \xi \ , \tag{4-153}$$

where ψ and ξ are dimensionless, turn equation (4–152) into

$$\frac{1}{\xi^2}\frac{d}{d\xi}\left(\xi^2\frac{d\psi}{d\xi}\right) = e^{-\psi} \ . \tag{4-154}$$

We now have to decide on the boundary conditions that have to be imposed on this differential equation. At the center of the cluster there are no forces and the first derivative of $\mathbb{V}$ or ψ must be zero. Because the potential can have an arbitrary additive constant, we can set the potential to be zero at the center. In terms of the new variables these two conditions are

$$\psi = 0 \qquad \text{and} \qquad \frac{d\psi}{d\xi} = 0 \qquad \text{at} \qquad \xi = 0 \,. \qquad (4\text{-}155)$$

Taken together with equation (4–154) they lead to a solution that has no closed form (Ch43)*.

PROBLEM 4–32. Show that in the limit of very small and of large ξ values the respective solutions are (Ch39)

$$\psi \sim \frac{1}{6}\xi^2 - \frac{1}{120}\xi^4 + \frac{1}{1890}\xi^6 + \cdots, \quad \xi \ll 1 \,, \qquad (4\text{-}156)$$

$$\psi \sim \ln\left(\frac{\xi^2}{2}\right), \quad \xi^2 > 2 \,. \qquad (4\text{-}157)$$

This can be verified by substitution in equations (4–154) and (4–155). From this, the radial density and mass distributions can be found (Ch43). The density distribution is plotted in Fig. 4.15.

One difficulty with this plot and with the asymptotic solution (4–157) is that the density $\rho_0 e^{-\psi}$ is proportional to ξ^{-2}. This causes the total mass integrated to large distances to become infinite. We therefore need a cut-off mechanism that will restrict the radius of a cluster of stars to a finite value. In Chapter 9 we will see how an external pressure can also lead to a finite size and to a structure for cold interstellar gas clouds consonant with observations.

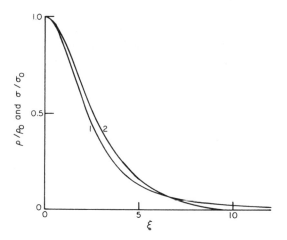

Fig. 4.15. Plot of density ρ, curve 1, and areal density σ, curve 2, against radial distance ξ from the center of a distribution. For a cluster, σ/σ_0 represents the star density drop with radial distance, as measured directly on a photographic plate. ρ_0 and σ_0 represent values at the center. $\rho/\rho_0 = \exp(-\psi)$, $\xi = (4\pi\rho_0 m G/kT)^{1/2} r$. (From *Principles of Stellar Dynamics* by S. Chandrasekhar (Ch43). Reprinted through permission of the publisher, Dover Publications, Inc., New York.)

Ionized Gases

It is interesting to compare these results to those obtained for ionized gases — *plasma* — in which both positive and negatively charged particles are present. The derivation of the Poisson–Boltzmann equation was in no way based on particle charges. It concerned itself only with an inverse square law force and a uniform mass distribution that could equally well have been a charge distribution. Using the density for an assembly of dissimilar particles (see equation (4–53)), the Poisson–Boltzmann equation can be written as

$$\frac{1}{r^2}\frac{d}{dr}r^2\frac{d\mathbb{V}}{dr} = -4\pi\sum_i n_{i0}q_i e^{-q_i\mathbb{V}/kT} \tag{4-158}$$

for plasma. Here q_i is the charge of particles of type i. If we restrict ourselves to large interparticle distances — a condition that holds in intergalactic, interstellar, and interplanetary space — then

$$q_i\mathbb{V} \ll kT , \tag{4-159}$$

and we can use the Taylor expansion

$$e^{-q_i\mathbb{V}/kT} = 1 - \frac{q_i\mathbb{V}}{kT} + \frac{1}{2}\left(\frac{q_i\mathbb{V}}{kT}\right)^2 + \cdots . \tag{4-160}$$

Neglecting quadratic and higher terms, the charge density on the right of (4–158) then becomes

$$\rho = \sum_i n_{i0}q_i - \frac{\mathbb{V}}{kT}\sum_i n_{i0}q_i^2 . \tag{4-161}$$

The first term vanishes because of charge neutrality for the bulk of the plasma. Note that this term does not vanish in the gravitational case; there it is dominant. The second term in (4–161) can be written as

$$\rho = -\frac{\mathbb{V}e^2}{kT}\sum_i n_{i0}Z_i^2 \qquad \text{where} \qquad q_i = eZ_i . \tag{4-162}$$

Substituting in (4–158) we have

$$\frac{1}{r^2}\frac{d}{dr}\left(r^2\frac{d\mathbb{V}}{dr}\right) = \frac{4\pi e^2\mathbb{V}}{kT}\sum_i n_{i0}Z_i^2 \tag{4-163}$$

or

$$\frac{d}{dr}\left(r^2\frac{d\mathbb{V}}{dr}\right) = r^2\mathbb{V}\left(\frac{1}{L^2}\right) , \tag{4-164}$$

where

$$\frac{1}{L^2} = \frac{4\pi e^2}{kT}\sum n_{i0}Z_i^2 . \tag{4-165}$$

L has the dimension of a length. It is called the *Debye shielding length* and is a distance over which a charged particle embedded in a plasma can exert an appreciable electrostatic field. Beyond that distance its electrostatic influence rapidly diminishes. For fully ionized hydrogen, $L = 6.90(T/n)^{1/2}$ cm.

One reason why the shielding length is of interest in astrophysics is because it points out the impossibility of maintaining an electric field over any large scale. A field cannot be influential over distances much larger than L. Even for tenuous interstellar gas clouds with $n_{i0} = 10^{-3}$ cm^{-3}, $Z_i = 1$, and $T = 100$ K, L turns out to be about 20 meters. This is completely negligible compared to typical interstellar distances ~ 1 pc.

Electrostatic forces may be important in large-scale processes, but only when they appear in conjunction with large-scale magnetic fields that can prevent the flow of charged particles along the electric field lines and therefore prevent the charge separation required for electrostatic shielding. The behavior of plasmas in the presence of magnetic fields is treated in the theory of magnetohydrodynamic processes (Co57), (Sp62). We will consider these processes in Chapters 6, 9, and 10.

Answers to Selected Problems

4–2. Suppose we take n_i steps of length λ_i. The mean square deviation then is $n_i \lambda_i^2$; and a similar result holds for all step sizes. Hence the final mean square deviation is $\sum_i n_i \lambda_i^2 = N \langle \lambda^2 \rangle$.

4–4. For escape the deviation has to be $\sim n^{1/2}$ steps of length L. Hence $n \sim R^2/L^2$.

4–5. For a given value of $R\alpha$, the value of $1/r$ averaged over all θ-values is

$$\langle \frac{1}{r} \rangle = \frac{\langle \sin \theta \rangle}{R\alpha} = \frac{2}{\pi R\alpha} .$$

If we also average over different values of $1/R\alpha$ we obtain

$$\left\langle \frac{1}{r} \right\rangle = \frac{2}{\pi} \left\langle \frac{1}{R\alpha} \right\rangle \quad \text{and} \quad \text{Total mass} = \frac{\langle v^2 \rangle}{G \langle 1/r \rangle} = \frac{3\pi \langle v_p^2 \rangle}{2G \langle (R\alpha)^{-1} \rangle} .$$

4–6.

$$\frac{nm \langle v^2 \rangle}{3} \sim 4 \times 10^{-17} \text{ dyn cm}^{-2} .$$

4–7. The collision probability per star pair in unit time is $nv\sigma$ s^{-1}. In P seconds the probability is $nv\sigma P$ per star pair, or about 1.5×10^{-11} that the Sun would have formed a planetary system in the time available. If there are 10^{11} stars altogether, 1.5 pairs, or 3 solar systems would have formed in this way in 5×10^9 yr. However, we know of more than 100 planetary systems within <30 pc of the Sun. Their number density already is too high to be accounted for by the Jeans hypothesis.

4–8. $P \sim 10^{-14}$ dyn cm^{-2}.

4–9. $v \sim (3kT/M)^{1/2} \sim 0.8 \, \text{cm s}^{-1}$ for $M = 4\pi\rho r^3/3 \sim 7 \times 10^{-14} \, \text{g}$.

4–10. The momentum transfer rate is $\pi r^2 n v^2 m = dp/dt = 3.3 \times 10^{-21} \, \text{g}$ cm s^{-2}. The mass of the grains is given in Problem 4–9 and, hence, $dv/dt = 5 \times 10^{-8} \, \text{cm s}^{-2}$. This gives the initial acceleration. However, as a grain gains velocity and mass, the acceleration decreases toward a value of zero, which is reached when the grain reaches velocity v.

4–11. The mass gain $dM/dt = \pi r^2 n v m = 3 \times 10^{-27} \, \text{g s}^{-1}$. At this rate the grain would gain 1% of its mass in $2 \times 10^{11} \, \text{s}$, a few thousand years.

4–12. $\langle v^2 \rangle = 3kT/m$; $v_P \sim 1.6 \times 10^6 \, \text{cm s}^{-1}$; $v_e \sim 6.7 \times 10^7 \, \text{cm s}^{-1}$;

4–13. $P \sim 1.4 \times 10^{-13} \, \text{dyn cm}^{-2}$.

4–14. If n is the number of photons passing through unit area per unit time, the pressure is

$$P = \frac{\int n(\nu) h\nu \, d\nu}{c} .$$

The force on the grain is its area, multiplied by P,

$$P = \frac{1.37 \times 10^6}{3 \times 10^{10}} = 4.6 \times 10^{-5} \, \text{dyn cm}^{-2} , \quad F = 3.6 \times 10^{-9} \, \text{dyn.} \quad (4\text{-}166)$$

4–15. For isotropic scattering we average the function $(1 - \cos\theta)$ over all solid angle increments to get the mean momentum transfer. Thus

$$P = \frac{\mathcal{E}}{3c} + \frac{2\mathcal{E}/3c}{4\pi} \int \int (1 - \cos\theta) \sin\theta \, d\phi \, d\theta ,$$

with θ chosen as zero in the forward scattering direction and $\mathcal{E}$ the energy incident on the grain per unit area and time. Thus $P = \mathcal{E}/c$ and the grain has a force of $1.4 \times 10^{-12} \, \text{dyn}$ acting on it at Earth's distance from the Sun. The gravitational force is $mM_{\odot}G/R^2$; here m is the particle mass, $M_{\odot}$ is the solar mass, and R is the distance of the Sun from Earth: $m = (4\pi\rho s^3)/3 = 2.5 \times 10^{-11} \, \text{g}$, the gravitational force is $1.5 \times 10^{-11} \, \text{dyn}$ and the radiative repulsion is about 10% of the gravitational attraction. The ratio of gravitational to radiative force remains constant because both diminish as the R^2.

4–16. From the equation for the period of a grain in an elliptic orbit

$$\tau = \frac{2\pi a^{3/2}}{(MG_{\text{eff}})^{1/2}} .$$

We note that the period depends on the square root of the effective gravitational constant. For $G_{eff} = \frac{2}{3}G$ the period of the grain will be $(\frac{3}{2})^{1/2}$ yr, that is, 1.22 yr. The orbital velocity will be $v_E/1.22$, where v_E is Earth's orbital velocity; that speed will be roughly $24 \, \text{km s}^{-1}$. The collision velocity of Earth with such a grain would therefore be $\sim 5 \, \text{km s}^{-1}$.

4–18. The thinness of the atmospheric layer implies that g is constant throughout. Hence the scale height h is determined by equation (4–51) and $h \sim 10^6 \, \text{cm}$.

4–19. Note that this involves an error function integral whose value is, for example, given later in this chapter in equation (4–110).

4–20. $m \sim 2 \times 10^{-24}$ g, and $v \sim 2.4\,\mathrm{km\ s^{-1}}$.

4–21. Consider Snell's law of optics, which states that radiation incident on the interface between vacuum and a medium with refractive index n obeys the relation $n \sin \theta_n = \sin \theta_0$, where θ_0 is the angle that a ray incident on the interface makes with the normal, and θ_n is the corresponding angle of the ray after it enters the medium. In thermal equilibrium, the amount of radiation crossing in each direction is the same. But radiation incident on the interface from the medium at angles $\theta_n > \sin^{-1}(1/n)$ is totally reflected and cannot exit. The rate at which radiation exits from the medium is, therefore, restricted both by this limiting angle and the speed c/n at which it is incident on the interface. The outgoing radiation rate corresponding to the incident rate given by (4–76) is

$$\frac{c}{n}\langle \cos \theta \rangle = \frac{c}{2\pi n} \int_0^{2\pi} \int_0^{\sin^{-1}(1/n)} \cos \theta \sin \theta d\theta d\phi = \frac{c}{2n^3} \ .$$

For balance between incident and exiting radiation, the radiation density in the medium must then be n^3 times higher than on the vacuum side.

4–23. For each of the three neutrino species

$$Z(p)\,dp = \frac{4\pi p^2\,dp}{h^3} \cdot V, \quad p = \frac{\mathcal{E}}{c}, \quad dp = \frac{d\mathcal{E}}{c} \ . \quad \therefore \quad Z(\mathcal{E})\,d\mathcal{E} = \frac{4\pi \mathcal{E}^2\,d\mathcal{E}}{h^3 c^3} V.$$

The total energy density and mass density for all three species, respectively, are

$$\frac{\mathcal{E}}{V} = 3 \int_0^{\Phi_\nu} \frac{\mathcal{E} Z(\mathcal{E})\,d\mathcal{E}}{V} = \frac{3\pi \Phi_\nu^4}{h^3 c^3} \quad \text{and} \quad \rho_\nu = \frac{\mathcal{E}}{c^2 V} = \frac{3\pi \Phi_\nu^4}{h^3 c^5} \ .$$

4–24. $kT \sim 655\,\mathrm{eV} \sim 10^{-9}$ erg,

$$T \sim \frac{10^{-9}\ \mathrm{erg}}{1.38 \times 10^{-16}\ \mathrm{erg\ K^{-1}}} \sim 8 \times 10^6\ \mathrm{K}.$$

We reason that $kT \sim$ excitation energy; the higher ionized state gives rise to a weak line because T is not sufficiently high to lead to frequent ionization to this level. That is, for the higher ionized state $kT <$ ionization energy. For the lower ionized states kT is probably more comparable to the ionization energy. Actually T is $\sim 1.5 \times 10^6$ K in the corona.

4–28. This can be demonstrated by writing $\langle \mathcal{E} \rangle$ as the first expression in (4–69), writing the analogous summation for $\langle \mathcal{E}^2 \rangle$, and taking the partial derivative of $\langle \mathcal{E} \rangle$ with respect to T. Although this holds for just one frequency ν, we can linearly sum mean square deviations as in (4–16) to show that (4–118) holds over the entire blackbody spectrum.

4–30. Equation (4–133) leads to $dS = (4a/3)d(VT^3)$, which gives the desired result with an additive constant of integration which needs to be zero to keep the entropy proportional to the volume.

5 Photons and Fast Particles

5:1 The Relativity Principle

In discussing Newton's laws of motion in Sections 3:4 and 3:8, we were careful to note that they held only under restricted conditions. All motions had to be described with respect to *inertial frames of reference* — frames at rest or moving at constant velocity with respect to the mean motion of ambient galaxies.

Under these conditions not only Newton's laws but all other laws of physics are obeyed. This general statement — first formulated by Einstein — is called the *principle of relativity*. It implies that an observer cannot determine the absolute motion of his inertial frame of reference — only its motion relative to some other frame. The principle also has many other important consequences which, taken together, form the basis of the theory of *special relativity* (Ei–a). As also mentioned in Section 3:8, Einstein broadened the concept of the inertial frame beyond Newton's scope of a frame moving at constant velocity with respect to fixed stars. He showed that we can include coordinate frames fixed in any freely falling, nonrotating bodies. Althougy such local inertial frames may accelerate with respect to frames that are far from any attracting massive objects, they are fully equivalent to them as far as the principle of relativity is concerned. Finally, Einstein postulated that the speed of light is the same in all reference frames, whether they move or are stationary. This actually is a consequence of the relativity principle. If this speed were not the same, an observer could determine whether he was at rest or moving.

We should note that the relativity principle is founded on observations. It could not have been predicted from logic alone.

In recent centuries, long before Einstein's birth, there has always been an awareness that some sort of relativity principle might exist. Even the wording of the principle was similar to Einstein's, though its implications were quite different. In Galileo's time, the speed of light was believed to be infinite as measured in any reference frame. This implied that the instantaneous transmission of signals and messages over large distances should be possible. Because any velocity added to an infinite velocity still gave infinite speed, it was clear that no matter how an observer moved he would always see light traveling at the same, infinite, speed. Similarly all other laws of physics seemed to hold identically in any of Newton's inertial frames.

Then in 1666 Ole Rømer discovered that the speed of light is finite, though large. This tended to detract from the *Galilean relativity principle* because it seemed that an observer moving into the direction of a light source would see the light wave moving faster than an observer moving away from the source. But at the end of the nineteenth century Albert A. Michelson and others discovered that the speed of light was identical in all the moving reference frames they were able to check. Independently Einstein postulated a new principle of relativity similar to Galileo's except that the speed of light now was finite and equal in all reference frames. To some extent this concept had already been present, for several decades, in the electromagnetic theory of James Clerk Maxwell, but the required constancy of the speed of light was considered a weakness of the theory, not its strength.

As we will see, Einstein's relativity principle also led him to conclude that no physical object could travel at a speed in excess of light, $c = 2.998 \times 10^{10}$ cm s^{-1}; the concept of an infinite velocity had no correspondence in physical moving objects.

The theory of relativity has the task of formulating the laws of physics in such a way that physical processes can be accurately described in any moving coordinate system. This study conveniently divides into two parts. The first theory is more restricted. It deals with physical processes as viewed from inertial reference frames and specifically excludes any consideration of gravity. This is *special relativity*. The second, more general theory incorporates not only special relativity but also the study of gravitational fields and arbitrarily accelerated motions. It is therefore called the *general theory of relativity*.

5:2 Relativistic Terminology

Suppose a physical process occurs in a system at rest with respect to some inertial reference frame K'. An observer O in some other inertial frame K views this process. If K and K' are moving at large velocities V, relative to each other, observer O will see events in K' distorted both in space and in time, but the special theory of relativity will allow him to reconstruct events as they would occur in his own system K. This is a very useful property of relativity theory. We will find many applications of it in astrophysics where high velocities are often encountered. The special theory, however, goes beyond this limited function of reconstructing clear pictures from apparently distorted observations. It gives new insight into the relation between time and space, and among momentum, energy, and mass; it justifies the impossibility of massive bodies moving through space at velocities exceeding the speed of light and yields many other new results.

To make full use of the theory, we will need to take a few preparatory steps. We must define new concepts and formulate them mathematically.

(a) To the extent that it is valid, the special theory abolishes an absolute standard for a state of rest. It states that there is no way of defining zero speed in an absolute way. Bodies may be at rest — but only *relative* to some other body or frame of reference.

We know that this statement need not be quite true. A preferred natural state of rest does exist for any locality in the Universe. It is the state of rest relative to the mean motion of ambient galaxies. This was not known at the time relativity theory was established. It tends to weaken the statement we formulated above, and allows us to state only that an absolute standard of rest is inconsequential to special relativity. The theory draws no distinction between absolute rest and constant velocity.

(b) In relativity we will talk about *events* that have to be described both by a place and a time of occurrence. We need four coordinates to define an event — three space coordinates and a time coordinate.

Correspondingly there exists a hypothetical four-dimensional space having spatial and time coordinates. In this space, events are represented by *world points* (x, y, z, t). Any physical process can be described as a sequence of events and can be represented as a grouping or continuum of world points in the four-dimensional space–time representation. Each physical particle can be represented by a *world line* in this four-dimensional plot. Even if its spatial coordinates (x, y, z) remain constant in a given reference frame, a particle's location in time will progress along its world line.

Einstein realized that the constancy of the speed of light c necessitated a revision of the concepts of *distance* and *time*. A mirror M_1, stationary with respect to an observer O, could be said to be at a distance $c(t_1 - t_0)/2$ if the reflection of a light pulse emitted by the observer at time t_0 reached him at time t_1. A second, stationary mirror M_2, then was equidistant from O if its reflection of the same light pulse also returned at identical instant t_1. Conversely, if two flashes of light reached O at the same instant t_1 from equidistant points S_1 and S_2, they could be said to have been emitted simultaneously. Einstein made this the definition of *simultaneity*.

Figure 5.1 shows an observer O at rest with respect to equidistant light sources at points S_1 and S_2. Another observer O' is at rest with respect to equidistant mirrors M_1' and M_2' and in motion with respect to O at relative velocity V. The straight line $S_1 S_2$ is collinear with the line $M_1' M_2'$ and the vector direction of V. Observers O and O' thus pass by each other and can synchronize their watches at the instant their positions coincide. They respectively designate this instant t_0 and t_0'.

When flashes of light are simultaneously emitted at S_1 and S_2 at time t_0, they will arrive simultaneously at observer O but not at O', who will have moved to the

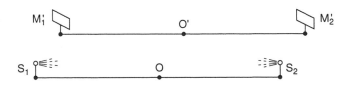

Fig. 5.1. Diagram to describe measurements carried out by observers O and O' in motion at relative velocity V. M_1' and M_2' are stationary mirrors equidistant from O'. Sources S_1 and S_2 are equidistant from O and emit flashes of light. S_1, S_2, M_1', M_2', and V all are collinear. O' is moving to the right relative to O.

right during the time it took light to cover the distance. O' will observe the flash of light from S_2 reaching him before the flash from S_1.

Suppose O' emits a series of carefully clocked light pulses that are reflected by mirrors M_1' and M_2'. One of these pulses reflected from M_2' returns to him simultaneously with the arrival of a flash from S_2. Somewhat later, another pulse, reflected by M_1' reaches him simultaneously with a flash emitted at S_1. Observer O notifies O' that the flashes from S_1 and S_2 were emitted simultaneously. Since O' knows M_1' and M_2' to be equidistant from him, he infers that M_2' passed S_2, before M_1' passed S_1. From this he concludes that the distance between S_1 and S_2 is shorter than the distance between M_1' and M_2'.

By symmetry, observer O concludes the exact opposite. Quite generally, the *lengths* of rods and distances between points along the direction of motion appear contracted when viewed by an observer in relative motion.

The same considerations also affect measures of time. Observers O' and O both see each other in relative motion at velocity V. Observer O' clocks point M_1' passing first by S_1 and then by S_2 at times separated by an interval $[S_1 S_2]/V$. Observer O measures the elapsed time between these events in the same way. But he judges the distance $[S_1 S_2]$ between light sources to be longer than that measured by O' and, therefore, the time elapsed appears to him longer than the time interval measured by O'. To O, the clock used by observer O' appears to run slower than his own. Observer O' again concludes the opposite. Time appears *dilatated* in a coordinate frame in motion relative to an observer.

Given these apparently conflicting measurements, which indicate that neither time intervals Δt nor lengths Δx have absolute values independent of the motion of an observer, Einstein pointed out how we might still make sense of our observations. The principle of relativity offers a simple guide. If the laws of physics are independent of the observer's inertial reference frame, then one consequence should be that the wave front from a flash of light propagating through vacuum should expand spherically, independent of the observer's motion. We will return to this expanding wave, below, to see how this helps us to make sense of measurements made by observers in relative motion. However, we first need to define a number of additional concepts.

(c) Two distinct events labeled a and b are separated by an *interval*, s_{ab}, of length

$$s_{ab}^2 = -[(x_a - x_b)^2 + (y_a - y_b)^2 + (z_a - z_b)^2 - c^2(t_a - t_b)^2] . \qquad (5\text{-}1)$$

This suggests that we could define a new coordinate $\tau = ict$, where i is the imaginary number, to obtain (5–1) in the form

$$s_{ab}^2 = -[(x_a - x_b)^2 + (y_a - y_b)^2 + (z_a - z_b)^2 + (\tau_a - \tau_b)^2] . \qquad (5\text{-}2)$$

This form brings out a symmetry between time and space coordinates. Equation (5–2) is just the Pythagorean expression for the separation of two points in a four-dimensional flat space. Such a space is also called a *Euclidean space*, and the particular four-dimensional space described in (5–1) is known as a *Minkowski space* (Mi08). Equation (5–2) helps to point out some of the properties of space and time coordinates. The time coordinate in the formulation (5–2) is an imaginary quantity, whereas the spatial coordinates are real. Unfortunately, the substitution $\tau = ict$ is not very useful. Special relativity, in its full form, deals with quantities that are best described in tensor notation. But that notation cannot be properly used if time is taken to be an imaginary quantity. Rather, as we will see, x, y, z, and ct should be considered to be components of a four-vector in a space that is said to have *signature* $(+ + + -)$, meaning that the Pythagorean expression for the square of the interval between events is the sum of the squares of the spatial components of the separation, with the square of the time increment subtracted.

(d) We can formulate equation (5–1) in differential form

$$ds^2 = -(dx^2 + dy^2 + dz^2 - c^2\, dt^2)\,, \tag{5-3}$$

where ds is called the *line element*.

(e) The interval between two events is said to be *timelike* if $s_{ab}^2 > 0$ and *spacelike* if $s_{ab}^2 < 0$. When s_{ab}^2 just equals zero, we see from either equation (5–1) or (5–3), that

$$v_x^2 + v_y^2 + v_z^2 \equiv v^2 = c^2\,. \tag{5-4}$$

The surface containing all intervals $s_{ab} = 0$, or line elements $ds = 0$ is called the *light cone*. It contains all trajectories going through a point (x, y, z, t) with the speed of light. A two-dimensional projection of this cone is shown in Fig. 5.2, where we have chosen coordinates $y = z = 0$ and the projection of the surface

$$x^2 + y^2 + z^2 = c^2 t^2 \tag{5-5}$$

now becomes $x = \pm ct$ with slope

$$\frac{dt}{dx} = \pm\frac{1}{c}\,. \tag{5-6}$$

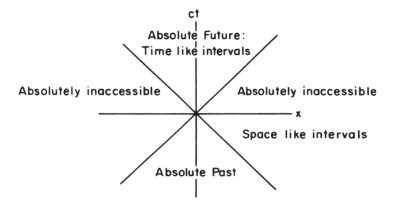

Fig. 5.2. *World diagram* to show the relation between different kinds of events.

Consider an observer placed at the origin of the coordinate system. All lines representing physical particles must indicate velocities $v < c$ and, therefore, are contained in that part of the light cone containing the t-axis. The lower half of the diagram represents the past. The upper half contains all world points lying in the future. The two parts of the diagram containing the x-axis are absolutely inaccessible in the sense that velocities greater than the speed of light would be required to reach them.

It is interesting that the concept of absolute past and future depends on the fact that the speed of light cannot be exceeded. If it could, we would be able to travel to a sufficiently distant point and "catch up" with light that had been emitted, say, in the supernova of 1054 A.D. With a sufficiently good telescope we could then "look back" and see the star just prior to explosion. The event could thus be brought into our "present," but it would still be inaccessible to us in the sense that we would not be able to influence the event in any way. This problem is looked at further in Section 5:12.

(f) The time read on a clock moving with the reference frame of an observer is called the *proper time* for that frame; and the length of an object measured in that frame is called the *proper length*.

5:3 Relative Motion

Let us now consider two inertial frames of reference K and K', whose axes x, y, z and x', y', z' are parallel (Fig. 5.3). Relative to K, K' moves with velocity V along the x-axis. An event has coordinates (x, y, z, t) as measured by an observer at rest in system K, and coordinates (x', y', z', t') as measured by an observer at rest in K'.

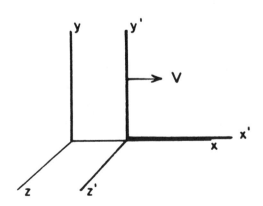

Fig. 5.3. Notation for moving coordinate frames.

At some time $t = t' = 0$, let the origins of the two reference frames coincide. The subsequent motion will not affect the identity of the y- and z-components: $y' = y$ and $z' = z$; but t and x will be related to t' and x' through a more complicated set of relations, the *Lorentz transformations*, which read (Lo–):

$$x = \frac{x' + Vt'}{\sqrt{1 - V^2/c^2}}, \quad y = y', \quad z = z', \quad t = \frac{t' + V(x'/c^2)}{\sqrt{1 - V^2/c^2}},$$

or   (5–7)

$$x_1 = (x'_1 + \beta x'_4)\gamma(V), \quad x_2 = x'_2, \quad x_3 = x'_3, \quad x_4 = (x'_4 + \beta x'_1)\gamma(V),$$

where we have set $x \equiv x_1$, $y \equiv x_2$, $z \equiv x_3$, $ct \equiv x_4$, $\beta \equiv V/c$, and $\gamma(V) \equiv (1 - \beta^2)^{-1/2}$. The second of the two formulations (5–7) shows the symmetry between space and time coordinates. $\gamma(V)$ is called the *Lorentz factor*.

These equations can be derived and follow directly from the principle of relativity and the constancy of the speed of light. Here we will only show that the equations are consistent with some of the predictions of the principle. For example, since the speed of light is the same in systems K and K', we would expect that a light wave emitted at $t = t' = 0$ — that is, when the origins of the coordinate systems coincide — would propagate spherically in both systems.

PROBLEM 5–1. Equation (5–5) describes the propagation of a spherical wave front in the coordinate system K. Show that according to (5–7), the corresponding equation describing the propagation of the wavefront in K' is

$$x'^2 + y'^2 + z'^2 = c^2 t'^2$$

making this wave appear spherical too. This is the validation of equations (5–7) along the lines we were proposing at the end of Section 5:2(b).

Another consequence of the relativity principle is that formulae expressing x', y', z', and t' in terms of x, y, z, and t can be obtained easily by changing V to $-V$.

PROBLEM 5–2. Show that this procedure is valid by actually solving equations (5–7) for x', y', z', and t'.

We also want to examine whether the speed of light will always appear to be c, viewed from any reference frame. We can answer this question by discussing how velocities transform according to equations (5–7). Let us write the expressions in differential form:

$$dx = (dx' + V\,dt')\gamma(V), \quad dy = dy', \quad dz = dz',$$ (5-8)
$$dt = \left(dt' + (V/c^2)\,dx'\right)\gamma(V).$$

This allows us to write the derivatives

$$v_x = \frac{dx}{dt} = \frac{dx' + V\,dt'}{dt' + \dfrac{V}{c^2}\,dx'} = \frac{v'_x + V}{1 + v'_x \dfrac{V}{c^2}} \,,$$

$$v_y = \frac{dy}{dt} = \frac{v'_y}{\left(1 + v'_x \dfrac{V}{c^2}\right)\gamma(V)} \,, \tag{5-9}$$

$$v_z = \frac{dz}{dt} = \frac{v'_z}{\left(1 + v'_x \dfrac{V}{c^2}\right)\gamma(V)} \,.$$

These equations prescribe the *composition* (addition) *of velocities.* If $v'_z = v'_y = 0$, and we write $v'_x = v'$, then equations (5–9) show that $v_y = v_z = 0$ and $v_x = v$ where

$$v = \frac{v' + V}{1 + \dfrac{v'V}{c^2}} \,. \tag{5-10}$$

When all motions are along the x-axis, a velocity measured as having a value v' in reference frame K', will appear to have velocity v in a frame K. The velocities v, v', and V are related by equation (5–10). V is the velocity of K' relative to K (Fig. 5.3).

Three cases are of interest:

(a) If $v' = V = c$, then substitution shows that $v = c$.

(b) If $v' < c$ and $V = c$, or if $v' = c$ and $V < c$, then $v = c$. This also can be shown by substitution in equation (5–10). It means that the speed of light is constant and has a value c in all inertial frames of reference.

(c) Finally, if $v' < c$ and $V < c$, then $v < c$.

PROBLEM 5–3. Show that the result (c) is always true by writing $v' = (1 - \delta)c$, $V = (1 - \Delta)c$ where $0 < \delta, \Delta < 1$.

PROBLEM 5–4. If the speed of light is infinite, Galilean relativity results. Give the transformations equivalent to (5–7) to (5–9) and obtain the law of composition of velocities. These expressions should be consistent with Newtonian physics.

Expression (5–10) is interesting because it also shows that a particle traveling at a speed less than the speed of light can never be accelerated to a speed equaling c. To see this, suppose that the particle initially was moving with velocity V. It is now given an extra velocity v' that is also less than c. From case (c), above, we see that the resultant velocity is always less than c. We can keep adding small increments to the particle's velocity, but to no avail. It will always move at a speed less than the speed of light. Highly energetic cosmic-ray particles travel at very nearly the speed of light. When accelerated, they move a little faster, but never faster than c.

The Lorentz transformation leaves the interval s between two events invariant, but this is done at the expense of changes in the apparent time and spatial separations of events. If a clock is at rest at position $x = 0$ in K, then the proper time for observer O at rest in frame K is given by t, whereas O$'$ at rest in K' measures

$$t' = \left(t - V\frac{x}{c^2}\right)\gamma(V)\bigg|_{x=0} = t\,\gamma(V)\,. \qquad (5\text{-}11)$$

Actually, we are not interested in an absolute time, only in time intervals $\Delta t = t_1 - t_2$ and $\Delta t' = t'_1 - t'_2$, where the equations (5–11) reduce to

$$\Delta t' = \frac{\Delta t}{\sqrt{1 - V^2/c^2}} \equiv \Delta t\,\gamma(V)\,. \qquad (5\text{-}12)$$

To the observer O$'$, O's clock appears to be going slower. He notes a *time dilatation* or *time dilation* in moving reference frames. The relation between Δt and $\Delta t'$ is independent of the choice of position, x. The choice $x = 0$ was not necessary.

In Problem 5–9 we will see that this time dilation can prolong the decay time of fast-moving, unstable, cosmic-ray particles by many orders of magnitude. The time dilation is a dominant effect for the decay of such particles.

We can similarly derive the change in spatial separation between simultaneously observed events. If the positions of two points at rest in the K system are x_a and x_b as measured by observer O at rest in K, the *proper length* of a line joining the two points is $\Delta x = x_b - x_a$. O$'$, the observer at rest in K', measures the separation of the two points at some given time t'. We use the equations

$$x_a = (x'_a + Vt')\gamma(V), \qquad x_b = (x'_b + Vt')\gamma(V)\,, \qquad (5\text{-}13)$$

where t' is the same in both expressions because O$'$ sees both points simultaneously. The spatial separation observed from the K' frame, then, is

$$\Delta x' = x'_a - x'_b = (x_a - x_b)\sqrt{1 - \frac{V^2}{c^2}} = \frac{\Delta x}{\gamma(V)}. \qquad (5\text{-}14)$$

Because the square root term is always less than unity the length measured by O$'$ is shorter than the proper length. We call this the *Lorentz contraction* (Lo–).

The Lorentz contraction is found only along the direction of motion, while the transverse dimensions y and z according to equations (5–7) remain unaffected. This could at first sight lead us to believe that a moving sphere should appear flattened into an oblate ellipsoid, and that a cube would appear distorted in some way dependent on its orientation with respect to the moving axes.

This view was held for more than half a century after the discovery of the special relativistic transformations by Lorentz and Einstein. But in 1959 Terrell (Te59) suggested that a sphere should always appear spherical, a cube cubical, and so on.

He showed that the Lorentz transformations, though producing some distortions, primarily act to change the apparent orientation of the object by effectively rotating it.

To see how this comes about, suppose that a cube is moving with velocity V along the x-direction. This motion is relative to an observer O′ who looks at the cube in a direction transverse to its motion.

We will be interested in the apparent length of the edges 1, 2, and 3. Let the length of each edge be L, as measured by an observer O at rest with respect to the cube, and let edge 1 be perpendicular both to the direction of motion and to the direction of the observer (Fig. 5.4).

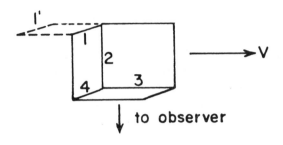

Fig. 5.4. The sides of a rapidly moving cube.

When observer O′ sees both edges 1 and 4 simultaneously — as she would when taking a photograph — she does not observe photons that were simultaneously emitted at these two edges. The light reaching her from edge 1 was emitted at a time L/c earlier than light arriving from edge 4. But at that earlier time, edge 1 occupied position 1′. A photograph will therefore show a view of the cube with the far edge occupying position 1′ and the near edge occupying position 4. The projected length of side 2 is the projected distance between 1′ and 4, namely Lv/c.

This factor does not enter in discussing the length of edge 4, because all points along this edge simultaneously emit those light rays that later are simultaneously observed. Side 4 is perpendicular to the direction of motion and its length is left unchanged by the Lorentz transformation; the Lorentz transformations also leave sides 1 and 2 unchanged. But side 3 is shortened by a factor $\sqrt{1 - (V^2/c^2)}$ (see equation (5–14)).

A photograph will show sides 1, 2, and 3 having lengths L, LV/c, and $L/\gamma(V)$, respectively. If we define an angle ϕ by $V/c = \sin\phi$, then it is easy to see that these sides have apparent lengths, L, $L\sin\phi$, and $L\cos\phi$. The cube appears rotated by an angle ϕ.

Although this is true for a small, distant cube at its point of nearest approach, there are added distortions if the same cube is seen, say, earlier in its trajectory.

Light arriving from the leading, nearest edge is then emitted later — when the cube is closer — than light from edge 4, the trailing edge. The near edge therefore appears disproportionately long. In general, the cube appears both distorted and rotated (Ma72a)*.

That the appearance of a sphere should remain spherical already follows from Problem 5–1. However, just as for the cube, a moving observer would see the sphere rotated.

5:4 Four-Vectors

Let us now turn to the relationship between the *world diagrams* of two observers O and O' moving with inertial frames K and K'. As in Fig. 5.3 we will take K' to be moving in the direction of K's positive x-axis with velocity V relative to K. The origin of coordinates will then have components $y = y' = 0$ and $z = z' = 0$ at all times.

Let us also choose the origins of K and K' to coincide at some time given by (5–11) as $t = t' = 0$. This means that $x = x' = 0$ at that time. As seen by O, the origin of K' then has the world line t', shown in Fig. 5.5. The line passes through

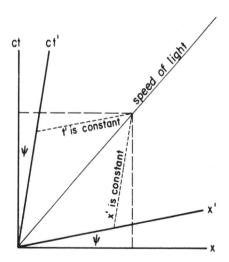

Fig. 5.5. Minkowski diagram showing characteristics of a moving inertial coordinate system K' as seen by another inertial observer.

the origin and has a slope

$$\frac{c\,dt}{dx} = \frac{c}{V}\,. \tag{5-15}$$

That t' actually is the time axis for O' follows from the first and last equations of the top row of expressions in (5–7) if we set $x' = 0$. Again, if we set $t' = 0$ in these two equations, we see that the slope of the x'-axis in O's world diagram must be $c\,dt/dx = V/c$. The angle ψ between the ct- and ct'-axes therefore equals the angle between the x- and x'-axes:

$$\psi = \tan^{-1}\frac{V}{c}\,. \tag{5-16}$$

The light cone bisects both the spaces K and K' in this diagram, often called the *Minkowski diagram* (Ro68)*.

A vector in the four-dimensional spaces K and K' appears equally long to both observers. If the vector joins events $(0,0,0,0)$ and (x_1, y_1, z_1, t_1) as seen by O, it will join $(0,0,0,0)$ and (x'_1, y'_1, z'_1, t'_1) as seen by O'. But because we can always choose the x-direction to coincide with the direction of motion, we can again set $y = y'$, $z = z'$, so that the lengths squared of the two vectors become

$$\begin{aligned}
L^2 &= \{c^2t^2 - (x^2 + y^2 + z^2)\} \\
&= \{[(ct' + Vx'/c)^2 - (x' + Vt')^2]\gamma^2(V) - (y'^2 + z'^2)\} \tag{5-17} \\
&= \{c^2t'^2 - (x'^2 + y'^2 + z'^2)\} = L'^2\,, \\
L &= L'. \tag{5-18}
\end{aligned}$$

The vector therefore has the same length, judged by either observer. Such a vector with components x, y, z, ct is called a *four-vector*. Four-vectors play a particularly important role in special relativity — first because the theory's natural setting is a *four-space*, and second because the length of four-vectors is *invariant* with respect to coordinate transformations. This means that one observer measures exactly the same vector magnitude as any other. But since relativity postulates that the laws of physics are invariant in all inertial frames, these invariant lengths assume a special significance in the formulation of the laws of physics.

We note that the length L specified here corresponds to the interval s defined in equation (5–1). This interval therefore is an invariant. If two events 1 and 2 occur in one and the same place for an observer O', we see that $\Delta s_{12}^2 = c^2\Delta t_{12}^2 - \Delta l_{12}^2 = c^2\Delta t_{12}'^2 > 0$. The square of the interval, Δs_{12}^2, is positive because the elapsed time $\Delta t_{12}'^2$ is a real quantity. In O's frame Δl_{12} is the spatial separation. We see that if an interval between events is timelike, there exists a frame in which the events occur in the same place. If the interval is spacelike we can similarly show that a frame exists in which the two events are simultaneous.

The general transformation of a four-vector with components A_1, A_2, A_3, and A_4 reads (see (5–7))

$$\begin{aligned}
A_1 &= [A'_1 + \beta A'_4]\gamma(V), \quad A_2 = A'_2, \quad A_3 = A'_3, \\
A_4 &= [A'_4 + \beta A'_1]\gamma(V)\,. \tag{5-19}
\end{aligned}$$

We will find, for example, that the *four-momentum* with components $(p_x, p_y, p_z, \mathcal{E}/c)$ transforms as a four-vector. So also does a four-vector (A_x, A_y, A_z, ϕ) having the electromagnetic vector and scalar potentials as components. We will encounter this later in Section 6:13. There are many other such four-vectors that correspond to useful physical parameters. Conversely, special relativity requires all physical entities to have a four-dimensional structure.

5:5 Aberration of Light

Next, we will want to use the Lorentz transformations to see how the measurement of angles depends on the relative motion of an observer. We will find here that the measurement of an angle — or rather the sine or cosine of an angle — does not at all involve the measurements of two lengths. Rather it requires the simultaneous measurement of two velocities. This comes about because a distant observer O must make his angular measurements using light signals received from an object, and the law of composition of velocities determines the angles these light rays subtend at the observer. Suppose a particle has a velocity vector that lies in the xy-plane. The velocity v has components $v_x = v \cos \theta$ and $v_y = v \sin \theta$ along the x- and y-axes of the reference frame K. Viewed by an observer at rest in the frame K', the velocity components are $v'_y = v' \sin \theta'$ and $v'_x = v' \cos \theta'$. The velocity transformation equations then allow us to write

$$\tan \theta = \frac{v_y}{v_x} = \frac{v' \sin \theta'}{[v' \cos \theta' + V] \gamma(V)} \ . \tag{5-20}$$

When we deal with a light ray, $v' = c = v$, and the angles subtended by the light ray transform as

$$\tan \theta = \frac{\sin \theta'}{[V/c + \cos \theta'] \gamma(V)} \ . \tag{5-21}$$

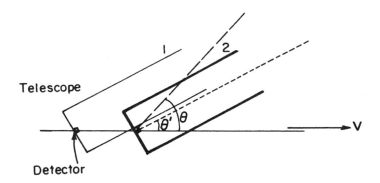

Fig. 5.6. Aberration of light. A telescope (stationary in some coordinate frame K') moves with velocity V, relative to a star. When the starlight enters, the telescope is in position 1. By the time the light has traveled the length of the instrument, the telescope has moved to position 2. All this time the telescope is pointed into direction θ', with respect to the x'-axis. But an observer whose telescope was at rest with respect to the star's reference frame K would have to point it at angle θ relative to x. Aberration is also found in Newtonian physics for a finite speed of light c. This is already indicated by the magnitude of the effect, which is of order V/c, whereas purely relativistic corrections are always of order V^2/c^2 and, therefore, are smaller. A relativistic correction needs to be applied to the Newtonian aberration in order to obtain the true aberration angle.

From (5–9) we see that for $v = v' = c$, v_x, and v_y alone lead to

$$\sin \theta = \frac{\sin \theta'}{\gamma(V)[1 + \beta \cos \theta']}, \qquad \cos \theta = \frac{\cos \theta' + \beta}{1 + \beta \cos \theta'} . \qquad (5\text{-}22)$$

When $V \ll c$ and the terms in β^2 are negligible, the sine equation becomes

$$[\sin \theta + \beta \sin \theta \cos \theta'] = \sin \theta' . \qquad (5\text{-}23)$$

The *aberration angle* $\Delta\theta = \theta' - \theta$ in the position of a star seen through a telescope is then

$$\Delta\theta \sim \beta \sin \theta \qquad \text{for} \qquad \Delta\theta \ll 1, \quad \beta \ll 1 . \qquad (5\text{-}24)$$

If light travels in a direction opposite to that in which the telescope moves, $\sin \theta'$ has a negative value, that is, $\theta' < \theta$, as shown in Fig. 5.6. This angle is of great practical importance in observational astronomy. If a star is to be observed at different times of the year, the direction in which the telescope must be aligned needs to be changed.

PROBLEM 5–5. The orbital velocity of Earth about the Sun is $30 \, \text{km s}^{-1}$, which means that the velocity of Earth changes by $60 \, \text{km s}^{-1}$ over a six-month interval. Taking $V \sim 60 \, \text{km s}^{-1}$ shows that for a star in the zenith this angle is

$$\Delta\theta \sim \frac{V}{c} \sim \frac{60}{3 \times 10^5} \sim 40'' \text{ of arc} ,$$

which is readily measured.

The aberration of light was first observed in 1728 by James Bradley. His finding gave conclusive proof that Nicolaus Copernicus had been right, in 1543, when he asserted that Earth orbits the Sun, rather than the Sun orbiting Earth.

5:6 Momentum, Mass, and Energy

The velocities we have discussed thus far are three-dimensional velocities. In relativity, however, the proper form to use is a four-vector because, as emphasized in Section 5:4, four-vectors have an *invariant magnitude*. We define a four-velocity with components

$$u_x \equiv \frac{dx}{ds}, \quad u_y \equiv \frac{dy}{ds}, \quad u_z \equiv \frac{dz}{ds}, \quad u_t \equiv c\frac{dt}{ds} . \qquad (5\text{-}25)$$

Because

$$ds = \sqrt{c^2 \, dt^2 - dx^2 - dy^2 - dz^2} = \frac{cdt}{\gamma(v)} , \qquad (5\text{-}26)$$

the four-velocities (5–25) become

$$u_x = \frac{dx}{dt}\frac{1}{c\sqrt{1-(v/c)^2}} = \frac{v_x\gamma(v)}{c}, u_y = \frac{v_y\gamma(v)}{c}, \quad u_z = \frac{v_z\gamma(v)}{c}, \quad u_t = \gamma(v)$$

or   (5-27)

$$u_i = [\gamma(v)\,dx_i/dt]c^{-1}\,.$$

The square of the magnitude of u is:

$$\left(\frac{cdt}{ds}\right)^2 - \left(\frac{dx}{ds}\right)^2 + \left(\frac{dy}{ds}\right)^2 + \left(\frac{dz}{ds}\right)^2 \equiv u_t^2 - (u_x^2 + u_y^2 + u_z^2)\,. \quad (5\text{-}28)$$

We see that u is a dimensionless quantity that does not have the units of velocity: cm s^{-1}. But we may introduce a new quantity $U_i \equiv cu_i$, which would have the more familiar dimensions and still preserve Lorentz invariance. For particles with *rest-mass* m_0 we can then write

$$\frac{m_0}{2}\left[U_t^2 - (U_x^2 + U_y^2 + U_z^2)\right] = \mathcal{L}\,. \quad (5\text{-}29)$$

Equations (5–28) and (5–29) exhibit an invariance property that we will need below. The magnitude of (5–28) is 1 for particles with rest-mass and 0 for electromagnetic radiation and gravitational waves. Corresponding to this, $\mathcal{L} = m_0c^2/2$ or 0, respectively. As can be seen from equations (3–109), equation (5–29) displays the properties of a *Lagrangian* $\mathcal{L}$ in the absence of a force field. Frequently we are interested in a particle's *momentum*. This can be written in the form

$$\mathbf{p} = m_0\mathbf{v}\gamma(v) \quad (5\text{-}30)$$

and involves three components that correspond to the quantities

$$-\frac{d\mathcal{L}}{dU_x} = m_0U_x = p_x\,, \quad -\frac{d\mathcal{L}}{dU_y} = m_0U_y = p_y\,, \quad -m\frac{d\mathcal{L}}{dU_z} = m_0U_z = p_z\,. \quad (5\text{-}31)$$

The linear momentum accounts for the first three components of a four-vector whose fourth component has the form m_0U_t. In relativity the fourth component is a measure of the particle energy $\mathcal{E}$

$$\frac{d\mathcal{L}}{dU_t} = m_0U_t = p_t = \frac{\mathcal{E}}{c}\,. \quad (5\text{-}32)$$

The complete relativistic momentum four-vector then has components

$$(p_x, p_y, p_z, \mathcal{E}/c). \quad (5\text{-}33)$$

It is clear that in the limit $v \ll c$, the first three components give the classical momentum $\mathbf{p} = m_0\mathbf{v}$. However, the energy takes on the new form

$$\mathcal{E} = \frac{m_0c^2}{\sqrt{1-\frac{v^2}{c^2}}} = m_0c^2\gamma(v)\,. \quad (5\text{-}34)$$

At zero velocity this reduces to

$$\mathcal{E} = m_0 c^2, \tag{5-35}$$

an expression now known to be accurate to at least one part in a million (Ra05), stating that mass and energy are equivalent (Ei–b)*. It is this equivalence that allows stars to radiate. The nuclear reactions that give rise to stellar radiation always involve a mass loss that liberates energy in the form of photons or neutrinos. As the star radiates it conserves mass–energy by becoming less massive. Interestingly, Einstein not only realized that a radiating body would lose mass, he also emphasized that radiation conveys mass from an emitting to an absorbing body. Even though electromagnetic radiation has no rest-mass, it does carry mass and is deflected by gravitational fields las are other massive bodies, as we will see in Section 5:14.

For small velocities equation (5–34) can be approximated by the expansion

$$\mathcal{E} = m_0 c^2 + \frac{1}{2} m_0 v^2 + \cdots, \tag{5-36}$$

where the second term represents kinetic energy. The next higher term would be of order $m_0 v^4 / c^2$. In mechanical or chemical processes m_0 remains essentially constant and we normally see changes only in the $m_0 v^2 / 2$ term. This is why that term has classically been so important even though it is far smaller than the energy contained in a particle's mass.

Equation (5–34) shows that $\mathcal{E} \to \infty$ as $v \to c$, which means that an infinite amount of work would be required to accelerate a particle to the speed of light. As with all special relativistic effects, this statement is valid in inertial frames but need not be true for others. This is why there is no conflict with the observations that distant galaxies travel at nearly the speed of light and that some may pass across the cosmic horizon when their speed, relative to our galaxy, exceeds the speed of light. Such a horizon is called an *event horizon* because, if there is some event that may be occurring in the galaxy just as it crosses the speed of light velocity threshold, this is the last event occurring in the galaxy that we shall ever witness. Because these distant galaxies are at rest in reference frames that are accelerated relative to ours, special relativity does not hold, and we can make no general statements about speed limitations unless we talk in terms of a less specialized theory, such as the general theory of relativity.

Two important relations should still be noted. First, equations (5–30) and (5–34) show that

$$\mathbf{p} = \frac{\mathcal{E}}{c^2} \mathbf{v} . \tag{5-37}$$

Second, writing the four-vector components as

$$p_i = m_0 c u_i, \tag{5-38}$$

we obtain the magnitude of the four-momentum vector

$$-(p_x^2 + p_y^2 + p_z^3 - p_t^2) = -p^2 + \frac{\mathcal{E}^2}{c^2} = m_0^2 c^2 \tag{5-39}$$

which again is invariant. Equation (5–39) can be rewritten as

$$\mathcal{E}^2 = p^2 c^2 + m_0^2 c^4 . \tag{5-40}$$

Since photons have zero rest–mass, and velocity c, (5–37) and (5–40) become

$$p = \frac{\mathcal{E}}{c} . \tag{5-41}$$

The relations (5–37) and (5–40) are of particular importance in cosmic-ray physics, where particle energies may be as high as $\sim 10^{20}$ eV. The rest–mass of a proton with this energy is only 931 MeV, but the total energy of a cosmic-ray proton can be $\sim 10^{11}$ times its rest-mass energy. This feature allows a cosmic-ray primary, incident on the top layers of the Earth's atmosphere, to undergo collisions that produce billions of shower particles whose total rest-mass exceeds that of the primary proton by many orders of magnitude. The classical concept of conservation of mass is violated here, but the more encompassing principle of conservation of mass–energy remains intact.

5:7 The Doppler Effect

Since energy is the fourth component of a four-vector $(\mathbf{p}, \mathcal{E}/c)$, it transforms, by equations (5–19) and (5–33), as

$$\mathcal{E} = \gamma(V)[\mathcal{E}' + V p_x'] \tag{5-42}$$

when the relative motion is along the x-direction.

If we wish to see how photon energies transform, we note that for a ray directed at an angle θ' with respect to the x'-axis (5–41) leads to

$$\mathcal{E} = \frac{\mathcal{E}' + (\mathcal{E}' V/c) \cos \theta'}{\sqrt{1 - (V/c)^2}} = \mathcal{E}'[1 + \beta \cos \theta'] \gamma(V) . \tag{5-43}$$

The angle θ' is that shown in Fig. 5.6, but we have to recall that the direction of the photon's travel is opposite to the viewing direction. We know from (4–40) that $\mathcal{E} = h\nu$. Using this in equation (5–43) gives,

$$\nu = \nu'(1 + \beta \cos \theta') \gamma(V) \tag{5-44}$$

which gives the Doppler shift in frequency for radiation emitted by a moving source (Fig. 5.7). In contrast to the classical prediction, we see from (5–22) that there is a red shift even when the motion of the source is purely transverse ($\cos \theta = 0$, $\cos \theta' = -V/c$). This corresponds to a time dilation — a frequency decrease. When the source radiates in a direction opposite to its direction of motion, $\beta \cos \theta' < 0$, and $\nu < \nu' \gamma(V)$. When it radiates in the forward direction, $\nu > \nu'$.

Quasars have large cosmological red shifts symbolic of their great distances across the expanding Universe and their correspondingly large recession velocities.

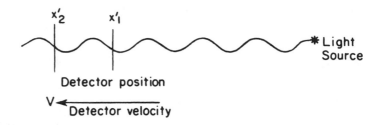

Fig. 5.7. The Doppler effect. A detector moves at velocity V, relative to a light source. It starts measuring the frequency of radiation at time t_1' and finishes at time t_2'. During this interval it is receding from the source, moving from position x_1' to x_2'. A wave that would just have reached x_1' by time t_2' is therefore not counted, nor are any waves lying between x_1' and x_2', at time t_2'. The detector therefore senses a lower frequency ν'. This explanation provides for the first-order Doppler shift proportional to V/c, which is also present classically. The correct relativistic expression contains an additional factor $(\sqrt{1 - V^2/c^2})^{-1}$ given in equation (5–44).

Their spectra generally exhibit *Lyman-α* emission lines at a strongly red-shifted frequency ν_0 and a series of absorption lines, for the same transition at higher frequencies $\nu_0+\Delta_1, \nu_0+\Delta_2, \ldots$ corresponding to spectral shifts at velocities of hundreds or thousands of kilometers per second. The clouds responsible for the absorption are called *Lyman-α* absorbers, and the densely packed series of narrow Lyman-α absorption lines sometimes seen in a quasar's spectrum, due to many absorbing clouds along the line of sight, are referred to the *Lyman-α forest*.

5:8 Poynting–Robertson Drag on a Grain

Consider a grain of dust in interplanetary space. As it orbits the Sun it absorbs sunlight, and re-emits this energy isotropically. We can view this two-step process from two different viewpoints.

(a) Seen from the Sun, a grain with mass m absorbs light coming radially from the Sun and re-emits it isotropically in its own rest-frame. A re-emitted photon carries off angular momentum proportional: (i) to its equivalent mass $h\nu/c^2$, (ii) to the velocity of the grain $R\dot{\theta}$; and (iii) to the grain's distance from the Sun R. Considering only terms linear in V/c, and neglecting any higher terms, we see that the grain loses orbital angular momentum L about the Sun at a rate

$$dL = \frac{h\nu}{c^2}\dot{\theta}R^2, \qquad \frac{1}{L}dL = \frac{h\nu}{mc^2}, \qquad (5\text{-}45)$$

for each photon whose energy is absorbed and re-emitted, or isotropically scattered, in the grain's rest-frame.

(b) Seen from the grain, radiation from the Sun arrives at an aberrated angle θ' from the direction of motion, instead of at $\theta' = 270°$ (see equations (5–22)). Hence,

$$\cos \theta = \frac{\cos \theta' + (V/c)}{1 + (V/c) \cos \theta'} = 0, \qquad \cos \theta' = -\frac{V}{c} . \tag{5-46}$$

Here V is $\dot\theta R$, the grain's orbital velocity, and the photon imparts an angular momentum $pR \cos \theta' = -(h\nu/c^2)R^2\dot\theta$ to the grain.

For a grain with cross-section σ_g

$$\frac{dL}{dt} = -\frac{L_\odot}{4\pi R^2} \frac{\sigma_g}{mc^2} L , \tag{5-47}$$

where $L_\odot$ is the solar luminosity.

Either way, the grain's velocity decreases on just absorbing sunlight. From the first viewpoint, this happens because the grain gains mass, which it then loses on re-emission; from the second, it is because the grain is slowed down by the transfer of angular momentum.

PROBLEM 5–6. A grain having $m \sim 10^{-11}$ g, $\sigma_g \sim 10^{-8}$ cm^2 circles the Sun at 1 AU. Calculate the length of time needed for it to spiral into the Sun — to reach the solar surface — assuming that the motion throughout is approximated by circular orbits.

PROBLEM 5–7. Suppose one part in 10^8 of the Sun's luminosity is absorbed or isotropically scattered by grains circling the Sun. What is the total mass of such matter falling into the Sun each second?

5:9 Motion Through the Cosmic Microwave Background Radiation

We can derive the apparent angular distribution of light emitted isotropically in the reference frame of a moving object. Let the object be at rest in the K' system. Then the intensity $I(\theta')$ has the same value I', for all directions θ' (Fig. 5.8). The energy radiated per unit time into an annular solid angle $2\pi \sin \theta' \, d\theta'$ is $2\pi I' \sin \theta' \, d\theta'$. In the K reference frame the intensity distribution is $I(\theta)$ and we would like to find the relation between $I(\theta)$ and I'.

The relativity principle requires that a body in thermal equilibrium in one inertial frame of reference also be in thermal equilibrium in all others. A blackbody radiator will therefore appear black in all inertial frames. From (4–72) and the definition of $I(\nu)$ (4–84),

$$I(\nu) = \frac{2h\nu^3}{c^2} \left[\frac{1}{e^{h\nu/kT} - 1} \right] . \tag{5-48}$$

For this to be true, we see that the ratios $I(\nu)/\nu^3$ and ν/T both must be invariant under a Lorentz transformation. The total intensity seen from an arbitrary direction

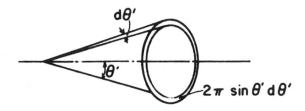

Fig. 5.8. Distribution of radiation, viewed in spherical polar coordinates.

θ and integrated over all frequencies then is $\int I(\nu)\,d\nu$. This is proportional to ν^4 and leads to

$$\frac{I(\theta)}{I'} = [(1 + \beta\cos\theta')\gamma(V)]^4, \quad \frac{T(\theta)}{T'} = (1 + \beta\cos\theta')\gamma(V) . \qquad (5\text{-}49)$$

The intensity of the radiation is proportional to the fourth power of the Doppler shift (5-44), and the temperature is directly proportional to the Doppler shift.

An isotropically radiating, fast-moving body appears to radiate the bulk of its energy in the forward direction ($\beta\cos\theta' \sim 1$), and only a small amount in the backward direction ($\beta\cos\theta' \sim -1$). From (5–22), we can obtain the expression

$$d\theta = d\theta'/[\gamma(V)(1 + \beta\cos\theta')] . \qquad (5\text{-}50)$$

This allows us to write

$$2\pi\int I(\theta)\sin\theta\,d\theta = 2\pi I'\int[(1+\beta\cos\theta')\gamma(V)]^2\sin\theta'\,d\theta' = 4\pi I', \quad (5\text{-}51)$$

which is important. It means that the total power radiated by a source is the same for any set of observers in inertial frames. We will make use of this fact in Section 6:21 to compute the total power emitted by a relativistic electron spiraling in a magnetic field.

The Universe is bathed by an isotropic flux of thermal radiation. This radiation field allows us to determine an absolute rest-frame on the basis of a local measurement. Such a frame in no way violates the validity of special relativity which, as stated earlier, does not distinguish between different inertial frames. Rather, the establishment of an absolute rest-frame emphasizes the fact that special relativity is really only meant to deal with small-scale phenomena and that phenomena on larger scales allow us to determine a preferred frame of reference in which cosmic processes appear isotropic. We will return to this question in Chapters 11 and 12.

The cosmic microwave background radiation has a blackbody spectrum (4–72) (Pe65, Fi96*). Equation (5–49) tells us that, as the Earth moves through the ambient radiation bath, the Doppler shift transforms the observed intensity and temperature. Measurements by the Cosmic Microwave Background Explorer, COBE, and the Wilkinson Microwave Anisotropy Probe, WMAP, show the cosmic background flux increasing slowly as a function of angle, starting from the direction trailing the Earth's apparent motion through the Cosmos, and reaching a maximum in the direction of motion. At each angle with respect to the direction of motion, we observe a blackbody spectrum, but the spectral temperature depends on the angle, as in (5–49). The observed WMAP peak flux is $\sim$3.35 mK higher than the average background temperature of 2.725 K, and directed toward Galactic coordinates $(\ell, b)263.85° \pm 0.1°, 48.25° \pm 0.04°$ (Be03*).

We are confident that the observed shift is due to the Sun's motion through the background radiation bath, because the angular dependence precisely matches (5–44). As we can see from (5–49), the velocity of the Sun through the microwave background implied by the observed temperature shift — $\sim$3.35 mK — is 369 km s^{-1}. This is a superposition of the Sun's motion about the Galactic Center, the Galactic Center's motion relative to the barycenter of the Local Group, and the Local Group's motion relative to the background. This latter motion is of order 630 km s^{-1} in a direction $(\ell, b) \sim (276°, 30°)$. The Local Group appears to be falling toward a region called the *Great Attractor*, which lies in the direction of the galaxy cluster Abell 3627. The mass of Abell 3627 is estimated at $5 \times 10^{15} M_\odot$ (Kr96).

Large-scale motions continue to persist even on considerably larger scales. Doppler shifts of galaxies in a spherical volume around the Galaxy reaching out to $\sim$100 Mpc in each direction, show that the entire region still exhibits a bulk motion of order 200 km s^{-1}, relative to the microwave background, directed very roughly toward $(\ell, b) \sim (304°, 25°)$ with angular uncertainties of order 15°. The bulk velocity of this larger region with respect to the background is thus considerably lower than that of the Local Group, but the directions are not far apart (da00). Figure 12.1 exhibits some of these considerations.

PROBLEM 5–8. The Lorentz contraction is an important effect for extreme relativistic cosmic-ray particles. To a proton with energy 10^{20} eV the disk of the Galaxy would appear extremely thin. If the width of the disk is of the order of 100 pc in the frame of an observer at rest in the Galaxy, show that, to an observer moving with the proton, this width would appear to be $\sim$3 $\times 10^9$ cm, comparable to the length of Earth's equator in our rest-frame.

PROBLEM 5–9. The time dilation factor similarly is important at cosmic-ray energies. Consider the decay time of a neutron that has an energy comparable to the 10^{20} eV energies observed for protons. How far could such a neutron move across the Galaxy before it beta decayed? In the rest-frame of the neutron the mean decay time is $\sim$ 885 s; but in the framework of an observer at rest in the Galaxy it would be much longer. Show that the neutron could more than traverse the Galaxy.

PROBLEM 5–10. If a cosmic gamma ray has sufficiently high energy it can collide with a low-energy photon and give rise to an electron–positron pair. Because of symmetry considerations, this electron–positron pair has to be moving at a speed equal to the center of momentum of the two photons. The pair formation energy is of the order of 1 MeV. The energy of a typical 2.73 K cosmic background photon is of the order of 10^{-3} eV. What is the energy of the lowest energy gamma photon that can collide with a background photon to produce an electron–positron pair? Show that, in the frame within which the pair is produced at rest, energy conservation gives

$$\frac{h\nu_1 \left(1 - (v/c)\right)}{\sqrt{1 - (v/c)^2}} + \frac{h\nu_2 \left(1 + (v/c)\right)}{\sqrt{1 - (v/c)^2}} = 2m_0 c^2, \tag{5-52}$$

and momentum conservation requires the two terms on the left to be equal. These two requirements give

$$(h\nu_1)(h\nu_2) = (m_0 c^2)^2 \sim 2.5 \times 10^{11} \, \text{eV} . \tag{5-53}$$

The intergalactic infrared radiation density is high enough at photon energies $h\nu_2 \sim 10^{-1}$ to 10^{-2} eV, that gamma-ray photons with energies $h\nu_1 \sim 10^{13}$ eV produce electron–positron pairs in collisions with the cosmic infrared background (see Figs. 1.15, 1.16). The cross-section for this process is sufficiently high so that no gamma rays with energies in excess of $\sim 2 \times 10^{13}$ eV reach us from some of the most energetic active galactic nuclei AGNs, like the galaxy Markarian 501 at a distance ~ 150 Mpc, despite indications that γ-rays at such high energies are produced in these nuclei. The radiation reaching us appears to be sharply truncated beyond about 10^{13} eV $\equiv$ 10 TeV.

5:10 Particles at High Energies

Cosmic rays are extremely energetic photons, nuclei, or subatomic particles that traverse the Universe. Occasionally such a particle or photon impinges on Earth's atmosphere, or collides with an ordinary interstellar atom. What happens in such interactions?

We have no experimental data on particles whose energies are as high as 10^{20} eV, because our laboratories can only accelerate particles to energies of the order of 3×10^{12} eV. However, the relativity principle permits us some insight into even such interactions. We ask ourselves, how 10^{20} eV protons would interact with low-energy photons in interstellar or intergalactic space. Such 2.73 K blackbody photons have a frequency $\nu \sim 3 \times 10^{11}$ Hz.

As seen by the proton, these millimeter-wavelength photons appear to be highly energetic gamma rays. This follows because $\gamma(v)$ must be $\sim 10^{11}$ for the proton, whose rest-mass is only 9.31×10^8 eV. By the same token the proton sees the photon Doppler shifted by a factor of 10^{11}. In the cosmic-ray proton's rest-frame, the photon appears to have a frequency of $\sim 3 \times 10^{22}$ Hz. This corresponds to a gamma photon with energy ~ 100 MeV, and the proton acts as though it were being bombarded by 100 MeV photons

Photons at 100 MeV can be produced in the laboratory; the main effect of photon–proton collisions at this energy is the production of π-mesons through the interactions

$$\mathcal{P} + \gamma \rightarrow \mathcal{P} + \pi^{\circ}, \quad \mathcal{P} + \gamma \rightarrow \mathcal{N} + \pi^{+}. \tag{5-54}$$

In the first reaction the proton–photon collision produces a neutral pion π° and a proton having a changed energy. The second reaction produces a neutron and a positively charged pion.

The cross-sections for these interactions can be measured in the laboratory, and the results are then immediately applicable to our initial query. The cross-sections are so large that the highest energy cosmic-ray protons whose energies range up to $\sim 3 \times 10^{20}$ eV, can probably not traverse intergalactic space over distances $\gtrsim 30$ Mpc through the 2.73 K microwave photon flux, as illustrated by Figure 1.16 (Bi95), (Gr66), (Bi97). This collisional destruction of the highest energy cosmic rays by the microwave background radiation is often referred to as the *Greisen–Zatsepin–Kuz'min cutoff*, after the three physicists who first noted its significance (Ta98a). Most of the cosmic-ray primaries at the highest energies appear to be protons, though the chemical composition is still uncertain at energies above $\sim 10^{14}$ eV (Di97), (O'H98).

Frequently a physical problem can be considerably simplified if we choose to view it from a favorable inertial frame. The relativity principle shows us how to do this and gives us many new insights into the symmetries of physical processes.

5:11 High-Energy Collisions

Consider the elastic collision of a low-energy particle with a similar particle initially at rest. If we view this interaction in the resting particle's frame, and both particles have mass m, then the center of mass will move with velocity $v/2$ as shown in Fig. 5.9.

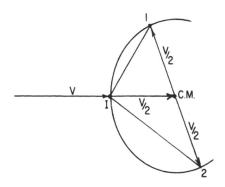

Fig. 5.9. Illustration of elastic collisions for identical particles.

For an initial approach velocity v of the moving particle, conservation of momentum requires that the two particles have velocities $v/2$ relative to the center of mass — after the collision, as well as before. For any time after the collision, a circle can be drawn through the impact point I and the particle positions 1 and 2 that define a diameter on the circle. This means that the particles always subtend a right angle at the impact point. So far our treatment has been nonrelativistic. In the relativistic case, the center of mass still lies on a line joining 1 and 2. Effectively, particles 1 and 2 are scattered away from the center of mass in opposing directions. Seen from a rest-frame, however, they will appear to be scattered predominantly into the forward direction. This is precisely the same concentration into the forward direction, which we saw for the rapidly moving light source that emits radiation isotropically in its own rest-frame (5–49).

When a cosmic-ray proton collides with the nucleus of a freely moving interstellar atom or with an atom that forms part of an interstellar grain, a fraction of the nucleus can be torn out. This may just be a proton or a neutron, or it could be a more massive fragment, say, a ^{3}He nucleus. Such *knock-on* particles always come

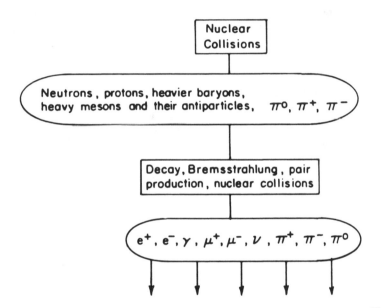

Fig. 5.10. Constituents of a cosmic-ray air shower. The primary particle, here shown as a proton, collides with the nucleus of an atmospheric atom, producing a number of secondary particles that suffer nuclear collision, decay, pair production, or *Bremsstrahlung* — a process in which a charged particle is slowed down by the emission of a gamma photon. A large succession of such events takes place. By the time the shower arrives at the surface of the Earth, most of the charged particles we observe are electrons, positrons, and muons. Although most of the primary nuclei are protons, several percent can be alpha particles (helium nuclei) and about one percent are heavier nuclei. Electrons and positrons also can be primary particles. The air showers are a prime example of the conversion of energy into rest–mass. On occasion, the energy of a single primary is sufficient to produce 10^9 shower particles.

off predominantly in the forward direction, close to the direction along which the primary proton was moving.

Similarly, when a cosmic-ray primary arrives at the top of the Earth's atmosphere after its long trek through space, it collides with an atmospheric atom's nucleus, giving rise to energetic secondary fragments, mesons, baryons, and their decay products. These decay into mesons, gammas, electron–positron pairs, or neutrinos, or they may collide with other atoms until a whole shower of particles rains down. Such a *cosmic-ray air shower* consisting of electrons, gamma rays, mesons, and other particles, even if initiated at an altitude of ten or more kilometers, often arrives at ground level confined to a patch no more than some hundred meters in diameter. The forward concentration is so strong that the showers are well confined even though they sometimes consist of as many as 10^9 particles.

This close confinement allows us to deduce the total energy originally carried by the primary; we need only sample the energy incident on a number of rather small detector areas. Most of our information about high-energy cosmic-ray primaries has come from just such studies made with arrays of cosmic-ray shower detectors. The *Akeno Giant Air Shower Array (AGASA)* in Japan covers an area of 100 km^2 sampled by 11 detectors, each with a collecting area 2.2 m^2 and separated by ~ 1 km from its nearest neighbor (Ta98a). The total energy in the shower can be determined from these samplings, and the time of arrival at each detector shows the direction from which the primary came. Figure 5.10 shows some of the constituents of cosmic-ray air showers.

5:12 Superluminal Motions and Tachyons

Some quasars are observed to periodically eject jets of relativistic particles that radiate at radio frequencies. These fast-moving jets, which at the distance of quasars typically extend no more than a tiny fraction of a second of arc, appear to exceed the speed of light. Similar *superluminal motions* are observed in jets streaming out of *microquasars*, stellar black holes surrounded by accretion disks from which matter is continually spiraling into the black hole. The appearance of a velocity higher than the speed of light, however, is only a projection effect.

Consider a quasar intermittently ejecting clouds of gas moving at relativistic velocities V, at an angle θ with respect to the line of sight to Earth, as measured by an observer on Earth (Figure 5.11). At some time t_1 radio waves are received from an ejected jet that radiates in the general direction of Earth. A time interval $\Delta t = t_2 - t_1$ later, the cloud is observed to have moved a distance $V \Delta t$ and continues to emit radio waves. Transverse to the line of sight to Earth it has traversed a distance $V \Delta t \sin \theta$; along the line of sight, a distance $V \Delta t \cos \theta$. As seen from Earth, the time interval Δt appears shorter than the time span Δ_a separating the two events at which the radiation was actually emitted because, while the first beam appearing to have been emitted at t_1 was traveling toward Earth, the cloud was moving closer to Earth at relativistic speed. As a result the actual line of sight distance between wave fronts, respectively arriving at t_2 and t_1, is $(c - V \cos \theta) \Delta t_a$ and

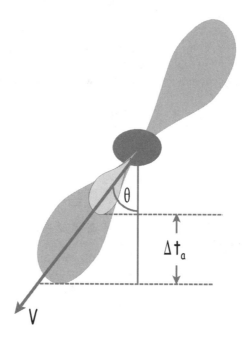

Fig. 5.11. Geometry of superluminal motion.

$\Delta t = (c - V \cos \theta) \Delta t_a / c$. The apparent transverse velocity of the cloud seen from Earth is $V \sin \theta / [1 - (V/c) \cos \theta]$. At sufficiently high velocities V directed close to the line of sight, where $\cos \theta$ approaches unity, the cloud will appear to have a transverse velocity exceeding the speed of light, giving rise to the name *superluminal velocities*. If the quasar also emits a cloud in the diametrically opposite direction it will appear to have a much smaller transverse velocity, $V \sin \theta / [1 + (V/c \cos \theta)]$. The actual measurement carried out by an observer at Earth, however, is a rate of angular displacement across the sky, rather than an actual velocity. This is $V \sin \theta / D$, where D is the distance to the emitting quasar and the jets in its immediate vicinity.

Galactic *microquasars*, black holes surrounded by compact accretion disks, often eject bi-lobed jets of gas in opposite directions. For a source at distance D, the proper motions of the approaching and receding clouds, respectively, μ_a and μ_r are

$$\mu_a = \frac{V \sin \theta}{[1 - (V/c) \cos \theta] D} \quad \text{and} \quad \mu_r = \frac{V \sin \theta}{[1 + (V/c) \cos \theta] D} . \qquad (5\text{-}55)$$

The ratio of Doppler-shifted wavelengths, respectively received from the approaching and receding jets, is then obtained from (5–44) as

$$\frac{\lambda_a}{\lambda_r} = \frac{1 - (V/c) \cos \theta}{1 + (V/c) \cos \theta} . \qquad (5\text{-}56)$$

Solving these three equations simultaneously yields the velocity V, angle θ, and distance D to the source. For the microquasar GRO J1655-40 a jet velocity $V \sim 0.92$ and a distance $D \sim 3$ kpc has been deduced this way (Mi98).

In quasars, the jets are observed primarily when they are pointed almost directly at Earth. The receding jet can then only be inferred, because the Doppler shift of recession not only red-shifts the radiation but, by (5–49) also diminishes its intensity to the point where it cannot be observed.

Although superluminal velocities do not involve velocities exceeding those of light, speculations occasionally arise about the possible existence of particles that do exceed the speed c. These have been called *tachyons*. When Einstein first discovered the special relativistic concept he clearly stated that matter could not move at speeds greater than the speed of light. He argued that the relation (5–34) between rest-mass and energy already implied that an infinite amount of energy was needed to accelerate matter to the speed of light. Particles with nonzero rest-mass therefore could never quite reach even the speed of light let alone higher velocities.

In recent years, this question has been re-opened by a number of researchers. They have argued that, while it certainly is not possible to actually reach the speed of light by continuous acceleration, this alone does not rule out the existence of faster-than-light particles created by some other means.

The basic argument in favor of even examining the possibility of tachyon existence is the formal similarity of the Lorentz transformations for velocities greater than and less than the speed of light, and the fact that the transformations taken by themselves say nothing that would rule out tachyon existence.

The similarity, of course, does not imply that particles and tachyons behave in precisely the same manner. If we look at expression (5–34), we note that for $V > c$ the denominator becomes imaginary. By choosing the mass of the tachyon to be imaginary, however, the energy $\mathcal{E}$ remains real, and so does the momentum, as shown by (5–37). Nevertheless, tachyons raise a disturbing difficulty. Special relativity shows that high-velocity tachyons should be able to influence the past, and thereby violate causality — relations between cause and effect. This does not entirely rule out tachyons below a certain speed limit, though it makes them problematic.

Thus far no tachyons have been detected, but further experimental investigations have been proposed (Ch96).

5:13 Strong Gravitational Fields

The introduction of gravitational fields requires a theory more general than the special theory of relativity, which restricts itself to inertial frames. For problems involving gravitation, the general theory of relativity (Ei16) or similar gravitational theories (Di67) have to be used. However, some simple gravitational results can be obtained without such theories if we remember that the set of inertial frames also includes freely falling frames of such small size that the gravitational field can be considered to be locally uniform. We will consider two such local inertial frames in a centrally symmetric gravitational potential Φ.

Consider an observer O′ at distance r from the central mass distribution (Fig. 5.12). We would like to know the form that the line element ds^2 would take in her

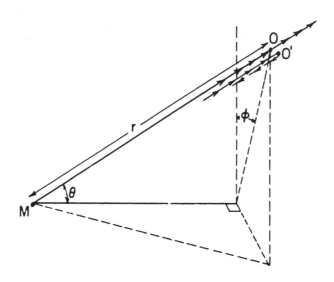

Fig. 5.12. Freely falling observer near a mass M.

frame of reference. We can suppose that O′ was initially moving outward from the star but at a speed less than the escape velocity. She only had enough kinetic energy to reach r. Here her velocity reached zero, and she is just beginning to fall back into the center. We see her when her velocity is zero.

Because O′ is falling freely, her line element will seem to her to have the form (5–3). In spherical polar coordinates this is

$$ds^2 = c^2 \, dt'^2 - r'^2(\sin^2 \theta' \, d\phi'^2 + d\theta'^2) - dr'^2 . \qquad (5\text{-}57)$$

We ask ourselves what ds^2 would seem to be, seen by an observer O far enough away from the mass distribution so that Φ essentially is zero or negligibly small. Φ as used here will be the negative of $\mathbb{V}(r)$ in (3–55).

We could of course assume that O gets all his information about O′'s system from light signals. But that is not necessary. The physical relationship between O and O′ is independent of how the observational information is conveyed.

Let us therefore suppose that O has taken a trip to find out for himself. We can suppose that he was near the central mass distribution, that he is now on his way out, and that he is in unpowered motion, freely falling radially outward, with just exactly enough energy to escape to infinity.

O goes through the radial distance r, close to O', just as O' passes through zero velocity and begins her infall. Since both observers are in inertial frames, the Lorentz transformations can be used to determine what O''s line element would seem like to O. Once again, the spatial components perpendicular to the direction of relative motion should be identical, so that

$$r^2(d\theta^2 + \sin^2\theta\, d\phi^2) = r'^2(d\theta'^2 + \sin^2\theta'\, d\phi'^2)\,. \tag{5-58}$$

The radial components, however, will appear changed because of the relative motion. If the gravitational potential is weak, the velocity of O relative to O' is immediately obtained from the fact that O just barely has enough kinetic energy to go to infinity, so that, equating kinetic and potential energy, per unit mass

$$\frac{1}{2}V^2 = \Phi. \tag{5-59}$$

Equations (5–12) and (5–14) then lead to

$$\Delta t = \frac{\Delta t'}{\sqrt{1 - (2\Phi/c^2)}}, \qquad \Delta x = \Delta x'\sqrt{1 - (2\Phi/c^2)}\,. \tag{5-60}$$

This is also the correct form for strong potentials Φ, where the classical concept of kinetic energy no longer has a clear meaning. We can therefore write that the line element (5–57) has the form

$$ds^2 = (c^2 - 2\Phi)\, dt^2 - r^2(\sin^2\theta\, d\phi^2 + d\theta^2) - \frac{dr^2}{1 - (2\Phi/c^2)} \tag{5-61}$$

as seen by O. This represents a measure of the clock rate and scale length in O''s frame as seen from O's coordinate system. O notes this down, and is able to convey these impressions when he reaches infinity. He has been traveling in an inertial frame all this time, and his results are therefore not suspect.

When the potential Φ is generated by a nonrotating mass M, we can rewrite (5–61) as

$$ds^2 = \left(c^2 - \frac{2MG}{r}\right)dt^2 - \left(1 - \frac{2MG}{rc^2}\right)^{-1}dr^2 - r^2(\sin^2\theta\, d\phi^2 + d\theta^2)\,. \tag{5-62}$$

The line element, or *metric* (5–62) is called the *Schwarzschild line element*. Karl Schwarzschild derived it within weeks after Einstein's publication in 1915 of his theory of general relativity (Sc16). To understand this expression's physical implications, it is important to carefully look at what we mean by the symbols r and t. Position and time can be defined in different ways, and we must take care to specify which definition we wish to use.

As used here, r is an *area coordinate* chosen to make the area of a sphere precisely $4\pi r^2$. We *define* $r \equiv (A/4\pi)^{1/2}$, where A is the surface area of a sphere centered on the mass M. As can be seen by setting (t, θ, ϕ) constant, the radial separation between two such concentric spheres with respective surface areas $4\pi r^2$ and

$4\pi(r+dr)^2$ is $dr/(1 - 2MG/rc^2)^{1/2}$. This is always greater than the *increment of coordinate length*, dr, measured by a stationary observer at position r.

The symbol t stands for *coordinate time*, marking the passage of time on the clock of a static observer whose radial coordinate r and angular coordinates (θ, ϕ) are constant and who is, therefore, moving along his own world line. Coordinate time is marked by synchronization of *clock rates* for all such observers. Each observer sees the clocks of all other observers running at the same rate as his own, though the times observed on these other clocks will generally differ from his. Coordinate time is not *proper time*, $ds/c = (1 - 2MG/rc^2)^{1/2}\, dt$, measured by a freely falling observer instantaneously at rest, $(r, \theta, \phi) = $ constant. The proper time interval is always shorter than the corresponding interval of coordinate time. This means that if we are to change the rates at which different clocks run, so as to synchronize them all with a standard clock at $r = \infty$, the individual rates have to be sped up by a factor of $(1 - 2MG/rc^2)^{-1/2}$. This creates a system that marks coordinate time t.

Three concepts underlie these definitions of space and time. Two of these are the *clock hypothesis* and *length hypothesis*, according to which two observers instantaneously at rest relative to each other, at some event (t, r, θ, ϕ) will make identical readings of all clocks and rulers, even though one may be falling freely, while the other is accelerated (or, equivalently, stationary in a gravitational field). The third concept is that of a *static field*, meaning that the metric coefficients in (5–62) are not only time-independent, making the field a *stationary field*, but also that all motions of particles and fields are time reversible. This latter criterion calls for a metric that lacks all time–space cross terms, $dt\, dr$, $dt\, d\theta$, or $dt\, d\phi$.

Any two clocks, A and B, can always be synchronized with a third clock C at infinity. This follows from the circumstance that the time taken for light to traverse distance CA remains constant in time, so that clocks with the same clock rates will always be out of phase by a constant amount. In addition, reversibility requires that the time taken for the light to transit the path CA equals the time taken to transit the reverse path AC. The same holds for paths CB and BC and paths AB and BA. We now see that the elapsed time required to traverse a path running along the three sides of the triangle $CABC$ is the same as the traversal time along the reverse path $CBAC$. This assures that the clock rates for any number of clocks can always be synchronized when the field is static. In particular, the field described by the Schwarzschild metric is static and permits synchronization of clock rates.

At the *Schwarzschild radius*

$$r_s \equiv \frac{2MG}{c^2} \tag{5-63}$$

something odd must take place. Here, according to (5–60), the clocks would appear to run infinitely slowly. A message emitted at some time t_0 would not arrive at larger radial distances until an infinite time later. Such a message would constitute the final event to ever reach an observer at infinity. The Schwarzschild radius of a black hole is, therefore, also its *event horizon*. Events occurring at $r < r_s$ can never be observed at infinity. r_s is often referred to as the *event horizon* because the passage of a particle through r_s is the last event that an observer at infinity can ever observe.

A massive object completely enclosed in r_s would appear invisible. Such objects have been called *black holes*. They are primarily detectable through the gravitational field they set up, but not through emitted radiation. A star could be orbiting about a black hole companion; its orbital motion about an apparently dark region in space would be a sign that a black hole might be there. Several examples of such binaries are now known. They generally involve X-ray novae.

X-ray novae occur where matter tidally torn out of a giant companion falls onto the surface of a compact star or an accretion disk around a compact star. Because black holes have no solid surfaces, an X-ray nova involving a black hole, of necessity, would require the black hole to be orbited by an accretion disk. The matter accumulates there until its hydrostatic pressure becomes so great that nuclear reactions set in and hydrogen explosively fuses into helium. This releases large amounts of energy in a nova outburst. The system then settles back to its earlier state and the accumulation of hydrogen by the disk begins all over again. During the quiescent periods between outbursts, the binary system's orbital characteristics can be monitored and the mass of the companion deduced. For the X-ray nova GS 2000 +25, the unseen companion's mass is of order $8M_\odot$. As we will see in Chapter 8, this is well above the mass that white dwarfs or neutron stars can maintain, and strongly suggests the presence of a star that has collapsed to form a black hole (Ca96). Hydrogen drawn out of the giant star evidently falls on an accretion disk encircling the black hole and accumulates until it explosively erupts.

For a solar mass, the Schwarzschild radius is $r_s \sim 3 \times 10^5$ cm. As we will see in Section 8:16, this is only a factor of order ~ 5 smaller than the radius of a neutron star. For an object with mass $M \sim 10^9 M_\odot$, $r_s \sim 3 \times 10^{14}$ cm or about 20 AU.

We will mention black holes again later. However, for the moment, it is still worth discussing two matters. First, as indicated in Table 1.6, stellar black holes appear to account for $\lesssim 3\%$ of all stellar mass (Fu04). With typical masses $\sim 7.5M_\odot$, they probably account for $\lesssim 0.2\%$ of all stars by number. Second, space travelers must be careful. Once they enter a black hole they can never return. The interior of such an object is as isolated from us as a separate universe.

5:14 Gravitational Time Delay; Deflection of Light

For light traveling radially from the surface of the Sun to Earth's orbit, we have $ds = 0$ and $d\phi = d\theta = 0$. Equation (5–62) then tells us that the traversal of an increment of radial distance dr requires an interval of coordinate time $dt = dr/c(1 - 2MG/rc^2)$. We can integrate this for the total distance from the Sun's surface at radius $R_\odot$ to Earth's orbital radius $R_\oplus$.

$$\int dt = \frac{1}{c} \int_{R_\odot}^{R_\oplus} \frac{dr}{(1 - 2MG/rc^2)} = \left(\frac{R_\oplus - R_\odot}{c}\right) + \frac{2MG}{c^3} \ln\left(\frac{R_\oplus - 2MG/c^2}{R_\odot - 2MG/c^2}\right).$$
$$(5\text{-}64)$$

The last term on the right gives the gravitational time delay in excess of the travel time $(R_\oplus - R_\odot)/c$ expected if the Sun had no mass. For a solar radius $R_\odot = 7 \times 10^{10}$ cm, the Earth's orbital radius $R_\oplus = 1.5 \times 10^{13}$ cm, and $2MG/c^2 = 3 \times 10^5$ cm, we obtain a gravitational delay of ~ 50 μs, or one part in $\sim 10^7$ of the total traversal time.

We can similarly consider light arriving at Earth from a distant pulsar that happens to lie in Earth's orbital plane. At one time of the year, Earth will be on the same side of the Sun as the pulsar. Six months later, Earth's trajectory will have taken it to the far side of the Sun so that the pulsar's radiation has to pass close by the Sun to reach Earth. Were it not for the gravitational time delay, the arrival times of the pulsar signals would be observed to be increasingly delayed by an amount $R_\oplus(1 - \cos \alpha)/c$, during this half year, as the pulsar–Sun–Earth angle α increased from 0 to π and the distance grew between the Earth and the pulsar. But, as α approaches π and the pulsar's radiation passes close by the Sun, the additional gravitational delay rises rapidly, reaching a maximum ~ 100 μs — twice the delay in the Sun–Earth travel time — as the radiation suffers first a delay on approaching the Sun and then a further delay on receding. Neglecting the term $2MG/c^2$, which is far smaller than $R_\odot$ or $R_\oplus$, the time delay becomes

$$\Delta t \sim \left(\frac{2MG}{c^3}\right) \ln\left(\frac{R_\oplus}{R_\odot}\right). \tag{5-65}$$

Strictly speaking, we have only calculated the gravitational time delay for radial infall to the Sun and transmission on to Earth. There is also a small additional delay due to the almost tangential passage of radiation past the Sun at closest approach. We can estimate this correction by setting $ds = 0$ and $dr = 0, \theta = \pi/2$ in equation (5–62). The increment of traversal time is now $dt = r\,d\phi/c(1 - 2MG/R_\odot c^2)^{1/2}$. Integrating over an angle $\Delta\phi \sim 1$ radian at closest approach to the Sun, we see that the additional increase in gravitational time delay will only be of order $\sim MG/c^3 \sim 5\,\mu$s for close passage by the Sun. The radial motion of light to and from the Sun accounts for all but a few percent of the total gravitational delay.

The time delay on transmitting radar signals from Earth to Venus and back, when the planet is on the far side of the Sun, has also been measured. The round trip gravitational delay when the planet lies at closest angular separation from the solar limb is $\sim 200\,\mu$s, corresponding to two full traversals past the Sun.

With the time delay established, we can estimate the radial deflection of light by the Sun. Light rays are gravitationally bent toward the Sun on close passage. Their small angular deflection is readily calculated from (5–64). Consider the wave front of radiation approaching the Sun. The portion passing nearest to the Sun suffers the greatest gravitational delay. The gradient in the delay as a function of impact parameter R follows from (5–64) and is

$$\frac{dt}{dR} = -\frac{d}{dR}\left[\frac{4MG}{c^3} \ln\left(R - \frac{MG}{c^2}\right)\right] \sim -\frac{4MG}{c^3 R}. \tag{5-66}$$

With this deflection the front undergoes a change in direction (Fig. 5.13).

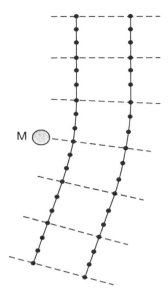

Fig. 5.13. Deflection of light and its wave front due to the gravitational time delay on passage close to a mass M. The wave front, shown by dashed lines, propagates from top to bottom in this figure.

$$\delta\phi \sim \frac{cdt}{dR} = \frac{4MG}{c^2 R} \sim 8.5 \times 10^{-6}\,\text{radians} \sim 1.8\,\text{arcsec}\,. \tag{5-67}$$

5:15 Gravitational Lenses

The Sun can be thought of as a weak lens, which bends a beam of light from a distant source and makes it converge to a focus. For the Sun, this focal length is its radius divided by the angular deflection (5–67). This amounts to $\sim$550 AU $\sim 10^{16}$ cm.

When radiation from a distant quasar passes by a massive galaxy its deflection again is of order $4MG/Rc^2$, where the mass and radius now refer to the galaxy, $M \sim 4 \times 10^{11} M_\odot$, $R \sim 3 \times 10^{22}$ cm $\sim 4 \times 10^{11} R_\odot$. This again corresponds to a deflection of about 2 seconds of arc. The focal length now is of order 3×10^{27} cm, or 10^3 Mpc. An observer stationed one focal length from the galaxy and on the straight line that runs from the quasar to the intervening galaxy and beyond, would see light from the quasar as a ring around the galaxy. Such rings are called *Einstein rings*. If the quasar, galaxy, and observer do not lie precisely on a straight line, so that the quasar sits somewhat off the axis joining the observer to the galaxy, only a fraction of an arc might be observed. Deviations from spherical symmetry in the galaxy's mass distribution may also break up the projected image of the quasar on the sky so that only a few patches of light are observed.

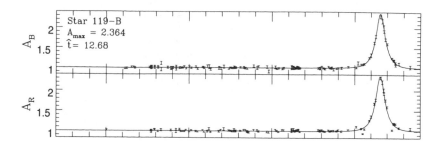

Fig. 5.14. Gravitational lensing of a background source by a Galactic halo star. Each marking on the horizontal scale represents a ten-day period. Note the symmetry of each curve — one obtained in blue light (B), the other in red (R) — their perfect coincidence in time, and their identical shapes. These three features characterize a gravitational lens (Aℓ97).

Gravitational lensing and Einstein rings or arcs are useful for assessing the total mass of matter that aggregates at the center of a cluster of galaxies. Such arcs can be discerned in Fig. 1.11. If the distances to the cluster and quasar are known from their red shifts, the angular position of the observed arcs immediately yields the value of the cluster mass through relation (5–67).

By observing a large number of faint stars in the bulge surrounding the Galaxy's central regions we occasionally register a rising signal from a distant star. It typically lasts a week. This is the lensing of a background source by a foreground star. The characteristic features of such an event, in contrast to a great many other variable stellar events, is the symmetry of the rise and fall of the curves and the lack of any difference in the ways that blue and red light curves vary. Figure 5.14 shows one such pair of light curves.

5:16 An Independent Measure of the Hubble Constant

The binary quasar 0957+561 has two principal components, A and B. They are lensed images of a single quasar at red shift $z = 1.41$, which is offset by a few seconds of arc from a massive foreground galaxy, G, at $z = 0.36$. Image A is displaced by 5.2 arcsec from G, while image B is only 1.034 arcsec away (Be97). Image A is less deflected, because it is seen at a larger angular distance from G, and we would expect light reaching us along this path to be less delayed than light more strongly deflected along path B. This is borne out by observations on the variability of the quasar. Variations in the flux received from A are replicated ∼420 days later by component B (Ku97b). From these data we can determine the Hubble constant (Re64), (Re64a).

We note from (5–64) that the difference in the time delays along the two light paths is

$$\Delta t_B - \Delta t_A = \frac{4MG}{c^3}\left(\ln\frac{D}{R_B} - \ln\frac{D}{R_A}\right) = \frac{4MG}{c^3}\ln\frac{\phi_A}{\phi_B}, \qquad (5\text{-}68)$$

where D is the distance to the galaxy G and the displacement of the images is $\phi_i = R_i/D$, with $i = A, B$. From (5–67) we also have

$$\phi_A - \phi_B = \frac{4MG}{c^2}\left(\frac{1}{R_A} - \frac{1}{R_B}\right) = \frac{4MG}{c^2 D}\left(\frac{1}{\phi_B} - \frac{1}{\phi_A}\right). \qquad (5\text{-}69)$$

PROBLEM 5–11. Show that equations (5–68) and (5–69) can be solved to yield a distance $D \sim 5.4 \times 10^{27}$cm, for the observational data cited, and that the derived Hubble constant is $cz/D \sim 60\,\mathrm{km\,s^{-1}\,Mpc^{-1}}$ Although this estimate is crude, it roughly confirms values obtained by entirely different methods cited in Chapter 2.

5:17 Orbital Motion Around a Black Hole

We now ask how a particle moves in a strong gravitational field, just outside the Schwarzschild radius of a mass M. To come to grips with this, we may invoke the *strong equivalence principle* which postulates that any physical law holding in special relativity will also hold in freely falling coordinate frames around arbitrary mass distributions. By *freely falling* we will mean that only gravitational forces are acting, and that electrical, magnetic, or other nongravitational forces are all absent.

To understand this principle, let us return to the man whom we encountered in his rocket ship in Section 3:8, and see what happens if his ship strays close to M in free fall. Within the small, freely falling ship the curvature of space can be considered negligibly small. Any particle in similar free fall along some arbitrary trajectory within the ship will traverse the space at constant velocity relative to a point fixed within the falling ship. The relative velocities of any two such particles passing through the ship, along quite arbitrary orbits, will also appear constant. Thus, the laws of special relativity apply. The same will hold true for photons which, as we saw in Sections 3:9, 5:14 and 5:15, also fall in a gravitational field. Coordinate frames attached to freely falling particles or photons within an infinitesimal freely falling locale are called *local inertial frames*. The strong equivalence principle states that all local inertial frames at the same *event* — i.e., coinciding in space and time — are in uniform relative motion, so that the laws of special relativity and Lorentz invariance apply. Local inertial frames that are far apart, however, may be in mutual acceleration.

The line element defining the space around M is given by equation (5–62). In analogy to equations (5–26) to (5–29) we can then rewrite equation (5–62) in terms of four-velocities,

$$2\mathcal{L} = \left(1 - \frac{2MG}{c^2 r}\right)\left(\frac{c^2 dt}{ds}\right)^2 - \left(1 - \frac{2MG}{c^2 r}\right)^{-1}\left(\frac{cdr}{ds}\right)^2$$

$$(5\text{-}70)$$

$$-r^2\left(\frac{cd\theta}{ds}\right)^2 - r^2\sin^2\theta\left(\frac{cd\phi}{ds}\right)^2,$$

where $\mathcal{L}$ is the Lagrangian per unit mass–energy. As indicated in Section 5:6, $2\mathcal{L} = c^2$ for massive particles of rest–mass $m_0 = 1$. The trajectories of photons are *geodesics*, the shortest distances between any two points traversed. This extremum implies that $ds = 0$ and $\mathcal{L} = 0$. For massive particles s may be identified with the proper time measured by a clock moving with the particle along the geodesics. Equation (5–70) defines the geodesics in space–time that govern the motions of particles and photons, as long as particle mass m_0 or photon energy $h\nu/c^2$ is negligibly small compared to the mass M of the black hole (Ch83)*.

We can denote differentiation with respect to s by a dot, writing $c\dot{t} \equiv c^2 dt/ds$, $\dot{r} \equiv cdr/ds$, $\dot{\theta} \equiv cd\theta/ds$, and $\dot{\phi} \equiv \phi/ds$. As in Section 5:6 we then obtain the momenta

$$p_t = \frac{\partial\mathcal{L}}{\partial\dot{t}} = \left(1 - \frac{2MG}{c^2 r}\right)\dot{t}, \qquad p_\phi = -\frac{\partial\mathcal{L}}{\partial\dot{\phi}} = \left(r^2\sin^2\theta\right)\dot{\phi},$$

(5-71)

$$p_r = -\frac{\partial\mathcal{L}}{\partial\dot{r}} = \left(1 - \frac{2MG}{c^2 r}\right)^{-1}\dot{r}, \qquad p_\theta = -\frac{\partial\mathcal{L}}{\partial\dot{\theta}} = r^2\dot{\theta}.$$

As we saw in Section 3:17, spherical symmetry confines the motion of an orbiting particle or photon to a plane, permitting us to arbitrarily assign a value of $\pi/2$ to θ throughout the particle's trajectory.

Because the metric (5–62) and the corresponding Lagrangian (5–70) are independent of time t, p_t is a constant of motion

$$p_t = \left(1 - \frac{2MG}{rc^2}\right)\dot{t} = \text{constant} = \frac{\mathcal{E}}{c}, \qquad (5\text{-}72)$$

where we have made use of (5–32). The energy $\mathcal{E}$ (normalized to unit m_0 or $h\nu/c^2$) is conserved.

Similarly, because (5–62) and (5–70) are independent of the angle ϕ,

$$p_\phi = r^2\sin^2\theta\,\dot{\phi} = \text{constant} = L, \qquad (5\text{-}73)$$

showing also the conservation of orbital angular momentum L correspondingly normalized.

With $\dot{t}$ and $\dot{\phi}$ given by equations (5–71) we can now express the Lagrangian as

$$\left(1 - \frac{2MG}{c^2 r}\right)^{-1}\left[\frac{\mathcal{E}^2}{c^2} - \dot{r}^2\right] - \frac{L^2}{r^2} = 2\mathcal{L} = c^2 \text{ or } 0, \qquad (5\text{-}74)$$

depending on whether we are dealing, respectively, with particles or photons. For particles this leads to

$$\dot{r}^2 = \frac{\mathcal{E}^2}{c^2} - \left(1 - \frac{2MG}{rc^2}\right)\left(c^2 + \frac{L^2}{r^2}\right) \equiv \frac{\mathcal{E}^2}{c^2} - \mathcal{V}^2(r), \qquad (5\text{-}75)$$

where we have defined a new quantity $\mathcal{V}^2(r)$ that has the characteristics of a potential and is termed the *effective potential*. When added to the left side of the equation, reminiscent of twice the particle's kinetic energy per unit mass, $\mathcal{V}^2(r)$ acts as twice the potential, and $\mathcal{V}^2(r) + \dot{r}^2$ is a constant of motion, $\mathcal{E}^2/c^2$.

Because the left side of (5–75) has to be positive or zero, the energy $\mathcal{E}/c$ must always exceed or equal $\mathcal{V}(r)$. In its own rest-frame, the particle's energy $\mathcal{E}$ is constant, which means that differentiating (5–75) with respect to proper time gives

$$2\dot{r}\ddot{r} = -\frac{d\mathcal{V}^2(r)}{dr}\dot{r} \qquad (5\text{-}76)$$

or

$$\ddot{r} = -\frac{1}{2}\frac{d}{dr}\mathcal{V}^2(r) = -\frac{MG}{r^2} + \frac{L^2}{r^3} - \frac{3MGL^2}{r^4c^2}. \qquad (5\text{-}77)$$

A circular orbit — an orbit for which the radius is constant — is, therefore, possible only at a maximum, minimum, or point of inflection of $\mathcal{V}^2(r)$, i.e., at

$$MGc^2r^2 - L^2c^2r + 3MGL^2 = 0, \qquad (5\text{-}78)$$

which is quadratic in r and can be solved for the two extrema

$$r = \frac{L^2}{2MG}\left[1 \pm \left(1 - \frac{12M^2G^2}{c^2L^2}\right)^{1/2}\right]. \qquad (5\text{-}79)$$

For a star like the Sun, $r_s \sim 3 \times 10^5$ cm = 3 km, which is minute compared to the Sun's radius $R_\odot \sim 7 \times 10^{10}$ cm or the Earth's orbital radius $\sim 1.5 \times 10^{13}$ cm. But for a neutron star, whose radius is only of order 12 km, $3r_s/2r$ can be significant for material orbiting close to the star's surface.

The expression for $\mathcal{V}(r)$ is plotted in Fig. 5.15 as a function of $r/(MG/c^2)$ for a variety of values of Lc/MG. If $L^2c^2 > 12M^2G^2$ there are two circular orbits, one corresponding to a minimum of $\mathcal{V}(r)$, the other to a maximum. If $L^2c^2 = 12M^2G^2$ there is only one solution, a point of inflection, shown in Fig. 5.15 for $cL/MG = \sqrt{12} = 3.464$. The only stable circular orbits are those for which $\mathcal{V}(r)$ is at a minimum. Where $\mathcal{V}(r)$ is at maximum, the circular orbit is unstable. The *innermost stable circular orbit* is the one where $\mathcal{V}(r)$ has a point of inflection and $cL/MG = \sqrt{12} = 3.464$.

PROBLEM 5–12. Show that the radius of the innermost stable circular orbit — frequently referred to as the *last stable circular orbit* r_{lsco} — is

$$r_{lsco} = 6MG/c^2 = 3r_s, \qquad (5\text{-}80)$$

and show that the effective potential $\mathcal{V}^2(r) = (8/9)c^2$ at this point.

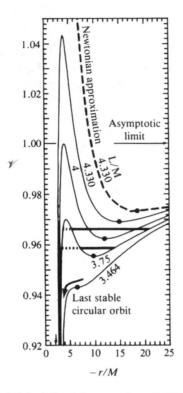

Fig. 5.15. The function $\mathcal{V}(r)$ defined by equation (5–75), plotted as a function of $r/(MG/c^2)$. Minima in the curves for different values of Lc/MG indicate stable circular orbits. Maxima indicate unstable circular orbits. The point of inflection shows where the last stable circular orbit occurs. Note: Specialists in general relativity like to express quantities in *geometrized units* where the speed of light and the gravitational constant are set equal to unity, $c = G = 1$. This is why the abscissa in this figure is labeled r/M instead of $r/(MG/c^2)$ and the values of the curves are shown in units of L/M rather than Lc/MG. It is worth remembering this notation when reading works on general relativity. (After S. Chandrasekhar, (Ch83)*).

PROBLEM 5–13. Derive the period of a particle orbiting a mass M in a circular orbit — for which the left side of equation (5–75) becomes zero. This can be done by again choosing $\theta = \pi/2$ and noting that $dt/d\phi = (1/c)(\dot{t}/\dot{\phi})$, as in equations (5–71) to (5–73). Equation (5–78) then provides the useful expression

$$L^2 = \frac{MGr}{1 - (3MG/c^2 r)} \quad \text{or} \quad r = \left(\frac{L^2}{MG}\right)\left(1 - \frac{3r_s}{2r}\right). \tag{5-81}$$

After some algebra this leads to $dt/d\phi = (r^3/GM)^{1/2}$. The time needed for the angle ϕ to change by 2π gives the period P

$$P = 2\pi \left(\frac{r^3}{GM}\right)^{1/2} \quad \text{and} \quad P_{lsco} = \frac{12\pi MG\sqrt{6}}{c^3} . \tag{5-82}$$

The period P derived in this way is measured in coordinate time, rather than proper time. But coordinate time conveniently corresponds to the time clocked by an observer at great distances, making P the period a terrestrial observer would measure.

We should still note that none of the central mass distributions M we have considered here have carried an electrical charge, nor have they carried angular momentum. A rotating, electrically charged black hole exhibits more complex behavior but, surprisingly, its properties are completely defined by just three parameters, its mass M, its net electric charge Q, and its spin angular momentum J. Black holes with $Q = J = 0$ are called *Schwarzschild black holes*. Spinning black holes are often called *Kerr black holes* after Roy Kerr who first set down the metric describing them.. For similar reasons, charged black holes are referred to as *Reissner–Nordström* black holes, and charged spinning black holes as *Kerr–Newman black holes*.

5:18 Advance of the Perihelion of Mercury

A particle orbiting close to a massive body no longer travels along Kepler's closed elliptic orbits. For most of the planets in the Solar System the deviation is too small to observe, but for Mercury, which is closest to the Sun, the effect is just barely detectable, and long constituted a puzzle. Mercury's perihelion point gradually shifts over the millennia, so that the entire orbital ellipse appears to slowly precess about the Sun, advancing in the sense of the planet's orbital motion. We talk about the *perihelion advance of Mercury*. For bodies orbiting other stars, the same effect is called the *advance of the periastron*.

We start by referring back to the dominant elliptical motion described by Kepler's laws in Section 3:5, and note that equations (3–30) and (5–77) resemble each other. If we neglect the last term on the right of equation (5–77), which is of order $(3r_s/2r)(L^2/r^3)$ and negligible compared to the second term if $r \gg r_s$, we see that (5–77) adopts the same form as equation (3–30). It, therefore, has a solution of the form of (3–33) as transformed by (3–36) or, more precisely, (3–57).

$$\frac{1}{r} = \frac{MG}{L^2}[1 + e\cos(\phi - \phi_0)] . \tag{5-83}$$

This is just the Newtonian elliptical orbit with eccentricity e. In order to incorporate the last term on the right of (5–77), which reflects the general relativistic effect, we adopt the same change of variables as in (3–31) and obtain the equation

$$\frac{d^2y}{d\phi^2} + y = \frac{MG}{L^2} + \frac{3MG}{c^2}y^2 . \tag{5-84}$$

PROBLEM 5–14. To find an approximate solution to (5–84) insert the trial function

$$y = \frac{MG}{L^2} \left\{ 1 + e \cos \left[\left(1 - \frac{3M^2G^2}{c^2L^2} \right) \phi - \phi_0 \right] \right\}, \tag{5-85}$$

in the equation and convince yourself that only terms lower than the leading terms by factors of c^{-2} or c^{-4} remain uncancelled.

The form of (5–83) tells us that the angle ϕ has to advance by $\Delta\phi = 2\pi(1 + 3M^2G^2/c^2L^2)$ from one perihelion passage to reach the next. The perihelion advances by an angle

$$\Delta\phi = \frac{6\pi M^2 G^2}{c^2 L^2} \tag{5-86}$$

per orbit. Using Table 1.4 and physical constants from Appendix B, convince yourself that this corresponds approximately to the measured *perihelion advance of Mercury* of 43 arc seconds per century.

For the binary pulsar PSR 1913 + 16, comprising two neutron stars orbiting each other only $\sim 2 \times 10^{11}$ cm apart, the corresponding *periastron advance* is $\sim 4.2°$ per year, an annual advance that is substantially larger (Ta82).

5:19 Accretion Disks Around X-ray Binaries

X-ray binaries are pairs of stars generally consisting of a white dwarf, neutron star, or black hole orbiting close to a larger star with lower surface gravity. The intense gravitational pull of the compact star tidally strips matter from its larger companion. If this material has too much angular momentum, it cannot directly fall onto the compact partner but, instead, accretes into a disk orbiting around it, called an *accretion disk*. Viscous friction transfers angular momentum from the faster rotating inner parts of this disk to its periphery. Deprived of angular momentum the inner portions move closer to the surface of the compact star, eventually reaching the last stable circular orbit and then spiraling onto the star.

PROBLEM 5–15. A neutron star has a mass $1.4M_\odot$. If the star's surface lies at $r = 10$ km, estimate the maximum temperature reached when protons tidally removed from a binary companion directly hit the star's surface $T \sim M_* m_P G/kr$, and estimate the maximum energy in electron volts of the X-rays or gamma-rays that could be emitted. Make a similar estimate for protons incident on an accretion disk at three times the stellar radius.

Many X-ray binaries exhibit millisecond flux modulations of their X-ray emission. The frequency of these oscillations changes in the course of the binary's orbital period. These kilohertz (kHz) X-ray modulations are called *Quasi Periodic Oscilla-tions, QPOs* and have frequencies ranging from several hundred cycles per second up to about 1300 Hz. The frequency of this X-ray modulation may be the frequency at which the inner edge of the accretion disk orbits the neutron star and may thus be associated with the orbital frequency at the last stable circular orbit. However, other explanations for the QPOs have also been advanced (Wa03).

PROBLEM 5–16. The radius of a neutron star can be no greater than the inner radius of its accretion disk. Show that a minimum stellar density can be inferred as $\rho_{min}(M) = 3\pi/P_{lsco}^2 G$, where P_{lsco} is given by (5–82).

Strong magnetic fields embedded in a neutron star or black hole can also have an effect on oscillation rates. In sorting out such different effects we are some-times helped by concentrating on phenomena that are common to both black holes and neutron stars, because we can then be sure that neither a solid surface, nor a Schwarzschild-radius *event horizon*, nor a nonaligned magnetic field can be in-volved. Because the magnetic field of a black hole is completely determined by its charge Q and its angular momentum L its magnetic field is always aligned with its spin axis. For neutron stars the magnetic field is normally not aligned with the spin axis.

PROBLEM 5–17. For a $5M_\odot$ black hole, show that $r_{lsco} \sim 44$ km, that the Schwarzschild radius is ~ 5 km, and that the orbital period P_{lsco} of a particle is only ~ 2.3 milliseconds. Compare these values to those for a neutron star with mass $1.4M_\odot$ and convince yourself that the corresponding radius and period, respectively are $r_{lsco} \sim 12.4$ km and 640 microseconds. Finally, for a $10^8 M_\odot$ black hole in the nucleus of a galaxy, show that these respective parameters would be ~ 6 AU and ~ 13 hours. Convince yourself also that the ratio of $2\pi r_{lsco}/P_{lsco} = c/\sqrt{6}$ regard-less of the mass of the compact object.

PROBLEM 5–18. The surface area of a Schwarzschild black hole of mass M is

$$A = 4\pi r_s^2 = 16\pi M^2 G^2/c^4 . \tag{5-87}$$

A particle of mass $m_0 \ll M$ is slowly lowered from afar and brought to rest at a small distance $b \ll r_s$ above surface A. Show that the potential energy of the particle at this point is $-Mm_0G/(b + r_s) = m_0c^2(b/r_s - 1)/2$. This means that a particle released from this radial distance with zero initial velocity, has a total mass–energy of only $m_0c^2b/2r_s$. Show that when the particle falls through the event horizon at r_s, the mass of the hole increases by $m_0b/2r_s$ and the area A of the event horizon increases by

$$\Delta A = 8\pi m_0 bG/c^2 , \tag{5-88}$$

independent of mass M.

5:20 The Smallest Conceivable Volume

We do not yet have a satisfactory quantum theory of space–time and gravitation, and do not know whether space–time is quantized. One argument, however, suggests that it could be. Equation (5–87) gives the surface area of a Schwarzschild black hole as $16\pi M^2 G^2/c^4$. We may now ask whether there is a minimum amount by which this area must increase when a particle of arbitrarily small mass m_0 falls into the hole. Equation (5–88) tells us that the area of the hole increases by $\Delta A = 8\pi m_0 b G/c^2$, when a particle at rest is dropped from a height b above the black hole surface. This is a minimum increase. Were the particle not at rest initially, it could only add more mass–energy to the black hole as it entered, and this would further increase the hole's surface area.

To minimize the increase ΔA in surface area, the product $m_0 b$ must be minimized. The smallest distance b from which a particle can be dropped is given by its proper radius. The Heisenberg uncertainty principle tells us that this must be $\sim \hbar/\Delta p \sim 2\hbar/m_0 c$. Here Δp is the root mean square uncertainty in the momentum along the radial direction given by (4–63). For a particle at rest this may be taken to be of order half the rest-mass momentum component $p_t = m_o c$ in equation (5–32). Minimizing the particle's radius thus implies increasing its mass. But increasing the mass indefinitely can also increase b because the Schwarzschild radius of m_0 then increases as $2m_0 G/c^2$. The smallest possible radius is obtained by making the two radii equal,

$$b = \frac{2m_0 G}{c^2} = \frac{2\hbar}{m_0 c} \quad \text{or} \quad m_0 \equiv m_P = \left(\frac{\hbar c}{G}\right)^{1/2} = 2.18 \times 10^{-5}\,\text{g}\,, \quad \text{(5-89)}$$

where m_P is called the *Planck mass*. This yields the Schwarzschild radius of the mass,

$$b = 2\left(\frac{\hbar G}{c^3}\right)^{1/2} \equiv 2\ell_P = 3.23 \times 10^{-33}\,\text{cm}\,, \quad \text{(5-90)}$$

where ℓ_P is called the *Planck length*, $\ell_P = 1.616 \times 10^{-33}$ cm. Substituting for the product $m_0 b$ in equation (5–88), we see that the area of a Schwarzschild black hole must increase by at least

$$\Delta A = 16\pi \left(\frac{\hbar G}{c^3}\right) = 4\pi \left(\frac{2m_0 G}{c^2}\right)^2 = 4\pi \ell_P^2\,. \quad \text{(5-91)}$$

whenever mass is added. This minimum increase just equals the minimum area enclosing a mass m_0 dropped into the black hole.

The increment of area ΔA in (5–91) appears to have a fundamental significance. It is possible to show that its value is independent of whether radiation or massive particles are absorbed by the black hole, whether the infalling matter is electrically charged or neutral, and that it is independent not only of the mass M of the black hole but also of its charge Q and spin angular momentum J (Be73). ΔA may, therefore, represent a quantum of area, and ℓ_P a corresponding quantum of length.

5:21 The Zeroth Law of Black Hole Dynamics

The *surface gravity* of a Schwarzschild black hole is $MG/r_s^2 = c^4/4MG$. Because such black holes are spherically symmetric, the surface gravity is identical over the entire surface of the event horizon. It can be shown that the surface gravity is also constant over the entire event horizon of any stationary black hole — one whose properties do not vary with time (Ba73). This constancy is referred to as the *zeroth law of black hole dynamics* which applies to spinning black holes of constant angular momentum J, with or without net electric charge Q. At first sight this appears surprising because the surface of a spinning black hole is not spherical. In the next section, we will see that black holes have a temperature that is directly proportional to the surface gravity. Because of this, the zeroth law is also referred to as the *zeroth law of black hole thermodynamics*.

5:22 Entropy and Temperature of a Black Hole

The First Law of Thermodynamics

In Chapter 4 we introduced the first law of thermodynamics,

$$dU = TdS - dW . \tag{5-92}$$

It tells us that, in going from an initial state A to some final state B, a system undergoes a change in internal energy $dU = U_B - U_A$ determined by the heat $đQ \equiv TdS$ that has been added to it, minus the work dW the system has done. In (4–119) the work dW consisted in expanding against an external pressure P by an increase in volume dV. For black holes dW could include increasing the rotational energy or the electric charge of the system. Though the heat input and the work done may vary in arbitrary ways, the change in internal energy depends only on the *difference* between added heat and work done.

The Second Law of Thermodynamics

The second law of thermodynamics tells us that the entropy S of a closed system is never observed to decrease, no matter what changes the system undergoes. Because matter entering a black hole remains trapped there, and because any addition of mass–energy always increases the surface area A, Jacob Bekenstein in 1973 proposed that A is a measure of the black hole's entropy (Be73).

For simplicity, let us consider a black hole into which we again drop an electrically neutral particle released at rest. It adds neither charge nor angular momentum to the black hole, and does no work on the hole because the black hole's surface expands without encountering an external pressure. We can then write the first law of thermodynamics as

$$\kappa TdA \equiv TdS = dU , \tag{5-93}$$

where dU is just the change in the black hole rest-mass energy $d(Mc^2)$. The constant κ must have dimensions erg K^{-1} cm^{-2} in order for the units on the left side of the equation to match those on the right, for any system — not just for Schwarzschild black holes. If we search for a product of fundamental constants with these dimensions we find a suitable candidate in $\kappa = k(c^3/4\hbar G)$, where k is the Boltzmann constant and $4\hbar G/c^3 = \ell_P^2$. With this substitution we obtain

$$T = \frac{1}{\kappa}\frac{dU}{dA} = \left(\frac{4\hbar G}{kc^3}\right)\frac{dU}{dA} = \frac{\hbar c^3}{8\pi k M G}. \tag{5-94}$$

The minimum increase in the entropy of a black hole then becomes

$$\Delta S = \kappa \Delta A = \left(\frac{kc^3}{4\hbar G}\right)\Delta A = 4\pi k, \quad \text{and} \quad S = \frac{kc^3}{4\hbar G}A, \tag{5-95}$$

where the expression on the right is obtained by integration of the equation on the left.

Specialists in general relativity often use units in which k, G, $\hbar$, and c all are arbitrarily set equal to unity. In such units, the entropy of a black hole can be seen to be one quarter its area: $S = A/4$.

5:23 The Third Law of Black Hole Thermodynamics

In addition to the zeroth, first, and second laws of thermodynamics, there is a third law. In nonrelativistic physics we have known for a long time that it is possible to lower the temperature of a system to approach absolute zero, but never to quite reach it. In black hole thermodynamics, where we have associated temperature with surface gravity, the temperature becomes zero for a black hole that is rotating so rapidly that it is on the verge of blowing apart and has no surface gravity.

A rigorous relativistic calculation finds that the maximum spin angular momentum that a black hole of any mass M can have is $J_{max} = GM^2/c$. An assembly of particles with angular momentum exceeding J_{max} would be unable to collapse to form a black hole.

PROBLEM 5–19. In Chapter 7 we will see that the rotational angular momentum states of any physical object are discrete and separated by increments $\hbar$. For a black hole, whose lowest angular momentum state is zero, the next highest angular momentum states are $\pm\hbar$. Consider a sphere whose mass m_0 is uniformly spread over its spherical surface of radius b. If it is rotating with equatorial velocity v, its angular momentum is $J < m_0 b v$. Show that, for the Planck mass (5–89), v would have to become relativistic to produce even the lowest excited rotational angular momentum states $J = \pm\hbar$.

From Problem 5–19, we see that the angular momentum of a Planck-mass black hole may be either $-\hbar$, 0 or $+\hbar$. However, the third law of thermodynamics tells us that the maximum rotation rate cannot be reached in a finite number of steps. We can, therefore, be all but certain that the particle of Planck mass m_P which we dropped from rest into the black hole, in Section 5:20, would have zero angular momentum.

PROBLEM 5–20. A mass m_P may be endowed with electric charge. Show that if the charge exceeds 11 electron charges, a black hole of mass m_P could not form, because the electrostatic repulsion would exceed the gravitational attraction of matter: $e^2/m_0^2 G > 1$.

5:24 Radiating Black Holes

For a long time, black holes were thought to absorb all incident particles and radiation without ever emitting any in return. This view has now changed. The idea that black holes might radiate was first proposed by Stephen Hawking who showed that these holes would emit blackbody radiation (Ha75).

The theory is complicated, but can be roughly understood in broad terms. Suppose we accept the assumption that a black hole emits blackbody radiation at the temperature given by (5–94). A blackbody radiates most of its power at wavelengths λ for which the product of wavelength and temperature is

$$T\lambda \sim 1/2 \text{ cm K} . \tag{5-96}$$

Dimensionally, we know that $kT \sim h\nu = hc/\lambda$, and we would be tempted to set $T\lambda = hc/k$. But numerically, we find that $hc/k \sim 1.44$ cm K, and so we get a better approximation to the temperature–wavelength product if we set

$$T\lambda \sim 2\hbar c/k . \tag{5-97}$$

The wavelength at which the black hole will radiate electromagnetically is determined by the motion of charges. The black hole tidally tears apart virtual electron–positron pairs in the ambient vacuum, where they are continually generated and recombine. The member of a pair that comes too close to the hole is gravitationally captured and separated from its partner, which will orbit above the Schwarzschild surface at a velocity somewhat below the speed of light until it finds another partner to recombine. The orbital frequency of this motion is $\nu \lesssim c/2\pi r_s$, where r_s is the Schwarzschild radius. The wavelength of radiation emitted at this frequency is $\lambda = c/\nu \gtrsim 4\pi MG/c^2$. With this wavelength we obtain a coarse estimate of the black hole temperature

$$T \lesssim \frac{\hbar c^3}{2\pi k MG} . \tag{5-98}$$

This roughly agrees with the actual temperature given in (5–94). ——————————

PROBLEM 5–21. Using the result (5–94) show that the temperature of a black hole is

$$T \sim 1.6 \times 10^{-7} \frac{M_\odot}{M} \text{ K}. \tag{5-99}$$

For black holes as massive as a star, the temperature is negligibly low and the black hole hardly radiates at all. But primordial black holes that might have been formed in the earliest moments of the Universe, with masses $\ll 10^{15}$ g, would long ago have radiated away their mass — slowly at first, then progressively faster. Initially, the radiation is confined to electromagnetic and gravitational waves and any other massless species. As the black hole loses mass, its temperature rises, it loses mass increasingly rapidly, and finally explodes in a flash of high-energy radiation. In the final moments, when the temperature is sufficiently high to produce electrons and positrons, and eventually the more massive mesons and baryons, such particles and their antiparticles can also be expelled.

PROBLEM 5–22. The luminosity of a sphere emitting blackbody radiation is given by equation (4–78). Assume that a black hole emits energy only in the form of blackbody radiation. Substituting the temperature (5–94) and the Schwarzschild radius into this expression, show that the luminosity $L_\bullet$ is given by

$$\frac{L_\bullet}{c^2} = -\frac{dM_\bullet}{dt} = \frac{4\pi\sigma}{c^2} \left(\frac{\hbar c^3}{8\pi k M_\bullet G} \right)^4 \left(\frac{4M_\bullet^2 G^2}{c^4} \right), \tag{5-100}$$

and that the time τ required to radiate away the entire mass is

$$\tau \sim 10^{-25} M_\bullet^3 \text{ s}, \tag{5-101}$$

where $M_\bullet$ is given in grams. Because the Universe is $\sim 4 \times 10^{17}$ s old, primordial black holes with an original mass $\sim 1.6 \times 10^{14}$ g would be approaching this explosion phase now. In its final moments, when the black hole mass had dropped to one ton (10^6 g), the hole would explode in one tenth of a microsecond, and emit enormously energetic particles.

When gamma-ray bursts were first observed, attempts were made to determine whether we were detecting the final outbursts of such primordial black holes. We now know that gamma bursts do not show increasingly high-energy gamma rays toward the end of a burst. This tells us that none of the bursts observed to date represents a primordial black hole. But more massive black holes could have been created in the early phases of the Universe. If they were substantially more massive than 10^{15} g — only a tiny fraction of the mass of the Earth — they would still be radiating so slowly as to be undetectable. Their presence would be observable only through their gravitational deflection of light or massive bodies.

Answers to Selected Problems

5–3. From equation (5–10) we obtain

$$v = \frac{c(2 - \delta - \Delta)}{(2 - \delta - \Delta + \Delta\delta)} < c.$$

For counterpropagating velocities we obtain an expression of the form

$$-c < v = \frac{(\Delta - \delta)c}{\delta + \Delta - \delta\Delta} = \frac{(\Delta - \delta)c}{(\Delta - \delta) + (2 - \Delta)\delta} < c.$$

5–4. Letting $c \to \infty$, $x = x' + Vt'$, $y = y'$, $z = z'$, $t = t'$, and

$$v_x = v'_x + V, \; v_y = v'_y, \text{ and } v_z = v'_z.$$

5–6.

$$\frac{dL}{dt} = -\frac{L}{R^2} \cdot \left(\frac{L_\odot}{4\pi}\right) \cdot \frac{\sigma_g}{(mc^2)}.$$

.

Because $L = mvR$ and $v^2/R = GM_\odot/R^2$,

$$\frac{dL}{L} = \frac{dR}{2R} \text{ and}$$

$$t = \left[\frac{2\pi}{L_\odot} \cdot \frac{mc^2}{\sigma_g}\right] \int_{R_\odot}^{1 \text{ AU}} R \, dR = 6 \times 10^3 \text{ yr}.$$

5–7. From Problem 5–6,

$$\frac{dL}{L} = -\frac{1}{m}\left(\frac{L_\odot \sigma_g}{4\pi R^2 c^2}\right) dt.$$

When the mass of the scattered radiation equals the grains' mass, $dL/L_0 = dR/2R = -dt$, and $R/R_0 = e^{-2(t-t_0)}$, where the subscript denotes, respectively, the initial orbital angular momentum, orbital radius, and time.

Because the light scattered in one second has a mass–energy of $10^{-8}L_\odot/c^2 \sim 4 \times 10^{-4}$ g, any grain suffers an exponential loss of its angular momentum and an exponential reduction in orbital radius per second, irrespective of its radial distance from the Sun. This provides a useful estimate of the rate at which matter falls into the Sun.

5–8. $\mathcal{E} = \gamma(V)m_0c^2$. Because $\mathcal{E} = 10^{20}$ eV, we see that $\gamma(V) = 10^{11}$ and $\Delta x' = \Delta x/\gamma(V) = 3 \times 10^{20}/10^{11} \sim 3 \times 10^9$ cm.

5–9. For a mean rest-frame life $\Delta t = 885$ s, $\Delta t' = \gamma(V)\Delta t \sim 10^{14}$ s for $\gamma(V) \sim 10^{11}$. At $v \sim c$, the distance the neutron can travel is $\sim 3 \times 10^{24}$ cm $= 1$ Mpc.

5–12. The first result follows from equation (5–79), which has a point of inflection with only a single solution with $12M^2G^2/c^2L^2 = 1$. The second result is derived from (5–75).

5–13. The orbital period in circular motion is $P = 2\pi(dt/d\phi) = (2\pi/c)(\dot{t}/\dot{\phi})$. From (5–71) to (5–74) we see that

$$P = 2\pi \frac{dt}{d\phi} = \frac{2\pi\mathcal{E}}{c^2} \frac{r^2}{L\left(1 - (2MG/rc^2)\right)} \ .$$

For circular motion the left side of equation (5–75) equals zero, and provides a value for $\mathcal{E}$ in terms of M and L which, on substitution into the equation for P, leads to (5–82).

5–15. The maximum kinetic energy the proton attains just before a direct impact onto the neutron star surface is $\sim$200 MeV. The maximum energy of any photons produced on impact must therefore lie at $\leq$200 MeV, equivalent to a temperature of $\sim$2.3 $\times$ 10^{12} K. The radiation emitted by material falling onto a neutron star or its accretion disk from a binary companion is generally observed at X-ray energies in the kilovolt range, i.e., at energies roughly four orders of magnitude lower. This indicates, in part, that the impact energy of infalling matter is distributed over a sizeable area before being radiated away, but also results from the material having too high an angular momentum to fall directly onto the neutron star. Instead it first falls onto the accretion disk orbiting the star, and then dissipatively spirals in, falling onto the star's surface the shorter distance from the inner edge of the disk.

5–18. For a mass $m_0b/2r_s$ added to an initial mass M, the area of the black hole increases to

$$A = 4\pi \left[\frac{2G}{c^2} \left(M + \frac{m_0b}{2r_s} \right) \right]^2 = 4\pi \left(r_s + \frac{m_0b}{M} \right)^2 ,$$

which leads to $A \sim 4\pi \left[r_s^2 + (2m_0Gb/c^2) \right]$ and $\Delta A = 8\pi m_0bG/c^2$.

6 Electromagnetic Processes in Space

6:1 Coulomb's Law and Dielectric Displacement

In earlier chapters we noted the similarities between Coulomb's law for the attraction of charged particles and Newton's law for the attraction of masses. Both are inverse square law forces. Coulomb's law states that the attraction between two charges q_1 and q_2 is proportional to the product of the charges, inversely proportional to the square of the distance between them, and lies along the direction separating the charges.[1]

$$\mathbf{F} = \left(\frac{q_1 q_2}{r^3}\right) \mathbf{r} \, . \tag{6-1}$$

The charges q can be either positive or negative. Where a large number of separate charges exert a force on a given charge q, the total force is the vector sum of individual terms of the form of equation (6–1).

$$\mathbf{F} = q \sum_i \left[\frac{q_i}{r_i^3}\right] \mathbf{r}_i \tag{6-2}$$

and we can define an *electric field* **E**,

$$\mathbf{E} = \frac{\mathbf{F}}{q} \, , \tag{6-3}$$

which can be considered the seat of the force. All this assumes that the charges q_i and q are at rest in vacuum. If the charges q_i are moving, the charge q will experience an additional magnetic force, and if the charges are not in vacuum but in a polarizable dielectric material, the material will adjust itself to cancel out some of the force. The actual force acting on q then becomes less than that given in equation (6–2).

To clearly specify this we define a vector *dielectric displacement* **D**, independent of the properties of the material in which the charges are embedded. **D** is strictly a geometric quantity and specifies the field that would be obtained if all charges were in a vacuum. In the presence of a uniform dielectric, equation (6–2) becomes

---

[1] A number of different conventions on units are in common use. Throughout, we will make use of electrostatic units of charge, esu, which set the proportionality constant in equation (6–1) equal to unity when the force is measured in dynes and the separation in centimeters.

$$\mathbf{F} = \frac{q}{\varepsilon} \sum_i \frac{q_i}{r_i^3} \mathbf{r}_i \; . \tag{6-4}$$

Equation (6–3) still holds true since it defines electric fields; but the dielectric displacement now becomes

$$\mathbf{D} = \varepsilon \mathbf{E} \; , \tag{6-5}$$

which is seen to be independent of ε and dependent only on the positions and magnitudes of the charges, that is, on the quantities q_i and $\mathbf{r}_i$ for most real materials. The *dielectric constant* ε, which is a dimensionless quantity, can be taken to be independent of $\mathbf{E}$ for field strengths below a critical value.

The displacement produced by a charge q_1 at distance r is

$$\mathbf{D} = \left(\frac{q_1}{r^3} \right) \mathbf{r} = \left(\frac{4\pi q_1}{4\pi r^3} \right) \mathbf{r} \; , \tag{6-6}$$

so that

$$\mathbf{D} \cdot \mathbf{n} = \left(\frac{4\pi q_1}{r A} \right) \mathbf{r} \cdot \mathbf{n} \; , \tag{6-7}$$

where $\mathbf{n}$ is the normal to the surface at point $\mathbf{r}$ and A is the total area of the enclosing surface. The dots denote a *scalar product*. If so many charges are involved that the charge distribution becomes continuous, a more general form of expression (6–7) is applicable:

$$\int \mathbf{D} \cdot ds = \int 4\pi \rho \, dV = \int \nabla \cdot \mathbf{D} \, dV \; , \tag{6-8}$$

where ρ is the *charge density*, and the last equality is obtained from *Gauss's theorem* on vector integration which states that for an arbitrary vector $\mathbf{X}$,

$$\int \mathbf{X} \cdot ds = \int \nabla \cdot \mathbf{X} \, dV \; , \tag{6-9}$$

where $\nabla\cdot$ is the *divergence operator*. The integral on the left is a *surface integral*, and ds is an element of the surface over which the integration takes place.

One may wonder why we emphasize the relation between $\mathbf{D}$ and $\mathbf{E}$ in such detail when we have set out to discuss electromagnetic processes in space. We might expect that the emptiness of the cosmos would assure that $\mathbf{D}$ and $\mathbf{E}$ are always identical. This is not quite true; much of our knowledge of the contents of interstellar space depends on small differences between $\mathbf{E}$ and $\mathbf{D}$. We define one more quantity that will be useful later. It is the *polarization field* $\mathbf{P}$, which is a measure of the difference between the displacement and electric fields:

$$\mathbf{P} = \frac{[\mathbf{D} - \mathbf{E}]}{4\pi} = \frac{(\varepsilon - 1)\mathbf{E}}{4\pi} \; . \tag{6-10}$$

The field set up through the rearrangement of charges in the polarizable material is $4\pi \mathbf{P}$. It tends to oppose the externally applied field, reducing its value from $\mathbf{D}$ to $\mathbf{E}$. The factor 4π introduced here is a matter of convention and has the following significance: . At a plane boundary, with charge density σ per unit area, **D** just equals $4\pi\sigma$. The polarization field, instead, will depend on σ', the induced charge density per unit area. Now, **P** is the electric dipole moment per unit volume. If this volume contains n dipoles having charge q and separation d, then $\mathbf{P} = nqd$. The charge density σ' then is nqd also, because we can visualize a cube of unit volume, made up of d^{-1} dipole layers each of thickness d and containing nd dipoles. This makes P numerically equal to σ' — no factor of 4π occurs.

Thus far we have acted as though static fields perhaps were important on a scale of cosmic dimensions. This probably is not true. In a near vacuum electric charges generally can quickly rearrange themselves into a configuration where all electric fields are neutralized, that is, into a charge-neutralized configuration where any small volume element contains essentially the same number of positive and negative charges. The dimensions of such volumes are given by the Debye shielding length discussed previously in Section 4:23. There is one exception to this general rule, and it is important. We will show in the next section that electric charges are tied to magnetic field lines in space. If an electric field is applied perpendicular to the direction of a cosmic magnetic field the charges cannot flow across the magnetic field lines to neutralize the electric field. Large-scale electric fields may then persist.

6:2 Cosmic Magnetic Fields

An electric charge q traveling through a cosmic magnetic field experiences a force **F** called the *Lorentz force*:

$$\mathbf{F} = \frac{q\mathbf{v} \wedge \mathbf{B}}{c}, \tag{6-11}$$

where **v** is the velocity of the charge, **B** is the *magnetic flux density* also called the *magnetic induction* and c is the speed of light.[2] The *cross product* in equation (6–11) shows that the force, and hence the acceleration experienced by the charge, is perpendicular to both the velocity and the direction of the magnetic field. The charge therefore spirals (see Fig. 6.1) along the magnetic field lines without changing energy. (To do work on the particle we would require a force that has some component along the direction of motion). In a constant magnetic field, the particle describes a helical motion with constant pitch. The velocity component v_z along the direction of the field is a constant of the motion, and the circular velocity v_c about the field lines then defines a *pitch angle* θ so that

$$\tan\theta = \frac{v_c}{v_z}. \tag{6-12}$$

The *gyroradius* or *Larmor radius*, R_L, of this motion is easily obtained by setting the magnetic force equal to the centrifugal force acting on the particle. If the particle has

---

[2] In empty space, the flux density **B** equals the *magnetic field* **H** defined, further on, by equation (6–17). Because of this equality, astrophysicists often refer to **B** as the "magnetic field." We will also do this but, strictly speaking, it is incorrect. The unit of flux density is the *gauss, G*.

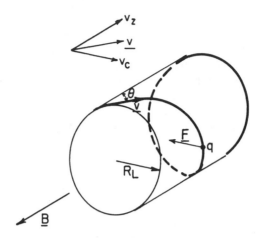

Fig. 6.1. Diagram to illustrate spiral motion in a magnetic field.

transverse momentum p_c and *gyrofrequency* $\omega_c = v_c/R_L$, the force has magnitude

$$\dot{p} = \frac{p_c v_c}{R_L} = \frac{qBv_c}{c}, \quad R_L = \frac{p_c c}{qB}, \quad \text{and} \quad \omega_c = \frac{v_c}{R_L} = \frac{qBv_c}{p_c c}. \qquad (6\text{-}13)$$

The gyrofrequency is sometimes also called the *cyclotron frequency*.

PROBLEM 6–1. Show that the Larmor radius of a proton moving at $10\,\text{km s}^{-1}$ through a field of $10^{-6}\,\text{G}$ is small compared to interstellar and even interplanetary distances. The unit of field strength, G, is the *Gauss* named after the nineteenth century German mathematician and scientist Karl Friedrich Gauss.

Because the Larmor radius is small compared to the expected dimensions of interstellar and interplanetary fields, charged particles moving with thermal velocities characteristic of cosmic gases are effectively tied to the magnetic field lines. They can move along the field lines but cannot cross them any appreciable distance. We say that the particles are "frozen" to the field and the motion of such particle-field combinations is called *frozen-in flow. Magnetohydrodynamics* is the subject that deals with problems arising from such flows (Co57).

We notice that the only way a charged particle can escape from being frozen to the lines of force is through an encounter with another particle. Each particle then assumes a completely new orbit. If such collisions are sufficiently frequent, the particles can diffuse across magnetic fields.

Inasmuch as cosmic magnetic fields have their origins in the organized motion of charged particles, the frozen-in flow is not only due to the presence of a magnetic field, but also maintains the field that causes it. This self-consistent motion of charges is not an obvious result, but magnetohydrodynamics shows that it is real.

The collisional processes just mentioned, therefore, conspire not only to prevent freezing-in, but also as a consequence tend to destroy the magnetic fields that are maintained by the frozen-in flow. For this reason, frozen-in fields cannot be maintained in dense gases where collisions are frequent. Collisions produce electrical resistance, dissipating particle motions and the energy resident in the magnetic field. Frozen-in flow therefore has a short life in a dissipative medium (Sp62).

Magnetohydrodynamics also tells us that the presence of a force, such as a gravitational or electrostatic force acting normal to the magnetic field, can produce a drifting motion in which charges move in directions perpendicular both to the applied force and to the magnetic field direction. Such particle drifts occur in the *Van Allen belts* of charged particles that constitute part of the Earth's *magnetosphere*. Drifts, however, do not directly act to dissipate cosmic magnetic fields unless the drifting particles suffer collisions.

6:3 Ohm's Law and Dissipation

A current generally consists of two types of terms. The first expresses the actual flow of charge in response to an applied electric field. The second corresponds to a virtual current representing a change in the applied field. This change gives rise to a magnetic field (see Section 6:5) just as a moving charge would. It is genuinely important. We write the electric current as

$$\mathbf{j} = \rho \mathbf{v} + \frac{1}{4\pi} \frac{\partial \mathbf{D}}{\partial t} \ . \tag{6-14}$$

The value of the velocity $\mathbf{v}$ in this equation is determined by two competing effects. The applied electric field seeks to continuously accelerate the charge, whereas distant collisions with other electric charges continually seek to slow the particle down. The resistivity of the medium is a measure of this slowing down. Its reciprocal is the *conductivity* σ (not to be confused with the charge density per unit area discussed in Section 6:1). In terms of $\mathbf{E}$ and σ, equation (6–14) can be written as

$$\mathbf{j} = \sigma \mathbf{E} + \frac{1}{4\pi} \frac{\partial \mathbf{D}}{\partial t} \ . \tag{6-15}$$

The dimension of σ is s^{-1}, as can be seen from (6–5) and (6–15), because ϵ is dimensionless. In general, the conductivity depends on the density of the gas, its temperature, state of ionization, and chemical composition. Distant collisions predominate in slowing down the motion of charged particles in space, and they determine the value of σ. This follows quite generally from Sections 3:13 and 3:14, but will be more explicitly shown in Section 6:18.

6:4 Magnetic Acceleration of Particles

One of Michael Faraday's nineteenth century contributions to electromagnetism was his discovery that a time-varying magnetic field gives rise to electric currents in a conducting medium encircling the field. The plane in which this current flows is perpendicular to the direction of the time-varying field component. In integral form *Faraday's law* is expressed as

$$\frac{1}{c}\frac{\partial}{\partial t}\int \mathbf{B} \cdot ds = - \oint \mathbf{E} \cdot d\mathbf{l}, \tag{6-16}$$

where the integral on the left is a surface integral over the area enclosed by the loop through which the current flows (Fig. 6.2 (a)). The integral on the right is a

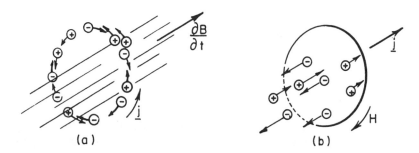

Fig. 6.2. Illustrations for Faraday's and Ampère's laws. (a) Faraday's law states that the current in a conducting loop, and the associated electric field, are determined by the rate of change in the number of magnetic *lines of force* enclosed by the loop (see equations (6–15) and (6–16)). The number of lines of force crossing unit area is proportional to the magnetic field, **B**. (b) Ampère's law states that the magnetic field integrated along a loop enclosing a current is determined by the total current crossing the enclosed area (see equation (6–17)).

line integral taken over that loop and the current observed by Faraday has been replaced by the electric field that gives rise to it in accordance with equation (6–15). We note now that, if any region of interstellar space should suddenly be subjected to a rising magnetic field, electric charges would experience an effective electrical field **E** proportional to the time rate of change of **B**. In the laboratory this effect is used to elevate charges to very high energies. The first device that successfully accomplished this acceleration was the betatron constructed by D. W. Kerst in 1940.

A rapid rise in magnetic field strength can result from the compression of a cosmic cloud in a direction perpendicular to its magnetic field. Such a compression can occur in the collision of interstellar clouds, either with one another, or with high-velocity gases ejected from hot stars or exploding supernovae. This process may produce low-energy cosmic rays, sometimes called *suprathermal particles*. It is not sufficiently powerful to produce extremely energetic particles. More effective mechanisms for accelerating particles are discussed in Section 6:6 below.

PROBLEM 6–2. Suppose the magnetic field in a region of space increases from 10^{-6} to 10^{-5} G over a period of 10^7 yr. To what energy would nonrelativistic electrons and protons be accelerated if they moved perpendicular to the field and suffered no collisions? How does the final energy depend on the initial energy? To do this, it is useful first to derive the energy–field relationship $d\mathcal{E}/\mathcal{E} = dB/B$ that follows from (6–13), and to recall the conservation of angular momentum of the circling particles as B changes.

6:5 Ampère's Law and the Relation Between Cosmic Currents and Magnetic Fields

In Section 6:2 we noted that cosmic magnetic fields exist by virtue of the gyrating electric charges frozen to the field. This idea is more precisely expressed by *Ampère's law* which states that a current produces an encircling magnetic field (Fig. 6.2 (b)):

$$\frac{4\pi}{c} \int \mathbf{j} \cdot d\mathbf{s} = \oint \mathbf{H} \cdot d\mathbf{l} . \tag{6-17}$$

Here again the left side of the equation is a surface integral taken over the entire surface encircled by the magnetic field in the line integral on the right.

We believe that cosmic magnetic clouds are configurations in which equation (6–17) is obeyed in every locale. The shapes of the magnetic fields and currents are therefore likely to be quite complicated. We can think of initial configurations called *force-free magnetic fields* in which the magnetic fields and the flow of charges are so arranged that no forces result to destroy the configuration. Such structures must have $\mathbf{j} \wedge \mathbf{B} = 0$ everywhere. Force-free configurations may well represent the structure of cosmic magnetic fields.

6:6 Magnetic Mirrors, Magnetic Bottles, and Cosmic-Ray Particles

In Section 6:4 we noted that a betatron accelerates charged particles. A different scheme for magnetically accelerating cosmic-ray particles was suggested by Enrico Fermi. In *Fermi acceleration* the cosmic-ray particles are thought to travel between cosmic gas clouds. Each cloud has an embedded magnetic field. When a particle approaches the cloud and enters its field perpendicular to the field direction, it is turned back by virtue of the magnetic force given by equation (6–11). For, after traveling in a semicircle, the particle once again finds itself at the edge of the cloud and headed into the direction from which it came.

As shown in Fig. 6.3, a similar reflection can occur for particles approaching a cloud along the lines of force. If the particle impinges on a cloud that is receding from it, the particle's momentum after encounter is smaller than before. If the particle impinges on an approaching cloud, its final momentum is higher than before the collision. In general the probability for collision is greater for an approaching than for a receding cloud. (This corresponds to everyday experience. On a highway

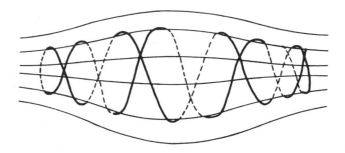

Fig. 6.3. Charged particle trajectories in a magnetic bottle. Light lines denote magnetic lines of force. A high density of field lines indicates a strong magnetic field.

we pass more cars going in the opposite direction than going along the direction in which we are traveling.)

Statistically, therefore, particles will derive increased momentum from encounters with clouds and can be accelerated to high energies. The process is similar to the acceleration of a ping-pong ball between two slowly approaching paddles. After the ball has made many bounces off each paddle, it is going far faster than either of these reflecting surfaces.

Generally, a charged particle moves along the lines of force of a magnetic field, spiraling as it goes. The pitch angle is given by equation (6–12). If the particle encounters a region of the magnetic field where the lines are more compressed, it experiences an increase in the field strength and by Faraday's law (6–16) its circular velocity v_c increases. However, because the field itself is not doing any work on the charge, the increase in v_c must be bought at the expense of kinetic energy initially resident in the longitudinal motion, that is, at the expense of a reduction in v_z. When the particle has advanced into the intense magnetic field to such a depth that all its kinetic energy is spent in circular motion, the pitch angle θ becomes $\pi/2$; the particle is reflected and spirals back out of the intense field.

As the particle first spirals into the intensifying magnetic field, its angular momentum about the axis of symmetry of the motion is conserved. Hence the *magnetic moment* **M:**

$$\mathbf{M} = \frac{\mathbf{j} \wedge \mathbf{r}}{2c}, \quad \mathbf{j} = q\mathbf{v} , \tag{6-18}$$

is also conserved. Substituting the gyroradius,, equation (6–13), for r we find

$$M = \frac{v_c p_c}{2B} \tag{6-19}$$

along the direction of the magnetic field. From this it follows that the transverse kinetic energy is directly proportional to the field B. If a particle has an initial pitch angle θ in a field B, it can only penetrate the field until it reaches a region where the field is B_0 and $\sin \theta = 1$:

$$B_0 = \frac{B}{\sin^2 \theta} . \tag{6-20}$$

Here it is reflected and spirals back out of the intense field.

A *magnetic bottle* consists of two such *magnetic mirrors* between which a particle is reflected going back and forth without possibility of escape. The Fermi mechanism ping-pong acceleration could involve a (shrinking) magnetic bottle in which the two magnetic mirrors approach.

We sometimes characterize cosmic-ray particles by a *magnetic rigidity BR_L* which equals pc/q for motion strictly perpendicular to the field (equation (6–13)). The rigidity has dimensions of energy per charge.

PROBLEM 6–3. Consider an interstellar cloud moving with velocity V. It acts as a magnetic mirror so that a particle suffers a change in speed $\Delta V = 2V$ added to its own initial velocity in any reflection off the cloud. With two approaching clouds a succession of collisions can occur. Using the law of composition of velocities compute how many collisions a proton with initial energy $\mathcal{E}$ would need in order to double its energy. Let $V = 7\,\text{km s}^{-1}$, typical of interstellar cloud velocities, let the distance between approaching clouds (magnetic mirrors) be of the order of 10^{17} cm, and let $\mathcal{E} = 10^{10}$ eV. How long would it take to double the particle's energy? Is this time appreciably different for protons and electrons? Is there a maximum to the energy increase this process permits?

A more likely variant of the Fermi process has come to be widely accepted. The particles are accelerated by successive reflections off a rapidly expanding supernova ejection shell. After each impact the accelerated particle returns — deflected back toward the shell by an external magnetic field into which the shell is expanding. A succession of reflections off the expanding shell progressively accelerates the particle to cosmic-ray energies. This variant has also come to be referred to as *Fermi acceleration*.

A cosmic-ray particle must eventually suffer destruction due to one of several competing processes — inelastic impact on another particle, loss from the Galaxy, and hence loss from contact with the accelerating clouds, and so on. The number of accelerating reflections between destructive events is thus limited. To reach truly high energies, therefore, cosmic-ray particles must be injected into the accelerating fields with rather high initial energies. Sufficiently energetic particles may be produced in supernova explosions. But the highest energy cosmic rays appear to reach the Galaxy from extragalactic sources.

The most energetic particles, having energies of order 10^{20} eV, have a gyroradius that exceeds the Galaxy's dimensions. These particles could therefore not be retained in the Galaxy for any length of time. If their origin were local they should appear to come primarily from the central regions of the Galaxy or the Milky Way plane. Instead, many come from other directions and are therefore believed to be extragalactic (Bi97).

We know that the heavy nuclei, which form an abundant part of the cosmic-ray flux, would suffer destructive collisions during the long stay in interstellar space required by the rather slow Fermi acceleration mechanism. Yet we find iron nuclei to be abundant at least up to energies of the order of 10^{12} to 10^{13} eV (Sw93). We are driven toward a mechanism that could accelerate these particles rapidly.

Currently, quasars, the active nuclei of galaxies, gamma-ray bursts, pulsars, magnetars and supernova explosions slamming into the interstellar medium, or a fast stellar wind, are all considered to be likely sources for producing cosmic rays at various energies. Further observations should lead to a clarification of these hypotheses. We can also hope that the study of solar flares, which are responsible for the solar cosmic-ray component, will give us a better understanding of at least one mechanism for accelerating energetic particles.

PROBLEM 6–4. Pulsars have been considered the source of at least some cosmic-ray particles. The particles are thought to be accelerated by magnetic lines of force that co-rotate with the central neutron star (Fig. 6.4). Suppose that the field velocity

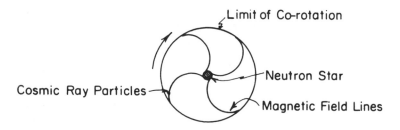

Fig. 6.4. Cosmic-ray acceleration near a neutron star.

is simply ωr, where ω is the star's angular velocity and r is the radial distance from the star. Consider charged particles to be dragged along, frozen to the magnetic field lines. What is the energy of the particles, then, as a function of radial distance if special relativistic physics is approximately valid in this problem? Beyond what radial distance can the particles and magnetic field not co-rotate? How does this compare to the neutron star radius?

6:7 Maxwell's Equations

Four equations of electromagnetism allow us to derive all classical electromagnetic effects. They are

$$\nabla \cdot \mathbf{D} = 4\pi\rho \quad \text{(see equation (6–8))}, \tag{6-21}$$

$$\nabla \wedge \mathbf{E} = -\frac{1}{c}\frac{\partial \mathbf{B}}{\partial t} \quad \text{(equivalent to equation (6–16))}, \tag{6-22}$$

$$\nabla \wedge \mathbf{H} = \frac{4\pi}{c}\mathbf{j} \quad \text{(equivalent to equation (6–17))}, \tag{6-23}$$

and finally

$$\nabla \cdot \mathbf{B} = 0 \quad \text{(isolated magnetic charges do not exist).} \tag{6-24}$$

Comparing (6–24) with (6–21) we can see that it is telling us that *magnetic monopoles*, isolated magnetic charges analogous to electric charges, do not exist in Nature. Only magnetic dipoles and higher multipole configurations occur. Despite this statement, a search for magnetic monopoles has gone on ever since Paul Dirac (Di31) pointed out that quantization of the electron's charge could be understood if a few or even only one such magnetic monopole existed in Nature. Thus far, no *Dirac monopoles* have been found. Monopoles may, however, have existed in the first few moments in the evolution of the Universe, when it was $\lesssim 10^{-35}$ s old.

The four Maxwell equations generally must be supplemented by four auxiliary expressions:

$$\mathbf{D} = \varepsilon \mathbf{E}\,, \tag{6-5}$$

$$\mathbf{B} = \mu \mathbf{H}\,, \tag{6-25}$$

$$\mathbf{j} = \sigma \mathbf{E} + \frac{1}{4\pi}\frac{\partial \mathbf{D}}{\partial t}\,, \tag{6-15}$$

$$\nabla \cdot \mathbf{j} = 0\,. \tag{6-26}$$

Equation (6–25) expresses a relation between the magnetic vectors $\mathbf{B}$ and $\mathbf{H}$ which is similar to that between $\mathbf{D}$ and $\mathbf{E}$. The *magnetic permeability* μ can have values greater than or less than unity, depending on whether the medium is *paramagnetic or diamagnetic*. In most cosmic gases $\mu = 1$, for all practical purposes, but paramagnetic grains in interstellar space appear to be responsible for the observed polarization of starlight (see Section 9:13). Equation (6–26) states that currents in the sense defined by equation (6–15) are continuous, having no sources or sinks.

6:8 The Wave Equation

From equations (6–22) and (6–23) and from the relations (6–15) and (6–25), we can obtain the expression

$$\begin{aligned}
\nabla \wedge (\nabla \wedge \mathbf{E}) &= -\frac{1}{c}\frac{\partial}{\partial t}\nabla \wedge \mathbf{B} \\
&= \frac{-4\pi\mu}{c^2}\frac{\partial}{\partial t}\left(\sigma \mathbf{E} + \frac{\varepsilon}{4\pi}\frac{\partial \mathbf{E}}{\partial t}\right),
\end{aligned} \tag{6-27}$$

provided the dielectric constant ε and permeability μ do not vary with time and μ is scalar. Both μ and ε generally can be tensor quantities but they frequently act as scalars. Let us use the identity

$$\nabla \wedge (\nabla \wedge \mathbf{E}) = \nabla(\nabla \cdot \mathbf{E}) - \nabla^2 \mathbf{E} , \qquad (6\text{-}28)$$

where the operator $\nabla^2 \equiv \nabla \cdot \nabla$ is called the *Laplacian*, sometimes written as Δ. If we consider only regions in which the space charge is neutral, then $\nabla \cdot \mathbf{E} = 0$ and

$$\nabla^2 \mathbf{E} = \frac{\mu\varepsilon}{c^2} \frac{\partial^2 \mathbf{E}}{\partial t^2} + \frac{4\pi\mu\sigma}{c^2} \frac{\partial \mathbf{E}}{\partial t} . \qquad (6\text{-}29)$$

In a nonconducting medium $\sigma = 0$ so that

$$\nabla^2 \mathbf{E} - \frac{\mu\varepsilon}{c^2} \frac{\partial^2 \mathbf{E}}{\partial t^2} = 0 . \qquad (6\text{-}30)$$

This is the equation for waves propagating with speed

$$V = \frac{c}{\sqrt{\mu\varepsilon}} . \qquad (6\text{-}31)$$

PROBLEM 6–5. Derive a similar expression for the magnetic field:

$$\nabla^2 \mathbf{H} - \frac{\mu\varepsilon}{c^2} \frac{\partial^2 \mathbf{H}}{\partial t^2} = 0 , \qquad (6\text{-}32)$$

paying particular attention to the limitations imposed on ε, μ, and σ in arriving at this result.

We have set $\sigma = 0$ in equations (6–30) and (6–32). But it is important to note that conductivity is a frequency-dependent quantity. At optical and even at radio frequencies σ usually is very low. Certainly, at optical frequencies the wavelength of the electromagnetic wave is short compared to the distance between charges, and the wave readily propagates through a vacuum. At the longest radio wavelengths a transition occurs: charges in the medium can respond to the electric and magnetic fields of a propagated wave, and the conductivity rises. When the second term on the right of equation (6–29) dominates, the expression assumes the form of a diffusion equation and the wave is *damped*.

The propagating waves are *transverse* (Fig. 6.5). If the direction of propagation for a *plane wave* is the x-direction, uniformity within its plane dictates that all partial derivatives with respect to y and z are zero. Moreover, the divergence relations give

$$\frac{\partial E_x}{\partial x} = 0 \quad \text{and} \quad \frac{\partial H_x}{\partial x} = 0 , \qquad (6\text{-}33)$$

and the curl equations (6–22) and (6–23) give

$$\frac{\partial E_y}{\partial x} = -\frac{1}{c} \frac{\partial H_z}{\partial t} , \qquad \frac{\partial H_y}{\partial x} = \frac{1}{c} \frac{\partial E_z}{\partial t} ,$$

$$\frac{\partial E_z}{\partial x} = \frac{1}{c} \frac{\partial H_y}{\partial t} , \qquad \frac{\partial H_z}{\partial x} = -\frac{1}{c} \frac{\partial E_y}{\partial t} , \qquad (6\text{-}34)$$

$$0 = \frac{\partial H_x}{\partial t} , \qquad\qquad 0 = \frac{\partial E_x}{\partial t} .$$

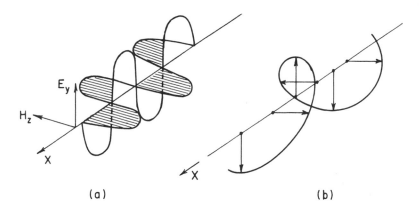

Fig. 6.5. Electromagnetic waves. (a) Wave propagating along the x-direction with electric field plane-polarized along the y-direction. (b) Circularly polarized wave propagating along the x-direction. For simplicity, only the electric field direction is shown here. The direction of the **E** vector rotates about the x-axis. The sense of rotation shown is said to be left-handed circularly polarized (LHP). Any electromagnetic wave can be constructed from a suitable superposition of left- and right-handed circularly polarized waves. Plane-polarized waves are obtained by superposing LHP and RHP waves of the same amplitude. Their relative phase determines the plane of the **E** vector (see Fig. 6.7).

If **n** is the unit vector along the direction of propagation, we see that (6–33) and equations (6–34) are satisfied by an expression of the form

$$\mathbf{H} = \mathbf{n} \wedge \mathbf{E}\,, \tag{6-35}$$

so that the **E** and **H** fields are always perpendicular and the solution of the *wave equation* (6–30) has the form

$$f_i = A\cos(2\pi\nu t - kx), \qquad i = y, z\,, \tag{6-36}$$

where $k = \sqrt{\mu\varepsilon}(\omega/c)$ and $\omega = 2\pi\nu$. Here, ν is the *frequency*, ω is the *angular frequency*, and k is the *wave number* of the wave — the number of waves per unit length along the direction of propagation. f_i represents electric and magnetic field components. The *wavelength*, $\lambda \equiv 1/k$.

6:9 Phase and Group Velocity

Let us write the equations for propagation of two waves f^- and f^+ that have angular frequencies $\omega - \Delta\omega$ and $\omega + \Delta\omega$, respectively:

$$f^- = A\cos[(\omega - \Delta\omega)t - (k - \Delta k)x]\,,$$
$$f^+ = A\cos[(\omega + \Delta\omega)t - (k + \Delta k)x]\,.$$

The superposition of these waves gives

$$f = f^- + f^+ = A\{\cos[(\omega - \Delta\omega)t - (k - \Delta k)x]$$
$$+ \cos[(\omega + \Delta\omega)t - (k + \Delta k)x]\} , \qquad (6\text{-}37)$$
$$f = 2A\cos(\omega t - kx)\cos[(\Delta\omega)t - (\Delta k)x] .$$

This means that there is a *carrier wave* frequency represented by $\cos(\omega t - kx)$ that is *amplitude modulated* by a wave $\cos(t\Delta\omega - x\Delta k)$. The carrier wave velocity is called the *phase velocity*:

$$V = \frac{\omega}{k} , \qquad (6\text{-}38)$$

and the velocity of the modulation is called the *group velocity*:

$$U = \frac{\partial\omega}{\partial k} . \qquad (6\text{-}39)$$

We will note later that U is the physically more interesting quantity. It represents the speed at which *information* can be conveyed or energy transported. As long as the medium is purely dispersive, that is, $\omega = \omega(k)$, there is no difficulty in defining U. But if the conductivity σ becomes appreciable, we have absorption, the amplitude A becomes complex, and U no longer has a clear physical meaning. For long wavelength cosmic radio waves, this kind of absorption prevents transit through the Earth's ionosphere. These long waves must then be observed from rockets or satellites taken above the atmosphere. At even longer wavelengths the interstellar medium absorbs, and such waves are not transmitted at all. We will return to this problem in Section 6:11.

6:10 Energy Density, Pressure, and the Poynting Vector

The scalar product of equation (6–22) with **H**, subtracted from the product of (6–23) with **E** is

$$\frac{1}{c}\mathbf{H} \cdot \frac{\partial\mathbf{B}}{\partial t} + \frac{\mathbf{E}}{c} \cdot \frac{\partial\mathbf{D}}{\partial t} + \frac{4\pi\sigma\mathbf{E} \cdot \mathbf{E}}{c} = -H \cdot \nabla \wedge \mathbf{E} + \mathbf{E} \cdot \nabla \wedge \mathbf{H} . \qquad (6\text{-}40)$$

Using the vector identity

$$\nabla \cdot (\mathbf{A} \wedge \mathbf{B}) - \mathbf{B} \cdot \nabla \wedge \mathbf{A} \quad \mathbf{A} \cdot \nabla \wedge \mathbf{B} \qquad (6\text{-}41)$$

we find that

$$\frac{1}{8\pi}\frac{\partial}{\partial t}(\varepsilon E^2 + \mu H^2) = -\sigma\mathbf{E}^2 - \nabla \cdot \mathbf{S} , \qquad (6\text{-}42)$$

where

$$\mathbf{S} \equiv (c/4\pi)\mathbf{E} \wedge \mathbf{H} . \qquad (6\text{-}43)$$

S is called the *Poynting vector*. If we apply Gauss's theorem (6–9) relating volume and surface integrals, (6–42) can be written as

$$\frac{\partial}{\partial t} \int \frac{\varepsilon E^2 + \mu H^2}{8\pi}\, dV = - \int \sigma E^2\, dV - \oint \mathbf{S} \cdot d\mathbf{s}\,. \tag{6-44}$$

Here the first term on the right is equivalent to the rate of change of kinetic energy of moving charges. It involves the scalar product of the force on the particles and their velocity, since $\sigma \mathbf{E}$ represents a current. Hence,

$$\int \sigma E^2\, dV \rightarrow \sum e\mathbf{v} \cdot \mathbf{E} = \sum \mathbf{v} \cdot \dot{\mathbf{p}}\,. \tag{6-45}$$

This is the time derivative of the kinetic energy summed over all particles. The other two terms in (6–44) represent the flow of electromagnetic energy. The term on the left of equation (6–44) is the rate of change of electromagnetic energy in the volume; $(\varepsilon E^2 + \mu H^2)/8\pi$ is the energy density of the fields. The second term on the right represents the flow of energy through the enclosing surface and S therefore is the *electromagnetic flux density*. Equation (6–44) states that the increase of electromagnetic energy in a volume equals the electromagnetic energy radiated away minus the increasing kinetic energy of charges enclosed in the volume.

Previously we found that the pressure P due to randomly oriented electromagnetic waves is just one-third the numerical value of the energy density. In Section 4:7 we determined this on kinetic grounds:

$$P = \frac{1}{3}\left[\frac{1}{8\pi}(\varepsilon E^2 + \mu H^2)\right]\,. \tag{6-46}$$

The case of static fields is similar except that a magnetic pressure can now exist without an accompanying electric pressure; the conductivity σ is high and a current $\sigma \mathbf{E}$ maintains the magnetic field. There exists a kinetic pressure due to the flow of charges, and this will depend on $\sigma \mathbf{E}$. The situation is further complicated because the magnetic pressure actually is a tensor quantity that depends on the orientation of the fields.

For a magnetic field there always exists a tension along the lines of force and an outward pressure perpendicular to the lines of force. We can see this in the following way. The magnetic energy density in a cube of unit dimension is $\mu H^2/8\pi$. If the cube is compressed an amount dl along a direction parallel to the field lines, the field strength remains constant but the volume decreases by dl. Because the energy density remains constant while the volume decreases, the total energy in the volume decreases by $(\mu H^2/8\pi)\, dl$. This means that the amount of work done to compress the cube is $-(\mu H^2/8\pi)\, dl$, and indicates that there is a pressure $-\mu H^2/8\pi$ along the field lines.

If the cube is compressed along a direction transverse to the field lines, the number of lines of force in the volume does not change, and a compression Δl increases the field strength to $H/(1 - \Delta l)$. The energy density now becomes $\sim(\mu H^2/8\pi)(1 + 2\Delta l)$, and because of the decrease in volume $(1 - \Delta l)$, the total energy change on compression is $\sim (\mu H^2/8\pi)\Delta l$. In this case an amount of work $(\mu H^2/8\pi)\Delta l$ must be done to compress the cube, and the pressure resisting compression is $\mu H^2/8\pi$. For a volume containing randomly directed bundles of field lines, the net effect of averaging over two transverse and one longitudinal direction is an overall outward pressure $P = (\mu H^2/8\pi)/3$.

This is the reason why difficulties arise in the problem of star formation in the presence of magnetic fields. It is relatively simple to see how matter can contract along the lines of force, but it is more difficult to understand how condensation takes place perpendicular to the direction of the field because the gases are frozen to the field lines and the pressure of the magnetic field attempts to resist any contraction. To see how severe this problem is, note that the transverse pressure is $H^2/8\pi$. Initially a typical field strength might be 10^{-6} G, so that $P_{\text{initial}} \sim 10^{-13}$ dyn cm^{-2}. As a protostar contracts from $\sim 10^{18}$ cm down to 10^{11} cm, conservation of the number of field lines requires that $H \propto r^{-2}$, so that $H^2 \propto r^{-4}$ and we would end up with a protostar having 10^8 G magnetic fields and 10^{15} dyn cm^{-2} magnetic pressures. The gravitational forces are far too weak to produce such high fields. We conclude that somehow we are looking at the problem in the wrong way. Stars manage to form despite these difficulties. We will return to this problem in Sections 10:5 and 10:6.

PROBLEM 6–6. The transverse pressure of a static magnetic field is $P_s = H^2/8\pi$; the magnetic part of the radiation pressure (6–46) is $P_r = \frac{1}{3}H^2/8\pi$. What is the significance of the factor $\frac{1}{3}$?

6:11 Propagation of Waves Through a Tenuous Ionized Medium

Consider an ionized medium without electric or magnetic fields. Let this medium be tenuous, so that collisions between ions and electrons are rare. Then, for small departures from equilibrium, electric fields in the electromagnetic wave accelerate the electrons in the medium relative to the more massive positive ions:

$$m\ddot{\mathbf{r}} = e\mathbf{E}(\mathbf{r}, t \ . \tag{6-47}$$

Here e and m are the charge and mass of the electron, and $\mathbf{E}$ is the field associated with the wave. Let the wave have the form

$$\mathbf{E}(r, t) = E_0(\mathbf{r})e^{i\omega t} \quad \text{(real part)} , \tag{6-48}$$

where only the real part will be considered. The displacement of the electron from its equilibrium position then is

$$\mathbf{r} = -\frac{e}{m\omega^2}\mathbf{E} \ . \tag{6-49}$$

This satisfies both equations (6–47) and (6–48). The displacement of the electrons effectively sets up a large number of dipoles which, as discussed in Section 6:1, give rise to a polarization field $\mathbf{P}$. If n is the number density of electrons, the polarization field is to be expressed as the sum of the individual dipole fields produced by the passing wave

$$\mathbf{P} = ne\mathbf{r} = -\frac{ne^2}{m\omega^2}\mathbf{E} \ . \tag{6-50}$$

The definition of the polarization field, equation (6–10), then tells us that the dielectric constant of the medium must be

$$\varepsilon = 1 - \frac{4\pi ne^2}{m\omega^2} \ . \tag{6-51}$$

Because the propagation phase velocity is inversely proportional to the refractive index at frequency ω, $n_\omega = \varepsilon^{1/2}$ (we can set $\mu = 1$ in all problems dealing with cosmic wave propagation), the *phase velocity* in a plasma will be greater than the speed of light! But no information and no energy is transmitted at this velocity. Therefore no violation of special relativity is involved. The more significant *group velocity* is always less than c.

If a wave propagates along the x-direction through the cosmic medium, the transverse $\mathbf{E}$ and $\mathbf{B}$ field components have the form (6–36):

$$f = f_0 \cos(kx \pm \omega t) \tag{6-52}$$

and

$$\omega^2 = \frac{k^2 c^2}{\varepsilon} = \frac{k^2 c^2}{1 - (4\pi ne^2/m\omega^2)} \ , \tag{6-53}$$

where equation (6–51) has been invoked with $\mu = 1$. This can be written as

$$\omega^2 = k^2 c^2 + \frac{4\pi ne^2}{m} \equiv k^2 c^2 + \omega_p^2 \ , \tag{6-54}$$

where

$$\omega_p \equiv \left(\frac{4\pi ne^2}{m}\right)^{1/2} \sim 5.6 \times 10^4 n^{1/2} \ , \mathrm{rad\ s}^{-1} \tag{6-55}$$

is called the *plasma frequency*. It is related to the Debye length L (see equation (4–165)) by $(mL^2/kT)^{1/2} = \omega_p^{-1}$, which is the time for an electron with a thermal velocity component $(kT/m)^{1/2}$ to cross a Debye length.

If $\omega < \omega_p$, the wave number k becomes imaginary and the wave will not propagate through the medium.

In radio-astronomy — as mentioned in Section 6:9 — observations at low frequencies cannot be carried out from below the ionosphere. Radio waves cannot be transmitted at frequencies below the ionospheric plasma frequency. This frequency varies because the electron density is not uniform. Typically, however, the cut-off is at frequencies of a few MHz. For the interstellar medium the cut-off is roughly an order of magnitude lower, as indicated in Figure 1.15.

When $\omega > \omega_p$, propagation can take place. The group velocity of the wave is

$$U = \frac{d\omega}{dk} = \frac{c}{\sqrt{1 + \omega_p^2/c^2 k^2}} \ . \tag{6-56}$$

The velocity of propagation therefore is frequency dependent. This phenomenon is important in the propagation of pulses emitted by a *pulsar* (He68a)*. If the emitted pulse contains a range of frequency components, the arrival time at the Earth will be delayed most at the lowest frequencies. We can write (6–56) as

$$U = \frac{c}{\sqrt{1 + \omega_p^2/(\omega^2 - \omega_p^2)}} .$$
(6-57)

The arrival time of a pulse that has traveled a distance D is D/U and the frequency dependence of the arrival time is

$$\frac{d(D/U)}{d\omega} \sim -\frac{D}{c}\frac{\omega_p^2}{\omega^3} \qquad \text{for} \qquad \omega \gg \omega_p .$$
(6-58)

Observations of pulsars show delays in arrival time taking the form (6–58). Because the propagated radiation frequencies must exceed the plasma frequency (6–55), they also put an upper limit on the number density of electrons in the dispersing medium. More important, however, is the conclusion that the frequency dependence of the time delay is directly proportional to Dn, the total number of electrons per unit cross-sectional area along the line of sight to the emitting object. This useful relation follows directly from (6–55) and (6–58). The integrated electron number density along the line of sight is called the *dispersion measure* $\mathcal{D}$:

$$\mathcal{D} \equiv \int_0^D n(\ell)\, d\ell = D\langle n \rangle .$$
(6-59)

If the mean number density of electrons in the interstellar medium is known, the dispersion measure (6–58) can give us the pulsar distance. Conversely, if the distance D is known from other observations, a mean value of n along the line of sight is obtained. The mean value estimated in this way would include a contribution due to any electrons surrounding the emitting region and part of the emitting object as well as true interstellar electrons. The dispersion measure within the pulsar would not be distance dependent, whereas the dispersion due to the interstellar medium would. On this basis we can distinguish the two contributions and find that the pulsar itself contributes negligibly. The dispersion measure along lines of sight leading to Galactic sources at known distances gives a mean interstellar electron density of about $\langle n \rangle = 0.03\,\mathrm{cm}^{-3}$. This value varies from source to source, depending on the number of bright, hot, ionizing stars along the line of sight. Using a mean density of electrons $\sim 0.03\,\mathrm{cm}^{-3}$, we find that the distribution of the nearer pulsars fits the distances of the nearer Galactic spiral arms (Da69). The pulsars also show a tendency to cluster close to the Galactic plane (Fig. 6.6).

Throughout this section we have assumed that the collision frequency ν_c between ions and electrons is low. However, when ν_c becomes high, energy losses through dissipation can no longer be neglected. This problem is treated in Section 6:18.

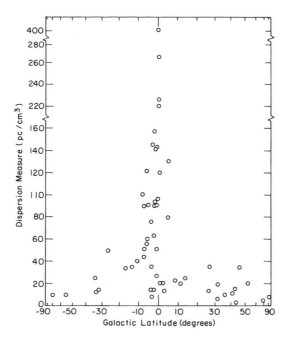

Fig. 6.6. Dispersion measure plotted against Galactic latitude for the 63 pulsars known in early 1972 (Te72). By 1997 more than 700 pulsars were known, and the number has continued to grow.

6:12 Faraday Rotation

Information about electron number densities in the cosmic medium can also be obtained from the *Faraday rotation* of a wave's plane of polarization. To understand this effect, consider an electron moving in a plane perpendicular to the direction of a magnetic field, **B**. It will be deflected by a force (6–11)

$$\mathbf{F} = \frac{e\mathbf{v} \wedge \mathbf{B}}{c} \ . \tag{6–11}$$

If the electron is also under the influence of an electromagnetic wave, it experiences a further force due to the wave's E field. Finally the gyrations under the combined influence of these fields must be balanced by an outward directed centrifugal force. The expression relating these three forces is

$$e\mathbf{E} \pm \frac{eB\omega\mathbf{r}}{c} = -m\omega^2\mathbf{r} \ , \tag{6-60}$$

where **E** is the component of the field vector perpendicular to the magnetic field, and the second term on the left has a negative sign when the electron rotates counterclockwise viewed along the direction of the **B** field. This is the motion induced by an electromagnetic wave with a right-handed circular polarization RHP propagating parallel to **B**. A left-handed circular polarization LHP gives rise to a force $+eB\omega r/c$ directed along the direction of displacement from the electron's equilibrium position. Note, however, that the value of e is negative for an electron. Solving for **r** gives

$$\mathbf{r} = -\frac{e}{m}\left(\omega^2 \pm \frac{eB\omega}{mc}\right)^{-1}\mathbf{E}.$$

$$(6\text{-}61)$$

The dielectric polarization, as in (6–50) becomes $\mathbf{P} = ne\mathbf{r}$, giving rise to a dielectric constant

$$\varepsilon = 1 - \frac{4\pi ne^2}{m\omega(\omega \pm \omega_c)}, \qquad \omega_c \equiv \frac{eB}{mc}.$$

$$(6\text{-}62)$$

Here ω_c is the gyrofrequency, or *cyclotron* frequency, (6–13). Since the index of refraction $\varepsilon^{1/2}$ is not the same, it follows that the left- and right-handed polarized radiation travel at different velocities through an ionized medium in a longitudinal magnetic field.

If a wave is initially plane-polarized with a given direction of polarization, the polarization angle can be expressed as a superposition of two circularly polarized waves of given phase, say θ_0, and equal amplitude. As the waves propagate, the phase relationship changes because one wave lags behind the other. The direction of polarization therefore rotates. Sometimes the **E** vectors will be in phase; at other times they will be out of phase.

Figure 6.7 shows two sets of superposed, opposite circularly polarized waves. The one on the left has $\theta_0 = 180°$. The one on the right has $\theta_0 = 90°$. The **E** vectors and their sums are shown at different times during the waves' period P. The sum of the vectors is indicated by the dashed line. We can see that the direction of the plane-polarized wave is given by an angle equaling half the phase lag. However,

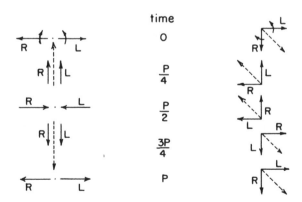

Fig. 6.7. Addition of circularly polarized waves, to give plane-polarized radiation. The relative phase delay at time $t = 0$ is θ_0.

the initial direction of one of the $\mathbf{E}$ vectors must also be specified. In Fig. 6.7, for example, we took the left-handed polarized $\mathbf{E}$ vector to point to the right at $t = 0$.

Turning now to the velocity of propagation and refractive index, we find the difference Δn between indexes n_L and n_R to be

$$n_L^2 - n_R^2 = \varepsilon_L - \varepsilon_R = 2n_\omega \Delta n . \tag{6-63}$$

We write

$$n_\omega \sim 1 - \frac{4\pi n e^2}{2m\omega^2} , \tag{6-64}$$

where (6–64) is obtained from equation (6–51) provided $n_\omega - 1 \ll 1$. Substituting the dielectric constants from (6–62) into (6–63) we obtain

$$\Delta n = \frac{\dfrac{4\pi n e^2 (2\omega_c)}{m\omega(\omega^2 - \omega_c^2)}}{2\left(1 - \dfrac{2\pi n e^2}{m\omega^2}\right)} . \tag{6-65}$$

If $\omega \gg \omega_c$ and $ne^2/m\omega^2 \ll 1$,

$$\Delta n = \frac{4\pi n e^2 \omega_c}{m\omega^3} . \tag{6-66}$$

This distance lag per unit time is $c\Delta n/n_\omega^2 \sim c\Delta n$. The phase lag of the LHP relative to the RHP wave therefore becomes $\omega\Delta n$, and the plane of polarization rotates through half this angle in unit time:

$$\Delta\theta \sim \frac{\omega\Delta n}{2} . \tag{6-67}$$

The difference in velocity of propagation, and hence the rate at which the polarization vector rotates, is therefore proportional to the number density n and to B. For a given velocity difference, the phase rotates at a rate inversely proportional to the wavelength λ, because the distance one wave has to lag behind the other becomes greater for longer wavelengths. On the other hand, the velocity difference between the waves is proportional to ω^{-3}, according to (6–66), and therefore is proportional to λ^3. Hence the angle $\theta(D)$ through which the plane of polarization is rotated over distance D is proportional to λ^2. In observing distant radio sources emitting polarized radiation we can determine the angle θ as a function of wavelength. This gives a value for the product of electron density n and the magnetic field component along the viewing direction (provided the path length is known). More correctly, because the rotation depends on the presence of both a properly oriented magnetic field and the local particle density at the field's position, the rotation actually gives a value of the product of particle density and magnetic field integrated along the line of sight.

If, as is sometimes supposed, the particles and fields actually do not occupy the same positions in space, but are physically separated from one another, then the Faraday rotation only produces a lower limit to the field strength and particle density. Passage of radiation through a magnetic field whose direction is scrambled by turbulence also yields just a lower limit.

For pulsars the dispersion measure tells us the mean number density of electrons along the line of sight (Section 6:11). The Faraday rotation can then be used to estimate the mean component of the magnetic field strength along the line of sight. This procedure has been followed in obtaining the local Galactic magnetic field (see Fig. 9.5). Because the field direction changes along this path, only a statistical estimate of actual field strength is obtained in this way.

PROBLEM 6–7. Suppose the field strength is B everywhere, that it varies randomly in direction from region to region, but that its direction is constant over any region of length L. If the source distance is NL, and the electron number density is n, show by a random walk procedure that

$$\theta \sim \sqrt{N} L \left(\frac{2\pi n e^3 B}{m^2 c^2 \omega^2} \right).$$

For simplicity assume that B always points directly toward or away from the observer.

6:13 Light Emission by Slowly Moving Charges

When an electric charge is set into accelerated motion it can emit radiation. If this motion is induced by an incident electromagnetic wave, we may find that the charge — or group of charged particles — absorbs or scatters the radiant energy. To see this consider the current associated with the accelerated charge. The current induces a magnetic field at some distance from the position of the charges, but the magnetic field strength variations will normally be somewhat out of phase with the variations in the current. This is due to the time delay involved in transmitting the information about the current strength from one position to another. That information can only be transmitted at the speed of light. For the moment we will regard the charges and currents as sources of electric and magnetic fields. If we use the Maxwell equations (6–22) and (6–23) for empty space where $\mathbf{E} = \mathbf{D}$ and $\mathbf{H} = \mathbf{B}$, and we can write $\mathbf{j}_c$ to symbolize the conduction current $\sigma \mathbf{E}$,

$$\nabla \wedge \mathbf{H} = \frac{4\pi}{c} \mathbf{j}_c + \frac{1}{c} \frac{\partial \mathbf{E}}{\partial t}, \tag{6-68}$$

$$\nabla \wedge \mathbf{E} = -\frac{1}{c} \frac{\partial \mathbf{H}}{\partial t}. \tag{6-69}$$

Now consider a *vector potential* $\mathbf{A}$ as giving rise to the magnetic field, while a *scalar potential* ϕ, together with $\mathbf{A}$, gives rise to the electric field; then we can write

$$\mathbf{H} = \nabla \wedge \mathbf{A} \tag{6-70}$$

and

$$\mathbf{E} = -\nabla\phi - \frac{1}{c}\frac{\partial \mathbf{A}}{\partial t}, \tag{6-71}$$

which are consistent with the Maxwell equations above. Separable equations, each depending on only one of these potentials, can then be obtained provided that

$$\nabla \cdot \mathbf{A} + \frac{1}{c}\frac{\partial \phi}{\partial t} = 0 \tag{6-72}$$

holds. Equation (6–72) is called the *Lorentz condition*.

PROBLEM 6–8. Check the validity of this statement by direct substitution of equations (6–70) to (6–72) into the Maxwell equations. In this way obtain

$$\nabla^2 \mathbf{A} - \frac{1}{c^2}\frac{\partial^2 \mathbf{A}}{\partial t^2} = -\frac{4\pi}{c}\mathbf{j}_c \tag{6-73}$$

and

$$\nabla^2 \phi - \frac{1}{c^2}\frac{\partial^2 \phi}{\partial t^2} = -4\pi\rho. \tag{6-74}$$

In empty space, equations (6–73) and (6–74) have the right side equal to zero; it is nonzero only at the actual location of charges and currents. Furthermore, in a static case where the time derivative vanishes, ϕ obeys the Poisson equation (4–151) which we used earlier in discussing plasmas. When we solve the Poisson equation, the potential is expressed in terms of an integral over the volume-distributed charges divided by the distance of the charges from the point at which the potential is evaluated. In view of this we write the potential as

$$\phi(\mathbf{R}_0, t) = \frac{1}{R_0} \int \rho\left(t - \frac{R_0}{c} + \frac{\mathbf{r}\cdot\mathbf{n}}{c}\right) dV, \tag{6-75}$$

where R_0 is the distance from the *center of charge*, and $\mathbf{r}\cdot\mathbf{n}$ is the projected distance of a point at $\mathbf{r}$ in the charge distribution (see Fig. 6.8). The unit vector along the direction from the charge distribution is $\mathbf{n}$. Equation (6–75) tells us that the potential at any given time is determined by the charge distribution at a time $R/c = (R_0 - \mathbf{n}\cdot\mathbf{r})/c$ earlier. The similarity between equations (6–73) and (6–74) suggests that we can also write

$$\mathbf{A}(\mathbf{R}_0, t) = \frac{1}{cR_0} \int \mathbf{j}_c\left(t - \frac{R_0}{c} + \frac{\mathbf{r}\cdot\mathbf{n}}{c}\right) dV. \tag{6-76}$$

Since a plane wave in vacuum obeys the relation (6–35)

$$\mathbf{H} = \mathbf{n} \wedge \mathbf{E} \tag{6-35}$$

(6–71) leads to

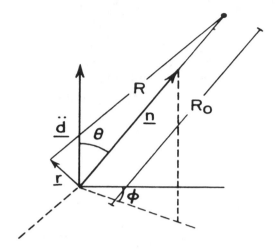

Fig. 6.8. Diagram to illustrate radiation by a dipole; see equations (6–75) to (6–85).

$$\mathbf{H} = \frac{1}{c}\dot{\mathbf{A}} \wedge \mathbf{n} \qquad (6\text{-}77)$$

as long as the magnetic field strength is measured at a large distance from the charge distribution so that $\nabla\phi$ can be neglected.

We can now determine the energy radiated away by the moving charges. The Poynting vector is immediately obtained from equations (6–35) and (6–43)

$$\mathbf{S} = \frac{c}{4\pi}H^2\mathbf{n} . \qquad (6\text{-}78)$$

For radiation by a *dipole* — that is, two slightly separated dissimilar charges — we can take the integral over the current distribution in (6–76) to be just the rate of change of the dipole moment

$$\mathbf{A} = \frac{1}{cR_0}\dot{\mathbf{d}} , \qquad (6\text{-}79)$$

where

$$\dot{\mathbf{d}} = \frac{d}{dt}\sum e\mathbf{r} . \qquad (6\text{-}80)$$

Here $\mathbf{d} = \sum e\mathbf{r}$ is the dipole moment of the charge distribution; the time derivative refers to a time $t' = t - (R_0/c)$, and the dimension of the dipole must be small compared to the radiated wavelength λ; for then

$$\frac{\mathbf{r}\cdot\mathbf{n}}{c} \ll \frac{\lambda}{c} = P \qquad (6\text{-}81)$$

and neglect of the term $\mathbf{r}\cdot\mathbf{n}/c$ in equation (6–76) involves a neglect only of a time increment small compared to the *period of oscillation* P. We now see from (6–77) and (6–35) that the field strengths at a distance R_0 from the dipole are

$$\mathbf{H} = \frac{1}{c^2 R_0} \ddot{\mathbf{d}} \wedge \mathbf{n} \,, \tag{6-82}$$

$$\mathbf{E} = \frac{1}{c^2 R_0} (\ddot{\mathbf{d}} \wedge \mathbf{n}) \wedge \mathbf{n} \,, \tag{6-83}$$

and the *intensity* of radiation dI directed into a solid angle $d\Omega$ is given by the Poynting vector integrated over that angle:

$$dI = \frac{1}{4\pi c^3} (\ddot{\mathbf{d}} \wedge \mathbf{n})^2 \, d\Omega \,. \tag{6-84}$$

Integrating over all angles $d\Omega = \sin\theta \, d\theta \, d\phi$,

$$I = \int \int \frac{\ddot{d}^2}{4\pi c^3} \sin^3 \theta \, d\theta \, d\phi \tag{6-85}$$

$$I = \frac{2}{3c^3} \ddot{d}^2 \,. \tag{6-86}$$

For two opposite charges e and $-e$, separated by a distance $\mathbf{r}$, the dipole moment is

$$\mathbf{d} = e\mathbf{r} \tag{6-87}$$

and the *total radiated energy* per second is

$$I = \frac{2e^2 \ddot{r}^2}{3c^3} \,. \tag{6-88}$$

PROBLEM 6–9. A magnetic dipole can be considered to consist of two fictitious magnetic charges q_m and $-q_m$ separated by a distance $\mathbf{a}$. The magnetic dipole moment would then be $\mathbf{M} = q_m \mathbf{a}$. (a) Show that the magnetic field along the axis of this configuration is $\mathbf{H} = 2aq_m/r^3$. (b) At the surface of a pulsar $H \sim 10^{12}$ G, $r \sim 10^6$ cm, $\omega \sim 10^2$ (rad/s)^2. By analogy to equation (6–88) show that, if H is the strength of a magnetic dipole field aligned perpendicular to the axis of the star's rotation, the intensity of radiation is (Pa68)

$$I = \frac{2}{3c^3} \ddot{\mathbf{M}}^2 \sim 10^{36} \text{ erg s}^{-1}. \tag{6-89}$$

We should still note that a system of charged particles all of which have the same charge-to-mass ratio cannot radiate as a dipole. The center of charge and the center of mass coincide for such a system; and if the center of mass $\sum m\mathbf{r}$ remains stationary, the derivatives $\ddot{d}$ all vanish:

$$\ddot{\mathbf{d}} = \sum e\ddot{\mathbf{r}} = \sum \frac{e}{m} m\ddot{\mathbf{r}} = 0 \,. \tag{6-90}$$

For such an assembly of charges we can still obtain *electric quadrupole radiation*, or radiation generated by higher electric or magnetic *multipole processes*. These processes depend on the inclusion of terms in $\mathbf{r} \cdot \mathbf{n}/c$ that we had previously neglected. The current j_c is now expressed as an expansion in $\mathbf{r} \cdot \mathbf{n}/c$:

$$j_c \left(t' + \frac{\mathbf{r} \cdot \mathbf{n}}{c} \right) = j_c(t') + \frac{\partial}{\partial t} \left(\frac{\mathbf{r} \cdot \mathbf{n}}{c} \right) j_c(t') + \cdots , \tag{6-91}$$

where again $t' = t - R_0/c$. If we only retain the first two terms of the expansion and sum over all charges, equation (6–76) yields

$$\mathbf{A} = \frac{\sum e\mathbf{v}}{cR_0} + \frac{1}{c^2 R_0} \frac{\partial}{\partial t} \sum e\mathbf{v}(\mathbf{r} \cdot \mathbf{n}) , \tag{6-92}$$

where the first term again is produced only by a time-varying dipole moment, and we now understand that $\mathbf{v}$ and $\mathbf{r}$ values are measured for time t', although all primes have been dropped for ease in writing. One can show (see La51) that this leads to

$$\mathbf{A} = \frac{\dot{\mathbf{d}}}{cR_0} + \frac{1}{6c^2 R_0} \frac{\partial^2}{\partial t^2} \mathbf{Q} - \frac{1}{cR_0}(\dot{\mathbf{M}} \wedge \mathbf{n}) , \tag{6-93}$$

where

$$\mathbf{M} \equiv \frac{1}{2c} \sum e\mathbf{v} \wedge \mathbf{r} \quad \text{and} \quad \mathbf{Q} \equiv \sum e(3\mathbf{r}(\mathbf{n} \cdot \mathbf{r}) - \mathbf{n}r^2) , \tag{6-94}$$

are the *magnetic dipole moment* and *electric quadrupole moment*, respectively. Note that the magnetic dipole term also vanishes when the *charge-to-mass ratio* is the same for all particles. This comes about because angular momentum is proportional to $\mathbf{M}$ and conservation of angular momentum implies $\dot{\mathbf{M}} = 0$. The second term on the right of equation (6–93) is called the *electric quadrupole term*.

The higher multipole terms are small compared to dipole radiation terms, since they effectively involve an expansion in v/c. As (6–81) shows, this is a small quantity when the dimensions of the system are small compared to the wavelength.

The considerations presented here in classical terms also apply in the *quantum theory of radiation*. Instead of talking about the intensity of radiation given off by a moving system of charges, we then talk about the probability for emission of radiation. Where differences in the charge-to-mass ratio do not vanish, the emission probability is normally much higher for electric dipole radiation than for multipole radiation. In systems with the same e/m ratio for all constituent particles, electric dipole radiation is *forbidden* by the quantum mechanical *selection rules* to be discussed in Section 7:6. For example, we now recognize large masses of interstellar molecular hydrogen H_2 in the Universe. But the presence of this gas for many years could not be established by observations of its mid-infrared spectrum, primarily because the symmetry of the hydrogen molecule forces us to look for lines that are emitted or absorbed only through the very weak electric quadrupole process. With improved infrared detectors these spectral lines are now detected, though they are often far fainter than radiation emitted by minor chemical constituents thousands of times less abundant.

This, incidentally, brings up one last important point: that *emission* of radiation is just the reverse of *absorption* and the probability for absorption in an atomic system is identical to the probability for *induced emission* (see Section 7:10). Induced emission is the process in which an atom or molecule emits radiation in response to stimulation by a light wave that has exactly the same frequency as the wave that the atom can emit. We then find that the stimulated and stimulating radiation have exactly the same characteristics; that is, the photons all belong to the same phase cell (see Section 4:11). This induced emission is different from the quantum mechanical *spontaneous emission*, which corresponds to emission by an unperturbed atom or molecule radiating on its own without any apparent external influence. The stimulus for spontaneous emission will become clearer from quantum mechanical considerations that we will take up in Section 7:10.

6:14 Gravitational Radiation

The general approach to radiative processes presented here is also relevant to *gravitational radiation*. As discussed in earlier sections, both gravitational forces and electrostatic forces diminish as the square of the distance respectively separating the masses or charges. This allows us to use a formalism somewhat similar to electromagnetic theory in dealing with gravitational radiation. One immediate consequence of such considerations is a statement about the strength of the expected gravitational radiation. Because the ratio of inertial to gravitational mass is constant for all matter, gravitational dipole radiation is not permitted. The much weaker quadrupole radiation is the first allowed multipole emission process. The magnitude of the expected radiation at any given multipole level will also be considerably smaller, simply because the ratio of gravitational mass to inertial mass is much smaller than the ratio of electric charge to mass. The ratio of intensities can therefore be expected to differ by factors of order $e^2/m^2G \sim 10^{42}$ if the electron charge-to-mass ratio is used. It is then clear that appreciable gravitational radiation can only be expected from large masses undergoing large accelerations. Ordinary binary stars are not sufficiently massive or compact to yield measurable amounts of gravitational radiation. However, the faint emission of gravitational radiation can be deduced from the gradual evolution of the orbits of binary pulsars orbiting each other at close range.

PROBLEM 6–10. (a) Using equation (6–77) together with expression (6–93) for the quadrupole moment, obtain a Poynting vector of the form (6–78) and show that the radiation intensity for quadrupole radiation is proportional to $(\dddot{Q})^2$ and c^{-5}. The dots indicate the third derivative with respect to time.

The actual intensity for gravitational quadrupole radiation (La51) is

$$I = \frac{G}{45c^5}\dddot{Q}^2 \,, \tag{6-95}$$

where Q is a tensor having the form of (6–94) but with mass replacing the electric charge e. For a mass M enclosed within radius a, this can be written as

$$I \sim \frac{GM^2 a^4 \omega^6}{c^5} f(\varepsilon^2) . \qquad (6\text{-}96)$$

(b) For an ellipsoid rotating with the mass symmetry axis perpendicular to the axis of rotation, the quadrupole moment is proportional to the square of the ellipticity, $\varepsilon \sim [1 - a_{min}^2/a_{max}^2]$; specifically, $f(\varepsilon^2) = \varepsilon^2/4$ (Ch70). Assume that $\varepsilon \ll 1$. For a pulsar with $a \sim 10^6$ cm, $M \sim 10^{33}$ g, $\varepsilon \sim 10^{-5}$, and $\omega \sim 10^2$ rad s^{-1} show that the intensity of gravitational radiation is $\sim 7 \times 10^{31}$ erg s^{-1}. This is smaller than the magnetic dipole radiation; but very early in the pulsar's career, when it spins with a period of the order of one millisecond, the ω^6 dependence and a possibly increased ε value may allow the gravitational radiation to equal or dominate the magnetic dipole radiation.

(c) For a binary pulsar — a pulsar orbiting another compact star — ε has to be taken as the orbital eccentricity defined in Section 3:1, and $f(\varepsilon) \sim (2/5)(1 + 73\varepsilon^2/24 + 37\varepsilon^4/96)(1 - \varepsilon^2)^{-7/2}$ (Pe64). The first binary pulsar to be discovered was PSR 1913+16. It has two neutron stars with nearly identical masses (Hu75), (Th93). The system's total mass is $M = 2.83 M_\odot$, its orbital eccentricity is $\varepsilon \sim 0.617$, its semimajor axis is $a = 1.94 \times 10^{11}$ cm, and its orbital period is 27,906 s. Using equation (6–96) show the rate of gravitational radiation to be $I \sim 1.9 \times 10^{31}$ erg s^{-1} and show that the resulting decline in the orbital period is of order $\dot{P}/P \sim 10^{-16}$ s^{-1}. This is in good agreement with the observed value, $\dot{P}/P \sim 8.2 \times 10^{-17}$ s^{-1} (Ta82).

Supernova explosions, and massive concentrations of matter falling into compact galactic nuclei are also expected to emit gravitational radiation.

6:15 Light Scattering by Unbound Charges

When a plane-polarized electromagnetic wave moving along the z-direction is incident on a charged particle having mass m and charge e, the particle is subjected to an electric field of form

$$\mathbf{E} = \mathbf{E_0} \cos(\mathbf{k} \cdot \mathbf{r} - \omega t + \alpha) . \qquad (6\text{-}97)$$

If the field is sufficiently weak so that the velocity imparted to the charge is always small — $v \ll c$ — then the force $e\mathbf{E}$ is always large compared to the force $e v \wedge \mathbf{H}/c$ acting on the particle. This is evident from (6–35). The acceleration experienced by the particle is given by

$$m\ddot{\mathbf{r}} = e\mathbf{E} \qquad (6\text{-}98)$$

and the dipole moment produced by the displacement of the charge, $\mathbf{d} = e\mathbf{r}$, has a second time-derivative

$$\ddot{\mathbf{d}} = \frac{e^2}{m}\mathbf{E} .\qquad(6\text{-}99)$$

We now see that equation (6–84) predicts a scattered light intensity per solid angle along direction $\mathbf{n}$:

$$dI = \frac{e^4}{4\pi m^2 c^3}(\mathbf{E} \wedge \mathbf{n})^2 \, d\Omega .\qquad(6\text{-}100)$$

We speak of a *differential scattering cross-section*, $\sigma(\theta, \phi)$,

$$d\sigma(\theta, \phi) \equiv \sigma(\theta, \phi)d\Omega = \frac{dI(\theta, \phi)}{S} = \left[\frac{e^2}{mc^2}\right]^2 \sin^2\theta \, d\Omega ,\qquad(6\text{-}101)$$

where θ is the angle between the scattering direction $\mathbf{n}$ and the electric field $\mathbf{E}$ of the incident wave, and S is given by (6–43). We note:

(a) That the frequency of the radiation is not changed by scattering;

(b) That the angular distribution of scattered light is not dependent on the frequency;

(c) That the total cross-section is not frequency dependent. It is obtained by integrating $d\sigma(\theta, \phi)$ over all angles θ, ϕ.

$$\sigma = \int_0^\pi \int_0^{2\pi} \sigma(\theta, \phi) \sin\theta \, d\phi \, d\theta \qquad(6\text{-}102)$$

For electrons

$$\sigma_e = \frac{8\pi}{3}\left(\frac{e^2}{mc^2}\right)^2 = 6.65 \times 10^{-25}\ \text{cm}^2 .\qquad(6\text{-}103)$$

This is called the *Thomson scattering* cross-section.

(d) The differential scattering cross-section is symmetrical in θ about $\theta = \pi/2$.

(e) σ is a factor of $(m_p/m_e)^2 \sim 10^6$ times less for protons than for electrons.

(f) No light can be scattered into the incident direction of polarization, where $\mathbf{E} \wedge \mathbf{n} = 0$. By the same token, light scattered at right angles to $\mathbf{E}$ and the incident wave is entirely polarized parallel to the $\mathbf{E}$ vector.

(g) If the initial wave incident on the particle is unpolarized, we obtain a scattering cross-section independent of Φ but dependent on the polar angle Θ included between the directions of the incident and scattered waves (Fig. 6.9). The angle θ is a function of the angles Φ and Θ, and $\cos\theta = \sin\Theta \cos\Phi$.

For a given angle Θ, therefore,

$$\langle \sin^2\theta \rangle = 1 - \sin^2\Theta\langle \cos^2\Phi \rangle = 1 - \frac{\sin^2\Theta}{2} = \frac{1}{2}(1 + \cos^2\Theta) ,\qquad(6\text{-}104)$$

where we have made use of the fact that $\langle \cos^2\Phi \rangle = 1/2$ when the average is taken over all angles Φ. For unpolarized radiation we can write

$$d\sigma = \frac{1}{2}\left(\frac{e^2}{mc^2}\right)^2 (1 + \cos^2\Theta) \, d\Omega .\qquad(6\text{-}105)$$

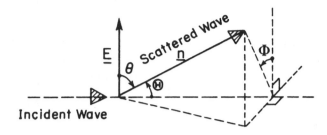

Fig. 6.9. Direction of incident and scattered waves (see equations (6–104) and (6–105)).

This yields the important result that:

(h) For unpolarized radiation the cross-section has peak values in the forward and backward directions. Most of the light is scattered forward along the direction in which the wave was initially moving, or backward into the direction from which the wave came.

PROBLEM 6–11. Show that there is a force component

$$F(\Theta) = (1 - \cos\Theta)\, d\sigma\, \frac{\mathbf{S}}{c}$$

acting on the scattering charge along the direction of propagation. Show that when this is averaged over all values Θ, we obtain a total force $\mathbf{F}$ along the direction of incidence

$$\mathbf{F} = \frac{2}{3}\left(\frac{e^2}{mc^2}\right)^2 E^2 = \frac{\sigma \mathbf{S}}{c}. \tag{6-106}$$

In the vicinity of bright hot stars this can be the dominant force acting on electrons. Much of the visible light reaching us from the solar corona seems to be due to scattering by electrons.

6:16 Scattering by Bound Charges

The *zodiacal light* — diffuse scattered sunlight in the ecliptic plane — is due to radiation scattered off small solid grains circling the Sun in the orbital plane of the planets. This glow extends into the corona and weakly contributes to its brightness. We can understand scattering by dust and by molecules, by considering a *harmonically bound charge* that would normally oscillate at a natural frequency ω_0. The electric field attempts to force the oscillator to vibrate at a different frequency ω instead. The equation of motion for this *forced oscillation* is

$$\ddot{\mathbf{r}} + \omega_0^2 \mathbf{r} = \frac{e\mathbf{E}}{m}. \tag{6-107}$$

If $\mathbf{E} = 0$, we obtain oscillation at frequency ω_0. If $\mathbf{E}$ has the form (6–97), equation (6–107) has the solution

$$\mathbf{r} = \frac{e\mathbf{E}}{m} \frac{1}{(\omega_0^2 - \omega^2)} \tag{6-108}$$

and

$$\ddot{\mathbf{d}} = \frac{e^2}{m}\mathbf{E}\left(\frac{1}{1 - (\omega_0^2/\omega^2)}\right) \tag{6-109}$$

(see equation (6–99)). The scattering cross-section then is

$$\sigma = \frac{\sigma_e}{(1 - \omega_0^2/\omega^2)^2} . \tag{6-110}$$

When $\omega \gg \omega_0$, the electron acts as though it were free and we again have $\sigma = \sigma_e$. If $\omega_0 \gg \omega$ we obtain

$$\sigma = \frac{\sigma_e \omega^4}{\omega_0^4} = \frac{8\pi}{3} \frac{e^4}{m^2 c^4} \frac{\omega^4}{\omega_0^4}, \tag{6-111}$$

called the *Rayleigh scattering cross-section*. Rayleigh scattering is responsible for the scattering of visible light in the daytime sky. The electrons are strongly bound to their parent molecules so that ω_0 is large compared to the frequency of visible light ω. For red light $(\omega/\omega_0)^4$ is smaller than for blue light by a factor close to $(2)^4 = 16$. Hence blue light is scattered most strongly. Red light passes more easily straight through the atmosphere without deflection, whereas blue light is scattered out of a straight path, and the sky appears blue when we look away from the Sun.

The scattering by fine dust grains is also of great interest in astronomy. For spherical grains with refractive index n, the cross section for scattering can be shown to be

$$\sigma = 24\pi^3 \left[\frac{n^2 - 1}{n^2 + 2}\right]^2 \frac{V^2}{\lambda^4}, \tag{6-112}$$

if the radius a of the sphere is much smaller than the wavelength λ. V is the volume $(4\pi/3)a^3$. We note that the factor λ^{-4} is reminiscent of Rayleigh scattering, showing that the two types of scattering are related. The differential cross-section has exactly the same angular dependence, (6–105), as Thomson and Rayleigh scattering.

This is an important property for the study of disks around young stars. The light from T Tauri stars becomes linearly polarized as it is scattered by a dusty disk surrounding the star. Even when such disks are too compact and too distant to be resolved, their orientation in the sky can be determined from the direction of polarization of the scattered light, which must be perpendicular to the orientation of the disk. Some T Tauri stars are binary systems. Judging by the polarization of the two stars, their disks are found to be roughly aligned along the direction separating the stars in the sky. Since these disks may eventually give rise to new planetary systems, it is interesting to see how these planets will be oriented relative to their parent stars. Because we have not yet devised sufficiently advanced instruments to image distant planetary systems directly, this is a particularly useful indicator of the potential structure of planetary systems around binary stars (Je04).

6:17 Extinction by Interstellar Grains

Interstellar grains absorb and scatter radiation so that starlight does not reach an observer directly. We talk about *extinction*. The term extinction refers to the fractional amount of light prevented from reaching us. It is a useful concept when we do not know how much of the radiation is scattered and how much is absorbed. The scattered radiation sometimes can be observed in *reflection nebulae* — clouds of dust grains illuminated by a bright star. These clouds show spectra remarkably similar to those of the illuminating star; the scattered portion of the radiation looks very much like light scattered off snow. The particles, in this sense, are white or gray. On the other hand, when we see starlight that has passed through a cloud, we find an amount of extinction that to first approximation is inversely proportional to the wavelength, λ. The data are shown in Fig. 6.10. Some λ^{-4} scattering undoubtedly also takes place; the grain size distribution appears to differ for different chemical constituents of the dust; the smallest grains certainly are small compared to the wavelength of visible light; but the overall effect is to produce a mean cross-section roughly proportional to λ^{-1}.

Thus far we have only talked about spherical grains that are purely dielectric. It is, however, also possible for grains to have a metallic character; they can then absorb and emit radiation — they do not merely scatter. For metallic grains the dielectric constant has an imaginary component and we talk about a *complex re-*

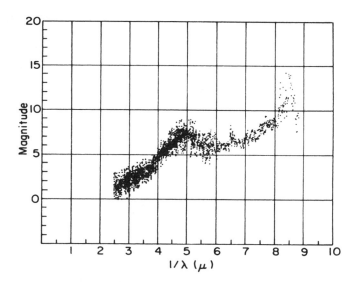

Fig. 6.10. Interstellar extinction curve showing magnitudes of extinction as a function of reciprocal wavelength. The data were obtained from observations of the stars ζ-Persei and ε-Persei, and have been normalized to an extinction difference $E(B - V)$ of one magnitude and $V \sim 0$. The curve would therefore roughly characterize extinction over a path length of order ~ 1 kpc through the galactic plane (after Stecher (St69)).

fractive index, m. In these terms, for a column density N of grains small compared to the wavelength, absorption dominates over scattering, and the extinction is given (Gr68)* by

$$\mathcal{E} = 6\pi N \left(\frac{m^2 - 1}{m^2 + 2} \right) \frac{V}{\lambda} \quad \text{(IP)} , \tag{6-113}$$

where $\mathcal{E}$ is the total light extinguished for unit incident energy, and the symbol (IP) means that the imaginary part of the expression in parentheses should be used.

For particles with dimensions comparable to the wavelength of the extinguished radiation, the expressions become quite complicated even for spherical particles; and for nonspherical grains the theory of extinction is extremely laborious.

We might be tempted to attribute interstellar extinction to metallic absorption alone, because then the $1/\lambda$ relation might be directly obtained. However, matters are not that simple. The refractive indices m and n are wavelength dependent — a dependence determined by the chemical makeup of the grains. The observed wavelength dependence of absorption and scattering by interstellar grains, and their ability to polarize light both through extinction and thermal radiation, put tight constraints on the sizes, chemical composition, shapes, and fluffiness of the grains (Dw97).

Radiation scattered by the processes mentioned in Sections 6:15 and 6:16 should lead to polarization that can be shown to have the Θ-dependence

$$P = \frac{\sin^2 \Theta}{1 + \cos^2 \Theta} . \tag{6-114}$$

In addition, polarization, both in absorption and in emission, can be produced by systematically oriented elongated or disk-shaped grains. The light that reaches us from distant stars located close to the plane of the Galaxy shows polarization believed due to this process. How small grains could be aligned to give consistently polarized radiation is discussed in Section 9:13. The interaction of grains with radiation is a complicated subject. Detailed discussions can be found in (Bo83)*.

6:18 Absorption and Emission of Radiation by a Plasma

In Section 6:11, we treated the propagation of radiation through a tenuous ionized medium. Sometimes, however, the plasmas are dense and collisions between ions and electrons have a high frequency of occurrence, ν_c. In this sense, a medium may be tenuous for high-frequency waves, but dense at lower frequencies. This is quite common. Because the nature of the transmission is so different in these two cases, the spectrum of radiation received from a source may also be quite different at high and low frequencies. A relatively abrupt change in spectrum taken together with other data then allows us to determine the collision frequency and, hence, the density of the medium. Radio-astronomy can, in this way, provide an extremely useful technique for measuring the density of interstellar ionized gases.

To show this we consider an ionized medium and define the *collision frequency* ν_c as the frequency with which an electron successively becomes deflected through a total angle of 90°, usually through a series of small collisions with ions. This angle of 90° is the deflection a particle has to suffer to give up all the directed momentum it had at some previous period, that is, to lose all sense of the direction into which it was accelerated by some previously applied force. We only consider collisions with ions because we will be interested in the dissipation of energy through collisions. When an electron collides with another electron the motion is that of a symmetric dipole; no energy is radiated away and these collisions can therefore be neglected.

We now consider an electron accelerated by the electric field component **E** of an electromagnetic wave. As the particle reaches appreciable velocity induced by the field, it suffers a collision and gives up all its directed momentum. This means that there is an acceleration by the electromagnetic wave traveling through the medium and deceleration through collisions. The combined forces on the electron then are

$$m\ddot{\mathbf{r}} = e\mathbf{E}(r, t) - m\dot{\mathbf{r}}\nu_c \,. \tag{6-115}$$

Here m is the reduced electron mass. The second term on the right shows a momentum loss equal to the instantaneous momentum $m\dot{\mathbf{r}}$ of the electron every time there is a collision, or ν_c times in unit time interval. This is just what we stated formally in defining ν_c. Our problem will be to calculate the actual value of ν_c. However, before we do that, we can proceed to solve equation (6–115) and obtain the transmission properties of the plasma in different frequency ranges relative to ν_c.

We have already noted that some of the momentum conferred on the electrons by the electromagnetic wave is lost in collisions. This means that the energy transferred from the wave to the particles becomes dissipated. Because the energy of the electromagnetic wave depends on the square of the wave amplitude, E^2, we can expect to find that E will decrease as the wave propagates through the medium. We will therefore make use of a function $\mathbf{E}(r, t)$ of the form

$$\mathbf{E}(r, t) = \mathbf{E}_0 e^{-\kappa x/2} \cos \omega t \tag{6-116}$$

in equation (6–115). Here κ is the *absorption coefficient*. The factor 2 in the exponent of this *damping term* is provided so that the energy in the wave, rather than its amplitude, may decay by a factor of $1/e$ in distance $x = 1/\kappa$. Note that the absorption coefficient always has units $(\text{length})^{-1}$.

Because the rate of energy loss from the wave will be determined by the total number of collisions per unit volume, we rewrite equation (6–115) as

$$nm\ddot{\mathbf{r}} + nm\nu_c\dot{\mathbf{r}} = ne\mathbf{E}_0 e^{-\kappa x/2} \cos \omega t \,, \tag{6-117}$$

where n is the electron density.

If we use a complex field **E** instead of a field with real values in equation (6–117), the solution of the differential equation becomes much simpler. However, in order to remember that only the real parts of the equation have physical significance, we add the annotation (RP):

$$nm\ddot{\mathbf{r}} + nm\nu_c\dot{\mathbf{r}} = ne\mathbf{E}_0 e^{-(\kappa x/2)+i\omega t} = ne\mathbf{E} \quad (RP). \tag{6-118}$$

PROBLEM 6–12. By substitution, show that a particular solution of (6–118) is

$$\mathbf{r} = -\left(\frac{e\mathbf{E}_0}{m\omega}\right) e^{(i\omega t - \kappa x/2)} \left[\frac{i\nu_c + \omega}{\nu_c^2 + \omega^2}\right]. \tag{6-119}$$

The current due to the n particles per unit volume can now be written as

$$\mathbf{j} = ne\dot{\mathbf{r}} = \frac{ne^2}{m}\left[\frac{\nu_c - i\omega}{\nu_c^2 + \omega^2}\right]\mathbf{E} \quad (RP). \tag{6-120}$$

As in equation (6–50), $ne\mathbf{r}$ is an induced polarization field, and the imaginary term in the brackets on the right of (6–120) is the induced polarization current

$$\frac{d\mathbf{P}}{dt} = i\omega\mathbf{P} = i\omega\frac{\varepsilon - 1}{4\pi}\mathbf{E} \quad (RP). \tag{6-121}$$

Here the imaginary number i enters as a consequence of the assumed field in (6–118). Equation (6–121) is then a direct consequence of the definition (6–10). The real term in the brackets of (6–120) determines the current $\sigma\mathbf{E}$ due to the flow of charge. We note two features of equation (6–120). The term proportional to ν_c on the right represents the dissipation of energy and is therefore directly related to the absorption coefficient κ. The second term, proportional to $i\omega$, depends on the dielectric constant in the medium, and hence yields the phase velocity $c\varepsilon^{-1/2}$ of the wave through the medium. Formally written:

$$\mathbf{j} = \left(\sigma + i\omega\frac{\varepsilon - 1}{4\pi}\right)\mathbf{E} \quad (RP) \tag{6-122}$$

with

$$\varepsilon = 1 - \frac{4\pi e^2 n}{m(\omega^2 + \nu_c^2)} \quad \text{and} \quad \sigma = \frac{e^2 n\nu_c}{m(\omega^2 + \nu_c^2)}. \tag{6-123}$$

If we write the imaginary and *complex dielectric constants* as

$$\varepsilon_i = -\frac{i4\pi\sigma}{\omega} \quad \text{and} \quad \varepsilon_c = \varepsilon + \varepsilon_i, \tag{6-124}$$

equation (6–122) can be written in a form characteristic of a pure dielectric. In fact, all of Maxwell's equations take on this form. This can be seen directly by noting that $\mathbf{j}$ appears only in equations (6–15) and (6–23) in the set of Maxwell's differential equations. For a complex field, as it appears in (6–118), a propagating wave has a form (see equation (6–36))

$$\mathbf{E} = \mathbf{E}_0 \exp\left[i\left(\omega t \pm \frac{\omega\varepsilon_c^{1/2}x}{c}\right)\right] \quad (RP), \tag{6-125}$$

so that (6–118) holds if

$$\frac{\kappa}{2} = \frac{i\omega}{c}\varepsilon_c^{1/2} \quad \text{(RP)}.$$

(6-126)

We can always write ε_c in the form

$$\varepsilon_c = (N + iQ)^2 ,$$

(6-127)

where N and Q are real quantities as long as we choose

$$\varepsilon_i = 2NQi = -\left(\frac{4\pi\sigma}{\omega}\right)i \quad \text{and} \quad \varepsilon = N^2 - Q^2 .$$

(6-128)

We are therefore interested in the quantity

$$\frac{\kappa}{2} = -\frac{\omega Q}{c} = \frac{4\pi\sigma}{2Nc} .$$

(6-129)

In practice $\omega \gg \nu_c$, ω_p, in all radio-astronomical processes, so that (6–123) and (6–128) give

$$|\varepsilon| \gg \frac{4\pi\sigma}{\omega} \quad \text{and} \quad N \sim \varepsilon^{1/2} .$$

(6-130)

With this same approximation we then also obtain κ at frequency ω,

$$\kappa(\omega) = \frac{4\pi(e^2 n/m\omega^2)\nu_c/c}{\sqrt{1 - 4\pi e^2 n/m\omega^2}} = \frac{\nu_c(\omega_p^2/\omega^2)/c}{\sqrt{1 - \omega_p^2/\omega^2}} ,$$

(6-131)

where ω_p is the plasma frequency (6–55).

We still need to calculate the collision frequency ν_c; but most of the work for this has already been done. Equation (3–74) gives the force acting on a particle of reduced mass μ deflected in the superposition of inverse square law fields produced by a density of n scattering centers per unit volume. To avoid confusion with the symbol μ used here for magnetic permeability, we will continue to use the symbol m for the electron's reduced mass. In equation (6–115) we had defined the drag force $m\dot{r}\nu_c$ and we now set this equal to the right side of (3–69), noting that since $\dot{r}$ is the velocity before collision, it plays the same role as v_0 in (3–69):

$$m v_0 \nu_c = m 2\pi n v_0^2 \int_{s_{\min}}^{s_{\max}} s(1 - \cos\Theta)\, ds .$$

(6-132)

For small deflections Θ

$$1 - \cos\Theta \approx 2\tan^2\frac{\Theta}{2} = 2\left[\frac{Ze^2}{v_0^2 sm}\right]^2 .$$

(6-133)

The second half of this relation is based on an analogy with equation (3–71) but with Coulomb forces replacing gravitational forces. Z is the typical charge on an ion. We therefore have

$$\nu_c = 4\pi n \left\langle \frac{1}{v_0} \right\rangle \frac{1}{\langle v_0^2 \rangle} \frac{Z^2 e^4}{m^2} \int_{s_{\min}}^{s_{\max}} s^{-1} \, ds,$$

(6-134)

$$= \frac{4\sqrt{2\pi}}{3} n \frac{Z^2 e^4}{\sqrt{m}(kT)^{3/2}} \ln \frac{s_{\max}}{s_{\min}},$$

where we have made use of (4–112) and (4–113). From (6–131) and (6–134)

$$\kappa(\omega) = \frac{32\pi^{3/2} e^6 n^2 Z^2}{3\sqrt{2} c \omega^2 (kTm)^{3/2}} \ln \frac{s_{\max}}{s_{\min}}.$$

(6-135)

Here we have assumed a succession of many weak deflections, meaning that the minimum impact parameter $s'_{\min}$ must be sufficiently large to give a potential energy small compared to the kinetic energy. Specifically, for $Z = 1$, $s'_{\min} \gg \{2e^2/mv_0^2\}$.

In the interstellar medium, s will rarely be less than 10^{-2} cm, whereas $2e^2 \sim 5 \times 10^{-19}$ (esu)^2; and typically mv^2 is 10^{-12} erg in ionized regions. This shows that the inequality for $s'_{\min}$ is usually well satisfied except in very rare chance collisions with a small impact parameter.

A second lower bound is given by the *de Broglie wavelength* of the electron $\lambda_e = h/mv$. At closer distances than this, the electron no longer behaves as a point charge, and we can use a lower limit $s''_{\min} > \lambda_e/2$. There also are two upper bounds we can set. First, we want a collision to appear instantaneous; that is, the time $1/\omega \gg s'_{\max}/v_0$. The time during which the electric field changes is long compared to the time in which the electron suffers a collision, or goes through its minimum approach. The second upper limit is $s''_{\max} = L$, the Debye length given by equation (4–165). Shielding by nearer particles screens out the effects of charges at distances greater than L. Using the limits $s'_{\max}$ and $s'_{\min}$ for the ionized interstellar matter, we can then write

$$\frac{s'_{\max}}{s'_{\min}} = \frac{mv_0^2}{2e^2} \frac{v_0}{\omega} \sim \frac{(2kT)^{3/2}}{2e^2 \omega m^{1/2}}.$$

(6-136)

The full expression for ionized hydrogen, $Z = 1$, reads

$$\kappa(\omega) = \frac{32\pi^{3/2}}{3\sqrt{2}} \frac{e^6 n^2}{c(mkT)^{3/2}\omega^2} \ln \left[\frac{1.32(kT)^{3/2}}{e^2 m^{1/2}\omega} \right],$$

$$\kappa(\nu) = \frac{8}{3\sqrt{2\pi}} \frac{e^6 n^2}{c(mkT)^{3/2}\nu^2} \ln \left[\frac{1.32(kT)^{3/2}}{2\pi e^2 m^{1/2}\nu} \right].$$

(6-137)

6:19 Radiation from Thermal Radio Sources

If we now look at the results obtained in the previous section, we note that we have an absorption coefficient $\kappa(\nu)$ that tells us the amount of absorption obtained per unit length of travel through an ionized medium. An electromagnetic wave traveling a distance $D = \int dx$ through the medium will encounter an *optical depth*

$$\tau(\nu) = \int \kappa(\nu) \, dx \, . \tag{6-138}$$

If the temperature throughout the region is constant, only the density will vary with position x and we find that (La74)

$$\tau(\nu) = F(T, \nu) \int n^2 \, dx \sim 8.2 \times 10^{-2} T^{-1.35} \nu^{-2.1} \int n^2 \, dx \, , \tag{6-139}$$

where the function F is $\kappa(\nu)/n^2$ from equation (6–137), and the integral

$$\mathcal{E}_m = \int n^2 \, dx = \langle n^2 \rangle D \tag{6-140}$$

is called the *emission measure*. It is a measure of the amount of absorption and emission expected along D. In radio-astronomy it is customary to express the electron number density n in terms of cm^{-3} and D, the path length covered, in parsecs. The emission measure then has units cm^{-6} pc.

The emission measure is just a measure that tells how frequently atomic particles approach each other closely along a line of sight through a given region. For this reason, such quantities as the number of atomic recombinations giving rise to a given emission line are also proportional to $\mathcal{E}_m$. When the recombination line strength $R(\nu_1)$ for a spectral line ν_1 is a known function $F_1(T)$ of the temperature,

$$R(\nu_1) = F_1(T)\mathcal{E}_m \, . \tag{6-141}$$

Hence, if we measure both the recombination line strength — possibly in the visible part of the spectrum — and also the radio thermal emission, both the emission measure and the temperature of the region can be determined. For this to be true, radio measurements are best taken at frequencies for which the region is optically thin, so that self-absorption of radiation by the cloud need not be considered.

An interesting feature of the self-absorption by an optically thin cloud is that the brightness should be independent of the frequency ν. This comes about because the absorption $\kappa(\nu)$ is inversely dependent on ν^2 — if we neglect the weak frequency dependence of the logarithmic term in (6–137). At the same time, the energy density of radio waves corresponding to a blackbody at gas temperature, T, would be

$$\rho(\nu) \sim \frac{8\pi k T \nu^2}{c^3}, \qquad h\nu \ll kT, \tag{4–83}$$

at very long wavelengths. The product of optical depth or effective emissivity for the gaseous region, and blackbody intensity $I(\nu) = \rho(\nu)c/4\pi$ is therefore frequency independent as long as the region is optically thin. At low frequencies, where $\tau(\nu) \gtrsim 1$, this behavior ceases to be true. The effective emissivity then remains close to unity, and the only frequency-dependent term is $I(\nu)$. This is the Rayleigh–Jeans limit, where a thermal source exhibits a spectrum proportional to ν^2.

For the flat part of the spectrum the product $S(\nu) = \tau(\nu)I(\nu)$ is proportional to $T^{-0.35}\mathcal{E}_m$; and this latter product can immediately be determined from a measurement of the surface brightness anywhere in this frequency range. In the steep part

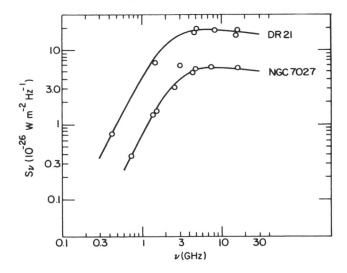

Fig. 6.11. Data obtained by a number of observers on the compact HII region DR21 and the planetary nebula NGC 7027 (see text). After P.G. Mezger (Me68). (From *Interstellar Ionized Hydrogen*, Y. Terzian, Ed., 1968, W.A. Benjamin, Inc. Reading, Massachusetts.)

of the spectrum, where the region is opaque, the measured surface brightness at frequency ν depends only on T (see equation (4–84)). These two sets of observations, taken together, provide data on both the temperature and the emission measure $\mathcal{E}_m$. Figure 6.11 shows spectra for some very compact ionized hydrogen regions and these clearly show the expected form. On this log–log plot the low-frequency spectrum has the expected slope of 2. At high frequencies it is flat.

NGC 7027 is a planetary nebula for which Mezger (Me68) — see the data of Fig. 6.11 — found an emission measure of 5.4×10^7 cm^{-6} pc and a temperature of $\sim 1.1 \times 10^4$ K. If the object is assumed to have a depth along the line of sight similar to the observed diameter, then an actual density can be computed. Mezger gave the value $n \sim 2.3 \times 10^4$ cm^{-3} for this object. From the density and the total volume, we can also obtain the nebular mass, which in this case is roughly $0.25 M_\odot$, or 5×10^{32} g. The observed diameter of NGC 7027 is about 0.1 pc.

Young compact HII regions, which may be found in the plane of the Galaxy in the vicinity of bright young stars, tend to have somewhat lower temperatures, about the same densities, but sometimes much greater masses — up to several stellar masses. These are believed to represent the remains of clouds from which massive stars were formed. When massive protostars light up as they approach the main sequence, they emit an intense ultraviolet flux that heats and ionizes the gas. Such HII regions will be discussed in Chapter 9.

6:20 Synchrotron Radiation

When a charged particle moves at relativistic velocity across a magnetic field, it describes a spiral motion. The axis of this spiral lies along the direction of the magnetic field and the acceleration experienced by the particle is along directions perpendicular to the field lines. As the particle moves, the direction of the acceleration vector continually changes.

We first consider the motion of a relativistically moving particle orbiting in a plane perpendicular to the magnetic field. This constitutes no restriction on generality because a constant velocity component along the magnetic field lines leaves the radiation rate unaffected.

If we recall that a force corresponds to a rate of change of momentum, we can use (6–11) to calculate the rate at which the particle is deflected. Consider the direction of motion in Fig. 6.12 to be the x-direction, and let the radial direction be the y-direction. In a time Δt_0, the momentum change, which is along the y-direction, amounts to

$$\Delta p_y = \frac{evB}{c} \Delta t_0 \ . \tag{6-142}$$

Since the initial relativistic momentum p_x is

$$p_x = \frac{m_0 v}{\sqrt{1 - v^2/c^2}} \tag{5–30}$$

we can see that the angular deflection during time interval Δt_0 is

$$\delta = \frac{\Delta p_y}{p_x} = \frac{eB}{m_0 c} \sqrt{1 - \frac{v^2}{c^2}} \Delta t_0 = \frac{eB \Delta t_0}{m_0 c \gamma(v)} \ , \tag{6-143}$$

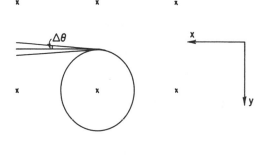

Fig. 6.12. Relativistic charged particle orbiting in a magnetic field. The direction of the field lines is into the paper. Because the acceleration of the particle has no z-component, radiation reaching an observer along the x-direction is linearly polarized along the y-direction — perpendicular to the magnetic field.

where m_0 is the particle's rest–mass. From this it follows that the time Δt_1, required for the particle to orbit one radian, $\delta = 1$, is

$$\Delta t_1 = \frac{m_0 c}{eB}\gamma(v) = \frac{1}{\omega_c} . \qquad (6\text{-}144)$$

The gyrofrequency ω_c given in equation (6–13) is the reciprocal of Δt_1; we see this if we use (5–30) to substitute $m_0\gamma(v)$ for p_c/v_c in (6–13). Having obtained the gyrofrequency of the particle, we might think that the problem is completely solved and that the particle radiates energy at that frequency. However, the spectrum radiated by the moving charge actually lies at frequencies often many orders of magnitude higher than ω_c. The reason for this is directly related to the strong concentration of emitted radiation into a narrow beam about the forward direction of motion. From equation (5–50), the angle $\Delta\theta$ in which radiation is received is related to the angle $\Delta\theta'$ into which the radiation is emitted in the coordinate system of the moving charge,

$$\Delta\theta \sim \sqrt{1 - \frac{v^2}{c^2}}\,\Delta\theta' = \gamma(v)^{-1}\,\Delta\theta' . \qquad (6\text{-}145)$$

Because of this, an observer is not properly oriented to receive radiation emitted by the particle except during a brief interval in each orbit during which the particle advances along its trajectory by about one radian

$$\Delta t_2 = \gamma(v)^{-1}\Delta t_1 = \frac{m_0 c}{eB} . \qquad (6\text{-}146)$$

But the radiation emitted during interval Δt actually arrives at the observer over an even smaller time span, because radiation emitted by the particle at the beginning of the interval Δt has a longer distance to travel to the observer than radiation emitted at the end of the interval, when the particle is nearer to him. If the particle travels a distance of length L during interval Δt, radiation emitted at the end of the interval will only arrive a time

$$\Delta t \sim -\left(\frac{L}{c} - \frac{L}{v}\right) \qquad (6\text{-}147)$$

later than radiation emitted at the beginning of the interval. Inasmuch as

$$L \sim v\Delta t_2 , \qquad (6\text{-}148)$$

we obtain

$$\Delta t \sim \left(1 - \frac{v}{c}\right)\Delta t_2 \sim \frac{m_0 c}{2eB}\left(1 - \frac{v^2}{c^2}\right) , \qquad (6\text{-}149)$$

because for highly relativistic particles

$$\left(1 - \frac{v}{c}\right) \sim \frac{1}{2}\left(1 + \frac{v}{c}\right)\left(1 - \frac{v}{c}\right) = \frac{1}{2}\left(1 - \frac{v^2}{c^2}\right) . \qquad (6\text{-}150)$$

We can define a radiation frequency ω_m that roughly corresponds to the reciprocal of twice this time interval:

$$\omega_m \equiv \frac{eB}{m_0 c}\left(1 - \frac{v^2}{c^2}\right)^{-1} = \gamma^2(v)\omega_c = \frac{eB}{m_0 c}\left(\frac{\mathcal{E}}{m_0 c^2}\right)^2 \sim \frac{1}{2\Delta t},\qquad (6\text{-}151)$$

where $\mathcal{E}$ is the total energy of the particle, $\mathcal{E} \gg m_0 c^2$, and ω_c is the radiation frequency of a nonrelativistic particle moving in a magnetic field. The significance of the factor 2 is explained in Section 6:21, immediately below. Because $(1 - v^2/c^2)$ is a very small number, it is clear that ω_m is many orders of magnitude greater than the gyrofrequency,

$$\omega_m \gg \omega_c = \frac{eB}{m_0 c}.\qquad (6\text{-}152)$$

Let us summarize what we have done:

(1) First, we computed the orbital frequency of a particle moving in a magnetic field.

(2) Next, we calculated the time in the observer's frame during which the particle was capable of emitting radiation into his direction.

(3) Finally, we computed the length of time elapsing between the arrival of the first and last portions of the electromagnetic wave train at the position of the observer. This elapsed time was very small compared to the period of the particle's gyration in the magnetic field, and the corresponding frequency $\omega_m \sim 1/2\Delta t$ was found to be $(\mathcal{E}/m_0 c^2)^2$ higher than the nonrelativistic gyrofrequency for this field.

6:21 The Synchrotron Radiation Spectrum

The actually expected synchrotron radiation spectrum obtained when the above-sketched calculations are done rigorously for monoenergetic electrons is a set of extremely finely spaced lines at high harmonics of the gyrofrequency. The peak of the spectral distribution function $p(\omega/\omega_m)$ occurs at a frequency $\omega = 0.5\omega_m$. But, as suggested by Fig. 6.13, roughly half of the emitted energy lies above and half below the frequency $2\omega_m$ corresponding to the reciprocal of the time interval Δt we derived in (6–149). Details of the theory are discussed in references (Gi64)* and (Sh60)*.

One can show that the energy actually radiated by a particle of energy $\mathcal{E}$ per unit time into unit frequency interval $d\nu$ is

$$P(\nu, \mathcal{E}) = 2\pi P(\omega, \mathcal{E}) = \frac{16 e^3 B}{m_0 c^2}\, p\left(\frac{\omega}{\omega_m}\right).\qquad (6\text{-}153)$$

In the limit of very high and very low frequencies the function $p(\omega/\omega_m)$ has asymptotic values

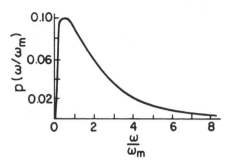

Fig. 6.13. Envelope of the narrowly spaced lines comprising the synchrotron spectrum of a particle radiating at peak emission frequency $\omega_m = (eB/m_0c)(\mathcal{E}/m_0c^2)^2$. Actual particle energies always vary to some extent and, hence, the finely spaced lines are never seen. One observes a continuum of the shape of the envelope (after I. S. Shklovskii (Sh60)).

$$p\left(\frac{\omega}{\omega_m}\right) = 0.256 \left(\frac{\omega}{\omega_m}\right)^{1/3} , \qquad \omega \ll \omega_m ,$$

$$p\left(\frac{\omega}{\omega_m}\right) = \frac{1}{16} \left(\frac{\pi\omega}{\omega_m}\right)^{1/2} \exp\left(-\frac{2\omega}{3\omega_m}\right) , \qquad \omega \gg \omega_m . \qquad (6\text{-}154)$$

PROBLEM 6–13. In Section 5:9 we saw that the power radiated by a body is independent of an observer's rest-frame as long as both the source and observer move in inertial frames of reference. Make use of this fact to obtain the total power radiated in the form of synchrotron radiation, by computing the emission of the spiraling charge as viewed from an inertial frame moving with the charge's instantaneous velocity. For a charge whose total energy is $\mathcal{E}$ this total power is

$$P(\mathcal{E}) = \frac{2}{3} \frac{e^4 B^2}{m_0^2 c^3} \left(\frac{\mathcal{E}}{m_0 c^2}\right)^2 \qquad (6\text{-}155)$$

for motion perpendicular to the magnetic field. For an electron

$$P(\mathcal{E}) = 1.58 \times 10^{-15} B^2 \left(\frac{\mathcal{E}}{m_0 c^2}\right)^2 \text{ erg s}^{-1}$$

$$= 2.48 \times 10^{-2} \left(\frac{B^2}{8\pi}\right) \left(\frac{\mathcal{E}}{m_0 c^2}\right)^2 \text{ eV s}^{-1}.$$

Verify that these expressions are at least approximately consistent with expressions (6–151), (6–153), and (6–154) by noting that $P(\omega, \mathcal{E})\omega_m$ roughly corresponds to $P(\mathcal{E})$, and that equation (6–153) with a numerical integration under the curve in Fig. 6.13 gives the same result.

PROBLEM 6–14. One astronomical object in which synchrotron radiation is important is the *Crab Nebula*. Take the magnetic field strength in some of the Crab's bright filaments to be of order 10^{-4} G and show that a classically moving electron would radiate at a frequency of about 300 Hz, independent of the energy. On the other hand, if the energy becomes 10^9 eV, some 2×10^3 times the rest-mass energy, the peak radiation will occur at about 600 MHz. If the energy becomes 10^{12} eV, the radiation peaks in the visible part of the spectrum at 6×10^{14} Hz.

The exact form of the observed spectrum depends on the energy spectrum of the radiating particles as well as on the function $P(\nu, \mathcal{E})$. If we integrate the radiation coming from different distances r along the observed line of sight out to some distance R, the resulting spectral intensity at frequency ν becomes

$$I_\nu \, d\nu = \int_0^{\mathcal{E}_{\max}} \int_0^R P(\nu, \mathcal{E}) n(\mathcal{E}, r) \, dr \, d\mathcal{E} \, d\nu \,, \tag{6-156}$$

where $n(\mathcal{E}, r)$ is the number density of particles of energy $\mathcal{E}$ at distance r.

Frequently, the electrons have an exponential spectrum with constant exponent $-\gamma$, $n(\mathcal{E}) \propto \mathcal{E}^{-\gamma}$. The intensity then obeys the proportionality $I_\nu \propto \nu^{-\alpha}$ where $\alpha = (\gamma - 1)/2$. To show this we note from Fig. 6.13 and from (6–153) that $P(\nu, \mathcal{E})$ is equal to $16e^3 B/m_0 c^2$ multiplied by an amplitude 0.1 and 2π, since the bandwidth is $\Delta\nu = \Delta\omega/2\pi \sim 3\omega_m/2\pi$.

Let us now suppose that every electron deposits its total radiated power at frequency ω_m. Equation (6–151) shows that $\mathcal{E} \propto \omega_m^{1/2}$ so that $\mathcal{E}^{-\gamma} \propto \omega_m^{-\gamma/2}$, $\Delta\mathcal{E} \propto \Delta\omega/\omega_m^{1/2}$, and the total radiated power in (6–155) obeys the proportionality

$$I(\nu)\Delta\nu \propto \omega_m \mathcal{E}^{-\gamma} \Delta\mathcal{E} \propto \omega_m^{(1-\gamma)/2} \Delta\omega \tag{6-157}$$

for a constant spectrum along the path of integration. Hence

$$I(\nu) \propto \nu^{-\alpha}, \qquad \alpha = \frac{(\gamma - 1)}{2}. \tag{6-158}$$

Here, α is called the *spectral index* of the source. To obtain this relationship between electron energy and electromagnetic radiation spectra, the source must be optically thin. Optically thick (self-absorbing) sources are discussed below. For a wide variety of nonthermal cosmic sources $0.2 \lesssim \alpha \lesssim 1.2$. For extragalactic objects indexes up to $\alpha = 2$ occur. In the frequency range below a few gigahertz, many quasars have $\alpha < 0.5$; but they often contain optically thick components. Most radio galaxies have $\alpha > 0.5$ (Co72). Spectra of some extragalactic sources are shown in Fig. 6.14. For the Galaxy, confirmation of these general concepts is good. We observe a Galactic cosmic-ray electron spectrum with $\gamma \sim 2.6$. This is measured at the Earth's position, but the electrons have reached us from great distances. Radio waves also show an overall Galactic spectrum with index $\alpha \sim 0.8$ — in agreement with (6–158).

Equation (6–158) is of great importance in astrophysics because it permits us to estimate the relativistic electron energy spectrum by looking at the synchrotron radio emission from a distant region. The total intensity of the radio waves is, however,

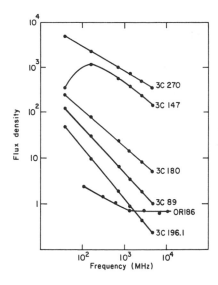

Fig. 6.14. Radio spectra of several extragalactic objects. Note that some have a constant slope, consistent with equation (6–158). Others show considerable curvature. (Compiled from Kellermann, Pauliny-Toth, and Williams (Ke69) and Jauncey, Neill, and Condon (Ja70). Flux density is measured in units of Janskys. 1 Jy $\equiv 10^{-26}$ Wm^{-2}Hz^{-1}. With the permission of the University of Chicago Press.)

not only a function of the total number of electrons along a line of sight, it also depends on the magnetic field strength in the region where the relativistic electrons radiate. Proton synchrotron radiation may also be important (Re68b).

PROBLEM 6–15. Show that $I(\nu)$ is proportional to $B^{(\gamma+1)/2}$. For a randomly oriented field, B^2 takes on the mean value of the component of the (magnetic field)^2 perpendicular to the line of sight. Hence

$$I(\nu) \propto B^{(\gamma+1)/2}\nu^{-(\gamma-1)/2}$$
$$\propto B^{\alpha+1}\nu^{-\alpha}. \tag{6-159}$$

Synchrotron emission can be accompanied by synchrotron absorption. Some strong extragalactic radio sources believed to radiate synchrotron emission exhibit a spectrum that is black in just those regions where synchrotron radiation would have the highest emissivity. This is interpreted (see Section 7:10) as meaning that the sources are opaque to their own radiation. Figure 6.14 shows that the flux for many nonthermal sources is high at low frequencies. On the other hand, at these low frequencies, the flux cannot exceed the flux of a blackbody

$$I(\nu)\,d\Omega = \frac{2kT}{c^2}\nu^2\,d\Omega\,, \tag{4-84}$$

whose temperature T is determined by the electron energy, $kT \sim \mathcal{E}$. Now, equation (6–151) gives the relation between the energy $\mathcal{E}$, the magnetic field in the source B, and the emitted frequency $\nu \sim \omega_m/2\pi$. Substituting for kT in (4–84), we then have

$$I(\nu)\, d\Omega = \left(\frac{8\pi\nu^5 m_0^3 c}{eB} \right)^{1/2} d\Omega \,, \qquad (6\text{-}160)$$

which expresses the magnetic field strength in the source in terms of the observed flux at frequency ν and the angular size of the source. The low-frequency spectrum then is no longer a blackbody spectrum because the energy of electrons decreases at lower radiated frequencies. Effectively the temperature of the electrons $\mathcal{E}/k$ is frequency dependent. The required data needed to compute the source magnetic field can be gathered with a radio interferometer (Ke71). The peaked spectrum of synchrotron self-absorption characterizes the radio source 3C147 (Fig. 6.14).

A synchrotron radiating electron has a lifetime (see equation (6–155))

$$\tau \sim \frac{\mathcal{E}}{P(\mathcal{E})} = \frac{3m_0^4 c^7}{2\mathcal{E}e^4 B^2} \,. \qquad (6\text{-}161)$$

For a 10^9 eV, electron this amounts to $\sim 10^5$ yr for $B \sim 10^{-3}$ G, a field that would be consistent with other observations on quasars. Yet, active galactic nuclei are sometimes observed to flare up on time scales much shorter than a year. These short-term variations can arise from synchrotron emission from superluminal jets. The Lorentz factor $\Gamma(v)$ arising from the bulk motion toward the observer can then reduce the observed time interval as indicated in (6–149). In addition, superluminal jets exhibit rapid lateral expansion as explained in Section 5:12. The expansion of the radiating surface further contributes to the increasing flux reaching an observer. And, finally, the changes in the magnetic field on expansion affect the flux reaching an observer at a fixed spectral frequency ν. Depending on the details of the jet expansion, the observed rise time of a flare can be reduced by many powers of the Lorentz factor. In one specific model this reduction is of order $\Gamma(v)^{7/2}$ (Re67). A Lorentz factor of ~ 35 could then reduce the observed rise time of a flare to well below a year.

We will see in Section 9:5 that highly relativistic expansion also plays an essential role in producing *gamma-ray bursts, GRBs*, which are often observed to persist for no more than a few seconds.

6:22 The Compton Effect and Inverse Compton Effect

When a high-energy photon impinges on a charged particle, it tends to transfer momentum to it, giving it an impulse with a component along the photon's initial direction of propagation. This is an effect that we had neglected in dealing with the low-energy Thomson scattering process. Although we talk about *Compton scattering* when we discuss the interaction of highly energetic electromagnetic radiation with charged particles, and *Thomson scattering* when lower energies are involved, the basic process is exactly the same; we are only talking about differences in the mathematical approach convenient for analyzing the most important physical effects in different energy ranges.

Corresponding to the Compton effect, there is an exactly parallel *inverse Compton effect*, in which a highly energetic particle transfers momentum to a low-energy photon and endows it with a large momentum and energy. These processes are exactly alike except that the coordinate frames from which they are viewed differ. To an observer O′ at rest with respect to the high-energy particle, the inverse Compton effect will appear to be ordinary Compton scattering. She would believe a highly energetic photon was being scattered by a stationary charged particle.

Because of this similarity we will only derive the expressions needed for the Compton effect, and then discuss the inverse effect in terms of a coordinate transformation. We set down four equations governing the interaction of a photon with a particle. The effect is more readily described in terms of photons rather than of electromagnetic waves, but this is only a matter of convenience and does not reflect a physical difference in the radiation involved. The considerations we have to take into account are:

(i) Conservation of mass–energy, given by

$$m_0 c^2 + h\nu = \mathcal{E} + h\nu' \,, \tag{6-162}$$

where ν and ν' are the radiation frequency before and after the collision, m_0 is the rest-mass, and $\mathcal{E}$ is the relativistic mass–energy of the recoil particle (Fig. 6.15).

(ii) The relation of $\mathcal{E}$ to m_0 is (5–34):

$$\mathcal{E} = m_0 c^2 \left(1 - \frac{v^2}{c^2}\right)^{-1/2} \equiv m_0 \gamma(v) c^2 \,. \tag{6-163}$$

(iii) Conservation of momentum along the direction of the incoming photon yields

$$\frac{h\nu}{c} = \frac{h\nu'}{c} \cos\theta + m_0 \gamma(v) v \cos\phi \,. \tag{6-164}$$

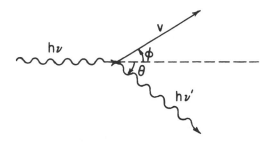

Fig. 6.15. Compton scattering.

(iv) The corresponding expression for the transverse momentum is

$$0 = \frac{h\nu'}{c}\sin\theta - m_0\gamma(\upsilon)\upsilon\sin\phi . \tag{6-165}$$

We now have four equations and four unknowns, υ, ν', θ, and ϕ.

PROBLEM 6–16. Show that these four equations can be solved to give the expression

$$\frac{c}{h}\left(\frac{1}{\nu'} - \frac{1}{\nu}\right) = \frac{1 - \cos\theta}{m_0\gamma c} . \tag{6-166}$$

By taking the wavelength of radiation to be $\lambda = c/\nu$, $\lambda' = c/\nu'$ we obtain

$$\lambda' - \lambda = 2\lambda_c \sin^2\frac{\theta}{2} , \tag{6-167}$$

where

$$\lambda_c \equiv \frac{h}{m_0\gamma c} \tag{6-168}$$

is called the *Compton wavelength* of the particle. For an electron $\lambda_c = 2.4 \times 10^{-2}$ Å or 2.4×10^{-10} cm. We note that for visible light, the change in wavelength amounts to only ~ 0.05 Å in 5000 Å, a nearly negligible effect. This is why momentum transfer could be neglected in Thomson scattering. However, in the X-ray region, say, at wavelengths of 0.5 Å, we encounter 10% effects; and at higher energies very large shifts can be expected, $(\lambda' - \lambda)/\lambda \gg 1$.

The cross-section for Compton scattering must be computed quantum mechanically and turns out to be dependent on the energy of the incoming photon. The expression for this cross-section (see Fig. 6.16) is known as the *Klein–Nishina formula*,

$$\sigma_c = 2\pi r_e^2\left\{\frac{1+\alpha}{\alpha^2}\left[\frac{2(1+\alpha)}{1+2\alpha} - \frac{1}{\alpha}\ln(1+2\alpha)\right] + \frac{1}{2\alpha}\ln(1+2\alpha) - \frac{1+3\alpha}{(1+2\alpha)^2}\right\} , \tag{6-169}$$

where r_e is the *classical electron radius* and α is the ratio of photon to electron energy. For an electron

$$r_e \equiv \frac{e^2}{m_0 c^2} = 2.82 \times 10^{-13}\,\text{cm}, \quad \text{and} \quad \alpha = \frac{h\nu}{m_0 c^2} . \tag{6-170}$$

In the extreme energy limits σ_c is approximated by

$$\sigma_L = \sigma_e\left\{1 - 2\alpha + \frac{26}{5}\alpha^2 + \cdots\right\}, \quad \alpha \ll 1 \quad \text{(low energies),} \tag{6-171}$$

$$\sigma_H = \frac{3}{8}\sigma_e\frac{1}{\alpha}\left(\ln 2\alpha + \frac{1}{2}\right), \quad \alpha \gg 1 \quad \text{(high energies) ,} \tag{6-172}$$

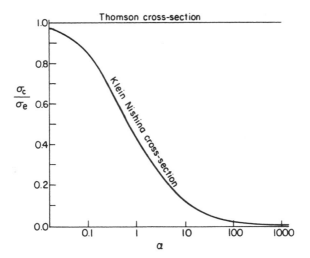

Fig. 6.16. Comparison of Compton and Thomson scattering cross-sections as a function of $\alpha = h\nu/m_0c^2$ (after Jánossy (Já50)).

where

$$\sigma_e = \frac{8\pi}{3}r_e^2 \tag{6-173}$$

is the Thomson scattering cross-section.

For a proton the cross-section would be smaller — inversely proportional to the mass. This means that Compton scattering is primarily an electron scattering phenomenon. Scattering by atoms takes place as though each atom had Z free electrons and the atomic scattering cross-section is just Z times greater than that for an individual electron. In Compton scattering the atomic binding energy is small compared to the photon energies and the electrons can be regarded as essentially free.

Let us still turn to the inverse Compton effect. Here, a highly relativistic electron collides with a low-energy photon and transfers momentum to convert it into a high-energy photon. This process can be followed from the point of view of an observer moving with the electron. He sees the incoming radiation blue-shifted (see equation (5–44)) to a wavelength

$$\lambda_D = \lambda\sqrt{\frac{c-v}{c+v}}. \tag{6-174}$$

Still in this frame of reference, the scattered wave has wavelength (6–167)

$$\lambda' = 2\lambda_c \sin^2\frac{\theta}{2} + \lambda_D , \tag{6-175}$$

since this is a simple Compton process to the observer initially at rest with respect to the electrons. For back-scattered radiation $\sin^2(\theta/2) = 1$.

Now, when this wave is once again viewed from the stationary reference system
— rather than from the viewpoint of the fast electron — a back-scattered photon is
found to have wavelength

$$\lambda_s \sim \lambda' \sqrt{\frac{c-v}{c+v}} \sim \lambda \left(\frac{c-v}{c+v}\right) + 2\lambda_c \left(\frac{c-v}{c+v}\right)^{1/2}. \tag{6-176}$$

This is the same transformation as (6–174); a stationary observer also sees back-
scattered radiation blue-shifted. Note that we have considered only direct back-
scattering and that this expression does not consider scattering at other angles. Quite
generally, however, it is clear that the wavelength of the photon becomes appreciably
shortened in the process and its energy is increased by factors of order

$$\frac{c+v}{c-v} \sim \frac{(1+v/c)^2}{(1-v^2/c^2)} \sim \frac{\mathcal{E}^2}{m_0^2 c^4}, \tag{6-177}$$

where $\mathcal{E}$ is the initial energy of the particle. As will be discussed in Section 9:8, the
total power radiated by an electron in inverse Compton scattering is closely related
to the power radiated in the form of synchrotron radiation. The total synchrotron
emission is proportional to the magnetic field energy density in space, $B^2/8\pi$. The
total inverse Compton scattering power loss for electrons is proportional to the electromagnetic radiation energy density in space. The proportionality constant for these
two processes is identical.

Regions emitting synchrotron radiation contain relativistic electrons capable of
inverse Compton scattering. These, as well as relativistic positrons and potentially
protons, are able to inverse Compton scatter synchrotron radiation to γ-ray energies.
γ-rays with energies ranging up to 10 TeV $= 10^{13}$ eV, emitted by the AGN Markarian
501 are believed to be produced in this *synchrotron-self-Compton process*.

Interestingly, a maximum brightness temperature of 10^{12} K can be set on optically thick synchrotron emitting sources. At this brightness the inverse Compton
scattering by the radiation emitted within the source quickly reduces the energy of
the relativistic electrons and thus bounds the brightness temperature.

6:23 The Sunyaev–Zel'dovich Effect

X-ray observations frequently detect intense emission from ionized gas at a temperature of $\sim 10^8$ K at the center of a cluster of galaxies. This gas not only radiates
through free–free emission, it also inverse Compton scatters the cosmic microwave
background radiation. The X-ray emission is proportional to the emission measure
(6–140) along the line of sight $\int n_e^2 \, dr$. The inverse Compton scattering, in contrast, is proportional to the column density of electrons $\int n_e \, dr$. It boosts the energy
of photons back-scattered toward an observer O, and diminishes the flux of background radiation reaching him from beyond the cluster. The overall effect is to raise
the observed flux at short wavelengths and diminish it in the long-wavelength tail of
the blackbody microwave background spectrum.

These two effects provide us with a measure of both $\int n_e^2\,dr$ and $\int n_e\,dr$, and thus permit us to solve for both the electron density and $\int dr$, the depth of the cloud along the line of sight. Assuming this depth to be the same as the transverse cloud diameter $D\theta$, where θ is the angular diameter of the X-ray emitting cloud and D is the distance to the cluster, we can solve for $D \sim \theta^{-1} \int dr$. The cluster's red shift z, taken together with this value of D, provides an independent measure of the Hubble constant, $H = zc/D$.

The change in the microwave background radiation is called the Sunyaev–Zel'dovich effect (Su72). For the cluster of galaxies CL 0016+16 the temperature decrement in the long-wavelength tail of the blackbody spectrum, at a wavelength of 1 cm is $\Delta T \sim 7 \times 10^{-4}\,K$ over a region that spans $\sim$75 arcsec (Ca96a). The countless distant clusters of galaxies in the sky are expected to show up as a slight patchiness in the microwave background on scales of roughly a minute of arc and at an apparent change in the measured background temperature of order $\delta T/T \sim 10^{-5}$ to 10^{-3}.

6:24 The Cherenkov Effect

We now come to a process that is primarily important for studying cosmic-ray particles — the *Cherenkov effect*. This effect dominates the interaction of incoming particles with the Earth's atmosphere. The Cherenkov effect causes these particles to decelerate radiatively and the light emitted can be used as a sensitive means for detecting the particles (See Sections 5:10 and 5:11).

To see how the effect works, we consider a highly relativistic particle entering the Earth's atmosphere. Because the particle is arriving from a region where the density has been very low and is entering a region of relatively high density, it finds that it has to make some adjustments. The presence of the electrically charged particle produces an impulse on atoms in the upper atmosphere and causes the atoms to radiate. The impulse comes about because the particle cannot move faster than the speed of light in this dense medium and must rapidly brake. The electric field due to the particle therefore appears to atmospheric atoms to be abruptly switched on; a rapidly time-varying field arises at the position of the perturbed atom. This is just the condition required to cause the atom to radiate. The relativistic particle will continue to affect atoms along its path in this way, as it keeps slowing down to the local speed of propagation of light. Once the deceleration is complete, the electric field changes produced in the vicinity of atoms take on a less abrupt character and the radiative effects are diminished.

There are many parallels between Cherenkov radiation and hydrodynamic shocks, which we will treat in Section 9:4. Just as a supersonic object that produces sonic booms and keeps on losing energy until it slows down to the local speed of sound, the cosmic-ray particle also keeps losing energy, through the Cherenkov effect, as it slows down to the local speed of light in the medium through which it is traveling.

For reasons given in Section 5:9, the radiation produced in this manner is emitted into a small forward angle of full width $\Delta\theta$ (see Fig. 6.17):

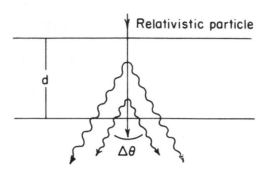

Fig. 6.17. Diagram to illustrate Cherenkov radiation.

$$\Delta\theta \sim 2\sqrt{1 - \frac{v^2}{c^2}} \tag{5-51}$$

just as in synchrotron emission or in any other relativistic radiation effect. The time of arrival of the radiation is also dictated by considerations similar to those found in synchrotron radiation. If the layer through which the radiation passes, before it has become sufficiently slowed down, is of thickness d, the time elapsing between the arrival of the first and the last photons of a wave train at the observer is, in analogy to equation (6–149),

$$\Delta t_c \sim \left(\frac{d}{v} - \frac{d}{c}\right) \sim \frac{d}{v}\left(1 - \frac{v}{c}\right) \sim \frac{d}{2c}\left(1 - \frac{v^2}{c^2}\right), \tag{6-178}$$

so that the corresponding frequency is of the order of

$$\omega_c = \frac{1}{\Delta t_c} \sim \frac{2c}{d}\left(1 - \frac{v^2}{c^2}\right)^{-1} \sim \frac{2c}{d}\left(\frac{\mathcal{E}}{m_0 c^2}\right)^2. \tag{6-179}$$

If the distance traversed in the upper atmosphere is of the order of $d \sim 10^6$ cm $= 10$ km, and a proton of energy 3×10^{14} eV is considered, $\mathcal{E}/mc^2 \sim 3 \times 10^5$ and $\omega_c \sim 6 \times 10^{15}$ or $\nu_c \sim 10^{15}$ Hz.

In many cases, an energetic primary produces a shower of secondary particles through collisions with atoms of the upper atmosphere. These secondaries can also give rise to Cherenkov radiation. The Cherenkov radiation spectrum does not normally peak at a frequency near ω_c. It depends more on atomic radiation properties of atmospheric gases and is relatively insensitive to the energy of the primary particle.

One interesting feature of Cherenkov detection is that it not only identifies the existence of cosmic-ray particles, but also gives the direction of arrival with reasonable accuracy; the uncertainty $\Delta\theta$ in the direction from which the particles appear to arrive is quite small.

Cherenkov radiation can be produced not only by high-energy cosmic-ray particles but also by high-energy gamma rays. Gamma rays with energies $\gtrsim 3 \times 10^{11}$ eV can be observed in this way. The detection process is indirect in this instance, and depends on the formation of very energetic secondary charged particles in the upper atmosphere. These, in turn, generate a visible light pulse through Cherenkov radiation. The direction of arrival indicates the direction of the emitting source.

Observations of powerful γ-ray flares from the blazars Markarian 421 and 501 — respectively at red shifts $z = 0.031$ and 0.034 — revealed not only γ-ray emission at energies exceeding ~ 10 TeV (1 TeV $\equiv 10^{12}$ eV), but also provided coarse spectra at these high energies (Kr01). By observing the high-energy spectral cutoff of the blazars at these very high energies, it has been possible to determine that the energy cut-off is due to the destruction of the γ-rays through collisions with intergalactic infrared photons to produce electron–positron pairs (See (5–52), (5–53)).

6:25 The Angular Distribution of Light from the Sky

In Section 6:8 we dealt with the Laplacian in Cartesian coordinates. For many problems, however, it is more convenient to use polar coordinates. This is particularly so for describing the surface brightness distribution across the sky.

PROBLEM 6–17. Show that the Laplacian expressed in Cartersian coordinates,

$$\nabla^2 = \frac{\partial^2}{\partial x^2} + \frac{\partial^2}{\partial y^2} + \frac{\partial^2}{\partial z^2} \tag{6-180}$$

can be rewritten in spherical polar coordinates as

$$\nabla^2 = \frac{1}{r^2} \frac{\partial}{\partial r} \left(r^2 \frac{\partial}{\partial r} \right) + \frac{1}{r^2 \sin\theta} \frac{\partial}{\partial \theta} \left(\sin\theta \frac{\partial}{\partial \theta} \right) + \frac{1}{r^2 \sin^2\theta} \frac{\partial^2}{\partial \phi^2} . \tag{6-181}$$

To do this it is helpful to recall the relation between Cartesian and spherical polar coordinates, $z = r\cos\theta$, $x = r\sin\theta\cos\phi$, $y = r\sin\theta\sin\phi$ and to apply the rule for partial differentiation of a function $f(r, \theta, \phi)$,

$$\frac{\partial}{\partial x} f = \frac{\partial r}{\partial x} \frac{\partial}{\partial r} f + \frac{\partial \theta}{\partial x} \frac{\partial}{\partial \theta} f + \frac{\partial \phi}{\partial x} \frac{\partial}{\partial \phi} f , \tag{6-182}$$

with similar expressions for partial differentiation with respect to y and z.

Equation (6–181) shows that $r^2 \nabla^2 f(r, \theta, \phi)$ separates into a purely radial and a purely angular term, so that by writing $f = R(r)Y(\theta, \phi)e^{i\omega t}$ we can separate the radial and angular parts of a wave equation such as (6–32)

$$\nabla^2 f = \alpha \frac{\partial^2 f}{\partial t^2} = -\alpha\omega^2 f , \tag{6-183}$$

where $\alpha = 1/c^2$ if we take $\epsilon = \mu = 1$. Then

$$\frac{1}{R}\frac{d}{dr}\left(r^2\frac{dR}{dr}\right) + r^2\frac{\omega^2}{c^2} = -\frac{1}{Y}\left[\frac{1}{\sin\theta}\frac{\partial}{\partial\theta}\left(\sin\theta\frac{\partial Y}{\partial\theta}\right) + \frac{1}{\sin^2\theta}\frac{\partial^2 Y}{\partial\phi^2}\right]. \quad (6\text{-}184)$$

Because the left side of this equation depends only on r, and the right side only on θ and ϕ, each side must equal some constant, λ. In particular, if we are interested solely in the angular part of the wave equation, we can write

$$\frac{1}{\sin\theta}\frac{\partial}{\partial\theta}\left(\sin\theta\frac{\partial Y}{\partial\theta}\right) + \frac{1}{\sin^2\theta}\frac{\partial^2 Y}{\partial\phi^2} + \lambda Y = 0. \quad (6\text{-}185)$$

We can separate this into two further equations, if we follow this procedure once more, by writing $Y(\theta,\phi) = \Theta(\theta)\Phi(\phi)$, which yields

$$\frac{d^2\Phi}{d\phi^2} + \nu\Phi = 0, \quad (6\text{-}186)$$

$$\frac{1}{\sin\theta}\frac{d}{d\theta}\left(\sin\theta\frac{d\Theta}{d\theta}\right) + \left(\lambda - \frac{\nu}{\sin^2\theta}\right)\Theta = 0. \quad (6\text{-}187)$$

The solution to (6–186) is obtained by choosing ν the square of an integer m that may be positive, negative, or zero.

$$\Phi_m(\phi) = \frac{1}{2\pi}e^{im\phi} \quad (\text{RP}). \quad (6\text{-}188)$$

This choice of ν assures that Φ will be single-valued over the entire range of ϕ. The multiplying constant $1/2\pi$ normalizes Φ to unity over the range $0 \leq \phi < 2\pi$. Substituting $\nu = m^2$ in (6—187) and, for convenience, replacing $\Theta(\theta)$ with a function $P(\cos\theta)$, one finds that the solutions to (6–187) diverge for values $\theta = 0$ or an integer multiple of π, unless λ has a special value $\lambda = \ell(\ell+1)$, ℓ is a positive integer, and $|m|$ is an integer or zero, with $|m| \leq \ell$.

The functions $P_\ell^m(\cos\theta)$ are called the *associated Legendre functions*, except when $m = 0$, in which case one uses the symbol P_ℓ and the corresponding function is called a *Legendre polynomial*. The polynomials can be written as,

$$P_\ell(x) = \frac{1}{2^\ell \ell!}\frac{d^\ell}{dx^\ell}(x^2 - 1)^\ell. \quad (6\text{-}189)$$

The associated Legendre functions are then formed through

$$P_\ell^m(x) = (1 - x^2)^{|m|/2}\frac{d^{|m|}}{dx^{|m|}}P_\ell(x). \quad (6\text{-}190)$$

We are now able to rewrite (6–185) in the form

$$\left[\frac{1}{\sin\theta}\frac{\partial}{\partial\theta}\left(\sin\theta\frac{\partial}{\partial\theta}\right) + \frac{1}{\sin^2\theta}\frac{\partial^2}{\partial\phi^2}\right]Y_{\ell m} = -\ell(\ell+1)Y_{\ell m}. \quad (6\text{-}191)$$

where the functions $Y_{\ell m}(\theta, \phi)$ are called *spherical harmonics* whose functional form is

$$Y_{\ell m}(\theta, \phi) = (\mp 1)^m \left(\frac{2\ell + 1}{4\pi} \frac{(\ell - |m|)!}{(\ell + |m|)!} \right)^{1/2} P_\ell^m(\cos\theta) e^{im\phi} , \qquad (6\text{-}192)$$

with $(\mp 1) = -1$ for $m \geq 0$ and $+1$ for $m < 0$. The functions $P_\ell^m(\cos\theta)$ obey

$$\int_0^\pi P_\ell^m(\cos\theta) P_{\ell'}^m(\cos\theta) \sin\theta d\theta = \frac{2}{(2\ell + 1)} \left[\frac{(\ell + |m|)!}{(\ell - |m|)!} \right] \delta_{\ell, \ell'} . \qquad (6\text{-}193)$$

The spherical harmonics form a complete set of *orthonormal* — orthogonal and normalized — functions, the lowest orders of which are illustrated in Figure 6.18,

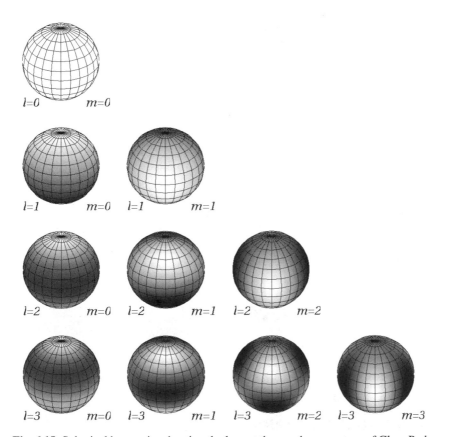

Fig. 6.18. Spherical harmonics showing the lowest three orders, courtesy of Clem Pryke.

$$\int_0^{2\pi} d\phi \int_0^\pi \sin\theta d\theta Y_{\ell'm'}^*(\theta, \phi) Y_{\ell m}(\theta, \phi) = \delta_{\ell'\ell} \delta_{m'm} , \qquad (6\text{-}194)$$

where the asterisk denotes the complex conjugate of the function, and the delta functions $\delta_{n'n}$ are unity when $n = n'$ and zero otherwise. The orthonormality of the spherical harmonics means that an arbitrary distribution of light across the celestial sphere can be described in terms of a superposition of these functions with different amplitudes $a_{\ell m}$, and relative phases (θ, ϕ). Note that the solution to the wave equation has a time dependence $e^{i(m\phi + \omega t)}$, indicating a rotation of the wave patterns in Fig. 6.18 about the vertical axis. The rotation is in the positive or negative direction, respectively, depending on whether m is negative or positive. When $m = 0$ the pattern pulsates. In Chapter 7 we will see that spherical harmonics are important for the description of angular momenta of quantized systems, such as rotating molecules or interstellar dust grains. In Chapter 13, we will again encounter these functions as we attempt to decipher the origins of the spatial inhomogeneities across the sky observed in the microwave background radiation.

The radial dependence, obtained by setting the left side of (6–184) equal to $\lambda = \ell(\ell+1)$, is a spherical Bessel function of the first kind $j_\ell(\omega r/c)$. Exact solutions for all values of $z \equiv \omega r/c$, respectively, for $\ell = 0$, 1 and 2, are

$$j_0(z) \equiv \frac{\sin z}{z}; \ j_1(z) \equiv \frac{\sin z}{z^2} - \frac{\cos z}{z}; \ j_2(z) \equiv \left(\frac{3}{z^2} - 1\right)\frac{\sin z}{z} - \frac{3}{z^2}\cos z,$$

$$(6\text{-}195)$$

as can be checked by substitution for R in (6–184). An asymptotic solution for all values of ℓ and r much greater than the wavelength $\omega/2\pi c$ is

$$j_\ell(z) = \frac{1}{z}\sin(z - \frac{\ell\pi}{2}) \quad \text{for} \quad z \to \infty. \tag{6-196}$$

ADDITIONAL PROBLEM 6–18. A rotating mass has energy $I\omega^2/2$, where I is the moment of inertia and ω is the angular frequency. Suppose that the rate of change of energy is proportional to the $n + 1$ power of ω:

$$\dot{\mathcal{E}} = K\omega^{n+1}. \tag{6-197}$$

Show that

$$n = \frac{\ddot{\omega}\omega}{\dot{\omega}^2}. \tag{6-198}$$

For the *Crab Nebula pulsar*, we observe that $n \sim 2.5$. Does this more nearly match the result expected for magnetic dipole or for gravitational radiation?

Answers to Selected Problems

6–1. $R_L = p_c c/qB = m_p v_p c/qB = 10^8$ cm .

6–2. For circular motion with a Larmor radius R_L:

$$\mathcal{E} = \frac{p_c v_c}{2} = \frac{qBv_c R_L}{2c},$$

and because angular momentum $R_L v_c$ is conserved,

$$\frac{d\mathcal{E}}{\mathcal{E}} = \frac{dB}{B} \ .$$

Hence the particles have a tenfold increase in energy.

6–3. Let $\mathcal{E}_i = \gamma(v_i)m_0c^2$ be the particles' initial energy ($\mathcal{E}_i = 10^{10}$ eV), where

$$\gamma(v_i) = \frac{1}{\sqrt{1 - v_i^2/c^2}}$$

and v_i is the initial velocity. v_f is the final velocity. Each interaction with the moving cloud imparts a velocity increment $\Delta V = 2 \cdot (7 \text{ km s}^{-1}) \sim 14 \text{ km s}^{-1}$ to the particles, as measured in their rest-frame. For protons, $m_0c^2 \sim 10^9$ eV, so that $\gamma(v_i) \sim 10$. If the energy is doubled $\gamma(v_f) \sim 20$, and

$$\left(\frac{v_i}{c}\right)^2 \cong (1 - 0.01) \Rightarrow \frac{v_i}{c} = 1 - 0.005,$$

$$\left(\frac{v_f}{c}\right)^2 \cong (1 - 0.0025) \Rightarrow \frac{v_f}{c} = 1 - 0.00125,$$

where v_f is the final velocity.

Hence $v_f - v_i = (0.0037)c = 1.1 \times 10^3 \text{ km s}^{-1}$.

By the law of composition of velocities (5–10), if v' is the velocity after one collision with a cloud:

$$v' \simeq (v_i + \Delta V)\left(1 + \frac{v_i \Delta V}{c^2}\right)^{-1} \sim v_i - v_i^2 \frac{\Delta V}{c^2} + \Delta V - 0\left[\frac{(\Delta V)}{c^2}\right],$$

$$v' - v_i = \Delta V\left(1 - \frac{v_i^2}{c^2}\right) = \frac{\Delta V}{\gamma^2(v_i)} \ .$$

The number of collisions needed to double the energy is then

$$\frac{1}{\Delta V}\int_{v_i}^{v_f} \frac{dv}{(1 - v^2/c^2)} \sim \frac{c}{2\Delta V}\ln\frac{\gamma_i^2}{\gamma_f^2} \sim 1.5 \times 10^4 \ . \tag{6-199}$$

If the clouds are initially separated by 10^{17} cm and approach each other at constant speed ΔV, they will collide after $\sim 7 \times 10^{10}$ s. Moving at relativistic speeds, the protons cover the mean distance between collisions $\sim 5 \times 10^{16}$ cm during this interval in an average time of $\sim 1.7 \times 10^6$ s. This can provide for as many as $\sim 4 \times 10^4$ collisions or for nearly an eightfold increase of energy before the clouds collide and the process comes to an end.

If the initial separation between clouds were to be increased, the number of collisions the relativistic particles could make would not increase, because the ratio of c to ΔV would remain constant. From this we see that the cloud collisions described here can produce no more than approximately an eightfold increase in proton energies.

Electrons have a higher γ value at corresponding energies but the time to double their energies, by (6–199), remains the same.

$6\text{–}4.\ \mathcal{E} = m_0 c^2 \gamma(wr), \qquad \gamma(wr) = \dfrac{1}{\sqrt{1 - w^2 r^2 / c^2}}.$

At $wr = c$, particles and field can no longer co-rotate. For a pulsar with a millisecond rotation rate, $w \sim 6 \times 10^3$ s^{-1}, and the maximum co-rotation radius is $\sim 5 \times 10^6$ cm, or about 5 stellar radii for the neutron star of Problem 5–15.

$6\text{–}6.$ The pressure exerted by a plane electromagnetic wave incident perpendicular to a surface is given by the magnitude of the Poynting vector (6–43) divided by the speed of light S/c, because the momentum per unit volume of such a wave is its energy density divided by c. Half of this pressure is due to the electric field component, the other half to the magnetic field. In terms of photons, the energy density of a number of photons per unit volume $n(\nu)$ at frequency ν is $nh\nu$, and the momentum carried by each photon is $h\nu/c$. The factor $\frac{1}{3}$ that enters the pressure relation in (4–42) is appropriate only when the photons or electromagnetic waves are incident on a surface at random angles, so that only a fraction of the momentum is transferred to the surface. In contrast, the longitudinal and transverse pressures we attributed to the static magnetic fields in this problem are selectively taken along the directions at which they take on their extreme values.

$6\text{–}7.$ Problem 4–3 gives the rms deviation of N steps of length L as $\sqrt{N}L$. In each step, the Faraday rotation angle is given by $\theta(L) = \frac{1}{2}(w/c)L\Delta n$. Substitution of (6–66) and the gyrofrequency from (6–62) gives the result.

$6\text{–}9.$ If we imagine fictitious magnetic charges, by analogy to electric charges,

$$\mathbf{H} = \frac{q_m}{(r - a/2)^2} - \frac{q_m}{(r + a/2)^2} = \frac{2q_m \mathbf{a}}{r^3} = \frac{2\mathbf{M}}{r^3}.$$

Hence $\mathbf{d}$ and $\mathbf{M}$ are analogous and by substitution we are led to the result (6–89).

$6\text{–}10.$ (a) $\mathbf{H} = \dfrac{\dot{\mathbf{A}} \wedge \mathbf{n}}{c} \quad$ and $\quad \mathbf{A} = \dfrac{1}{6c^2 R_0} \dddot{\mathbf{Q}},$

$$\mathbf{H} = \frac{1}{6c^2 R_0} \frac{\dddot{\mathbf{Q}} \times \mathbf{n}}{c},$$

$$\mathbf{S} = \frac{c}{4\pi} H^2 \mathbf{n} \propto \frac{(\dddot{\mathbf{Q}})^2}{c^5}.$$

(c) Equation (3–42) tells us that the total energy of the system is $\mathcal{E}\mu = -M\mu G/2a \sim 3.5 \times 10^{47}$ erg, where $\mu \sim M/4$ is the reduced mass for identical neutron stars. As gravitational radiation reduces the system's energy, the period also diminishes. Equation (3–47) shows that P is proportional to $\mathcal{E}^{-3/2}$, so that $\dot{P}/P = 3I/2\mathcal{E}\mu \sim 8 \times 10^{-17}$ s^{-1}.

$6\text{–}13.$ To show this, use the total radiated power (6–86):

$$\frac{2}{3c^3}\ddot{d}^2 = \frac{2e^2}{3c^3}\left(\frac{\dot{p}}{m_0}\right)^2 .$$

Equations (6–13) and (5–40) then give

$$\dot{p} = \frac{eB}{m_0 c}p_c = \left(\frac{eB}{m_0 c}\right)\left(\frac{\mathcal{E}}{c}\right) .$$

Substitute, to get the result (6–155).

6–14. $B = 10^{-4}$ G,

$$\nu_c = \frac{\omega_c}{2\pi} = \frac{eB}{2\pi m_0 c} = 300\text{ Hz} ,$$

$$\omega_m = \gamma^2 \omega_c = 1200\text{ MHz} \quad \text{for } \gamma = 2 \times 10^3 .$$

The peak lies at $\omega = \omega_m/2 = 600$ MHz.

6–15. By (6–156)

$$I(\nu)\Delta\nu \propto \int_0^{\mathcal{E}} P(\mathcal{E})n(\mathcal{E})\,d\mathcal{E} ,$$

where, by (6–155)

$$P(\mathcal{E}) \propto \mathcal{E}^2 B^2 .$$

$$\therefore\ I(\nu)\Delta\nu \propto B^2 \int_0^{\mathcal{E}} \mathcal{E}^2 \mathcal{E}^{-\gamma}\,d\mathcal{E} \propto K B^2 \mathcal{E}^{3-\gamma} .$$

But by (6–151)

$$\mathcal{E} \propto \left[\frac{\omega_m}{B}\right]^{1/2}$$

$$\therefore\ I(\nu)\Delta\nu \propto B^2 B^{(\gamma-3)/2}\omega_m^{(3-\gamma)/2} ,$$

$$I(\nu) \propto B^{(\gamma+1)/2}\nu^{(1-\gamma)/2} .$$

6–16. Squaring (6–164) and (6–165) and adding gives

$$-2h^2\nu\nu'\cos\theta + h^2(\nu^2 + \nu'^2) = (m_0 v\gamma c)^2 = m_0^2 c^4(\gamma^2 - 1)$$

Squaring (6–162) gives

$$m_0^2 c^4(\gamma^2 - 1) = h^2(\nu^2 + \nu'^2 - 2\nu\nu') + 2hm_0\gamma c^2(\nu - \nu') .$$

Equating these two expressions we obtain $h\nu\nu'(1 - \cos\theta) = (\nu - \nu')m_0\gamma c^2$ which is equivalent to (6–166).

6–17. We can write

$$r = (x^2 + y^2 + z^2)^{1/2},\ \theta = \cos^{-1}(z/(x^2 + y^2 + z^2)^{1/2}),\ \phi = \tan^{-1}(y/x) ,$$

to obtain

$$\frac{\partial r}{\partial x} = \sin\theta\cos\phi, \quad \frac{\partial\theta}{\partial x} = \frac{\cos\theta\cos\phi}{r}, \quad \frac{\partial\phi}{\partial x} = -\frac{\sin\phi}{r\sin\theta}, \tag{6-200}$$

$$\frac{\partial r}{\partial y} = \sin\theta\sin\phi, \quad \frac{\partial\theta}{\partial y} = \frac{\cos\theta\sin\phi}{r}, \quad \frac{\partial\phi}{\partial y} = \frac{\cos\phi}{r\sin\theta}, \tag{6-201}$$

$$\frac{\partial r}{\partial z} = \cos\theta, \quad \frac{\partial\theta}{\partial z} = -\frac{1}{r}\sin\theta, \quad \frac{\partial\phi}{\partial z} = 0. \tag{6-202}$$

Then, apply the rule (6–182) with (6–200) to obtain

$$\frac{\partial f}{\partial x} = \sin\theta\cos\phi\frac{\partial f}{\partial r} + \frac{\cos\theta\cos\phi}{r}\frac{\partial f}{\partial\theta} - \frac{\sin\phi}{r\sin\theta}\frac{\partial f}{\partial\phi}. \tag{6-203}$$

The application of (6–182) to $\partial f/\partial x$ in similar fashion yields

$$\frac{\partial^2 f}{\partial x^2} = \left(\frac{\partial r}{\partial x}\right)\frac{\partial}{\partial r}\left(\frac{\partial f}{\partial x}\right) + \left(\frac{\partial\theta}{\partial x}\right)\frac{\partial}{\partial\theta}\left(\frac{\partial f}{\partial x}\right) + \left(\frac{\partial\phi}{\partial x}\right)\frac{\partial}{\partial\phi}\left(\frac{\partial f}{\partial x}\right), \tag{6-204}$$

the right side of which, through the use of (6–200) and (6–203), can be expressed entirely in terms of spherical polar coordinates. With similar efforts to obtain $\partial^2 f/\partial y^2$ and $\partial^2 f/\partial z^2$, and considerable care in collecting the many cancelling or simplifying terms, one finally arrives at the equivalence of the Laplacians (6–180) and (6–181).

6–18.
$$\dot{\mathcal{E}} = K\omega^{n+1} = I\omega\dot{\omega},$$

$$\therefore \dot{\omega} = \frac{K}{I}\omega^n, \quad \ddot{\omega} = \frac{K}{I}n\omega^{n-1}\dot{\omega}, \quad \text{and} \quad \ddot{\omega}\omega = n\dot{\omega}^2.$$

Equations (6–89) and (6–96) show that $n + 1 = 4$ for the magnetic dipole, and $n + 1 = 6$ for the gravitational quadrupole radiation. The current data therefore are in closer agreement with a magnetic dipole mechanism. Since pulsars may emit predominantly gravitational radiation when they are first formed, observations leading to a value $n = 5$ could be obtained right after the formation of such an object. This would be interesting because the observations could be made in the radio domain but would give evidence of gravitational radiation that might be too difficult to directly detect.

7 Quantum Processes in Astrophysics

7:1 Absorption and Emission of Radiation by Atomic Systems

In Chapter 6, we considered the means by which particles could absorb or emit radiation. But we restricted ourselves to processes in which the Maxwell field equations of classical electrodynamics could be applied. These equations break down on the scale of atomic systems. The electron bound to a positively charged nucleus does not lose energy because of its accelerated motion, although the classical theory of radiation predicts that it should. Instead, the ground state of hydrogen, or any atom, is stable for an indefinitely long period. Moreover, when energy is actually radiated from excited states, and the atomic ground energy state is finally reached, we always find that only discrete amounts of energy have been given off in each transition. Again, this is at variance with classical predictions.

Since the interpretation of astronomical observations often depends on an understanding of transitions that occur between atomic levels, we will consider how they take place and what can be learned from them.

Almost everything we know about stars or galaxies is learned through spectroscopic observations. Our ideas of the chemistry of the Sun and of the chemical composition of other stars is based entirely on the interpretation of line strengths of different atoms, ions, or molecules. Our understanding of the temperature distribution in the solar corona is based on the strength of transitions observed for several highly ionized atoms, notably iron. Our picture of the distribution of magnetic fields across the surface of the Sun is founded on the interpretation of the splitting of atomic lines by magnetic fields in the solar surface. What we know about the motion of gases and their temperatures at different heights above the solar surface, again, is largely based on spectroscopic information. Small shifts in line positions and the shape and width of the lines yields much of the information we need. Some idea of the number densities of atoms, ions, or electrons at different levels of the solar atmosphere can also be obtained from line width and shape.

Even more information can be obtained spectroscopically because the Sun is nearby and can be clearly resolved. We can map photospheric velocities in fine detail over the entire solar surface by observing Doppler-shifted spectral lines to detect a large number of oscillatory modes. The Sun "rings" like a spherical bell. Some of the observed modes probe surface layers while others test the deep interior. We can thus conduct a seismological survey to obtain densities, temperatures, and rotation rates throughout the Sun. Even for more distant objects like emission-line stars or quasars, where detailed resolution does not appear possible, new information can still only be gained using spectroscopic techniques. Spectroscopy and a knowledge of quantum physics provide essential insight into astronomical processes.

7:2 Quantization of Atomic Systems

Classical theories of physics no longer apply on atomic scales, but a number of important features are shared by quantum and classical theory. Thus, in a closed system we find that:

(a) Mass–energy is always conserved.

(b) Momentum and angular momentum are always conserved.

(c) Electric charge is always conserved.

On the atomic scale, these conservation principles take on a form that deviates somewhat from classical formulations. However, where these differences are important, we can still be sure that:

(d) As the size of an atomic system grows, features predicted by quantum theory approach those calculated on the basis of classical physics. This is called the *correspondence principle*.

However, three major differences must also be noted:

(a') *Action*, a quantity that has units of (energy × time) or (momentum × distance), is quantized. The unit of action is $\hbar$. By this we mean that in a bound atomic system action can only change by integral amounts of Planck's constant h divided by 2π; $h/2\pi \equiv \hbar$.

(b') Even if they existed — and they do not — states of an atomic system whose characteristic action differs by an amount less than $\hbar$, cannot be distinguished. This is *Heisenberg's uncertainty principle*.

(c') Two particles having half-integral spin cannot have identical properties in the sense of having identical momentum, position, and spin direction. This is *Pauli's exclusion principle* (see Section 4:11).

The three statements (a'), (b'), and (c') are not axioms of quantum mechanics. Rather, they can be considered to be useful rules that emerge from the theory and make quantitative predictions about the behavior of electrons, atoms, and nuclei.

The concept of action is not as familiar as the idea of *angular momentum*, which has the same units and is subject to quantization in an identical way. We might therefore take a brief look at how angular momentum changes occur in atoms.

In any bound atomic system a change of angular momentum along any given direction in which we choose to make a measurement will always have a value $\hbar$ or an integer multiple of $\hbar$. The direction of this angular momentum is important.[1] We shall therefore talk about a *measured angular momentum component* with the

---

[1] Although the angular momentum along the measured direction can only change in steps whose size is $\hbar$, we have no such definite prescription for the changes that can simultane-

understanding that we have a definite direction in mind whenever we make a measurement.

Angular momentum quantization can be understood in more basic terms. All the fundamental particles involved in building up atoms have well-defined *spins*. For electrons, protons, and neutrons the values are $\pm\hbar/2$. A change from one spin orientation of an atomic electron, over to another orientation amounts to a change of one unit of $\hbar$ in the measured angular momentum component. Such a change can be brought about by the absorption or emission of a single photon. Photon spin angular momenta $\pm\hbar$ are invariably collinear with the direction of motion — a Lorentz-invariant property of the photon. In quadrupole and higher-order multipole radiation photons, respectively, carry away angular momenta of $2\hbar$ or higher, comprising both spin and orbital angular momentum; but only the spin component is collinear with the direction of propagation. All the different states of an atom can be reached from any other state through a succession of photon absorption or emission processes, or through a set of spin–flip transitions of the electrons or within the nucleus.

However, quantization is intrinsic to atoms even without this argument concerning photons. We can therefore be sure that if an atomic system has a state of zero angular momentum, then all other states must have integral values of the angular momentum. Similarly, if the lowest angular momentum state has a value $\hbar/2$, then all other states must have half integral angular momenta (see also Section 7:6).

This is one way in which statement (a'), above, provides insight into the structure of quantized systems. In addition, principle (b') gives us some general quantitative information. Let us consider the simplest atom, hydrogen, in terms of this principle. The energy of the lowest state can be estimated directly. The smallest possible size of an electrostatically bound atom must be related to the uncertainty in momentum through

$$p^2 r^2 \sim \langle \Delta p^2 \rangle \langle \Delta r^2 \rangle \sim \hbar^2. \tag{7-1}$$

Here we have taken the mean squared value of the radial momentum and the radial position as being equal to the uncertainty in these parameters. Through the virial theorem applied to a system bound by inverse square law forces, we can write the energy of a state either as half the electrostatic potential energy for the interacting proton and electron, or as the negative of the system's kinetic energy (3–85). Thus the lowest energy state is

$$\mathcal{E}_1 = -\frac{Z}{2}\frac{e^2}{r} = -\frac{p^2}{2\mu} \sim -\frac{\hbar^2}{2\mu r^2} , \tag{7-2}$$

where μ is the reduced mass of the electron. Here we have made use of equation (7–1) and the extreme right side of (7–2), which also shows that

---

ously take place in the transverse angular momentum. The uncertainty principle precludes a simultaneous definitive measurement of the longitudinal and transverse angular momentum components. All that we can say is that there exist *selection rules* that also specify allowed changes of the *total angular momentum* of a system, as we will see in Section 7:6. They state that the magnitude of the angular momentum squared J^2 changes by integral amounts of a basic step size $\hbar^2$. These integral amounts depend on the initial value of J characterizing the system, and on the multipole considerations mentioned in Section 6:13.

$$\mathcal{E}_1 = -\frac{Z^2 \mu e^4}{2\hbar^2} , \qquad (7\text{-}3)$$

$$r = \frac{\hbar^2}{Z\mu e^2} . \qquad (7\text{-}4)$$

This root mean square radius r is called the *Bohr radius* of the atom and $\mathcal{E}_1$ is the atom's *ground state energy*. For hydrogen $Z = 1$ and $\mathcal{E}_1 = -13.6\,\text{eV}$, $r \sim 5.29 \times 10^{-9}\,\text{cm}$. We have proceeded here on the assumption that the electrostatic potential confines the electron to a limited volume around the proton, and have derived a solution consistent with the uncertainty principle. Nothing has been assumed about possible orbits that the electron might describe around the proton; the very act of setting the mean squared value of position and momentum equal to the mean squared value of their uncertainties, implied that the electron is to be found in the whole volume, not just in a well-defined orbit having a narrow range of r or p values. Neither, however, does the electron spend equal amounts of time throughout the volume. Its dwell time in different locations has a probability distribution generally approximated by a standing wave that has zero amplitude beyond the zone of confinement. The wavelength of this standing wave is called the *de Broglie wavelength*, λ. Its associated momentum is $p = h/\lambda$. Most quantum systems permit a confined particle to be in one of several possible states defined by energy, angular momentum, or ambient electric or magnetic fields.

PROBLEM 7–1. If we wanted to distinguish successive states having different radial positions and momenta, the product pr for these states would have to differ by $\hbar$; otherwise, they would not be distinguishable in Heisenberg's sense. Setting $p_n r_n = n\hbar$, show that

$$\mathcal{E}_n = -Z^2 \frac{\mu e^4}{2n^2}\hbar^{-2}. \qquad (7\text{-}5)$$

If the nth radial state of the atom has a phase space volume $(4\pi p_n^2 \Delta p_n)(4\pi r_n^2 \Delta r_n)$, show that the number of possible states with principal quantum number n is proportional to n^2. Show also that the Bohr radius of the nth state is $r_n = n^2\hbar^2/Z\mu e^2$.

In order to find the actual number of states corresponding to the quantum number n, we still have to invoke the Pauli exclusion principle — statement (c′) above. We know that the state $n = 1$ corresponds to a single cell in phase space. Accordingly there can only be two states, one in which the nuclear spin and electron spin are parallel and the other in which they are antiparallel. Using the result of Problem 7–1, we then see that the nth radial state comprises $2n^2$ different substates, all having the same energy to the approximation considered here. The factor 2 enters because the electrons can be in either of two spin states.

We see from this that just the most basic concepts (a′), (b′), and (c′) suffice to tell a great deal about the structure of hydrogen and hydrogenlike atoms such as singly ionized helium, five times ionized carbon, or any other bare nucleus orbited by just one electron.

We should, however, not pretend that all problems of atomic structure can be handled as simply. Equation (7–2), for example, makes use of Newtonian mechanics and electrostatic interactions alone. We have neglected all relativistic effects and all interactions of the spins of particles that constitute the atom. In dealing with such features, or with the interactions between particles and various types of fields, we need to use the full mathematical structure provided by quantum mechanics. At the basis of any such structure, however, are the elementary principles (a) to (d) and (a') to (c'). We shall make much use of them in the next few sections.

PROBLEM 7–2. We can show that the principles (a') to (c') also permit a determination of the size of the atomic nucleus. To see this consider the *nucleons* — protons and neutrons — to be bound to each other by a short-range attractive potential

$$\mathbb{V} = -\mathbb{V}_0 \quad \text{if} \quad r < r_0, \qquad \mathbb{V} = 0 \quad \text{if} \quad r \geq r_0 . \qquad (7\text{-}6)$$

Each nucleon is tied to the nucleus by a binding energy $\mathcal{E}_b$, so that, while in the nucleus, it has kinetic energy $\mathbb{V}_0 - \mathcal{E}_b$. Using equation (7–1) show that

$$r \sim \frac{\hbar}{[2M(\mathbb{V}_0 - \mathcal{E}_b)]^{1/2}} . \qquad (7\text{-}7)$$

where M is the nucleon mass. If $\mathbb{V} \sim 2\mathcal{E}_b$, and $\mathcal{E}_b$ is roughly 6 MeV, show that a typical nuclear radius is of order 10^{-13} cm. This gives a characteristic interaction cross-section of $\sim 10^{-26}$ cm^{-2} for nucleons. We will find this to be of interest in Chapter 8, where we will examine nuclear processes in stars.

Note that the nuclear radius is quite insensitive to the value of r_0. The depth of the potential well and the binding energy determine the size of the nucleus.

7:3 Atomic Hydrogen and Hydrogenlike Spectra

We next turn to the spectra of atomic hydrogen, the most abundant element in the Universe (He44), (He67)*. The energy of a spectral line seen in absorption or emission is the energy difference of the atomic levels between which the transition occurs.

To start with one of the simplest concepts, we notice that in (7–3) and (7–5) the energy depends on the reduced hydrogenic mass. It therefore has a somewhat different value for normal hydrogen, which has only a proton in its nucleus, and for deuterium which has a nucleus composed of a neutron and a proton, making the nucleus about twice as massive. The reduced mass, μ_D, of an electron orbiting a deuterium nucleus accordingly is

$$\mu_D = \frac{m_e m_D}{m_e + m_D} \sim \frac{2 m_e m_P}{m_e + 2 m_P} \sim m_e \left(1 - \frac{m_e}{2 m_P}\right), \qquad (7\text{-}8)$$

whereas

$$\mu_P \sim m_e \left(1 - \frac{m_e}{m_P}\right). \tag{7-9}$$

Subscripts e, D, and P, here represent electrons, deuterons, and protons. Equations (7–5), (7–8), and (7–9) lead us to expect that deuterium energy levels will lie further apart than levels of ordinary hydrogen by about one part in $2m_P/m_e \sim 3700$. In the visible portion of the spectrum this corresponds to a line shift of the order of 1.5 Å toward shorter wavelengths. Such a spectral shift is readily determined. But for many years, despite much searching, no deuterium was ever detected in any astronomical object at levels comparable to the terrestrial abundance of deuterium, D/H $\sim 2 \times 10^{-4}$. In the interstellar medium, atomic D/H ratios are now reliably measured at $\sim 1.5 \times 10^{-5}$ (Oℓ03); and, as Fig. 12.5 indicates, the best estimates for the cosmic abundance of deuterium give D/H $\sim 2 - 3 \times 10^{-5}$. The high terrestrial deuterium abundance appears to be an anomaly due to chemical fractionation. Deuterium and hydrogen bond to other molecules with different bond strengths and, at identical temperatures and densities, tend to react differently to form quite different molecules. In the interstellar molecular cloud NGC 1333, where temperatures are estimated to be as low as 10 K, the ratio of deuterated to normal ammonia is an astonishingly high $ND_3/NH_3 \sim 10^{-3}$ (vd02). Such measurements remind us that, to avoid false conclusions in the compilation of tables of cosmic abundances such as Table 1.1, it is important to check for the prevalence of elemental and isotopic species in a variety of different atomic, molecular, and ionic states that may be favored at select temperatures and densities.

The reduced mass just mentioned also helps to distinguish hydrogenic transitions from spectral lines of ionized helium. Singly ionized helium HeII has one electron surrounding a nucleus with charge $Z = 2$. According to equation (7–5) the energy of any given state should therefore be just four times as great as the corresponding hydrogenic energy. This integral relation would sometimes lead to an exact identity of line energies for transitions involving *principal quantum numbers*, n, that were twice as great for helium as for hydrogen. The difference in reduced mass, however, shifts these lines sufficiently, so that ambiguities can often be avoided in astronomical observations. When the Doppler line shift for a moving source is not known, identification on the basis of one or two lines may, however, not be possible, and a search may have to be made for lines of other well-known atoms or — for helium — lines that are not common to the hydrogen spectrum, that is, transitions involving one level with an odd principal quantum number, and one level with an even value of n.

Although we have presented this similarity of spectra as though it were a matter of difficulty, it is often a great help. After years of theoretical work that explains many fine details, we understand the hydrogen spectrum well. Whenever it becomes possible to relate properties of complex atoms to specific similar properties of hydrogen, a whole body of theoretical knowledge becomes available at once, and this can lead to a better understanding of the more complex system.

Until recent decades, interest in hydrogenic spectra centered largely around transitions in which at least one of the states had a low principal quantum number, say, $n \lesssim 5$. Not much thought was given to very high lying states, and transitions to be found between such states were always thought to give rise only to very weak spectral lines. It therefore came as a surprise that transitions involving states with $n = 90, 104, 159, 166$, and many others in this same range were observable in radio-astronomy (Hö65). Not only these, but also the correspondingly excited helium states could be identified and again distinguished on the basis of reduced mass differences. Transitions between such highly excited states are sometimes referred to as *Rydberg transitions*. The spectral lines emitted in these transitions have permitted the observation of ionized hydrogen regions over great distances in the Galaxy. Ionized regions that are not detectable in the visible range now are readily accessible because radio waves are transmitted through dust clouds that extinguish visible light from all but the nearest portions of the Galaxy. Often these regions had been previously known, because dense ionized plasmas emit readily measured thermal continuum radiation (Section 6:19). But the discovery of the line radiation from hydrogen and hydrogenlike ions in these incompletely ionized regions allowed us to deduce the radial velocity of the region and its distance within the Galaxy calculated on the basis of differential rotation models (Section 3:12).

For completeness we should still present some of the terminology used in discussions of the hydrogen spectrum. Transitions involving lower states $n = 1, 2$, 3, and 4, respectively, are members of the *Lyman, Balmer, Paschen*, and *Brackett* spectral series (see Fig. 7.1). The line with the longest wavelength in each of these spectral series is termed α; the second line is called β, and so on. Thus the transition $n = 4 \rightarrow n = 2$ gives rise to the Balmer-β line in emission. Members of the Balmer

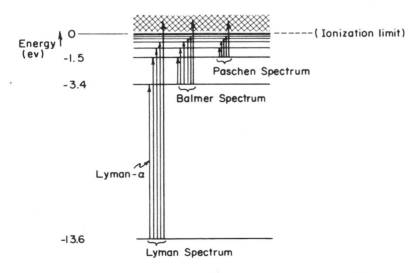

Fig. 7.1. Energy-level diagram of atomic hydrogen.

spectrum sometimes are written out as Hα, Hβ, and so on. Lyman spectral lines are generally written as Ly-α, Ly-β, and so on.

Problem 7–1 showed that the nth energy level of the hydrogen atom comprises $2n^2$ distinct quantum states. Each of the n^2 electronic states has two distinct components, corresponding to the two different orientations of the electron's spin relative to the nuclear spin direction. These two configurations have slightly different energies and hence a transition from the higher to the lower state can occur spontaneously.

The energy separation is called the *hyperfine splitting*, shown for the hydrogen ground state in Fig. 7.2. The split represents an energy difference between atoms

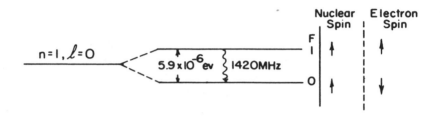

Fig. 7.2. Energy-level diagram for the *hyperfine splitting* of the ground state of the hydrogen atom. The state in which the electron and proton spins are aligned has a slightly higher energy. The transition frequency 1420 MHz corresponds to a radio wavelength of 21 cm.

having electron and nuclear spins opposed (total spin angular momentum quantum number $F = 0$) or parallel ($F = 1$). This splitting is present at all levels n and assures a total multiplicity of $2n^2$ states at any given level.

In radio-astronomy this transition within the state $n = 1$ has played a leading role. It occurs at a frequency of 1420 MHz, corresponding to a wavelength of 21 cm and an energy difference of $\sim 6 \times 10^{-6}$ eV — less than one part in two million of the binding energy of the atom in its ground state ~ 13.6 eV. The distribution of hydrogen in our galaxy was first mapped by means of 1420 MHz observations. This was possible because, as already stated, radio waves are not absorbed by the dust that extinguishes visible light. It is interesting that we now have rather good maps showing the distribution of gas in the Galaxy, but no comparable map showing the distribution of stars. Stars do not emit sufficiently in the radio or infrared parts of the spectrum and, across the span of the Galaxy, their visible light is totally extinguished by intervening dust clouds.

When we take a look at the first excited electronic state of the hydrogen atom, $n = 2$, we encounter two types of sublevels whose energies happen to be close. First, just as in the ground state, there are again two hyperfine states in which the electron has zero orbital angular momentum about the nucleus. A transition from these two states to the ground state through the emission of a photon is forbidden because the orbital angular momentum of the atomic system would have to remain unchanged in such a transition. This is not possible because the photon involved always carries off angular momentum. In tenuous ionized regions of interstellar space, the lifetime of atoms in such a state of $n = 2$ can therefore be very long. Eventually, the atom can revert to the ground state through the emission of two photons rather than just one; but the lifetime against this two-photon decay is of the order of 0.12 s (Nu84), in contrast to the usual 10^{-8} s required for normal allowed transitions. Metastable helium atoms similarly have a two-photon decay time measured as $\sim 2 \times 10^{-2}$ s (Va70).

We might still ask whether the angular momentum criterion in such transitions could be satisfied if the electronic transition was accompanied by a spin flip from a parallel to an antiparallel configuration. The coupling between the electron spin and the electromagnetic radiation, however, is low and does not suffice to make that transition as probable as the two-photon decay.

The second set of levels, within the state $n = 2$, all have an *orbital angular momentum quantum number* $l = 1$, with a corresponding (total angular momentum)^2 of $l(l + 1)\hbar^2 = 2\hbar^2$. The (angular momentum)^2 does not have the value $l^2\hbar^2$, because, aside from a well-defined angular momentum component about one (arbitrarily) chosen axis, there always remains an uncertain angular momentum about two orthogonal axes, which adds an amount $l\hbar^2$ to the (angular momentum)^2 (see Section 7:6). Corresponding to $l = 1$, there are three sublevels, each split into two further, hyperfine states. One of these sublevels has an angular momentum component $\hbar$ along some given direction, the second has a component 0, and the third has a component $-\hbar$ along that direction. These three components are labeled $m = 1$, 0, and -1. The label m is called the *magnetic quantum number* because the states have different energies when a magnetic field is applied to the atom. In the absence of a magnetic field, there is a splitting of the order of 10^{-5} eV between some of these states. This is called the *fine structure* of the atom and is shown in Fig. 7.3 [2]

The excited state $n = 3$ again has a hyperfine split sublevel of zero angular momentum, $l = 0$. There are three such pairs of states with $l = 1$, and five pairs with $l = 2$, corresponding to magnetic quantum numbers $m = 2, 1, 0, -1$, and -2. Under normal conditions, these levels are *degenerate*, meaning that they have precisely the same energy. In an applied magnetic field, however, the energy of the states is shifted somewhat and the energy separation between states becomes proportional to the field strength H for low values of H. This splitting is called *Zeeman splitting*.

Zeeman splitting can be understood in the following way. The orbital angular momentum of the electron implies a loop current that has an associated magnetic dipole field. Depending on whether this dipole, respectively, is aligned along, per-

---

[2] For atoms other than hydrogen, the labeling of states does not proceed in this particular way because the spins and orbital angular momenta of the electrons interact through their magnetic moments. However, the enumeration of the different quantum states still proceeds in terms of their distinguishing characteristics — that is, in terms of the Heisenberg and Pauli principles.

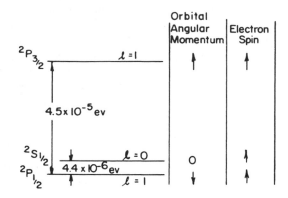

Fig. 7.3. Energy-level diagram showing the fine structure of the $n = 2$ level of hydrogen. The labeling in the left-hand column has the following significance. The letters S and P denote the total orbital angular momenta 0 and 1, respectively. The right lower index gives the total angular momentum resulting from a vectorial addition of the electron and orbital angular momenta. The left upper index is the *multiplicity* $(2S + 1)$ of the term, where S now is the total electron spin. (This double meaning of S sometimes leads to confusion.) As an example, the $^2P_{3/2}$ state has $l = 1$; the orbital and electron spins are parallel, giving a total spin $3/2$; and, because the spin for a single electron has magnitude $1/2$, the left superscript is 2.

pendicular to, or opposed to the field, we have an atomic state with decreased, unaltered, or increased energy.

Quantitatively, the orbital angular momentum of the electron about the nucleus gives rise to a magnetic dipole moment with components along the field direction of

$$\mu_B m_i = \frac{e\hbar}{2mc}m_i, \qquad i = 0, \pm 1, \pm 2, \ldots, \tag{7-10}$$

where μ_B is called the *Bohr magneton*. The energy of a state in a magnetic field is then

$$\mathcal{E} = \mathcal{E}_0 + \mu_B H m_i = \mathcal{E}_0 + \hbar \omega_L m_i, \qquad \omega_L = \frac{eH}{2mc}. \tag{7-11}$$

The state with the smallest energy has its angular momentum antiparallel to the field direction. This is the configuration in which the quantum number m_i has its lowest value. The *Larmor frequency*, ω_L, should be compared to the gyrofrequency (6–13) which is twice as large: $\omega_c = 2\omega_L$.

We note that the classical energy of a magnetic dipole in a magnetic field would be $\mathbf{M} \cdot \mathbf{H}$. But when this expression is used with equation (6–18), and we seek to find the energy of a magnetic dipole aligned with the field, we obtain

$$\mathcal{E} = \mathbf{M} \cdot \mathbf{H} = \frac{e(\mathbf{v} \wedge \mathbf{r}) \cdot \mathbf{H}}{2c} = \frac{eLH}{2mc} = \frac{\omega_c L}{2}. \tag{6-18a}$$

Here we have made use of equation (6–13) to see the classical energy dependence on ω_c. We can now see why ω_L is only half the gyrofrequency. On the other hand, if

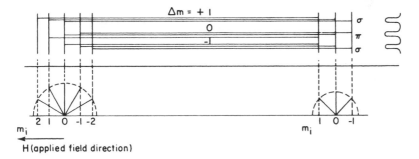

Fig. 7.4. Zeeman splitting: Transitions between energy states shifted through the application of an external magnetic field. The figure shows both the orientation of the angular momentum components relative to the direction of the applied field, and the shifts in the transition energy. We note that the angular momentum components along the field direction always have values that are integral multiples of $\hbar$. The total length of the angular momentum vector is $[l(l + 1)]^{1/2}$, that is, nonintegral. Symbols σ and π denote the states of polarization of the emitted radiation (see text). A large horizontal distance between levels indicates a large transition energy between two states.

we make use of the Larmor frequency in (7–11) we preserve an analogy to photons in that, for $m_i = 1$, the magnetic energy becomes $\mathcal{E} - \mathcal{E}_0 = h\nu_L = \hbar\omega_L$.

Figure 7.4 shows the splitting in the energy levels corresponding to quantum numbers $l = 2$ and $l = 1$ and gives the spectral lines that arise from a transition between such states.

Zeeman splitting provides us with useful information about the magnetic fields on the surface of the Sun and distant stars. In some strongly magnetic stars of spectral type A, fields higher than 30,000 G have been recorded. The general dipole field for the Sun has a value of the order of a gauss, but the local field strength varies greatly. In sunspots, fields of 3×10^3 G are not uncommon.

The determination of magnetic field strengths from the spectra alone would normally be difficult, because the lines are broad and often overlap since the splitting is small. Fortunately, however, the lines marked σ in Fig. 7.4 have a different polarization from the line marked π. The *magnetographic method* makes use of this polarization to separate out the different components through use of analyzers sensitive only to light of a given polarization. By carefully measuring the line centers of the variously polarized components, we can obtain the energy splitting between states, even when the lines are strongly broadened through disturbing effects. For solar work, spectral lines of iron or chromium are often used. The energy splitting gives H directly through equation (7–11).

Interstellar magnetic fields have been measured in a similar way by means of radio observations. The principle of this technique is identical to that used in solar work.

In neutral hydrogen regions of interstellar space, magnetic fields split the 21 cm line of atomic hydrogen. Three different transitions are expected, corresponding to $\Delta m = 0, \pm 1$. Viewed along the direction of the magnetic field, only two lines appear, respectively, at frequencies $\nu = \omega/2\pi$ (see equation (7–11))

$$\nu = \nu_0 \pm \frac{eH}{4\pi mc}, \qquad \nu_0 = 1420\,\text{MHz}. \tag{7-12}$$

These two components are circularly polarized in opposite senses (Fig. 7.5). They

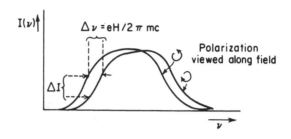

Fig. 7.5. Zeeman shift of the positions of the two circularly polarized components of the 1420 MHz hydrogen line viewed along the direction of the magnetic field.

are called the σ-components (Fig. 7.4) and appear linearly polarized when viewed normal to the field direction; their direction of polarization is at right angles to the direction of the field lines.

There is also an unshifted component, the π-component, which appears at frequency ν_0 when viewed normal to the direction of the field; it is linearly polarized with the direction of polarization parallel to the field. Viewed along the field lines, this component does not appear at all.

The observation of Zeeman splitting is made difficult by the rapid motion of the interstellar gas atoms. This produces a Doppler broadening of the lines (Section 7:7). A $1\,\text{km\,s}^{-1}$ random motion gives a frequency shift of order $\delta\nu/\nu \sim 3 \times 10^{-6}$. At the 21 cm line frequency of 1420 MHz, this corresponds roughly to 4×10^3 Hz. In contrast, the frequency split $\Delta\nu$ due to the magnetic field is $2.8 \times 10^6 H$ Hz between the two σ-components. This means that a field of order 10^{-5} G, only gives a split of $\Delta\nu \sim 30$ Hz.

Normally such a splitting would be all but impossible to observe in the presence of the overriding Doppler broadening. The saving feature is the difference in polarization. By working at the edge of the line, where the slope is steep, the difference in intensity ΔI of the two polarized components can be accentuated (Fig. 7.5). This technique has established the existence of fields of order $\sim 2 \times 10^{-5}$ G in interstellar clouds.

7:4 Spectra of Ionized Hydrogen

(a) Positive Ions

Hydrogen can become ionized through the absorption of a photon whose energy is 13.6 eV or higher. Once the minimum energy required to loosen the electron from the proton is reached, the excess energy can always be absorbed in the form of translational kinetic energy of the electron and proton. This feature is important in determining the appearance of very hot stars.

We might think that the recombination of electrons with protons would similarly regenerate ultraviolet photons. However, this occurs only part of the time. Frequently the recombination leaves the atom in one of its excited states, with a subsequent cascade through lower states down to the ground state. In this process, a number of less energetic photons are created. At energies somewhat higher than the ionization limit, the absorption cross-section is of order 10^{-17} cm^2. Hence, for typical interstellar densities of order 1 atom cm^{-3}, the mean free path for absorption is only of order 10^{17} cm, in contrast to standard interstellar distances of order 3×10^{18} cm or more. In a random walk, an ionizing photon would then have 10^3 opportunities for ionization and recombination in crossing 3×10^{18} cm. The probability that an ionizing photon would penetrate the full 3×10^{18} cm without ever ionizing a single hydrogen atom is of order $e^{-30} \sim 10^{-13}$. As a result, we seldom observe ultraviolet photons beyond the ionization limit — the *Lyman limit*.

The small region of the Galaxy in which the Sun happens to be located, however, has unusually low interstellar gas density. This low density region, referred to as the *Local Bubble* has made it possible to observe nearby white dwarfs even at extreme ultraviolet wavelengths.

(b) Negative Ions

A hydrogen atom can become ionized not only through the loss of an electron, but also by gaining one to become a negative ion H$^-$, a *hydride ion*.

$$\mathrm{H} + e \rightarrow \mathrm{H}^- + h\nu. \tag{7-13}$$

The structure of this ion is somewhat similar to that of the neutral helium atom in having two electrons bound to a nucleus. The second electron is only weakly bound, because the first electron is quite effective in screening out the nuclear charge. The binding energy is 0.75 eV and there is only one bound state. All transitions from or to this state involve the continuum, that is, a neutral hydrogen atom and a free electron (Fig. 7.6).

Because its 0.75 eV binding energy is so low, H$^-$ ions readily absorb visible starlight from cool stars like the Sun. This absorption continues out to a wavelength of 1.65 μm where photon energies drop below the bound–free transition energy. However, even at longer wavelengths absorption can take place, since the H$^-$ ion also can have free–free transitions — absorption in which energy is taken up by

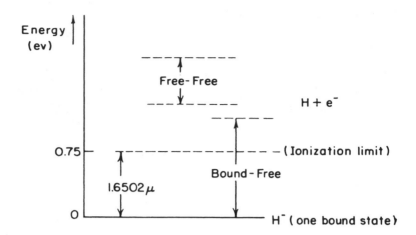

Fig. 7.6. Energy-level diagram of the H⁻ ion. There is only one bound state with binding energy 0.75 eV. All radiative transitions must be either between the bound state and a state with a free electron, or else between two free states.

the translational energy of an unbound electron in the presence of a hydrogen atom. This is no small effect. The H⁻ ion plays an important role in the transport of energy through the solar atmosphere (Ch58).

We should still explain how H⁻ even happens to exist in the atmosphere of cool stars. It survives because metal atoms such as sodium, calcium, and magnesium, which have low ionization potentials, can be easily ionized even by the light of cool stars. Some of the electrons generated in this way attach themselves to hydrogen atoms to form the H⁻ ions.

Absorption due to H⁻ is always accompanied by subsequent re-emission. Most of the light we receive from the Sun is due to a continuum transition in which atoms and electrons recombine to form the hydride ion, H⁻.

Many other elements of course also have ions that play an important role in astrophysics. Molecules can be ionized too and molecular ions have characteristic spectra of their own. They are often observed in the ionized gaseous tails of comets, where we detect OH^+, CO^+, H_2O^+, H_3O^+, and other ions spectroscopically or by means of a mass spectrograph aboard a space probe sent into the comet's tail.

7:5 Hydrogen Molecules

Molecules can have three types of quantized states. First, there is the possibility that atoms in the molecule vibrate relative to each other. In that case the *vibrational energies* are quantized. Second, there is the possibility of quantized rotation. This means that the angular momentum is quantized. Third, just as in atoms, there exist different quantized electronic states.

The binding energy between atoms is relatively weak. By this we mean that the energy required to separate two atoms that have formed a molecule is normally smaller than the energy required to ionize an atom. Only some large alkali atoms have ionizing energies lower than the highest molecular binding energies. Correspondingly, the radiative transitions between excited vibrational states also tend to occur at lower energies than those found for the transitions involving the lower electronic levels of most atoms. Characteristically, we tend to find that electronic transitions, that is, transitions between electronic excited states in atoms or molecules, occur in the visible and ultraviolet part of the spectrum. Vibrational transitions occur in the near infrared part of the spectrum, roughly at wavelengths between 1 and 20 μm, and rotational transitions occur in the far infrared $\lambda \gtrsim 20\,\mu$m and microwave spectrum.

This is just a rule of thumb and is not strictly obeyed. We already know that hydrogen atoms can have electronic transitions that reach way out to the very lowest energies associated with radio wavelengths. We discussed those in Section 7:3. However, pure vibrational transitions do not often occur in the visible spectra of astrophysically important substances, and pure rotational spectra are not expected at wavelengths shorter than a few microns.

For many purposes, the vibrations of two atoms relative to one another can be treated as quantized harmonic oscillations. Figure 7.7 shows the vibrational energy levels for hydrogen H_2 in the lowest electronic state of the molecule. If the vibrations become too violent, the molecule dissociates into two separate atoms. The

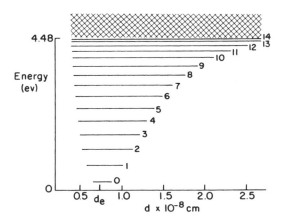

Fig. 7.7. Vibrational energy levels in the hydrogen molecule H_2 ground electronic state. The lengths and positions of the lines along the abscissa indicate the range of separations d between nuclei in the molecule. The equilibrium separation is denoted by d_e. There are only 14 vibrational states below 4.48 eV, the energy at which H_2 dissociates into two atoms. Each of the vibrational states comprises a number of possible substates having rotational angular momenta $J \geq 0$.

dissociation energy is 4.48 eV — slightly higher than the energy of the 14th excited vibrational state.

The ground state of the molecule is not at zero energy. Rather, analogous to photon energies (Section 4:13, equation (4–71)), the lowest vibrational state lies at roughly half the energy difference between the ground state and first excited state. This displacement is characteristic of ground vibrational levels.

Hydrogen molecules are a major constituent of interstellar space and are the predominant constituent of dark clouds in the Galactic plane. Unfortunately, the total mass present is only indirectly inferred. First, there is the difficulty of detecting the gas. Hydrogen is a symmetric dipole molecule, and as we already saw in Section 6:13, symmetric configurations can at best radiate if they have a quadrupole moment. Moreover, the probability for this type of transition is many orders of magnitude lower than for the more usual dipole radiation of asymmetric molecules. In addition, however, even the lowest rotational states (Fig. 7.8) can be collisionally excited only at gas temperatures above $\sim$100 K, while interstellar molecular clouds are normally far cooler. Once excited, the lifetime before radiating to a lower state is measured in years if not centuries. Pure rotational emission of H_2 is, therefore, rarely observed. Instead, the presence of the molecular hydrogen is inferred from the observation of the lowest rotational transitions of carbon monoxide, CO, which is far less abundant than H_2 but radiates readily. These transitions, however, are not entirely reliable tracers of H_2 because the emitted radiation is readily reabsorbed by layers of CO through which the transiting CO emission must pass to escape a cloud.

The vibrational spectrum of H_2 can be excited in shocked regions where temperatures rise to $\sim$2000 K. At these temperatures, colliding molecules excite each

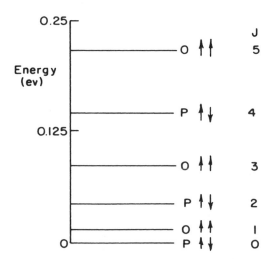

Fig. 7.8. Rotational energy-level diagram of H_2 for the ground vibrational state. Alternate states have nuclear spins aligned: parallel, ortho (O), or antiparallel, para (P).

other vibrationally. Once excited, H_2 can undergo a radiative transition to a lower vibrational state, with probability 10^{-6} s^{-1}, meaning that the molecule typically remains excited for a week or two before radiating its excitation energy away. The photon energy then corresponds to the energy difference between two of the levels in Fig. 7.7 and is well defined. The spectrum of the radiation observed identifies the molecular hydrogen uniquely, and also tells us the state of excitation of the gas (He50), (He67)*.

Though molecular hydrogen can be readily dissociated by ultraviolet photons insufficiently energetic to ionize hydrogen atoms and, therefore, able to pass through neutral atomic gas without much hindrance, H_2 is well shielded inside dark clouds by an abundance of interstellar dust. The dust absorbs the ultraviolet photons and thus shields the hydrogen molecules.

7:6 Selection Rules

In Section 7:2 we saw that the interaction of an atomic system with radiation obeys certain conservation principles. Compliance with these principles requires that some transitions be forbidden while others are allowed. The rules that tell us about the permitted transitions are called *selection rules*. We had already seen roughly how these rules come about. Here we will examine the question in somewhat greater depth.

When any two quantum systems combine to form a larger system, the addition of angular momenta takes place so that, along any arbitrarily chosen direction z, the final angular momentum J_{zf} is the sum of the two initial angular momenta along that direction,

$$J_{zf} = J_{z1} + J_{z2}, \tag{7-14}$$

where subscripts 1 and 2 refer to the two original systems.

Because the precise measurement of the z-component precludes a simultaneous precise measurement of the transverse components, equations of the form (7–14) do not exist for the orthogonal x- and y-directions. But equation (7–14) holds whether we interpret the symbols J_{zi} as the z-components of the angular momentum or only as the quantum number for this angular momentum component which, when multiplied by $\hbar$, represents the actual angular momentum component. Below we will interpret the symbol J_{zi} as a quantum number. The values that these quantum numbers can assume are zero, half-integer, or integer.

A second statement can be made about the addition of the squares of the angular momenta. The (angular momentum)^2 is also a precisely measurable quantity. The selection rules permit it to be measured simultaneously with J_{zi}. The allowed values of (angular momentum)^2 always take on numerical values

$$\text{(angular momentum)}_i^2 = J_i(J_i + 1)\hbar^2, \qquad i = 1, 2, \ldots, f. \tag{7-15}$$

The right side of this expression resembles the right side of equation (6–191). To understand why, it is useful to remember from Section 7:2 that quantization restricts angular momentum to values that are integer or half-integer multiples of $\hbar$. Classically, we also know that an orbital angular momentum L is proportional to the cross-product of a radial distance and a linear momentum, $\mathbf{L} = \mathbf{r} \wedge \mathbf{p}$. But because $\hbar$ already has the correct units for angular momentum, and the radial distance $\mathbf{r}$ is a vector quantity that has units of length, the momentum $\mathbf{p}$ must be the product of $\hbar$ and a vector quantity that has units of (length)$^{-1}$. The quantum mechanical formulation accomplishes this in representing linear momentum by an operator $-i\hbar\nabla$ — proportional to the gradient of the state ψ of the quantum system under consideration.

We may write the square of the angular momentum as $\mathbf{L} \cdot \mathbf{L}^*$, where the asterisk denotes the complex conjugate. This can now be rewritten as

$$\mathbf{L}^2 = -\hbar^2 (\mathbf{r} \wedge \nabla) \cdot (\mathbf{r} \wedge \nabla)^* = -\hbar^2 r^2 \nabla^2 . \qquad (7\text{-}16)$$

Because radial components of gradients do not contribute to $\mathbf{r} \wedge \nabla$, ∇^2 here is restricted to the tangential components, represented by the second and third terms on the right of equation (6–181). This tells us that $\mathbf{L}^2$ has precisely the angular dependence of (6–185)

$$\hbar^2 \left[\frac{1}{\sin\theta} \frac{\partial}{\partial\theta} \left(\sin\theta \frac{\partial\psi}{\partial\theta} \right) + \frac{1}{\sin^2\theta} \frac{\partial^2\psi}{\partial\phi^2} \right] = -\mathbf{L}^2\psi , \qquad (7\text{-}17)$$

and, correspondingly, $\mathbf{L}^2\psi$ should have values $\ell(\ell+1)\hbar^2\psi$. Here, we have simply interchanged symbols ℓ for J_i.

The relationship between J_i and J_{zi} requires that J_{zi} take on values $J_{zi} = J_i, J_i - 1, J_i - 2, \ldots, 1 - J_i, -J_i$. This too is reminiscent of the result we cited, in Section 6:25, that well-behaved solutions of (6–185) required that m take on positive or negative integer values with $|m| \leq \ell$, except that for angular momentum both integer and half-integer values are permitted. The half-integer values generally are associated with spins of electrons, protons or other fermions.

Since the maximum value that J_{zi} can assume is J_i, we note that, for this particular value, equation (7–15) tells us that there is an additional amount of angular momentum $(J_i\hbar^2)^{1/2}$ to be associated with the transverse angular momentum components. These components can therefore never be zero, unless J_i itself is zero also; and in that case they *must* be zero. The transverse angular momentum contribution comes about because the uncertainty principle does not permit a simultaneous precise measurement of two or more angular momentum components. But when the (angular momentum)^2 is zero, all angular momentum components must be zero, because each individual component of (angular momentum)^2 must be ≥ 0.

We said that the z-direction can be arbitrarily chosen. Let us choose it to represent the direction along which a photon approaches the atomic system. In the scheme used here we can assign subscript 1 to the photon, 2 to the initial atomic state, and f to the atomic state after photon absorption. Lorentz invariance requires the photon spin angular momentum to lie along the direction of propagation, $J_{z1} = 1$, and

$$J_{zf} = J_{z2} \pm 1. \qquad (7\text{-}18)$$

This tells us that an atomic system with half-integer values of J_{zi} must have J_i half-integer, and that no transitions to integer values of J_{zi} or J_i are possible through photon absorption or emission. Similarly a system with integer angular momentum quantum numbers always maintains those properties under photon absorption.[3]

We note that the quantum numbers J_{zf} and J_{z2}, alternatively, may be taken to correspond to the magnetic quantum numbers we previously labeled m. Equation (7–18) then gives a selection rule that states $\Delta m = \pm 1$ when the direction of photon emission is along the magnetic field lines. This is why only two Zeeman shifted lines are observed along that direction. Along a direction of emission perpendicular to the field, equation (7–18) is still true, but the association of J_{z2} with m is then no longer valid. A division of photons into groups having $J_{z1} = \pm 1$, that is, in terms of left- or right-handed polarized light, then mixes the contributions from different magnetic energy levels m. This happens because the photons from the various levels m, viewed along that direction, are plane-polarized. In Section 6:12 we saw that plane-polarized light can be considered to be a superposition of left- and right-handed polarized components. Hence, viewed along the direction perpendicular to the field, Δm may have values 0, as well as ± 1, even though (7–18) is still obeyed.

Equation (7–18) leads to one other important selection rule. It is impossible for an atomic system to undergo transitions between angular momentum states whose values are zero through absorption or emission of one photon. It is easy to see that this must be true. If $J_2 = 0$, then $J_{z2} = 0$ also, and similarly if $J_f = 0$, $J_{zf} = 0$. But by (7–18) both these z-components cannot be zero simultaneously. Hence the selection rule as stated must be true. This rule is absolutely inviolable, no matter whether electronic or vibrational transitions are involved. It is always true!

$$J = 0 \not\leftrightarrow J = 0. \tag{7-19}$$

In quantum mechanics selection rules such as these are linked to the symmetry of the atomic system. When the system has sufficiently complex symmetries, the selection rules also become correspondingly complex. We have only shown one or two of the simplest selection relations here, but it is worth remembering that even the more complex appearing rules actually are basic symmetry statements, which become relatively straightforward when viewed in terms of the appropriate symmetry. The angular momentum selection rules discussed here are based on the rotational symmetry of atomic systems.

Equation (7–19) holds true only for transitions involving a single photon. It is possible, however, though with low probability, for two-photon transitions to occur between states $J = 0$; the angular momentum carried off by the individual photons is then oppositely directed. Such transitions are possible in tenuous nebulae in interstellar space where atoms in an excited state of zero angular momentum can exist

---

[3] This is true even for photons produced in the laboratory that carry not only spin but also orbital angular momentum along the direction of propagation. The orbital angular momentum component of these photons is always an integer multiple of $\hbar$ along the direction of propagation, and so is their total angular momentum component J_{z1} (Aℓ99).

undisturbed for long periods (Va70), (Sp51b). In laboratory systems, where pressures are higher, such excited states normally become de-excited through atomic collisions.

An interesting feature of angular momentum quantization is that the existence of quantized states for all matter implies the impossibility of interacting with radiation having nonquantized angular momentum. Whatever fields, electric, magnetic, weak or strong nuclear, gravitational, or others that may exist in the Universe, should therefore have radiation that is quantized in terms of half-integer spin angular momentum $\hbar$, or multiples thereof if it is to interact with atomic matter. When gravitational waves are discovered, we are confident we will find them to have quantized spin angular momentum. The current prediction is that their spin should be $2\hbar$ — twice that of photons (Gu54).

PROBLEM 7–3. If a system of particles has a *moment of inertia*

$$I = \sum_j m_j r_j^2 \tag{7-20}$$

about its axis of rotation, show that the energy and angular momentum are related by

$$\mathcal{E} = \frac{\omega L}{2} = \frac{L^2}{2I}, \tag{7-21}$$

provided we are talking about a classical, rigid, nonrelativistic rotator whose angular frequency of rotation is ω. From this, show that the energy carried away by each quantum of radiation would have to be

$$\delta\mathcal{E} = \hbar\omega \tag{7-22}$$

due to quantization $\hbar$ of angular momentum. Equation (7–15) gives the square of the total angular momentum as $\hbar^2[J(J+1)]$. Show that the energy of each state is

$$\mathcal{E} = \hbar^2 \frac{J(J+1)}{2I} \tag{7-23}$$

and the energy of the quantum released in the transition $J \to J - 1$ is

$$\delta\mathcal{E} = \hbar^2 \frac{J}{I} . \tag{7-24}$$

For massive objects rotating rapidly, show that this is equivalent to the classical formula.

PROBLEM 7–4. If an interstellar molecule has a rotational energy kT in thermal equilibrium with surrounding gas (Section 4:19), say $T \sim 100\,\mathrm{K}$, what is the range of frequencies it will radiate if typical values of atomic weights are 10^{-23} g and typical molecular radii are 2 Å? Make use of some of the expressions obtained in the previous problem to show that radiation may be expected in the far infrared or submillimeter region.

PROBLEM 7–5. Set up an expression for the probability that a molecule with moment of inertia I about a given spin axis will be found in an excited rotational state J when it is in thermal equilibrium with gas at temperature T. From this convince yourself that a molecule in a cool interstellar cloud ($T \sim 100\,$K) cannot be excited into very high rotational states.

PROBLEM 7–6. If interstellar grains are in thermal equilibrium with the surrounding gas ($T \sim 100\,$K), and have typical radii, 10^{-5} cm with typical masses 10^{-15} g, at what frequency could they be expected to radiate away angular momentum? This process is possible because small inhomogeneous grains can be expected to have appreciable electric dipole moments. Observations would, however, have to be made from above the atmosphere, and interstellar plasma absorption might interfere.

PROBLEM 7–7. It has been suggested that massive cosmic objects might exist having such high angular momenta that contraction to high density becomes impossible. Such an object might slowly cool down without ever becoming a star because its central temperature does not get sufficiently high to sustain nuclear reactions. It might, however, be capable of losing angular momentum through systematic emission of circularly polarized radiation carrying spin angular momentum $\hbar$.

An object like this might also emit gravitational radiation. Because the graviton carries away twice the spin angular momentum of photons, see how the formulas of Problem 7–3 would change if applied to gravitons. Try the same thing for neutrinos that carry away only half the spin angular momentum of photons. Under normal conditions, in any of these cases, the probability for the emission of a quantum is quite small as the discussion of transition probabilities in the next sections will show. If electromagnetic radiation were actually emitted it would not be transmitted by the interstellar medium because the expected rotational frequency ω for massive objects would be small. Why not? What would happen to the energy (Re71)?

7:7 The Information Contained in Spectral Lines

Left to itself an excited atomic system will spontaneously jump into a lower energy state. The mean time required for such a transition varies from one system to another and depends on the symmetries of the states, the selection rules discussed in Section 7:6, and other factors. If the system has a total probability P of leaving the excited state in any unit time interval, its total life in the excited state is $\delta t = 1/P$. The energy of the state therefore cannot be determined with arbitrarily great accuracy. The limited time in the excited state implies that the energy can only be determined to an accuracy $\delta \mathcal{E}$ given by the uncertainty principle

$$\delta\mathcal{E} = \frac{\hbar}{\delta t}. \tag{7-25}$$

Improbable transitions therefore have a narrow natural line width. The accuracy to which the transition energy between two states i and k can be determined depends on the lifetime of both the upper and the lower states; the total frequency width of the line $\delta\nu$, then, is the sum of the widths of the two levels. This total width is usually denoted by

$$\delta\omega = 2\pi\delta\nu = \frac{(\delta\mathcal{E}_i + \delta\mathcal{E}_k)}{\hbar} \equiv \gamma; \tag{7-26}$$

γ is called the *natural line width* for the transition.

In Section 7:8 we will find that a spontaneously decaying atom emits radiation with a spectral distribution or *line shape*

$$I(\omega) = I_0\frac{\gamma}{2\pi}\frac{1}{(\omega - \omega_0)^2 + \gamma^2/4}. \tag{7-27}$$

The intensity drops to half the maximum value at frequencies $\omega_0 \pm \gamma/2$.

In astronomical sources the natural line width is seldom directly observed, but deviations from this width can give us a great deal of information. Let us list the various line broadening effects.

(a) Doppler Broadening

This effect is due to random motions of emitting or absorbing atoms or molecules. For small velocities the frequency shift of the radiation is roughly proportional to the line of sight velocity component v_r (see equation (5–44)):

$$\Delta\omega = \omega_0\frac{v_r}{c}, \qquad v_r \ll c. \tag{7-28}$$

Two kinds of motion can contribute to Doppler broadening: the thermal velocities of emitting atoms within a cloud and the turbulent velocities peculiar to the clouds superposed along a line of sight. Sometimes these two effects can be resolved.

We can, for example, observe the absorption of stellar radiation by interstellar sodium atoms. Sodium absorbs strongly in the yellow part of the visible spectrum in a pair of lines known as the sodium D lines at 5890 and 5896 Å. If these lines are examined at very high resolution, so that shifts of the order of one part in 10^6 are detected, velocity differences as low as $\sim 3 \times 10^4$ cm s^{-1} can be distinguished. What we then observe is a series of discrete broadened absorption lines due to individual clouds absorbing light along the line of sight, and a definite characterizing width for individual lines representing a given cloud. Whether this individual line width is entirely due to thermal motions, or also partially due to turbulent motions on a smaller scale within each cloud, cannot be immediately recognized (Ho69).

PROBLEM 7–8. If atoms of mass m move with velocities determined by Maxwell–Boltzmann statistics, the probability of observing a given line of sight velocity v_r

is proportional to $\exp(-v_r^2 m/2kT)$ (see equation (4–56)). Show that the line shape for thermal Doppler broadening therefore should be of the form

$$I(\omega)\,d\omega = I_0 \exp\left[-\frac{mc^2(\Delta\omega)^2}{2\omega_0^2 kT}\right] d\omega. \tag{7-29}$$

See also Problem 4–27 and show how the line width can be immediately related to the temperature through equation (7–28).

In general the width at half maximum for Doppler broadening

$$\delta = \omega_0 \left[\frac{2kT(\ln 2)}{mc^2}\right]^{1/2} \tag{7-30}$$

is much greater than the natural line width γ. However, it drops off exponentially and, therefore, much faster than the natural line width. The observed wings of very strong lines, for example, *Lyman-α* in interstellar space, generally are due to natural line width and to the other causes listed below (Je69).

(b) Collisional Broadening

In the relatively dense atmospheres of stars, atoms or ions often suffer a collision while they are in an excited state. Because any given collision may induce a transition to a lower state, the collective effect of such collisions is to increase the total transition probability. Thus, if the spontaneous transition rate were γ and the number of collision-induced transitions in unit time is γ_c,

$$\frac{\delta\mathcal{E}}{\hbar} = \gamma + \gamma_c. \tag{7-31}$$

The emitted line has the spectral intensity distribution of the natural line shape, except that γ is replaced by $\gamma + \gamma_c$.

(c) Other Types of Broadening

There are other effects due to interactions with neighboring atoms that can cause shifting and splitting of states; through the influence of electric fields — the *Stark effect*; through resonance coupling between atomic systems; and other effects. These processes all lead to line broadening but at low densities their effects are small.

PROBLEM 7–9. To obtain a better feel for the relative importance of the Doppler and collision line widths for visual spectra obtained from stellar atmospheres, show that: if the atmosphere has a density n, thermal velocities $(3kT/M)^{1/2}$, and collision cross-section σ,

$$\frac{\gamma_c}{\delta} \sim \frac{n\sigma c}{\omega_0}, \tag{7-32}$$

where factors of order unity are neglected. For visible light $\omega_0 \sim 10^{15}\,\text{s}^{-1}$, and collision cross-sections have typical dimensions of atoms, $10^{-16}\,\text{cm}^2$. In normal stellar photospheres $n \ll 10^{21}\text{cm}^{-3}$ and hence $\gamma_c \ll \delta$.

7:8 Absorption and Emission Line Profile

In estimating the amount of neutral interstellar hydrogen along the light path between an ultraviolet emitting star and Earth, we need to know both the shape and the total strength of the Lyman-α absorption line. These two pieces of information permit us to determine the amount of hydrogen in terms of the observed absorption line width. Intrinsically the calculation of line strength and shape is a quantum mechanical problem. Classical theory permits us to calculate the line shape on the basis of a harmonic oscillator model. The model also yields the right order of magnitude for the line strength. However, we must take care not to take the classical model too seriously; by itself it does not lead to quantized energy states.

We will first derive an expression for the emission line profile using semiclassical methods. We start with equation (6–88) for the total energy radiated per second by a charged oscillator, $I = 2e^2\ddot{r}^2/3c^3$. We can see that this intensity corresponds to a force

$$\mathbf{F} = \frac{2}{3}\frac{e}{c^3}\dddot{\mathbf{d}} \tag{7-33}$$

because the average work done by that force, in unit time, is then

$$\langle \mathbf{F} \cdot \dot{\mathbf{r}} \rangle = \left\langle \frac{2e}{3c^3}\dddot{\mathbf{d}} \cdot \dot{\mathbf{r}} \right\rangle = \left\langle \frac{2}{3c^3}\frac{d}{dt}(\dot{\mathbf{d}} \cdot \ddot{\mathbf{d}}) - \frac{2}{3c^3}\ddot{\mathbf{d}}^2 \right\rangle = I. \tag{7-34}$$

Here the term containing $\dot{\mathbf{d}} \cdot \ddot{\mathbf{d}}$ vanishes because $\dot{d}$ and $\ddot{d}$ are exactly out of phase in simple harmonic motion. The *damping force* F is small compared to the harmonic force; the oscillation in other words lasts over many cycles. We can therefore write the equation of motion as

$$m\ddot{\mathbf{r}} = -m\omega_0^2\mathbf{r} + \frac{2}{3}\frac{e^2}{c^3}\dddot{\mathbf{r}}. \tag{7-35}$$

This equation is very much in the spirit of (6–107), except that we have a damping (instead of a harmonic driving) force here. Because the damping is weak, the motion is almost harmonic and we can approximate

$$\dddot{\mathbf{r}} = -\omega_0^2\dot{\mathbf{r}}. \tag{7-36}$$

We then rewrite (7–35) as

$$\ddot{\mathbf{r}} = -\omega_0^2\mathbf{r} - \gamma\dot{\mathbf{r}} \quad \text{with} \quad \gamma = \frac{2}{3}\frac{e^2\omega_0^2}{mc^3} \ll \omega_0 . \tag{7-37}$$

Since $\gamma \ll \omega_0$ this has the approximate solution

$$\mathbf{r} = \mathbf{r}_0 e^{-\gamma t/2} e^{-i\omega_0 t} . \qquad (7\text{-}38)$$

The oscillating dipole thus sets up an oscillatory field of the form

$$\mathbf{E}(t) = \mathbf{E}_0 e^{-\gamma t/2} e^{-i\omega_0 t} \ (\text{RP}), \qquad (7\text{-}39)$$

where only the real part enters into physical consideration. This is not monochromatic any more because it changes with time; and only time-invariant oscillating fields can be strictly monochromatic. A time-dependent change in intensity affects the frequency spectrum.

The total field is now written in terms of an integral over the entire range of frequency components:

$$\mathbf{E}(t) = \int_{-\infty}^{\infty} E(\omega) e^{-i\omega t} \, d\omega. \qquad (7\text{-}40)$$

By a theorem from *Fourier theory*, an integral of this form can be inverted to give

$$E(\omega) = \frac{1}{2\pi} \int_{-\infty}^{\infty} E(t) e^{i\omega t} \, dt. \qquad (7\text{-}41)$$

If we now introduce the field (7–39) into this equation and note that $E(t)$ is defined only for time $t \geq 0$, we can readily integrate to obtain

$$E(\omega) = \frac{1}{2\pi} \frac{E_0}{i(\omega - \omega_0) - \gamma/2} \ (\text{RP}) . \qquad (7\text{-}42)$$

We then obtain the spectral line intensity (see Fig. 7.9):

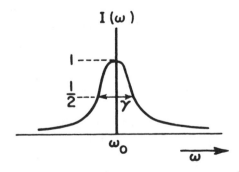

Fig. 7.9. Natural width of a spectral line. The curve is normalized to peak intensity $2I_0/\pi\gamma$ (see equation (7–42)).

$$I(\omega) = |E(\omega)|^2 = I_0 \frac{\gamma}{2\pi} \frac{1}{(\omega - \omega_0)^2 + \gamma^2/4} \,,$$

$$I(\nu) = I_0 \left(\frac{\Gamma}{2\pi}\right) \left[(\nu - \nu_0)^2 + \frac{\Gamma^2}{4}\right]^{-1} , \qquad \gamma \equiv 2\pi\Gamma , \qquad (7\text{-}43)$$

where I_0 is the total intensity integrated over all frequency space and $\omega = 2\pi\nu$, $\omega_0 = 2\pi\nu_0$:

$$I_0 = \int_{-\infty}^{\infty} I(\omega)\, d\omega = \int_{-\infty}^{\infty} I(\nu)\, d\nu \,. \qquad (7\text{-}44)$$

This type of line shape is sometimes called the *Lorentz profile*, and Γ and γ are called the *natural line width*. This is the full frequency width at half maximum. We have not yet shown that the absorption and emission profiles are the same. This question is taken up next, in Section 7:9.

7:9 Quantum Mechanical Transition Probabilities

Much astrophysical information can be obtained from the intensity of absorption or emission lines. The strengths of the lines define the number densities of atoms, ions, or molecules in a source or along the line of sight to a source, and the ratio of the strengths of lines can be used to determine the excitation temperature of gases through application of the Saha equation (Section 4:16).

In order to obtain this information in useful form, we must first relate the intensity of the spectral absorption or emission lines to the number density of atoms or ions in different energy levels; and we can only do that if we know the transition probability between states of the system.

Very roughly the transition probability depends on three factors: (a) on the symmetry properties of the atomic system, (b) on its size in relation to the wavelength to be absorbed or emitted, and (c) on the statistics of the radiation field. The first of these factors includes the selection rules discussed in Section 7:6, the statements about charge-to-mass ratio made in Section 6:13, and other similar restrictions. The second factor represents the relative probability for dipole, quadrupole, or higher multipole radiation; this is size dependent as seen, for example, in (6–93). The third factor, to be discussed now, depends only on the radiation field and is quite general for any transition regardless of the atomic or nuclear system involved.

The probability that an atomic system will undergo a transition from some state i to another state j is proportional to the number of ways in which a change in the photon field can occur. For example, the probability for emission of a photon with radial frequency ω is (see equation (4–65)) proportional to

$$\frac{\omega^2 \, d\omega \, d\Omega}{(2\pi c)^3} = \frac{\nu^2 \, d\nu}{c^3} \, d\Omega \qquad (4\text{-}65)$$

for photons polarized in one sense. Here $d\Omega$ is the increment of solid angle. This factor, considered in isolation, makes transitions in the optical domain, where

$\omega \sim 3 \times 10^{15}$ rad s^{-1}, much more probable than, say, in the radio region at $\omega \sim 3 \times 10^9$ rad s^{-1}.

The emission probability is also proportional to $n_\omega + 1$, where n_ω is the ambient density of photons per phase cell that already have the momentum and polarization characterizing the photon to be emitted. The newly emitted photon then has precisely the same frequency, direction of propagation, and polarization as the stimulating photon or photons and we speak of *stimulated* or *induced emission*. Stimulated emission is the exact opposite of ordinary *photon absorption*. The number of absorptions is again proportional to $n_\omega + 1$, if n_ω is taken to be the number density per phase cell of photons left after the atom has reached the upper state. We therefore see that the *transition probability* $P(\omega, \theta, \phi)$ per unit solid angle and frequency range obeys the relation

$$P(\omega, \theta, \phi)\, d\Omega\, d\omega \propto [n(\omega, \theta, \phi) + 1] \frac{\omega^2}{(2\pi c)^3}\, d\Omega\, d\omega . \qquad (7\text{-}45)$$

Here $n(\omega, \theta, \phi)$ is the probability per unit frequency range that a photon state is occupied when the atomic system is in its upper energy state and $\nu = 2\pi\omega$ is the mean spectral frequency. Let us now return to the factors (a) and (b) mentioned earlier. These have to be evaluated quantum mechanically to yield a *matrix* with elements U_{ij} giving the *transition amplitude* between any two states i and j of the atomic system. The actual transition probability between these two states is proportional to $|U_{ij}|^2$. The prescription for obtaining the transition probability per unit solid angle, then, is to multiply the product $|U_{ij}|^2 P(\omega, \theta, \phi)$ by the numerical factor $2\pi/\hbar^2$. Thus the transition probability per unit time becomes

$$\frac{2\pi}{\hbar^2} |U_{ab}|^2 [P(\omega, \theta, \phi)]\, d\Omega = \frac{2\pi}{\hbar^2} |U_{ab}|^2 [n(\omega, \theta, \phi) + 1] \frac{\omega_{ab}^2}{(2\pi c)^3}\, d\Omega . \qquad (7\text{-}46)$$

Since the energy of the states is quite narrowly defined $n(\omega, \theta, \phi)$ normally does not change appreciably over the bandwidth of the line. The matrix elements already include the integration over the frequency bandwidth that appears explicitly in equation (7–45). The transition probability (7–46) thus includes an integration over the entire frequency range ω and, specifically, includes consideration of strongly absorbed or emitted photons at the line center as well as the less readily absorbed and emitted photons in the line wings. More precisely stated, it includes consideration of the line shape (7–43).

We still need to relate the quantum mechanical transition probability to equation (6–86) which expressed the intensity I absorbed by an oscillating dipole in terms of the second time derivative of the dipole moment d.

$$I = \frac{2}{3c^3} \ddot{d}^2 = \frac{2e^2 \ddot{r}^2}{3c^3} . \qquad (6\text{-}86)$$

Because **r** has the time dependence (7–38), equation (6–86) is readily rewritten as

$$I = \frac{2e^2 \omega^4}{3c^3} \langle r^2 \rangle = \frac{32\pi^4}{3c^3} e^2 \nu^4 \langle r^2 \rangle, \qquad (7\text{-}47)$$

where the brackets $\langle \rangle$ indicate a time average. The intensity I is related to the spontaneous transition probability of quantum mechanics; and we must therefore set $n = 0$ in equation (7–46) if a comparison with (7–47) is to be made. In the dipole approximation the matrix element U_{ab} makes a contribution

$$|U_{ab}|^2 = 2\pi\hbar\omega_{ab}e^2|r_{ab}|^2 \sin^2\theta, \qquad (7\text{-}48)$$

where an integration has been carried out over the possible directions of polarization. The physical meaning of $e^2|r_{ab}|^2$ will be discussed below. The total intensity is now given by the product of the transition probability (7–46) and the photon energy $\hbar\omega_{ab}$:

$$I\,d\Omega = \hbar\omega_{ab} \cdot \frac{2\pi}{\hbar^2} \frac{\omega_{ab}^2}{(2\pi c)^3} \cdot 2\pi\hbar\omega_{ab}e^2|r_{ab}|^2 \sin^2\theta\,d\Omega, \qquad (7\text{-}49)$$

where θ represents the angle between the vector $\mathbf{r}$ and the direction of propagation of the emitted radiation. Integrating over all angles of emission we obtain the total intensity of spontaneously emitted radiation

$$I = \frac{4}{3}\frac{e^2}{c^3}\omega_{ab}^4|r_{ab}|^2 = \frac{64}{3}\pi^4\frac{e^2}{c^3}\nu_{ab}^4|r_{ab}|^2. \qquad (7\text{-}50)$$

We see that the formula obtained quantum mechanically is almost the same as the classical expression (7–47). We only have to replace the time-averaged $\langle r^2 \rangle$ by $2|r_{ab}|^2$ if we wish to obtain identical forms. This connection is consistent with the correspondence principle. We note, however, that $e^2|r_{ab}|^2$ is not an exact quantum mechanical analogue to the mean square dipole moment. For each individual state of the atomic system, a or b, the dipole moment would be given by expressions involving the diagonal matrix elements er_{aa} or er_{bb}, respectively. The quantities er_{ab}, instead, denote a property that is influenced both by the initial and the final state of the system. They have no exact classical analogue and we therefore need not be surprised that a factor 2 appears in equation (7–50). We had no reason to expect complete identity of classical and quantum mechanical forms. After all, the quantum theory of radiation is supposed to go a step beyond the classical results in order to provide an understanding of abrupt transitions. Its results must therefore differ in some essential form from those of classical theory.

Thus far we have only a formal solution that does not yet allow us to estimate the strength of an emission or absorption line. We can, however, still make use of equation (7–37) for this purpose. We note that γ^{-1} is a time constant, so that γ taken by itself is equivalent to a transition probability. By setting the value for γ equal to the transition probability (7–46), we are able to estimate U_{ab}, and also an absorption cross section for radiation. We write

$$\gamma = \frac{2}{3}\frac{e^2\omega_{ab}^2}{mc^3} = \frac{2\pi}{\hbar^2}\frac{\omega_{ab}^2}{(2\pi c)^3}\int |U_{ab}|^2[n(\omega,\theta,\phi)+1]\,d\Omega. \qquad (7\text{-}51)$$

For spontaneous emission, $n(\omega,\theta,\phi)$ can be set equal to zero. The value for γ that is used here has been derived on the basis of a dipole radiator model, and the integral on the right-hand side of equation (7–51) therefore contains the same factor $\frac{2}{3}$ that already came up in the evaluation of the classical expressions (6–85) and (6–86).

$$\int |U_{ab}|^2 \, d\Omega = \frac{2}{3} |U_{ab}|^2 4\pi \ . \tag{7-52}$$

Hence

$$|U_{ab}|^2 = \frac{\pi e^2 \hbar^2}{m} \ . \tag{7-53}$$

The absorption for a constant level of continuum radiation $I(\omega)$ is

$$\sigma I(\omega) = \int \sigma(\omega) I(\omega) \, d\omega \ , \tag{7-54}$$

where $\sigma(\omega)$ is the cross-section at frequency ω and has the dimensions of area, say, cm^2. The integrated cross-section σ has corresponding dimensions cm^{2}Hz. If an atomic system is surrounded by an isotropic photon gas of density $n'(\omega, \theta, \phi)$, the total number of photons absorbed can be expressed as

$$n'(\omega, \theta, \phi)\sigma c \, d\Omega = \frac{2\pi}{\hbar^2} |U_{ab}|^2 [n(\omega, \theta, \phi) + 1] \frac{\omega_{ab}^2}{(2\pi c)^3} \, d\Omega \ . \tag{7-55}$$

Here the left side represents the number of photons per unit frequency range of a continuous spectrum intercepted by the cross-sectional area in unit time, and the right-hand side gives the probability for absorption of a photon as expressed in (7–46). We can cancel the photon densities in equation (7–55) if we follow the procedure of letting $n(\omega, \theta, \phi)$ stand for the fractional number of photon states occupied when the atomic system is in its upper state. Because we have taken n' to represent the number density of photons per unit frequency band present before absorption, that is, when the atomic system still is in its lower state, we can see that

$$n'(\omega, \theta, \phi) = [n(\omega, \theta, \phi) + 1] \frac{\omega_{ab}^2}{(2\pi c)^3}. \tag{7-56}$$

The factor $\omega_{ab}^2/(2\pi c)^3$ appears because n is a number density per phase cell whereas n' is a density per unit volume of normal, three-dimensional configuration space. From (7–53) and (7–55) it then follows that

$$\sigma = \frac{2\pi^2 e^2}{mc} = 2\pi^2 r_e c \ , \tag{7-57}$$

where

$$r_e \equiv \frac{e^2}{mc^2} \ . \tag{6-170}$$

The cross-section (7–57) has precisely the value we would obtain by classical means if we modified equation (6–107) to include a radiative reaction force (see equation (7–33)) representing the force on the moving charge due to its emission of radiation:

$$F_{\rm rad} = \frac{2}{3}\frac{e^2}{c^3}\overset{...}{r}\,. \tag{7-58}$$

A series of remarks is now in order:

(1) The cross-section obtained here holds only for atomic systems for which an oscillating charged dipole represents a satisfactory description. This must be strongly emphasized! Each type of atom or molecule has its own structure and therefore will interact with photons in its own way. However, an essential feature shared by many atomic systems is that electrons are bound to a nucleus or core. In a stable quantum state the electron then resists the efforts of an applied electromagnetic field to move it from its equilibrium position or, more accurately, from its equilibrium orbital distribution within the atomic system.

In this respect the electron behaves as though it were harmonically bound to the more massive core. This justifies the use of the classical dipole approximation as a guide to the quantum treatment. However, it does so only for atoms or molecules having a dipole moment and for wavelengths long compared to the atomic dimensions. The limitations that held for classical radiators hold equally well in the quantum limit. This was already pointed out in Section 6:13, but it is worth stating again.

(2) No atom behaves precisely like a classical harmonic oscillator. Its cross-section is not precisely that given by (7–57). We can define an *oscillator strength f* that represents the actual absorption strength of a given line in units of $2\pi^2 e^2 (mc)^{-1}$. A value $f = 1$ represents an absorption equal to that of the classical dipole.

(3) As already noted in equation (7–54), the cross-section of the atomic system varies with frequency. Its frequency distribution is of the form (7–43).

PROBLEM 7–10. Show that if the absorption cross-section is

$$\sigma_{ab}(\omega) = \frac{2\pi e^2}{mc} f_{ab} \frac{\gamma/2}{(\omega - \omega_{ab})^2 + (\gamma/2)^2}\,, \tag{7-59}$$

$$\sigma_{ab}(\nu) = \frac{2\pi e^2}{mc} f_{ab} \frac{\Gamma}{2}\left[(\nu - \nu_{ab})^2 + \left(\frac{\Gamma}{2}\right)^2\right]^{-1}\,, \qquad \Gamma = \frac{\gamma}{2\pi}\,,$$

the total cross-section obtained in (7–57), multiplied by an oscillator strength f_{ab}, is obtained on integrating over all frequencies.

---

(4) The identity of absorption and emission cross-sections is evident from the form of equation (7–46). This is true both of the magnitude of the absorption and its spectral distribution. When no radiation field at all is present, that is, $n(\omega, \theta, \phi) = 0$, we still have the vacuum field or zero-level photon population which induces the spontaneous emission of radiation to be discussed in more detail in Section 7:10.

(5) The actual magnitudes of various kinds of transitions are also of interest. The absorption cross-section (7–57) corresponding to unit oscillator strength has a value

$\sigma \sim 0.17\,\text{cm}^2\,\text{s}^{-1}$. When multiplied by the radiative flux *in unit frequency interval*, σ gives the total amount of radiation absorbed by an atom.

PROBLEM 7–11. Show that the maximum absorption cross-section has the value $\sigma(\omega_{ab}) = (3\lambda^2/2\pi)f_{ab}$, so that for $f_{ab} \sim 1$ the apparent size of the atom at resonance is roughly a factor of two lower than λ^2, the wavelength of the radiation squared.

PROBLEM 7–12. What fraction of the radiation in an emission or absorption line lies within the bandwidth γ defined by the natural line width?

PROBLEM 7–13. Show that the spontaneous transition probability is roughly $\gamma \sim (5\lambda^2)^{-1}$ in cgs units. It therefore has a value of $10^8\,\text{s}^{-1}$ for visible light, $\lambda \sim 5 \times 10^{-5}$ cm.

In a different spirit we can use (7–50) to write the transition probability w as

$$w \sim \frac{e^2}{c^3}\frac{\omega_{ab}^3}{\hbar}|r_{ab}|^2 \sim \frac{e^2}{c^3\hbar}\left(\frac{me^4}{\hbar^3}\right)^2\left(\frac{\hbar^2}{me^2}\right)^2\omega_{ab} \sim \left(\frac{e^2}{c\hbar}\right)^3\omega_{ab} \sim \frac{1}{(137)^3}\omega_{ab},$$
(7-60)

where we have made use of equation (7–3) for a hydrogenlike atom to roughly estimate the radiated frequency ω_{ab} and have set the Bohr radius (7–4) equal to $|r_{ab}|$. The *fine structure constant*

$$\alpha = \frac{e^2}{\hbar c} \sim \frac{1}{137}$$
(7-61)

taken to the third power then appears in the last element of equation (7–60). The transition probability for visible radiation is of order $10^8\,\text{s}^{-1}$ and, correspondingly, we can see from (7–60) that it should be of order $10^{11}\,\text{s}^{-1}$ for X-rays, $\sim 10^{14}\,\text{s}^{-1}$ for γ-radiation, and $\leq 10^4\,\text{s}^{-1}$ for radio waves. Interestingly, (7–60) is independent of the mass of the emitting particle. It does not have to be an electron, but can be an ion. The lifetime of the state is just the reciprocal of the transition probability.

Oscillator strengths can vary greatly. For the hydrogen Lyman series we have values Ly-α(0.42), Ly-β(0.08), Ly-γ(0.03), Ly-δ(0.01), and so on. Occasionally f values are slightly larger than 1.0. At the other extreme, values of 10^{-10} or even less can also occur. The oscillator strengths must therefore be evaluated individually for any given atom or molecule and depend strongly on the structural properties of the atomic system.

The f values for different transitions in an atom or molecule are not independent. In particular, a given atom cannot have an arbitrarily large number of strong absorption or emission lines. If we sum f values for all possible transitions, between all possible states in an atom or ion, we should obtain a number equal to the total number of electrons in the atom. If the atom has strongly bound inner electrons, then the sum should equal the number of the more weakly bound valence electrons, i.e., electrons in the incompletely filled outer shell of an atom. This is the *Thomas–Kuhn sum rule* (Aℓ 63)*. For hydrogen the sum of all f values should equal unity.

PROBLEM 7–14. For many years astronomers believed that atoms could be repelled by sunlight to a sufficient extent to account for the long comet tails we observe.

For a small molecule having an oscillator strength $f = 1$ and a mass $m = 5 \times 10^{-23}$ g, calculate the ratio of solar radiative repulsion to gravitational attraction. For comet tails the observed repulsive acceleration corresponds to an effective ratio of the order 10^2 to 10^3. Assume that all of the sunlight is roughly evenly distributed between 4×10^{-5} and 7×10^{-5} cm wavelengths. Does it appear likely that radiation produces this repulsion?

Magnetohydrodynamic forces are currently thought primarily responsible for the acceleration of tail constituents. In contrast, as seen from Problem 4–14, radiation pressure does exert major forces on interplanetary dust grains and is responsible for propelling grains away from the comet nucleus into a dust tail.

---

(6) Atomic systems that do not have a dipole moment can at best undergo transitions through quadrupole or magnetic dipole radiation. The transition probabilities for such processes are of order $(r/\lambda)^2$ smaller, where r is a typical dimension of the atomic system. This is consistent with our finding in Section 6:13, because $r/\lambda \sim r\omega/c \sim v/c$. From (7–60) we also see that $(r\omega/c)^2 \sim (1/137)^2$ for atoms. In rough agreement with these estimates, we find that actual transition probabilities for magnetic dipole and for quadrupole transitions are, respectively, of order 10^2 to 10^3 and 0.01 to 0.1 s^{-1} (He50).*

7:10 Blackbody Radiation

Two basic requirements need to be satisfied to obtain blackbody radiation. First, the temperature of absorbing particles must be constant in the vicinity of the emitting surface so that the photons emanating from the surface are in thermal equilibrium at a well-defined temperature; second, for equilibrium to become established, we require that the assembly of absorbing particles be large enough so that a succession of absorptions and re-emissions occurs before energy escapes from the surface of the assembly.

The number density $n(\mathcal{E})$ of particles in an excited energy state $\mathcal{E}$ is given by the Boltzmann factor (4–47) in terms of n_0, the number density in a lower state:

$$n(\mathcal{E}) = n_0 e^{-\mathcal{E}/kT}. \tag{7-62}$$

Let us look for the conditions under which the number of photons absorbed by the assembly of particles just equals the number emitted. We may restrict our discussion to transitions occurring between two specific energy states of the assembly. The presence of other states will not alter the conclusions.

A particle in the lower of the two states can absorb a photon and transit to the upper energy state. In the upper state, it can either spontaneously emit a photon or else be induced to emit one through stimulation by radiation of appropriate spectral frequency. In equilibrium, the sum of the induced and spontaneous downward transitions must equal the number of upward transitions. Let the probability of emitting a photon of frequency ν in unit time interval be $A(\nu)$. Let the corresponding probability of absorbing a photon be $n(\nu, T)cB(\nu)$, where $n(\nu, T)$ is the photon density at temperature T and frequency ν, and $B(\nu)$ is a transition cross-section at frequency ν; then the probability for stimulated emission for a given excited particle equals the probability for absorption by some other particle in the lower state. As illustrated in Fig. 7.10, this is a consequence of time-reversal symmetry that holds

Fig. 7.10. In part (a) a photon of frequency ν stimulates the emission of a similar photon, while the particle energy drops by an amount $\mathcal{E} = h\nu$. In (b), the time-reversed process takes place. The particle transits to the higher energy state $\mathcal{E}$ by absorbing a photon. As discussed in the text, spontaneous emission of radiation from a singly excited radiation oscillator (c) corresponds to the time-reversed process of absorption of energy (d) by a particle in its ground state.

for all electromagnetic processes.

We are now ready to write the equation for equilibrium between absorption and emission of photons in the frequency range $d\nu$ around ν.

$$n(\nu, T)B(\nu)n_0c\, d\nu = [A(\nu) + n(\nu, T)cB(\nu)]n(\mathcal{E})\, d\nu . \qquad (7\text{-}63)$$

Combining this with equation (7–62) gives

$$n(\nu, T)\, d\nu = \frac{A(\nu)/[B(\nu)c]}{e^{h\nu/kT} - 1}\, d\nu . \qquad (7\text{-}64)$$

If we consider spontaneous transitions to be stimulated by the ground state of the radiation field discussed in Section 4:13, equation (4–72), we can set $A(\nu)$ equal to the number density of radiation oscillators multiplied by the transition probability per unit time $B(\nu)$. This is equivalent to stating that all emission processes are induced, that the probability for emission is proportional to the sum of all populated radiation oscillator states, that the ground state of each oscillator ($n = 1$) is always populated, and that some radiation oscillators containing what we have called photons are in higher states n. From the phase-space enumeration of the number density of radiation oscillators at frequency ν (equation (4–65)) we then see that

$$A(\nu) = \frac{8\pi\nu^2}{c^2} B(\nu) .$$

(7-65)

To relate these coefficients to earlier work, we note that (7–50) gives

$$\frac{I}{h\nu_{ab}} = \frac{64}{3h}\pi^4 \frac{e^2\nu_{ab}^3}{c^3}|r_{ab}|^2 = \int A(\nu)\, d\nu = A_{ab} .$$

(7-66)

Here, $A(\nu)$ is dimensionless, $B(\nu)$ is an area, and A_{ab} has units of frequency, Hz.

PROBLEM 7–15. According to the correspondence principle, the transition probability should be related to the classical radiation intensity in the limit of large atomic systems. In the ionized regions of interstellar space, transitions often occur between highly excited states of atomic hydrogen (Ka59), (Hö65). Show that the correspondence argument leads to

$$\frac{d\mathcal{E}}{dt} = h\nu A_{n,n-1} = \frac{\omega^4 e^2 r_n^2}{3c^3},$$

(7-67)

where r_n is the Bohr radius in the nth state. Show that this gives

$$A_{n,n-1} = \frac{64\pi^6 m_e e^{10}}{3c^3 h^6 n^5} = \frac{5.22 \times 10^9}{n^5}.$$

(7-68)

PROBLEM 7–16. Show that $B(\nu)$ differs from $\sigma_{ab}(\nu)$ in equation (7–59) only by a factor of 2. Derive a relation between A_{ab} and f_{ab}.

We see now, that

$$n(\nu, T)\, d\nu = \frac{8\pi\nu^2}{c^3} \frac{d\nu}{e^{h\nu/kT} - 1} .$$

(7-69)

This corresponds to equation (4–72) for blackbody radiation and shows that the blackbody process depends heavily on the concept of stimulated emission.

The process we have described is stable and self-regulating. If $n(\nu, T)$ is lower than the value given in equation (7–69), spontaneous emission will exceed the sum of absorption and stimulated emission; the population of photons will then increase until it reaches the value given by (7–69). Conversely, if $n(\nu)$ is too high, absorption will lower it back to the equilibrium value.

We should still note that the *Einstein coefficients* $A(\nu)$ and $B(\nu)$ are sometimes defined somewhat differently from the way we have — for example, in terms of emitted or absorbed energy, rather than photons. Throughout this section we also have taken the statistical weights g_n, g_{n-1}, discussed in Section 4:16 to be unity. If this is not so, equations (7–62) and (7–63) must be modified but (7–69) remains unaltered.

7:11 Stimulated Emission and Cosmic Masers

Let us ask what would happen if the relationship between n_0 and $n(\mathcal{E})$ were not given by the Boltzmann relation (7–62). For small deviations from thermal equilibrium the photon bath would tend to cause n_0 and $n(\mathcal{E})$ to come back to equilibrium. But if the number of particles in the upper state starts to exceed the number in the lower state, then an entirely different process comes into play. Clearly, this situation can never come into existence under conditions of thermal equilibrium, because $\exp(-\mathcal{E}/kT)$ is always less than unity for positive values of $\mathcal{E}$ and T. A *population inversion*, $n(\mathcal{E}) > n_0$, can therefore only be brought about by an artificial process. Sometimes we describe a population inversion as a state of negative temperature, for then the exponential term in equation (7–62) can exceed unity. However, this is primarily a descriptive device and does not define any physical process.

Under population inversion, the probability for stimulated emission always exceeds the probability of absorption. In any given transition, a radiation oscillator is therefore more likely to rise to a higher energy state than to a lower one. As the radiation propagates through the assembly of particles, it is amplified. Moreover, because the emitted photons have the same characteristics as the stimulating photons, the amplified radiation is coherent. In the laboratory the process described here corresponds to a *maser*. On a cosmic scale we therefore talk about maser processes. A maser that operates at optical frequencies is called a *laser*.

Cosmic maser action is maintained provided that the pumping of energy into the assembly of particles — the rate of excitation of the upper levels — keeps up with the downward spontaneous and induced transitions. The density of particles in the upper energy state $n(\mathcal{E})$ must remain greater than that in the lower state n_0.

The pumping process can take several forms. We might have a very energetic photon excite particles into an energy state $\mathcal{E}'$ from which the transition probability to the ground state is low and the probability for transition to a metastable state with energy $\mathcal{E}$ is high. This type of maser is called a *three-level maser*. Another means for producing a population inversion can come about chemically. Suppose that a molecule is formed in the interaction of two atoms and that it is formed in a high energy state. If the formation rate is sufficiently high, a population inversion can be maintained and maser action can set in.

In select circumstellar clouds and interstellar regions, OH radicals and/or water vapor molecules H_2O have certain energy states pumped up to population inversion. It is curious that different regions of space show a variety of different OH levels inverted. We therefore seem to have a number of different pumping mechanisms that evidently come into play under different conditions (Me97).

Cosmic masers emit extremely intense coherent radiation. Because all induced photons travel along the same direction, they appear to come from an improbably compact region (Fig. 7.11). The radiation reaching our telescopes arrives contained in a well-defined, extremely small, solid angle.

The smallness of the observed solid angle is misleading. It may not represent the actual size of the cloud, but might represent only the dimension of the region in

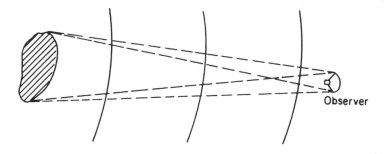

Fig. 7.11. Maser emission. The observer may not see the entire cloud of particles (dashed line) in coherent radiation. He may see only that portion in which the coherent wave originated. (See text.)

which the coherent radiation originated. This volume can be seen, from phase-space arguments, to equal h^3/p^3, or $c^3/\nu^3 = \lambda^3$ where λ is the wavelength of light.

Several types of masers are known. There are OH and H_2O masers whose positions coincide with those of Mira variable stars. Assuming that it emits isotropically, such a maser's luminosity is roughly $10^{-4}L_\odot$. It shows variability, over periods of months, synchronized with the star's pulsations. We also recognize OH and H_2O masers associated with dust clouds in or near HII regions. The H_2O masers can have isotropic luminosities up to $\sim L_\odot$, and show variability over periods of weeks. A number of other molecules such as SiO, also form interstellar masers. Some types of masers may be pumped by a strong infrared radiation flux. For other masers the pumping mechanism is not known. Megamasers, extremely powerful emitting regions, are also found around nuclei of active galaxies. A water vapor megamaser with isotropic luminosity as high as $23,000L_\odot$ has been detected in an AGN at red shift $z = 0.66$ (Ba05a).

One characteristic of the interstellar masers is that the radiation is highly polarized. As already explained, stimulated emission always involves formation of photons with the same sense of polarization as the stimulating photon. This means that all the photons derived from a given progenitor will have identical polarization and the radiation is 100% polarized.

To realize how quickly the intensity of a beam increases as it traverses a cloud in which the population is inverted, we note that for a gain g per interaction, and for an optical depth n in the cloud, the outgoing beam will have an intensity of

$$N = g^n.$$

Suppose that the gain is 1.1 — that is, the probability for stimulated emission is 10% higher than for absorption. After 100 mean free paths, the total number of photons in the beam will be 10^4. After 200 such successive absorption–emission processes, the number would have reached 10^8, and so on. An emitting region, therefore, need not have an opacity in excess of several hundred in order to emit extremely bright maser radiation. Maser radiation is concentrated into a very narrow spectral bandwidth and in this narrow spectral range it often has the equivalent brightness of a fantastically hot body, with T up to 6×10^{13} K (Ra71)*. This temperature can be considerably higher than that of the brightest synchrotron sources mentioned at the end of Section 6:22, because no free electrons are involved and hence no inverse Compton scattering to limit the brightness.

7:12 Stellar Opacity

In the interior of a star, where matter is highly ionized, the interaction of photons with ions often determines the rate at which energy is transported through the star. We will be interested in four distinct types of interaction of radiation with matter: (a) Thomson or Compton scattering of radiation by free electrons (Sections 6:15, 6:22), (b) *free–free absorption* or emission (Section 6:18), (c) *bound–free interactions* in which an electron undergoes a transition between a bound and a free state, and (d) *bound–bound* transitions, as in the excitation or de-excitation of atoms or ions by photons.

In order to compute the mean opacity of stellar matter we proceed in three steps. First, we need to know the interaction cross-section of radiation with matter for each of the four processes. This gives us the opacity due to the individual interactions through a simple proportionality. However, the total opacity of stellar matter is not just the sum of the individual opacities. A suitably chosen mean opacity must be computed, properly weighting the individual contributions made by processes (a) to (d) and also taking induced emission into account. Stimulated emission decreases the opacity because the energy transport rate is increased.

The contributions of the various processes to the opacity depend strongly on the temperature. In the cool surface layers of a star, where atoms are only partially ionized, the opacity may be dominated by bound–bound and bound–free transitions. At high temperatures where ionization may be nearly complete, the opacity due to free–free interactions becomes dominant. At the highest temperatures where induced emission reduces the opacity due to factors (b) through (d), electron scattering plays a dominant role, because Thomson scattering is frequency independent.

We will let *extinction* denote the amount of radiation eliminated from a beam of light through absorption or scattering. We can then define the extinction $\mathcal{E}$ of a slab of matter of unit thickness through which radiation passes at normal incidence as

$$\mathcal{E} \equiv \kappa \rho, \tag{7-70}$$

where the *opacity* of the medium is denoted by the symbol κ and ρ is its density. The opacity for radiation at a particular spectral frequency ν is denoted by $\kappa(\nu)$. Summing over the opacity contributions of processes (a) to (d) at any given frequency, we can write a total opacity $\kappa^*(\nu)$ as

$$\kappa^*(\nu) = \kappa_e + [\kappa_{ff}(\nu) + \kappa_{bf}(\nu) + \kappa_{bb}(\nu)][1 - e^{-h\nu/kT}], \tag{7-71}$$

where the subscripts, respectively, represent electron scattering and free–free, bound–free, and bound–bound transitions. κ^* is the true opacity with induced emission taken into account.

The proper averaging of $\kappa^*(\nu)$ over the entire range of frequencies depends on our purpose. In the case of stellar energy transport, which will be discussed in Chapter 8, we need to know the mean free path of the radiation as it travels through the star. Since this is inversely proportional to the opacity, we need to average $1/\kappa^*(\nu)$ over the entire spectral range. This average, however, must still take into consideration that the radiation spectrum is not flat and that the energy transport rate will therefore also depend on the radiation spectrum defined by the local temperature at any given point of the star. We will consider this later in Chapter 8. For now, we show only how $\kappa^*(\nu)$ depends on the individual opacities, and how these are determined by atomic interaction cross-sections for radiation.

(a) Scattering by Free Electrons

At temperatures sufficiently low for photon energies to lie well beneath the electron rest–mass energy, that is, $T \ll mc^2 k^{-1} \sim 10^{10}$ K, relativistic effects can be neglected and the scattering cross-section is simply the Thomson cross-section

$$\sigma_e = \frac{8\pi}{3}\left(\frac{e^2}{mc^2}\right)^2 = 6.65 \times 10^{-25}\,\text{cm}^2\ . \tag{6–103}$$

This is frequency independent. At the centers of highly dense stellar cores the temperature may become large enough so that the Klein–Nishina cross-section for Compton scattering (6–169) gives a more accurate representation, and a frequency dependence then does exist. At the mean density of the Sun, $\rho \sim 1\,\text{g cm}^{-3}$, the number of electrons per cubic centimeter is of order 10^{24}, so that the mean free path of radiation between electron scattering events is only of the order of 1 cm. If n_e is the number of electrons in unit volume the opacity for scattering is given by

$$\kappa_e \rho = \sigma_e n_e\ . \tag{7-72}$$

(b) Free–Free Interactions

This process was discussed for tenuous plasmas in Section 6:18. The same theory also describes the denser plasmas inside stars. We note that the classical expression (6–137) must unwittingly contain the induced emission factor $[1 - \exp(-h\nu/kT)]$ which at long wavelengths approaches $h\nu/kT$. If we, therefore, divide (6–137) by this factor and also by the number densities, we obtain an absorption coefficient per ion for unit density of ions and electrons. It has the form

$$\alpha_{ff} = \frac{8}{(6\pi)^{1/2}}\frac{Z^2 e^6}{cm^2 h\nu^3 v}\,\ln[\ldots]\ , \tag{7-73}$$

where we have set $v = (3kT/m)^{1/2}$ and have assumed that the argument of the logarithmic function has the same character as in (6–137). The actual, quantum mechanically correct, result is (To47)

$$\alpha_{ff} = \frac{4\pi e^6}{3\sqrt{3}chm^2} \frac{Z^2}{v} \frac{g_{ff}}{v^3},$$ (7-74)

where Z is the effective charge of the ion considered and g_{ff} is called the *Gaunt factor*. It contains the logarithm in (6–137) and is of order unity for most cases of interest.

(c) Bound–Free Absorption

Quantum mechanically, one can also compute an absorption coefficient for bound–free transitions, when only one electron per atom is active in absorbing radiation. This has the form (Cℓ68)*

$$\alpha_{bf} = \frac{64\pi^4 m e^{10} Z^4}{3\sqrt{3}ch^6 n^5} \frac{g_{bf}}{v^3},$$ (7-75)

where n is the principal quantum number and the Gaunt factor, g_{bf}, again is of order unity and only mildly dependent on n and v. This equation only holds when the photon energy exceeds the ionization energy χ_n in the nth state

$$h\nu > \chi_n \sim \frac{2\pi^2 m e^4 Z^2}{n^2 h^2}.$$ (7-76)

(d) Bound–Bound Transitions

These have cross-sections already discussed in Section 7:9. They depend strongly on the actual structure of the individual atom, and do not give rise to a continuum absorption cross-section as do the factors (a) to (c). As discussed in the next section, these cross-sections play an important role in determining the radiative transfer rate through a stellar atmosphere; they do not play a significant role in the stellar interior, where processes (a), (b), and (c) dominate radiative transfer rates (Chapter 8).

The opacity of low-density ionized matter also is a measure of the radiant power emitted from unit volume at given temperature T. If the chemical composition of the plasma corresponds to the cosmic abundance (Table 1.1), Fig. 7.12 shows the radiated power; it assumes that self-absorption by the plasma can be neglected. At high densities where that assumption no longer is valid the plasma would radiate with a brightness characteristic of any blackbody at temperature T.

For these low-density plasmas, forbidden line emission dominates. *Forbidden transitions* are those for which dipole (and sometimes higher multipole) radiative transitions are not allowed by the *selection rules*. When a plasma is at sufficiently low densities so that collisions are rare, a forbidden radiative transition with a correspondingly low transition probability may imply a metastable lifetime of seconds

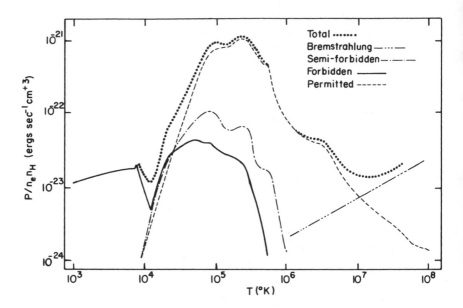

Fig. 7.12. Radiated power for unit volume of a low-density ionized gas in collisional equilibrium. The power indicated is for electron and hydrogen concentrations of $n_e = n_H = 1\,\mathrm{cm}^{-3}$. n_H represents the total number density for hydrogen atoms and protons. To obtain the total power radiated from unit volume, the ordinate would have to be multiplied by $n_e n_H$. The plasma considered has a chemical composition typical of cosmic sources (see Table 1.1) (after D. P. Cox and E. Daltabuit (Co71a)).

or years for the excited metastable atom. Because collisions are rare in the low-density medium, these transitions may take place nevertheless. As the gas density increases, collisions not only excite higher energy states, they also start dominating the de-excitation of atoms, whose energy is transferred to the colliding particles, and the forbidden lines disappear. If we know the lifetimes of a number of different metastable atoms in a hot interstellar nebula we can often conclude a great deal about temperature and density conditions by studying the forbidden lines.

In stars and dense stellar atmospheres collisions between atoms and ions are frequent and no forbidden lines are expected; but even the Earth's upper atmosphere is sufficiently tenuous that forbidden oxygen lines appear in auroral spectra.

7:13 Chemical Composition of Stellar Atmospheres — The Radiative Transfer Problem

In order to determine the abundance of various chemical elements in the atmospheres of stars we must be able to correctly interpret their absorption and emission spectra. This is complicated and depends on the correct choice of a model of the stellar atmosphere. By this we mean that we have to choose an effective temperature T_e, a value for the star's surface gravity, and a parameter ξ_t representative of the turbulent velocity in the atmosphere.

In interpreting individual *Fraunhofer lines* — absorption lines — in terms of the number density n_i of a given atom or ion in an excited state i, we find that the theory always yields expressions proportional to $n_i f g$, where f is the oscillator strength of the transition and g represents the statistical weight of the lower energy level — the level from which the transition takes place. For hydrogen, helium, and other one- and two-electron ions, the f values can be quantum mechanically computed. For such complex spectra as those of iron, however, f values must be obtained through laboratory experiments.

Another important parameter required for an abundance determination when the absorption line is very strong is the *damping constant γ*, which represents the broadening of the line due to the intrinsically finite lifetime of the states and due to the shortening of this life through collisions.

We can define an *equivalent width W_λ* of a Fraunhofer line. It represents the total energy absorbed in this line, divided by the energy per unit wavelength emitted by the star in its continuum spectrum around wavelength λ. Figure 7.13 shows this relationship.

For very weak lines the amount of radiation absorbed, and hence the equivalent width, depends linearly on the abundance n_i and on the product gfn_i. As W_λ approaches the Doppler width due to thermal and turbulent motion, the absorption line becomes saturated and the *curve of growth* (Fig. 7.14), which represents the growth

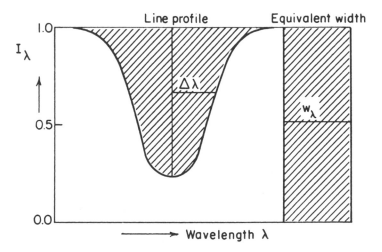

Fig. 7.13. Profile and equivalent width W_λ of a Fraunhofer line. The intensity of the continuum has been set equal to 1. The area under the line profile is equal to that of a completely "black" strip in the spectrum of width W_λ, usually measured in milliangstroms (after A. Unsöld (Un69)).

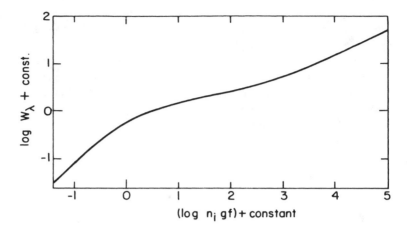

Fig. 7.14. Curve of growth showing increasing absorption with increasing amount of material traversed in a stellar atmosphere. The equivalent width W_λ is plotted against $n_i g f$ where n_i is the abundance of an element in a particular energy state, f is the oscillator strength of the Fraunhofer line, and g is the statistical weight of the absorbing state (after A. Unsöld (Un69)).

of W_λ with increasing material traversed, flattens out. For still stronger lines absorption in the wings of the lines becomes possible. Here the parameter γ (equation (7–51)) determines the amount of radiation that is absorbed.

In determining the abundance of various chemical elements in stars we have to keep in mind that the population of different atomic or ionic energy states depends quite critically on the atmospheric temperature and also to some extent on the surface gravity that determines the pressure. The Boltzmann and Saha equations are applied in these computations, on the assumption that the atmosphere is in thermodynamic equilibrium. Often, the f values for a given transition are not well enough understood, but in some instances we can at least obtain an idea of the relative abundance of an element in a given star compared to its abundance in the Sun.

To relate quantitatively the total flux from a star or nebula to its chemical and physical properties we proceed in the following way. The *intensity* $I(\nu)$ of radiation at spectral frequency ν changes as it crosses a layer of matter of thickness dx. There is a loss of intensity through absorption and a corresponding gain through emission. For normal incidence on the layer the total intensity change is

$$\frac{dI(\nu)}{dx} = -\kappa(\nu)\rho I(\nu) + j(\nu)\rho \,, \tag{7-77}$$

where the first term represents extinction (see equation (7–70)) and $j(\nu)$ is the spontaneous *emission coefficient*, whose units are erg g^{-1} s^{-1} sterad^{-1} Hz^{-1}. The function $j(\nu)$ may strongly depend on the radiation intensity itself as in the case of strong scattering or induced emission. When $j(\nu)$ is entirely due to induced emission the opacity $\kappa^*(\nu)$ in (7–71) is the difference between $\kappa(\nu)$ and $j(\nu)/I(\nu)$.

Equation (7–77) can be rewritten as

$$\frac{1}{\rho\kappa(\nu)}\frac{dI(\nu)}{dx} = -I(\nu) + J(\nu), \quad \text{where} \quad J(\nu) \equiv j(\nu)/\kappa(\nu). \tag{7-78}$$

$J(\nu)$ is called the *source function* and (7–78) is the *transfer equation*. Sometimes one sees the expression $j(\nu)\rho$ in (7–77) written simply as $j(\nu)$ with correspondingly changed units erg cm^{-3} s^{-1} sterad^{-1} Hz^{-1} (Ry71*), (Ch50*). The definition of the source function in (7–78), however, remains unchanged in either notation.

In Sections 8:7 and 8:8 we will discuss the transfer of radiation from the center of a star to its periphery. It will then be necessary to consider not only normal incidence on a layer, but also incidence at other azimuthal angles θ. For arbitrary angles of incidence (θ, ϕ) we can express the energy density of radiation as

$$\int \rho(\nu)\, d\nu = \frac{1}{c}\int\int I(\nu, \theta, \phi)\, d\Omega\, d\nu, \tag{7-79}$$

where $I(\nu, \theta, \phi)$ is the *specific intensity* or *brightness* in the direction (θ, ϕ). Its units are erg cm^{-2} s^{-1} sr^{-1} Hz^{-1}.

The *integrated flux*, measured in units of erg cm^{-2} s^{-1}, depends on the intensity as

$$F = \int F(\nu)\, d\nu = \int\int I(\nu, \theta, \phi)\cos\theta\, d\Omega\, d\nu, \tag{7-80}$$

where $F(\nu)$ is called the *net flux*, whose units are erg cm^{-2} s^{-1} Hz^{-1}. If we consider $I(\nu, \theta, \phi)$ to be a function that specifies the angular distribution of radiation at frequency ν, then $\rho(\nu)$ and $F(\nu)$ involve the zeroth and first moments of this function. The second moment leads to the radiation pressure

$$P = \int P(\nu)\, d\nu = \frac{1}{c}\int\int I(\nu, \theta, \phi)\cos^2\theta\, d\Omega\, d\nu, \tag{7-81}$$

as follows from the discussion of Sections 4:5 and 4:7. The radiation pressure is important in the theory of stellar structure, where hydrostatic equilibrium requires a balance between gravitational forces and pressure gradients. In some stages of stellar evolution, notably in stages leading to planetary nebulae, these gradients depend more strongly on radiant than on kinetic gas pressures. Radiant pressures also play a role in determining the atmospheric structure particularly of giant and supergiant stars.

Let us still describe the factors that determine the shape of a spectral absorption or emission line seen in a star's atmosphere. We have already discussed factors that lead to broadening of a line. However, we still should mention that for gas at a given temperature T, the emission line intensity $I(\nu)$ will normally not exceed the blackbody intensity at that temperature and at frequency ν. Stimulated and spontaneous emission will therefore tend to increase the brightness of an emission line in its wings, as radiation is transferred through the star's atmosphere. The center of the line may already have become saturated — reached its peak intensity — close to the surface of the star. This effect will lead to emission line broadening. Similarly, absorption lines become broadened on passage through the cool outer portions of a star's atmosphere because absorption in the wings becomes increasingly probable as more matter is traversed. We talk about a *curve of growth* for a spectral line. By this we mean a plot of the equivalent width W_λ (Fig. 7.13) against the product $n_i f$ — the number of atoms n_i in a column of unit area through the atmosphere, the *column density*, and the oscillator strength of the transition f. Sometimes, as in Fig. 7.14, the curve of growth is a plot of W_λ not against $n_i f$ but against some other specified function, in this case $\log n_i g f$, where g is the statistical weight of the absorbing state.

7:14 A Gravitational Quantum Effect

While we know that the behavior of electrons in an electromagnetic field is responsible for the quantized structure of atoms, we now ask whether particles in a gravitational field would also have quantized states. Gravitational experiments to detect quantization are difficult because the gravitational attraction between an electron and a proton is $\sim 10^{40}$ times weaker than the electrostatic force between them, so that even the dipole–dipole interactions between atoms can completely mask any gravitational effects. Neutrons, however, carry no electric charge, and experiments show their dipole moments, if not zero, to be lower than 10^{-25} e cm – less than 10^{-12} electron charges across a neutron diameter of 10^{-13} cm. These properties make them ideal for testing quantum effects of gravity.

Quite generally, a neutron trapped between two reflecting surfaces should exhibit quantized energy states and obey all the rules of quantum mechanics outlined in Section 7:2. In the Earth gravitational field, we could place these reflecting surfaces horizontally. If the plates initially are touching, and we slowly move them apart to where the de Broglie wavelength λ of a low-energy neutron is just twice the separation Δ, a standing wave would be set up with nodes at both upper and lower plates. The neutron momentum would then be $p = h/\lambda$, as discussed in Section 7:2. By "low energy", we mean a neutron that has just enough kinetic energy to reach the upper plate against the gravitational pull, namely, $\mathcal{E} = m_n g \Delta = m_n g \lambda / 2$, where $g = 980$ cm s^{-2} is the surface gravity on Earth. We can now write

$$\mathcal{E} = \frac{p^2}{2m_n} = \frac{h^2}{2\lambda^2 m_n} = \frac{m_n g \lambda}{2} \quad \text{and} \quad \lambda = 2\Delta = \left(\frac{h^2}{m_n^2 g} \right)^{1/3} . \qquad (7\text{-}82)$$

This corresponds to a separation between plates of only $\sim 1.3 \times 10^{-3}$ cm, a very small height indeed, but the experiment has been carried out with a beam of cold neutrons projected almost horizontally between two plates. Only the bottom plate in this experiment was reflecting. The top plate was an absorber; but because the low energy neutrons naturally reached only a limited height in Earth's gravitational field, the top plate was needed solely to weed out excessively energetic neutrons. Initially, as the separation between the two plates was gradually increased from zero, the almost horizontally directed beam was totally blocked from transmission. Only when a separation of order 1.5×10^{-3} cm was reached, could the first neutrons in the beam pass through. They passed horizontally between the two plates by rebounding several times off the bottom plate, reaching a height Δ before falling back for the next bounce. Higher quantized transmission levels were observed as Δ was further increased, though these were less reliably detected in what already was a technically difficult experiment (Ne02). The experiment, however, clearly showed that energy states in a gravitational field are quantized.

Answers to Problems

7–1. From the virial theorem (3–85) and in analogy to equation (7–2),

$$E_n = -\frac{p_n^2}{2\mu} = -\frac{n^2\hbar^2}{2r_n^2\mu} = -\frac{Ze^2}{2r_n} .$$

From the last two equalities we can solve for the Bohr radius r_n and obtain the energy,

$$E_n = -\frac{Z^2 e^4 \mu}{2n^2\hbar^2} .$$

Using (7–1), together with the values of p_n and r_n just derived,

$$\text{the number of states} = \frac{\text{phase-space volume}}{\text{volume of unit cell}} = \frac{16\pi^2 n^2 \hbar^3}{\hbar^3} = 16\pi^2 n^2 .$$

7–2.

$$r \sim \frac{\hbar}{\sqrt{p^2}} \quad \text{where} \quad \frac{p^2}{2M} = \mathbb{V}_0 - \mathcal{E}_b .$$

7–3. For rigid rotation at angular frequency ω,

$$\mathcal{E} = \frac{1}{2}\sum_i m_i v_i^2, \quad L = \sum m_i v_i r_i = 2\mathcal{E}/\omega ,$$

$$\text{and } I = \sum m_i r_i^2 \quad \therefore \ \mathcal{E} = \frac{I\omega^2}{2}, \ L = I\omega, \text{ and } \mathcal{E} = \frac{\omega L}{2} = \frac{L^2}{2I} .$$

Because angular momentum is quantized, $\delta L = \hbar$ and $\delta\mathcal{E} = (L/I)\delta L = \omega\hbar$. Since $L = \hbar\{J(J+1)\}^{1/2}$,

$$\mathcal{E}_J = \frac{L^2}{2I} = \frac{\hbar^2(J+1)J}{2I} \quad \text{and} \quad \delta\mathcal{E} = \mathcal{E}_J - \mathcal{E}_{J-1} = \hbar^2\frac{J}{I} .$$

For rapidly rotating massive objects $\hbar J \sim L$ and $\delta\mathcal{E} = L\delta L/I$.

7–4. $\frac{1}{2}I\omega^2 = \frac{3}{2}kT$, and $T = 100$ K. If the molecule is roughly spherical,

$$I \sim \frac{2}{5}mr^2 \sim \frac{2}{5}(10^{-23}\,\text{g})(2 \times 10^{-8}\text{cm})^2 = 1.6 \times 10^{-39}\,\text{g cm}^2.$$

Hence $\omega \sim 5 \times 10^{12}\,\text{s}^{-1}$, $\quad \nu = \dfrac{\omega}{2\pi} \sim 8 \times 10^{11}\,\text{Hz}$, and $\lambda \sim \dfrac{3}{8}$ mm.

7–5. From Section 4:15 we see that the excitation probability is proportional to a Boltzmann factor. Its form is

$$\exp\left(-\frac{\hbar^2 J(J+1)}{2IkT}\right),$$

which is small for large J and low T. For the molecule of Problem 7–4, the exponent is of order $-(0.025)J(J+1)$, so that rotational states far above $J = 6$ are unlikely to be excited.

7–6. For two rotational degrees of freedom (Section 4:19),

$$E = \frac{1}{2}I\omega^2 = kT \sim 1.4 \times 10^{-14}\,\text{erg}, \quad\text{and}\quad I = \frac{2}{5}(10^{-15})(10^{-10})\,\text{g cm}^2.$$

$\therefore\ \omega \sim 6 \times 10^5\,\text{s}^{-1}$, and $\nu \sim 10^5$ Hz, which is slightly below the AM radio band.

7–7.

$$\delta\mathcal{E}_{\text{grav}} = 2\hbar\omega, \quad \delta\mathcal{E}_{\text{neutrino}} = \frac{1}{2}\hbar\omega,$$

and results of Problem 7–3 apply. For electromagnetic radiation the emitted frequency would lie far below the interstellar plasma frequency (6–55) and the radiation would be rapidly absorbed by ambient gas.

7–8. Equation (4–56) gives the velocity distribution function

$$f(v_r) = \left(\frac{m}{2\pi kT}\right)^{3/2}\exp\left(-\frac{mv_r^2}{2kT}\right),$$

and (7–28) gives

$$v_r = \frac{\Delta\omega}{\omega_0}c.$$

Combining these yields

$$I(\omega) = I_0\exp\left(-\frac{mc^2\Delta\omega^2}{2\omega_0^2 kT}\right).$$

From Problem 4–27

$$\langle v_r^2\rangle = \frac{kT}{m} = \frac{\langle\Delta\omega^2\rangle}{\omega_0^2}c^2, \quad\text{so that}\quad T = \frac{m\langle\Delta\omega^2\rangle}{k\omega_0^2}c^2.$$

7–9. With δ from (7–30) and $\gamma_c = n\sigma\langle v\rangle = n\sigma(3kT/m)^{1/2}$, we obtain $\gamma_c/\delta = (n\sigma c/\omega_0)[3/(2\ln 2)]^{1/2}$.

7–10. Making use of the definite integral

$$\int_{-\infty}^{\infty} \frac{(\gamma/2)\, d\omega}{(\omega - \omega_{ab})^2 + (\gamma/2)^2} = \pi$$

in equation (7–59) yields

$$\sigma = \int_{-\infty}^{\infty} \sigma_{ab}(\omega)\, d\omega = \frac{2\pi^2 e^2}{mc} f_{ab} \ .$$

7–11. Equation (7–59) shows that at maximum

$$\sigma_{\max} = \frac{2\pi e^2}{mc} f_{ab} \frac{1}{\gamma/2}$$

and (7–51) gives

$$\frac{\gamma}{2} = \frac{e^2 \omega_{ab}^2}{3c^3 m} \ ,$$

where $\omega_{ab} = 2\pi c/\lambda$, so that

$$\therefore \ \sigma_{\max} = \frac{3}{2} \frac{\lambda^2}{\pi} f_{ab} \ .$$

7–12. The fraction we are seeking is given by the ratio of the integrals I_1 and I_2 :

$$I_1 = \frac{I_{ab}}{\pi} \int_{\omega_{ab}-\gamma/2}^{\omega_{ab}+\gamma/2} \frac{(\gamma/2)\, d\omega}{(\omega - \omega_{ab})^2 + (\gamma/2)^2} = \frac{I_{ab}}{2} \ ,$$

$$I_2 = \frac{I_{ab}}{\pi} \int_{-\infty}^{\infty} \frac{(\gamma/2)\, d\omega}{(\omega - \omega_{ab})^2 + (\gamma/2)^2} = I_{ab} \ ,$$

$$I_1/I_2 = 1/2 \ .$$

7–13. From (7–51) and the definition of ω_{ab},

$$\gamma = \frac{2}{3} \frac{e^2}{mc^3} \omega_{ab}^2 \sim \frac{2}{3} \frac{e^2}{mc^3} \frac{4\pi^2 c^2}{\lambda^2} \sim \frac{1}{5\lambda^2} \ .$$

7–14. As Problem 7–12 shows, half the absorbed light lies within a bandwidth γ. Hence, effectively all the light absorbed lies within a bandwidth 2γ. If the total bandwidth of light from the Sun resides in a bandwidth $\Delta = (c/\lambda_{min}) - (c/\lambda_{max})$, the absorbed fraction is $2\gamma/\Delta$. For a molecule with absorption cross-section $\sigma(\omega)$, the radiative repulsion from the Sun becomes $F_r = 2L_\odot \gamma \sigma(\omega)/4\pi R^2 c\Delta$, while the gravitational attraction is $F_g = MmG/R^2$. This results in

$$\frac{F_r}{F_g} = \frac{L_\odot \gamma \sigma(\omega)}{2\pi MmGc\Delta} \ .$$

7–15. By (7–47) and (7–66)

$$I = \frac{d\mathcal{E}}{dt} = \frac{e^2\omega^4\langle r^2\rangle}{3c^3} = \hbar\omega A_{n,n-1} \, ,$$

and from (7–5)

$$\mathcal{E}_n = \hbar\omega = \frac{\mu}{2}\frac{e^4}{\hbar^2}\left\{\frac{1}{(n-1)^2} - \frac{1}{n^2}\right\} .$$

For n large $\omega \sim \mu e^4/\hbar^3 n^3$, and from (7–2) and (7–5), $r_n^2 = (n^2\hbar^2/\mu e^2)^2$ to obtain $A_{n,n-1} = 64\pi^6\mu e^{10}/3c^3 n^5 h^6$.

7–16. In equations (7–63) and (7–65) we called the number of photons of a given polarization absorbed per particle per second $n(\nu, T)cB(\nu)$. In defining σ, we talked about photons of either polarization, so that $\sigma(\nu) = 2B(\nu)$.

$$\frac{I}{h\nu} = \int A(\nu)\, d\nu = A_{ab}, \tag{7–66}$$

$$I(\omega) = \frac{I_{ab}}{\pi}\frac{\gamma/2}{(\omega - \omega_{ab})^2 + (\gamma/2)^2} \, , \tag{7–43}$$

$$\sigma(\omega) = \frac{2\pi e^2}{mc}f_{ab}\frac{\gamma/2}{(\omega - \omega_{ab})^2 + (\gamma/2)^2} \, . \tag{7–59}$$

$$\therefore A(\omega) = \frac{1}{\pi}\frac{A_{ab}(\gamma/2)}{(\omega - \omega_{ab})^2 + (\gamma/2)^2}, \quad A(\nu) = \frac{8\pi\nu^2}{c^2}B(\nu) \, .$$

$$\therefore A(\omega) = \frac{2\omega^2}{\pi c^2}B(\omega) \, .$$

$$= \frac{\omega^2}{\pi c^2}\sigma(\omega) \, ,$$

$$\therefore A_{ab} = \frac{8\pi^3 e^2\nu^2}{mc^3}f_{ab} \, .$$

8 Stars

8:1 Observations

We do not really know how stars are formed, nor just how they die. But we think we understand the structure of the most commonly found stars and the mechanisms that generate the energy we see as starlight.

How sure can we be of this understanding? How correct is the theory of stellar structure? Such questions are difficult to answer. Many of the most important stellar processes take place deep in a star's interior, whereas the observations available to us mainly register surface characteristics. Conditions in the star's central regions are inferred, and our evidence is indirect.

Generally, the merit of a theory is judged by the number of unrelated observations it can explain; when the observations are indirect, a larger than usual body of data is desirable. For the theory of stellar structure and evolution we have several different classes of observations:

(a) We have measurements on the masses of a variety of stellar types. However, the number of precision measurements is small. Each such measurement involves the detailed analysis of a stellar binary system (Section 3:5), and binary stars may be atypical.

(b) We can determine the luminosity of a star quite accurately provided the star is near the Sun where its distance is easily measured and interstellar extinction is negligible. In the past few decades, bolometric magnitudes have become well established as satellite observations have increasingly provided far-infrared, far-ultraviolet, and X-ray fluxes. These observations suggest frequent sizeable deviations from blackbody behavior.

(c) The surface temperature of stars can be obtained in three different ways.

(i) We can define a color temperature, using equation (4–77),

(ii) We can determine an effective temperature if the star's angular diameter is known from (4–78),

(iii) We can observe the strengths of spectral lines representing transitions between various excited atomic states. Because the relative population of excited states is governed by the (temperature-dependent) Boltzmann factor, the temperature can be computed directly, provided the relevant transition probabilities are known. These probabilities can be calculated or, preferably, measured in the laboratory.

These three independent techniques yield a satisfactory estimate of stellar surface temperature, but the temperature in the interior of stars remains unobserved.

(d) We can determine the angular diameter of a star by interferometric means (Section 4:12). Occultation of a star by the Moon or by a planet can also be used, with accurate timing giving the angular diameter in terms of the occulting body's motion across the star's surface (Ev96). Eclipse data from binaries can also yield stellar diameters.

The luminosity, diameter, and surface temperature of a star are related. They involve no more than two independent parameters, provided the star's spectrum is sufficiently simple to allow the assignment of a single representative temperature.

(e) The chemical makeup of stars is spectroscopically determined. The abundance of the different elements obtained in this way refers only to the surface layers of the stars. Using current observational techniques, we cannot directly verify conjectures about the composition inside a star.

We find that the abundance of elements on the surfaces of stars varies. Normally, hydrogen is by far the most abundant constituent. By mass its concentration lies between 68 and 76%. Helium, the next most abundant element, has a concentration of 24 to 30% by weight. Oxygen, carbon, nitrogen, and neon follow in order of decreasing abundance; together they account for <0.01 to 2% of the total mass. Magnesium, silicon, iron, and sulfur are next; each of these has an approximate abundance of the order of one part per thousand, by mass. Stars that appear to have formed long ago, when the Galaxy was young, have the lowest abundances of heavy elements.

One task of a theory of stellar evolution must be the correct prediction of the abundance of chemical elements in different types of stars. The interior of stars is the most plausible place for heavier elements to be formed from the relatively pure hydrogen–helium mixture that was the main constituent of the Galaxy at the time the earliest stars formed in globular clusters.

The relative abundances of the various isotopes of different elements are repeatedly found in similar ratios in the Earth's crust, in meteorite fragments, and in the interstellar medium and many but not all stars in our neighborhood of the Galaxy. The similarity of these ratios cannot be accidental and should be explained by a comprehensive theory of stellar evolution.

(f) For a limited number of stars we have measurements on the surface magnetic fields. Peculiar stars of spectral type A have fields of order 10^4 G. White dwarfs have fields $B \sim 10^5$ to 10^8 G. Neutron stars generally have magnetic field strengths around 10^{12} G, but the *magnetars* among them have field strengths ranging up to $\sim 10^{15}$ G. Current theories of stellar evolution have not yet incorporated magnetic effects to any great extent.

(g) Although projection effects preclude an analysis of the rotational velocity of individual stars, statistical studies indicate that young O and B stars have extremely high rotational velocities, and that these velocities progressively diminish with spectral type from O to M (see Table A.4). We are only just starting to consider effects of rotation on stellar evolution.

(h) Similarly, a statistical study of stellar velocities in the Galaxy (see Table A.6) tells us that the history of stars of different spectral types must be dissimilar. Presumably these stars originated at different epochs in the Galaxy's evolution. A final theory of stellar evolution will have to consider such age differences for stars, and will have to take into account the extent to which the chemical composition of the interstellar gas from which stars are formed has changed as the Galaxy aged.

(i) For a number of different stars, notably K-giants, O-stars, and nuclei of planetary nebulae, we now have data on mass loss. For the first two of these spectral types, the information comes from spectral measurements of gas outflow. In the case of planetary nebulae, we actually see the accumulation of ejected gas. Such evidence is important both for studying unstable states of stars, and for understanding chemical changes in the interstellar medium when material which has undergone nuclear transformation in stars is returned to interstellar space. Nova and supernova ejecta also yield important data in this respect. Unfortunately we do not as yet have enough information to judge the extent to which violent but infrequent explosive events contribute to the cycling of matter between stars and interstellar space. Nor do we know the fraction of explosively ejected matter escaping a galaxy.

(j) For the Sun, we also have data on internal rotation rates from helioseismology and oblateness studies (Di86); and we know the rate of neutrino emission (Sc02). For other stars such data are completely lacking. Similarly, we have information on solar cosmic-ray and X-ray emission; X-ray data by now also exist for an appreciable number of other stars. In each of these cases, however, it is unclear how the circumstellar regions from which this radiation reaches us are affected by conditions prevailing inside the stars.

(k) Finally, a very important body of statistical information is contained in the Hertzsprung–Russell and color-magnitude diagrams (Figs. A.2 and 1.3 to 1.7).

The confinement of stars to quite narrow ranges on an H–R diagram sets a condition that must be met by any theory of stellar structure and evolution. The theory must prohibit the appearance of stars in empty regions, and account for the relative density of stars in populated portions of the H–R diagram. An acceptable theory must explain the significance of the main sequence, the existence of the red-giant and horizontal branches, the variable turn-off point that has the main sequence joining the red-giant branch at different locations in different groups of stars, and other features.

Stellar masses and diameters, the Hertzsprung–Russell diagrams available for a large number of different stellar groups and populations, and the tables of chemical abundances compiled for many astronomical objects, provide a wealth of observational detail against which to gauge the merit of our theories.

The purpose of the present chapter is to outline the main ideas involved in current theories of stellar structure and to show the extent to which these theories fit observations (Bu57)*.

8:2 Sources of Stellar Energy

We have shown how one determines the overall characteristics of stars — their radii, masses, and luminosities. But what is the source of energy that heats a star and replenishes the energy it loses in radiated starlight?

Before this problems can be discussed, we may want to answer a somewhat different question. How much energy does a typical star radiate away in the course of its life? Here we may want to know the average luminosity of the star and its age at death — whatever form that death might take. It is not easy to decide how old a given star is, because stellar ages are far greater than the few hundred years during which reliable astronomical observations have been carried out; but two pieces of information are useful.

First, stars like the Sun occupy positions on the lower main sequence of the Hertzsprung–Russell diagram, and are found not to have noticeably changed either in brightness or in color since photographic techniques became well established around the turn of the nineteenth to twentieth century.

Second, the Sun must be older than the Earth which, as judged from the abundance of the radioactive uranium isotope ^{238}U and its decay products is $\sim$4.7 Gyr old. From paleontological evidence, we surmise that the temperature of the Sun cannot have varied a great deal in the past $\sim$1.9 $\times$ 10^9 yr during which life is definitely known to have existed on Earth (Mo05). Fossil remains that we find today indicate that liquid water must have been present on Earth during this entire interval. Other indicators suggest that life may have existed significantly earlier (Section 14:4). Had the Sun been somewhat cooler or hotter during these epochs the oceans might have frozen or evaporated away, and the observed early forms of life would have died out.

We may therefore assume that, to rough approximation, the Sun has radiated at its present rate for $\sim$5 Gyr. Because its luminosity is $L_\odot = 4 \times 10^{33}$ erg s^{-1}, the total radiated energy emitted thus far is $\sim$6 $\times$ 10^{50} erg. Because the solar mass is $M_\odot = 2 \times 10^{33}$ g, this amounts to an energy-to-mass ratio of 3 $\times$ 10^{17} erg g^{-1}.

Could this much energy have been provided by chemical reactions, or else through slow gravitational contraction which, as seen from equation (4–141), yields radiant energy of the order of the potential energy released?

Neither of these sources turns out to be adequate. The energy yield of chemical reactions, including the burning of fossil fuels, normally does not exceed 100 kilocalories, or $\sim$4 $\times$ 10^{12} erg g^{-1}. If the Sun had depended on chemical sources it could have continued to shine no more than $\sim$5 $\times$ 10^4 yr, a factor of 10^5 too short.

If we assume, for purposes of a rough estimate, that the Sun has the same density throughout, the total potential energy released to date would be

$$\mathbb{V} = -\int_0^{R_\odot} \left(\frac{4\pi}{3}\right) \rho r^3 \frac{G}{r} (4\pi\rho r^2) \, dr = -\frac{3}{5} \frac{M_\odot^2 G}{R_\odot} \, , \qquad (8\text{-}1)$$

which amounts to $\sim$2 $\times$ 10^{48} erg. This corresponds to 10^{15} erg g^{-1}, still two or three orders of magnitude short of the required energy. Even a hundredfold density increase at the center of the star could not change this result significantly. We cannot rule out, without further evidence, a much denser central core with $\rho \sim 10^{15}$ g cm^{-3} and radius $R \sim 10^5$ cm. Approximately the right amount of gravitational energy would then be available. But whereas this source of energy seems to be important for very compact stars, it appears to play no significant role in normal stars.

The only remaining source of energy involves nuclear reactions. The high abundance of hydrogen and helium in the Universe suggest that hydrogen may be transmuted into helium at the centers of stars. We note that the mass difference between four hydrogen atoms and one atom of helium is

$$4m_H - m_{H_e} = 0.029 m_H \, . \tag{8-2}$$

The transmutation of hydrogen into helium therefore includes a mass loss of the order of 7×10^{-3} g for each gram of converted hydrogen. Since the energy given off in the annihilation of mass m is mc^2, this amounts to an energy liberation of 6×10^{18} erg g^{-1} — ample compared to the amount required, even if only a fraction of a star's hydrogen content is converted into helium (Be39).

If we now ask about the life span of stars on the main sequence and about the rate at which stars are born, we can proceed in the following way. Let us first assume that we know the life span τ_i for a given type, i, of main sequence star. Let the number density in the Galaxy be ϕ_i for this kind of star. We can then define a birthrate function — usually called the *Salpeter birthrate function* ψ_i

$$\psi_i = \frac{\phi_i}{\tau_i} \, , \tag{8-3}$$

giving the rate of star formation in unit volume of the Galaxy (Sa55) (Fig. A.6). For disk population (Population I) stars the formation rate will of course be high only in and near the Milky Way disk, while the birthrate will be negligible in the halo.

We can also obtain an estimate of the age of a star as it moves off the main sequence. Suppose that a fraction of the stellar mass $f(M)$ needs to be exhausted of hydrogen before the star moves onto the red-giant branch. If the initial composition of the stellar material contains a fraction (by mass, not by number of atoms) X, in the form of hydrogen, the energy $\mathcal{E}$ liberated by the star while it still resides on the main sequence is

$$\mathcal{E} = f(M)X \left(\frac{0.029 m_H}{4 m_H} \right) M c^2 = 6.4 \times 10^{18} f(M) X M \text{ erg.} \tag{8-4}$$

The numerical factor gives the energy in ergs liberated by one gram of hydrogen converted into helium. The time taken to expend this energy is just the energy $\mathcal{E}$ divided by the star's luminosity L. Now, Fig. 8.1 shows that the mass–luminosity relation for main sequence stars is roughly $L = L_\odot (M/M_\odot)^a$, where $3 \lesssim a \lesssim 4$. Taking $a \sim 3.5$, the star's life τ on the main sequence becomes

$$\tau = \frac{\mathcal{E}}{L} = 6.4 \times 10^{18} X f(M) \left(\frac{M_\odot}{M} \right)^{5/2} \frac{M_\odot}{L_\odot} \quad \text{seconds.} \tag{8-5}$$

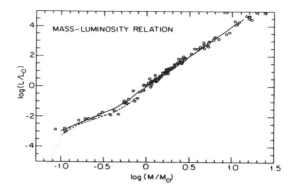

Fig. 8.1. Mass-luminosity diagram for main sequence stars. The dashed and solid lines are different theoretical fits to the data represented by solid points (Re87). (Reprinted with permission from the Royal Astronomical Society.)

$f(M)$ is of order 15% for stars with solar composition for which $X \sim 0.7$. Inserting these two numerical values, making use of the mass–luminosity relation once more, and converting the time scale to years we find, very roughly, that

$$\tau \sim \left(\frac{L_\odot}{L}\right)^{5/7} \times 10^{10}\,\mathrm{yr}. \tag{8-6}$$

The Sun should therefore have a total life span of order 10^{10} yr, whereas O stars, which are some ten thousand times more luminous and ten times more massive, survive only a few million years.

8:3 Requirements Imposed on Stellar Models

Granted that sufficient energy is available from *hydrogen burning* (8–2) and perhaps from other nuclear reactions, we still need to investigate whether the hypothesis of nuclear energy conversion also fits all the other observations. These are:

(a) Conditions inside stars must be compatible with adequate nuclear reaction and energy generation rates to match the observed luminosity of stars. The energy released at the star's surface must further be predominantly in the form of visible, ultraviolet, or infrared radiation since most of a normal star's radiation is observed at these wavelengths. If a predominant fraction of the generated nuclear energy were channeled elsewhere, say into neutrino emission, then we would still be faced with the problem of accounting for the visual starlight.

(b) Nuclear reaction rates depend on the temperature, density, and chemical composition of the matter in stars. The values of these parameters required to maintain the star's luminosity must remain compatible with stable stellar structure.

Pressure equilibrium, for example, must be maintained throughout the star. This is determined by two factors. First, the pressure in any region is determined by the local temperature, density, and chemical composition. The relationship among these quantities is summarized in an equation of state such as the ideal gas law or some similar expression. Second, the local pressure must be just able to support the weight of material lying overhead — matter at larger radial distance from the center of the star. This is called the condition of *hydrostatic equilibrium*.

If the temperature and density are too high, the local pressure becomes too large and the star expands. If the pressure is too low, the star will contract. We will see that any appreciable deviation from pressure equilibrium leads to a readjustment that takes no more than about an hour. A star that lives for many æons must therefore be very close to pressure equilibrium throughout, unless it pulsates.

(c) The energy generated at the center of the star must be able to reach the surface within a time small compared to its evolutionary age; otherwise, the whole life of the star would have to be described by transient conditions, and the stable characteristics of main sequence stars could not be explained.

(d) The temperature at any given distance from the center of the star must not only lead to the correct pressure (condition (b)), it must also be compatible with adequate energy transfer rates to assure that the luminosity just equals the rate of energy generation (condition (c)).

(e) At the center of the star the *luminosity* must be zero. This means that there is no finite outflow of energy, no mysterious source pouring out energy from an infinitesimal volume about the center of the star.

At the same time, there can be no more than an infinitesimal mass enclosed in an infinitesimal volume about the center of the star. These two requirements impose boundary conditions on the differential equations implied by requirements (a) to (d).

(f) At the surface of the star the pressure and temperature can usually be taken to be very small compared to values found in the central regions. This follows from the equation of state and from condition (b) which required pressure balance throughout the star. It is a statement of the fact that stars have high internal pressures and that the boundary between star and surrounding empty space is relatively sharp. Nevertheless, some caution has to be observed in applying this last condition; and differences will arise between *early spectral-type stars* — where energy is transported through the surface layers primarily by radiation — and *late spectral types* whose surfaces are convective.

8:4 Mathematical Formulation of the Theory

The requirements described above can be summarized in a number of differential equations. In giving this formulation we will find it convenient to follow a procedure slightly different from that of Section 8:3.

(a) The change of pressure dP on moving a distance dr outward from the center of a star is

$$dP = -\frac{\rho G M(r)}{r^2} \, dr, \tag{8-7}$$

where ρ is the local density and $M(r)$ is the mass enclosed by a surface of radius r. This increment of pressure is produced by the gravitational attraction between $M(r)$ and the mass $\rho \, dr$ enclosed in the incremental volume of height dr and unit base area. G is the gravitational constant.

(b) The change of mass $dM(r)$ on moving a distance dr outward from the center is

$$dM(r) = 4\pi r^2 \rho \, dr. \tag{8-8}$$

(c) The change of luminosity $L(r)$ within an increment dr at distance r from the center of the star is

$$dL(r) = 4\pi r^2 \rho \varepsilon \, dr, \tag{8-9}$$

where ε is the energy generation rate per unit mass.

(d) In general, this generation rate is a function of the local density ρ, temperature T, and the mass concentrations X_i of elements i. Hydrogen and helium mass concentrations are usually labeled X and Y, respectively:

$$\varepsilon = \varepsilon(\rho, T, X, Y, X_i) \quad \text{where} \quad i = 1, \ldots, n, \tag{8-10}$$

when n elements other than hydrogen and helium are present in significant amounts.

(e) The local pressure is related to the temperature, density, and chemical composition. We will find it convenient to write this in the form

$$P = P(\rho, T, X, Y, X_i), \qquad i = 1, \ldots, n, \tag{8-11}$$

because it will facilitate comparison of pressures derived from expressions (8–7) and (8–11). The right side of this equation is a general form for an equation of state which often is well approximated by Dalton's law (4–38).

(f) Next, the temperature gradient must be related to the parameters that assure a stable luminosity profile throughout the star. Two possibilities arise here:

(i) If the star has a low opacity κ,

$$\kappa = \kappa(\rho, T, X, Y, X_i), \qquad i = 1, \ldots, n, \tag{8-12}$$

light can travel long distances within the star before being absorbed or scattered, and no large temperature gradients arise. In this case the transfer of energy is achieved by radiation alone. The photons are emitted, scattered, absorbed, and re-emitted many times; and their energy and number density changes as they diffuse through the star in a complex random walk that eventually takes them from the center to the star's surface. There they start on their long journey through space.

(ii) If the opacity is high, this random walk may be excessively slow. The center of the star then becomes too hot and the stellar material starts to convect. A convective pattern of heat transfer sets in and, as we will see below, the temperature gradient is given by the so-called *adiabatic lapse rate* which depends on the ratio of heat capacities of the material, $\gamma = c_p/c_v$ (see Section 4:19).

Corresponding to these two alternatives, we can derive temperature gradients of the form

$$\frac{dT}{dr} = F_1[\kappa, L(r), T, r] \quad \text{for radiative transfer,} \tag{8-13}$$

or

$$\frac{dT}{dr} = F_2(T, P, r, \gamma) \quad \text{for convective transfer.} \tag{8-14}$$

(g) The two boundary conditions implied by (e) and (f) in Section 8:3 are:

(i) at $r = 0$, $M(r) = 0$, and $L(r) = 0$, (8–15)

(ii) at $r = R$, $T \ll T_{\text{central}}$ and $P \ll P_{\text{central}}$, (8–16)

where R is the star's radius. For purposes of computing hydrostatic pressures, the relations (8–16) are tantamount to writing

$$T(R) \approx 0, \qquad P(R) \approx 0 \tag{8-17}$$

Equations (8–7) to (8–17) constitute the foundations of the theory of stellar structure.

One point of particular interest should still be mentioned. The equations presented state nothing about the physical source of the generated energy. The overall structure and appearance of the star can therefore give no clue about whether nuclear reactions indeed are responsible for stellar luminosities, or which particular reactions predominate at any given evolutionary stage. We have to derive this information by indirect means — mainly by looking at the debris ejected from stars that become unstable or by spectrally analyzing stars whose surfaces become denuded to expose matter previously evolved at the center.

8:5 Relaxation Times

Suppose we could artificially perturb the temperature or pressure within a star. After this perturbation stopped, the star would again relax to its initial temperature and pressure equilibrium. We will find that the *relaxation time* in response to a pressure change is very much faster than the time required to re-establish temperature equilibrium.

(a) We first wish to estimate the time required to reach pressure equilibrium. Let the perturbed pressure $P_p(r)$ differ from the equilibrium pressure $P(r)$ by a fractional amount f

$$P_p(r) - P(r) = fP(r) . \tag{8-18}$$

This pressure acts on a mass $M - M(r)$ lying at radial distance greater than r with a force $F = 4\pi r^2 fP(r)$. As a result this material moves with an acceleration

$$\ddot{r} = \frac{4\pi r^2 fP(r)}{M - M(r)} . \tag{8-19}$$

We suppose that a displacement Δr amounting to a fraction g of the total radius R is required to relieve the pressure difference

$$\Delta r = gR \, . \tag{8-20}$$

Then the time required to obtain this displacement with the acceleration given in equation (8–19) is of the order of

$$\tau_p \sim \left(\frac{2\Delta r}{\ddot{r}} \right)^{1/2} = \left[\frac{gR[M - M(r)]}{2\pi r^2 f P(r)} \right]^{1/2} . \tag{8-21}$$

Let us compute the approximate value of τ_p. We can estimate $P(r)$ and $M(r)$ by assuming a uniform density throughout the star and considering a star with one solar mass $M = M_\odot = 2 \times 10^{33}$ g contained in one solar radius $R = R_\odot = 7 \times 10^{10}$ cm. The density then is $\rho \sim 1$, and from (8–7),

$$P(r) = - \int_R^r \frac{4\pi}{3} \rho^2 r G \, dr = \frac{2\pi}{3} \rho^2 G (R^2 - r^2). \tag{8-22}$$

Let us choose $r \sim R/2$; then

$$P \left(\frac{R}{2} \right) \sim 10^{15} \, \text{dyn cm}^{-2} \, ,$$

$$M(R) - M \left(\frac{R}{2} \right) \sim 2 \times 10^{33} \, \text{g} \, ,$$

and

$$\tau_p \sim 5 \times 10^3 \sqrt{\frac{g}{f}} \, \text{s}^{-1} \, .$$

For small perturbations, g/f can be expected to be of order unity and the relaxation time is of the order of an hour.

In Section 9:3, we will see that the *speed of propagation of pressure information*, the speed of sound, is roughly $(P/\rho)^{1/2}$, which is also roughly the speed of ideal gas particles (4–31). This speed is of order $3 \times 10^7 \, \text{cm s}^{-1}$ in the Sun. Pressure information can therefore be conveyed over distances $R_\odot$ in $\sim 2 \times 10^3$ s, a time comparable to the pressure adjustment time.

PROBLEM 8–1. (a) Show that the temperature $T(R/2)$ under the conditions assumed above is $\sim 10^7$ K.

(b) One difference between a planet and a star is that for planets, Coulomb forces on electrons and ions are more important than gravitational forces (Sa70a). The opposite is true of a star. Let $\mathcal{E}_c$ be a typical Coulomb binding or repulsive interaction energy. Show that the planet's mass M_p is of order

$$M_p \lesssim \frac{1}{\rho^{1/2}} \left[\frac{|\mathcal{E}_c|}{GAm_H} \right]^{3/2} \sim \frac{R|\mathcal{E}_c|}{GAm_H}, \tag{8-23}$$

where A is an average atomic mass measured in atomic mass units, and m_H is the mass of the hydrogen atom.

(c) If $\mathcal{E}_c$ is of the order of the binding energy $\sim 10^{-11}$ erg per *chemical bond*, show, by referring to Table 1.4, that Jupiter, which consists largely of hydrogen, lies near the upper limit of the mass range for planets.

(b) Next we wish to estimate the time taken to transport heat from one point within the star to another. If the transport process is radiative, the time can be computed from a random walk model; we only need to know the mean free path of radiation, given by the opacity of the material κ (see Section 7:12). When a beam of n photons passes through a layer of thickness dl, a fraction dn will be absorbed or scattered by the material. The loss of photons from the beam can then be expressed as

$$dn = -n\kappa\rho\,dl, \tag{8-24}$$

where ρ is the density of the material. We note that we have not gone into detail about the scattering process. Some processes strongly scatter light into a forward direction. Such scatterers make a medium much less opaque than isotropic scattering centers. We will assume here that the scattering is isotropic. Alternatively, we could count a photon as being lost from the beam only after a large number of collisions has increased its angle with respect to the original direction of propagation significantly, say to $90°$, so that all memory of the original direction is lost. We made a similar assumption about electron scattering in Section 6:18. We wish to calculate the mean free path of the photons under such conditions. Integrating equation (8–24) we obtain

$$n = n_0 e^{-\kappa\rho l}. \tag{8-25}$$

The mean distance $\langle l \rangle$ traveled by a photon before it is absorbed or strongly scattered is then

$$\langle l \rangle = -\frac{\int_0^{n_0} l\,dn}{n_0} = \int_0^{\infty} l\kappa\rho e^{-\kappa\rho l}\,dl = \frac{1}{\kappa\rho}. \tag{8-26}$$

For a star like the Sun $\kappa\rho$ is of order unity, and the mean free path is of the order of a centimeter. To traverse a distance of the order of the solar radius $R \sim 10^{11}$ cm, we would require 10^{22} steps, which would cover a total distance $\sim 10^{22}$ cm. The total time taken is $\sim (R^2\kappa\rho/c)$ when we do not count the time required between absorption and re-emission. The time constant therefore is at least of the order of 10^{11} s, several thousand years.

(c) Energy can also be transported by convection, where sufficiently high thermal gradients exist. A buoyancy force then accelerates a hot blob of matter upward and returns cooler material down toward the center of the star. For nondegenerate matter we can take $\Delta\rho$, the density difference between hot and cold material, to be

$$\Delta\rho \sim \frac{\rho}{T}\Delta T. \tag{8-27}$$

The upward force on unit volume of the hotter material is

$$F(r, \rho, \Delta T) = \frac{M(r)G}{r^2} \frac{\rho}{T} \Delta T \,, \tag{8-28}$$

which leads to a convective motion accelerated at a rate

$$\ddot{r} = \frac{M(r)G}{r^2} \frac{\Delta T}{T} \,. \tag{8-29}$$

If the blob travels a distance of the order of one-tenth of a solar radius the time required is

$$t = \left[\frac{2R_\odot}{10\ddot{r}} \right]^{1/2} \sim 3 \times 10^6 \, \mathrm{s} \sim 1 \, \mathrm{month} \tag{8-30}$$

for $M_r \sim 10^{33} \, g$, $T \sim 10^7 \, \mathrm{K}$, and $\Delta T \sim 1 \, \mathrm{K}$. By (8–27) the density gradient is then $\Delta\rho/\rho \sim 10^{-7}$, vindicating our assumption of a constant density in applying (8–22) to estimate the pressure deep in the Sun. Equation (8–30) implies a rather fast transport rate. It sets in whenever radiative heat transfer is too slow to maintain thermal equilibrium within a star. We will return to this stability problem in Section 8:9, where we will also justify the choice of $\Delta T \sim 1 \, \mathrm{K}$.

(d) If the electrons at the center of a star are degenerate, they can readily transport heat at a rate much faster than is possible by either radiative or convective means. Degenerate electrons cannot readily transfer their energy to other electrons because all the lower electron energy states already are filled and there is no space for another electron that is about to lose energy. The mean free path for electrons therefore becomes extremely long, and heat transport proceeds swiftly. In the limiting case an electron could traverse the entire degenerate region and not lose energy until it reached the nondegenerate surroundings. If the span in question amounts to a distance of the order of $R_\odot/10^2$, the traversal times at $T \sim 10^7 \, \mathrm{K}$ would be of the order of one second. This represents the thermal relaxation time for the degenerate core of a star.

8:6 Equation of State

The equation of state needed to define the pressure in terms of temperature, density, and composition depends on whether: (i) conditions at the center of a star are nondegenerate or degenerate; and (ii) the temperature is sufficiently high to involve relativistic behavior.

(a) Nondegenerate Plasma

At the high temperatures found within stars all but the heaviest elements are completely ionized. Electrons and ions are far apart compared to their own radii since electrons and bare nuclei have radii of order 10^{-13} cm. The ideal gas law can therefore be expected to hold:

$$P = nkT \,, \tag{4–38}$$

Table 8.1. .

	Number of Ions	Number of Electrons
Hydrogen	$\dfrac{X\rho}{m_H}$	$\dfrac{X\rho}{m_H}$
Helium	$\dfrac{Y\rho}{4m_H}$	$\dfrac{Y\rho}{2m_H}$
Others	$\dfrac{Z\rho}{\langle A\rangle m_H}$	$\dfrac{Z\rho}{2m_H}$

where n is the number of particles in unit volume.

In Table 8.1 we enumerate the contribution to the number density by the various particles. The symbols X, Y, Z, represent the concentration, by mass, of hydrogen, helium, and heavier elements, respectively. $\langle A \rangle$ is the mean atomic mass of the heavier elements. In the last column of the table the number of electrons contributed by the heavier elements is given on the hypothesis that the number of electrons per atom is $\langle A \rangle /2$. This is a fairly good approximation for the less massive elements. The number of ions contributed by the heavier elements amounts to a negligibly small fraction of the total population — only about one part per thousand. The total number density of particles to be inserted in the ideal gas or Dalton's law relation (4–38), therefore, is roughly

$$n = \frac{\rho}{m_H}\left[2X + \frac{3}{4}Y + \frac{1}{2}Z\right] , \tag{8-31}$$

and the equation of state reads

$$P = \frac{\rho kT}{m_H}\left[2X + \frac{3}{4}Y + \frac{1}{2}Z\right] . \tag{8-32}$$

At first we might think that P represents the total pressure; but it is only the pressure contributed by particles. A further pressure due to electromagnetic radiation must be added to yield the total pressure. This is true for both nondegenerate and degenerate matter.

We had already found in Section 4:7 that the radiation pressure has a value numerically equal to one-third of the energy density. Inside the star that density is aT^4; the refractive index is practically unity, and the relationship of Problem 4–21 reduces to equation (4–74). Hence

$$P_{\text{Rad}} = \frac{aT^4}{3} . \tag{8-33}$$

The equation of state for nondegenerate matter then reads

$$P_{\text{Total}} = \frac{\rho kT}{m_H}\left(2X + \frac{3}{4}Y + \frac{1}{2}Z\right) + \frac{aT^4}{3} . \tag{8-34}$$

(b) Degenerate Plasma

For a momentum p_0 corresponding to the Fermi energy, equation (4–65) tells us that the maximum number of electrons that can occupy unit volume is

$$n_e = \frac{8\pi}{3} \frac{p_0^3}{h^3} . \tag{8-35}$$

The number density of electrons can also be written as

$$n_e = \left(X + \frac{1}{2}Y + \frac{1}{2}Z \right) \frac{\rho}{m_H} \equiv \frac{1}{2}(1+X)\frac{\rho}{m_H} , \tag{8-36}$$

because

$$X + Y + Z = 1 . \tag{8-37}$$

The pressure contribution due to isotropically moving electrons is then given through equations (4–27), (4–28), and (4–30), as

$$P_e = \int_0^{p_0} \int_0^{2\pi} \int_0^{\pi/2} 2n_e(p)p\cos\theta v \cos\theta \sin\theta \, d\theta \, d\phi \, dp , \tag{8-38}$$

$$= \frac{1}{3} \int_0^{p_0} \frac{8\pi p^2}{h^3} pv \, dp , \tag{8-39}$$

where $n_e(p)$ is the number density of electrons having momenta in the range p to $p + dp$.

(i) In the nonrelativistic case $v = p/m_e$ and the electron pressure is

$$P_e = \frac{8\pi}{15} \frac{p_0^5}{m_e h^3} . \tag{8-40}$$

Substituting for p_0 from (8–35) and (8–36), equation (8–40) becomes

$$P_e = \frac{h^2}{20m_e m_H} \left(\frac{3}{\pi m_H} \right)^{2/3} \left(\frac{(1+X)}{2}\rho \right)^{5/3} . \tag{8-41}$$

(ii) In the relativistic case $v \sim c$, and equation (8–38) integrates to

$$P_e = \frac{2\pi c}{3h^3} p_0^4 . \tag{8-42}$$

Using (8–35) and (8–36) again to eliminate p_0, we obtain

$$P_e = \frac{hc}{8m_H} \left(\frac{3}{\pi m_H} \right)^{1/3} \left(\frac{1+X}{2}\rho \right)^{4/3} . \tag{8-43}$$

To obtain the total pressure we need to add the pressures P_i contributed by individual ions. These normally are nondegenerate, as was pointed out in Section 4:15.

$$P_i = \left(X + \frac{1}{4}Y \right) \frac{\rho kT}{m_H} . \tag{8-44}$$

Finally, we have to add the radiation pressure from equation (8–33) to obtain

$$P_{\text{Total}} = P_e + P_i + P_{\text{Rad}} . \tag{8-45}$$

8:7 Luminosity

We have estimated the time required for a star to recover from a thermal perturbation when a range of different conditions prevails within the star. However, we still have to ask ourselves, "When does each of these different conditions predominate? Under what circumstances is radiative heat transfer dominant? When is convection a major contributor, and what are the conditions that favor degeneracy?" These are the questions we must examine next. When we obtain an answer we will also be able to quantitatively express the rates of energy transfer that add to give the bolometric luminosity of a star. Luminosity is expressed in units of power – energy emitted per unit time.

The total flux at radial distance $r = r_0$ from the star's center is the difference between the outward and inward directed energy flow through a surface that can be assumed to be plane because the radiation mean free path is very small compared to r_0. Let the temperature at r_0 be T_0. Radiation passes through a surface r_0 at all values of azimuthal angle θ (Fig 8.2). The unattenuated flux originating at a distance

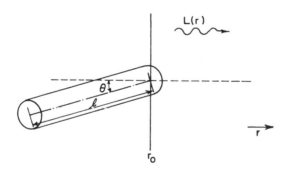

Fig. 8.2. Illustration to show the relation between luminosity and temperature gradient.

l along direction θ, which passes through the surface in unit time, is

$$aT^4(l, \theta) \cdot c \cos \theta \cdot \frac{2\pi \sin \theta \, d\theta}{4\pi} , \tag{8-46}$$

where

$$T(l, \theta) = T_0 - \frac{dT}{dr} l \cos \theta , \tag{8-47}$$

$c \cos \theta$ represents the cylindrical volume from which radiation crosses unit area of the surface in unit time, and $2\pi \sin \theta \, d\theta / 4\pi$ gives the fraction of the total solid angle at (l, θ) containing the radiation that will pass through the appropriate unit area, at r_0.

In actuality, radiation from (l, θ) does not reach r_0 unattenuated. A photon originating at l only has a probability $\pi(l)$ of reaching r_0:

$$\pi(l) = \kappa \rho e^{-\kappa \rho l} \ . \tag{8-48}$$

This follows from (8–24) and also gives the proper normalization

$$\int_0^\infty \pi(l) \, dl = 1 \ . \tag{8-49}$$

The radiant flow $F(r_0)$ through unit area is now formally given as

$$F(r_0) = \int_0^\infty \int_0^\pi a \left(T_0 - \frac{dT}{dr} l \cos \theta \right)^4 \cdot c \cos \theta \cdot \frac{2\pi \sin \theta \, d\theta}{4\pi} \pi(l) \, dl \ . \tag{8-50}$$

However, to obtain the actual radiative flux we must decide what value of κ to use in equation (8–48). Equation (8–24) did not take into account stimulated emission which, as explained in Section 7:12, is important. On the other hand, if we use $\kappa^*(\nu)$ from (7–71) we have to average properly over all frequencies to arrive at a suitable mean opacity. We will do this in Section 8:8 below.

PROBLEM 8–2. The luminosity $L(r)$ at any radial distance within the star is

$$L(r) = 4\pi r^2 F(r) \ . \tag{8-51}$$

Show by integration that to first order

$$L(r) = -\frac{16\pi a c}{3\kappa \rho} r^2 T^3 \frac{dT}{dr} \ . \tag{8-52}$$

8:8 Opacity Inside a Star

In Section 7:12 we had discussed the four sources of opacity: electron scattering, free–free transitions, free–bound transitions, and bound–bound transitions. However, we have not yet indicated how to compute the mean opacity obtained from these four contributing factors. It is this opacity that has to be used in expression (8–52).

Two factors enter. First, we have to average over all radiation frequencies; but clearly, if the opacity is to give an accurate assessment of the radiative transfer rate, those frequencies at which the radiation density gradient is greatest should receive greater weight in the averaging process. Second, those frequency ranges in which the opacity is smallest potentially make the greatest contribution to energy transport. We therefore will be more interested in averaging $1/\kappa(\nu)$ rather than $\kappa(\nu)$.

Let us write (8–52) in its more fundamental form, involving energy density $\rho(\nu)$ of radiation at frequency ν and temperature T (see equation (4–72)).

$$L(r, \nu) = \frac{-4\pi r^2}{3\rho \kappa^*(\nu)} c \frac{d\rho(\nu)}{dr} \ . \tag{8-53}$$

Here we have defined a contribution $L(r, \nu)$, at frequency ν, to the total luminosity $L(r)$ at r; and we have set the total energy density U equal to the blackbody energy density. $\kappa^*(\nu)$ is the opacity at frequency ν that takes account of stimulated emission:

$$L(r) = \int_0^\infty L(r, \nu)\, d\nu \quad \text{and} \quad U = \int_0^\infty \rho(\nu)\, d\nu = aT^4 . \qquad (8\text{-}54)$$

We can neglect bound–bound transitions since they play a negligible role in the stellar interior. Equation (7–71) therefore simplifies to

$$\kappa^*(\nu) = [\kappa_{bf}(\nu) + \kappa_{ff}(\nu)](1 - e^{-h\nu/k}) + \kappa_e , \qquad (8\text{-}55)$$

and we can define a mean opacity

$$\frac{1}{\kappa} = \frac{\displaystyle\int_0^\infty \frac{1}{\kappa^*(\nu)} \frac{d\rho(\nu)}{dT} \frac{dT}{dr}\, d\nu}{\displaystyle\int_0^\infty \frac{d\rho(\nu)}{dT} \frac{dT}{dr}\, d\nu} , \qquad (8\text{-}56)$$

called the *Rosseland mean opacity*, in which (4–72) can be used to obtain $d\rho(\nu)/dT$. As can be seen from (8–53), the Rosseland mean opacity does indeed favor the frequencies important to the transfer process by using the energy density gradient $d\rho(\nu)/dr$ as a weighting function for $1/\kappa^*(\nu)$, which is a measure of the mean free path at frequency ν. The opacity at any frequency is the sum of contributions from bound–free (bf) and free–free (ff) transitions, and from electron scattering (e).

$\kappa_{ff}(\nu)$ and $\kappa_{bf}(\nu)$ themselves are sums over the opacity contributions of the individual states of excitation n of the various atoms and ions A present at radial distance r in the star,

$$\kappa_{ff}(\nu)\rho = \sum_A \int \alpha_{ff} \frac{\rho X_A}{A m_H} n_e(v)\, dv , \qquad (8\text{-}57)$$

$$\kappa_{bf}(\nu)\rho = \sum_{A,n} \alpha_{bf} \left(\frac{\rho X_A}{A m_H} \right) N_{A,n} . \qquad (8\text{-}58)$$

Here $\rho X_A / A m_H$ is the number density of atoms of kind A, X_A is the abundance by mass of atoms or ions with mass number A, m_H is the mass of a hydrogen atom, and $N_{A,n}$ is the fraction of these atoms or ions in the nth excited state. $n_e(v)$ is the number density of electrons in a velocity range dv around v. The quantities α_{ff} and α_{bf} are the atomic absorption coefficients defined in (7–74) and (7–75). As shown in (7–72),

$$\kappa_e \rho = \sigma_e n_e , \qquad (8\text{-}59)$$

where the right side is the product of the electron number density and the Thomson (or — at high energies — the Compton) scattering cross-section.

To evaluate $N_{A,n}$ we make use of the *Saha equation* (4–107), which for high ionization leads to

$$N_{A,n} = n^2 \left[n_e \frac{h^3}{2(2\pi m_e kT)^{3/2}} e^{\chi_n/kT} \right], \qquad (8\text{-}60)$$

where we have considered that most of the ions are in the $(r + 1)$st ionization state. We can understand this equation in the following way.

χ_n is the energy needed to ionize the atomic species A from the nth excited state; m_e is the electron mass. Using a *Bohr atom* model this energy is (7–5, 7–76)

$$\chi_n \sim \frac{2\pi^2 e^4 m_e}{h^2} \frac{Z'^2}{n^2}, \qquad (8\text{-}61)$$

where Z' is the effective charge of the ion considered. Equation (8–61) assumes that all the excited atoms of a given species A will be in the same state of ionization at radial distance r from the star's center. In our present notation this means that in equation (4–107), $n_r/n_{r+1} = N_{A,n}$. We note that $N_{A,n}$ is proportional to n^2. This is because the statistical weight g_r — the number of sublevels — of the nth excited state is $2n^2$ (see Problem 7–1). From Section 4:16 we also have $g_e = 2$. Similarly, the ion can also exhibit two spin states $g_{r+1} = 2$. But for any given final state there are only two possible combinations of spin, $g_{r+1} g_e = 2$.

Making use of equation (7–75) for α_{bf}, with χ_n from (8–61) substituted into this expression, we can now obtain

$$\kappa_{bf}(\nu) = \frac{2}{3} \sqrt{\frac{2\pi}{3}} \frac{Z'^2 e^6 h^2 \rho (1 + X) Z}{c A m_H^2 m_e^{1.5} (kT)^{3.5}} \left[\frac{1}{n} \frac{\chi_n}{kT} e^{\chi_n/kT} \left(\frac{kT}{h\nu} \right)^3 g_{bf} \right]. \qquad (8\text{-}62)$$

Here Z is the metal abundance by fraction of the total mass. We have summed (8–58) only for these constituents, because hydrogen and helium do not contribute significantly to the bound–free transitions. The summation over states has been neglected, since the lowest state n usually contributes most. We also have used an electron density from Table 8.1:

$$n_e = \frac{1}{2}(X + 1)\frac{\rho}{m_H}. \qquad (8\text{-}63)$$

Equation (8–62) can be considerably simplified if approximate values of the opacity suffice. For example, we can restrict our attention to those levels for which $\chi_n/kT \sim 1$, $h\nu/kT \sim 1$, because this makes use of the frequencies and ionization potentials that will contribute most to the opacity. Constituents which would be ionized at lower temperatures, $\chi_n \ll kT$, are already almost fully ionized and have too few bound electrons to be effective, while those with higher χ_n values absorb too few of the photons present. Similarly the photons of frequency $\nu \sim kT/h$ are weighted most favorably by the Rosseland mean.

For most elements we can also choose a typical value $Z'^2/A \sim 6$.

With these approximations we obtain *Kramer's Law of Opacity* for bound–free absorption:

$$\kappa_{bf} = 4.34 \times 10^{25} Z(1 + X)\frac{\rho}{T^{3.5}} \frac{\langle g_{bf} \rangle}{f} \quad \frac{\text{cm}^2}{\text{g}}, \qquad (8\text{-}64)$$

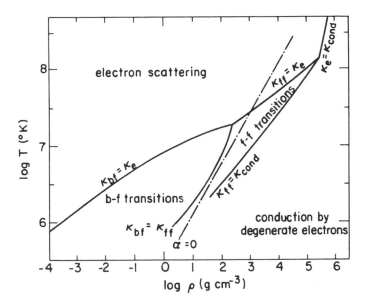

Fig. 8.3. Opacity as a function of density and temperature in a star of population I composition. The diagram is divided into four regions characterized by different mechanisms of energy transport. The sources of opacity that dominate these mechanisms are electron scattering, bound–free transitions, free–free transitions, and the effective opacity that would describe the energy transport by degenerate electrons. The dashed line shows where the degeneracy *parameter* α (see equation (4–94)) equals zero (after Hayashi, Hoshi, and Sugimoto (Ha62).

where $\langle g_{bf} \rangle$ is the mean *Gaunt factor* (7–75), which is always of order unity, and f contains correction factors — all of order unity also — which arise because of the approximations we have made. For free–free transitions, we can similarly obtain expressions (Sc58b)*:

$$\kappa_{ff} = \frac{2}{3}\sqrt{\frac{2\pi}{3}} \frac{e^6 h^2 (X+Y)(1+X)\rho}{cm_H^2 m_e^{1.5}(kT)^{3.5}} \frac{g_{ff}}{196.5}$$

(8-65)

$$= 3.68 \times 10^{22} \langle g_{ff} \rangle (X+Y)(1+X)\frac{\rho}{T^{3.5}} \quad \frac{cm^2}{g} \ ,$$

where $\langle g_{ff} \rangle$ is the mean Gaunt factor (7–74). We note that if we had taken $\kappa(\nu)$ in equation (6–137) and substituted into (8–56) for $\kappa^*(\nu)$, we would have obtained an opacity expression proportional to $e^6 n^2 c^{-1} (m_e kT)^{-1.5}$ and a weighted mean function proportional to ν^{-2} that would be proportional to $h^2(kT)^{-2}$. This is just the dependence found in (8–65). For electron scattering, (8–59) combined with the number of free electrons (8–63), yields

$$\kappa_e = \frac{4\pi}{3}\frac{e^4}{c^4 m_H m_e^2}(1+X) \sim 0.19(1+X) \quad \frac{cm^2}{g} .$$

(8-66)

Electron scattering is the main contributor to the opacity at low densities and high temperatures, where the interaction between electrons and ions is weakened.

Figure 8.3 shows the relative importance of scattering and absorption for different densities and temperatures. At high densities, where electrons become degenerate, heat is transferred most rapidly through conduction by these electrons.

Figure 8.4 shows the opacity as a function of temperature in stars whose com-

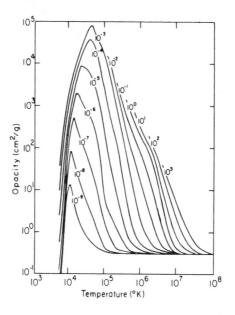

Fig. 8.4. Opacity for stars whose composition is similar to that of the Sun. Each curve represents a different density value ρ, measured in g cm^{-3}. (After Ezer and Cameron (Ez65). With permission of the editors of *Icarus. International Journal of Solar System Studies*, Academic Press, New York.)

position is similar to that of the Sun.

Thus far we have discussed radiative transfer only in the interior of a star. However, the equations of radiative transfer also play a dominant role in the transport of energy through stellar atmospheres (Section 7:13).

PROBLEM 8–3. Using equations (8–7), (8–52), and the ideal gas law, show that the luminosity of stars should be roughly proportional to M^3.

We find, in reality, that main sequence stars more nearly obey the *mass-luminosity relation* (Fig. 8.1):

$$L \propto M^a, \qquad 3 \lesssim a \lesssim 4. \tag{8-67}$$

Presumably this relation holds in main sequence stars because radiative transfer dominates there, while convective transfer (Section 8:9 below) is more important in the giants, and degenerate electron transfer dominates in compact stars and in compact stellar cores.

Radiative transfer is always present, even when other processes dominate. The total energy transfer rate is the sum of all the different rates.

8:9 Convective Transfer

Let us establish the conditions under which the temperature gradient becomes so large that the medium starts to convect and the spherically symmetrical temperature distribution about the stellar center becomes unstable.

Consider an element of matter at some density ρ_1' and pressure P_1' surrounded by a region with exactly the same characteristics (ρ_1, P_1) (see Fig. 8.5):

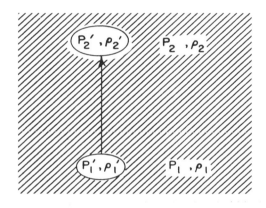

Fig. 8.5. Convective outward motion of a low-density "bubble." When thermal gradients become too high, convective motion sets in and becomes the dominant vehicle for heat transport.

$$\rho_1' = \rho_1, \qquad P_1' = P_1. \tag{8-68}$$

The element is then moved to a new position, subscript 2, where its final pressure P_2' equals the ambient pressure P_2:

$$P_2' = P_2. \tag{8-69}$$

Using (8–30) we found that a convective motion of this kind is fast compared to the time required for radiative heat transfer. We can therefore consider the process to be adiabatic. Equations (4–129) then imply that

$$\rho_2' = \rho_1' \left(\frac{P_2}{P_1} \right)^{1/\gamma}. \tag{8-70}$$

For the highly ionized plasma in a star, the ratio of heat capacities is $\gamma = c_p/c_v = \frac{5}{3}$.

If the initial displacement of the element was upward, and we find that $\rho_2' > \rho_2$, then the element will be forced down toward its initial position, 1; the medium is then stable. However, if $\rho_2' < \rho_2$, then the initial displacement leads to further motion along the same direction — upward — and the medium is unstable. Convection sets in.

The condition for stability therefore is

$$\rho_2 < \rho_2' = \rho_1' \left(\frac{P_2'}{P_1'}\right)^{c_v/c_p} = \rho_1 \left(\frac{P_2}{P_1}\right)^{c_v/c_p}, \tag{8-71}$$

where we have made use of expressions (8–68) to (8–70). This can be rewritten as

$$\frac{d\rho}{\rho} < \left(\frac{P+dP}{P}\right)^{c_v/c_p} - 1 = \frac{c_v}{c_p}\frac{dP}{P}. \tag{8-72}$$

In terms of radial gradients this becomes

$$\frac{1}{\rho}\frac{d\rho}{dr} < \frac{c_v}{c_p P}\frac{dP}{dr}, \tag{8-73}$$

which, with the ideal gas equation (4–37), leads to the stability condition

$$\frac{dT}{dr} > \frac{T}{P}\left(1 - \frac{c_v}{c_p}\right)\frac{dP}{dr}. \tag{8-74}$$

Both dP/dr and dT/dr have negative values. The right side of (8–74) is called the adiabatic temperature gradient, and we conclude that stability will prevail when the absolute value of the temperature gradient dT/dr is less than that of the adiabatic gradient. When the absolute value of dT/dr becomes larger than the absolute value of the adiabatic gradient, instability sets in and heat is transferred by convection.

To compute the heat transfer rate we have to know four quantities: the velocity v of the buoyant element, its heat capacity C, its density, and the temperature differential ΔT between the element and the surroundings to which it finally imparts this temperature. The heat transport rate per unit area is then

$$H = C\rho v \Delta T. \tag{8-75}$$

Here C is the heat capacity under the assumed adiabatic conditions. Using the acceleration given in equation (8–29) and assuming transport over a distance one-tenth of a solar radius, the mean velocity v is of order $[\ddot{r}R_\odot/10]^{1/2}$

$$H \sim C\rho \left[\frac{GM(r)}{Tr^2}\frac{R_\odot}{10}\right]^{1/2} (\Delta T)^{3/2}$$

$$\tag{8-76}$$

$$\sim C\rho \left[\frac{GM(r)}{Tr^2}\right]^{1/2} \left(\frac{d\Delta T}{dr}\right)^{3/2} \left(\frac{R_\odot}{10}\right)^2.$$

The distance $R_\odot/10$ is chosen somewhat arbitrarily since convective theories are currently not clear on how to estimate the convection cell size — the distance ℓ over which an element is transported before it diffusively mixes with its surroundings and ceases to exist. This *mixing length* ℓ is sometimes taken to be proportional to the local pressure scale height but without specifying the proportionality constant. A more accurate estimate appears to be that ℓ is simply equal to a buoyant element's distance beneath the outer surface of the convection zone, suggesting that material from all layers beneath this surface strives to deposit its energy at the outermost radial distance it can convectively reach (St97).

We take $d\,\Delta T/dr$ to be the difference between the actual and the adiabatic gradient. The equations we have obtained hold equally well for the upward convection of hot matter and downward convection of cool material. For a given gradient $d\,\Delta T/dr$ we can now obtain the order of magnitude of H if the heat capacity is known. We have not yet discussed the equation of state, although we have assumed an ideal gas law above. For a completely ionized plasma the heat capacity is roughly known, even though the process described here proceeds neither at constant pressure nor at constant volume. We will, however, not be far wrong in taking $2RT$ per gram, where R is the gas constant (see equation (4–34)).

We now wish to see at what gradients the convective flux exceeds radiative transfer. This can be done by checking the value of $d\Delta T/dr$ at which the total convective flux equals the luminosity. With $r \sim R_\odot/2$, $M(r) \sim M/2$, $\rho \sim 1\,\mathrm{g\ cm^{-3}}$, $C \sim 2\times10^8\ \mathrm{erg\ g^{-1}\ K^{-1}}$, $T \sim 10^7\,\mathrm{K}$, and $L \sim 10^{34}\ \mathrm{erg\ s^{-1}}$, we have

$$L = 4\pi R_\odot^2 H \sim 4\pi R_\odot^2 C\rho \left(\frac{GM(r)}{Tr^2}\right)^{1/2}\left(\frac{R_\odot}{10}\right)^2\left(\frac{d\,\Delta T}{dr}\right)^{3/2},$$

$$\frac{d\,\Delta T}{dr} \sim 10^{-10}\ \mathrm{K\ cm^{-1}}. \tag{8-77}$$

The average temperature gradient for a star is of order $T_c/R \sim 10^7/10^{11} \sim 10^{-4}$ K cm^{-1}, where T_c is the central temperature. The required excess gradient is only of the order of one millionth of the total gradient. Over a distance $\Delta r \sim R_0/10$, the excess temperature drop corresponds to ~ 1 K, the figure we had previously used in establishing the time constant for convective transport in equations (8–29) and (8–30).

We have now dealt with all the differential equations discussed in Section 8:4; but we still have to derive the energy generation rate through nuclear reactions at the center of a star. This is done in the next section.

8:10 Nuclear Reaction Rates

The nuclear reactions that take place in stars are largely reactions in which two particles approach to within a short distance, become bound to each other, and at the same time release energy. These *exergonic* processes are the ultimate source of energy for the star.

Let us look into the various factors that determine the reaction rate. We will assume that two kinds of particles are involved and will label them with subscripts 1 and 2, respectively. The reaction rate then is proportional to:

(i) The number density n_1 of nuclei of the first kind;

(ii) The number density n_2 of nuclei of the second kind;

(iii) The frequency of collisions, which depends on the relative velocity v with which particles approach each other; and

(iv) The velocity-dependent interaction cross-section $\sigma(v)$ which normally is proportional to $1/v^2$. However, in order for a reaction to occur, the Coulomb barrier, which bars positively charged particles from approaching a nucleus, must be penetrated.

This makes the reaction rate proportional to

(v) The probability $P_p(v)$ for penetrating the Coulomb barrier; this has an exponential form

$$P_p(v) \propto \exp\left[-\frac{4\pi^2 Z_1 Z_2 e^2}{hv}\right] . \qquad (8\text{-}78)$$

Here $Z_1 e$ and $Z_2 e$ are the nuclear charges.

Once the nuclear barrier has been penetrated there is a probability P_N for nuclear interaction. This is insensitive to particle energy or velocity but does depend on the specific nuclei involved.

We therefore introduce a factor proportional to

(vi) P_N the probability for nuclear interaction. For the interaction of two protons this interaction is known from theory. For all other reactions laboratory data have to be used to evaluate the probability.

The rate of the process further is proportional to

(vii) The distribution of velocities among particles. This can be assumed to be Maxwellian because the nuclei normally are not degenerate. Equation (4–59) gives

$$D(T, v) \propto \frac{v^2}{T^{3/2}} \exp\left[-\frac{1}{2}\frac{m_H A' v^2}{kT}\right] , \qquad (8\text{-}79)$$

where $A' = A_1 A_2/(A_1 + A_2)$ is the reduced atomic mass, measured in atomic mass units.

We can now write the overall reaction rate in unit volume as

$$r = \int_0^\infty n_1 n_2 v \sigma(v) P_p(v) P_N D(T, v)\, dv . \qquad (8\text{-}80)$$

This integral is readily evaluated because of the narrow range of velocities in which the product of P_p and D is high. Outside this velocity range the integrand is too small to make a significant contribution to the integral. We proceed in the following way. The integral in equation (8–80) has the form

$$\int_0^\infty v \exp\left[-\frac{a}{v} + bv^2\right] dv . \qquad (8\text{-}81)$$

The integrand has a sharp maximum at the minimum value of the exponent. We take the derivative of the exponent with respect to v and set this to zero. This gives the value v_m

$$v_m = \left(\frac{a}{2b}\right)^{1/3} = \left(\frac{4\pi^2 Z_1 Z_2 e^2 kT}{hm_H A'}\right)^{1/3} . \tag{8-82}$$

To evaluate the integral, however, we still need to estimate the effective velocity range over which the integrand is significant. For order of magnitude purposes it will suffice to take a range between points where the value of the integrand has dropped a factor of e. This happens at v values for which

$$\left(\frac{a}{v} + bv^2\right) - \left(\frac{a}{v_m} + bv_m^2\right) = 1 . \tag{8-83}$$

Because the deviation from v will be small, we set

$$v = v_m + \Delta , \tag{8-84}$$

and substitute in equation (8–83). Terms linear in Δ drop out through use of (8–82), but the quadratic terms yield

$$\left(\frac{a}{v_m^3} + b\right)\Delta^2 = 3b\Delta^2 = 1 , \tag{8-85}$$

$$\Delta = \pm\sqrt{\frac{1}{3b}} = \pm\sqrt{\frac{2kT}{3A'm_H}} . \tag{8-86}$$

The integral (8–80) can now be readily evaluated. First, however, we would like to lump all the proportionality constants into a single constant B and relate velocity to temperature everywhere.

We note that

$$\Delta \propto T^{1/2} \tag{8-87}$$

and that the integrand is proportional to

$$v_m \cdot \frac{1}{v_m^2} \cdot \frac{v_m^2}{T^{3/2}} = \frac{v_m}{T^{3/2}} = T^{-7/6} , \tag{8-88}$$

where we have made use of the relation (8–82). This means that $r \propto T^{-7/6}\Delta \propto T^{-2/3}$

We can set

$$n_1 = \frac{\rho_1}{m_1} = \frac{\rho}{m_1}X_1 \quad \text{and} \quad n_2 = \frac{\rho}{m_2}X_2 , \tag{8-89}$$

where X_1 and X_2 are the concentrations, and m_1 and m_2 the masses of nuclei of species 1 and 2. Absorption of factors m_1 and m_2 into the proportionality constant B then yields the reaction rate

$$r = B\rho^2 X_1 X_2 T^{-2/3}\exp\left[-3\left(\frac{2\pi^4 e^4 m_H Z_1^2 Z_2^2 A'}{h^2 kT}\right)^{1/3}\right] . \tag{8-90}$$

Thus far we have developed an estimate of reaction rates without much thought about the individual reactions involved, the required temperatures and densities, and the resulting energy liberation rate. We now turn to these.

We first ask ourselves how much energy would be needed for two particles to interact. It is clear that a nuclear reaction can only take place if the particles approach to within a distance of the order of a nuclear diameter $D \sim 10^{-13}$ cm. However, because both nuclei are positively charged they tend to repel. The work required to overcome the repulsion is

$$E = \frac{Z_1 Z_2 e^2}{D} \sim 2 \times 10^{-6} Z_1 Z_2 \,\text{erg} \sim Z_1 Z_2 \,\text{MeV} . \tag{8-91}$$

This might lead us to think that temperatures of the order of 10^{10} K would be required for nuclear reactions to proceed. This is far higher than the 10^7 K temperature we had estimated in Problem 8–1.

Two factors allow the actual reaction temperature to be so low. First, a small fraction of the nuclei with thermal distribution $D(T, v)$ has energies far above the mean (Fig. 8.6). Second, two particles have a small but significant probability of ap-

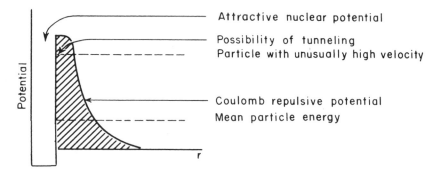

Fig. 8.6. Energies involved in nuclear reactions.

proaching each other by tunneling through the Coulomb potential barrier rather than going over it. This probability is quantum mechanically determined and is included in the function $P_p(v)$.

These two factors suffice to allow nuclear reactions to proceed at mean energies some 10^3 times lower than those employed to produce nuclear interactions in the laboratory. The main difference is that in the laboratory speed is essential, whereas the star is in no hurry. In the laboratory we want high reaction rates so that results may be obtained within a few minutes or, at most, hours. In contrast, a reaction having a probability of transmuting a given particle in the course of ten billion years is sufficiently fast to produce the luminosities characterizing many stars like the Sun. This extension of the available time by a factor of $\sim 10^{14}$ is the prime difference that permits low-temperature generation of energy and transmutation of the elements at the centers of stars at cosmically significant rates.

8:11 Particles and Basic Particle Interactions

A number of basic particles are involved in most nuclear reactions in stars. We list their properties in Table 8.2.

Table 8.2. Particles That Take Part in Many Stellar Nuclear Reactions.

Particle	Symbol	Rest–Mass g	Rest–Mass MeV	Charge esu	Spin	Class
Photon	γ	0	0	0	1	Photon
Neutrino	ν	$\lesssim 4 \times 10^{-34\,(b)}$	$\lesssim 2.5 \times 10^{-7}$	0	$\frac{1}{2}$	Lepton
Antineutrino	$\bar{\nu}$	$\lesssim 4 \times 10^{-34}$	$\lesssim 2.5 \times 10^{-7}$	0	$\frac{1}{2}$	Antilepton
Electron	e	9×10^{-28}	0.511	-5×10^{-10}	$\frac{1}{2}$	Lepton
Positron	e^+	9×10^{-28}	0.511	$+5 \times 10^{-10}$	$\frac{1}{2}$	Antilepton
Proton	$\mathcal{P}$	1.6×10^{-24}	938.256	$+5 \times 10^{-10}$	$\frac{1}{2}$	Baryon
Neutron[a]	$\mathcal{N}$	1.6×10^{-24}	939.550	0	$\frac{1}{2}$	Baryon

[a] Half-life 614 ± 1 s   [b] (Sp03).

The spin of a particle tells us the type of statistics it obeys. Integral spin implies obedience to the *Bose–Einstein* statistics, and half-integral spin labels a particle as a *fermion*.

A number of basic conservation laws govern all nuclear reactions:

(a) Mass–energy must be conserved (Section 5:6).

(b) The total electric charge of the interacting particles is conserved.

(c) The number of particles and antiparticles must be conserved. A particle cannot be formed from an antiparticle or vice versa. But a particle–antiparticle pair may be formed or destroyed without violating this rule. In particular:

(d) The difference between the number of *leptons* and *antileptons* must be conserved (*conservation of leptons*); and

(e) The difference between the number of *baryons* and *antibaryons* must be conserved (*conservation of baryons*).

With these rules in mind we enumerate some of the most common nuclear reactions in stars.

(i) Beta Decay

A *neutron*, as a free particle or as a nucleon inside an atomic nucleus, can decay giving rise to a proton, an electron, and an *antineutrino*. This reaction often is *exergonic* and can proceed spontaneously. A proton inside a nucleus, however, can also combine with an electron in *inverse beta decay*, giving rise to a nuclear neutron and the emission of a neutrino:

$$\mathcal{N} \to \mathcal{P} + e^- + \bar{\nu}, \quad \mathcal{P} + e^- \to \mathcal{N} + \nu \,. \tag{8-92}$$

(ii) Positron Decay

Here a proton gives rise to a neutron, positron, and neutrino. This process is *endergonic* — requires a threshold input energy — because the mass of the neutron and positron is considerably greater than the proton mass.

$$\mathcal{P} \to \mathcal{N} + e^+ + \nu \,. \tag{8-93}$$

In principle all these reactions could go either from left to right or right to left; but normally the number of available neutrinos or antineutrinos is so low that only the reaction from left to right need be considered.

(iii) $(\mathcal{P}, \gamma)$ Process

In this reaction a proton reacts with a nucleus with charge Z and mass A, to give rise to a more massive particle with charge $(Z + 1)$. Energy is liberated in the form of a photon, γ:

$$^{A}Z + \mathcal{P} \to \ ^{A+1}(Z + 1) + \gamma \,. \tag{8-94}$$

A typical reaction of this kind involves the carbon *isotope* ^{12}C and nitrogen isotope ^{13}N,

$$^{12}\text{C} + \ ^{1}\text{H} \to \ ^{13}\text{N} + \gamma \,. \tag{8-95}$$

(iv) (α, γ) and (γ, α) Processes

In processes of this kind an *α-particle* — helium nucleus — is added to the nucleus or ejected from it. The excess energy liberated in adding an alpha particle is carried off by a photon. The energy required to tear an alpha particle out of the nucleus also can be supplied by a photon. These two processes are particularly important in nuclei containing an even number of both protons and neutrons — the *even–even nuclei*. These nuclei are especially stable and play a leading role in processes that determine the formation of heavy elements.

(v) $(\mathcal{N}, \gamma)$ and $(\gamma, \mathcal{N})$ Processes

Such processes involve the addition or subtraction of a neutron. The nucleus emits or absorbs a photon in the reaction to assure energy balance.

8:12 Energy-Generating Processes in Stars

A variety of different energy-generating processes can take place in stars. We will enumerate them in the succession in which they occur during a star's life.

(a) When the star first forms from the interstellar medium, it contracts radiating away gravitational energy. During this stage no nuclear reactions take place.

(b) When the temperature at the center of the star becomes about a million degrees, the first nuclear reactions set in. From the discussion of Section 8:10 it is clear that these reactions do not switch on sharply as some given temperature value is exceeded. The temperature is not a threshold in this sense, even though threshold energies are involved. Instead, we can think of a critical temperature T_c at which reactions will proceed at a certain rate. We may choose to define the critical temperature as that temperature at which the mean reaction time becomes as short as five billion years. Because of the rapid increase in reaction rates with temperature, the nuclear reactions would be quickly exhausted (on the scale of a billion years) if T_c were significantly exceeded.

The first reactions to occur in a star are those that destroy many of the light elements initially in the interstellar medium and convert them into helium isotopes. We list the reactions and the energy released (Sa55a). Note that this energy is carried away by photons or neutrinos, but we have not specifically shown this here:

$$
\begin{aligned}
{}^2D + {}^1H &\rightarrow {}^3He, & 5.5\,\text{MeV} & \\
{}^6Li + {}^1H &\rightarrow {}^3He + {}^4He, & 4.0 & \\
{}^7Li + {}^1H &\rightarrow 2\,{}^4He, & 17.3 & \qquad (8\text{-}96)\\
{}^9Be + 2\,{}^1H &\rightarrow {}^3He + 2\,{}^4He, & 6.2 & \quad\left.\begin{array}{c}\\ \\ \end{array}\right\}\ \text{two}-\text{step}\\
{}^{10}B + 2\,{}^1H &\rightarrow 3\,{}^4He, & 19.3 & \qquad\text{reactions}\\
{}^{11}B + {}^1H &\rightarrow 3\,{}^4He, & 8.7 &
\end{aligned}
$$

These reactions have lifetimes of the order of 5×10^4 yr at respective temperatures $\sim 10^6$, 3×10^6, 4×10^6, 5×10^6, 8×10^6, and 8×10^6 K. At temperatures ranging from about half a million degrees to five million degrees, these reactions take place rapidly as the star contracts along the *Hayashi track* (Fig. 1.5) — a fully convective stage in the star's pre-main sequence contraction. The elements burn up everywhere including the surface layers of the stars, where they can be destroyed because convection circulates the material between the interior and the surface.

Brown dwarf stars are stars more than ten times as massive as Jupiter, but with $0.01 M_\odot \leq M < 0.08 M_\odot$. They are not sufficiently massive to ignite hydrogen fusion in their centers (Bo96). However, they can overcome the Coulomb barrier that prevents planets from indefinitely contracting (Problem 8–1). Hydrostatic equilibrium in the core of a brown dwarf is maintained by electron degeneracy pressures. Regardless of brown dwarf mass, this equilibrium sets in when the brown dwarf has contracted to a radius roughly equal to that of Jupiter. Their elevated central temperature permits brown dwarfs to derive nuclear energy through the deuterium-, lithium-, and beryllium-burning reactions listed in (8–96). Once these light elements are exhausted, the dwarfs slowly contract, feebly radiating until degeneracy pressure (8–41) stops further contraction.

With few exceptions these light elements are only found in small concentrations in the surface layers of stars. However, large concentrations of, say, lithium can be found in some stars; these are called lithium stars and constitute a puzzle (De02). The theory of stellar structure and evolution seeks to explain both such anomalies and the more general origin of the chemical elements.

None of the reactions listed in (8–96) contribute a large fraction of the total energy emitted by stars during their lifetime. However, they are of interest in connection with the theory of the formation of elements; and the low abundance of these chemical elements in Nature provides one test of the accuracy of our notions.

(c) When the temperature at the center of the star reaches about ten million degrees, hydrogen starts burning (Be39). The reactions and mean reaction times for any given particle are given below for $T = 3 \times 10^7$ K. The amount of energy liberated in each step is also given.

$$
\begin{array}{llll}
^1\text{H} + {}^1\text{H} \rightarrow {}^2\text{D} + e^+ + \nu, & 1.44\,\text{MeV}, & 14 \times 10^9\,\text{yr}, \\
^2\text{D} + {}^1\text{H} \rightarrow {}^3\text{He} + \gamma, & 5.49\,\text{MeV}, & 6\,\text{s}, & (8\text{-}97) \\
^3\text{He} + {}^3\text{He} \rightarrow {}^4\text{He} + 2\,{}^1\text{H}, & 12.85\,\text{MeV}, & 10^6\,\text{yr}.
\end{array}
$$

The first and second reactions have to take place twice to prepare for the third reaction. Not all of the energy liberated contributes to the star's luminosity. Of the energy generated in the first step, 0.26 MeV is carried away by the neutrino and is lost. The total contribution to the luminosity is therefore 26.2 MeV for each helium atom formed. This set of reactions is the main branch of the *proton–proton reaction*. Other branches are shown in Fig. 8.19.

Hydrogen burning can also take place in a somewhat different way, making use of the catalytic action of the carbon isotope ^{12}C. This set of reactions comprises the *carbon cycle* or a more elaborate scheme sometimes called the *CNO bi-cycle*, since carbon, nitrogen, and oxygen are all involved; the CN portion is energetically the more significant (Cℓ68)*. The reaction times are given for 15×10^6 and 20×10^6 K.

Reaction	MeV	15×10^6 K	20×10^6 K
$^{12}\text{C} + {}^1\text{H} \rightarrow {}^{13}\text{N} + \gamma$	1.94	$\sim 10^6$ yr	$\sim 5 \times 10^3$ yr
$^{13}\text{N} \rightarrow {}^{13}\text{C} + e^+ + \nu$	2.22	15 min	
$^{13}\text{C} + {}^1\text{H} \rightarrow {}^{14}\text{N} + \gamma$	7.55	2×10^5 yr	2×10^3 yr
$^{14}\text{N} + {}^1\text{H} \rightarrow {}^{15}\text{O} + \gamma$	7.29	2×10^8 yr	10^6 yr
$^{15}\text{O} \rightarrow {}^{15}\text{N} + e^+ + \nu$	2.76	3 min	
$^{15}\text{N} + {}^1\text{H} \rightarrow {}^{12}\text{C} + {}^4\text{He}$	4.97	10^4 yr	30 yr
$^{15}\text{N} + {}^1\text{H} \rightarrow {}^{16}\text{O} + \gamma$	12.1	4×10^{-4} of $^{15}\text{N}(\mathcal{P}, \alpha)^{12}\text{C}$ rate	
$^{16}\text{O} + {}^1\text{H} \rightarrow {}^{17}\text{F} + \gamma$	0.60	2×10^{10} yr	5×10^7 yr
$^{17}\text{F} \rightarrow {}^{17}\text{O} + e^+ + \nu$	2.76	1.5 min	
$^{17}\text{O} + {}^1\text{H} \rightarrow {}^{14}\text{N} + {}^4\text{He}$	1.19	2×10^{10} yr	10^6 yr

$$(8\text{-}98)$$

The second part of the cycle occurs about 4×10^{-4} times as often as the first, because the $^{15}N(\mathcal{P}, \alpha)^{12}$C reaction is about 2.5×10^3 times more probable than the $^{15}N(\mathcal{P}, \gamma)\,^{16}$O reaction.

In the decay of the ^{13}N particle, 0.71 MeV is carried off by the neutrino; and in the ^{15}O decay 1.00 MeV is similarly lost on average. The total energy made available to the star per helium atom formed is therefore only 25.0 MeV, slightly less than the energy available from the proton–proton reaction. The reaction rates given here are for total concentrations of the carbon and nitrogen isotopes amounting to $X_{CN} \sim 0.005$. The relative predominance of the proton–proton reaction and the carbon cycle as a function of temperature is given in Fig. 8.7. The ^{13}C formed in the

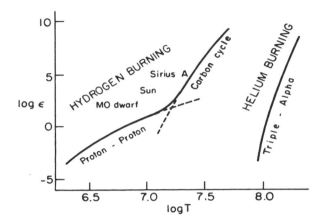

Fig. 8.7. Nuclear energy generation rate as a function of temperature (with $\rho X^2 = 100$, $X_{CN} = 5 \times 10^{-3} X$ for the p–p reaction and carbon cycle, but $\rho^2 Y^3 = 10^8$ for the triple-α process) (Sc58b). (From Martin Schwarzschild, *Structure and Evolution of the Stars*, ©1958 by Princeton University Press, p. 82.)

CN cycle can act as a source of neutrons as can other particles with mass number $4n + 1$. We will see this in reactions (8–103) and (8–104) below.

The hydrogen-burning reactions we have discussed contribute the energy given off by the star during its long stay on the main sequence. Once the hydrogen at the center of the star is largely depleted, helium burning can set in as described immediately below. In general, hydrogen burning continues in a shell surrounding the depleted core.

(d) When the hydrogen-burning phase of a star is completed, no further nuclear energy-generating processes may be available in the core for some time, and the central portions of the star slowly contract. Hydrogen burning may, however, continue in a shell surrounding the central core. (Figs. 8.8, 8.9). Meanwhile, the star's central temperature rises as potential energy lost in contraction is converted into heat. At about 10^8 K, helium burning sets in (Sa52). In this *triple-alpha* process three α-particles are transmuted into a carbon nucleus. Two steps are involved:

$$
\begin{aligned}
^4\mathrm{He} + {}^4\mathrm{He} &\rightarrow {}^8\mathrm{Be} + \gamma, \quad -95\,\mathrm{keV}, \\
^8\mathrm{Be} + {}^4\mathrm{He} &\rightarrow {}^{12}\mathrm{C} + \gamma, \quad +7.4\,\mathrm{MeV}.
\end{aligned}
\tag{8-99}
$$

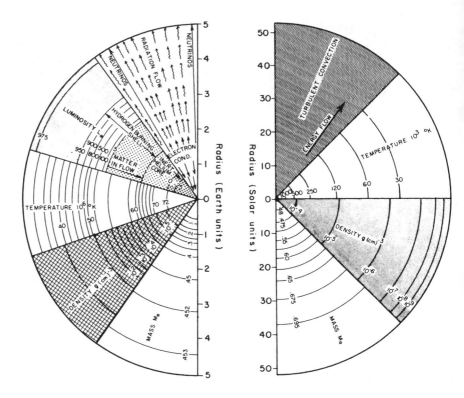

Fig. 8.8. Shell structure of a red giant in whose central regions hydrogen has become depleted (after Iben (Ib70)). The section on the left is a blown-up version of the tiny disk in the center of the drawing at the right. ("Globular Clusters Stars" ©1970 by Scientific American, Inc. All rights reserved.)

The first reaction is *endergonic*. Energy has to be supplied to make it proceed. The ^{8}Be nucleus is unstable and decays back into two alpha particles. An equilibrium is set up between alpha particles and ^{8}Be nuclei in which the concentration of ^{8}Be is quite small, of the order of 10^{-10} times the concentration of alpha particles. This particular abundance is determined by the lifetime of the metastable ^{8}Be, the density and energy (temperature) of the helium, and by the magnitude of the (negative) binding energy, $-95\,$keV.

(e) The star's core does not stay long in the helium-burning phase because the available amount of energy is small ($\sim$10%) compared to the energy generated in the hydrogen-burning phase. At higher internal temperatures a succession of (α, γ) processes may set in to form ^{16}O, ^{20}Ne, and ^{24}Mg. This type of process is called the α-*process*. After depletion in the core, helium burning may continue in a shell surrounding the depleted core. This shell is surrounded by a hydrogen-burning shell.

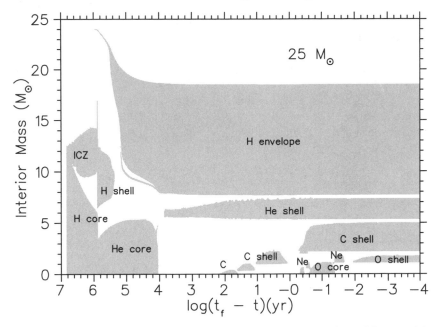

Fig. 8.9. The change of the internal structure as a function of time of a $25 M_\odot$ star with an initial elemental composition like that of the Sun (Tables 1.1, 1.2). The time sequence starts 10^7 yr before final collapse and progresses from left to right, each vertical slice through the star representing its state at a given time. The sequence terminates at the extreme right, about an hour before the star undergoes a pre-supernova collapse. Shaded regions indicate the evolving convective zones and are labeled according to their physical origin. *ICZ* stands for an intermediate convective zone, separating interior zones from an external nonconvective region. The star undergoes an initial mass loss that lowers its mass to $\sim 19 M_\odot$ about 10^4 yr before collapse. By then the initial hydrogen core has burned up and hydrogen burning continues in a shell; helium burning that replaced hydrogen burning in the core has also been exhausted but continues in a shell interior to the hydrogen-burning shell. A short-duration carbon-burning phase in the core is succeeded in turn by core neon and oxygen burning, with both elements subsequently continuing to burn in inner shells. (Courtesy of M. F. El Eid, B. S. Meyer and L.-S. The (Eℓ04).)

(f) At higher temperatures yet, 10^9 K, reactions may take place among the ^{12}C, ^{16}O, and ^{20}Ne nuclei. At this stage there would be no supply of free helium, but these particles can be made available through a (γ, α) reaction. The densities at this stage are of the order of $\rho \gtrsim 10^6$ g cm^{-3}. A typical reaction is

$$2\,^{20}\text{Ne} \rightarrow\,^{16}\text{O} +\,^{24}\text{Mg}, \quad 4.56\,\text{MeV} . \tag{8-100}$$

^{24}Mg can capture alpha particles to form ^{28}Si, ^{32}S, ^{36}A, and ^{40}Ca.

That this *equilibrium* or *e-process* actually takes place may be partly confirmed by the relatively large natural abundance of these isotopes compared to isotopes of the same substances, or neighboring elements in the periodic table. The even–

even isotopes of magnesium, silicon, sulfur, argon, and calcium, account for 79, 92, 95, 84, and 97%, by number, of all the isotopes of these respective elements (An89). That these elements also are abundant relative to neighboring elements is evident from Table 1.1. The α and e processes probably occur rapidly — perhaps explosively.

These processes eventually terminate in the iron group of nuclei, the most stable of all the elements, because their masses per nuclide are at a minimum. By the time an equilibrium concentration is reached between these even–even nuclei, the expected temperature and density have risen to

$$T \sim 4 \times 10^9 \, \text{K}, \quad \text{and} \quad \rho \sim 10^8 \, \text{g cm}^{-3} \, . \tag{8-101}$$

(g) In a second generation star — one that has formed from interstellar gases containing appreciable amounts of the heavier elements — we may find that ^{21}Ne is produced

$$^{20}\text{Ne} + {}^{1}\text{H} \rightarrow {}^{21}\text{Na} + \gamma, \ 2.45 \, \text{MeV}, \ 10^9 \, \text{yr at } 3 \times 10^7 \, \text{K},$$
$$^{21}\text{Na} \rightarrow {}^{21}\text{Ne} + e^+ + \nu, \ 2.5 \, \text{MeV}, \ 23 \, \text{s} \, . \tag{8-102}$$

At high temperatures in the helium core we can then have the exergonic reaction

$$^{21}\text{Ne} + {}^{4}\text{He} \rightarrow {}^{24}\text{Mg} + \mathcal{N}, \quad 2.58 \, \text{MeV}. \tag{8-103}$$

Similarly from the carbon cycle there will be some ^{13}C available and we may have the reaction

$$^{13}\text{C} + {}^{4}\text{He} \rightarrow {}^{16}\text{O} + \mathcal{N}, \quad 2.20 \, \text{MeV}. \tag{8-104}$$

These neutrons are preferentially captured by the heavy nuclei, particularly those in the iron group, and these can then be built up into heavier elements yet. There are of the order of a hundred ^{13}C and ^{21}Ne nuclei available for each iron group element; hence, an abundance of neutrons is at hand. Elements as heavy as ^{209}Bi can be built up in this way. The chain only ends at ^{210}Po, which α-decays with a half-life of only 138d. In addition, light nuclei such as ^{22}Ne can also be built up and, with the exception of the even–even nuclei, most particles with $24 \leq A \leq 50$ are believed to have been built up through neutron capture. This neutron process is *slow*; it is therefore called the *s-process*. Neutron capture at this stage typically requires several years to several thousand years. This time scale is slow compared to beta decay rates, and only those elements can be built up that involve the addition of neutrons to relatively stable nuclei.

Evidence for a stage with abundant neutrons comes from peaks in the abundance curve at mass numbers $A \sim 90$, 138, and 208. These nuclei have closed shells of neutrons with $N = 50$, 82, and 126.

The *s*-process occurs after the star has traversed the horizontal branch of the H–R diagram during hydrogen shell and helium burning, and returns to the asymptotic giant branch (AGB) portion of the diagram shown in Fig. 1.7. The *s*-process is set in motion by a series of helium shell flashes that convectively mix both portions of the outermost hydrogen-rich layer and parts of the helium-rich shell into the carbon-rich core. At the $o(10^8)$ K core temperatures involved, ^{12}C and ^{1}H (see reactions (8–98)) can rapidly yield ^{13}C which produces the required neutrons through interaction with ^{4}He as in reaction (8–104) (Ri99), (He04).

(h) In addition to the slow neutron process, neutrons can also be added to heavy nuclei in a *rapid process* (*r-process*) that takes place at least in some stars. Some such process is required to explain the existence of elements beyond ^{210}Po.

If a star runs out of all energy sources, a rapid implosion can take place on a free-fall time scale which, as we saw earlier, corresponds to times of the order 1000 s. Extremely high temperatures then set in and iron group nuclei can be broken up into alpha particles and neutrons; for, in ^{56}Fe there are 4 excess neutrons for 13 alphas. All this takes place at temperatures of order 10^{10} K and with neutron fluxes of order 10^{32} cm^{-2} s^{-1}. The *r*-process can build up elements to about $A \sim 260$ where further neutron exposure induces fission that cycles material back down into the lower mass ranges.

In the breakup of iron group elements into helium, the ratio of specific heats γ becomes less than $\frac{4}{3}$ so that an implosion occurs (Section 4:20). This is accompanied by γ-photon production, pair formation, and electron–positron pair annihilation which at these high pressures gives rise to large neutrino fluxes. The neutrino flux lifts off the outer layers and the rapid neutron process then takes place while the star is again expanding explosively. It is believed that this is the process involved at least in some types of supernova explosions. A neutron star forms from the central imploding core. Prominent among the ejecta is ^{56}Ni, which decays with a mean life of 8.8 days into ^{56}Co, which in turn decays with a mean life of 111.3 days into ^{56}Fe. This sequential decay powers the optical emission of supernovae for the many weeks during which they can be seen far across the Universe (Bu00).

Detailed computations based on neutron capture cross-sections and nuclear decay times, both measured in the laboratory, show that many features of the abundance curve in the region between $A = 80$ and $A = 200$ can be explained by the *r*-process. This leads us to believe that the sequence of events described above is at least roughly correct. In particular, as Figure 8.10 indicates, the very earliest, most metal-poor stars, formed from nearly pure hydrogen–helium mixtures early in the evolution of the Universe, appear to have undergone *r*-process nucleosynthesis. We will discuss these Population III stars further in Chapter 13.

Two notes may be added.

(i) Proton-rich isotopes are relatively rare although they can be produced in $(\mathcal{P}, \gamma)$ processes (sometimes called *p-processes*) or in a $(\gamma, \mathcal{N})$ reaction. Such nuclei could be produced if hydrogen from the outer layers of a star could come into contact with hot material from the core in convective processes. Generally, however, the *r*-process can account semiquantitatively for the abundance ratios of many of the heavier elements.

(ii) The uranium isotopes ^{235}U and ^{238}U might be expected to arise in roughly equal abundances in the *r*-process. Their present ratio as found in the Earth is of the order of 0.0072. This would be expected on the basis of their respective half-lives of 0.71 and 4.5 Gyr if we assume a common time of formation $\sim$6 Gyr ago.

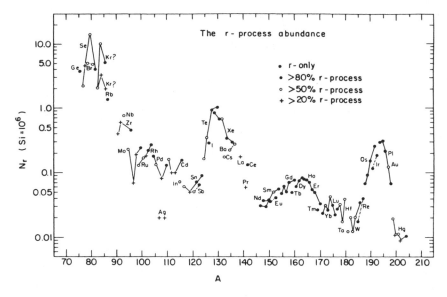

Fig. 8.10. r-process abundances estimated for the Solar System by subtracting the calculated s-process contribution from the total observed abundances of nuclear masses. Isotopes of a given element are joined by lines, dashed lines for even Z and solid lines for odd Z. All abundances are normalized to silicon, whose abundance would be 10^6 on the same scale. Three peaks, and a broad rare earth bump, are the most prominent features of this plot. The question marks indicate a high uncertainly about the relative roles of the r- and s-processes in producing krypton (after Seeger, Fowler, and Clayton (Se65)).

Radioactive dating of terrestrial and meteoritic materials, however, still faces many uncertainties, some of which are illustrated by Problem 13–16 in Section 13:36.

A brief stage in which carbon, oxygen, and silicon are successively burned at progressively higher temperatures during the explosion of a star in the mass range of 20 to 40 $M_\odot$ has been suggested (Ar70) as responsible for the observed abundances of elements in the range $20 \le A \le 64$. Initially carbon in the helium-depleted core of a star would undergo fusion reactions of the kind:

$$
\begin{aligned}
{}^{12}\mathrm{C} + {}^{12}\mathrm{C} \rightarrow\ & {}^{23}\mathrm{Na} + \mathcal{P}, && 2.238\,\mathrm{MeV}, \\
& {}^{23}\mathrm{Mg} + \mathcal{N}, && -2.623\,\mathrm{MeV}, \\
& {}^{20}\mathrm{Ne} + \alpha, && 4.616\,\mathrm{MeV}.
\end{aligned}
\tag{8-105}
$$

These reactions would take place at a temperature of 2×10^9 K. The initial density would be of order 10^5 g cm^{-3}. The reactions are assumed to last for about one-tenth of a second, after which the explosive expansion has cooled matter enough to stop the processes. If the star collapses further, however, and reaches higher temperatures, 3×10^9 K, oxygen also burns and thereafter silicon ^{28}Si disintegrates. In this latter process, the silicon splits into seven α's, which are absorbed by other ^{28}Si nuclei to form increasingly massive nuclei in the range up to ^{56}Fe, as indicated in Figure 8.11.

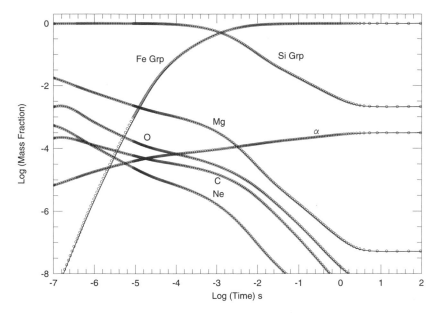

Fig. 8.11. The evolution of nuclear species at the center of a massive star, under constant thermodynamic conditions at a central stellar temperature $T = 5 \times 10^9$ K and density $\rho = 10^9$ g cm^{-3}. These calculations track the evolution from helium burning to statistical equilibrium and show the evolution of nuclear abundances under the α- and e-processes to the final production of the iron peak nuclei. Curves for the individual isotopes shown are for α-particles, ^{12}C, ^{16}O, ^{20}Ne, and ^{24}Mg. The silicon group mass fraction is the sum of the mass fractions of ^{28}Si, ^{32}S, ^{36}Ar, ^{40}Ca, and ^{44}Ti. The iron group mass fraction represents the sum of the mass fractions of the more massive even–even nuclei, ^{48}Cr, ^{52}Fe, ^{56}Ni, and ^{60}Zn. The species shown represent the complete set of even–even nuclei from ^{4}He out to the iron peak nuclei. Silicon burning to the iron group does not proceed directly through capture of heavy nuclei, but rather through photodisintegrations and lighter particle captures. The initial abundances at the start of the calculations are shown at the left edge of the diagram, a time 10^{-7} seconds after the onset of the assumed temperature and density. After only one microsecond, the iron group abundance has significantly risen, while C, O, Ne, and Mg have all declined and the α-particle mass fraction has gradually risen. These trends continue with a steepening decline in the silicon group mass fraction after a millisecond. By the time a second has passed, final equilibrium is all but reached. These plots show the very rapid evolution of nuclear species at the center of a massive collapsing star and explain the predominant formation of iron group nuclei in stars that collapse just before exploding as supernovae. Courtesy of W. Raph Hix, J. Craig Wheeler, and their colleagues (Hi98).

These processes will occur if ignition of the nuclear fuel takes place at a temperature $\sim$5 $\times 10^9$ K and peak density 2 $\times 10^7$ g cm^{-3} in the helium depleted core, and give an r-process isotope distribution in accord with solar values, provided a neutron excess

$$\eta = \frac{(n_\mathcal{N} - n_\mathcal{P})}{(n_\mathcal{N} + n_\mathcal{P})} \qquad (8\text{-}106)$$

of order ~ 0.002 is assumed. This excess, however, appears unnecessary at higher central temperatures and densities, such as those cited in Fig. 8.12, where statistical

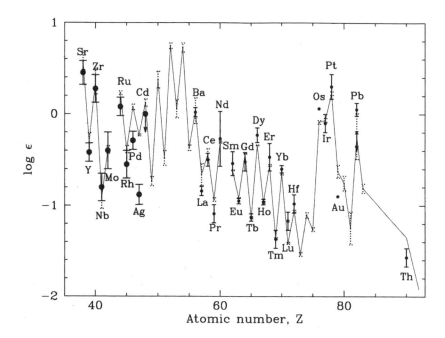

Fig. 8.12. Neutron-capture elements observed in the low-mass ultra-metal-poor star CS 22892-052, whose abundance of iron relative to hydrogen is $<10^{-3}$ times that of the Sun. The elements shown are all formed in the r-process. The observed abundances and their error bars are shown in bold. Calculated r-process abundances for the Sun are shown joined by thin lines, culminating in error bars plotted with dotted lines. These calculated values are scaled to the observed abundances of strontium (Sr), yttrium (Y), zirconium (Zr), niobium (Nb), and molybdenum (Mo) in CS 22892-052, but the absolute abundance of those metals in the Sun is much higher. The Sun also exhibits appreciable abundances of s-process isotopes produced in and ejected from relatively low-mass stars. These are absent in the ultra-metal-poor halo population stars, which appear to have formed early in the Galaxy's evolution, in the wake of the earliest, rapidly evolving, massive stars that ended life as supernovae of type II (SNe II). The r-process elements ejected in those explosions were apparently incorporated in this next generation of ultra-metal-poor stars, whose low masses and luminosities have permitted them to shine to this day. For the Sun, the r-process elemental abundances are calculated by subtracting the more readily computed s-process abundances from the total observed metallicities. (Ar70, Co04b). Courtesy John Cowan (Tr04).

equilibrium between nuclear species sets in (Me02a). It appears as though a variety of different explosive conditions can provide abundance ratios roughly compatible with observed values, and that several of these may have been active over the æons.

The central structure of a star may also alter the outcome of an explosion. Neutron stars that are formed in the course of a supernova explosion occasionally are observed moving at high velocities relative to the center of the expanding, explosively ejected supernova remnant. This suggests that the explosion was asymmetric and that the star's original structure could not have been concentric.

8:13 Compact Stars

Thus far we have dealt with stars whose densities are roughly comparable to the Sun's, except late in life when their central cores become very compact. We now turn to stable stars that are orders of magnitude more compact, throughout: the *white dwarfs* and *neutron stars*. The structure of these stars can be considered from the same general viewpoint that allowed us to understand processes in the interior of ordinary stars. Before proceeding in this direction we should, however, review one particular argument that we had brought out to demonstrate the importance of nuclear reactions in stellar interiors. In (8–1) we had shown that the potential energy per unit mass of stellar substance is $\sim 3MG/5R$, whereas the available nuclear energy is of the order of $10^{-2}c^2$, if matter–antimatter annihilation is ruled out. It follows that very compact stars may be able to liberate amounts of gravitational energy in excess of the normal nuclear energies available. This happens when

$$R \lesssim \frac{MG}{10^{-2}c^2} \sim 10^7 \, \text{cm} = 100 \, \text{km} \qquad (8\text{-}107)$$

for stars of one solar mass. This radius is still appreciably larger than R_s, the Schwarzschild radius:

$$R_s = \frac{2MG}{c^2} . \qquad (8\text{-}108)$$

Because white dwarfs typically have masses of the order of 10^{33} g, the corresponding Schwarzschild radius would be

$$R_s \sim 1.5 \times 10^5 \, \text{cm} = 1.5 \, \text{km},$$

which is small compared to the white dwarf radii $\lesssim 10^4$ km noted below in Section 8:14, but only an order of magnitude lower than the neutron star radii ~ 15 km, considered in Section 8:16.

8:14 White Dwarf Stars

We previously suggested that a compact star can be so dense that matter becomes degenerate in its interior. At the surface of a white dwarf the density and its outer layers are not degenerate. However, the nondegenerate layer is so thin that we can treat the star as completely degenerate throughout.

We first note that most of the pressure in the star's interior must be provided by the degenerate electrons. The more compact the star is, the higher the Fermi energy of the electrons and the higher the electron gas pressure. The partial pressure of the nuclei is negligibly low, because, as discussed in Section 4:15, the nucleons are nondegenerate. This permits us to set the total pressure equal to that of the electrons alone, $P = P_e$.

In Section 8:6 we gave the electron pressure for a nonrelativistic and for a completely relativistic degenerate electron gas, respectively, as

$$P = \frac{h^2}{20 m_e m_H} \left(\frac{3}{\pi m_H} \right)^{2/3} \left(\frac{(1+X)}{2} \rho \right)^{5/3} \quad \text{nonrelativistic} \qquad (8\text{–}41)$$

and

$$P = \frac{hc}{8 m_H} \left(\frac{3}{\pi m_H} \right)^{1/3} \left(\frac{(1+X)}{2} \rho \right)^{4/3} \quad \text{relativistic.} \qquad (8\text{–}43)$$

In general, there will exist an important transition region where the gas is neither highly relativistic nor completely nonrelativistic. In that region the pressure can be shown (Problem 8–4) to take the form:

$$P = \frac{8 \pi m_e^4 c^5}{3 h^3} f(x), \qquad \rho = \mu_e \frac{8 \pi m_H m_e^3 c^3}{3 h^3} x^3 , \qquad (8\text{-}109)$$

where

$$x = \frac{p_0}{m_e c} , \qquad (8\text{-}110)$$

$$\mu_e = \frac{2}{1+X} . \qquad (8\text{-}111)$$

The function $f(x)$ is

$$f(x) = \frac{1}{8} [x(2x^2 - 3)(x^2 + 1)^{1/2} + 3 \sinh^{-1} x] . \qquad (8\text{-}112)$$

PROBLEM 8–4. For a degenerate relativistic gas all momentum states (4–65) are filled, and equation (5–30) holds. Using equation (4–27) show that

$$P = \frac{1}{3} \int_0^{p_0} p v(p) n_e(p) \, dp \qquad (8\text{-}113)$$

$$= \frac{8 \pi}{3 m_e h^3} \int_0^{p_0} \frac{p^4 \, dp}{[1 + (p/m_e c)^2]^{1/2}} . \qquad (8\text{-}114)$$

Setting $\sinh u = p/m_e c$, show that

$$P = \frac{8\pi m_e^4 c^5}{3h^3} \int_0^{u_0} \sinh^4 u \, du \, . \tag{8-115}$$

We see that (8–115) has the same coefficient as (8–109). We can also show that integration of (8–115) yields the expression (8–112) for $f(x)$.

Small values of x, $x \ll 1$, correspond to the lower density, outer regions of the star, where the gas is nonrelativistic, while high x-values correspond to the frequently relativistic central portions of the star.

PROBLEM 8–5. Evaluate $f(x)$ in the limits $x \ll 1$ and $x \gg 1$ and show that equations (8–41) and (8–43) are obtained.

Equation (8–109) is computed on the basis of statistical mechanics and involves no assumptions concerning stars. It is an *equation of state* for a partially relativistic degenerate gas no matter where it may be found. We should note that the pressure is temperature independent in this equation. It only depends on x, which is a measure of the momentum at the Fermi energy of the electron gas; this is only density dependent. The mathematical problem of computing conditions at the center of the star can therefore be separated into two parts: one hydrostatic, the other thermodynamic.

The hydrostatic equilibrium conditions are the same as those obtained earlier:

$$\frac{dP}{dr} = -\rho \frac{GM(r)}{r^2}, \qquad \frac{dM(r)}{dr} = 4\pi r^2 \rho \, . \tag{8-7}, \quad (8-8)$$

To integrate these equations, we assume that we know what the chemical composition of the white dwarf is, because that composition determines the value of μ_e in equation (8–111). We next choose an arbitrary central density ρ_c for the star, and then integrate the hydrostatic equations outward from the star's center until we reach a radius r where the pressure has dropped to zero. In this computation that radius represents the surface of the star. The value of $M(r)$ at this radial distance corresponds to the total mass of the star and the value of r represents the actual stellar radius.

This procedure can be repeated for a range of different central densities, and we therefore obtain a whole family of stellar models with varying central densities. Similarly, we can obtain a new family of models having different chemical compositions; but no recomputation is required here, because a change of chemical composition is mathematically equivalent to a simple change of variables.

Let us see why.

If the initially computed values are denoted by primes, and new variables — corresponding to a new chemical composition — are denoted by unprimed symbols, we find that the required relations are

$$P = P',$$
$$\rho = \frac{\mu_e}{\mu_e'} \rho',$$

$$M(r) = \left(\frac{\mu'_e}{\mu_e}\right)^2 M(r') , \qquad (8\text{-}116)$$

$$r = \left(\frac{\mu'_e}{\mu_e}\right) r' .$$

We can readily see that substitution of these expressions into equations (8–109), (8–7), and (8–8) leaves their form unchanged; a change in the chemical composition is therefore equivalent to a change in central density. Consequently we are dealing with a one-parameter family of models. Everything about a given star can be described entirely in terms of an equivalent central density. We present the results of the described computations in the form of Table 8.3.

Table 8.3. Central Densities, Total Mass, and Radius of Different White Dwarf Models, Taking $\mu_e = 2$ (Negligible Hydrogen Concentration).[a]

$\log \rho_c$	$M/M_\odot$	$\log R/R_\odot$
5.39	0.22	−1.70
6.03	0.40	−1.81
6.29	0.50	−1.86
6.56	0.61	−1.91
6.85	0.74	−1.96
7.20	0.88	−2.03
7.72	1.08	−2.15
8.21	1.22	−2.26
8.83	1.33	−2.41
9.29	1.38	−2.53
∞	1.44	−∞

[a] See text for comments. (After M. Schwarzschild (Sc58b), from *Structure and Evolution of the Stars* ©1958 by Princeton University Press, p. 232.)

We note that:

(1) The larger the mass of the white dwarf, the smaller is its radius.

(2) For masses comparable to the Sun, the white dwarf's radius is a factor of $\sim 10^2$ smaller than $R_\odot$.

(3) There exists an upper limit to the mass — the *Chandrasekhar limit* — above which no stable white dwarf configuration exists — infinite central pressure would be needed to keep the star from further collapse. This limit is $\sim 1.4 M_\odot$.

The reasons for the limit are apparent if we consider that there are quite different relations between central pressure and density in the relativistic and nonrelativistic extremes. From (8–41) and (8–43):

$$\text{nonrelativistically}: \quad P \propto \rho^{5/3}, \quad \frac{dP}{dr} \propto \rho^{2/3}\left(\frac{d\rho}{dr}\right) , \qquad (8\text{-}117)$$

$$\text{relativistically}: \quad P \propto \rho^{4/3}, \quad \frac{dP}{dr} \propto \rho^{1/3} \left(\frac{d\rho}{dr} \right). \tag{8-118}$$

For both the gravitational pressure gradient given by (8–7) and (8–8) is

$$\frac{dP}{dr} \propto -\frac{\rho(r)}{r^2} \int_0^r \rho(r') 4\pi r'^2 \, dr' . \tag{8-119}$$

In the crudest approximation, we can set the density equal to the stellar mass divided by the cube of the radius R so that

$$\begin{aligned}
\text{nonrelativistically}: \ & dP/dr \propto M^{5/3}/R^6 \ , \\
\text{relativistically}: \quad & dP/dr \propto M^{4/3}/R^5 \ , \\
\text{gravitational}: \quad & dP/dr \propto M^2/R^5 \ .
\end{aligned} \tag{8-120}$$

We note that the dependence of the relativistic pressure gradient on radius has the same power as the gravitational force. Both increase as R^{-5} as the star contracts. This means that once a relativistic white dwarf core is forced to contract by hydrostatic pressure, the counterforce produced through contraction increases at the same rate as the gravitational attraction, and that tends to compress the star even further. There is, therefore, no way in which the star can come into equilibrium. On the other hand, a nonrelativistic gas at the center of a white dwarf can always adjust itself through contraction, until the gravitational forces compressing the star are countered. We then have the following situation. For small stellar masses, the central pressure is determined more nearly by the nonrelativistic approximation, and the star can find a stable equilibrium position. For more massive objects, the central density becomes so high during contraction that the relativistic regime is reached and further contraction no longer leads to equilibrium. The Chandrasekhar limit is therefore symptomatic of the transition from a predominantly nonrelativistic to a predominantly relativistic central core (Ch39)*. Rather similar arguments set roughly similar limits to the masses of neutron stars, as we will see in Section 8:16 and Figure 8.17.

8:15 Stellar Evolution and The Hertzsprung–Russell Diagram

At this stage it is useful to draw a connection between events in the interior of a star and the star's appearance at its surface. As described in Chapter 1, a star leaves the main sequence of the Hertzsprung–Russell diagram on exhausting the hydrogen content of its central core. The theory of stellar evolution allows us to determine the ages of stars at this stage. Using plausible stellar models, we can compute what the hydrogen-burning time scale should be for stars having different masses. A *globular cluster* contains stars with a variety of masses; but at any given time only stars having a particular color and magnitude are about to leave the main sequence to embark onto the red-giant branch. Since stellar luminosities and masses are related (see Problem 8–3), we can use the *turn-off point* where stars leave the main sequence as

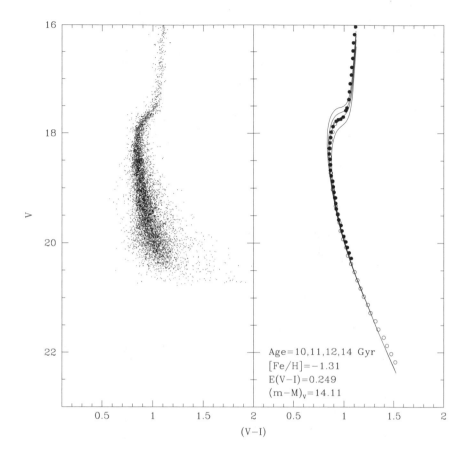

Fig. 8.13. Visual magnitude V plotted against the visual-infrared (V−I) color for the globular cluster M12. The plot in the panel on the left shows the selection of stars from which the age of the cluster was estimated. In the panel on the right the best fit to these data is shown by the filled circles. Additional data at fainter magnitudes (vB02), are shown as open circles. The theoretical curves are *isochrones* – snapshots of the trajectories of stars of different masses through the (V, V−I) color-magnitude diagram at different epochs. From upper left to lower right the isochrones refer to cluster ages 10, 11, 12, and 14 Gyr, the least massive stars taking the longest time to turn off the main sequence. Reflecting the great age of the cluster, whose stars formed when the abundance of heavy elements in the Universe was still low, the isochrones assume the abundance of iron relative to hydrogen, indicated by [Fe/H] = −1.31 — the notation is logarithmic — to be ∼20 times lower than the solar abundance. Reddening of the starlight by interstellar dust needs to be taken into account and is one of the main uncertainties in producing such a plot. In deriving the isochrones the authors concluded that the intrinsic (V−I) color of M12 is E(V−I) = 0.249 magnitudes lower than shown in this plot. The absolute magnitude M_v is 14.11 magnitudes lower than the apparent magnitude $V \equiv m_v$. Although the isochrone fit to the data is not perfect at the high luminosity end of the plots, the indicated age of ∼12 Gyr is consistent with a variety of other cosmological age measures. (Courtesy of J. R. Hargis, E. L. Sandquist, and M. Bolte (Ha04).)

an indicator of the masses of stars that are just completing the hydrogen-burning phase. The age of these stars can then be calculated and this defines how long ago the stars must have been formed — assuming that the cluster stars were all born at roughly the same time. The right-hand portion of Fig. 8.13 shows the calculated tracks that stars with ages in the range 10 to 14 Gyr would follow near the turn-off point, and show that the age of the globular cluster M12 must be about 12 Gyr. Interestingly, the ages of different clusters computed in this way vary considerably, indicating that they were formed at different epochs. They may have originally formed in smaller galaxies that eventually were tidally captured and merged with the Milky Way.

Figure 8.14 indicates the path through the Hertzsprung–Russell diagram following evolution off the main sequence. Lifetimes here are far shorter. For stars of 15 and 120 $M_\odot$, the hydrogen-burning phase, respectively, ranges from roughly 12 to 3×10^6 yr, and the corresponding helium-burning phase lasts only 2 to 0.5×10^6 yr (Sc92). The detailed evolution depends not only on the mass of the star but also on its initial chemical composition. When the Universe was young, metal abundances were orders of magnitude lower than today, and helium abundances were ~20% lower. The impact of these differences on a star's evolution is reflected by the differences in the two panels of Fig. 8.14.

Let us still consider the appearance of white dwarfs in the Hertzsprung–Russell diagram. Once a white dwarf's interior becomes degenerate no further contraction takes place. The star then gradually cools down over a long period, but no further nuclear reactions take place. This determines the star's location in the H–R diagram. We recall the definition of the effective temperature of a star

$$L = \sigma T_e^4 (4\pi R^2) \qquad (4\text{--}78)$$

and rewrite this in terms of the solar luminosity and surface temperature:

$$\log \frac{L}{L_\odot} = 4\log \frac{T_e}{T_{e\odot}} + 2\log \frac{R}{R_\odot}. \qquad (8\text{-}121)$$

If we then use the white dwarf radii and masses from Table 8.3, we can obtain plots of L against T_e as a function of different mass values. Choosing five different representative masses, we obtain the curves shown in Fig. 8.15.

The agreement with observations is satisfactory, and we can feel reasonably confident that the discussion pursued here is at least roughly correct. This is important. In our neighborhood of the Galaxy white dwarfs have a number density of ~5 × 10^{-3} per cubic parsec, corresponding to ~3.5 × 10$^{-3} M_\odot$ pc^{-3} (Ho02). The local overall mass density is ~$0.1 M_\odot$ pc^{-3}. We should come to understand white dwarfs well if we are to learn how low-mass stars die.

Stars more massive than the Chandrasekhar limit cannot maintain themselves as white dwarfs and collapse to become even more compact.

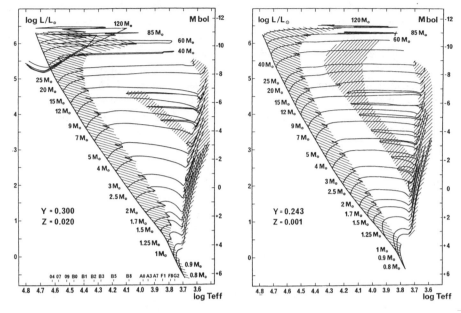

Fig. 8.14. Theoretical post-main-sequence evolutionary tracks through the Hertzsprung–Russell diagram for stars of different masses. The left panel shows plots for an initial solar composition, $X = 0.68$, $Y = 0.30$, $Z = 0.02$. The right panel shows plots for an initially primordial composition with almost no heavy elements, $Z = 0.001$, a helium abundance $Y = 0.243$, and a hydrogen abundance $X = 0.756$. Slow phases of nuclear burning, where a star spends appreciable time, are indicated by hatched areas. For stars more massive than $7M_\odot$ the tracks show evolution through the end of carbon burning. For stars with $2 \leq M/M_\odot \leq 5$ the tracks show the stars evolving to the early phase on the asymptotic-giant branch (AGB) on the right. For stars with $M \leq 1.7M_\odot$ the evolution is traced up to the helium flash. For stars $M > 15M_\odot$ and $Z = 0.001$, helium burning takes place mainly in the blue supergiant region, whereas for solar-abundance stars with $15 < M/M_\odot < 25$, most of the helium burning occurs in the AGB phase. Stars with $M > 40M_\odot$ for $Z = 0.02$, and $M > 85M_\odot$ for $Z = 0.001$, are Wolf–Rayet stars with low surface-hydrogen abundances. During helium burning they undergo mass losses at rates $\geq 10^{-4} M_\odot \, \mathrm{yr}^{-1}$. At the end of carbon burning, stars with $Z = 0.02$ are very blue and have a lowered luminosity. (After Schaller *et al.* (Sc92).)

8:16 Supernovae, Neutron Stars, and Black Holes

Stars of intermediate mass develop a core of densely packed neutrons. The masses of relatively few *neutron stars* are known. They are derived from observed orbital motions of binary pulsars and, as seen in Figure 8.16, cluster closely around $1.4M_\odot$ (Th93).

We can imagine the evolution toward this state in the following way (Sa67)*. As a star evolves, the value of μ_e in the equations of Section 8:14 changes. Initially, as hydrogen is depleted, μ_e assumes a value of 2. This holds, for exam-

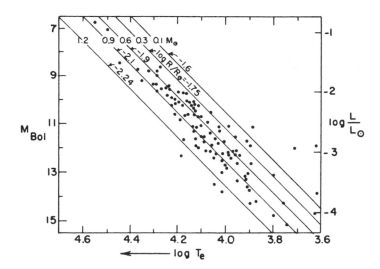

Fig. 8.15. White dwarf Hertzsprung–Russell diagram. Lines of constant radius are shown. Also shown are the masses based on completely degenerate core models containing elements having $\mu_e = 2$. (After Weidemann (We68). Reprinted with permission from *Annual Review of Astronomy and Astrophysics*, Vol. 6, ©1968 by Annual Reviews, Inc.)

ple, for a star in which the major constituent is ^{12}C. But as the chemical elements evolve toward the more neutron-rich species, equations (8–36) and therefore (8–111) no longer hold, and for a star rich in ^{56}Fe we find $\mu_e = 2.15$. Now, the Chandrasekhar limiting mass is proportional to μ_e^{-2}, as seen from (8–116), so that we can draw a number of curves of mass against central density — as in Fig. 8.17. In these curves we assume the lowest possible temperature and we show plots for stars of various chemical compositions. Corresponding to each chemical composition we have a different Fermi energy E_F for the electrons at the center of the star — a direct consequence of the star's changed central density. As the central density increases, for a given composition, the electron Fermi energy always increases up to the point where inverse beta decay takes place and drives the electrons into the nuclei. This is what produces the increasingly neutron-rich elements cited in Table 8.4. The symbolic reaction is

$$\mathcal{P} + e \rightarrow \mathcal{N} + \nu. \tag{8-122}$$

The reverse reaction cannot take place if the Fermi energy is sufficiently high because all the electron states into which the radioactive nucleus might decay are already occupied. This gives otherwise unstable nuclei an environmentally induced stability. The neutrinos diffuse to the star's surface and escape; the star loses energy. This process is called *neutronization*.

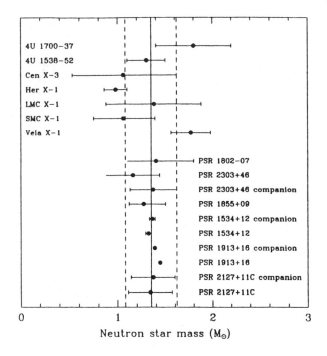

Fig. 8.16. Masses of 17 neutron stars. Stars found in massive X-ray binaries are shown at the top; radio pulsars and their companions are shown at the bottom (Th93).

Table 8.4. Density and Electron Fermi Energy at Which Inverse Beta Decay Becomes Energetically Favorable (after E. E. Salpeter (Sa67)).[a]

	^{12}C	^{32}S	^{56}Fe	^{120}Sn
$\log \rho \, (\mathrm{g\,cm^{-3}})$	10.6	8.2	9.1	11.5
E_F (Mev)	13	1.7	3.7	24

[a]Reprinted with permission of the publishers of The American Mathematical Society, from *Lectures in Applied Mathematics*, ©1967, Vol. 10, Part 3, "Stellar Structure Leading up to White Dwarfs and Neutron Stars," p. 34.

The value of μ_e, which is an effective nuclear mass per free electron, also increases during contraction. When the Fermi energy reaches 24 MeV, the density is $\rho \sim 10^{11.5} \, \mathrm{g\,cm^{-3}}$ and $\mu_e \sim 3.1$. At this stage free neutrons become energetically favorable so that a further increase in density leads to an increased partial density of neutrons, a practically constant density of ions, and a constant electron Fermi energy of 24 MeV.

As the density increases, E_F increases to the point where reaction (8–122) proceeds rapidly and the electrons are driven into the nuclei causing the collapse of

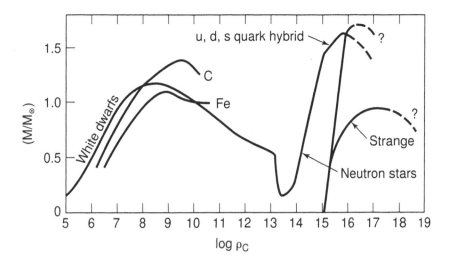

Fig. 8.17. Mass of a cold star, as a function of central density. The full curve is for a representative initial chemical composition and assumes relativistic hydrostatic equilibrium. The curves labeled C and Fe assume pure carbon and iron composition. Negative slopes on the curves indicate regions where there are no hydrostatically stable configurations. The density is given in units of g cm^{-3}. The dashed portion of the curve near central densities of the order 10^{18} g cm^{-3} is very uncertain because the physical state of matter at these densities is uncertain. In addition to white dwarfs and neutron stars, hypothetical strange stars are also shown (Sa67, Gℓ97). (Reprinted with permission of the publishers of The American Mathematical Society, from *Lectures in Applied Mathematics*, ©1967, Vol. 10, Part 3, "Stellar Structure Leading up to White Dwarfs and Neutron Stars," p. 33.)

the central core because the electron pressure no longer increases at a sufficient rate during the contraction.

In Fig. 8.17 the curves for stars containing ^{12}C and ^{56}Fe show a maximum mass at ρ_c values where inverse beta decay first sets in and μ_e increases. To the right, the curve for free neutrons is shown. It has a maximum just beyond $\rho_c \sim 10^{15}$ g cm^{-3}.

The reason for this maximum is relatively easy to understand if we compute the mass expected, respectively, on the basis of a nonrelativistic neutron gas and an extreme-relativistic gas.

The virial theorem gives the ratio of pressure P to density ρ in terms of stellar mass as

$$3\left\langle \frac{P}{\rho} \right\rangle \sim \frac{GM}{R} \propto M^{2/3}\langle n^{1/3}\rangle, \tag{8-123}$$

where mean values are denoted by brackets, and n is the number density of neutrons. As can be seen from (8–35), (8–40), and (8–42),

$$P \propto n^{5/3} \quad \text{nonrelativistic,} \tag{8-124}$$

$$P \propto n^{4/3} \quad \text{extreme relativistic.} \tag{8-125}$$

Similarly, the ratio of mass density to number density is

$$\langle\rho\rangle/\langle n\rangle \sim m_{\mathcal{N}} \quad \text{nonrelativistic,} \tag{8-126}$$

$$\langle\rho\rangle/\langle n\rangle \sim \frac{E_F}{c^2} \quad \text{extreme relativistic,} \tag{8-127}$$

because in the extreme case, the rest-mass energy can be neglected. But because $E_F \propto n^{1/3}$, we then have

$$\langle\rho\rangle \propto \langle n\rangle^{4/3} \quad \text{extreme relativistic,} \tag{8-128}$$

and from equation (8–123) we then find

$$M^{2/3} \propto 3 \left\langle \frac{P}{\rho} \right\rangle \langle n^{1/3} \rangle^{-1} \propto \begin{cases} \langle n^{1/3} \rangle & \text{nonrelativistic,} \\ \langle n^{1/3} \rangle^{-1} & \text{extreme relativistic.} \end{cases} \tag{8-129}$$

This means that the mass first increases as $\langle n\rangle^{1/2}$ and then decreases as $\langle n\rangle^{-1/2}$ as the density continues to increase.

The masses of neutron stars have been ascertained only from observations of pulsars in binary systems. Double neutron star binaries appear to have masses of $\sim 1.4 M_\odot$ (Fig. 8.16), often assumed to also be typical of isolated neutron stars. But other binaries that include neutron stars yield masses ranging up to $\sim 2 M_\odot$ (Ni05). If this higher mass is characteristic of isolated neutron stars, it may be providing clues about the nature of the collapse that generated these compact objects.

The cores of massive neutron stars may contain baryonic matter that is liquid or crystalline and contain drops, rods, or slabs, of quarks or hadrons. *Quarks* are the constituent particles of nucleons and mesons. *Nucleons* — protons or neutrons — contain three quarks; mesons contain two. The term *hadron* comprises nucleons or mesons in any of their ground or excited states. *Hyperons* also play a role in such models. Hyperons are *strange* particles that decay into pairs of hadrons — pions and/or nucleons. The physics of liquid and solid quark, hadronic, or hyperon structures is not yet understood, making the equations of state of matter difficult to compute. This makes the rigidity of the cores quite uncertain. An indication of the complexity of neutron star structures is given in Fig. 8.18. At the high densities involved, general relativistic effects must also be taken into consideration. This is the domain that is of greatest interest, because potentially it is in these last stages that a star can give off by far its greatest amount of energy by converting a large fraction of its mass into radiation, neutrinos, and perhaps gravitational waves. The star then turns into a *black hole*.

The neutrino pressures resulting from the collapse of a massive star are believed responsible for exploding away the star's outer layers in a supernova explosion. Computation of the various stages in this hydrodynamic process are difficult, and we do not know just how such explosions take place. Nevertheless, a central factor

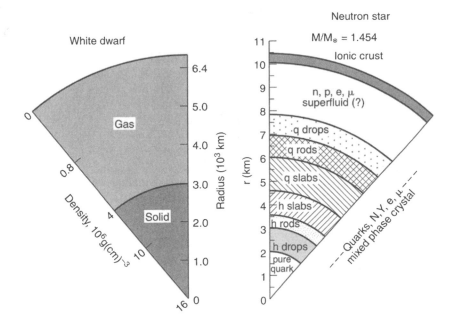

Fig. 8.18. Density and structure of white dwarf and neutron stars. The diagram on the left is a white dwarf model. The shell structure on the right represents a neutron star. Note that the white dwarf radius is 6400 km and that of the neutron star is only 10.5 km. The neutron star contains an inner sphere of pure quark matter surrounded by a crystalline region of mixed quark and hadronic matter. Various kinds of phases are shown for h(adronic) and q(uark) matter. Only the dominant constituent is given in the labels. Particles in these regions include quarks, nucleons (N), hyperons (Y), and leptons. A liquid of neutron star matter, containing nucleons and leptons surrounds these mixed phases. A thin outer crust of heavy ions provides the star's surface composition (after Ruderman (Ru71) and Glendenning (Gℓ97). Reprinted with permission from Ruderman, "Solid Stars," ©1971 by Scientific American, Inc. All rights reserved.

in the collapse to the neutron star and black hole stages is the liberation of gravitational energy and the formation of neutrinos, many of which escape into space. A glance at Table 1.6 shows that the energy of neutrinos emitted in stellar core collapse actually exceeds the total radiant, or electromagnetic radiation released in stars. Although neutron stars and black holes are rare, accounting for only ~5% of the stellar population, by mass, the gravitational energy that can go into neutrino production during the final stages of stellar collapse is extremely high. In the spirit of (8–1) it amounts to $3M_n G/5r_n \sim 1.2 \times 10^{20}$ erg g^{-1} for a neutron star mass $M_n = 1.4 M_\odot$ and radius $r_n = 15$ km. This contrasts to 6×10^{18} erg g^{-1} released in hydrogen burning.

   This ultimate form of stellar death is made particularly complex by a number of important considerations that we have so far neglected to mention. Not only is it essential to know whether neutrons, quarks, hadrons, or hyperons are present at any particular depth in the star, we also need to consider whether these particles will arrange themselves as fluids or superfluids, crystals of various kinds, or phases that have no analogue in the laboratory. These will have to be understood more thoroughly before we can claim to understand the structure of neutron stars.

We also do not know much about the formation of black holes. We believe that a neutron star that is sufficiently massive undergoes further collapse to from a black hole where, as discussed in Chapter 5, it no longer can emit any radiation that will be received outside the star. But what happens to the magnetic field when a black hole is formed, and to the electric charge? How can we talk about conservation of baryons in the Universe when all these particles disappear without a trace, giving off only photons or gravitational radiation and neutrinos as the collapse takes place? What are the correct dynamic equations in such a collapse? Are they, as we tend to assume, the equations of general relativity? Again a thorough discussion is needed of this problem, which already has a long and distinguished history (Op39a), (Op39b), (Gℓ97).

8:17 Pulsars, Magnetars, and Plerions

We know that an ordinary star which suddenly collapses has to rotate rapidly if angular momentum is to be conserved. The most rapidly spinning neutron star discovered todate rotates 716 times per second (He06). However, most neutron stars exhibit spin periods ranging from $1/60$ to 8.5 s. In a rapidly rotating neutron star, the co-rotating magnetic field of order 10^{12} gauss can accelerate charged particles to cosmic-ray energies. Relativistic charged particles intensely radiate electromagnetic waves, most often radio waves, directed along the magnetic poles. As the star rotates about its spin axis, which generally is not aligned with the magnetic poles, this radiation can be directed toward Earth, once per rotation, and appears pulsed: hence the name *pulsar*. In time, the loss of angular momentum carried away by the radiation and cosmic-ray particles slows down the rotation, and after $\sim 10^7$ yr the spin rate diminishes to the point where relativistic particles no longer are created.

Some of the most rapidly rotating pulsars may have also received an initial kick from the supernova explosion that gave them birth. The kick makes itself apparent through the high velocities, about 1000 km s^{-1}, with which certain pulsars are seen streaming through the Galaxy — speeds exceeding the escape velocity from the Galaxy's gravitational pull. Some neutron stars have magnetic fields ranging to $\gtrsim 10^{14} - 10^{15}$ gauss. Their strong magnetic fields have led to the name *magnetar*. The strengths of the magnetic fields spin the stars down on time scales of $\sim 10^5$ yr. A few of these also exhibit γ-ray and X-ray emission, respectively, the *soft-γ-ray repeaters (SGRs)* and the *anomalous X-ray pulsars (AXPs)* (Ka03).

SGRs are observed to periodically emit a burst of X-ray and low-energy gamma rays. The burst may be due to the release of internally trapped magnetic fields as they break through the solid crust of the neutron star. For SGR 1806-20 the emitted X-rays showed a spectral absorption feature at $h\nu \sim 5$ keV $\sim 8 \times 10^{-9}$ erg corresponding to a spectral frequency $\nu \sim 1.2 \times 10^{18}$ s^{-1} suggestive of the *cy-*

clotron frequency $\nu_c = \omega_c/2\pi = eB/2\pi/m_p c$ to be expected from a proton in a field $B \sim 10^{15}$ gauss (Ib03).

AXPs, whose slow rotation period, like those of SGRs are in the 6–12 s range, appear to emit considerably more X-ray energy than can be accounted for by the star's spin-down rate. This suggests that the particles emitting the X-rays are accelerated by the strong electric fields generated as the star's magnetic field decays and equation (6-16) comes into play. AXPs and SGRs may be closely related types of magnetars. None of the observed AXPs show any sign of being binary stars. This is in contrast to *binary X-ray pulsars*. These emit X-rays as material from a close binary companion falls directly onto a compact star's surface or onto its orbiting compact accretion disk. The accreting star is spun up to a high rotational speed through absorption of angular momentum from material falling onto its surface, and has spin periods measured in milliseconds.

A few pulsars appear embedded in an expanding bubble of relativistic particles and magnetic fields powered by the pulsar. These bubbles, called *plerions* cool through synchrotron emission and adiabatic expansion, as surmised from their radio and X-ray spectra. They appear to be found only within the youngest Galactic supernova remnants, and therefore probably do not survive more than a few thousand years.

8:18 Hypernovae and Gamma-Ray Bursts

Gamma-ray bursts (GRBs) reaching Earth from remote sources last anywhere from 0.01 to 100 s and during this period are the brightest γ-ray sources in the sky. For more than two decades after the first report of their discovery (Kℓ73), there was no indication of their nature or origin. Then, in the late 1990s, it became clear that GRBs with durations exceeding 2 s appear to originate in hypernova explosions in distant galaxies.

Hypernovae are extremely energetic supernovae. In contrast to more common supernovae that eject material at a few thousand kilometers per second, hypernovae explosively eject matter at initial velocities in excess of 30,000 km s^{-1}. A current model for long-duration GRBs and hypernova explosions postulates that they result from the core collapse of an isolated massive star — a *core collapse supernova* or *collapsar*. As the star collapses a black hole embedded in an accretion disk is formed in its center. Neutrinos emitted from the accretion disk exert an outward pressure on the collapsing star, through absorption, scattering, and the formation of electron/positron pairs and photons in neutrino–antineutrino annihilation. Neutrino annihilation occurs mainly in the innermost regions of the disk, above and below the black hole, and along the rotation axis of the disk (Fr03).

The neutrino scattering and absorption opacities, κ_{sc} and κ_{abs}, respectively, are

$$\kappa_{sc} \sim \frac{5\alpha^2 + 1}{24} \frac{\langle \epsilon_\nu^2 \rangle \sigma_0}{(m_e c^2)^2} \frac{\rho}{m_u}(Y_n + Y_p) \qquad \text{and} \qquad (8\text{-}130)$$

$$\kappa_{abs} = \frac{3\alpha^2 + 1}{4} \frac{\langle \epsilon_\nu^2 \rangle \sigma_0}{(m_e c^2)^2} \frac{\rho}{m_u} (Y_n + Y_p) \,. \tag{8-131}$$

Here, $\alpha = -1.26$ is a coupling constant, ϵ_ν is the neutrino energy, $\sigma_0 = 1.76 \times 10^{-44}$ cm^2, ρ_0 is the density of stellar material along the rotation axis above the disk, $m_u = 1.67 \times 10^{-24}$ g is the atomic mass unit, and $Y_n \equiv (n_n/n_b) \sim 0.5$ and $Y_p \equiv n_p/n_b \sim 0.5$ are the number fractions of neutrons and protons, with respect to the total number density of baryons n_b.

The neutrinos in the disk are largely produced through capture of degenerate electrons and inverse beta decay (8–92), as high electron-degeneracy-pressures, produced in the collapse drive electrons into nuclei. However, antineutrinos can also be produced. The mutual annihilation produces electron–positron pairs and photons which are readily absorbed by the surrounding medium. Onset of the explosion along the rotation axis occurs when the inward-directed gravitational acceleration of the black hole on matter in the collapsing star $a_g = -M_{BH}G/r^2$ is exceeded by the outward-directed acceleration due to neutrino scattering, absorption, and annihilation, $a_g + a_{\nu,sc} + a_{\nu,abs} + a_{\nu,\bar{\nu}} > 0$.

For a star with an initial mass in the range $\sim$40 to $60 M_\odot$, the rate at which the central black hole accretes mass has been estimated to be between $\sim 0.5 - 1 M_\odot$ s^{-1}, and the mass of the final black hole is thought to be around 14 to 23 $M_\odot$ (Fr03). The accretion energy released as mass from the disk spirals into the black hole, is converted into the explosive energy of a bi-lobed, highly-relativistic jet beamed along the star's poles. As would be expected from the discussion of Section 5:9, the high *Lorentz factor* $\Gamma(v) \equiv [(1 - (v/c)^2]^{-1/2} \sim 100$ of the ejected jet accounts for a confined angle within which the γ-emission appears particularly strong. More important, the high value of $\Gamma(v)$ accounts for two other factors — the high energy of the individual highly Doppler-shifted γ-photons received, and the short duration of the GRB. The Doppler shift (5–44) tells us that the energy of the photons arriving at Earth will be Doppler shifted by a factor $\Gamma(v)$. And, as equations (6–149) and (6–150) tell us, if the duration of emission in the rest-frame of the jet is Δt_{rest}, the duration of the γ-emission for an observer at whom the jet is directed will be $\Delta t_{obs} \sim (1 - v/c)\Delta t_{rest} \sim (1/2)\Gamma(v)^{-2}\Delta t_{rest}$. With a Lorentz factor as high as $\Gamma(v) \sim 100$, γ-rays emitted by the jet in the course of two days would arrive at Earth within an interval of 10s. However, for distant GRBs the cosmic red shift z increases this duration by a factor of $z + 1$. GRB 050904 with red shift z=6.295 had a duration of 225 s (Ka06). As we will see in Chapter 11, such a high red shift indicates that stars sufficiently massive to explode as hypernovae already existed when the Universe was merely 1 Gyr old.

More recently, GRBs lasting less than $\sim$0.1 s in the higher energy 100-400 keV range have been identified as due to the merger of neutron-star-, black-hole-, or mixed neutron-star/black-hole-binaries to form a single black hole (Ge05, Vi05). Such mergers provide a way in which the vast amounts of energy liberated in a GRB can be produced in such short bursts. GRB 050509 and 050709, respectively observed on May 9 and July 9, 2005, were bursts emanating from galaxies at red shifts $z = 0.225$ and 0.16, with explosive energies in the $10^{48} - 10^{50}$ erg range, if their emission was isotropic. The emission, however, is likely to be strongly beamed. The total emitted energy released in such mergers could potentially be as high as of order $M_\odot c^2 \sim 10^{55}$ erg if the radiation efficiency were high, though current expectations are that this efficiency is likely to be low. Further observations will undoubtedly clarify all these questions.

8:19 Microquasars

Microquasars are black holes of stellar mass. On a far smaller scale, they mimic quasars, which are black holes with masses ranging up to $\sim 10^9 M_\odot$; but because microquasars can be locally found in the Galaxy they can be studied more readily. Microquasars are believed to be spinning black holes with masses of order $10 M_\odot$, surrounded by an accretion disk roughly 10^3 km across. The accretion disk continues to be fed mass from a close binary companion star. As discussed in Problem 5–15, X-rays are emitted as matter is tidally stripped from the star and crashes onto the disk. Along the accretion disk axis, jets of relativistic particles stream out at velocities that appear to exceed the speed of light. These *superluminal velocities*, however, are only apparent as we saw in Section 5:12 (Mi98).

8:20 Vibration and Rotation of Stars

We know from the virial theorem for nonrelativistic systems that the absolute value of the potential equals twice the kinetic energy per unit mass. In a star this kinetic energy is represented by the thermal motions of the atomic particles whose speeds are of order of the speed of sound v_s. We can therefore write

$$\frac{GM}{R} \sim v_s^2, \tag{8-132}$$

where G is the gravitational constant, M is the mass, and R is the radius of the star. We can now estimate a very rough order of magnitude of the stellar vibration frequency, by noting that the period $P_{\rm vib}$ should be comparable to the time it takes to transmit information about pressure changes across the entire dimension of the star. This time equals $2R/v_s$, and we can write

$$P_{\rm vib}^{-1} = \nu_{\rm vib} \sim \frac{v_s}{2R} \sim \sqrt{\frac{GM}{4R^3}} \sim \sqrt{G\rho}, \tag{8-133}$$

where ρ is the density of the stellar material.

Accordingly, the vibrational periods for neutron stars ought to be about one-tenth of a millisecond, and the vibration period for white dwarfs should be of the order of a second. These stars can have a wide range of densities, and only representative periods are given in Table 8.5. These can vary by over an order of magnitude.

The maximum rotational frequency is determined by the equilibrium between centrifugal and gravitational forces, since at higher frequencies the star would be torn apart:

$$R\nu_{\text{rot, max}}^2 = \frac{GM}{(2\pi R)^2} \qquad (8\text{-}134)$$

and

$$\therefore P_{\text{rot, min}}^{-1} \sim \frac{1}{\pi}\sqrt{G\rho}. \qquad (8\text{-}135)$$

Table 8.5. Approximate Relation Between Stellar Density, Vibrational Period, and Minimum Rotational Period.

Star	Density ρ	P_{vib}	$P_{\text{rot,min}}$
Neutron star	$10^{15}\ \text{g cm}^{-3}$	$10^{-4}\ \text{s}$	$3 \times 10^{-4}\ \text{s}$
White dwarf	10^7	1	3
Central core of Sun	150	$10^{2.5}$	10^3
Brown dwarf	10	10^3	$10^{3.5}$
RR Lyrae star	10^{-2}	$10^{4.5}$	10^5
Cepheid variable	10^{-6}	$10^{6.5}$	10^7

When the Crab Nebula pulsar was discovered to have a period of only 33 ms, it became clear that white dwarfs could not be considered to play a role in the pulsar phenomenon because the vibration frequencies of white dwarfs would be too low. On the other hand, the rotational period of a neutron star could well be several milliseconds. The discovery that the period of the Crab Nebula pulsar is increasing, so that it could have been closer to the minimum expected rotational period of a neutron star shortly after the supernova explosion in the year 1054 AD, further supported the theory that pulsars are rapidly rotating neutron stars that are losing angular momentum and slowing down as time goes on. Because magnetic pressures would tend to disrupt rapidly rotating neutron stars, the actual minimum rotation period will be somewhat greater than shown in Table 8.5.

Table 8.5 also shows that RR Lyrae variables and Cepheid variables have periods consistent with the very simple vibration picture discussed in the present section. Such stars are indeed pulsating. This can be judged by the periodic Doppler line shifts and color temperature changes. The observed periodicity is the pulsation period. The nova remnant DQ Her (Nova Herculis, 1934) has a 71 s period, and periodic behavior has been seen in some white dwarfs, although these periods are too long to represent fundamental pulsations. For the Sun the vibrational period observed in solar oscillations fits nicely into Table 8.5. The rotation period, $\sim 10^6$ s in the interior is well above the minimum.

In recent years, an increasing volume of information on the Sun has become available through *helioseismology*, the study of the Sun's vibrations. Section 6:25 taught us to think of the waves permeating a medium in terms of a superposition of spherical harmonics. In the Sun the waves are acoustic and have both surface components like those sketched in Fig. 6.18 and radial components, respectively corresponding, very roughly, to the functions Y and R in Section 6:25. The function R, however, becomes more complex, taking account of the Sun's gravitational potential, density, and pressure at each radial distance from the center. Other factors enter as well. We will deal more quantitatively with such waves in Section 13:8, where we shall be concerned with acoustic waves in primordial density fluctuations in the early Universe.

By conducting careful studies of brightness variations across the solar surface, over a wide range of both spatial and temporal scales, it has become possible to determine the magnitudes of many of the harmonics constituting the Sun's oscillations. These yield information on the density profile of the Sun, the rotation in its interior, the structure of the convective zone, and many other characteristics. Much of the knowledge previously gained through studies of stellar nucleosynthetic models alone has now been independently analyzed, partly confirmed, and partly revised and enriched. Efforts to bring the tools of helioseismology to bear on observations of other stars are still quite preliminary. *Stellar seismology* is difficult, because surface features on stars cannot be resolved with nearly the same resolution with which solar features are studied.

8:21 Solar Neutrino Observations

Thus far we have described the currently expected sequence of nuclear processes that may be taking place in the interior of stars, and have identified these with phases in the life of a star as it journeys through the Hertzsprung–Russell diagram. Because the sequence of events, the variety of possible reactions, and the number of assumptions required are so numerous, direct verification of the postulated nuclear reactions would be highly desirable.

The most promising observations that can be made in this respect are measurements on neutrinos emitted in the nuclear reactions. As already indicated *neutrinos* carry off somewhere around 2 to 6% of the hydrogen-to-helium conversion energy, depending on whether the proton–proton reaction or the CN cycle predominates. The neutrinos can escape from a star, virtually without hindrance, because in a $1 \, \mathrm{cm}^2$ column, one stellar radius deep, the neutrino would typically encounter only $\rho R/m_H \sim 10^{35}$ nuclei, where $\rho \sim 1 \, \mathrm{g \, cm^{-3}}$, $R \sim 10^{11} \, \mathrm{cm}$, and $m_H \sim 10^{-24} \, \mathrm{g}$. Because the neutrino interaction cross-section with nuclei normally is of order $10^{-45} \, \mathrm{cm}^2$, only one neutrino in 10^{10} would be intercepted on its way out of the star.

Early attempts to directly observe neutrinos from the Sun were carried out by Raymond Davis, Jr. and co-workers at the Brookhaven National Laboratory (Da68). Their experiment was based on the large absorption cross-section for neutrinos exhibited by the chlorine isotope $^{37}\mathrm{Cl}$ in the reaction

$$^{37}\mathrm{Cl} + \nu \rightarrow \ ^{37}\mathrm{Ar} + e^-. \tag{8-136}$$

This reaction requires a minimum neutrino energy of 0.81 MeV. The argon isotope ^{37}Ar is radioactive and makes itself evident through a 34-day half-life, 2.8 keV *Auger* (*X-ray*) transition which can be recorded. The reaction cross- section for this process is only large for high-energy neutrinos, however, and so not all of the neutrinos emitted by the Sun could be counted in this way. The experimenters figured they would only be able to observe the neutrinos from the decay of the boron isotope ^{8}B, which is formed in very small quantities according to the predictions of nuclear theory. The neutrino given off in that decay can have energies as high as 14 MeV. The scheme, which first gives rise to boron, is shown in Fig. 8.19, together with the

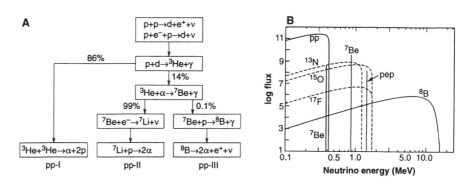

Fig. 8.19. (A) Nuclear reactions in the Sun, showing relative probabilities for the different reactions and the energy of neutrinos produced. The branching ratios are temperature dependent. The ratios shown are expected at temperatures of $\sim 1.5 \times 10^7$ K at concentrations $X = 0.726$, $Y = 0.26$, $Z = 0.014$ (Ba72). (B) Neutrino spectrum expected from the Sun in the absence of *neutrino oscillations* (Ra95). Reprinted with permission from *Science* ©1995 American Association for the Advancement of Science.

probability for the occurrence of competing reactions. This pioneering experiment consistently measured neutrino fluxes that were roughly one-quarter of the flux predicted by solar models, and constituted a major puzzle for several decades, until the possibility that neutrinos might have rest–mass came to be taken seriously.

Neutrinos come in three varieties called *flavors*, ν_e, ν_μ, and ν_τ, respectively associated with electrons, muons, and τ leptons. Neutrinos with mass oscillate among themselves, meaning that an electron neutrino can convert itself into a muon- or tau-neutrino on its way out from the Sun, and before arrival at Earth, through the so-called Mikheyev–Smirnov–Wolfenstein (MSW) effect (Ra95). Because the chlorine detectors were set up to detect solely electron neutrinos, they were detecting only a fraction of the electron neutrinos actually emitted.

By now, several different types of neutrino observatories designed to detect neutrinos in different energy ranges are in operation. Referring to the three channels shown in Fig. 8.19(a), the results from these observatories confirm a proton–proton solar neutrino flux 1.01 ± 0.02 times the theoretical predictions; a ^{7}Be neutrino flux measured at $0.97^{+0.28}_{-0.54}$ the predicted level; and a ^{8}B flux at 1.01 ± 0.06 times that predicted by theory (Ba03).

Experiments conducted with antineutrinos traveling of the order of 200 km from nuclear power reactors to an antineutrino detector in the Kamioka mines, near Tokyo in Japan, have shown that roughly 60% of the generated antineutrinos fail to arrive (Eg03). Oscillations are favored when the mass differences between the different antineutrino (or equivalently the different neutrino) flavors are small. The best estimates of the differences in masses of any two flavors, say i and j are $\Delta m_{ij} \equiv |m_i^2 - m_j^2| \sim 7 \times 10 - 5$ eV^2. The absolute masses of the neutrinos are still unknown. Upper limits based on cosmological observations discussed in Section 13:18 are $m_\nu \lesssim 0.23$ eV $\sim 4 \times 10^{-34}$ g.

PROBLEM 8–6. A supernova exploded in the Large Magellanic Cloud on February 23, 1987. After traversing the distance of 50 kpc, nine neutrinos were detected at Earth within an interval lasting 12.5 s. Their energies ranged from $\sim$35 MeV to $\sim$6 MeV (Hi87). What is the upper limit to the neutrino mass?

Additional Problems

The following set of problems uses greatly simplified stellar models, mainly to show that we can obtain reasonable, order-of-magnitude estimates for stellar luminosities and lifetimes even without using sophisticated computing methods.

Reaction rates for the proton–proton reaction and the carbon cycle have the form given in equation (8–90). Schwarzschild (Sc58b) gives the energy generated per unit time and unit mass of matter for the proton–proton reaction as

$$E_{pp} = 2.5 \times 10^6 \rho X^2 \left(\frac{10^6}{T}\right)^{2/3} \exp\left[-33.8 \left(\frac{10^6}{T}\right)^{1/3}\right] \text{ erg g}^{-1} \text{ s}^{-1} , \quad (8\text{-}137)$$

and Clayton (Cℓ68) gives a similar expression for the CN cycle as:

$$E_{CN} = 8 \times 10^{27} \rho X X_{CN} \left(\frac{10^6}{T}\right)^{2/3} f \exp\left[-152.3 \left(\frac{10^6}{T}\right)^{1/3}\right] \text{ erg g}^{-1} \text{ s}^{-1} ,$$

$$(8\text{-}138)$$

where f is an electron screening factor, $f \sim 1$.

For Problems 8–7 to 8–9, consider an initial concentration $X_{CN} = 0.005X$ and $Y = 0.24$.

8–7. Consider the following model of a B1 star, that is, a massive young star. Its mass is $M = 10 M_\odot$ and its radius $R = 3.6 R_\odot$. Its central density is 10 g cm^{-3} and the bulk of energy generation takes place within a radial distance $0.15R$. Assume constant density for this region and a temperature 2.7×10^7 K. Determine the rate of energy generation by the star and its surface temperature assuming that the star radiates like a blackbody at its given radius. How long can the star exist in its present state without being appreciably changed by nuclear burning; that is, how soon will it use up the hydrogen in its central core at the present rate?

8–8.   Repeat this for a star like the Sun when it first starts life on the main sequence. In this *zero-age main sequence ZAMS* phase assume that it is burning hydrogen in the central region out to $0.22R$. Take a central density of $\sim$75 g cm^{-3} and a central temperature 10^7 K. Assume that the star's radius at that epoch is only $R = 0.87R_\odot$ – 87% of the Sun's current radius (Bℓ99).

8–9.   At present the concentration of hydrogen, by mass, at the center of the Sun is about 70% throughout a central region out to roughly $0.122R$. The density is about 155 g cm^{-3} in the center. Take the mean temperature throughout the region to be 1.1×10^7 K and recalculate the parameters for this solar model (Bℓ99).

8–10.   A red-giant star whose radius is $100\,R_\odot$ is in an evolutionary state where the inner hydrogen has all been exhausted but helium burning has not yet set in. The main energy source is hydrogen burning that takes place in a shell surrounding the inert helium core (Figs. 8.8, 8.9). Let hydrogen burning take place at a radial distance ranging from 1.8 to 2×10^9 cm. The mean density in this layer is about 50 g cm^{-3}. The temperature is 5×10^7 K. Calculate the above parameters. Take $X_{CN} = 10^{-3}X$, $X \sim 0.5$.

8–11.   A star composed of hydrogen alone is limited to a maximum brightness. When its luminosity-to-mass ratio exceeds $4 \times 10^4 L_\odot/M_\odot$, its surface layers are blown off. This *Eddington limit* is exceeded in the ejection of *planetary nebula* envelopes. (a) Show that this occurs when the radiative repulsion of electrons exceeds the star's gravitational attraction for hydrogen atoms so that ionized hydrogen no longer is gravitationally bound. (b) Show that for a star predominantly composed of ^{4}He, ^{12}C, ^{16}O, or ^{28}Si the luminosity-to-mass ratio can be as high as $8 \times 10^4 L_\odot/M_\odot$ before an outer envelope is blown off.

8–12.   A white dwarf star is thought to shine by virtue of its stored thermal energy. Assuming its mass to be $0.45M_\odot$, its radius $0.016R_\odot$, and its density roughly uniform throughout (however, see Fig. 8.18), calculate how long the star could radiate at its present rate if its luminosity is $10^{-3}L_\odot$.

8–13.   For a star in which radiative transfer dominates, the energy density at each point is very nearly the blackbody radiation density (at all frequencies ν) even though the radiant flux $F(r)$ is predominantly outward-directed from the center of the star. Show how closely the radiation density ρ really equals the blackbody radiation density for a given opacity value $\kappa(\nu)$, by considering the higher-order terms in equation (8–50).

8–14.   In Section 8:2 we argued that the Sun's energy must be nuclear, because equation (8–1) did not provide enough potential energy to account for the total solar luminosity over the past æons. Show that this argument could be refuted, by working out the potential energy of a structure in which roughly half the Sun's mass is uniformly distributed throughout a sphere of radius $R_\odot$ and half the mass is con-centrated in a core of radius 10 km. Although the observed oscillations of the Sun rule out such a model, it is possible that some evolved stars have a dense, degenerate neutron core. Such stars are known as Thorne–Żytkow stars (Th75).

8–15. The *X-ray source* Sco X-1 is believed to be a neutron star . A plasma with electron density $\sim 10^{16}$ cm^{-3} radiates at an apparent temperature of 5×10^7 K from a volume of radius 10^9 cm that encompasses the star's surface and accretion disk. Matter is tidally stripped from a companion star to fall onto the disk from which it spirals onto the surface of the neutron star, giving up potential energy at each step as it approaches the star. To account for the fast accretion rate of matter and the high radiation intensity it produces, this matter would have to be rapidly siphoned off the binary companion. Compute the infall rate required, and convince yourself that the energies produced are of the right order of magnitude by comparing gravitational to thermal energies (Ka97).

8–16.   Figure 8.16 indicates that neutron stars appear to have masses of order $M_{ns} = 1.4 M_\odot$, as judged by the gravitational potential they exert in a binary sys-tem. Assuming that such a star consists entirely of neutrons, show that the number of neutrons in the star appreciably exceeds M_{ns}/m_N, where m_n is the neutron rest–mass, i.e., that the *baryonic mass* of the star appreciably exceeds its *gravitational mass* because the potential energy of the star is substantial (Aℓ04). Following this argument to its extreme, why does the gravitational mass of a black hole not vanish entirely?

Answers to Problems

8–1. (a) $P = nkT$

$$P\left(\frac{R}{2}\right) \sim 10^{15} \, \text{dyn cm}^{-2}, \qquad n = \frac{\rho}{m_H} = \frac{1}{1.67 \times 10^{-24}} \text{cm}^{-3},$$

$$T\left(\frac{R}{2}\right) = \frac{10^{15} \times 1.67 \times 10^{-24}}{1.38 \times 10^{-16}} \sim 10^7 \, \text{K}.$$

(b) The gravitational potential energy at the planet's surface is

$$\frac{GM_p m}{R} \le |\mathcal{E}_c| \,,$$

where m is the mass of an atom, $m = Am_H$. For domination by Coulomb forces

$$M_p \lesssim \frac{|\mathcal{E}_c| R}{GAm_H} \,.$$

But $M_p \sim \rho R^3$

$$\therefore \; M_p^{2/3} \lesssim \frac{|\mathcal{E}_c|}{GA\rho^{1/3} m_H} \,, \qquad M_p \lesssim \left(\frac{|\mathcal{E}_c|}{GAm_H}\right)^{3/2} \frac{1}{\rho^{1/2}} \,.$$

(c) $\mathcal{E}_c \sim 10^{-11}$ erg, $R_J = 7.1 \times 10^9$ cm, $\left.\begin{array}{l} \\ \end{array}\right\}$ see Table 1.4.

$\qquad M_J = 2 \times 10^{30}$ g,

Assuming a constant density throughout, this yields $\rho_J \sim 1.344$ g cm^{-3}, and from (8–22) a central pressure $P \sim 1.3 \times 10^{13}$ dyn cm^{-2}. If the main constituent at the planet's center were atomic hydrogen, the ideal gas law would require $kT = Pm_H/\rho \sim 1.6 \times 10^{-11}$ erg, high compared to the dissociation energy of H_2 and comparable to the ionizing energy for hydrogen atoms.

8–2.

$$F(r) = -\int_0^\infty \kappa\rho e^{-\kappa\rho l}\, dl \int_0^\pi a\left(T - \frac{dT}{dr}l\cos\theta\right)^4 \frac{c}{2}\cos\theta\, d(\cos\theta)\,.$$

Expanding, we have to first order in dT/dr,

$$\simeq \frac{4ac\kappa\rho}{3}\frac{dT}{dr}T^3\int_0^\infty e^{-\kappa\rho l}l\, dl\,.$$

And using (8–26) and (8–51) we have

$$L(r) = \frac{16\pi acr^2}{3\kappa\rho}T^3\frac{dT}{dr}\,.$$

8–3. $dP \propto \rho\dfrac{M(r)\,dr}{r^2}$ (8–7),

$P = nkT$ so that, for roughly constant $\rho(r)$, $dP \propto \rho\, dT$,

$\therefore \dfrac{dT}{dr} \propto M(r)/r^2$ and $T(r) \propto M(r)/r$,

$L(R) \propto \left.\dfrac{R^2T^3}{\rho}\dfrac{dT}{dr}\right]_{r=R} \propto \dfrac{M^4}{\rho R^3} \propto M^3$.

8–4. The number of particles incident on a hypothetical surface in unit time and solid angle is

$$n(\theta, \phi, p)v\cos\theta\, d\Omega\,.$$

From (4–27), the pressure then is

$$P = \int_0^{p_0}\int_0^{2\pi}\int_0^{\pi/2} 2\cdot p\cdot\cos\theta\cdot v\cos\theta\cdot n(\theta, \phi, p)\, d\Omega\, dp\,.$$

If the gas is isotropic,

$$n(\theta, \phi, p) = \frac{n(p)}{4\pi}\,,\qquad P = \frac{1}{3}\int_0^{p_0} pvn(p)\, dp\,,$$

where $n(p)$ is given by (4–65) since all states are filled. Equation (5–30) tells us that $p = mv\gamma(v)$ which can be solved for v:

$$v = \frac{p}{m\sqrt{1 + p^2/m^2c^2}} \, ,$$

$$\therefore \ P = \frac{8\pi}{3mh^3} \int_0^{p_0} \frac{p^4}{\sqrt{1 + p^2/m^2c^2}} dp \, .$$

If $p/mc = \sinh u$, $\quad dp/du = mc \cosh u$, so that

$$P = \frac{8\pi m^4 c^5}{3h^3} \int_0^{u_0} \sinh^4 u \, du \, .$$

8–5. Equations (8–112) and (8–110), respectively, read

$$f(x) = \frac{1}{8}\left[x(2x^2 - 3)(x^2 + 1)^{1/2} + 3 \sinh^{-1} x\right], \qquad x = \frac{p_0}{m_e c} \, .$$

If $x \ll 1$, $\quad \sinh^{-1} x = x - \frac{x^3}{6} + \frac{3}{40}x^5 - \cdots$.
On expanding

$$8f(x) \approx x(2x^2 - 3)\left(1 + \frac{x^2}{2} - \frac{x^4}{8}\right) + 3x - \frac{x^3}{2} + \frac{9}{40}x^5 - \cdots$$

$$\approx \frac{8}{5}x^5, \qquad \text{for } x \ll 1 \, .$$

Substitution of $f(x)$ and (8–111) into (8–109) then gives (8–41).
If $x \gg 1$, $\quad \sinh x \sim e^x/2$, and $\quad \sinh^{-1} x = \ln(2x)$.

$$\therefore \ f(x) \sim \frac{x^4}{4}, \quad x \gg 1 \, ;$$

substitution in (8–109) leads to (8–43).

8–6. Let the energies of arriving neutrinos range from

$$\mathcal{E}_1 = \frac{m_\nu c^2}{(1 - (v_1^2/c^2))^{1/2}} \quad \text{to} \quad \mathcal{E}_2 = \frac{m_\nu c^2}{(1 - (v_2^2/c^2))^{1/2}} \, .$$

Then

$$m_\nu^2 c^6 \left| \frac{1}{\mathcal{E}_1^2} - \frac{1}{\mathcal{E}_2^2} \right| = |v_2^2 - v_1^2| \sim 2c\Delta v \, ,$$

where $\Delta v \equiv |v_2 - v_1|$. This velocity difference is related to the neutrino arrival times by $\Delta v = (D/t_2 - D/t_1) = [(t_1 - t_2)/t^2]D = c^2\Delta t/D$.

$$\therefore \ m_\nu^2 c^4 \left| \frac{1}{\mathcal{E}_1^2} - \frac{1}{\mathcal{E}_2^2} \right| = \frac{2c\Delta t}{D} \, .$$

For $\mathcal{E}_2 \gg \mathcal{E}_1 = 6$ MeV, $D = 50$ kpc, and $\Delta t = 12.5$ s, we then have $m_\nu \sim \mathcal{E}_1(2\Delta t/Dc^3)^{1/2} \sim 13.5 eV \sim 2.4 \times 10^{-32}$ g. This is an upper limit because the period over which neutrinos are emitted could last several seconds.

8–7. We use a temperature 2.7×10^7 K and density $\rho = 10\,\mathrm{g\,cm^{-3}}$ in equations (8–137) and (8–138):
This gives

$$E_{pp} = 20\,\mathrm{erg\,g^{-1}s^{-1}}, \qquad E_{CN} \sim 2.3 \times 10^3\,\mathrm{erg\,g^{-1}s^{-1}}.$$

$\therefore$ The total energy generated is $(4\pi/3)\rho E_{CN}(0.54R_\odot)^3 \sim 5 \times 10^{36}\,\mathrm{erg\,s^{-1}}$. The total surface area of the star $4\pi(3.6R_\odot)^2 \sim 8 \times 10^{23}\,\mathrm{cm^2}$. Hence the flux crossing unit area is $6.4 \times 10^{12}\,\mathrm{erg\,cm^{-2}\,s^{-1}}$ and, using the blackbody law, $T^4\sigma$ equals flux across unit area, $T \sim 18,000$ K.

The total available energy per hydrogen atom is about 10^{-5} erg with $n \sim 6 \times 10^{24}\,\mathrm{cm^{-3}\,s^{-1}}$. Hence the total energy available per cubic centimeter is about 6×10^{19} erg. The total time during which the star's core can supply energy, therefore, is about 9×10^7 yr.

8–8. With a central temperature 10^7 K, density $75\,\mathrm{g\,cm^{-3}}$, and hydrogen burning out to $0.22 \times 0.87R_\odot$, we find that the proton–proton reaction predominates and yields $260\,\mathrm{erg\,cm^{-3}}$. This leads to a surface temperature of about 5600 K, obtained by dividing the star's luminosity by its surface area and the Stefan–Boltzmann constant, and taking the fourth root.

8–9. Using $X = 0.7$ and a central density $\rho_c = 155\,\mathrm{g\,cm^{-3}}$ with $T_c = 1.1 \times 10^7$ K, we again find that the proton–proton reaction predominates, yielding $\sim 1500\,\mathrm{erg\,cm^{-3}\,s^{-1}}$. The total energy generated is $\sim 3.8 \times 10^{33}\,\mathrm{erg\,s^{-1}}$. This gives a surface temperature of $T = [(4\pi/3)\rho_c(0.122R_\odot)^3 E_{pp}/(S_\odot\sigma)]^{1/4} \sim 5800$ K, where σ is the Stefan–Boltzmann constant and $S_\odot$ is the surface area $4\pi R_\odot^2$.

8–10. At 5×10^7 K the carbon cycle predominates. If we assume burning in a thin shell ranging from 1.8×10^9 to 2×10^9 cm, we obtain a carbon-cycle energy generation rate of $4 \times 10^8\,\mathrm{erg\,cm^{-3}\,s^{-1}}$ throughout a volume of $10^{28}\,\mathrm{cm^{-3}}$. The luminosity therefore is $\sim 4 \times 10^{36}\,\mathrm{erg} \sim 10^3 L_\odot$ and the surface temperature at $R = 7 \times 10^{12}$ cm is $T = (L/4\pi R^2\sigma)^{1/4} = 3200$ K.

8–11. The radiative repulsion of an electron at distance R from a star is $L\sigma_e/4\pi R^2 c$, where σ_e is the Thomson cross-section (6–103). The gravitational attraction for a hydrogen atom is $m_H(MG/R^2)$. For the Sun, this ratio of repulsion to attraction is

$$\frac{\sigma_e L_\odot}{4\pi c M_\odot G m_H} \sim 3 \times 10^{-5}.$$

For a star with $L/M \sim 3 \times 10^4$ greater than the Sun, electrons are repelled and pull the protons along. For ^{4}He, ^{12}C, and other fully stripped heavy ions, each electron pulls along a proton plus a neutron, and twice this luminosity-to-mass ratio is needed to exceed the Eddington limit.

8–12. A white dwarf can only radiate away the kinetic energy of its ions. Because the electrons are degenerate, they cannot lose energy and provide the main support against hydrostatic pressures. Just before electron degenerate pressures start pre-dominating, the kinetic energy of the ions is about one-tenth of the white dwarf's potential energy. The virial theorem would predict that half of the energy should be kinetic energy, but there will be two electrons or more for each ion present, and the electrons will have higher energies because of the onset of partial degeneracy. The total available ion energy therefore is of order $0.1 M^2 G/R$, and the lifetime $\sim 0.1 M^2 G/RL$. It may still be worth stating how the luminosity can be derived. The nondegenerate outer layers of the white dwarf permit radiative transfer. Using equation (8–52) with $T \sim 0.1 MGm_i/kR$ and $dT/dr \sim T/R$, we obtain the lu-minosity if the opacity is known. The opacity can be computed from the Kramers' expressions, although a look at Fig. 8.4 shows that at high densities the opacity is nearly independent of density and has a very approximate value $\sim 10^8/T$ in the tem-perature range of interest. The cooling time can therefore be expressed solely as a function of the star's mass, radius, and chemical composition (ion mass). If the lu-minosity for a star of mass $0.45 M_\odot$ is $\sim 10^{-3} L_\odot$, and its radius is 1.1×10^9 cm, $\tau \sim 0.1 M^2 G/RL \sim 1.2 \times 10^{18}$ s, and the cooling time is of the order of 40 Gyr. This problem is discussed more rigorously in (Sc58b).

8–13. In Problem 8–2 we derived the expression equivalent to

$$F(r) = \frac{4ac}{3\kappa\rho} T^3 \frac{dT}{dr}$$

by neglecting higher-order terms, justified since $(dT/dr)x \ll 1$. The next order term in the expansion is

$$\frac{4ac\kappa\rho}{5} \int_0^\infty e^{-\kappa\rho l} T \left(\frac{dT}{dr}\right)^3 l^3 \, dl$$

$$= \frac{4ac}{(\kappa\rho)^3} \frac{T}{5} \left(\frac{dT}{dr}\right)^3 \int_0^\infty e^{-y} y^3 \, dy = \frac{24Tac}{5(\kappa\rho)^3} \left(\frac{dT}{dr}\right)^3 .$$

Integrating this over all radii r gives the additional energy increment

$$\rho(\nu) = \int_0^r 4aT \left(\frac{dT}{dr}\right)^3 \frac{18}{5(\kappa\rho)^2} \, dr .$$

This is the departure from blackbody energy density in a star. Hence

$$\frac{d\rho}{dr} = 4a \left[T^3 \frac{dT}{dr} + T \left(\frac{dT}{dr}\right)^3 \frac{18}{5} \frac{1}{(\kappa\rho)^2} \right]$$

$$= \frac{d\rho(\nu)}{dr} + \frac{d\rho^*(\nu)}{dr} .$$

8–14. For a uniformly dense, spherical central core, the potential energy is given by (8–1):

$$V = \frac{3}{5}\frac{GM^2}{R} \sim 4 \times 10^{52}\,\text{erg} \quad \text{for} \quad R = 10^6\,\text{cm}, \quad M = 10^{33}\,\text{g}\,.$$

Such a star could have been shining since the first stars formed $\sim 1.3 \times 10^{10}$ yr ago with a luminosity of 10^{35} erg s^{-1} $\sim 25 L_\odot$ without running out of gravitational energy.

8–15. The energy of a proton freely falling onto a neutron star's accretion disk is $mMG/R \sim 10^{-7}$ erg for $R \sim 10^9$ cm. This is $\sim 10^5$ eV and X-rays up to this energy can therefore be given off by the protons. The observed X-ray energy at 5×10^7 K is 5×10^3 eV. If the gas were to fall directly onto the neutron star, whose radius is only of order 10^6 cm, it could radiate at higher temperatures. However, the material slowly spirals in, giving up energy more gradually. When the hydrogen eventually reaches the neutron star surface, it builds up a layer that progressively thickens to about one meter before undergoing a thermonuclear flash, observable as a huge outburst of X-rays.

8–16. The gravitational mass of a neutron star containing N_n neutrons is

$$M_{ns} = N_n m_n + \epsilon/c^2\,,$$

where ϵ is the sum of the star's (negative) potential energy, internal energy, and rotational energy. Among these, the potential energy tends to dominate. To lowest order, given by the nonrelativistic equation for a star of uniform density, (8–1), it is of order $-3(N_n m_n)^2 G/5R_{ns}$. Since the neutron star radius R_{ns} exceeds the Schwarzschild radius $r_s = 2N_n m_n G/c^2$ by only a factor of order 2, we see that $\epsilon/c^2 \sim -0.3 N_n m_n$. The internal energy for a degenerate neutron star is roughly determined by the mean volume occupied by each neutron, which yields the Fermi energy $\mathcal{E}_F$. For radii of order $2r_s$, $\mathcal{E}_F$ is a small fraction of the rest–mass. Even the most rapidly rotating neutron stars do not have rotational energies coming close to the potential energy. The gravitational mass of the star may therefore be no more than $\sim 70\%$ of the rest–mass of the contributing nuclei, meaning that a neutron star with mass $\sim 1.4 M_\odot$ may contain $N_m \sim 2M_\odot/m_n$ neutrons. Without an equation of state for neutron star matter, one cannot say much more. A high central density leads to a rapid increase in internal energy, or equivalently P/ρ, which reduces the absolute value of ϵ but also brings the star closer to gravitational collapse, as indicated by equations (8–124) to (8–129).

Black holes differ from cold neutron stars in that the constituents are not at rest and need not have given up any potential energy. In principle, a black hole could be formed through a spherically symmetric collapse of photons with a total energy E into a volume of radius $r < 2GE/c^4$. Though the photons would have given up no potential energy in this *adiabatic advection*, they would be irreversibly trapped by their mutual gravitational pull.

9 Cosmic Gas and Dust

The word *astrophysics* implies a study of stars. But the past decade has clearly brought out that most of the baryonic — i.e., atomic — matter in the Universe is gaseous, permeating the vast spaces between galaxies. Within galaxies, gas clouds are less prevalent by mass but they, and the dust grains swept along by them play a crucial role in the formation of stars and planets.

In this chapter we establish a common framework within which processes in clouds of gas and dust can be understood. We start with a brief phenomenological description that makes use of many of the properties of radiation and matter derived in Chapters 6 and 7. With this depiction in hand, we develop the dynamics governing the evolution of gaseous bodies, in order to prepare ourselves for later chapters in which we will examine how stars and planetary systems form and how galaxies originated early in the evolution of the Universe.

9:1 Observations

(a) The Intergalactic Medium

(i) Radiation

Some of the most carefully conducted observations in hand tell us that the early Universe was hot and completely ionized. Matter and radiation were in thermal equilibrium. As the Cosmos expanded, it eventually cooled to 4000 K; electrons and protons combined to form neutral hydrogen atoms. Thermal radiation that had been intimately coupled to matter through Thomson scattering became free to traverse the Universe. The continuing expansion progressively shifted radiation to longer wavelengths. Persuasive evidence for this history is recorded in the blackbody microwave background radiation permeating the Universe today at a temperature of 2.725 ± 0.001 K (Fi02). The radiation field is isotropic, but a direction-dependent Doppler shift in its spectrum tells us that the Sun is moving through the radiation with a velocity of 371 ± 1 km s^{-1} in a direction marked by Galactic coordinates $(l, b) = (263.85° \pm 0.1, 48.25° \pm 0.04)$. In this coordinate system the Galactic center lies at $(l, b) = (0, 0)$ and the Galactic plane is the plane $b = 0$. Fluctuations of the radiation temperature at a level of one part in $\sim 10^5$ mapped across the sky tell us how matter was distributed at early times, and how it gravitationally collapsed to form the clusters of galaxies ubiquitous today. We will pursue this history in greater detail in Chapter 13. For now, we merely note it as the setting within which all other processes need to be viewed.

(ii) Hydrogen

The Thomson scattering optical depth through which the microwave background has passed on its way to Earth is estimated from the polarization that scattering produces, and is found to be $\tau = 0.17 \pm 0.04$. The scattering originates in gas that became at least partially reionized once the first massive stars were formed when the Universe was about 100 million years old (Be03). Another $\sim 10^9$ yr later , after the first quasars were born, the intense ionizing quasar emission fully reionized intergalactic space. We know this, because we see no traces of spectral absorption features due to atomic hydrogen anywhere along the line of sight toward the most distant quasars, except in the immediate environs of the quasars themselves, or in isolated Lyman-α absorbing clouds, at least some of which lie in galaxies along the line of sight. The lack of any indication of absorption — the absence of this so-called *Gunn–Peterson effect* to be discussed in Section 13:28 — sets an upper limit of order 10^{-12} neutral hydrogen atoms per cubic centimeter in intergalactic space today (Gu65, So95).

(iii) Helium

In contrast to neutral hydrogen, singly ionized intergalactic helium does give rise to absorption. He^+ has a spectrum similar to that of atomic hydrogen but shifted to roughly four times shorter wavelengths (Section 7:3). The He^+ line corresponding to Ly-α lies at 304 Å; it gives rise to an absorption dip shortward of a wavelength of $304(z + 1)$ Å, where z is the red shift of the quasar against which the absorption is observed (Da96). Whether this[1] HeII is diffusely spread throughout intergalactic space or clumped in gaseous clouds is still uncertain, but the optical depth appears to be of order unity at red shifts $z = 3$. The oscillator strength for this helium line is $f \sim 0.55$ corresponding to a total absorption cross-section $\sigma \sim 0.01$ cm^2 over unit frequency interval (see Section 7:9). The line frequency is $\sim 10^{16}$ Hz, so that the column density of singly ionized helium in intergalactic space must be of order $\sim 10^{18}$ cm^{-2}. Because the distances traversed by radiation reaching us from distant quasars is $o(10^{28})$ cm the HeII number density appears to be of order 10^{-9} cm^{-3} at $z = 3$, and a factor $(z + 1)^3 \sim 64$ lower today. At these densities the recombination time would far exceed the age of the Universe, and we conclude that the intergalactic HeII density must be of order 10^{-11} cm^{-3} today. This low abundance makes it likely that the bulk of the helium is doubly ionized and resides in the Warm-to-Hot Intergalactic Medium (WHIM) discussed below.

(iv) Magnetic Fields

Upper limits on the extragalactic magnetic field can be derived from Faraday rotation measures. However, these assume a uniformly directed field (Section 6:12). If

---

[1] In this notation, neutral helium is denoted by HeI, singly ionized helium by HeII, and doubly ionized helium by HeIII. Ionized states of other atoms follow the same convention.

the field is randomly oriented over short distances, its strength could be substantially higher. A coarse upper limit — probably far too high — for a randomly directed field is $B \lesssim 10^{-6}$ G, though fields of this strength may actually exist in the dense central regions of clusters of galaxies. For a field that might be systematically aligned over large cosmological distances the upper limit is far more stringent, $\lesssim 10^{-9}$ G for fields stretching over regions of the order of a few megaparsec (Kr94).

(b) Intracluster Gas and the Warm-to-Hot Intergalactic Medium, WHIM

The intergalactic medium is not homogeneous. We observe massive, hot, X-ray-emitting gaseous clouds trapped in the gravitational potential wells of large clusters of galaxies. Their free–free emission, Sunyaev–Zel'dovich effect (Section 6:23) and X-ray spectra tell us that their temperatures lie in the 10^7–10^8 K range, and their iron abundances, relative to hydrogen, are roughly one third as high as seen in the Sun. The intracluster gas, however, appears to constitute only a small portion of the entire extragalactic gaseous component. Ten times more prevalent is the WHIM.

Absorption by six-times-ionized oxygen OVII observed in the X-ray domain indicates the existence of massive, filamentary intergalactic clouds at temperatures of $\sim 10^6$ K, a warm-to-hot intergalactic medium stretching over distances of the order of megaparsecs. The observed spectral features indicate that these filaments have a fractional oxygen abundance, relative to hydrogen, about a factor of 30 lower than in the Sun (Ni05). These filaments, if ubiquitous, embrace a total baryonic mass exceeding that of all stars in the Universe by about an order of magnitude (Fu04). They indicate that the ionized hydrogen density averaged over all intergalactic space amounts to $n_P = n_e \sim 2.5 \times 10^{-7}$ cm^{-3} today, or $\sim 4 \times 10^{-31}$ g cm^{-3}. Given the highly ionized state of oxygen, we must conclude that most of the helium is fully ionized, unable to exhibit spectral lines.

(c) Lyman-α Absorbers

Optical spectra obtained along the line of sight to distant quasars show an abundance of red-shifted Lyman-α absorption lines, often referred to as the *Lyman-α forest*. The inferred column densities of hydrogen atoms derived from the depth of the absorption lines range from $\lesssim 10^{12}$ to 10^{22} cm^{-2} per absorbing cloud. The number of high-density clouds observed declines roughly in proportion to column density (So95). The clouds are appreciably more abundant at high red shifts than in our local vicinity and may be protogalactic or represent early galaxies. They can exhibit strong MgII and SiII absorption. Absorption by deuterium atoms has been observed in clouds red-shifted to $z \sim 3$. Their abundances relative to hydrogen range from 2×10^{-5} to 2×10^{-4} (Sc96), (We97).

(d) Quasars, Blazars, and Active Galactic Nuclei, AGN

Quasi-Stellar Objects, QSOs — quasars and BL Lacertae objects — are highly luminous, compact, extragalactic sources. Quasars exhibit strong optical emission lines whereas BL Lacertae sources have a bland continuum spectrum. The quasar optical-line emission comes from highly excited ions in nebulosity with an electron density of order $3 \times 10^6 \, \text{cm}^{-3}$ and a radius of the order of 1 pc. Although all quasars appear to be associated with host galaxies, the quasar emission is so luminous that the relatively faint host galaxies are detected with difficulty. A quasar's radio emission can be strong, suggesting synchrotron radiation by relativistic particles in a core with high magnetic field strength (Section 6:21). A number of galaxies, such as the *Seyfert galaxy* NGC 4151, the radio source Centaurus A, the BL Lac object 3C 279, and the quasar 3C 273, all are powerful sources of X-ray emission (Fig. 9.1). So

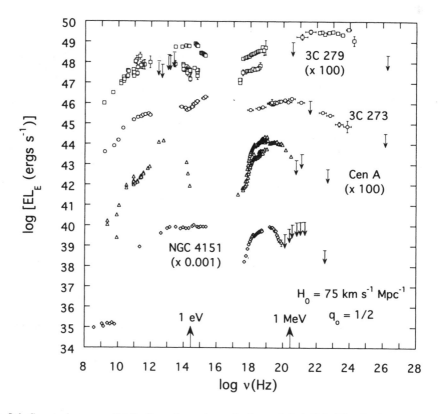

Fig. 9.1. Spectral energy distributions for active galactic nuclei, AGNs. The Seyfert galaxy NGC 4151, the powerful radio galaxy Cen A, the quasar 3C 273, and the blazar 3C 279, all have active galactic nuclei, AGNs (De95). The ordinate gives the deduced luminosity per natural logarithmic energy interval; the abscissa indicates the spectral frequency observed.

also is the elliptical galaxy M87, the first X-ray galaxy to be discovered. The X-ray flux from NGC 4151 amounts to $\sim 10^{43} \, \text{erg s}^{-1}$, and 3C 273 emits $1.5 \times 10^{46} \, \text{erg}$ s^{-1}, while 3C 279 can emit $\geq 3 \times 10^{47} \, \text{erg s}^{-1}$ at X-ray frequencies and, like NGC 4151, is highly variable. These objects emit as much energy at X-ray frequencies as at all others combined, and the flux from NGC 4151 is comparable to the visible flux emitted by normal spirals in the form of starlight. Many quasars and Seyfert galaxies are also powerful emitters of infrared radiation. Their primary radiation appears to be absorbed by dense ambient dust clouds and re-emitted at infrared wavelengths.

These galaxies and quasi-stellar objects, all of which have active nuclei apparently housing a massive black hole, are complex sources that may radiate by means of a variety of different mechanisms. Figure 9.1 shows representative spectra. Whether a given component of their radiation is due to thermal emission (Section 6:19), inverse Compton scattering (Section 6:22), synchrotron radiation (Sections 6:20 and 6:21), or some other mechanism, can be difficult to determine. Some of the sources show significant variability on a time scale of hours. Others vary little, or less rapidly. As discussed in Section 6:21 some mechanisms of emission have intrinsically faster onset or decay times than others, and the fluctuation rate may therefore permit us to decide which spectral ranges of emission are connected by one and the same emission mechanism and what that mechanism might be.

(e, f) Galaxy Spiral Arms and the Interarm Medium

In the Galaxy and in some extragalactic objects, neutral hydrogen can be observed through absorption or emission of atomic hydrogen at a wavelength of 21 cm. We can also observe Ly-α absorption in light emitted by O and B stars. These two types of data do not always agree, but the indications are that neutral hydrogen number densities in our part of the Galaxy are of order 0.1 to 0.7 cm^{-3}. Between the spiral arms, the density is lower (Je70), (Ke65). Molecular hydrogen is most readily detected in shocked regions, where temperatures are sufficiently high for molecules to be collisionally excited into higher rotational and vibrational states, from which they return to the ground state by emitting radiation at well-defined infrared spectral wavelengths (Section 7:5). In colder molecular clouds these molecules cannot be excited; their presence is inferred from carbon monoxide emission. The $J = 1$ rotational state of CO is collisionally excited at temperatures only a few degrees Kelvin above absolute zero, and can serve as a tracer of molecular hydrogen on the assumption that the two gases are everywhere mixed in roughly constant abundance ratios.

The electron number density is determined from the dispersion measures (6–58) of pulsar radiation. The electron density thus obtained is $\sim$0.03 cm^{-3} averaged over the arm and interarm domains in our locale of the Galactic disk.

(g, h) HII Regions and Planetary Nebulae

In fully ionized gases the electron temperatures and densities are readily determined. Free–free emission observed in the radio domain provides us with the temperature and (density)^2 integrated along the line of sight (Section 6:18). Visual and radio recombination-line data provide complementing information. The recombination rate can be computed using the bound–free absorption coefficient, α_{bf} (7–75).

The recombination cross-section for an electron at temperature T is then written as $Q_n(T)$ where n is the principal quantum number of the final state in the hydrogen-like ion.

Consider an idealized thermal equilibrium in which the number of ionizations to the nth level equal the number of recombinations. In velocity range dv, the recombination rate per electron is proportional to the electron's velocity $v \sim (3kT/m)^{1/2}$, to the cross-section $Q_n(v)$, and to the number densities $n_e(v)$ of electrons and n_{r+1} of ionized atoms. The ionization rate is proportional to the speed of light c, to α_{bf}, to the number density of atoms in the lower ionization state n_r, and to the number density of photons at a frequency ν sufficiently high to produce both ionization and an electron ejection velocity v. If χ_r is the ionizing energy,

$$\nu = \frac{1}{h}\left(\frac{m}{2}v^2 + \chi_r\right). \tag{9-1}$$

We can then write the equilibrium condition between ionization and recombination, very roughly, as

$$n_e(v)n_{r+1}vQ_n(v)\,dv = c \cdot n_r \cdot \frac{8\pi}{c^3} \frac{\nu^2 \alpha_{bf}(\nu)}{(e^{h\nu/kT} - 1)}\,d\nu, \tag{9-2}$$

by making use of the blackbody spectrum (4–72) for the number density of photons $\rho(\nu)/h\nu$. Use of the Saha equation (4–107) and the absorption coefficient given by expression (7–75) then leads to the relation

$$\frac{g_{r+1}g_e}{g_r} \frac{[2\pi mkT]^{3/2}}{h^3} e^{-\chi_r/kT} Q_n(v)v\,dv = \frac{8\pi}{c^3} \frac{64\pi^4 me^{10}Z^4}{3\sqrt{3}h^6 n^5} g_{bf} \frac{d\nu}{\nu[e^{h\nu/kT} - 1]}. \tag{9-3}$$

Using the relationship between variables v and ν in (9–1) and knowing that $g_r \propto n^2$ (see Problem 7–1) we obtain a relation for the recombination rate α_n for unit electron and ion density:

$$\alpha_n = \int_0^\infty vQ_n(v)\,dv \tag{9-4}$$

$$= \frac{g_r}{g_e g_{r+1}} \int_0^\infty \frac{2^9 e^{10}\pi^5 Z^4 me^{\chi_r/kT}}{c^3 h^3 n^5 [6\pi mkT]^{3/2}[e^{(\chi_r + mv^2/2)/kT} - 1]} \frac{d\left(mv^2/2\right)}{(mv^2/2 + \chi_r)},$$

where the fraction outside the integral is $\sim n^2$ (see Section 8:8).

For visible radiation equation (9–4) is considerably simplified since kT is small compared to χ_r so that the exponential dependence on χ_r can be neglected. The integral can then be expressed in approximate form (Za54):

$$\alpha_n = \frac{2.08 \times 10^{-11}}{T^{1/2}}\phi(T) \quad \text{cm}^3\,\text{s}^{-1}, \tag{9-5}$$

where $\phi(T)$ does not rapidly change with temperature. It has a value of 3.16 at $1,580\,\text{K}$ and 1.26 at $7.9 \times 10^4\,\text{K}$.

Although the thermal velocities of ions already lead to broadening of spectral lines, we can still determine bulk velocities of turbulent motion when these are high enough to lead to an actual split appearance of spectral lines, or if the thermal broadening contribution can be computed from independent temperature data.

Dust densities in such clouds can be determined by measuring the continuum radiation from the cloud in the visible or infrared part of the spectrum. Much of this radiation is likely to be starlight respectively scattered or absorbed and re-emitted by dust. Some assumption about particle size must then still be made, and ideally we should also know the chemical composition and physical structure of the grains. Judging from their spectra, likely candidates for grain composition are silicates, iron-containing minerals, or graphite grains.

(i) Supernova Remnants

The diameters of these remnants can stretch across many tens of parsecs and often display a circular arc structure. Doppler velocities indicating high expansion rates can be measured by spectral observations. For the Crab Nebula an actual expansion can be observed by a comparison of present-day and decades-old images. The expansion velocity for the Crab is of the order of 10^8 cm s^{-1}. Its radiation is strongly polarized along a direction perpendicular to the length of the continuum emitting wisps. Assuming that this comes from synchrotron radiation emitted by highly relativistic electrons spiraling magnetic field lines that run the length of the wisps, we can make an estimate of the magnetic field strength, $\sim 10^{-4}$ G. Lower temperature plasma emits the Hα spectral line clearly apparent in the red part of the visual spectrum.

(j, k) Galactic Molecular and Atomic Clouds

Atomic and molecular clouds in the Galaxy are largely confined to an extremely thin disk. Although consisting primarily of H_2, the molecular clouds are more readily observed through their carbon monoxide (CO) absorption or emission, and through the presence of other molecular lines. Although CO is orders of magnitude less abundant than H_2 its dipole emission is far stronger than the rotational quadrupole emission of H_2 expected from cold molecular clouds, many of which exhibit temperatures in the 10–20 K range. Molecular clouds appear to have a fractal structure, i.e.. a structure that has significant components on all logarithmic scales. Judging from the Doppler shifts of their spectral lines, the clouds exhibit turbulent velocity components restricted to a few kilometers per second. These low velocities keep the clouds from rising appreciable distances above or below the Galaxy's central plane. Maximum heights reached are of order 50 pc before the gravitational tug of the central plane pulls them back.

Atomic hydrogen clouds are most readily detected through their 21 cm line absorption or emission. Atomic sodium and singly ionized carbon, silicon, and other abundant atomic species embedded in these clouds also are readily observed through their spectral features. Like molecular clouds the atomic clouds remain closely confined to the Galactic plane.

(l, m) Stellar Winds

The stellar wind in O and B stars can be detected by observing the Doppler-shifted lines of highly excited ions. These can be observed in the ultraviolet part of the spectrum through observations from space. Doppler shifts indicate outflow velocities of order 100 km s^{-1}. Assuming that solar abundances also characterize the surface matter in these stars, one can interpret the observed line strengths to obtain the mass of ejected matter. Typically, a massive O star ejects matter at a rate of a solar mass in 10^5 to 10^6 yr. During the star's lifetime, this amounts to an appreciable fraction of the star's total mass.

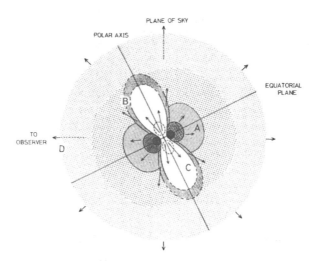

Fig. 9.2. A model of the protoplanetary nebula CRL 2688, evolving from the asymptotic giant branch, AGB, phase on its way to becoming a planetary nebula. The structure is rotationally symmetric about its polar axis. Four regions of the molecular envelope are labeled with letters. Region A is the equatorial disk. Region B represents expanding shells elongated along the polar directions. Regions C are produced by winds emanating from the star. Region D is a cold expanding envelope. The arrows show the presumed radial outflow of molecular gas. Regions enclosed by dashed lines are optical reflection nebulae (Ka87).

When a massive star reaches the asymptotic giant branch (AGB) stage it ejects dense dusty clouds of molecular hydrogen. Spectral observations of H_2O, CO, and other molecular constituents show the gas to be streaming out at velocities of tens of kilometers per second, apparently accelerated by the star's light pressure propelling the grains outward. Judging by the strengths of the observed spectral lines and the ejection velocities, the mass loss rates can be as high as $10^{-4} M_\odot$ yr^{-1}. This phase evidently can last no more than $\sim 10^5$ yr before the star exhausts its ejectable mass. Once a star has evolved through the AGB phase it can enter a *protoplanetary nebula stage*, that ultimately leads to it becoming a planetary nebula. Figure 9.2 depicts the star CRL 2688 undergoing this transition.

(n) Solar Wind

The solar-wind density, velocity, and variability is sampled by interplanetary probes placed at distances where the Earth's magnetosphere no longer interferes with observations. Magnetic fields are measured by magnetometers carried on these spacecraft. Considerable wind variations between quiet and active periods on the Sun are observed. At quiet times only a few electrons and protons are detected per cubic centimeter, and the velocity does not vary greatly from a mean value around $400 \, \text{km s}^{-1}$. Following a solar flare the density can rise by an order of magnitude and wind velocities reach values of $\sim 1000 \, \text{km s}^{-1}$. This fast stream of gas coming from the Sun interacts with the Earth's *magnetosphere* to give rise to a wide variety of effects, ranging from the colorful *aurora borealis* to the nuisance of poor radio wave propagation in the broadcast band. The solar wind also sweeps cometary ions into a comet's long straight tail extending into the antisolar direction.

(o) Comets

Comets have three distinct parts: a roughly spherical head, a long straight tail, and a shorter curved tail. The head, frequently of order 10^{10} cm in diameter, contains H_2O, C_2, C_3, CN, NH, CH, OH, and NH_2 among other molecules and radicals, as well as such ions as OH^+, CH^+, and CO^+. Typical molecular densities range around $10^4 \, \text{cm}^{-3}$. An extended H_2O and atomic hydrogen envelope around the head has also been discovered through observations from spacecraft. The velocities of the molecules in the head can be measured by Doppler shifts and broadening and amount to a few km s^{-1}. The head also includes a solid nucleus that may be a few to a few tens of kilometers in diameter, usually too small to be directly resolved, except from fly-by spacecraft.

The long straight tails seen in comets sometimes stretch over a distance larger than an astronomical unit. They are the most extended objects in the Solar System but their densities are low and the total mass contained is minute. Solely ions — no neutral molecules — are seen in these tails. The number density of the ions is deduced from the intensity of the molecular lines. The f values for excitation are known from computations, so that the observed brightness of the emission lines can be related to the rate at which the ions are excited by sunlight and thus to the number of molecular ions along the line of sight.

The shorter curved tails in comets reflect sunlight and exhibit its *Fraunhofer (absorption) line spectrum*. Because the solar lines do not appear broadened in this reflection, the scattering particles must be slowly moving dust and cannot be electrons, whose Thomson-scattered radiation would exhibit thermal broadening of

Table 9.1. Rough Characterization of Gas and Dust Aggregates.

	Representative Object	Density of Hydrogen $n_H, n_p, n(H_2)$ (cm^{-3})	Electron Density n_e (cm^{-3})	Dimensions (cm)
a. Intergalactic Medium		$n_p \sim 10^{-7}$	$\sim 10^{-7}$	
b. WHIM	filaments	$n_p \sim 10^{-5}$	$\sim 10^{-5}$	10^{24}
c. Ly-α Absorbers		$10^{12} < N_H < 10^{22}$ cm^{-2}		$\lesssim 10^{23}$
d. AGNs	Quasar 3C 273		3×10^6	3×10^{18}
e. Spiral Galaxy Arm:	Galaxy	0.1 to 0.7 $\Big\}$	0.03	3×10^{20} thick; $\Big\}$
f. Interarm Medium:		$\lesssim 0.05$		disk span 10^{23}
g. HII Region	Orion Nebula		10^4	5×10^{18}
h. Planetary Nebula	NGC 6543		6×10^3	10^{17}
i. Supernova Remnant	Crab Nebula		40	5×10^{18}
j. HI Cloud	Heiles Cloud I	40 to 125	~ 0.3	10^{19}
k. Molecular Clouds		$n(H_2) \sim$ 10^4 to 10^8 $\Big\}$		$10^{18} - 10^{19}$
l. Hot Cloud Cores	Orion IRc 2	$n(H_2) \gtrsim 10^6$		$\gtrsim 10^{17}$
m. Hot Stellar Wind	O star δ Ori	0.14	10^8 (1 AU from star)	
n. AGB Wind	NML Cyg	$n(H_2) \sim 10^8$ (100 AU from star) $\Big\}$		
o. Solar Wind			2 (1 AU from Sun) $\Big\}$	
p. Comet Head	Halley	$n_{molecules}$ $\sim 10^4$ $\Big\}$		10^{10} cm
q. Comet Dust Tail	Halley			10^{12} length
r. Comet Ionized Tail	Halley		$n_{ion} \sim 2$ (at 10^{11} cm) $\Big\}$	5×10^{12} length

a. n_e and T inferred from He$^+$ absorption.

b. Warm-Hot Intergalactic Medium (Ni05). The densities are averaged over all space.

d. Active Galactic Nuclei, AGNs: Quasars, Blazars, Seyfert Galaxies.
   Velocities from multiple absorption spectra.

e. Ly-α and 21 cm n_H data differ (Je70, Ke65).

i. (Co70), (Wo57).

j. (He69), (He68).

Table 9.1. Rough Characterization of Gas and Dust Aggregates (cont.)

Magnetic Field (G)	Turbulent or Bulk Velocity (cm s⁻¹)	Temperature (K)	Number Densities and Radii of Grains		Remarks
			n_g (cm⁻³)	a_g (cm)	
$\lesssim 10^{-9}$?		$> 2 \times 10^4$	$< 10^{-15}$	at 10^{-5}?	a.
		$\sim 10^6$			b.
					c.
$\sim 10^5$?	10^8 to 10^9	17,000			d.
$\sim 10^{-5}$	10^6		10^{-13}	$\sim 10^{-5}$	e.
					f.
	4×10^6	10^4	10^{-9}	$\sim 10^{-5}$	g.
		8400	$\sim 3 \times 10^{-10}$	$\sim 10^{-5}$	h.
3×10^{-4}	10^8	$< 17,000$			i.
$\lesssim 10^{-5}$	10^4 to 10^6	$\sim 10^2$	$\sim 10^{-9}$	$\sim 10^{-5}$	j.
					k.
	10^6	100	$\gtrsim 10^{-5}$	$\sim 10^{-5}$	l.
	1.4×10^8	10^4			m.
	2×10^6	300 (100 AU from star)	10^{-9} (100 AU from star)	$\sim 10^{-5}$	n.
3×10^{-5}	4×10^7	10^4 to 10^5	$\sim 10^{-13}$ (1 AU from Sun)	5×10^{-5}	o.
	2×10^5				p.
	10^6		$\sim 10^{-7}$	5×10^{-5}	q.
$\sim 3 \times 10^{-5}$	10^7				r.

l. (Pa01).

m. (Mo67).

n. Asymptotic Giant Branch (AGB) Stars, (Zu04).

o. Wind terminates at ~ 100 AU; grains orbit the Sun and do not move with the wind.

p. Atoms and radicals are: H, C, O, ..., CN, C_2, C_3, CH, OH, NH, ...,
   Ions: OH^+, CH^+, CO^+, H_2O^+, ...

r. Mainly CO^+ ions, CO_2^+, N_2^+, OH^+, ...

$\sim 10^8$ cm s^{-1}. The prevalence of dust is also corroborated by the shape of the lagging tail. It is curved because the repulsion of the dust by sunlight and the requirement for constant orbital angular momentum about the Sun produces an increasing lag for the repelled grains. The sizes of the dust grains can be roughly determined by the rate at which solar radiation pressure pushes them away from the head of the comet. Grains of assorted sizes follow different paths because the radiative repulsion varies, and it is possible to derive rough estimates of grain sizes at different locations across the width of the tail. The smallest grains lie closest to the radius vector pointing away from the Sun. The largest grains are most distant from this axis. From a rough estimate of grain sizes, we can compute the number density of grains as judged from the total scattered sunlight.

Table 9.1 summarizes some of the information on individual diffuse objects in the Solar System, in the Galaxy, and beyond. Within each class, variations in size, density, and other characteristics amounting to orders of magnitude are not uncommon. We have to be careful not to assume that different members of a class have identical properties.

9:2 Strömgren Spheres

In 1939, Bengt Strömgren considered the interaction of a very young star with the interstellar medium (St39). To make matters simple he made two assumptions. First, that the star lights up rapidly to full strength; and second, that the surrounding medium is homogeneous throughout. These two assumptions permitted him to draw a simple picture of the development of ionized hydrogen regions around massive stars emitting ultraviolet radiation.

If the star emits a number of photons dN_i capable of ionizing the surrounding gas, the number of electrons that are stripped off the atoms over the same time interval will also be dN_i if equilibrium is maintained. In practical cases this assertion is always true because the cross-section for ionization by energetic photons is of the order $\sigma \sim 10^{-17}$ cm^2 and typical gas densities in the vicinity of young stars might be of order $n_H \sim 10^3$ cm^{-3}. At these densities a photon can only travel a distance of order $(n_H \sigma)^{-1} \sim 10^{14}$ cm through the neutral medium before it ionizes an atom. But this is only ~ 6 AU, a distance small compared to the radii of ionized regions, which range from $\sim 10^{16}$ to 10^{20} cm. Hence, practically no ionizing photons can escape through the gas without becoming absorbed.

Although an energetic photon can travel only a short distance in the neutral medium, its path through the ionized gas is very long. It is occasionally scattered; but the scattering cross-section is relatively small — the Thomson cross-section is only $\sim 6.7 \times 10^{-25}$ cm^2 (6–103). We can assume that the ionizing photons proceed undisturbed through the ionized gas immediately surrounding the star until they hit the boundary region where the neutral gas commences. The thickness of the interface between ionized and neutral domains is of the order of the mean free ionizing path:

$$\delta = (n_H \sigma)^{-1}. \tag{9-6}$$

Neutral clouds are called Hɪ regions, ionized clouds, Hɪɪ regions.

We picture an ionizing star as embedded in an Hɪɪ region, which is separated from a surrounding Hɪ domain by a thin layer δ shown in Fig. 9.3. If the gas is

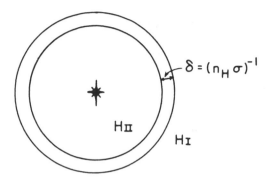

Fig. 9.3. Schematic diagram of a Strömgren sphere. Hɪɪ is the ionized gas, Hɪ is the neutral region, and δ is the thickness of the separating layer.

homogeneous, the separating boundaries are spherical, and the sphere containing the ionized gas is called the *Strömgren sphere*.

We now ask ourselves how quickly the sphere becomes established. To do this, we note that the number of atomic particles in a shell of radius R and thickness dR is $4\pi R^2 n_H\, dR$. When the star emits dN_i ionizing photons, the radius of the region grows by an amount dR given by

$$\frac{dN_i}{dt} = 4\pi R^2 n_H \frac{dR}{dt} . \tag{9-7}$$

Here we have gone through the additional step of formally dividing both sides by dt, to obtain the rate of development.

Equation (9–7), however, is only applicable during the initial stages of growth. It neglects the competing recombination of ions and electrons in the Strömgren sphere. For, if an electron and ion recombine to form an atom, a new ionizing photon will be required to separate the two particles. This new photon never reaches the boundary R and will therefore not contribute to the growth of the region. The recombination rate per unit volume is proportional to the product $n_e n_i$, since each electron has a probability of colliding that is proportional to the number of ions it encounters. In addition, the effective recombination rate per unit volume is also proportional to the recombination factor $\alpha_{\mathrm{eff}} \sim 3 \times 10^{-13}\ \mathrm{cm}^3\ \mathrm{s}^{-1}$. This depends, through (9–5), on the temperature of the ionized gas, normally $\leqslant 10^4$ K, and represents the sum over recombination factors leading to all states $n > 1$ (see column 5 of Table 9.2).

Direct recombination to $n = 1$ produces a photon capable of ionizing another hydrogen atom and does not contribute to α_{eff}. The full equation satisfied by the gas is therefore

Table 9.2. Absorption and Recombination Coefficients for Hydrogen and Helium.[a]

Atom	Term	a_{ν_0}	f	α_n	Q_n
				\multicolumn{2}{}{10,000 K}	
		10^{-18}		10^{-14}	10^{-22}
		cm^2		$\text{cm}^3\,\text{s}^{-1}$	cm^2
HI	1s	6.3	0.436	15.8	32
	2s	15	0.362	2.3	4.7
	2p	14	0.196	5.3	11
	3s	26	0.293	0.8	1.6
	3p	26	0.217	2.0	4.1
	3d	18	0.100	2.0	4.1
	4s	38	0.248	0.4	0.7
	4p	40	0.214	1.0	2.0
	4d	39	0.149	1.0	2.0
	4f	15	0.057	0.6	1.2
	Total			43	88
HeI	$1s^2\,{}^1S$	7.6	1.50	15.9	33
	$1s2s\,{}^3S$	2.8	0.25	1.4	3
	$1s2s\,{}^1S$	10.5	0.40	0.6	1
	Total			43	88
HeII	1s	1.7	0.42	70	140

[a] a_{ν_0} is the absorption cross-section at the ionization limit; the oscillator strength f is defined in Section 7:9; the recombination coefficients α_n and Q_n are defined by (9–4). (After Allen (Aℓ73). With the permission of Athlone Press of the University of London, 3rd ed. © C. W. Allen 1973.)

$$\frac{dN_i}{dt} = 4\pi R^2 n_H \frac{dR}{dt} + \frac{4\pi}{3} R^3 n_i n_e \alpha_{\text{eff}}. \tag{9-8}$$

During the late developmental stages this simple model would predict that dR/dt eventually becomes zero as the sphere grows so large that the star emits photons only just fast enough to keep up with the total number of recombinations. This will happen at an equilibrium radius

$$R_s^3 = \frac{3}{4\pi n_i n_e \alpha_{\text{eff}}} \frac{dN_i}{dt}. \tag{9-9}$$

Equations (9–7) and (9–9) are the extreme cases covered by equation (9–8). They describe the initial growth and final equilibrium value of the radius of the Strömgren sphere as long as the simplest assumptions are retained.

A number of comments are needed.

(1) Equation (9–8) has to be used in conjunction with some model for ultraviolet emission by stars. If the star's luminosity and temperature are known, then we can readily estimate N_i from the Planck blackbody relation (4–72). Actually, the ultraviolet spectrum of a very hot star does not closely approximate a blackbody

because absorption by the star's outer atmosphere changes the spectrum of the escaping radiation. This is called the *blanketing effect*. Despite this effect, however, the blackbody approximation gives roughly the correct magnitude for the number of ionizing photons to be inserted in equation (9–8).

Table 9.2 gives the ionization cross-section of hydrogen in its ground state as $a_{\nu_0} \sim 6.3 \times 10^{-18}$ cm^2 at the ionization limit, $\lambda = 912$ Å, $\nu_0 = 3.29 \times 10^{15}$ Hz. This cross-section declines as ν^{-3} at higher spectral frequencies. As a result, photons with energies much higher than ν_0 can penetrate great distances through a circumstellar cloud, while wavelengths just short of 912Å are strongly absorbed.

(2) As already stated, the recombination of an electron with an ion can yield a photon that is still capable of ionizing another atom. Even atoms recombining to $n \geq 2$ may be quickly reionized by intense Balmer, Paschen, Brackett, or higher-level continuum radiation (see Section 7:3). For this reason the second term on the right side of equation (9–8) is an upper limit on the loss of ionizing photons through recombination. Similarly the radius R_s of equation (9–9) is a lower limit for an equilibrium value. The effect turns out to be more important in very dense regions than in tenuous gases surrounding a star. For very dense regions the true R_s value may be more than ten times greater than that given by (9–9). For values of n_H around 10^4 to 10^5 cm^{-3}, typical of the denser ionized regions normally encountered, the radius R_s is a factor of 2 to 3 higher than predicted by (9–9).

(3) A quick consideration shows that equations (9–8) and (9–9) cannot be completely correct because they neglect the problem of pressure equilibrium. This is readily seen. The ionized region must have at least twice as many particles per unit volume as the neutral surrounding region because it contains at least one ion and one electron for each ionized atom. This means, according to the ideal gas law (4–37), that the pressure on the inner side of the boundary separating ionized from neutral regions would be at least twice as great as the pressure on the outside, and that, only if the temperature were the same on both sides. In practice, the temperature of the HI region is likely to be of order 70 K whereas the temperature of the HII region normally amounts to $\sim$7000 K. The total pressure inside the separating boundary is therefore of order 200 times greater than the pressure outside, and the HII region must rapidly expand.

(4) If we were to draw the very simplest picture of an expansion, we would proceed by visualizing the process in terms comparable to the inflation of a balloon. If the mass of the surrounding HI region is M, the mass per unit area at the separating surface is $M/4\pi R^2$. The pressure inside the HII region is $2n_i kT_i$. Here T_i is the temperature of the ionized region, and the factor 2 reflects that the number of ions n_i closely approximates the number of electrons. Neglecting the small gas pressure on the outside of the sphere, we obtain the outward acceleration of the boundary as

$$\ddot{R} = \frac{2n_i kT_i}{(M/4\pi R^2)} \, . \tag{9-10}$$

This can be integrated if we first multiply both sides by $\dot{R}$. Then

$$\frac{\dot{R}^2}{2} = \frac{8\pi}{3} \frac{n_i k T_i}{M} R^3,$$

(9-11)

which leads to a development time scale of order

$$t \sim \left(\frac{3M}{4\pi n_i k T_i R}\right)^{1/2}.$$

(9-12)

If we take M roughly equal to one solar mass $M_\odot \sim 2 \times 10^{33}$ g, $n_i \sim 10^4$ cm^{-3}, $T_i \sim 10^4$ K, and $R \sim 10^{17}$ cm, we find that

$$\dot{R} \sim 3 \times 10^5 \text{ cm s}^{-1}, \qquad t \sim 5 \times 10^{11} \text{ s}.$$

(9-13)

This velocity has to be compared to the random speed of atoms in the cool medium; this is only $\sim (3kT/m_H)^{1/2} \sim 1.5 \times 10^5$ cm s^{-1} at the low temperature of the HI region. The correct dynamics, therefore, cannot be described by equations (9–10) through (9–13) because pressure is normally propagated at the speed of sound, roughly the random speed of the atoms in HI regions. If the expansion on the inner edge of the HI region proceeds faster than the speed of sound, the outer portions of the region will not be aware that a pressure is being exerted at the inner boundary, and will therefore not move. As a result, the quantity of material actually accelerated at any given instant will be significantly less than the mass M used in the subsonic approximation (9–10), and the actual velocity $\dot{R}$ will be considerably higher. The equations of supersonic hydrodynamics must therefore be used. These will be derived in Section 9:4 below.

(5) Before proceeding to the dynamical treatment of expanding HII regions, it is interesting to point out that equation (9–7) may still hold well for extremely early stages of development, because dR/dt is then so high that the *ionization front*, that is, the region separating ionized and neutral regions, proceeds into the medium at velocities that can be orders of magnitude higher than the speed of sound in the medium. There is then no possibility at all for major instantaneous adjustments of density in response to pressure differences between ionized and neutral regions. This will also be discussed in Section 9:4.

(6) The expansion produced by gas pressure reduces the density of ionized material in the HII region and therefore decreases the recombination rate per unit volume. The factor $n_e n_i \alpha_{eff}$ of the second term in equation (9–8) decreases as R^{-6}, because both n_i and n_e decrease as R^{-3} when only expansion due to excess pressure (in contrast to expansion through further ionization) is involved. This means that the second term on the right of (9–8) is always reduced by pressure-induced expansion, thereby giving rise to a higher value for the expansion velocity $\dot{R}$ of the boundary.

(7) Finally, it is important that our whole concept of the development of a Strömgren sphere has been based on a picture in which the central star suddenly brightens and produces ionizing radiation. This, however, does not at all correspond to the development of massive stars shown in Fig. 1.5. The figure's caption notes that a massive O or B star takes some 6×10^4 yr to contract to the main sequence, and for a good fraction of that time is luminous without emitting much ionizing

radiation. Davidson (Da70) has argued that during the contraction stage, light pressure pushes gas and dust away from the star. This happens because a dust grain with radius a, accelerated by light pressure to a velocity v with respect to the gas, suffers collisions at a rate $n_H v \pi a^2$ with the atoms and, hence, suffers a drag (momentum loss) amounting to a deceleration

$$\dot{v}_d = -\frac{n_H m_H v^2 \pi a^2}{(4\pi/3)a^3 \rho} , \qquad (9\text{-}14)$$

where ρ is the grain's density.

For a star of luminosity L, the grain's radiative acceleration is

$$\dot{v}_r = \frac{L}{4\pi c R^2} \frac{\pi a^2}{(4\pi/3)a^3 \rho} \qquad (9\text{-}15)$$

so that equilibrium is established at a velocity

$$v \sim \left[\frac{L}{4\pi c R^2 n_H m_H}\right]^{1/2} , \qquad (9\text{-}16)$$

which has a value of $\sim 1.5 \times 10^6$ cm s^{-1} for $L \sim 10^{38}$ erg s^{-1}, $R \sim 10^{17}$ cm, and $n_H \sim 10^4$ cm^{-3}. This velocity is set up in a time

$$\tau \sim \frac{v}{\dot{v}} = \frac{(4/3)\rho a}{\sqrt{(L/4\pi c R^2)n_H m_H}} . \qquad (9\text{-}17)$$

If $\rho \sim 3$ g cm^{-3} and $a \sim 10^{-5}$ cm, $\tau \sim 2.5 \times 10^9$ s. From this we see that grains reach equilibrium velocity in a matter of a century. In contrast, the contraction of the star to the main sequence takes tens of thousands of years.

The grains drag the gas out to quite large distances through this process. A radiative pressure $(L/cR^2 4\pi)$ acting, say, on a column of length R and hence of mass $n_H m_H R$, would produce a mean acceleration of order

$$\dot{v} \sim \frac{L}{4\pi c n_H m_H R^3} \qquad (9\text{-}18)$$

and for the same conditions chosen above, but with $R \sim 3 \times 10^{17}$ cm, $\dot{v} \sim 10^{-6}$ cm s^{-2}. In 3×10^4 yr a distance of order R would be covered. Davidson therefore argued that when the star begins copious emission of ionizing radiation, most of the gas already has been pushed to large distances. The ionization of course still occurs, but it takes place at the edge of the low-density cavity in which the star now finds itself. What may happen then is that the newly ionized gas flows inward to the star, rather than outward away from it.

9:3 Pressure Propagation and the Speed of Sound

To understand the dynamics of gaseous regions, let us examine pressure waves that propagate at the speed of sound.

Atoms and molecules in a gas collide and transmit pressure. An impulse generated in the vicinity of a star can propagate across the interstellar medium to compress a distant region. When the impulse is weak, it generates a small density perturbation ρ_1 that propagates at velocity $\mathbf{v}_1$ through the ambient medium with mean density ρ_0. We may consider an infinitesimal volume at rest, through which this inhomogeneous gas flows. The rate at which the density in this volume changes is determined by two factors. The first is the difference in the velocity at which gas enters and exits the volume across different faces — i.e., the divergent gas velocity across the volume. The second is the rate at which density gradients are transported into and out of the volume. Both are spelled out in the *continuity equation*, which reads

$$\frac{\partial \rho}{\partial t} = -\nabla \cdot (\rho \mathbf{v}) = -(\rho \nabla \cdot \mathbf{v} + \mathbf{v} \cdot \nabla \rho) . \tag{9-19}$$

Here the density at any given point $\mathbf{r}$ is $\rho(\mathbf{r}) = \rho_0 + \rho_1(\mathbf{r})$, and similarly $\mathbf{v}(\mathbf{r}) = \mathbf{v}_0 + \mathbf{v}_1(\mathbf{r})$. We will consider a fluid at rest $\mathbf{v}_0 = 0$. Then the continuity equation can be written

$$\frac{\partial \rho_1}{\partial t} = -\nabla \cdot (\rho \mathbf{v}_1) = -(\rho_0 \nabla \cdot \mathbf{v}_1 + \mathbf{v}_1 \cdot \nabla \rho_1) . \tag{9-20}$$

In the absence of external forces, the acceleration of an element of the fluid is given by the pressure gradient across it divided by its density. The pressure $P(\mathbf{r})$ may be divided into an unperturbed part P_0 and a perturbed part $P_1(\mathbf{r})$, $P(\mathbf{r}) = P_0 + P_1(\mathbf{r})$. The accelerated flow through our incremental volume then is

$$\frac{d\mathbf{v}}{dt} = -\frac{1}{\rho_0}\nabla P , \quad \text{or} \quad \frac{d\mathbf{v}_1}{dt} \equiv \mathbf{v}_1 \cdot \nabla \mathbf{v}_1 + \frac{\partial \mathbf{v}_1}{\partial t} = -\frac{1}{\rho_0}\nabla P_1 . \tag{9-21}$$

where $d\mathbf{v}_1/dt$ is an exact differential. This is the *Euler equation* named after Leonhard Euler, who discovered the relation in the eighteenth century and made it the basis for all subsequent studies of fluid flow. For an ideal gas the pressure gradient in an adiabatic compression, i.e., a compression at constant entropy S, is then

$$\nabla P_1 = \frac{\partial P}{\partial \rho}\bigg]_S \nabla \rho_1 = \frac{\gamma P_0}{\rho_0}\nabla \rho_1 , \tag{9-22}$$

where we have made use of equation (4–129), noting that ρV is constant. Taking the divergence of the first and third terms, and making use of (9–20) to (9–22), though with neglect of second-order terms, we obtain

$$\frac{\partial^2 \rho_1}{\partial t^2} = \frac{\gamma P_0}{\rho_0}\nabla^2 \rho_1 . \tag{9-23}$$

This is a *wave equation*, similar in form to (6–30) and (6–32) governing the propagation of electromagnetic waves, except that it is longitudinal. The velocity vectors $\mathbf{v}_1$ of an *acoustic wave* lie along the direction of propagation, whereas, in electromagnetic waves the vectors $\mathbf{E}$ and $\mathbf{H}$ are transverse to the direction of propagation.

The acoustic density disturbance propagates as a plane wave

$$\rho_1 = A \exp[2\pi i(x/\lambda - \nu t)] \qquad \text{(RP)}, \qquad (9\text{-}24)$$

where the designation (RP) again indicates that we only take the real part of the expression. Inserting this solution into the wave equation, we obtain the speed of propagation, given by the wavelength λ divided by the time ν^{-1} required to traverse λ,

$$c_s = \lambda\nu = \left(\frac{\partial P}{\partial\rho}\right]_S\right)^{1/2} = \left(\frac{\gamma P_0}{\rho_0}\right)^{1/2}. \qquad (9\text{-}25)$$

c_s is the speed of sound under adiabatic conditions.

Quite generally $\frac{4}{3} \lesssim \gamma \lesssim \frac{5}{3}$, as already discussed in Chapter 4; the adiabatic speed of sound, therefore, always exceeds the root mean square velocity component along the direction of propagation, $v_{rms} \sim (kT/m)^{1/2}$ of the atoms or molecules. This can be understood by considering an element of mass enclosed in a layer of thickness Δx along the direction of propagation of the wave. An area of this layer having unit cross-section transverse to Δx contains an amount of mass $\delta\rho_1\Delta x$. The pressure increase in the acoustic wave, δP_1, does an amount of work $\delta P_1 \Delta x$ on this volume. We now take Δx to be the range across which the gradients in equation (9–22) are applied. This equation then tells us that $\partial P_1/\partial\rho_1$ is greater by a factor of γ than kT/m, the mean square of the velocity component along the direction of propagation that would have been set up by an increment of pressure δP under isothermal conditions described by (9–21). This is just the factor γ by which c_s exceeds the root mean square speed of atoms or molecules along a given direction.

9:4 Shock Fronts and Ionization Fronts

In Section 9:2 we gave one example of supersonic flow in the vicinity of hot stars. There are many others. Stellar *winds*, or streams of gas that continuously send stellar material out into surrounding space, blow at supersonic velocities ranging from a few thousands of kilometers per second for the hottest O stars down to speeds of the order of $400\,\text{km s}^{-1}$ for stars like the Sun.

Supersonic phenomena are encountered in stellar eruptions of all kinds, from the small outbursts that regularly occur in flare stars to the explosion of supernovae and the explosive ejection of gas from the nuclei of galaxies. They are more the norm than the exception in astrophysics. In this section, we will be concerned with the equations that describe the interaction of an HII region with the surrounding neutral medium; but the treatment is general and can be applied to many other supersonic phenomena.

Let us assume that a star has suddenly undergone an increase in brightness, that it rapidly ionizes the surrounding medium, and that a *shock front*, or an *ionization front*, or both, move outward into the cool HI region at supersonic speed.

There are two ways of considering a front — or dividing region — between the expanding ionized gas and the still unperturbed neutral hydrogen region. We can either consider the front as moving out into neutral gas at some velocity v, or else we can pretend that the neutral gas is moving into a stationary front at velocity, $-v$. After passing through the front, the gas is compressed and possibly ionized; there will also be energy changes — mainly heating.

Let us adopt this second point of view — that the front between the two regions is stationary — and make a number of demands on the gas flowing through the front, Fig. 9.4.

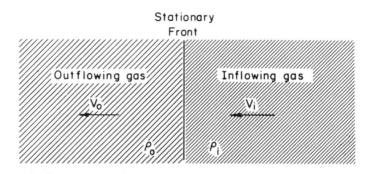

Fig. 9.4. Conditions on the two sides of a shock or ionization front.

(a) We require that the mass flow into the front equals the mass flowing out. This is just the continuity condition. If the inflow density and velocity are ρ_i and v_i and the outflow density and velocity are ρ_o and v_o, this requirement reads

$$\rho_i v_i = \rho_o v_o \equiv I. \tag{9-26}$$

Here, I is the *mass flow* through unit area in unit time.

(b) We can consider the front to be a surface that absorbs inflowing gas and emits outflowing gas. The pressure of inflowing material would be $\rho_i v_i^2$ even if the inflowing gas had no intrinsic pressure due to random motion of the atoms. The momentum transferred to the surface per unit area and unit time through absorption of the inflowing particles is $\rho_i v_i^2$ — which is what we mean by pressure. Similarly the back pressure due to the seemingly emitted outflowing gas would be $\rho_o v_o^2$. Since (9–26) has to be satisfied and since v_o generally differs from v_i, these two pressures will not normally be equal and opposite. We still have the two thermal pressures due to the random thermal motions of the gas atoms adding to the overall pressure acting on each side of the front. If the front is not to be accelerated — and we have assumed that there is a constant inflow and outflow velocity here — then momentum conservation requires that the overall pressures on the two sides of the front be equal and oppositely directed

$$P_o + \rho_o v_o^2 = P_i + \rho_i v_i^2 \ . \tag{9-27}$$

This is the condition of steady flow.

(c) On passing through the front, the energy content of the gas changes. A number of different sources contribute to the overall energy. For the inflowing gas there is:

(i) Kinetic energy due to the bulk flow, $v_i^2/2$ per unit mass flowing into the front;

(ii) Internal energy per unit mass U_i (see Section 4:19);

(iii) The work done on unit volume as it flows into the surface. Since we are picturing the gas as being stopped — absorbed — as it approaches the front.

The work done per unit area and time then involves a volume equal to the velocity v_i,

$$\text{For unit area}: \quad \text{work}/\text{time} = P_i v_i. \quad (9\text{-}28)$$

The product on the right of (9–28) is called the *enthalpy* of the gas. Reduced to unit volume, the enthalpy numerically just equals the pressure P_i. Reduced to unit inflow mass, it is P_i/ρ_i.

As it crosses the front, energy Q per unit mass may be fed into the fluid, so that the actual energy gain per unit mass of inflowing material is Q.

In unit time, a mass $\rho_i v_i$ of inflowing material crosses the front. This must contain energy equal to the energy contained in the gas flowing out of the front in unit time except that the outflow energy can be greater by an amount Q. The outflow energy consists of terms similar to those described under (i), (ii), and (iii). We therefore have the energy conservation equation

$$Q + \left(\frac{v_i^2}{2} + U_i + \frac{P_i}{\rho_i}\right) - \left(\frac{v_o^2}{2} + U_o + \frac{P_o}{\rho_o}\right) = 0, \quad (9\text{-}29)$$

where we have again used subscripts o to denote outflow. Equations (9–26), (9–27), and (9–29) relating inflow to outflow are sometimes called *the jump conditions*, because there is a discontinuity, a sudden jump, across the front.

PROBLEM 9–1. Referring to Sections 4:19 and 4:20, show that the internal energy per unit mass can be written as

$$U = c_v T = \frac{R}{\gamma - 1}\frac{P}{R\rho} = \frac{P}{(\gamma - 1)\rho}. \quad (9\text{-}30)$$

This leads to

$$\left[\frac{v_i^2}{2} + \left(\frac{\gamma_i}{\gamma_i - 1}\right)\frac{P_i}{\rho_i}\right] - \left[\frac{v_o^2}{2} + \left(\frac{\gamma_o}{\gamma_o - 1}\right)\frac{P_o}{\rho_o}\right] = -Q. \quad (9\text{-}31)$$

Because gases consisting of neutral atoms and gases containing only electrons and atomic ions both have γ values of $\frac{5}{3}$, we can finally write (9–31) in the form

$$\left[\frac{v_i^2}{2} + \frac{5}{2}\frac{P_i}{\rho_i}\right] - \left[\frac{v_o^2}{2} + \frac{5}{2}\frac{P_o}{\rho_o}\right] = -Q. \quad (9\text{-}32)$$

Equations (9–26), (9–27), and (9–32) describe the motion of a front into a monatomic medium.

Let us first examine the structure of an *ionization front*. We will assume that J ionizing photons are incident on the front per unit area in unit time. The mean energy of these photons is $\chi_r > \chi_o$ where χ_o is the energy required for ionizing atoms (Ka54)*. Since J is the number of ionizing photons incident on the front that divides the ionized from the neutral gas, it follows that J atoms are flowing into the front in unit time and J ions plus J electrons are streaming out.

By referring to (9–26) we see that the mass flow across the front is related to the flux of ionizing photons through the relation

$$I = mJ, \tag{9-33}$$

where m is the mass of the neutral atoms. For a pure hydrogen cloud $m = m_H$. We now define the ratio of densities

$$\frac{\rho_o}{\rho_i} \equiv \Psi. \tag{9-34}$$

Then, by (9–26) and (9–27)

$$P_o = P_i - \frac{\rho_i v_i^2 (1 - \Psi)}{\Psi}. \tag{9-35}$$

From (9–26), (9–32), and (9–35) we then obtain

$$\left[5\frac{P}{\rho} + v^2 + 2Q \right] \Psi^2 - 5 \left[\frac{P}{\rho} + v^2 \right] \Psi + 4v^2 = 0, \tag{9-36}$$

where we have dropped all subscripts, but the pressure, density, and velocity all refer to the inflowing material. We note that the energy supplied at the ionizing front goes partly into ionization and partly into kinetic energy of the particles. The part that goes into kinetic energy on average is $(\chi_r - \chi_o)$ per ion pair. Per unit mass of ionized material this relationship corresponds to a mean square velocity

$$u^2 = \frac{2(\chi_r - \chi_o)}{m} = 2Q. \tag{9-37}$$

Note that Q represents only the heating energy, not the energy needed to overcome atomic binding. This binding energy has not been specifically included in the formalism presented here. We have been able to neglect it by concentrating only on those photons having $\chi_r > \chi_o$. For a higher binding energy, fewer photons are available to ionize material.

We also note that the speed of sound depends on the ratio of heat capacities γ defined in (4–125),

$$c_s = \left[\frac{\gamma P}{\rho} \right]^{1/2} = \left[\frac{5P}{3\rho} \right]^{1/2}, \tag{9-38}$$

so that equation (9–36) can now be written entirely in terms of the three velocities v, u, and c_s, and in terms of the ratio Ψ,

$$[3c_s^2 + v^2 + u^2]\Psi^2 - [3c_s^2 + 5v^2]\Psi + 4v^2 = 0. \tag{9-39}$$

This is a quadratic equation in Ψ that can have a pair of coincident roots, two positive roots, or a pair of complex roots depending on whether

$$(3c_s^2 + 5v^2)^2 \gtrless 16v^2(3c_s^2 + v^2 + u^2) \tag{9-40}$$

or on whether

$$9(c_s^2 - v^2)^2 \gtrless 16v^2 u^2. \tag{9-41}$$

There are real roots under two conditions. The first is that

$$3(c_s^2 - v^2) \leq -4vu , \tag{9-42}$$

which means that v is greater than some critical speed v_R:

$$v \geq v_R = \frac{1}{3}\left(2u + \sqrt{4u^2 + 9c_s^2}\right). \tag{9-43}$$

This requires that the flux of ionizing photons be larger than a critical value J_R. By (9–26) and (9–33),

$$J \geq J_R = \frac{n}{3}\left(2u + \sqrt{4u^2 + 9c_s^2}\right), \tag{9-44}$$

where n is the initial number density of atoms in the neutral medium. The subscript R stands for *rarefied*.

The second condition under which real roots exist is if

$$3(c_s^2 - v^2) \geq 4vu , \tag{9-45}$$

which implies velocities of the front less than a critical value v_D and an ionizing flux below J_D, where D stands for *dense*:

$$v \leq v_D = \frac{1}{3}\left(-2u + \sqrt{4u^2 + 9c_s^2}\right), \tag{9-46}$$

$$J \leq J_D = \frac{n}{3}\left(-2u + \sqrt{4u^2 + 9c_s^2}\right). \tag{9-47}$$

The two critical speeds v_R and v_D correspond, for a given ionizing flux, to densities

$$\rho_R = \frac{I}{v_R} \quad \text{and} \quad \rho_D = \frac{I}{v_D} . \tag{9-48}$$

If the gas ahead of the ionizing front has a density ρ_R or ρ_D, only one possible value of ρ_o can result, apiece; the density in the ionized medium behind the front then has a fixed value. Although (9–39) is quadratic in $\Psi = \rho_o/\rho$, so that, for $\rho < \rho_R$ and also for $\rho > \rho_D$, there are two different density ratios that sustain identical inflow velocities v, equation (9–36) shows that there is only one positive value of v corresponding to any given value of Ψ. The density ratio across the front determines the sustainable inflow velocity, not vice versa.

For intermediate values of the initial density, $\rho_R < \rho < \rho_D$, there is no permissible value. This means that the ionization front cannot be in direct contact with the undisturbed HI region. We therefore have the following development of an ionized hydrogen region around a star that suddenly flares up and emits ionizing radiation. Initially, the interface between the ionized and neutral region is very close to the star; the flux J is still very high and well above the critical value J_R. We then have what is called the *R-condition*. The rate at which the front moves into the neutral medium (or vice versa according to the formalism used here) is $v = J/n$. This is just what equation (9–7) stated. However, as the ionization front moves farther from the star, the value of J decreases and the front slows down until the critical velocity v_R is reached. This velocity has the approximate value

$$v_R \sim \frac{4}{3}u \qquad (9\text{-}49)$$

because the mean energy of the photons is so high that the excess energy carried off by the ionized particles makes them move at velocities much higher than the speed of sound in the undisturbed neutral medium — the temperature in the ionized medium is much higher than in the neutral gas. Typical temperatures in ionized regions of interstellar space are between 5000 and 10,000 K, whereas HI regions have temperatures a factor of $\sim 10^2$ lower.

When the critical velocity v_R is reached, the ionization front no longer has direct contact with the undisturbed medium. It is now moving so slowly that a shock front signaling the impending arrival of the ionized region precedes the ionization front, and in so doing compresses the medium to a density greater than that of the undisturbed state.

Essentially, this just means that the ionization heats the gas that then expands into the neutral medium fast enough so that a compression wave travels into the neutral gas at a speed exceeding the local speed of sound. A shock front therefore precedes the ionization front into the neutral medium and modifies the density in this medium so that the boundary conditions (9–26), (9–27), and (9–29) once again are satisfied at the ionization front. As the ionization front moves still farther from the star, the velocity with respect to the undisturbed neutral medium drops below the lower critical value v_D. Here a gradual expansion is going on, no shock is propagating into the neutral medium, and the ionization front once again is in direct contact with the undisturbed medium. This is called the *D-condition*.

The boundary between interstellar space and the Solar System's sphere of influence, the *heliosphere*, is a particularly fascinating region. Because the Solar System moves supersonically through the ambient interstellar medium, a *bow shock* similar to the shock ahead of a supersonic aircraft is expected to develop where the interstellar gas first encounters the *heliosheath*, an outer region of the heliosphere. But the interstellar gas does not mingle with the solar wind at this surface; that is expected to mainly occur further in at the *heliopause*. Meanwhile, the supersonic *solar wind* moving outward from the Sun produces a *terminal shock* at the interface with the heliosheath, where the magnetic field strength B suddenly jumps by roughly a factor of 2, but where the outflow still does not yet mingle with the interstellar medium until the heliopause is reached further out in the heliosheath, the region separating the bow shock from the termination shock. Late in December 2004, after 27 years of travel, the spacecraft Voyager I launched in 1977 located the anticipated termination shock at 94 AU from the Sun and crossed into the heliosheath beyond. The predicted heliopause and bow shock, and details of the physical processes characterizing these surfaces await discovery as Voyager I moves on in its long journey to leave the Solar System and explore the regions beyond (Fi05).

PROBLEM 9–2. Show that

$$v_D \sim 3c_s^2/4u. \tag{9-50}$$

We should still note three factors:

(a) The conditions at a normal shock front are identical to those across an ionization front except that the ionization energy is not supplied, that is, $Q = 0$.

(b) When the neutral gas is molecular rather than atomic, a dissociation front normally precedes the ionization front into the neutral medium. The dissociation energy of hydrogen molecules is only 4.5 eV, far lower than the ionization energy of hydrogen atoms, 13.6 eV. Starlight photons with energies $4.5 \lesssim h\nu \lesssim 13.6$ eV readily pass through the HII region to create a *photodissociation region, PDR*, ahead of the ionization front. Such regions are somewhat interchangeably also called *photon-dominated regions* designated by the same acronym, *PDR*. The equations that govern conditions at the interface between the cool molecular gas and the PDR are precisely the same as those we derived above for the interface between the neutral and ionized regions, except that now χ_o represents the energy required to dissociate a molecule, and γ_i has a value of $\frac{7}{5}$ appropriate for molecular hydrogen at a temperature sufficiently high to excite rotation. The equations derived here therefore have a wide range of applications.

(c) Usually a magnetic field is present and the energy balance and pressure conditions must then also include magnetic field contributions. *Hydromagnetic shocks* are particularly significant because under conditions where collisions between particles are rare, the magnetic fields are the main conveyors of pressure throughout the medium. Pressure equilibrium between gas particles is established through mutual interaction via magnetic field compression. The speed at which information on pressure differences is conveyed hydromagnetically is called the *Alfvén velocity*, which can be shown to be $v_A \equiv (B^2/4\pi\rho)^{1/2}$; in magnetized plasmas it replaces the speed of sound.

In weakly ionized cool clouds, pressure can be transmitted to the ions at speeds v_A that may be much greater than the speed of an approaching collisional shock. The ions then become accelerated even before the collisional shock arrives. Since the number density of ions is low, they can only gradually accelerate the ambient neutral atoms through collisions. Nevertheless, the cold medium may reach appreciable velocities before the collisional shock arrives. Such *continuously* accelerated shocked regions are called *C-shocks*, while shocks that entail the previously derived *jump conditions* (9–26), (9–27), and (9–29) are called *J-shocks*.

At the interface between HII and HI regions we sometimes see bright rims that outline the dark, dust-filled regions not yet ionized. The bright rims generally are located at the edge of the nonionized matter and appear pointed toward the direction of a relatively distant ionizing star normally of spectral type earlier than O9. It is possible that these rims occur when ionizing radiation arriving at the HI region satisfies the D-condition and sets up an ionization front that moves into the neutral gas without being preceded by a shock wave.

When clouds collide at relativistic speeds, a *relativistic shock* ensues. The basic principles of the interaction remain unchanged, but the details of the jump conditions change. Continuity of mass flow into and out of the front, and conservation of energy, must now be replaced by a continuity of mass–energy flow, because new particles may be generated or destroyed at the front. We can think of relativistic shocks as having some of the features of the cosmic-ray air showers discussed in Sections 5:11 and 6:24, where a highly relativistic particle penetrating a stationary medium from outside creates large numbers of new particles. The only difference is that in relativistic shocks entire clouds of particles enter what amounts to a stationary medium at relativistic speeds. If the medium is magnetized, some of the entering particles or their secondaries may circle back to hit the approaching shock front again and again, to be Fermi-accelerated at the front to extreme energies, as already described in Section 6:6.

9:5 Gamma-Ray Bursts, GRB

In Section 8:18 we discussed core-collapse supernovae. When a collapsar explodes with an energy of order 10^{51} erg a high-temperature, optically thick electron–positron plasma, called a *fireball* is created. It expands ultrarelativistically into the ambient medium creating a *forward shock* in the medium, while, by symmetry, a *reverse shock* plows into the fireball. The ultrarelativistically expanding front of the fireball is often referred to as a *blast wave*.

A variety of processes spring into action at the shock front, including Fermi-acceleration of particles, synchrotron radiation by electrons, and inverse Compton production of high energy γ-rays created as ambient radiation is scattered off high-energy cosmic-ray electrons. If the shock front moves relativistically in the direction of the observer, the γ-rays produced at the expanding front are observed to arrive during a highly contracted period. The effect is identical to that described in Section 6:21 where we discussed superluminal jets. In the case of gamma-ray burst, however, Lorentz factors may be of order $\Gamma(v) \sim 100 - 1000$ (To02, Da05). In the rest-frame of the expanding front, the photons may be created over the course of a few days, but at the observer, they arrive within a span of seconds.

Gamma-ray bursts are rare events. Most of them are observed to come from highly red-shifted galaxies. Were it not for their immensely energetic outbursts, and the narrow beams into which the γ-rays are confined — the cone angle of the emitted beam may span no more than a few degrees — they would probably be missed. As the expansion of the fireball slows down, a lower-energy X-ray and optical afterglow persists, apparently beamed into a considerably wider angle. A search is afoot to see how many such afterglows might be found that have not been preceded by an observable GRB. This would help to define the beam open angle, and thus would lead to a more accurate estimate of the total energy emitted in the course of the explosion. Afterglows lacking an observed GRB have come to be known as *orphans* (To02).

9:6 Origin of Cosmic Magnetic Fields

Magnetic fields are known to exist in stars and in the interstellar medium. Stars like the Sun have typical surface magnetic fields of the order of 1 G, but in some A stars the surface fields can reach ~40,000 G. The fields in the interstellar medium are much weaker, typically of the order of 10^{-5} G. But there are wide variations. In some regions of the Galaxy no magnetic fields at all have been determined in measurements that should have detected fields of strength 10^{-6} G, while at other locations, quite strong fields exist. In the Crab Nebula supernova remnant, for example, the field strength can be as high as 10^{-4} G, and in compact HII regions and other clouds with densities as high as $n \geq 10^7 \, \mathrm{cm}^{-3}$ fields reach milligauss levels over regions that have dimensions ranging from 10^{13} to 10^{17} cm (Kr94), (Zw97).

Figure 9.5 shows that the observed Galactic magnetic field direction generally runs along the local spiral arm though its strength is not constant everywhere.

Where does this field come from? Is its origin primordial, dating back to some early stages of the Universe? We do not know.

If magnetic fields are not primordial, two alternatives suggest themselves:

(1) Magnetic fields are formed in the interstellar or intergalactic medium and find their way into stars as stars are formed from interstellar material; or

(2) Magnetic fields are formed in stars, possibly by dynamo mechanisms of the type discussed in Section 9:7, below, and the field is then introduced into the interstellar medium as mass is ejected from the stars. This might be consistent with the high strength of the Crab Nebula field. It would also be consistent with the observation that the solar wind carries along magnetic fields. Whether some portion of this field becomes detached from the Sun and strays out into the interstellar medium is not known. But many stars have much more massive winds than the Sun, and the outflow of magnetic fields may be a customary accompaniment to the outflow of mass.

Once a magnetic field exists in very weak form, it can be amplified by turbulent motions of the medium in which the fields are embedded. The net magnetic flux crossing any given fixed surface cannot be increased in this way, but by folding the field direction many times, local fields of greatly increased strength can be formed without an accompanying high net flux (Fig. 9.6). Turbulent motion therefore obviates the need for strong initial fields. Small, seed magnetic fields can be amplified by turbulent stretching and folding of the field lines.

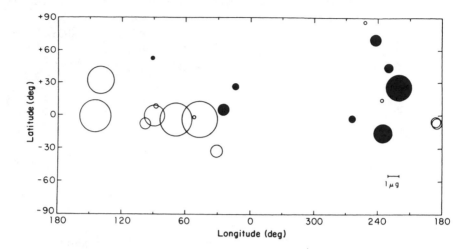

Fig. 9.5. Magnetic field direction at the Sun's Galactic location. These data actually are mean line-of-sight magnetic field components for pulsars as judged from their rotation and dispersion measures. For fields greater than 0.3 μG the circle diameter is proportional to the field strength. When the field has a direction toward the observer (positive rotation measure) the circles are filled. When they are away from the observer they are empty. The diameter for 1 μG is indicated in the figure. The observations are consistent with a relatively uniform field of about 3.5 μG directed along the local spiral arm. Note that the directions of greatest field strength are toward longitudes $\sim$60° and $\sim$240°, although there are large variations. These are also roughly the directions of the local spiral arm (Bo71). (From Manchester (Ma72b). With the permission of the University of Chicago Press.)

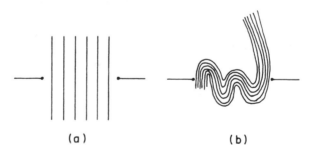

Fig. 9.6. Two magnetic field configurations with the same net flux. Configuration (a) has low field strength everywhere. Configuration (b) has high field strength in some places. In this figure, the field strength is taken proportional to the number of lines crossing unit length of the abscissa. This would be representative for field lines embedded in sheets normal to the plane of the paper.

Let us see how big this effect could be. If a field B_0 initially existed in some location within the Galaxy, the flow of gas at a velocity v could have stretched out field lines maximally at that velocity. Folding the field back on itself also could maximally occur at velocity v, so that the ability of a turbulent motion to amplify the field is limited by the speed of the motions. The amplification of the field through stretching and folding is given by the ratio of the initial volume V_0 containing the seed fields to the final volume V_f that would have been obtained through stretching the region through a rectilinear motion at velocity v

$$\therefore \frac{B_f}{B_0} = \frac{V_f}{V_0} . \tag{9-51}$$

Here B_f is the final magnetic field strength obtained through stretching and folding in a constant volume V_0.

Within the Galaxy explosive velocities of order $10^4 \, \text{km s}^{-1}$ are observed in supernova ejecta. We can choose this to represent the maximum turbulent velocity. The initial dimension of the Galaxy is ~ 30 kpc along a diameter and ~ 100 pc perpendicular to the disk. If the stretching motion were to go on for 10^{10} yr at $10^3 \, \text{km s}^{-1}$, a distance of 10 Mpc would be covered, and a turbulent folding would increase the magnetic field strength respectively by a factor of 300, or 10^5 depending on whether the turbulent motion took place predominantly within the Galactic plane or perpendicular to it.

Since the field in the Galaxy is estimated to have a strength of order 3×10^{-6} G, at the present epoch, the initial seed fields must have had strengths at least of the order of 3×10^{-11} G. There seems no way to escape this conclusion.

A primordial field of this magnitude must therefore have been present initially, or else some mechanism must have existed for producing this field. A number of processes have been suggested for setting up such a *seed field* that later could grow in strength through turbulent motion.

As illustrated in Fig. 9.7, the Poynting–Robertson effect, which slows down electrons orbiting a luminous source while leaving protons almost unaffected, can produce a current to set up a weak magnetic field (Ca66). Some, or perhaps even most of the energy may, however, end up in some form other than magnetic energy. A complete analysis of such effects is complicated and depends in detail on the interaction of the electrons with protons, on the resulting tendency for positive and negative charges to slightly separate along a radial direction from the light source, and so on.

This type of effect, which applies different forces to electrons and protons, acts like a battery and is referred to as a *battery effect*. Whether battery effects contribute significantly to the generation of interstellar magnetic fields is uncertain, but the effect discussed here should serve as an illustration of the type of mechanism that could perhaps be effective.

We note that the Poynting–Robertson drag on an electron is large because the Thomson cross-section (see equation (6–103)) is $(m_P/m_e)^2 \sim 3 \times 10^6$ times larger for electrons than for protons. Here m_e and m_P are the electron and proton mass.

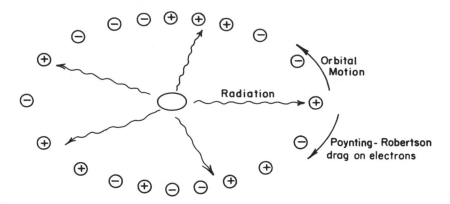

Fig. 9.7. Poynting–Robertson drag on electrons produced by a luminous body acting on an orbiting plasma. The drag on protons is much weaker; a small net current can therefore be induced.

The deceleration is even stronger for electrons, being a factor $(m_P/m_e)^3$ times larger than for protons, because for electrons the drag force acts on a smaller mass. From (5–47) we see that the orbital angular momentum L, for an electron in a circular orbit, changes at a rate

$$\frac{1}{L}\frac{dL}{dt} = \frac{L_s}{4\pi R^2}\frac{\sigma_e}{m_e c^2} , \qquad (9\text{-}52)$$

where L_s is the source luminosity, R is the distance from the source, and σ_e is the Thomson cross-section. The work dW/dt done on an electron in unit time is equal to the force F acting on it multiplied by the distance through which it moves at its orbital velocity v,

$$\frac{dW}{dt} = Fv = \left(\frac{1}{R}\frac{dL}{dt}\right)\frac{L}{Rm_e} = \frac{L_s \sigma_e L^2}{4\pi R^4 m_e^2 c^2} . \qquad (9\text{-}53)$$

The largest number of electrons that could be slowed down in this way would be $N \sim 4\pi R^2/\sigma_e$, because if we had more electrons than this, some would shadow the others. Hence the maximum total work that can be done on the clouds is

$$\frac{N dW}{dt} \sim \frac{L_s}{R^2}\frac{L^2}{m_e^2 c^2} = L_s \frac{v^2}{c^2} . \qquad (9\text{-}54)$$

This gives the maximum work that can go into building up a magnetic field over volume V:

$$NW \sim V \int \frac{d}{dt}\left(\frac{B^2}{8\pi}\right) dt \sim \frac{B_f^2}{8\pi}V , \qquad (9\text{-}55)$$

where $B^2/8\pi$ (see Section 6:10) is the instantaneous magnetic field energy density. For the Galaxy, gas is contained in a disk volume $V \sim 3 \times 10^{66}\,\text{cm}^{-3}$ and B_f is the final magnetic flux density, or field strength, $B_f \sim 3 \times 10^{-6}\,\text{G}$.

$$\therefore NW = \frac{L_s v^2}{c^2}\tau \sim 10^{54}\,\text{erg}\,. \tag{9-56}$$

Let us first see whether the total field could have been produced had the Galaxy at one time been as bright as a quasar for, say, 3×10^6 yr. Taking typical quasar internal velocities $v \sim 10^8\,\text{cm s}^{-1}$, $\tau \sim 3 \times 10^6$ yr $\sim 10^{14}$ s; we would require $L_s \sim 10^{45}\,\text{erg s}^{-1}$ at peak efficiency to produce a field of $\sim 3 \times 10^{-6}$ G.

This does not seem too unreasonable for producing the entire flux, so that perhaps we would not even need the subsequent turbulent amplification. However, if we want to do the same thing in our Galaxy right now, we find that there are so few electrons that only $\sim 10^{-5}$ of the total flux would be scattered by electrons near the center where velocities are $\sim 3 \times 10^7\,\text{cm s}^{-1}$. There $L_s \lesssim 10^{43}\,\text{erg s}^{-1}$, so that the overall rate of work done on electrons is decreased by $\sim 10^8$. In 10^{14} s a seed field of $\sim 3 \times 10^{-10}$ G could be formed for the Galaxy — in 10^{10} yr, a field of $\sim 2 \times 10^{-8}$ G. Some form of battery effect producing a seed field that might subsequently be amplified by turbulence appears to be one option for accounting for present-day magnetic fields (Ku97).

We should still note that the battery effect could also act to destroy magnetic fields. If the orbiting plasma contains an initial magnetic field, the drag acting on the electrons could be in a direction producing a current that reduces the magnetic field.

9:7 Dynamo Amplification of Magnetic Fields

Dynamos can be effective in amplifying weak magnetic fields once they exist. If electric currents can be set up perpendicular to the direction of a seed field, the field can grow. The question is whether cosmic magnetic fields can grow sufficiently fast by dynamo action to account for the field strengths observed today (Bh95), (Ch97).

We saw in equation (6–11) that the force on an electron moving in a magnetic field is $e\mathbf{v} \wedge \mathbf{B}/c$. This force gives rise to a current just as an electric field would. Equation (6–15) for the current density in the absence of varying displacement fields, therefore, becomes

$$\mathbf{j} = \sigma\left(\mathbf{E} + \frac{\mathbf{v} \wedge \mathbf{B}}{c}\right), \qquad \mathbf{E} = \left(\frac{\mathbf{j}}{\sigma} - \frac{\mathbf{v} \wedge \mathbf{B}}{c}\right). \tag{9-57}$$

Inserting this into (6–22) yields

$$\frac{\partial \mathbf{B}}{\partial t} = \left(\nabla \wedge \mathbf{v} \wedge \mathbf{B} - \frac{c\nabla \wedge \mathbf{j}}{\sigma}\right). \tag{9-58}$$

With (6–23) this reduces to

$$\frac{\partial \mathbf{B}}{\partial t} = \left(\nabla \wedge \mathbf{v} \wedge \mathbf{B} - \frac{c^2 \nabla \wedge \nabla \wedge \mathbf{B}}{4\pi\sigma\mu}\right), \tag{9-59}$$

and with (6–24) and the identity (6–28) we finally obtain

$$\frac{\partial \mathbf{B}}{\partial t} = \left(\nabla \wedge \mathbf{v} \wedge \mathbf{B} + \frac{c^2 \nabla^2 \mathbf{B}}{4\pi\sigma\mu} \right). \tag{9-60}$$

From (9–59) we see that the magnetic field grows only if the velocity transverse to the field is appreciable

$$v \geq c^2/4\pi\sigma\mu L), \tag{9-61}$$

where L is the dimension over which the magnetic field changes significantly.

To note what this means, we need to evaluate the conductivity σ. The acceleration an electron experiences in an electric field $\mathbf{E}$ is $\mathbf{E}e/m_e$. The acceleration produces an increase in the current, but only for a time equal to ν_c^{-1}, the reciprocal of the collision frequency derived in equation (6–134). The mean distance a charge can travel in this time interval is $\mathbf{E}e/2m_e\nu_c^2$, so that the mean velocity is $\mathbf{E}e/2m_e\nu_c$. For a charge density ne, the current density then becomes

$$\mathbf{j} = \sigma\mathbf{E} = \frac{\mathbf{E}ne^2}{2m_e\nu_c}, \tag{9-62}$$

so that

$$\sigma = \frac{ne^2}{2m_e\nu_c}. \tag{9-63}$$

For fully ionized hydrogen we obtain ν_c from equation (6–134), and

$$\sigma \sim \frac{3.5 \times 10^7 T^{3/2}}{\ln(s_{\min}/s_{\max})} \; \mathrm{s}^{-1}. \tag{9-64}$$

For a region of dimensions $L \sim 10^{18}$ cm and a temperature of order 100 K, we then find that the systematic electron velocity relative to the field needs to be of order 10^{-8} cm s^{-1} to make interstellar magnetic fields grow; otherwise they will decay. This seems like a ridiculously low velocity, and we might expect that magnetic fields would grow rapidly by means of dynamo processes. However, we do not yet quantitatively understand the means by which a systematic electron drift could be produced relative to the magnetic fields, without current-neutralizing protons drifting along as well — though, as we saw in the previous section, radiation pressure can produce such currents. Consequently, we do not yet know to what extent dynamo effects are significant on stellar or interstellar scales. Perhaps, once we better understand the nature of turbulent magnetohydrodynamic flows, dynamo action will also become better understood.

9:8 Cosmic-Ray Particles in the Interstellar Medium

Cosmic-ray particles, mainly high-energy electrons and protons, contribute an energy density of about 10^{-12} erg cm^{-3} to the interstellar medium. This compares to a mean starlight density of $\sim 7 \times 10^{-13}$ erg cm^{-3} and a kinetic energy of gas atoms, ions, and electrons ranging from about 10^{-13} erg cm^{-3} in the low-density cool clouds, to roughly 10^{-9} erg cm^{-3} in high-density HII regions.

The interaction between the cosmic rays, the gas, and the radiation field is quite strong. It usually involves an energy loss for cosmic-ray particles. Such losses can be divided in the following way (Gi64), (Gi69).

(a) Highly relativistic electrons having energies $\mathcal{E} \gg mc^2$, lose energy to the interstellar medium through a number of different processes that sometimes are collectively referred to as *ionization losses*. They comprise: (i) the ionization of atoms and ions; (ii) the excitation of energetic atomic or ionic states; and (iii) production of Cherenkov radiation. These effects are not always separable. Their relative strengths are determined in part by the electron energy and in part by the nature of the medium. Neutral and ionized gases give rise to different loss rates. Table 9.3 gives expressions for these and other cosmic-ray losses discussed below.

(b) Ultrarelativistic electrons can also suffer *Bremsstrahlung* losses. These occur when electrons are deflected by other electrons or nuclei. The deflection amounts to an acceleration that causes the particle to radiate. Again the loss rates differ for ionized plasma and for a neutral gas.

(c) Synchrotron and Compton losses (see Sections 6:20 to 6:22) are related loss rates, respectively, proportional to the energy density of the magnetic and radiation fields. That these two processes can be considered to be similar can be seen from a simplified argument. Imagine two electromagnetic waves — photons — traveling in exactly opposing directions in such a way that their magnetic field vectors are identical in amplitude and frequency and their electric fields are exactly opposite in amplitude but again at the same frequency. The electric field and the Poynting vector **S** both cancel for these two waves at certain times, and we are left with a pure magnetic field whose energy density is equal to the total energy in the radiation field. At this point, synchrotron loss should be equivalent to the losses from inverse Compton scattering off the two photons of equivalent energy.

(d) For cosmic-ray protons and nuclei we have ionization losses again given in Table 9.3. Synchrotron and Compton losses should be less than those of electrons by the ratio of masses taken to the fourth power $\sim 10^{13}$. There are also a variety of interactions between cosmic-ray nuclei and the nuclei of the interstellar gaseous medium and grains. Table 9.4 gives these interactions for several different groups of nuclear particles interacting with an interstellar gas composed of 90% hydrogen and 10% helium by number of atoms. The mean free path Λ gives the distance traveled between nuclear collisions. Essentially, a proton travels until it has passed through an effective layer thickness containing 72 g of matter per cm^2. In a cool cloud with density of order 10^{-23} g cm^{-3} this amounts to a distance of order 2 Mpc. Since the cosmic-ray particles describe spiral paths in the Galaxy's magnetic field, they traverse such a distance in about 6×10^6 yr. The more massive cosmic-ray nuclei suffer collisions more rapidly. The absorption path length $\lambda = \Lambda/(1 - P_i)$ (where P_i is the probability that the collision will again yield a nucleus belonging to the same

Table 9.3. Energy Losses of Cosmic-Ray Particles in the Interstellar Medium (after Ginzburg (Gi69)).[a]

(a) Ionization Losses[b] for

Electrons with $\mathcal{E} \gg mc^2$:

in a fully ionized plasma

$$-\frac{d\mathcal{E}}{dt} = \frac{2\pi e^4 n}{mc}\left\{\ln\frac{m^2c^2\mathcal{E}}{4\pi e^2 n\hbar^2} - \frac{3}{4}\right\} = 7.62\times10^{-9}n\left\{\ln\left(\frac{\mathcal{E}}{mc^2}\right) - \ln n + 73.4\right\} \text{eV s}^{-1}$$

in a neutral gas

$$-\frac{d\mathcal{E}}{dt} = \frac{2\pi e^4 n}{mc}\left\{\ln\frac{\mathcal{E}^3}{mc^2\chi_0^2} - 0.57\right\} = 7.62\times10^{-9}n\left\{3\ln\left(\frac{\mathcal{E}}{mc^2}\right) + 20.2\right\} \text{eV s}^{-1}$$

Electron Radiation Losses:

for plasma

$$-\frac{d\mathcal{E}}{dt} = 7\times10^{-11}n\left\{\ln\left(\frac{\mathcal{E}}{mc^2}\right) + 0.36\right\}\frac{\mathcal{E}}{mc^2} \text{eV s}^{-1}$$

for neutral gas

$$-\frac{d\mathcal{E}}{dt} = 5.1\times10^{-10}n\frac{\mathcal{E}}{mc^2} \text{eV s}^{-1}$$

Electron Synchrotron and Compton Losses[c]:

$$-\left[\left(\frac{d\mathcal{E}}{dt}\right)_s + \left(\frac{d\mathcal{E}}{dt}\right)_c\right] = 1.65\times10^{-2}\left[\frac{H^2}{8\pi} + \rho_{ph}\right]\left(\frac{\mathcal{E}}{mc^2}\right)^2 \text{eV s}^{-1}$$

(b) Losses for Nuclei:

in neutral gas

$$-\frac{d\mathcal{E}}{dt} = 7.62\times10^{-9}Z^2n\left\{4\left[\ln\left(\frac{\mathcal{E}}{mc^2}\right)\right] + 20.2\right\} \text{eV s}^{-1}, \text{ if } Mc^2 \ll \mathcal{E} \ll \left(\frac{M}{m}\right)^2 Mc^2$$

$$-\frac{d\mathcal{E}}{dt} = 7.62\times10^{-9}Z^2n\left\{3\left[\ln\left(\frac{\mathcal{E}}{mc^2}\right)\right] + \ln\frac{M}{m} + 19.5\right\} \text{eV s}^{-1}, \text{ if } \mathcal{E} \gg \left(\frac{M}{m}\right) Mc^2$$

for fully ionized plasma

$$-\frac{d\mathcal{E}}{dt} = 7.62\times10^{-9}Z^2n\left\{\ln\left(\frac{W_{max}}{mc^2}\right) - (\ln n) + 74.1\right\} \text{eV s}^{-1}$$

where $W_{max} = 2mc^2\left(\dfrac{\mathcal{E}}{mc^2}\right)^2$ if $Mc^2 < \mathcal{E} \ll \left(\dfrac{M}{m}\right) Mc^2$

$$= \mathcal{E} \text{ if } \mathcal{E} \gg \left(\frac{M}{m}\right)Mc^2$$

[a] Reprinted with the permission of Gordon and Breach Science Publishers, Inc., New York.

[b] Ionization losses comprise ionization and excitation of atoms and Cherenkov radiation including plasma oscillations.

[c] Magnetic field and photon energy densities, $H^2/8\pi$ and ρ_{ph}, are in erg cm^{-3}. M is the nuclear mass, Z is the nuclear charge, $\mathcal{E}$ is the particle energy, χ_0 is the ionization energy, and n is the number density of electrons in the medium.

Table 9.4. Cross-Sections, Mean Free Paths Λ, and Absorption Paths λ.[a]

Cosmic-Ray Particle	Cross-Section for Collision	Λ Mean Free Paths	λ Absorption Path
$\mathcal{P}$	3×10^{-26} cm^2	72 g cm^{-2}	— g cm^{-2}
α	11	20	34
Li, Be, B	25	8.7	10
C, N, O, F	31	6.9	7.8
$Z \geq 10$	52	4.2	6.1
Fe	78	2.8	2.8

[a] For cosmic-ray particles in different groups of elements interacting with an interstellar medium which consists of 90% hydrogen and 10% helium (in number density of atoms)(see text) (after Ginzburg (Gi69)). Reprinted with the permission of Gordon and Breach Science Publishers, Inc., New York.

initial cosmic-ray group) is somewhat longer than the mean free path, as shown in column 4 of Table 9.4.

PROBLEM 9–3. If the energy loss per collision of a cosmic-ray nucleon with a nucleus of the interstellar medium leads to a loss comparable to the total energy of the nucleon

$$\left(-\frac{d\mathcal{E}}{dt} \right)_{\text{nucl}} = cn\sigma\mathcal{E} , \tag{9-65}$$

show with the help of Table 9.4 that this loss dominates the other processes listed in Table 9.3 for cosmic-ray nuclei. Here σ is the collision cross-section and n is the number density of interstellar atoms.

PROBLEM 9–4. Using the above loss rate for protons and using the loss rates from Table 9.3 for cosmic-ray electrons having the spectrum shown in Fig. 9.8, calculate roughly how fast the electron and the proton cosmic-ray components lose energy, and estimate how fast the interstellar medium is being heated by cosmic rays if their energy density is of order 10^{-12} erg cm^{-3}. This cosmic-ray heating is important in dense molecular clouds where light does not penetrate.

The observed flux of cosmic-ray protons and alpha particles incident on the Earth's atmosphere is shown in Fig. 9.9. Similar data exist for many other elements. Roughly 90% of the nuclear component of the cosmic-ray flux at the top of the atmosphere consists of protons. Alpha particles make up ∼9%, and the remaining particles are heavier nuclei. Curiously, there is a great excess of Li, Be, B, and ^{3}He, despite their low overall cosmic abundance. All these constituents are easily destroyed at temperatures existing at the center of stars (Section 8:12). We can account for the presence of these elements if they are produced through collisions of

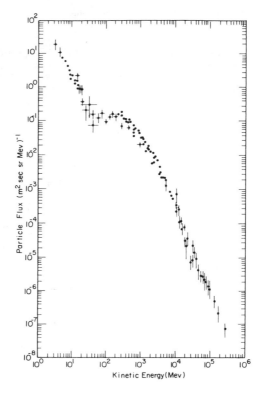

Fig. 9.8. Cosmic-ray electron spectrum at the Earth. Compiled from various sources by P. Meyer (Me69). Electron and positron abundances are comparable. 1 GeV $\equiv 10^9$ eV.

carbon, nitrogen, and oxygen cosmic-ray particles with hydrogen nuclei in the interstellar medium. The lighter elements are then the *spallation products* of the more massive parent particles. To obtain the amount of these low mass elements observed and also to obtain the correct ^{3}He/^{4}He ratio, cosmic-ray particles with energies in excess of 1 GeV would have had to pass through $\sim$3 g cm^{-2} of matter (Re68c). This suggests an age of about 2×10^6 yr if the particles have been spiraling within the Galaxy all this time, and may represent the mean time taken for cosmic-ray particles to be lost from — potentially diffuse out of — the Galactic disk.

We also find that the heavy elements are represented far more abundantly in the cosmic-ray flux than in meteorites or in the solar atmosphere. This suggests that these highly energetic particles originate in supernova explosions, pulsars, or white dwarfs, where high concentrations of heavy elements will have been produced during advanced stages of stellar evolution (Co71b).

The cosmic-ray flux in the Galaxy appears to be quite steady. Meteorites and lunar surface samples have been analyzed for tracks left by heavy nucleons. The total flux, as well as the relative abundance of heavy nuclei, cannot have changed drastically over the past 5×10^7 yr.

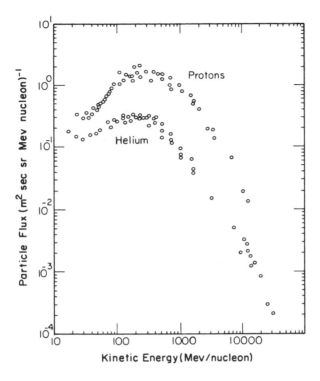

Fig. 9.9. Proton and alpha particle cosmic-ray flux at the Earth. At any given energy the proton flux is about one hundred times as intense as the electron flux. (Compiled from various sources by P. Meyer (Me69). Reprinted with permission from *Annual Review of Astronomy and Astrophysics*, Vol. 7, ©1969 by Annual Reviews, Inc.) (Error bars have been omitted.) At higher energies the flux J continues to drop, obeying a power law $dJ/dE \propto E^{-\gamma}$ with $\gamma \sim 3$ to energies of order 10^{20} eV.

Electrons and the somewhat less abundant positrons have fluxes which, at any given energy, amount to about 1% of the proton flux shown in Fig. 9.9. The spectrum of the diffuse X-ray flux arriving at the Earth (Fig. 9.10) is roughly similar to that of the cosmic-ray electrons. This suggests that the X-rays and γ-rays are formed by inverse Compton scattering in active galactic nuclei, AGNs.

9:9 Formation of Molecules and Grains

Interstellar grain material appears to be quite varied. Large organic molecules, and graphite and silicate grains are prominent constituents. The graphite grains appear to originate in the dense atmospheres of *carbon stars*. Water and other volatiles freeze out on these grains in the cold interior of dense clouds, where the grains are shielded from heating by star light (see Fig. 9.11).

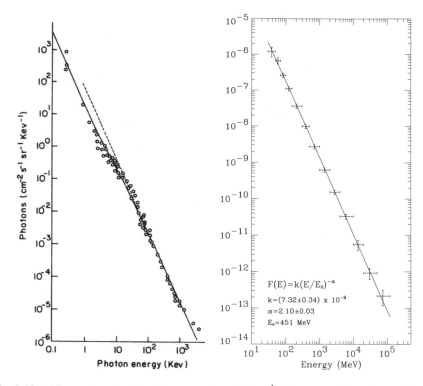

Fig. 9.10. Diffuse extragalactic electromagnetic radiation (a) X-ray spectrum observed from above the Earth's atmosphere. Compiled from various sources by A.S. Webster and M.S. Longair (We71). With the permission of the Officers and Council of the Royal Astronomical Society. (b) Diffuse extragalactic gamma-ray spectrum. Compiled by Sreekumar et al. Courtesy Stanley D. Hunter (Sr98). Both diffuse components may be due to the cumulative emission from distant active galactic nuclei, AGNs

Densities in interstellar space are so low that grain formation appears impossible there. To see this, consider the growth rate of a grain. Let its radius at time t be $a(t)$. Interstellar atoms and molecules impinge on the grain with velocity v. If the number density of heavy atoms having mass m is n, the growth rate of the grain is

$$4\pi a^2 \frac{da}{dt} = \frac{\pi a^2 nmv}{\rho}\alpha_s \,, \tag{9-66}$$

where ρ is the density of the interstellar atoms after they have become deposited on the grain's surface, α_s is the sticking coefficient for atoms impinging on the grain, and the left side of (9–66) represents the rate at which the grain's volume grows. Taking $v \sim \sqrt{3kT/m}$, with $T \sim 100\,\mathrm{K}$ and $m \sim 20\,\mathrm{amu} \sim 3 \times 10^{-23}\,\mathrm{g}$, $\rho \sim 3\,\mathrm{g\ cm}^{-3}$, and $n \sim 10^{-3}\,\mathrm{cm}^{-3}$, and the maximum value $\alpha_s = 1$, we obtain

$$\frac{da}{dt} \sim \frac{n\sqrt{3kTm}}{4\rho}\alpha_s \sim 10^{-22}\,\mathrm{cm\,s}^{-1}. \tag{9-67}$$

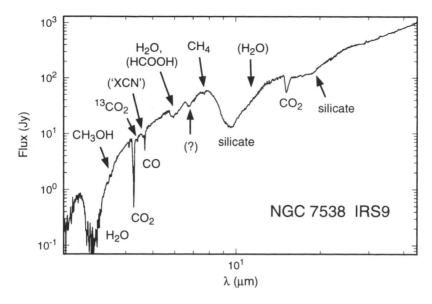

Fig. 9.11. Spectrum of the cold interstellar molecular cloud NGC 7538 IRS9, from 2.5 to 45 μm. Absorption features due to various solid constituents of interstellar grains are identified (Wh96).

To grow to a size of 10^{-5} cm, the available time would have to be 10^{17} s $\sim$ 3×10^9 yr. With more realistic values of α_s, the required length of time increases to an age greater than that of the Galaxy. Here we have neglected the deposition of hydrogen on a grain, because pure hydrogen would normally evaporate rapidly.

Of course, there exist regions in space where the number density of atoms such as oxygen, nitrogen, carbon, and iron is ~ 1 cm^{-3}. The Orion region is about this dense. If there were no destructive effects there, grains could perhaps form in a time $\sim 3 \times 10^6$ yr, if $\alpha_s \sim 1$. Furthermore, if the temperature in a dense cool cloud could become low enough so that hydrogen could solidify on the grains without rapid re-evaporation, the growth rate could still be higher by two or three orders of magnitude.

Thus far we have neglected destructive effects. In H II regions, radiation pressure often accelerates small grains to higher velocities than large grains. Collisions among grains can then take place at velocities as high as $\gtrsim 1$ km s^{-1}, to vaporize both of the colliding grains. The vapors then have to recondense. In addition, *sputtering* by fast-moving protons can knock atoms off a grain's surface after they have become attached. Such destructive effects tend to reverse the growth implied by equation (9–66), or at least will decrease the growth rate. This destruction would be stronger for frozen water (ice) where molecules are bound weakly, than for strongly bound substances like silicates or graphite. All three of these solids are constituents of interstellar grains.

Another, less catastrophic, destructive effect is evaporation, characterized by the *vapor pressure* of the material from which the grains are formed. In thermal equilibrium at a temperature T the vapor pressure, P_{vap}, gives the rate at which molecules or atoms evaporate from a grain's surface. The equilibrium vapor pressure is that pressure of ambient vapor at which the growth rate is just equal to the evaporation rate. This pressure is related to the vapor density through the equation of state. If Dalton's law (4–38) holds, the mass impact rate of molecules is

$$nmv \sim \left(\frac{m}{kT}\right)^{1/2} P_{vap} \tag{9-68}$$

per unit area. This then must also represent the evaporation rate from the surface. We see that the pressure in the ambient space must exceed the vapor pressure if the grain is to grow. Hydrogen has a vapor pressure of about 10^{-7} torr at 4 K (Table 10.2). This amounts to about 10^{11} molecules cm^{-3}. This density is far higher than any expected for interstellar space. On the other hand, grains are never likely to be cooler than 4 K. This means that hydrogen cannot very well remain on grains, unless it is chemically bound by the presence of other substances or else adsorbed on the basic grain material. For other substances this would not be true. Silicates and graphite have such low vapor pressures that no appreciable evaporation off grains would occur in periods of the order of the life of the Galaxy. Solid carbon monoxide, CO, or carbon dioxide CO_2 grains have been detected in the shielded interior of dusty molecular clouds, where temperatures as low as 12–15 K have been observed (La96). In bulk form these substances rapidly sublimate, respectively, at temperatures of order $\sim$20–25 K and $\gtrsim$50 K at the low pressures found in the interstellar medium, but the sublimation temperatures could be even lower for small grain sizes (Sc97).

For ice, the situation is somewhat more complicated. On approaching close to an individual star, H_2O molecules could evaporate as a grain's temperature rose. This is what happens when a comet approaches the Sun from the outer portions of the Solar System. The surface warms until water, ammonia, and other ices sublimate. For ice grains, sputtering through collisions with atoms is generally the dominant destructive mechanism; only on very close approach to a star can evaporation be significant.

Grains appear to form primarily in the atmospheres of cool giant stars or Mira variables; both eject gas into interstellar space. The atmospheres of these stars are dense so that n in (9–66) is high. Formation must, however, take place in a period as short as a year after material is blown beyond the photosphere. After that, the outflowing gas becomes too rarefied. With $n \sim 10^6$ cm^{-3} for the heavier atoms, $T \sim 2 \times 10^3$ K, $m \sim 12$ for carbon, it might be possible to form a graphite grain at a rate $da/dt \sim 3 \times 10^{-13}$ cm s^{-1}; a grain with $a \sim 10^{-5}$ cm would form in a year. This would be a nucleus that could subsequently still grow, possibly because radiation pressure would keep it moving slightly faster than the gas flowing out from the star (see equation (9–15)).

Mira variables and cool giants often emit an excess of infrared radiation. This is due to the formation dust in their atmospheres and its subsequent ejection into the interstellar medium.

The infrared emission from planetary nebulae also indicates the presence of dust. The grains could have been produced in an earlier red-giant phase of the star, in a circumstellar shell expanding with the velocity of the original red-giant stellar wind.

The novae Ser 1970, Aql 1970, and Del 1967 were found to grow brighter in their infrared emission as the visible part of their spectrum dimmed, suggesting that dust played an important role in their emission. In contrast, supernovae, which eject far more massive amounts of heavy elements into their surroundings, may expel most of it into extragalactic space where densities are too low for grains to form.

The formation of molecules presents problems similar to dust formation. In the past few decades, microwave techniques have uncovered an increasing number of interstellar molecules, NH_3 (ammonia), CO, H_2O, HCN (hydrogen cyanide), H_2CO (formaldehyde), CN (cyanogen), HC_3N (cyanoacetylene), the hydroxyl radical OH, and many others. Deuterated versions of hydrogen-containing molecules also abound. More than one hundred interstellar molecular species are now known (Oi97).

A number of destructive effects should still be mentioned. Molecules can be destroyed through dissociation — often as a consequence of absorption of, or ionization by, ultraviolet photons. Calculations indicate that galactic starlight may destroy molecules such as CH_4, H_2O, NH_3, and H_2CO in times of the order of a hundred years, unless the molecules are well shielded from light inside strongly absorbing dust clouds (St72). Ionization by energetic electrons or cosmic-ray particles could produce similar effects. Collisions among interstellar clouds that accelerate particles to elevated energies, as described in Section 6:6, could therefore destroy molecules. Conversely, the high densities at the contact face of two colliding clouds could also lead to more rapid formation of molecules. Competing formation and destruction rates must be compared in specific settings to see which predominates.

9:10 Formation of Molecular Hydrogen, H_2

Hydrogen atoms can combine to form molecular hydrogen by two different processes. The first of these is slow, but predominates when no other constituents except hydrogen itself are present, early in the evolution of the Universe, before the heavier chemical elements have been produced in stars. We will return to this process in Chapter 13, where we will see the central role that H_2 formation plays in the formation of the first stars. In today's universe, however, the formation of hydrogen molecules can progress more rapidly wherever cold dust grains are abundant (Ho71).

At typical grain temperatures of order 15 K, hydrogen atoms can diffuse over grain surfaces until two atoms find themselves at a catalytic site that facilitates the bonding of two atoms to form a molecule.

$$H + H \rightarrow H_2 + 4.5 \, \text{eV} \, . \qquad (9\text{-}69)$$

The release of the bond energy $E = 4.5$ eV propels the newly formed molecule off the surface. Because the mass M of the grain is so much larger than the mass of the molecule m_{H_2}, conservation of momentum tells us that most of the energy is carried off by the molecule, whose velocity will be

$$v \sim (2E/m_{H_2})^{1/2} \sim 2 \times 10^6 \, \text{cm s}^{-1} \, . \qquad (9\text{-}70)$$

Conservation of momentum, nevertheless, implies that the grain receives a significant kick, a velocity change of order $\Delta V \sim v(m_{H_2}/M)$ shared between translational and rotational velocities.

9:11 Polycyclic Aromatic Hydrocarbons

Hydrocarbons are molecules consisting primarily of carbon and hydrogen. They are some of the simplest *organic* — meaning carbon-based — molecules. A particularly stable form of hydrocarbon molecule is a benzene ring, an assembly of six carbon atoms forming a ring to which hydrogen and other atoms, ions or radicals may be attached. Molecules based on this type of structure are called *aromatic* hydrocarbons. When a molecule comprises several benzene rings, it is said to be a *polycyclic aromatic hydrocarbon*, abbreviated PAH. Naphthalene is made up of two benzene rings, anthracene of three, and literally millions of far more complex aggregates are known in organic chemistry.

The infrared emission from interstellar space frequently exhibits spectral characteristics reminiscent of various types of PAHs. An exact identification has not been made, reflecting the possibility that PAHs are highly varied. These macromolecules absorb visible starlight and redistribute the absorbed energy into *phonons* — vibrational and torsional modes within the molecule. The PAH acts as an assembly of coupled springs responding to an impulse. Small amounts of energy can also produce rotation. The excited modes quickly re-radiate energy in a cascade of infrared photons whose energies correspond to the chemical makeup of the radiating PAH and provide clues to its dominant chemical bonds. As the cascade progresses, the PAH returns to its ground state configuration, where it remains until again excited by an arriving photon.

Unlike larger grains, PAHs never attain an equilibrium temperature with starlight. Their infrared emission is strictly a short-term phenomenon that follows the absorption of a single photon. Visible and near ultraviolet starlight has an energy of several electron volts while the various vibrational and torsional modes of these large molecules are excited by energies well below an electron volt. This makes PAHs remarkably efficient converters of optical radiation into infrared. They absorb energetic photons and re-emit many quanta of lower energy.

9:12 Infrared Emission from Galactic Sources

Stars

Cool supergiants are often shrouded by dust that absorbs much of their visible radiation, re-emitting the energy primarily in the near infrared.

The near infrared emission from circumstellar dust is readily explained. Equation (4–81) specifies the temperature that a grain will assume at a given distance R from a star. Some uncertainty prevails because the absorptivity ε_a and re-radiation emissivity ε_r of the grains remains largely unknown.

$$T = \left(\frac{\varepsilon_a}{\varepsilon_r} \frac{L_\odot}{16\pi\sigma R^2} \right)^{1/4} , \qquad (4\text{--}81)$$

Because the grains are distributed over a range of distances from the parent star, we also expect a rather broad emission spectrum corresponding to emission by grains at different temperatures.

Ionized regions

At longer wavelengths, principally beyond 10 μm, Galactic HII regions strongly radiate. The Galactic center also is a powerful infrared source at wavelengths peaking around 100 μm.

The infrared emission from planetary nebulae and HII regions involves relatively little dust. Often the obscuration within the ionized region is not noticeable, and yet strong infrared emission is observed. How can dust be responsible?

To explain this phenomenon we have to return to the discussion of ionization equilibrium in Section 9:2. We had noted there that, for an equilibrium Strömgren sphere, the recombination rate equals the ionization rate. Each time an electron recombines with a proton to form a hydrogen atom we have two possibilities. Either the recombination leaves the atom in the first excited state $n = 2$, or else it leaves it in a higher level from which a cascade of photon emission eventually places the atom into state $n = 2$ or else $n = 1$. Any photon emitted through a transition from $n = 2$ to $n = 1$, or from any higher excited state down to $n = 1$, has a very high probability of being reabsorbed by another hydrogen atom in the $n = 1$ state within the HII region. The photon then wanders through the HII region in a random walk, as in Problem 4–4, though with a variable mean free path.

A photon, successively absorbed and reradiated by atoms moving with thermal velocities, v, eventually becomes Doppler-shifted by $\Delta\nu \gtrsim \nu\langle v^2 \rangle^{1/2}/c$, large compared to the natural line width, γ, in equation (7–59). The mean free path then increases, permitting rapid escape from the region. Nevertheless, the Ly-α photon's trajectory out of the region is typically several times longer than the cloud radius, and dust has a correspondingly greater chance of absorbing the radiation. We have acted here as though all of the Lyman spectrum photons were Ly-α radiation. This is almost true. A Ly-β photon emitted by one hydrogen atom is likely to be absorbed and to give rise to an Hα photon in an atomic transition $n = 3$ to $n = 2$, succeeded by a Ly-α transition $n = 2$ to $n = 1$.

As a rule of thumb, we can state that for each ionizing photon, the HII region eventually must produce one recombination. That recombination, followed by a succession of emission, reabsorption, and re-emission processes, eventually has to give rise to one photon of the Balmer spectrum, and one Ly-α photon. The oscillator strength for absorption of the Ly-α photon is near unity so that the effective cross-section per hydrogen atom is large. Even if the fractional neutral hydrogen density is very low, Ly-α absorption usually is large enough to trap the Ly-α photons in the nebula.

PROBLEM 9–5. For an HII region in which each ionizing photon has unit optical depth across a distance equal to the radius R, relate n_H, the number density of neutral atoms, to the absorption coefficient $a_{\nu o}$ listed in Table 9.2.

(a) With reference to Problem 7–11 determine the mean free path at the line center for Ly-α absorption and compare this to R.

(b) The absorption bandwidth, γ, usually is small compared to the Doppler frequency shift $\Delta\nu \sim \nu \langle v^2 \rangle^{1/2}/c$ where $\langle v^2 \rangle^{1/2}$ is the root mean square velocity in the HII region. For $\langle v^2 \rangle^{1/2} \sim 30\,\mathrm{km\,s^{-1}}$, what is the Ly-$\alpha$ absorption mean free path (see Problems 7–10, 7–11, and 7–13)?

(c) Successive absorption and isotropic re-emission of Ly-α in randomly moving atoms produces radiation whose central frequency undergoes a random walk in the spectral frequency domain. Eventually, an emitted Ly-α photon may be so far out in the spectral line wings that most atoms will only have a low absorption cross-section for it. The photon can then more readily escape the HII region. Show that this will happen after $(n\gamma/\Delta\nu)^2$ absorptions and re-emissions, for a frequency shift $n\gamma$ off line center and $n\gamma \gg \Delta\nu$.

Because each ionizing photon gives rise to a Ly-α quantum of radiation, and because this radiation is likely to be absorbed by a grain before it can escape to the boundary of the HII region, we might expect that all the radiation converted into Ly-α would (Kr68) eventually be absorbed by a grain and the energy would be thermally radiated in the far infrared. Most of the ionizing photons given off by a star have an energy less than twice the Ly-α energy. Hence we conclude that more than half the ionizing energy emitted by a star would eventually find its way into infrared radiation. This seems to be at least approximately true.

We can also compare the infrared emission and the free–free emission from ionized regions observed in the radio domain. This can be directly related to the expected recombination line intensities through equations such as (9–5) or (6–141). We can therefore derive a proportionality between the expected free–free radio emission from the HII region and the far infrared flux from grains.

The correlation between these two turns out to be so good that a systematic process connecting infrared and radio fluxes has been sought. An intriguing suggestion is that electric dipole emission from rapidly rotating $\sim 10^{-7}$ cm dust grains, rather than free–free emission, could be responsible for radio emission, at least in the 5 to 50 GHz frequency range, and that grains in different size ranges could be responsible both for the infrared and the radio emission (Dr98), (Fi02a).

Photodissociation Regions, PDR

Occasionally, grains also can heat ambient gas. In the photodissociation regions mentioned in Section 9:4 starlight has been stripped of all its ionizing photons. This deprives it of the ability to heat the neutral atoms through which it passes as it crosses from an HII domain surrounding an ionizing star and streams out into a molecular cloud. The primary heating mechanism for such regions, therefore, appears to be heating by electrons torn out of grains by the photoelectric effect.

Electrons are easily removed from solids that have a low *work function* — a low ionization potential. For interstellar grains, photons with energies of a few electron volts may suffice to eject an electron. Starlight incident on grains will then charge them up to the point where the work function plus the electrostatic potential at the surface of the grain equals the highest incident photon energy. For a grain with a low work function and radius r, the energy required to remove an electron is Ze^2/r, where Z is the number of electrons removed. With $r = 10^{-5}$ cm, this energy is $2.3 \times 10^{-14} Z$ erg. The charge distribution on grains, however, is also determined by competing collisions with electrons and ions, and generally depends upon ambient conditions and on grain size and composition (Ce95).

Photons escaping the HII region have energies at least as high as 11 eV — well below the ionization potentials of the most abundant species, hydrogen, helium, oxygen, nitrogen, or carbon. If the grain work function is of order 5 eV, this still permits the grain to be charged up to a surface potential of 6 volts, where the additional energy required to overcome the electrostatic attraction is $\sim 10^{-11}$ erg and $Z \sim 400$.

Photoelectric removal of electrons can therefore free up to several hundred electrons per grain. Each of the initially emitted electrons can carry away an energy of up to a few electron volts. The electron quickly gives up this energy to nearby atoms, principally through *charge exchange*, which has a high cross-section $\sim 10^{-15}$ cm^2. In charge exchange the free electron displaces a bound electron in an atom and transfers a fraction of its energy to the atom. Eventually a low-energy electron is attracted to and captured by the charged grain, only to be recycled through the process of photoelectric emission, to again transfer starlight energy to ambient atoms. This cycle is referred to as *photoelectric heating* by grains.

If the grains at the cloud surface facing the star become fully charged, so that no further photoemission can take place, the grains can still absorb incident optical and ultraviolet radiation. The grain temperature will then rise. Most of the absorbed heat will be radiated away at far infrared wavelengths and escape the cloud. A small amount of heat can also be transferred to colliding gas atoms and contribute to heating the gas, though generally at an insignificant level as seen in Fig. 10.4.

Molecular Clouds

Molecular clouds at extremely low temperatures are primarily cooled through dust emission at submillimeter wavelengths. Figure 9.12 shows the far infrared emission

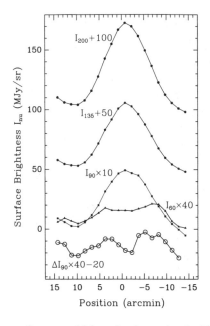

Fig. 9.12. Far infrared scans of a very cold dust cloud associated with molecular gas. The dust temperature indicated by the relative emission at 60, 90, 135, and 200 μm lies in the range of 12 to 15 K (La96).

by grains in a cloud at 10 to 15 K. Shielded from starlight by high ambient dust densities, grains in the interior of these clouds are primarily heated by cosmic rays. Part of the cosmic-ray energy loss, and part of the destruction of cosmic rays discussed in Section 9:5, can be attributed to this interaction with grains.

Large interstellar grains may have dimensions of the order of 10^{-5} cm, permitting them to radiate more effectively at long wavelengths. In the hot cores of giant molecular clouds, they are heated by frequent collisions with atoms and re-emit this energy in the far infrared, typically at wavelengths of the order of 100 μm. The infrared emission observed in star-forming regions is produced in this fashion. Grains are responsible for the bulk of protostellar cooling that permits these clouds to release gravitational energy as they slowly contract to form stars.

We will return to cooling by grains in Section 10:9.

Relativistic Processes

In Seyfert galaxies, where strong infrared emission from active nuclei is sometimes observed, infrared emission may arise through processes such as synchrotron radiation or inverse Compton scattering. Seyfert galaxies can be strong emitters of X-rays so that highly energetic electrons are likely to be present. In these galaxies it is often difficult to determine the primary mechanism responsible for infrared emission.

9:13 Orientation of Interstellar Grains

Starlight that has traversed long distances through the Galactic plane tends to be both reddened and polarized. The transmitted starlight is predominantly polarized parallel to the plane (Fig. 9.13), meaning that its oscillating electric field vector $\mathbf{E}$ is parallel to the plane. We therefore assume that light polarized perpendicular to the plane is eliminated — absorbed or scattered — by grains oriented with their long axes perpendicular to the Galactic plane, as discussed in Section 6:17. The grains absorb, emit, or scatter light, somewhat like a radio antenna, parallel to their long axes along which electrons can most readily flow in response to an oscillating $\mathbf{E}$ field. Tiny antennas operating at optical wavelengths have been tested in the laboratory and obey the same principles as radio antennas (Mü05).

How can grains become aligned in this way? In Problem 7–6, we saw that grains have angular frequencies of the order of 10^5 Hz. It therefore makes no sense to talk about the orientation of a stationary particle. There is, however, a different kind of orientation involving preferred directions of a grain's angular momentum vector.

In Section 9:9 we saw that the formation of molecular hydrogen on grains can give a grain a significant kick, as a newly formed H_2 molecule is ejected at a velocity $v \sim 2 \times 10^6$ cm s^{-1}. If half the recoil momentum from the molecule goes into the grain's translational motion and half into rotation, the angular momentum kick the grain receives should be of order

$$\delta L \sim \frac{1}{2} m_{H_2} va \sim 3 \times 10^{-23} \left(\frac{v}{2 \times 10^6 \text{cm s}^{-1}} \right) \left(\frac{a}{10^{-5}} \right) \text{g cm}^2 \text{ s}^{-1} , \quad (9\text{-}71)$$

where a is the grain radius.

On a typical grain, there may only be a few catalytic sites where molecules can form. Because the grains are small and rapidly spinning, as we saw in Problem 7–6, molecules ejected from the catalytic sites are likely to be ejected into random spatial directions around the grain. The grain's translational velocity should then increase as a random walk — as the square root of the number of molecules ejected — and be damped by collisions with ambient gas molecules. The grain's angular momentum, however, may increase more nearly in direct proportion to the number of ejected molecules. Because of the low expected number of catalytic sites on a grain, the many molecules ejected from a given site may produce a systematically anisotropic impulse. The grain will then be spun up to much higher angular momenta than

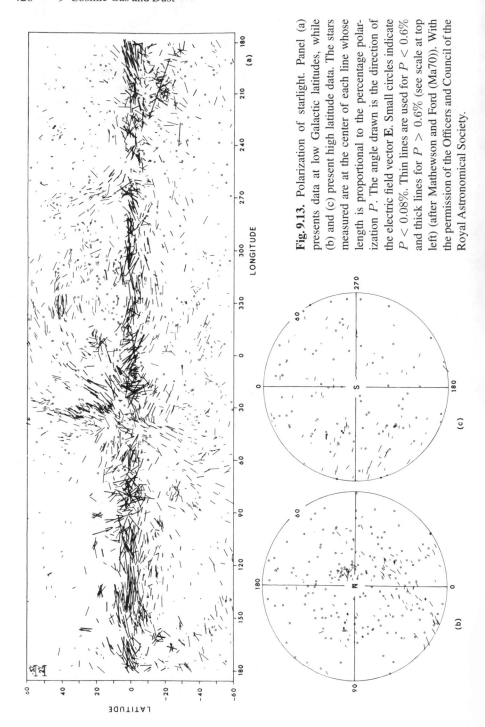

Fig. 9.13. Polarization of starlight. Panel (a) presents data at low Galactic latitudes, while (b) and (c) present high latitude data. The stars measured are at the center of each line whose length is proportional to the percentage polarization P. The angle drawn is the direction of the electric field vector $\mathbf{E}$. Small circles indicate $P < 0.08\%$. Thin lines are used for $P < 0.6\%$ and thick lines for $P > 0.6\%$ (see scale at top left) (after Mathewson and Ford (Ma70)). With the permission of the Officers and Council of the Royal Astronomical Society.

would otherwise be expected, particularly for a grain in thermal equilibrium with its surroundings at the low temperatures of interstellar space. This high spin-up rate is referred to as *suprathermal spin-up*.

If as many as, say, $N \sim 300$ molecules are produced at a catalytic site before the grain surface becomes altered, possibly in response to a cosmic-ray impact, and some other site on the grain becomes active, we might expect angular momenta as high as $L \sim 10^{-20}$ g cm s^{-1}.

PROBLEM 9–6. Show that the angular velocity of the grain is $\omega \sim 10^6$ rad s^{-1} in this case, and that $L \sim 10^{-20}$ g cm^2 s^{-1} also represents the thermal equilibrium value at a temperature of 100 K.

Since we know angular momentum to be quantized in units of $\hbar$ we see that the angular momentum quantum number will typically have a value

$$J \sim 10^7 \hbar . \tag{9-72}$$

A number of other forces illustrated in Fig. 9.14 also affect the grain's angular

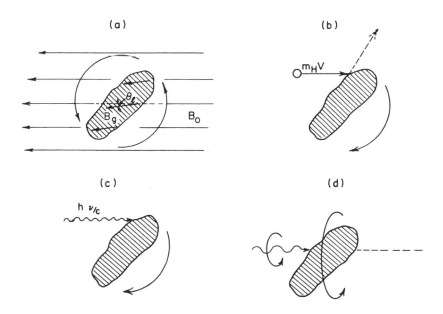

Fig. 9.14. Alignment mechanisms for interstellar grains: (a) process of paramagnetic relaxation; (b) alignment by streaming through gas, or (c) through a photon field. In process (c) the photon's linear momentum causes the grain to spin; in process (d) the photon's intrinsic spin angular momentum is of importance (see text).

momentum. Figure 9.14(b) shows ambient hydrogen atoms incident on the grain applying a torque. However, because the random gas velocities and the translational velocities of the grains through the gas generally are much lower than the speeds at which hydrogen molecules are ejected from the grain's surface after formation, the torques due to these collisions are likely to be negligible compared to those due to the formation and ejection of H_2 molecules. A similar effect due to the impact of starlight on the grain is illustrated in Fig. 9.14(c). Starlight, however, also has an intrinsic spin angular momentum $\hbar$, parallel to the direction of incidence, as shown in Fig. 9.14(d). The respective magnitudes of the angular momenta acquired by a grain per collision with atoms or photons are:

$$\delta L_H \sim \frac{mva}{\sqrt{3}}, \quad \delta L_e \sim \frac{h\nu a}{\sqrt{3}c} = \frac{ha}{\sqrt{3}\lambda}, \quad \text{and} \quad \delta L_i \sim \hbar. \tag{9-73}$$

where λ is the wavelength of incident radiation. These are the angular momenta induced, respectively, by collisions with hydrogen atoms, due to a photon's extrinsic angular momentum (Fig. 9.14(c)), and due to the intrinsic photon angular $\hbar$ (Fig. 9.14(d)). For small grains, the last term predominates over the second

$$\frac{a}{\sqrt{3}} \lesssim \frac{\lambda}{2\pi}. \tag{9-74}$$

It also predominates over atomic collisions if $\sqrt{3}n_\nu c\hbar > nmv^2 a$ or if $n_\nu h\nu > nmv^2$. Within the Galaxy, $n_\nu > 0.02\,\mathrm{cm}^{-3}$ so that for atomic densities $n_H \sim 10\,\mathrm{cm}^{-3}$, and $v \sim 10^5\,\mathrm{cm\ s}^{-1}$, the photon intrinsic effect will dominate.

A more important randomizing effect, however, is due to thermal emission by the grains (Ha70). Equation (4–74) gives this rate as $\sim 1.5 \times 10^{11}T^3\,\mathrm{cm}^{-2}\,\mathrm{s}^{-1}$. But the efficiency of grain emission is roughly proportional to a/λ, and, for thermal radiation the peak wavelength is of order $\lambda \sim 0.36/T$. This leads to a very crude estimate of the emission rate of photons per grain $\dot{n}_{\nu,gr}$ as

$$\dot{n}_{\nu,gr} = 1.5 \times 10^{11}(\pi a^2)T^3 \left(\frac{aT}{0.36}\right) \sim 130 \left(\frac{a}{5 \times 10^{-6}\mathrm{cm}}\right)^3 \left(\frac{T}{30\mathrm{K}}\right)^4 \mathrm{s}^{-1}. \tag{9-75}$$

This is four orders of magnitude higher than the impact rate $n_\nu c\pi a^2 \sim 1.5 \times 10^{-2}\,\mathrm{s}^{-1}$ of Galactic photons on such grains.

We next consider magnetic alignment of dust grains (Da51). In this process, a grain is set spinning about an arbitrary axis, for example, by the ejection of a newly formed high-velocity hydrogen molecule. We can now postulate that the grain material is paramagnetic. Such materials when placed into a magnetic field set up an internal field whose direction is (Fig. 9.14(a)) parallel to the external field. If the grain rotates with angular velocity ω about a direction perpendicular to the magnetic field, the internal field is forced — again at frequency ω — to change its direction relative to an axis fixed in the grain. The internal field, however, cannot change instantaneously. A slight misalignment of the internal and external field arises as shown in Fig. 9.14(a). The interaction of the induced internal field with the externally applied field attempts to compel parallelism by opposing the rotational motion. This drag torque is proportional to the external field **B**, to the internal field that also is proportional to **B**, to the grain volume, V, and to ω,

$$\text{Torque} = KVB^2\omega. \tag{9-76}$$

This holds for a grain spinning about an axis perpendicular to the direction of **B**. When the grain spins about an axis parallel to **B**, the induced field does not need to change its direction relative to the external field and no drag force arises. This damping process is called *paramagnetic relaxation.*

For an arbitrary spin direction, that component of the spin whose axis is perpendicular to the field will therefore be damped out in a time

$$\tau = \frac{I}{KVB^2}, \tag{9-77}$$

where I is the moment of inertia about the spin axis; the spin component whose axis is parallel to **B** remains undamped.

An aspherical grain whose spinning motion is slowed tends to align itself with its *axis of greatest moment of inertia* parallel to the angular momentum axis. This inertia axis is perpendicular to the long axis of an elongated grain. The net effect of paramagnetic relaxation is to align elongated grains with their long axes perpendicular to the magnetic field. There are indications that interstellar grains may be exceptionally paramagnetic, due to a feature termed *superparamagnetism* (Jo67, Go95, Dr99).

9:14 Acoustic Damping and The Barnett Effect

Paramagnetic substances are characterized by unpaired electron spins. If a rotating paramagnetic grain initially has an equal number of electron spins directed parallel and antiparallel to the angular velocity vector ω, it can dissipate its kinetic energy without loss of angular momentum by transferring some of its angular momentum to the electron spin system, in effect by flipping some of the electron spins. This partially aligns the electron spins and their spin magnetic moments, giving rise to a net magnetization. Dissipation is built into this system since, for a grain rotating about some direction other than its principal moment of inertia axis, the rotation axis continually shifts with respect to coordinates fixed within the grain. Stresses produced by centrifugal forces will similarly shift within the grains at the grains' spin frequency, giving rise to acoustic waves. Laboratory studies of various low-temperature materials indicate that most substances are sufficiently inelastic to damp out these shifts and waves in a period of a few years and, with that, also damp out any rotational component not aligned with the principal moment of inertia axis. A second dissipation mechanism arises from the magnetization of a paramagnetic grain as the electron spins flip. The magnetization vector then also precesses in a coordinate system fixed within the spinning grain. This also dissipates energy and ultimately results in the alignment of the magnetic moments of the unpaired spins, as well as the rotational axis, with the grain's principal moment of inertia axis. Both electron and nuclear spins contribute to this effect, which damps out any misalignment of the rotation and principal moment of inertia axes on time scales of $\sim 10^5$ s, roughly a day. An effect of this general class is referred to as a *Barnett effect* (Pu79, La99).

The systematic suprathermal spin-up due to the formation of hydrogen molecules at a small number of catalytic sites on a grain produces relatively steady rotational orientations not readily disrupted by interactions with ambient gas or emitted radiation. Barnett damping then ensures that paramagnetic alignment by interstellar magnetic fields inexorably results in the alignment of both the rotational and the principal moment of inertia axes with the direction of the local interstellar magnetic field.

Observations at far-infrared wavelengths permit us to determine the linear polarization of radiation emitted by grains deep inside a dust cloud. Because the grains radiate preferentially along their longest axes, the observed far-infrared polarization allows us to deduce the direction of the magnetic field deep in the cloud's interior (Hi96).

9:15 Stability of Isothermal Gas Spheres

The plane of the Milky Way is laced with molecular clouds, so dense with gas and dust that they eclipse all visible light from stars at greater distances. Only infrared radiation at progressively longer wavelengths can readily pass through. Some of these clouds are well isolated, roughly spherical, and have well-defined boundaries. Many were originally studied by Bart Bok and are called *Bok globules*. About one third of the globules contain young stars evidently formed within the globules (Aℓ01). Others are devoid of stars. How do such clouds maintain their structure for any length of time? Why do they not collapse under their own gravity to form stars?

To answer these questions we first need to understand the structure of a self-gravitating cloud of gas, its dependence on its temperature, and the influence of chemical composition. For the moment let us restrict ourselves to isothermal clouds at some temperature T.

Instead of phrasing the problem in terms of forces and potentials, as we did in Section 4:23, we start with two related expressions, (8–7) and (8–8), respectively dealing with pressure gradients and distribution of mass,

$$\frac{dP(r)}{dr} = -\frac{GM(r)\rho(r)}{r^2}; \quad \frac{dM(r)}{dr} = 4\pi r^2 \rho(r) . \tag{9-78}$$

These two combine to yield the equation for hydrostatic equilibrium,

$$\frac{1}{r^2}\frac{d}{dr}\left(\frac{r^2}{\rho(r)}\frac{dP(r)}{dr}\right) = -4\pi G\rho(r) , \tag{9-79}$$

which has an obvious similarity to equation (4–151), and shows how the gradient of the gravitational potential is related to the density and pressure gradient. For the ideal gas law (4–37) we have $P = \rho kT/m$, where m is the mass of the hydrogen atoms or molecules in the gaseous sphere. Inserting this into (9–79) gives

$$\frac{1}{r^2}\frac{d}{dr}\left(\frac{r^2}{\rho}\frac{d\rho}{dr}\right) = -\frac{4\pi Gm\rho}{kT}. \tag{9-80}$$

With the substitutions

$$\rho = \rho_0 e^{-\psi} \quad \text{and} \quad r = \left(\frac{kT}{4\pi Gm\rho_0}\right)^{1/2}\xi \tag{9-81}$$

we recover the result (4–154):

$$\frac{1}{\xi^2}\frac{d}{d\xi}\left(\xi^2\frac{d\psi}{d\xi}\right) = e^{-\psi}. \tag{9-82}$$

As we saw in Section 4:23 the sphere described by this equation extends out to infinity and contains an infinite amount of mass. Because the globules we are studying are well confined, we can model them more accurately if we truncate the assumed spherical gas cloud at some radius R_s. To replace the hydrostatic pressure P_s exerted by the gas stretching from R_s out to infinity, we postulate that there is some other external pressure, P_s, due to ambient radiation or tenuous, high-temperature ambient gas, acting on the truncated surface.

We next ask how big and how dense a spherical cloud of temperature T can be before it collapses under its own gravitational pull. If the density is low and the temperature high, thermal pressure will prevent the cloud from collapsing. But if we begin to compress the cloud, keeping its mass M and temperature T constant, the gravitational pull on its surface layers increases while a counteracting thermal pressure acts to resist this pull. We need to see which of these two ultimately prevails.

The analysis is simplified if we concentrate on a combination of high central density and low temperature. This choice is most likely to lead to collapse and implies a large value of ξ and a correspondingly simple density distribution given by equation (9–81).

PROBLEM 9–7. Show that for large values of ξ expressions (9–81) and (4–157) lead to a density $\rho = (kT/2\pi mGr^2)$. Show also that integration of the expression on the right of (9–78), out to a radius R_s, at this density, leads to a total mass $M = 2kTR_s/mG$. In turn, this allows the density to be more simply expressed as

$$\rho(r) = M/4\pi R_s r^2 \tag{9-83}$$

for large central densities and low temperatures. Note that, though the density becomes infinite at the cloud's center, the mass within any volume enclosing the center always remains finite as does the potential.

In this limit the gravitational pull on a volume element at the surface R_s becomes $-MG\rho(R_s)/R_s^2 = -M^2G/4\pi R_s^5$, and the thermal pressure amounts to

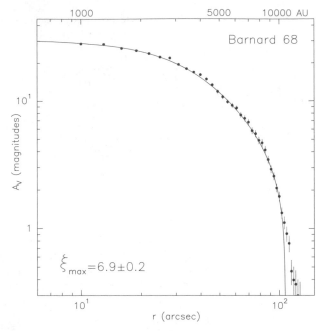

Fig. 9.15. The extinction of radiation by the roughly spherical gas cloud Barnard 68 translated into magnitudes of visual extinction A_v. Data points are shown with error bars; the best fit to produce the observed extinction with an isothermal model of a gas sphere is shown by the solid line. The theoretical fit closely follows observations to nearly the surface of the globule, where deviations from sphericity are significant (Aℓ01).

$\rho(R_s)kT/m = MkT/4\pi mR_s^3$. This clearly shows that gravitation must win out over a resisting thermal pressure as a cloud of mass M is isothermally compressed into progressively smaller volumes $4\pi R_s^3$.

The point at which gravity wins out can be found by solving equation (9–82). This is somewhat laborious and permits no analytical solution, but a sphere can be shown to become unstable at $\xi_s \sim 6.5$ or equivalently $R_s \sim 6.5(kT/4\pi\rho_0mG)^{1/2}$ (Bo56). Early work on isothermal gas spheres was done by R. Ebert and W. Bonnor, and these spheres have come to be known as *Bonnor–Ebert spheres* (Bo56, Eb55).

The collapse radius R_s depends not only on the temperature and central density, but — through the value of m — also on whether the gas is molecular or atomic hydrogen. This is not surprising since the thermal pressure of atomic hydrogen is twice that of the same mass density of molecular hydrogen. Chemical composition plays such an important role in interstellar processes that we speak of *chemodynamics* to indicate the close interdependence of chemical and dynamic processes.

The Bok globule Barnard 68 is an extremely dark molecular cloud. Its density profile has been probed by observing distant stars whose light has traversed the cloud to reach us. The cloud's radial distribution of gas and dust is determined by measuring the absorption of infrared radiation at several different wavelengths as a function of angular distance from the center. Figure 9.15 shows the deduced absorption profile for the cloud, and the corresponding absorption for an isothermal sphere having identical absorption per unit volume and $\xi = 6.9 \pm 0.2$. The value of ξ suggests that the globule is marginally unstable, and should at some time collapse to form stars. The cloud lies at a distance of ~ 100 pc, has a temperature $T \sim 10\,\mathrm{K}$ determined from the relative populations of different excited levels both of $^{12}\mathrm{CO}$ and ammonia molecules, a central density $n_{\mathrm{H_2}} = 3.3 \times 10^5$ cm^{-3}, and radius $R_s \sim 1.75 \times 10^4$ AU. The mass content of the cloud is $\sim 2 M_\odot$. The surface pressure required by the fit to an isothermal distribution is of order 10^{-11} dyn cm^{-2}, which is in rough agreement with other estimates of the ambient gas pressure in that part of the Galaxy. The close correspondence of the observed extinction profile to that of a Bonnor–Ebert sphere suggests that Barnard 68 is an isothermal pressure-confined, self-gravitating cloud (Aℓ01).

9:16 Polytropes

Although the good fit to Barnard 68 provided by an isothermal cloud of gas and dust appears convincing, there are many other interstellar clouds, including Bok globules where conditions may be more complex. In some cases a central isothermal globule or core may be embedded in a surrounding envelope in which conditions are adiabatic (Cu00). Turbulent gas clouds permeated by shocks and magnetic fields and *photon-dominated regions* at ionization fronts are also likely to adhere to adiabatic rather than isothermal conditions, as we saw in Section 9:4.

Where adiabatic conditions prevail expressions (4–129) play a central role. To the extent that the chemical composition of a cloud can critically alter the ratio of heat capacities, γ, cloud chemistry can again be decisive in determining the stability conditions of Section 4:22. Turbulent velocities in such clouds play a role similar to those of thermal velocities and may be assigned a commensurate temperature.

Replacing the volume V in expressions (4–129) by its reciprocal, the density ρ, leads to

$$P = K\rho^\gamma \equiv K\rho^{(n+1)/n} , \tag{9-84}$$

where n is a pure number, a characterizing constant called the *polytropic index*. The ratio $(n + 1)/n$ is just the ratio of specific heats γ in (4–125). The polytropic index $n = \frac{3}{2}$ is particularly important since it corresponds to $\gamma = \frac{5}{3}$, which applies to regions consisting of atomic hydrogen, molecular hydrogen at low temperatures, or fully ionized hydrogen. It also applies to nonrelativistic degenerate fermions and, as discussed in Sections 8:14 and 8:16, to the physics of white dwarfs and neutrons stars.

We now assume that the density takes the form $\rho = \lambda\theta^n$, where λ is a constant and θ, called the *polytropic temperature*, is a function of r. The equation of hydrostatic equilibrium (9–79) then becomes

$$\left[\left(\frac{(n+1)K}{4\pi G}\right)\lambda^{-(n-1)/n}\right]\frac{1}{r^2}\frac{d}{dr}\left(r^2\frac{d\theta}{dr}\right) = -\theta^n . \tag{9-85}$$

PROBLEM 9–8. By substitution, show that (9–85) has a solution of the form $\theta = Ar^q$, with $q = -4$ for $n = \frac{3}{2}$. This yields a density proportional to r^{-6}, which makes both the density and mass infinite at $r = 0$. This is not a physically meaningful solution and we therefore need to seek another possibility.

Let us choose a new variable for insertion in (9–85)

$$\xi = \frac{r}{\alpha}, \quad \text{with } \alpha = \left[\left(\frac{(n+1)K}{4\pi G} \right) \lambda^{-(n-1)/n} \right]^{1/2} . \tag{9-86}$$

This leads to

$$\frac{1}{\xi^2} \frac{d}{d\xi} \left(\xi^2 \frac{d\theta}{d\xi} \right) = -\theta^n , \tag{9-87}$$

called the *Lane–Emden equation*. Note the resemblance of this expression to equation (9–82), which arose from similar physical considerations but assumed an isothermal distribution of particle energies rather than adiabatic conditions.

PROBLEM 9–9. By substituting

$$\theta = 1 - \frac{1}{6} \xi^2 + \frac{n}{120} \xi^4 - \frac{1}{42} \left(\frac{n^2}{120} + \frac{n(n-1)}{72} \right) \xi^6 + \cdots \tag{9-88}$$

into (9–87) show that this expression provides a solution with finite central density, since θ approaches unity at $\xi = 0$, and that the proportionality constant λ is just the central density ρ_0. Note resemblances to, and differences from, the solution (4–156) given in Problem 4–32. With a more precise approximation one can also show that, for $n = \frac{3}{2}$, θ drops to a first zero at a value of $\xi \equiv \xi_R = 3.65375$ (Ch39). We can consider this to be the radius R of a finite gravitationally bound mass distribution whose central density is ρ_0.

PROBLEM 9–10. By integrating the second of equations (9–78) and substituting for r and ρ show that the total mass of the aggregate enclosed within radius ξ_R is

$$M = -4\pi \left(\frac{5K}{8\pi G} \right)^{3/2} \left(\xi^2 \frac{d\theta}{d\xi} \right)\bigg]_{\xi=\xi_R} \rho_0^{(3-n)/2n} . \tag{9-89}$$

A detailed computation using the expansion (9–88) shows that for $n = \frac{3}{2}$, $\xi^2(d\theta/d\xi)$ has a value of 2.71406 at $\xi = \xi_R$, so that

$$M_R = -4\pi \left(\frac{5K}{8\pi G} \right)^{3/2} (2.71406)\rho_0^{1/2} . \tag{9-90}$$

9:17 The Nature of Dark Matter

A wide variety of observations gathered over recent decades have indicated the existence of *dark matter* that has, to date, made itself known solely through its gravitational attraction for matter and radiation.

Dark matter appears to dominate the gravitational potential of clusters of galaxies. X-ray emitting intracluster gas has such high temperatures that it could not stay gravitationally bound unless cluster masses were far higher than the detected baryonic mass. Similar conclusions arise from observations of the apparently random velocities of galaxies in a cluster (Sections 3:15 and 4:4) and from the gravitational time delay and bending of light by galaxies and clusters (Section 5:14).

Large quantities of dark matter are also inferred from the distribution of hot gases surrounding some elliptical galaxies. These gaseous coronae appear to be in hydrostatic equilibrium, so equation (8–7) can be applied. Measurement of the X-ray surface brightness of the gas, taken together with its temperature derived from its X-ray free–free emission — *Bremsstrahlung* — tell us both the local gas pressure and density in an ideal gas approximation. The derived total mass gravitationally attracting this gas is estimated to be from 1 to $5 \times 10^{12} M_\odot$ out to 100 kpc in some of the best-studied cases (Fa85). These elliptical galaxies appear to have a *mass–luminosity ratio* of $\sim 100 \, M_\odot/L_\odot$. The dark matter clearly is far fainter than mass found in solar-mass stars.

Let us still look at how much of the dark matter may be distributed in a spiral galaxy's disk, as compared to its spherical halo.

PROBLEM 9–11. A measure of the disk mass is obtained from the vertical distribution of stars above and below the plane of the Milky Way. Consider stars at height z above or below the plane and picture the plane as a disk of uniform areal density σ and radius R. For a disk thin in comparison to the heights z to which the stars rise, show that the force acting on unit stellar mass located along the disk axis is (Fig. 9.16)

$$F(z) = - \int_0^R \left(\frac{2\pi r \sigma G}{(r^2 + z^2)} \right) \frac{z}{(r^2 + z^2)^{1/2}} \, dr = -2\pi \sigma G \quad \text{for } R \gg z. \quad (9\text{-}91)$$

From this show that for a large disk the *scale height* is given by

$$h = \frac{v^2}{4\pi \sigma G}, \quad (9\text{-}92)$$

where v is a measure of the vertical component of stellar random velocity in the Galaxy's plane.

For a galactic disk having mass $10^{10} \, M_\odot$, radius 10 kpc, and hence a uniform areal density 7×10^{-3} g cm^{-2}, the scale height would be 56 pc for a random velocity component of 10 km s^{-1}. More rigorous calculations and observations of the

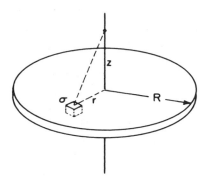

Fig. 9.16. Diagram to illustrate the Galaxy's scale height for stars passing through the plane.

distribution and velocities of stars in the Galaxy's disk show that, although the disk appears to be more massive than can be accounted for by the areal density of stars, gas and dust, dark matter appears to no more than double the total disk mass (Ba84). Most of the gravitational mass appears to be in a more spherical halo distribution.

Figure 1.13 shows the rotation curves of two typical spiral galaxies . They reflect the apparent existence of gravitating mass not accounted for either in luminous matter — stars — or as gas and dust. The rotational velocities $v(r)$ lie in the range 150 to $300\,\mathrm{km\,s}^{-1}$ and remain nearly constant out to large radii r from a galaxy's center. Faint traces of gas in a galaxy's disk, orbiting at far larger radii than normally associated with the visible mass, still show the mass $M(r)$ enclosed within a radius r to still be rising.

If the velocities are Keplerian and truly constant, equation (3–44) would imply that the gravitational mass $M(r)$ enclosed within a radius r from a galaxy's center is proportional to r,

$$M(r) = \frac{[v(r)]^2 r}{G} .$$ (9-93)

The observed velocities imply a gravitational mass $M(r)$ considerably in excess of the mass of observed stars or interstellar gas and dust. For a spherical mass distribution throughout the galaxy we have

$$M(r) = 4\pi \int_0^r \rho(r) r^2 dr = \frac{[v(r)]^2 r}{G}, \qquad \rho(r) = \frac{[v(r)]^2}{4\pi G r^2} ,$$ (9-94)

implying that the density drops as $1/r^2$. Although this derived density profile suffices for present purposes, a rather better radial fit is presented in Section 13:23.

A number of explanations for the discrepancy between the mass of observed stars, gas, and dust and the derived gravitational mass have been postulated:

1. The most radical suggestion is that general relativity does not correctly represent the laws of gravitation on scales exceeding $\sim 10^{21}$ cm. Mordehai Milgrom

has for many years championed a modified Newtonian dynamics (MOND) that describes the observed dynamics without assuming the existence of unobserved mass (Mi95a). Though we might ultimately be driven to such a new dynamics, it could require a massive re-examination of all of astrophysics. Giving up general relativity, our best current theory of gravitation, would be a drastic step. Most astronomers have preferred to look at other alternatives.

2. Cool stars, particularly low-mass brown dwarfs were long considered possible candidates for dark matter, but the *Massive Compact Halo Object (MACHO)* survey, that searches for gravitationally lensed background objects rules these out. As Fig. 5.14 shows, distant stars and quasars occasionally are observed to suddenly flare up for a period of days, as they pass behind a faint foreground star or planet. Although an appreciable number of such lensing phenomena have been observed, the survey has all but ruled out that sufficient dark matter to account for the dominant gravitational mass of the Galaxy's halo could be found in objects with masses in the range $\gtrsim 3 \times 10^{-8} M_\odot$ (Aℓ96, Yo04). The observed amount of infrared radiation due to stars in the mass range $0.1 M_\odot$ can also account for only $\sim$3% of the halo's mass. Stars of even greater mass are also ruled out because, unless they were faint white dwarfs or neutron stars, they would easily have been detected by now through their optical and near-infrared emission (Bo95).

3. Neither can the dark matter be *baryonic*, as we will see in Section 12:13, and in Fig. 12.5; otherwise the amount of ^{4}He and ^{7}Li formed in the first few minutes of cosmic evolution would have been higher than observed. The helium content of the Universe is quite sensitive to the total density of baryons at that time.

Neutrinos also appear ruled out, because they are fermions with too low a rest–mass. Current laboratory experiments place an upper limit of a few electron volts and possibly $\lesssim$0.5 eV on the mass of the electron neutrino (No97). Upper limits to neutrino masses derived from microwave background polarization measurements suggest masses $\leq$0.23 eV (Sp03). To contribute a significant amount of mass to a galaxy, the density of each of the three neutrino species and their antiparticles, all of whose rest–masses are expected to be close to equal, would have to be of order $\rho_\nu \sim 2 \times 10^{-25}$ g cm^{-3}. A rest–mass m_ν then implies a number density $n_\nu = \rho_\nu / m_\nu$. But by equation (7–1) the neutrino momenta would then have to be $m_\nu v_\nu \gtrsim n_\nu^{1/3} \hbar$ and

$$m_\nu \gtrsim [\hbar \rho_\nu^{1/3} / v_\nu]^{3/4} . \tag{9-95}$$

To keep the neutrino velocities below escape velocity from the galaxy, $v \sim 500$ km s^{-1}, their masses would have to be $\gtrsim 6 \times 10^{-33}$ g ~ 3.5 eV, which appears to be observationally ruled out.

4. Primordial black holes discussed in Section 5:24 can also be largely ruled out as a dominant source of dark matter. They have been searched for by the gravitational lensing they might produce, and are ruled out by the MACHO surveys within the same mass limits placed on planetary and stellar masses. Black holes of very low mass can also be ruled out. As discussed in Section 5:24, if their masses were well below 10^{15} g they would by now have exploded. Nevertheless, the existence of dark matter in the form of well-isolated primordial black holes in the mass range

10^{15} g $\lesssim M \lesssim 6 \times 10^{25}$ g cannot yet be ruled out. Larger masses than this would presumably lead to noticeable disruption of globular clusters and molecular clouds.

5. A variety of exotic *weakly interacting massive particles, WIMPS*, could also account for the missing mass. Physicists interested in fundamental particles have taken an especial interest in these, and active searches for them are under way in several laboratories (Wa97). Among them are *axions* of masses 10^{-6} to 10^{-7} eV and a neutralino of mass 10 to 1000 GeV predicted by some supersymmetric theories (Tu97). Whatever the nature of the dark matter particles, they cannot be relativistic. Their velocities v must be sufficiently low for them to remain gravitationally bound to a galaxy. In this sense they are cold, and we speak of *cold cark matter*, often referred to by the initials *CDM*.

Additional Problems

9–12. The *solar wind* is produced by the high temperature $\sim 2 \times 10^6$ K of the solar corona. The wind velocities, which are ~ 400 km s^{-1}, are higher than the thermal velocities of protons ~ 130 km s^{-1} in ionized hydrogen at that temperature. Show that this difference can be partly accounted for because randomly directed velocities in a confined gas all become projected onto a radial direction when the gas freely expands into a much larger volume where collisions between particles no longer are important. A similar process probably plays a role in all stellar winds. Assume equipartition of energy among protons, electrons, and magnetic fields in the corona.

9–13. In a *comet*, ionized gas is propelled into a straight tail pointing away from the Sun. Molecules such as CO, initially in the comet's head, suffer charge exchange with protons of the solar wind. In this process, which has a high cross-section $\sim 10^{-15}$ cm^2, an electron is transferred from the CO molecule to the proton. The newly formed CO$^+$ ion is now swept along by the magnetic field embedded in the solar wind. The magnetic field is predominantly transverse to the wind direction. If the solar wind velocity is 400 km s^{-1} compute the velocity of ions in the comet's tail if roughly 10% of the protons on any given magnetic line of force undergo charge exchange.

Answers to Selected Problems

9–1. By (4–124), $c_v = R/(\gamma - 1)$. The ideal gas law for unit mass states $T = P/R\rho$. This gives the result (9–30).

9–2. Because $u \gg c_s$, we can write

$$v_D = \frac{1}{3}\left(-2u + 2u\sqrt{1 + \frac{9c_s^2}{4u^2}}\right) \sim \frac{2u}{3}\left(-1 + 1 + \frac{9c_s^2}{8u^2}\right).$$

9–3. Table 9.4 shows a collision mean free path of order $20\,\mathrm{g\,cm^{-2}}$, which for an interstellar matter density of $10^{-23}\,\mathrm{g\,cm^{-3}}$ gives a path of order $2 \times 10^{24}\,\mathrm{cm}$ and a life of $\sim 6 \times 10^{13}\,\mathrm{s}$.

$$\therefore \quad \frac{d\mathcal{E}}{dt} \sim 1.6 \times 10^{-14}\,\mathcal{E}\,\mathrm{s^{-1}}.$$

The loss rates in Table 9.3(b) are of order $10^{-6}Z^2\,\mathrm{eV\,s^{-1}}$, so that collisional losses dominate for all nuclei with energies in excess of $\sim 10^9\,\mathrm{eV}$.

9–4. For electrons typical losses from Table 9.3(a), for the most significant part of the energy range covered in Fig. 9.8, are $10^{-6}\,\mathrm{eV\,s^{-1}}$. At $\sim 10^8\,\mathrm{eV}$ the life is $\sim 10^{14}\,\mathrm{s}$. Taken together with the result of Problem 9–3, this indicates a cosmic-ray energy loss to the interstellar medium of $10^{-12}\,\mathrm{erg}\ \mathrm{cm^{-3}}/10^{14}\,\mathrm{s}$ or $10^{-26}\,\mathrm{erg\,cm^{-3}\,s^{-1}}$.

If integrated over a gas-containing volume of $10^{66}\,\mathrm{cm^3}$ in the Galaxy, this would indicate an eventual radiation loss of order $10^{40}\,\mathrm{erg\,s^{-1}}$. The total luminosity of the Galaxy is 10^3 to 10^4 times higher; but only about 10% of this luminosity may contribute to heating the interstellar medium; and in the darkest clouds, where radiation does not readily penetrate, cosmic-ray heating is the dominant factor.

9–5. Our assumption is that $n_H R a_\nu = 1$. Because most atoms are in the lowest state, we use $a_{\nu o} \sim 6.3 \times 10^{-18}\,\mathrm{cm^2}$ so that

$$R \sim \frac{1.6 \times 10^{17}}{n_H}\,\mathrm{cm}\ .$$

In actuality, the ionization cross-section drops as $a_\nu \sim a_{\nu o}(\nu/\nu_o)^3$ for $\nu > \nu_o$, which lengthens the mean free path for higher-energy ionizing photons.

(a) For Ly-α, the oscillator strength f cited in Table 9.2 is 0.436 so that

$$\sigma \sim \frac{3\lambda^2 f}{2\pi} \sim 3 \times 10^{-11}\,\mathrm{cm^2}.$$

This would give an absorption distance

$$\frac{1}{\sigma n_H} \sim \frac{3 \times 10^{10}}{n_H}\,\mathrm{cm} \sim 2 \times 10^{-7}\,R.$$

However:

(b) A Doppler shift away from the central absorption frequency reduces the ability of an atom to absorb radiation. The mean absorption cross-section therefore is $(c\gamma)^2 \sigma_{max}/16\pi^2\nu^2\langle v^2\rangle$ and, for $\langle v^2\rangle^{1/2} = 30\ \mathrm{km\ s^{-1}}$, the mean free path is

$$\sim \frac{16\pi^2\nu^2\langle v^2\rangle}{(c\gamma)^2\sigma_{max}n_H} \sim \frac{16\pi^2\langle v^2\rangle}{(\lambda)^2(25\lambda^4)^{-1}(3\lambda^2/2\pi)f n_H} \sim \frac{7.4 \times 10^{16}}{n_H}\,\mathrm{cm}.$$

where σ_{max} is the cross-section at line center.

(c) To move a frequency interval $n\gamma$ off line center, in random walk steps of frequency shift $\Delta\nu$, would require $N = (n\gamma/\Delta\nu)^2$ steps (see equation (4-11)).

9–6. For a gas at $T \sim 100\,\mathrm{K}$, $kT \sim 1.4 \times 10^{-14}\,\mathrm{erg}$, and the grain that has moment of inertia $I \sim 10^{-26}\,\mathrm{g\,cm^2}$ has $\omega \sim 10^6\,\mathrm{rad\,s^{-1}}$ to make $kT \sim I\omega^2$. The angular momentum $L \sim I\omega$ therefore has a thermal equilibrium value that is also $L \sim 10^{-20}\,\mathrm{g\,cm^2\,s^{-1}}$. This is no coincidence. The "random" angular momentum acquired in $N = M/m$ collisions becomes the "systematic" angular momentum to be altered by the next generation of N collisions. These collisions endow the grain with a random angular momentum of the same magnitude as its initial value but oriented in some other arbitrary direction.

9–7. The two expressions in (9–81) give

$$\frac{4\pi Gm\rho_o r^2}{kT} = \frac{4\pi Gm\rho e^{\psi} r^2}{kT} = \xi^2 \ .$$

Taking logarithms of both sides and using (4–157) gives

$$\ln(4\pi Gm\rho r^2/kT) + \psi = \psi + \ln 2$$

or $\rho = kT/2\pi Gmr^2$. This leads to $M(R_s) = 2kTR_s/Gm$ and $\rho = M/(4\pi R_s r^2)$.

9–11. The force per unit mass integrated over the whole disk is

$$F(z) = -\int_0^R \left(\frac{2\pi r\sigma G}{r^2 + z^2} \right) \left(\frac{z}{(r^2 + z^2)^{1/2}} \right) dr = \frac{2\pi\sigma zG}{R} - 2\pi\sigma G \ ,$$

where the first term in parentheses is the force along the direction to a particular surface element and the second term in parentheses provides the component perpendicular to the plane. For $R \gg z$ this gives the desired result on integration. For a large disk the force changes only along a direction perpendicular to the plane near the disk axis and is $-2\pi\sigma Gz$. The height $h = (v^2/2)/(2\pi\sigma G)$ is the distance from the plane to which stars or gas clouds with kinetic energy per unit mass $v^2/2$ can rise against this force.

9–12. The total energy per proton in the corona is $3kT/2$. Because of the magnetic field, protons and electrons will be moving together in the solar wind expansion and the energy of random motion can be transferred into expansion velocity. The total magnetic energy can also decrease, at the expense of particle velocity, because $B \propto r^{-2}$ and the energy that is proportional to $B^2 r^3$ will be proportional to r^{-1}. The three sources of coronal energy — protons, electrons, and magnetic fields — provide an energy of $9kT/2$ for each hydrogenic mass m_H streaming away from the Sun. For a coronal temperature $T = 2 \times 10^6\,\mathrm{K}$, we have $v \sim (9kT/m_H)^{1/2} \sim 4 \times 10^7\,\mathrm{cm\,s^{-1}}$.

9–13. If 10% of the protons undergo charge exchange, then the momentum carried by the others must be shared with the captured CO^+ ions. These are 28 times as massive as a proton. Very little momentum is transferred by the proton that exchanges charge so that only the remaining protons supply momentum. This brings about a velocity reduction to a fraction $nM_H/[(n-1)m_H + M_{CO}] \sim 0.27$ of the original solar wind velocity. For an initial velocity of $400\,\mathrm{km\,s^{-1}}$ for protons, the final ion velocities would be $\sim 100\,\mathrm{km\,s^{-1}}$.

10 Formation of Stars and Planetary Systems

The Milky Way may contain as many as a hundred billion stars. Yet we have only the most incomplete notions of how all these stars formed. Within the past decade, we have also become aware of the many planetary systems orbiting all these stars. But we are still uncertain even about how our own system of planets condensed. In this chapter we examine the formation of stars and planetary systems, establish some of the mechanisms that appear to be at work, and describe ways in which further insight may be gained.

10:1 Star Formation

Three primary steps are required for a star to form: (1) a protostellar cloud of gas and dust has to radiate away energy to become increasingly compact; (2) it must reduce its angular momentum from the high values seen in diffuse hydrogen clouds to the low values observed in stars; and (3) it must dissipate its magnetic field, identified by Faraday rotation and Zeeman splitting, to yield the relatively low values observed in stars. To these three requirements we may add one more that apparently is met rather easily: (4) subjected to rapid compression — often induced by supernova explosions or the rapid expansion and intense radiation pressures of ambient ionized hydrogen regions — a cloud's contraction must be sufficiently inelastic, and accompanied by sufficiently rapid cooling and energy loss, to prevent expansion back to an appreciable fraction of the cloud's initial size.

Shocks around HII regions can rapidly compress cool clouds as we saw in Section 9:4. The shocked clouds continue to contract through radiative cooling via grain emission or radiation by H_2, H_2O vapor, CO, atomic oxygen, or singly ionized carbon, depending on the temperature, density, and ionization fraction of the gas. Criteria (1) and (4) therefore appear to be met. The shedding of angular momentum (2) and dissipation of magnetic fields (3), however, still pose unsolved problems. We will only be able to see one or two possible ways out of these difficulties.

Let us consider the angular momentum problem first. Differential rotation about the center of a galaxy induces a velocity shear in an interstellar cloud. This has an associated angular momentum about the cloud center. Contraction of the cloud preserves this angular momentum.

One way in which a cloud might contract, and still keep the angular momentum due to differential rotation low, might be to preferentially gather material with low angular momentum about the protostellar cloud center. For gas in circular motion about a galaxy's center this could be material gathered from a ring at some fixed radius R_c from the center. As equation (3–44) tells us, the angular momentum per unit mass about the center is $\sqrt{MGR_c}$, where M is the mass of the galaxy enclosed within radius R_c. If contraction of the cloud brings equal amounts of mass from regions at $(R_c - \Delta)$ and $(R_c + \Delta)$ to the cloud center at R_c a net amount of angular momentum per unit mass

$$L = \left(\frac{1}{2}\right) \left| \sqrt{MG(R_c - \Delta)} + \sqrt{MG(R_c + \Delta)} \right| \sim \frac{1}{8} \left(\frac{\Delta}{R_c}\right)^2 \sqrt{MGR_c} \quad (10\text{-}1)$$

would accrue. At the Sun's distance from the Galactic center, $R_c \sim 8$ kpc, and for $M \sim 2 \times 10^{44}$ g and $\Delta/R_c \sim 1.5 \times 10^{-6}$, or $\Delta \sim 3.6 \times 10^{16}$ cm, this amounts to $L \sim 1.5 \times 10^{17}$ cm^2 s^{-1} per unit mass, roughly comparable to the angular momentum per unit mass of the Solar System. The ring can therefore only have a total width $W \sim 3\Delta \sim 10^{17}$ cm, before the average angular momentum of the contracting cloud becomes excessive. Now, the areal mass density in the Galactic plane, i.e., the mass contained in a column perpendicular to the Galactic plane, is of order $\sigma \sim 10^{-4}$ g cm^{-2}. To gather a solar mass of 2×10^{33} g, the length ℓ of the circular segment from which the star forms would need to be $\ell \sim M_\odot/W\sigma \sim 2 \times 10^{20}$ cm or ~ 70 pc along an arc of radius $R_c \sim 8$ kpc.

Whether such a contraction could occur along an arc two thousand times longer than it is wide is doubtful. Although the main star-forming region in the Orion complex does exhibit long tubular structures of dense material at low relative velocities (Wi96), gravitational contraction is most effectively sustained if the contracting volume is roughly spherical (Eb55). For then a small compression in volume can amount to a relatively large change in potential energy, and rapid loss of energy through cooling will assure that the region remains collapsed — resistant to disruption by external turbulence. But this requires an efficient means for shedding angular momentum. While turbulence could introduce sufficient viscosity to transport away angular momentum, we know too little about turbulent transport to reliably estimate this effect.

Problems caused by the compression of embedded magnetic fields also remain unresolved in a contraction of the kind just described. If the magnetic field lines were largely normal to the radius vector R_c — which might be favored by shear produced through differential rotation — then contraction along the magnetic field would not be resisted by the field, but contraction perpendicular to the field direction would still necessitate compression of field lines. For a typical galactic magnetic field $B \sim 10^{-6}$ G, compression from a cross-section $HW \sim 10^{36}$ cm^2, where H is the scale height of the column of gas above and below the galactic plane, down to Solar System dimensions $\sim 10^{30}$ cm^2 would produce a field of the order of one gauss. Further compression of the material into an object the size of the Sun would result in a field of order 10^8 G. This magnetic field would have to be dissipated before a star like the Sun could form with a field of 1 G.

We will discuss these questions in greater detail in this chapter, but will find that the answers to many questions remain poorly understood.

10:2 Gravitational Condensation of Matter

Let us first consider the contraction of an isolated cloud unburdened by angular momentum or embedded magnetic fields. Initially it is neither expanding nor contracting; but if it is sufficiently cool it will begin to collapse under its own gravitational attraction. Consider the density of the cloud to be uniform throughout a sphere of radius r_0. Then the forces on matter at the surface of the sphere will produce an acceleration

$$\ddot{r} = -\frac{GM}{r^2} .$$

(10-2)

Integration with respect to time yields

$$\frac{\dot{r}^2}{2} = \frac{4\pi}{3} r_0^2 \rho_0 G \left(\frac{r_0}{r} - 1\right),$$

(10-3)

where ρ_0 is the initial density.

PROBLEM 10–1. Show that this equation can be integrated a second time, for example, through a substitution of a new variable u given by $r = r_0 \sin^2 u$, to give the free-fall collapse time

$$t_{\text{ff}} = \sqrt{\frac{3\pi}{32G\rho_0}} .$$

(10-4)

Note that this expression is independent of the size of the cloud. A second point of interest is that the free-fall expression is unaffected by any spherical distribution of matter that lies outside the cloud. This can be seen by viewing the potential within a spherical shell of radius R due to a mass distribution whose surface density is $M/4\pi R^2$ everywhere on the sphere.

PROBLEM 10–2. Show that the potential anywhere within this sphere is

$$V = -\frac{MG}{R} ,$$

(10-5)

and further show that the potential due to any spherical distribution of matter outside an empty sphere has a value that is constant throughout the sphere.

For any spherically symmetric distribution of matter in spherically symmetric motion, the dynamics within a central sphere always remain unaffected by the distribution outside. This result, which is also valid in general relativity and has the most wide-ranging consequences, is attributed to George Birkhoff, who first showed its generality in what has come to be known as *Birkhoff's theorem* (Bi23). We will discuss this theorem further in the context of galaxy formation in Chapter 13.

10:3 Jeans Criterion

Let us recall the acoustic waves discussed in Section 9:3. When a gas cloud is subject to its own gravitational potential ϕ_1, not only the pressure but also gravitational gradients have to be taken into account since the gas is accelerated by both. Equation (9–21), the Euler equation, then becomes

$$\frac{\partial \mathbf{v_1}}{\partial t} + \mathbf{v_1} \cdot \nabla \mathbf{v_1} = -\frac{1}{\rho_0} \nabla P_1 - \nabla \phi_1 . \tag{10-6}$$

The potential ϕ_1 is given by the Poisson equation (4–151)

$$\nabla^2 \phi_1 = 4\pi G \rho_1. \tag{10-7}$$

Using (9–22) and (9–25) we now write (10–6) as

$$\frac{\partial \mathbf{v_1}}{\partial t} + \mathbf{v_1} \cdot \nabla \mathbf{v_1} = -\frac{c_s^2 \nabla \rho_1}{\rho_0} - \nabla \phi_1 . \tag{10-8}$$

Taking the divergence of both sides of this equation and neglecting second-order terms, we obtain the analogue of (9–23)

$$\frac{\partial^2 \rho_1}{\partial t^2} = 4\pi G \rho_1 \rho_0 + c_s^2 \nabla^2 \rho_1, \tag{10-9}$$

where we have used Poisson's equation to substitute for $\nabla^2 \phi_1$. This yields a density wave of form

$$\rho_1 = A_g \exp[i(2\pi x/\lambda - \omega t)] \qquad \text{(RP)}, \tag{10-10}$$

where only the real part is considered. The angular frequency of the wave is

$$\omega^2 = \left(\frac{2\pi c_s}{\lambda}\right)^2 - 4\pi G \rho_0 \tag{10-11}$$

and the propagation velocity is

$$c_g = \frac{\lambda \omega}{2\pi} = \left(c_s^2 - \frac{G\lambda^2 \rho_0}{\pi}\right)^{1/2} = c_s \left(1 - \frac{G\lambda^2 \rho_0}{\pi c_s^2}\right)^{1/2}. \tag{10-12}$$

From this we see that, whenever the wavelength exceeds

$$\lambda_J = c_s \left(\frac{\pi}{G\rho_0}\right)^{1/2}, \tag{10-13}$$

the disturbance no longer propagates as a wave but grows exponentially as substitution into equation (10–11) shows.

The *Jeans length* λ_J is named for James Jeans, who first noted the tendency for a self-gravitating cloud to collapse under its own weight when perturbed at sufficiently long wavelengths. Shorter wavelengths are propagated away at the speed of sound. The Jeans instability is believed to play a role in the contraction of cool gaseous clouds to form new stars. In order to contract, the cloud has to be large, dense, and cool, to maximize $\lambda^2 \rho_0 / c_s^2$.

One note may still be of interest: at several places in this book we have encountered similarities in the behavior of gases subjected to Coulomb forces as compared to Newtonian gravitational forces. These similarities arose in the scattering of particles (Section 3:13), in the behavior of ionized gases and assemblies of stars (Section 4:23), as well as in the absorption of radiation in a medium (Section 6:18). Here we have considered the evolution of a disturbance in a neutral cloud in which only gravitational forces are at play. A comparison shows that results already derived in Section 6:18 can be directly adopted. We recall that we were concerned there with a wave propagating through an ionized medium. By analogy, we consider here a neutral medium in which small density perturbations have arisen. The perturbations $\delta\rho$ can have either a positive or a negative sign. A region of higher density than the average is attracted toward a massive center, while a region of low density is buoyantly repelled. In equation (4–147) we pointed out that gravitational behavior could be derived from electrostatic analogues if we replace the product of charges $Q_1 Q_2$ by $-G m_1 m_2$, where G is the gravitational constant and m_1 and m_2 are interacting masses. In this spirit we turn to Section 6:11 and ask how a somewhat perturbed medium acts under its own gravitational influence. We then find that a density perturbation of the form (6–52), $f = f_0 \cos(kx \pm \omega t)$, in a medium of mean density ρ obeys a *dispersion relation* — relation between frequency and wave number — of the form of (6–54) which, with the help of (4–147), becomes

$$\omega^2 = k^2 c_s^2 - 4\pi G \rho_0. \tag{10-14}$$

Here ω is imaginary for wave numbers k below a critical value

$$k_J = \left(\frac{4\pi G \rho_0}{c_s^2} \right)^{1/2} = \frac{2\pi}{\lambda_J}. \tag{10-15}$$

This is the same result as (10–13).

The negative sign of the gravitational term in equation (10–11) implies an exponential growth of the disturbance with an e-folding time

$$\tau = \frac{2\pi}{i\omega} = \frac{2\pi}{c_s(k_J^2 - k^2)^{1/2}} \quad \text{for} \quad |k| < k_J. \tag{10-16}$$

The *Jeans mass*, the mass enclosed in a sphere of radius $\lambda_J/2$, is

$$M_J = \frac{4\pi}{3} \rho_0 \left(\frac{\lambda_J}{2} \right)^3 = \frac{\pi^{5/2} c_s^3}{6 G^{3/2} \rho_0^{1/2}}. \tag{10-17}$$

We note that expression (8–1) can give us the total potential energy within a sphere of radius $\lambda_J/2$. When this is divided by the total kinetic energy for the gas we obtain from equation (10–17)

$$\frac{\text{potential energy}}{\text{kinetic energy}} \sim \frac{(3/5)M_J^2 G/(\lambda_J/2)}{M_J(3c_s^2/2\gamma)} \sim \frac{2\pi^2\gamma}{15} \geq 1, \qquad (10\text{-}18)$$

where γ is the ratio of heat capacities

Very roughly, the Jeans criterion tells us that contraction can occur when the potential energy due to self-gravitation exceeds the kinetic energy of random thermal motion in the gas.

10:4 Hydrostatics of Gaseous Clouds

From a somewhat different perspective, we may assume that the formation of stars takes place in clouds of gas that somehow have condensed, as galaxies have formed. The formation of stars then takes place in substantially denser, usually cooler regions where the density ρ is high and the speed of sound c_s is low. Instead of talking of the Jeans criterion, we may then return to the concept of hydrostatic equilibrium. Here, we return to equation (8–7) and note that a spherical cloud will contract if the pressure at its center is not sufficiently high to withstand compression from matter gravitationally bearing down on it from larger radial distances. For a cloud of uniform density, the pressure at the center is obtained by integrating (8–7):

$$dP = -\left\{\frac{\rho G M(r)}{r^2}\right\} dr = -\left[\frac{4\pi\rho^2 G r}{3}\right] dr \qquad (10\text{-}19)$$

and the cloud will collapse unless the central gas pressure $\rho k T/m$ exceeds the value

$$P = \frac{\rho k T}{m} > \frac{\rho G M}{2R}, \qquad (10\text{-}20)$$

where M is the mass contained within radius R, and m is the mass of a hydrogen atom or molecule. If the temperature of a molecular cloud is $20\,\text{K}$, collapse would require $R \sim 10^{18}\,\text{cm}$ for $M = 2 \times 10^{34}\,\text{g}$. However, while the temperatures in molecular clouds often appear to reach such low values and hence would appear to promise that stars of several solar masses could easily form, thermal velocities rarely predominate. At $20\,\text{K}$, the velocities of typical molecules are $[3kT/m]^{1/2}$ which, for $m = 3.2 \times 10^{-24}\,\text{g}$, are about $0.5\,\text{km}\,\text{s}^{-1}$, whereas observed turbulent velocities in such clouds often exceed a few kilometers per second. These random bulk velocities, as well as rotational velocities that could be of the same order, keep the cloud from collapsing.

Note that if we use (9–25) to approximate P by $c_s^2\rho/\gamma$ in (10–20), with $\gamma \sim \frac{5}{3}$, and set $M = (4\pi/3)\rho R^3$, equation (10–13) leads to the requirement for collapse, $R \gtrsim (18/5)^{1/2}(\lambda_J/2\pi)$ — showing the relationship between pressure equilibrium and the Jeans criterion.

10:5 Magnetic Reconnection

Let us now return to the need for shedding magnetic fields if a protostellar cloud is to collapse. Oppositely directed magnetic fields threading partially or fully ion-

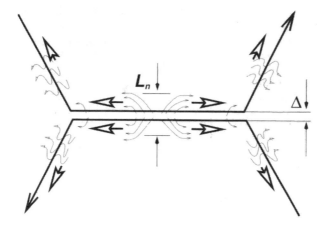

Fig. 10.1. Schematic diagram of magnetic reconnection in a partially ionized medium. The tranverse magnetic pressures in two regions of opposite magnetic polarity drive the gas at their interface into a reconnection layer of width Δ separating the two regions. The magnetic fields are imagined to be roughly horizontal but, as the ordinary arrow heads indicate, along opposite directions in the upper and lower halves of the figure. Where the sheets of oppositely directed fields meet they annihilate. To the left and right of the annihilation region, the magnetic field lines curve apart. The energy dissipated in the annihilation of the fields accelerates the ionized plasma, channeling it out of the annihilation region, parallel to the confining field lines, as indicated by hollow arrow heads. Because neutral gas can cross the field lines it diffuses in and out of the annihilation layer and is dragged along primarily through collisions with ions. This produces a thicker layer L_n, marked with thin lines, in which neutral gas is propelled out of the annihilation region. Courtesy Ethan Vishniac (Vi99).

ized gases can rapidly annihilate wherever they come into contact. This is amply demonstrated by observations of magnetic field annihilation on the surface of the Sun. Magnetohydrodynamic theories, however, still have difficulties explaining the extreme speed of the observed annihilation.

Figure 10.1 sketches how magnetic field annihilation is believed to take place. Along the vertical line of symmetry in the figure, where the horizontal coordinate takes on the value $x = 0$, the sum of magnetic and gas pressures, $B^2/8\pi + P$, is approximately constant. In the mid-plane of the figure, where we set the vertical coordinate $y = 0$, the magnetic field vanishes and, for an incompressible fluid, the total pressure $\rho v_x^2 + P$ is also approximately constant. Let the horizontal outflow velocity be $v_x = 0$ at the center of symmetry, $(x, y) = (0, 0)$, where the flow divides — part of it going to the right and part to the left. Let us further suppose that the gas pressure drops to zero at the extreme right and left of Fig. 10.1, at points $(\pm L_x, 0)$, where the magnetic field lines curve away from the central plane because the magnetized fluid there is not being driven toward the plane. Then the outflow velocity where the gas pressure drops to zero, becomes

$$v_x \sim B/\sqrt{8\pi\rho} = v_A/\sqrt{2}, \quad \text{where} \quad v_A \equiv B/\sqrt{4\pi\rho} . \tag{10-21}$$

v_A is the *Alfvén velocity*, previously mentioned in Section 9:4. Continuity requires that the mass influx into the reconnection region, $2\rho L_x v_y$ be balanced by mass outflow $2\rho v_x \Delta$. The mean inflow velocity is then given by

$$v_y \sim (\Delta/L_x) v_A / \sqrt{2}. \tag{10-22}$$

So far, we have neglected any Ohm's law dissipation. We can incorporate this through use of (9–57). Since there is no applied electric field, the current is not affected by the term $\partial \mathbf{D}/\partial t$. However, the motion of the plasma with respect to the magnetic field, leads to an effective electric field component $\mathbf{v} \wedge \mathbf{B}$ that adds on to the field component associated with the current $\mathbf{j}$, so that

$$\mathbf{E} = \left(\frac{\mathbf{j}}{\sigma} - \frac{\mathbf{v} \wedge \mathbf{B}}{c} \right). \tag{10-23}$$

which leads to (9–60)

$$\frac{\partial \mathbf{B}}{\partial t} = -c\nabla \wedge \left(\frac{\mathbf{j}}{\sigma} - \frac{\mathbf{v} \wedge \mathbf{B}}{c} \right) = \left(\frac{c^2}{4\pi\mu\sigma} \right) \nabla^2 \mathbf{B} + \nabla \wedge (\mathbf{v} \wedge \mathbf{B}). \tag{10-24}$$

For $v_y = 0$, (10–24) has the form of a diffusion equation. The quantity $\eta \equiv c^2/(4\pi\mu\sigma)$ is correspondingly called the *ohmic diffusion constant*.

For the flow in Fig. 10.1 we are primarily interested in the rate at which the magnetic field can annihilate, i.e., the velocity v_y at which the field B_x is driven into the recombination zone.

$$\frac{\partial \mathbf{B}_x}{\partial t} = \eta \frac{\partial^2 B_x}{\partial y^2} - \frac{\partial v_y B_x}{\partial y}. \tag{10-25}$$

In a steady state, $\partial B_x/\partial t = 0$, and so the velocity of the magnetic field must be of order $v_y \sim \eta/\Delta$. Substituting this into (10–22) and neglecting the factor $\sqrt{2}$, this leads to a reconnection speed

$$V_r \sim v_y \sim \left(\frac{\eta v_A}{L_x} \right)^{1/2} \sim v_A \left(\frac{\eta}{v_A L_x} \right)^{1/2} \equiv \frac{v_A}{\mathcal{R}_B^{1/2}}, \tag{10-26}$$

often named the *Sweet–Parker recombination rate* after P. A. Sweet and F. N. Parker, who first proposed it (Sw58), (Pa57). $\mathcal{R}_B$ is called the *magnetic Reynolds number*. Because $\mathcal{R}_B \gg 1$ in most astrophysical settings, the reconnection speed should usually be orders of magnitude slower than the Alfvén speed, which tends to be of the order of turbulent velocities. On the other hand, observations of magnetic fields in the solar corona and chromosphere appear to imply reconnection at speeds of order $\sim 0.1 v_A$, and it is not clear how the theory can permit such high speeds. *Magnetohydrodynamics* is a complex field that still awaits full development (Vi99)*.

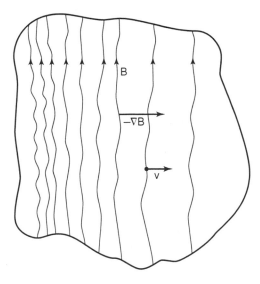

Fig. 10.2. Magnetic field gradients and ambipolar diffusion. Regions in which magnetic field lines are highly compressed tend to expand into domains where the magnetic field strength and pressure are lower. Ions circling the field lines are pulled along by the expansion, and drift through the neutral gas. Their drift velocity is determined by a balance between magnetic pressure gradients and resistive drag due to collisions with neutral atoms and molecules.

10:6 Ambipolar Diffusion

While reconnection is one process that permits a region to shed its magnetic fields, another process that operates in weakly ionized interstellar clouds is *ambipolar diffusion*, a drift of ions with respect to neutrals.

A gradient in the magnetic field strength exerts a force on electrons and ions; neutral atoms are affected only through their collisions with these charged particles. The magnetic fields and ions that circle the field lines can, therefore, drift through the neutral gas. For an interstellar cloud of radius R, whose central field strength B tapers off to zero at the cloud's surface, the transverse pressure gradient is $\sim B^2/8\pi R$ (Section 6:10). This gradient causes the ions and field lines to drift at a velocity v through the ambient neutral gas (Fig. 10.2). The magnetic field moves with the ions, beause the ions are confined to stay within a Larmor radius from a given line of force. The drift velocity causes the ions to collide with neutral atoms or molecules and give up their drift momentum to them. Let the collision cross-section for momentum loss be σ_c, the number density of neutrals be n_n, and the ion mass and number density be m_i and n_i. Then the friction force per unit volume of ionized material due to the drift is given by the mass of the ions in that volume $n_i m_i$, and the rate at which they lose their drift momentum, $n_i m_i v$, due to collisions.

Each ion suffers a momentum-transferring collision in a distance $(\sigma_c n_n)^{-1}$, which is covered in a time $(v\sigma_c n_n)^{-1}$. The frictional force per unit volume, therefore, is $n_n n_i m_i v^2 \sigma_c$, and the drift velocity in equilibrium is

$$v \sim \left(\frac{B^2/8\pi R}{n_n n_i m_i \sigma_c} \right)^{1/2}. \tag{10-27}$$

In a typical molecular cloud with radius $R \sim 10^{18}$ cm, we might find $B \sim 4 \times 10^{-6}$ G, $n_n \sim 10^6$ cm^{-3}, $n_i/n_n \sim 10^{-6}$, $m_i \sim 1.6 \times 10^{-24}$ g, and $\sigma_c \sim 10^{-16}$ cm^2. This yields a drift velocity $\sim$60 cm s^{-1}. At that rate, expulsion of the field from the cloud would take approximately 5×10^8 yr. Such long diffusion times constitute a difficulty that does not appear to be reduced by turbulent motions that might occasionally bring regions permeated by intense magnetic fields closer to the periphery of the cloud, thus shortening the path across which the fields would have to drift.

For a given degree of ionization the drift velocity is inversely proportional to density. For a more tenuous cloud with $n_n \sim 10^4$ cm^{-3} and $n_i/n_n \sim 10^{-6}$ the drift velocity is 6000 cm s^{-1} and the time for a magnetic field of 4 μG to drift out of a cloud with $R = 10^{18}$ cm would only be 5×10^6 yr. Ambipolar diffusion may therefore be active during the initial stages of contraction in star formation. It could continue to produce appreciable drifts only if the magnetic fields grew substantially as the density increased. Although observations tend to rule out fields greater than a few microgauss in most dense molecular clouds, important exceptions can be found. The Sagittarius B2 region near the Galaxy's center has fields of the order of 1 mG, and a circumnuclear disk at the Galaxy's center has fields as high as $\sim$2 mG (De97). In these regions the magnetic diffusion times could be as short as a million years even for dense clouds $n_n \sim 10^6$ cm^{-3} of radius 10^{18} cm. Ambipolar diffusion, therefore, appears to be a promising way of ridding protostellar material of magnetic fields.

10:7 Triggered Collapse

We now ask what kind of trigger might initiate the collapse of a protostellar cloud. If random bulk motions prevent collapse, then some process must be found that initiates contraction. For star formation in ordinary spiral galaxies the distribution of young stars along spiral arms provides one hint. Another comes from luminous galaxies in which star formation appears to be particularly active; those galaxies generally also appear disturbed and exhibit high-velocity gas flows. These factors all point to shock compression as a trigger initiating the collapse of a cloud. In the neighborhood of young stars, compression can occur at the edge of expanding ionization fronts or at shocks preceding an ionization front into a dark cloud. In regions where a burst of star formation has already taken place, explosions from supernovae evolving from young stars also may be providing the required shock-compression that triggers the formation of a next generation of stars.

PROBLEM 10–3. In equation (9–34) we defined a compression ratio $\Psi = \rho_o/\rho_i$ corresponding to the increased density of gas flowing out of a shock relative to the inflow density. Show that, even when extremely high shock velocities are invoked, the compression ratio in an adiabatic shock will not exceed

$$\Psi = \frac{\gamma_o + 1}{\gamma_o - 1} \, , \tag{10-28}$$

where γ_o is the ratio of heat capacities in the outflowing gas. When the flow is not adiabatic and an amount of heat $-Q$ per unit mass is used in exciting or dissociating the gas, show that the compression becomes

$$\Psi = \frac{\gamma_o \pm \{2(1 - \gamma_o^2)Q/v_i^2 + 1\}^{1/2}}{(2Q/v_i^2 + 1)(\gamma_o - 1)} \, . \tag{10-29}$$

For molecular hydrogen at very low temperatures the ratio of heat capacities is $\gamma = \frac{5}{3}$. At temperatures between $100\,\mathrm{K}$ and $300\,\mathrm{K}$, as rotationally excited states of the molecules become populated (see Problem 7–5), γ drops from $\frac{5}{3}$ to $\frac{7}{5}$ as further internal degrees of freedom add to the heat capacity of the gas. At even higher temperatures, vibrational degrees of freedom become invoked and the ratio of heat capacities can drop toward $\frac{9}{7}$ where, as Section 4:22 indicates, instability can set in. However, for cool post-shock gases, $\gamma = \frac{5}{3}$, and the compression attained in an adiabatic shock cannot exceed $\Psi = 4$. The compression rises to $\Psi = 6$ as rotational states become increasingly populated. But as vibrational states become excited and dissociation tends to set in, the flow no longer is adiabatic because Q is no longer zero and we then need to use the more general expression (10–29). We can see that the compression then becomes large, provided $-Q \sim v_i^2/2$. This, however, occurs only at rather high shock velocities, at several tens of kilometers per second. As already indicated in Section 9:4, all these conditions are appreciably altered by the presence of magnetic fields. The hydrodynamics then are fairly involved, though the general approach taken here still leads to useful insights.

10:8 Energy Dissipation

The compressive shocks just described can trigger collapse, but they do not guarantee permanent compression. Unless a compressed cloud can also dissipate energy, it will rebound elastically. In order for the cloud to remain compressed it must cool itself on a time scale comparable to the compression time. Energy must be radiated away, preferably altogether beyond the borders of the cloud. This is often a two-step process: an atom or molecule is first collisionally excited; in a second step it then radiates the excitation energy away. If this process is repeated sufficiently often, the energy drain on the cloud is appreciable, and it cools even as it is compressed. For excitation to take place the translational energy of the gas constituents in a shock must approach or exceed the excitation energy for low-lying atomic or molecular levels. Once this threshold is exceeded, the collisional excitation cross-sections, for virtually all atoms or molecules found in galactic clouds, tend to be of order 10^{-16} cm^2. In that respect there is rather little difference between the various atomic and molecular constituents.

The efficiency with which these constituents radiate, however, varies enormously. Neither atomic nor molecular hydrogen radiates efficiently below 1000 K. At these temperatures, the only available atomic hydrogen transition is the 21 cm, hyperfine transition corresponding to an energy jump of merely 6×10^{-6} eV. Because the Einstein spontaneous decay coefficient for this transition is only $A = 2.87 \times 10^{-15}$ s^{-1}, the cooling rate through 21 cm emission could maximally be of the order of 1 K in a hundred million years. In contrast, cloud collapse is believed to occur during tens of thousands of years and involves temperatures of hundreds of degrees Kelvin. Similarly, molecular hydrogen, a symmetric dipole molecule, can radiate only through quadrupole emission which, as discussed in Section 6:13, is an inefficient process. Where other coolants are entirely absent, as in the early Universe, before carbon, oxygen, and other heavy elements were synthesized in stars, or in regions where other molecular species are largely absent, molecular hydrogen can dominate cooling. But in low-temperature molecular clouds, we find that cooling largely depends on impurity constituents, such as CO and H_2O which, though low in abundance, are readily excited through collisions and rapidly radiate this energy away only to be collisionally excited again to repeat the cycle.

While collisions largely excite these molecules into rotational states, they elevate atomic impurities such as oxygen, O, carbon, C, or singly ionized carbon, C$^+$, to low-lying fine-structure levels within their ground electronic states. A *fine-structure transition* involves a change in electron-spin angular momentum relative to orbital angular momentum.

To obtain a quantitative estimate of cooling rates, we consider a cloud with density n and an impurity concentration X of a species of atom or molecule with a low-lying level that can be collisionally excited by hydrogen molecules of mass m. The excitation cross-section is σ. If the Einstein coefficient for spontaneous emission of a photon with energy ε is A and the gas temperature is T, then the cooling rate per unit volume is

$$L = Xn^2 \left(\frac{3kT}{m}\right)^{1/2} \sigma\varepsilon , \qquad (10\text{-}30)$$

provided the collisional excitation rate is far slower than the spontaneous emission rate

$$n \left(\frac{3kT}{m}\right)^{1/2} \sigma \ll A . \qquad (10\text{-}31)$$

Otherwise collisional excitation can be followed by de-exciting collisions and radiation becomes less efficient. Equation (10–30) can be rewritten, in terms of frequently encountered cloud parameters, as

$$L = 1.5 \times 10^{-18} \left(\frac{X}{10^{-4}}\right) \left(\frac{n}{10^6 \text{ cm}^{-3}}\right)^2 \left(\frac{T}{70 \text{ K}}\right)^{1/2} \qquad (10\text{--}32)$$
$$\times \left(\frac{\sigma}{10^{-16} \text{ cm}^2}\right) \left(\frac{\varepsilon}{10^{-3} \text{ eV}}\right) \text{ erg cm}^{-3} \text{ s}^{-1} ,$$

which we note is independent of A. In employing this format for writing an equation, we are stating effectively that $X \sim 10^{-4}$, $n \sim 10^6 \, \mathrm{cm}^{-3}$, and so on, are typical values for impurity concentrations, hydrogen density, and other parameters, and that L scales in proportion to X, to n^2, and so forth.

The cooling rate L should still be compared to the heat content per unit volume, H,

$$H \sim nkT = 10^{-8} \left(\frac{n}{10^6 \, \mathrm{cm}^{-3}} \right) \left(\frac{T}{70 \, \mathrm{K}} \right) \mathrm{erg \, cm}^{-3} . \tag{10-33}$$

The ratio of these two quantities gives the cooling time

$$t_{\mathrm{cool}} \sim \frac{H}{L} = 6.7 \times 10^9 \left(\frac{10^{-4}}{X} \right) \left(\frac{10^6 \, \mathrm{cm}^{-3}}{n} \right) \left(\frac{T}{70 \, \mathrm{K}} \right)^{1/2} \tag{10-34}$$

$$\times \left(\frac{10^{-16} \, \mathrm{cm}^2}{\sigma} \right) \left(\frac{10^{-3} \, \mathrm{eV}}{\varepsilon} \right) \mathrm{s} .$$

This cooling time is roughly 200 yr for the parameters assumed and corresponds to the time required by a shock at speed $15 \, \mathrm{km \, s}^{-1}$ to cross a distance of $10^{16} \, \mathrm{cm}$. This suggests that turbulent motions at supersonic velocities should be rapidly damped and cannot long persist in a cloud. We may still compare the cooling time to the free-fall time for a spherical cloud given by equation (10–4),

$$t_{\mathrm{ff}} \sim \left(\frac{3\pi}{32\rho G} \right)^{1/2} \sim 40,000 \left(\frac{10^6}{n} \right)^{1/2} \mathrm{yr} . \tag{10-35}$$

The collapsing cloud can cool itself far more rapidly than free fall. Infrared observations at wavelengths of several microns have revealed tantalizing protostellar candidates deep inside dusty clouds. These are understandably rare because the collapse phase is so fleeting. The free-fall time is a hundred times shorter than the main sequence lifetime of the most massive, shortest-lived stars. For other stars, it represents an even smaller fraction of the total lifetime. In contrast to the total number of stars, very few protostars exist at any given epoch.

The restriction (10–31) on the Einstein coefficient can be rewritten as

$$A \gg 10^{-5} \left(\frac{n}{10^6 \, \mathrm{cm}^{-3}} \right) \left(\frac{T}{70 \, \mathrm{K}} \right)^{1/2} \left(\frac{\sigma}{10^{-16} \, \mathrm{cm}^2} \right) \mathrm{s}^{-1}. \tag{10-36}$$

This needs to be compared to the Einstein A values for potential coolants. For CO the Jth rotational state has a coefficient $A_J \sim 1.118 \times 10^{-7} J^3 \, \mathrm{s}^{-1}$. The wavelengths for these transitions lie at $\lambda_J \sim (2600/J) \, \mu\mathrm{m}$ equivalent to a temperature for state J, $T_J = hc/2k\lambda_1 J(J+1) \sim 2.77 \, J(J+1) \, \mathrm{K}^{1}$. Atomic oxygen has a fine-structure transition at $63 \, \mu\mathrm{m}$ with $A = 9 \times 10^{-5} \, \mathrm{s}^{-1}$. We see that both CO at $J > 7$ and atomic oxygen meet the requirements of (10–36). Water vapor also has a large number of lines with high A values. The relative cooling capacities of CO and H_2O depend on density and temperature (Fig. 10.3).

---

[1] $1\mu\mathrm{m} \equiv 1 \, micron \equiv 1 \, micrometer = 10^{-6}$ meters

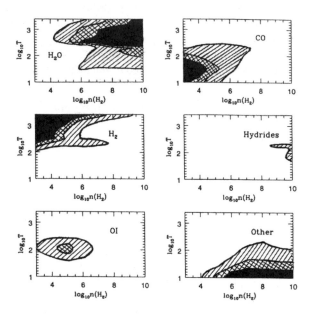

Fig. 10.3. The fractional cooling rate due to radiation emitted by different coolants in a molecular cloud as a function of temperature T and molecular hydrogen density $n(H_2)$. The coolants shown are H_2O, CO, H_2, neutral atomic oxygen (OI), hydrides other than H_2O, and other molecules. The radiated energy is proportional to the abundance of each species, determined by the chemical equilibrium at each temperature. Blank areas correspond to a fractional cooling power of < 0.2, hatched areas to 0.2–0.5, cross hatched areas to 0.5–0.7, and black areas to > 0.7 of the total cooling rate (from Neufeld, Lepp, and Melnick (Ne95)).

The cooling rates derived, assume that the optical depth of the cloud is low. If the column density rises to the point where self-absorption dominates, radiative cooling by atoms and molecules becomes progressively less efficient. A typical value for the *Einstein absorption coefficient* within the spectral line envelope is $B(\nu)$ at line frequency ν,

$$B(\nu) = \frac{c^2}{8\pi\nu^2}\frac{A}{\Delta\nu},\tag{10-37}$$

where $\Delta\nu$ is the Doppler width. The *column density* N_X of an atomic or molecular species X cannot exceed

$$N_X = [B(\nu)]^{-1}\tag{10-38}$$

without appreciable *line trapping* within the cloud and diminished cooling. We see this for a photodissociation region, PDR, interleaved between a circumstellar HII region and an adjacent cold gas cloud (Fig. 10.4).

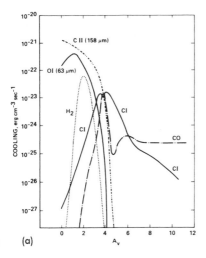
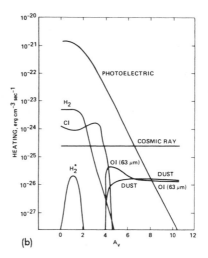

Fig. 10.4. Cooling and heating rates in a photodissociation region, PDR, as a function of visual extinction, A_V, on penetrating a cloud. The photon flux is 1.6 erg cm^{-2} s^{-1}; atomic densities are 10^3 cm^{-3}. (a) The curve labeled CII refers to fine-structure-line cooling at 158 μm by singly ionized carbon atoms; CI represents fine-structure-line cooling at 370 and 609 μm by neutral carbon atoms; OI represents similar cooling by oxygen atoms at 63 μm. H$_2$ and CO are the cooling rates due to rotational and vibrational transitions of these molecules. (b) Here "Photoelectric" stands for photoelectric heating by grains (Section 9:12); "Dust" means heating of the gas through collisions with warm dust; "Cosmic Ray" stands for cosmic-ray heating; H$_2$* and OI refer to collisional de-excitation, respectively, of radiatively pumped H$_2$ and oxygen atoms; H$_2$ and CI indicate heating through photodissociation of H$_2$ and photoionization of carbon atoms (Ho91).

10:9 Cooling of Dense Clouds by Grain Radiation

Once an interstellar cloud becomes sufficiently dense to absorb spectral line radiation emitted by its principal atomic and molecular constituents, grain emission may begin to dominate cooling. However, grains can cool the gas only as rapidly as atoms or molecules transfer their energy to the dust. If the grains are taken to be roughly spherical with radius a, the rate of heat transfer to a grain is

$$\frac{dQ_{\rm gr}}{dt} = n_2 \pi a^2 \left(\frac{3kT_2}{m_2}\right)^{1/2} \frac{(T_2 - T_{\rm gr})}{T_2} \alpha \left(\frac{c_v T_2}{\mathcal{N}}\right). \qquad (10\text{-}39)$$

Here, the expression on the left is the heat $dQ_{\rm gr}$ transferred to a grain in time dt, and we see that this is proportional to the number of hydrogen molecules per unit volume, n_2, assuming H$_2$ to be the dominant gas constituent; the grain collision cross-section for gas impact, πa^2; the speed $(3kT_2/m_2)^{1/2}$ with which the molecules of mass m_2 travel at the gas temperature T_2; the fractional difference in gas and grain temperature $(T_2 - T_{\rm gr})/T_2$, indicating the fraction of the molecular energy that can be transferred; an efficiency factor α denoting the efficiency with which impacts transfer energy from a molecule to a grain; and, finally, the energy the molecule has that it could transfer, $c_v T_2/\mathcal{N}$, where $\mathcal{N}$ is Avogadro's number and c_v is the heat capacity per mole of gas. The notation used here is the same as in Section 4:19.

The number density of grains n_{gr} is determined by the fraction by mass X_{gr} of matter in the form of grains and by the mass density ρ_{gr} of the grain material. Per unit volume in space we then have

$$n_{gr} = \frac{n_2 X_{gr} m_2}{\rho_{gr}(4\pi a^3/3)} = 3.8 \times 10^{-9} \left(\frac{n_2}{10^4\,\text{cm}^{-3}}\right)\left(\frac{X_{gr}}{10^{-3}}\right)$$

$$(10\text{-}40)$$

$$\times \left(\frac{2\,\text{g cm}^{-3}}{\rho_{gr}}\right)\left(\frac{10^{-5}\,\text{cm}}{a}\right)^3 \text{cm}^{-3},$$

where the numerator of the intermediate expression gives the mass of grains in unit volume of space, and the denominator is the mass per grain. The expression on the right exhibits representative values for typical parameters characterizing molecular clouds. We can now multiply the expressions for dQ_{gr}/dt and n_{gr} to obtain the cooling rate per unit volume of space L_{gr},

$$L_{gr} = \frac{3}{4}n_2^2 X_{gr}(3m_2 k)^{1/2}(T_2 - T_{gr})\alpha c_v T_2^{1/2}(\mathcal{N}\rho_{gr}a)^{-1}$$

$$= 2 \times 10^{-24}\left(\frac{n_2}{10^4\,\text{cm}^{-3}}\right)^2\left(\frac{X_{gr}}{10^{-3}}\right)\left(\frac{\alpha}{1/3}\right)\left(\frac{2\,\text{g cm}^{-3}}{\rho_{gr}}\right)\left(\frac{10^{-5}\,\text{cm}}{a}\right)$$

$$\times \left(\frac{c_v/\mathcal{N}k}{3/2}\right)\left(\frac{T_2}{50\,\text{K}}\right)^{3/2}\left(\frac{T_2 - T_{gr}}{30\,\text{K}}\right)\frac{\text{erg}}{\text{cm}^3\,\text{s}}. \qquad (10\text{-}41)$$

This cooling rate must be compared to the radiative cooling rate for grains at temperature T_{gr} radiating with efficiency η,

$$L_{rad} = 4\pi a^2 n_{gr}\sigma T_{gr}^4 \eta = \frac{3 n_2 X_{gr} m_2 \sigma T_{gr}^4 \eta}{\rho_{gr}a}$$

$$= 4.4 \times 10^{-21}\left(\frac{n_2}{10^4\,\text{cm}^{-3}}\right)\left(\frac{X_{gr}}{10^{-3}}\right)\left(\frac{T_{gr}}{20\,\text{K}}\right)^4\left(\frac{\eta}{10^{-4}}\right)$$

$$\times \left(\frac{10^{-5}\,\text{cm}}{a}\right)\left(\frac{2\,\text{g cm}^{-3}}{\rho_{gr}}\right)\frac{\text{erg}}{\text{cm}^3\,\text{s}}. \qquad (10\text{-}42)$$

Here σ is the Stefan–Boltzmann constant and η has been chosen as 10^{-4}, roughly corresponding to a radiation efficiency comparable to the ratio of grain radius to wavelength, a/λ, for grains radiating at wavelengths just short of one millimeter. Even when this efficiency is an order of magnitude lower, $\eta \sim 10^{-5}$, the grains can still radiate rapidly enough to keep the gas temperature below $50\,\text{K}$, and we would have $L_{rad} > L_{gr}$, meaning that the rate of radiation by grains is at least as rapid as the rate at which the gas can heat the grains.

As the cooling cloud contracts, the density rises and L_{gr} increases as the square of the density, while L_{rad} only grows linearly with density. With the above parameters we would have equality of grain heating and cooling, $L_{gr} \sim L_{rad}$, at densities $n_2 \sim 5 \times 10^5\,cm^{-3}$; but that number depends quite critically on the difference between gas and grain temperatures. As the density increases, the temperature difference between gas and grains tends to decline. If the grains become hotter they begin emitting at shorter wavelengths where their efficiency $\eta \sim a/\lambda$ increases roughly in proportion to T_{gr}, since a thermally emitting body shifts its peak emission frequency in proportion to T, as is evident from (4–72), so that $\eta \propto T_{gr} \propto \lambda^{-1}$. As a result L_{gr} rises roughly in proportion to T_{gr}^5.

Grains play a major role in the cooling of a contracting cloud that radiates away energy gained as gravitational potential energy is released. The predominance of grains in cooling is due to the broad wavelength range over which they emit and is not greatly affected by their low efficiency in radiating. This lack of efficiency implies that energy radiated by a given grain in the cloud will not be readily absorbed by another grain and will therefore escape. Per unit volume the opacity is

$$\kappa_{gr} = \pi a^2 n_{gr} \eta = \frac{3 n_2 X_{gr} m_2 \eta}{4 \rho_{gr} a} = 1.2 \times 10^{-22}\,cm^{-1} \tag{10-43}$$

for the same parameters used in equation (10–42). Since the opacity can only change in response to an increase in density or radiating efficiency, and since η is unlikely to exceed 10^{-3} at temperatures characteristic of protostellar clouds, n_2 can rise to a value of $10^9\,cm^{-3}$ and still leave the opacity as low as $\sim 10^{-16}\,cm^{-1}$. At those densities a sphere of radius $10^{16}\,cm$ would just barely have unit opacity, but the total mass encompassed would exceed $1 M_\odot$. Once the cloud becomes opaque, perhaps at the time its radius is of order of a few hundred astronomical units, it can at best only emit as a blackbody. In thermal equilibrium, rapid contraction requires an increase in temperature to make up for the decreasing surface area available for radiation.

Finally, we may want to examine the relationship, at this late stage of collapse, between the cooling time and the free-fall time. The temperature at the surface of the nebula can be estimated from the virial theorem (3–85), by setting $3kT = m_i MG/R$, where m_i is the prevalent atomic or molecular mass. At uniform density the potential energy of the cloud is given by equation (8–1). The cooling time is therefore

$$t_{cool} \sim \frac{3M^2 G}{5R}\left(\frac{1}{4\pi R^2 \sigma T^4}\right) = \frac{243 k^4 R}{20 \pi m_i^4 M^2 G^3 \sigma}, \tag{10-44}$$

where k is Boltzmann's constant. Taking $m_i = m_H$, we find the cooling time to be roughly one year. In contrast, the free-fall time is $(3\pi/32 G n_2 m_2)^{1/2}$ as seen from equation (10–4); for a protostellar nebula of constant density this would amount to about 3000 yr. The nebula can cool itself faster than it would collapse through free fall, unless excessive angular momentum that cannot be shed or excessive internal magnetic pressures halt the contraction.

Gas swirling around a star generally has a net angular momentum that prevents it from falling directly toward the star. It gathers into a disk that orbits the star in differential rotation; the outer parts of the disk have an angular velocity lower than the inner portions. A number of different types of accretion disks are significant. Protostellar gas clouds contracting to form a star form a disk around the gravitating center. These disks can remain even after the star forms. T Tauri stars are embedded in the disks. Later, as the star evolves, such disks may give rise to planetary systems that maintain the disk's original angular momentum but allow the cooling gas to form planetary bodies.

When planet formation has progressed to a point where much of the disk material has aggregated into planets or asteroids, destructive collisions between these bodies can result in the production of enormous amounts of debris. The *debris disk*, heated by starlight, tends to prominently radiate at infrared wavelengths. Generally, such a disk persists for less than a hundred million years. During this time the debris is partly swept up and captured by the larger planets and partly dragged into the star by the Poynting–Robertson effect.

Protostellar gas normally has an embedded magnetic field, and as the gas contracts toward the star, it preferentially follows the field lines. This leads to the lowest compaction of the field lines and avoids the buildup of magnetic pressures that would resist further contraction.

Magnetic field lines anchored in the rotating star may extend out into the accretion disk, and wind up until magnetic pressures become too high. Twisted magnetic loops are then thought to break out of the disk with current sheets developing where oppositely directed magnetic fields annihilate, as in Fig. 10.1.

Plasma becomes sufficiently hot to emit X-rays and a large X-ray flare results. Protostars still embedded in the cloud of gas and dust from which they formed frequently flare up in X-ray emission (Ne97).

Once the new star is fully formed and begins burning nuclear energy, it starts to shed its outer envelope in a *stellar wind*. This begins to blow away some of the accretion disk, but much of the escaping gas is initially funneled out, escaping along the system's polar axis.

Figure 10.5 shows the complexity of interplay between evolving stars and the interstellar medium. It indicates why the question of star formation has no single easy resolution: the formation of stars may proceed along varieties of different lines.

10:10 Condensation in the Early Solar Nebula

We now turn to another source of information on star formation — remnants from the earliest phases of the protosolar nebula, left over from a time when the Sun and the planets were just forming. These remnants are the earliest-formed meteorites. In order to decipher the message they have preserved and brought down to us over the æons, we must first go back to a number of thermodynamic considerations. We do this now and return to the meteoritic data in Section 10:11.

Let us ask how gases in the early Solar Nebula may have condensed into the solids that ultimately went into the formation of planetary bodies. Clearly, the first

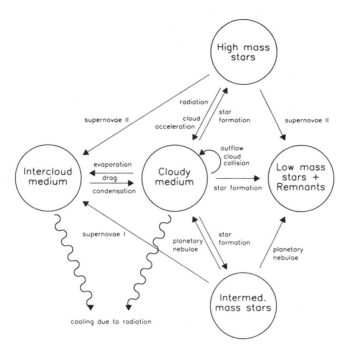

Fig. 10.5. The complex interplay between evolving stars and the interstellar medium (Sa97a).

law of thermodynamics — energy conservation — must hold. We start with equation (4–119)

$$\partial Q = dU + P\,dV \tag{10-45}$$

and note that the second law requires

$$T\,dS \geq \partial Q \;, \tag{10-46}$$

meaning that entropy S either increases or, at best, remains constant in any physical process as long as we deal with the entire system, generally the entire volume within which the process takes place. Equality in this relation holds only when the process occurs reversibly. Relatively few genuinely reversible processes exist, but sublimation and condensation are among them. Typically, a closed vessel kept at constant temperature and pressure near the sublimation point of a substance will see the growth of some crystals at the expense of others, while the vapor pressure in the vessel remains constant. No net work is done in this equilibrium state since the pressure and volume remain constant, but heat is transferred from growing crystals to sublimating crystals. This is the latent heat of evaporation per mole of substance, λ. We, therefore, see the entropy change for a mole of condensing material to be

$$\Delta S = -\frac{\lambda}{T} \;. \tag{10-47}$$

The latent heat λ depends on both the pressure and temperature, but the equilibrium vapor pressure rapidly rises near one particular temperature — the sublimation temperature. Below this narrow temperature range one finds the bulk material largely equilibrated in the condensed phase; above this range it is almost exclusively gaseous.

We now turn to another thermodynamic function — the *Gibbs free energy* named after the nineteenth-century American thermodynamicist J. Willard Gibbs. It is defined as

$$G \equiv U + PV - TS . \tag{10-48}$$

This is a function that describes the state of the system without regard to the ways in which it originated. G can be differentiated exactly and we write

$$dG = dU + P\,dV + V\,dP - S\,dT - T\,dS . \tag{10-49}$$

Applying equations (10–45) and (10–46) we obtain

$$dG \leq V\,dP - S\,dT . \tag{10-50}$$

For changes occurring at constant temperature and pressure, $dG \leq 0$, and the free energy either decreases or remains constant. A constant free energy requires a reversible process such as the condensation or sublimation just discussed.

Let us consider two phases designated by v for vapor and s for solid. In equilibrium we have

$$V_s\,dP - S_s\,dT = V_v\,dP - S_v\,dT \tag{10-51}$$

or

$$\frac{dP}{dT} = \frac{S_s - S_v}{V_s - V_v} = \frac{\Delta S}{\Delta V} = \frac{-\lambda}{T\Delta V} . \tag{10-52}$$

For condensation, $V_s \ll V_v$ and $\Delta V = -V_v = -RT/P$, where R is the gas constant defined in Section 4:6. Hence

$$\frac{dP}{dT} = \frac{P\lambda}{RT^2} , \tag{10-53}$$

from which we obtain

$$\ln P = \frac{-\lambda}{RT} + \text{constant} . \tag{10-54}$$

This reciprocal relationship between the logarithm of the vapor pressure and the temperature is shown by the straight lines of Fig. 10.6. Its plot shows the rapid drop in vapor pressure for elements of interest in the condensation of the early Solar Nebula. Table 10.1 shows the condensation temperatures of a number of pure elements as well as compounds at two different total pressures in the early Solar Nebula, respectively, 1 and 6.6×10^{-3} atm (1 atm = 760 torr = 1.01×10^6 dyn cm^{-2}).

We notice first that the elements which are most volatile — those with the highest vapor pressure at a given temperature — tend to be clustered in the right half of the periodic table, as shown in Fig. 10.7.

Table 10.1. Condensation Temperatures for Compounds and Elements in the Early Solar Nebula, at Two Different Total (Hydrogen) Gas Pressures, P_τ. (Reprinted with permission from Larimer (La67).)

$P_\tau = 1$ atm		$P_\tau = 6.6 \times 10^{-3}$ atm	
Compound or Element	$T(^\circ K)$	Compound or Element	$T(^\circ K)$
$MgAl_2O_4$	2050	$CaTiO_3$	1740
$CaTiO_3$	2010	$MgAl_2O_4$	1680
Al_2SiO_5	1920	Al_2SiO_5	1650
Ca_2SiO_4	1900	Fe	1620
$CaAl_2Si_2O_8$	1900	$CaAl_2Si_2O_8$	1620
$CaSiO_3$	1860	Ca_2SiO_4	1600
Fe	1790	$CaSiO_3$	1580
$CaMgSi_2O_6$	1770	$CaMgSi_2O_6$	1560
$KAlSi_3O_8$	1720	$KAlSi_3O_8$	1470
Ni	1690	$MgSiO_3$	1470
$MgSiO_3$	1670*	SiO_2	1450
SiO_2	1650	Ni	1440
Mg_2SiO_4	1620*	Mg_2SiO_4	1420
$NaAlSi_3O_8$	1550	$NaAlSi_3O_8$	1320
$MnSiO_3$	1410	$MnSiO_3$	1240
Na_2SiO_3	1350	MnS	1160
K_2SiO_3	1320	Na_2SiO_3	1160
MnS	1300	K_2SiO_3	1120
Cu	1260	Cu	1090
Ge	1150	Ge	970
Au	1100	Au	920
Ga	1015	Ga	880
Sn	940	Zn_2SiO_4	820
Zn_2SiO_4	930	Sn	806
Ag	880	Ag	788
ZnS	790	ZnS	730
FeS	680	Fes	680
Pb	655	Pb	570
CdS	625	CdS	570
Bi	620	$PbCl_2$	535
$PbCl_2$	570	Bi	530
Tl	540	Tl	475
In	400	Fe_3O_4	400
Fe_3O_4	400	In	360
H_2O	260	H_2O	210
Hg	196	Hg	181

* The condensation temperatures for $MgSiO_3$ and Mg_2SiO_4 are somewhat uncertain.

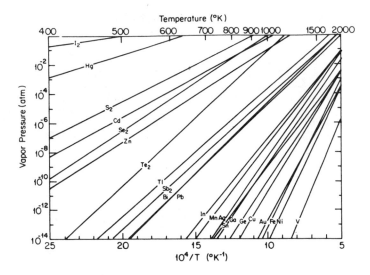

Fig. 10.6. Vapor pressures of elements found in meteorites. (Reprinted with permission from Larimer (La67).)

If the early Solar Nebula had cooled down slowly, we would expect that the first condensates would have contained only refractory materials, materials that evaporate at high temperatures. As the nebula cooled, more volatile substances would have condensed out. The condensation sequence with declining temperature would then have followed the order indicated in Table 10.1.

10:11 The Evidence Provided by Meteorites

At least some types of meteorites are believed to be remnants of the earliest stages of evolution of the Solar Nebula. They are thought to have condensed well before the first planets formed. Meteorites generally are classed either as *iron* or *stony*. The iron meteorites are metallic and rich in iron. The stony meteorites can take on different forms. Of particular interest to studies of the formation of the Solar System are *chondrites*, stony meteorites containing *chondrules*. Chondrules, in turn, are millimeter-sized silicate spherules that look as though they might have been droplets frozen from a melt. They consist largely of *olivine*, a mineral whose chemical makeup is $(Mg, Fe)_2SiO_4$, *pyroxine* $(Mg, Fe)SiO_3$, and *plagioclase feldspar*, which is a solid solution of $CaAl_2Si_2O_8$ and $NaAlSi_3O_8$. Table 10.1 shows that all of these minerals condense out at temperatures $\geq 1240\,K$ in the pressure range shown. The chondrules are embedded in a *matrix*, a more finely ground mass, generally of the same composition. In the matrix we also find millimeter-sized particles of nickel–iron with a nickel content ranging from about 5 to 60%. *Troilite*, whose chemical composition is FeS, is also present.

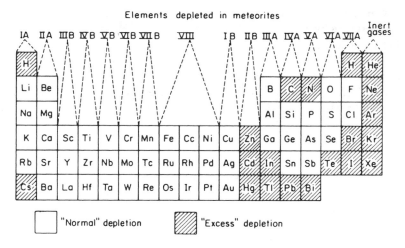

Fig. 10.7. Elements found depleted in most chondritic meteorites, relative to type I carbonaceous chondrites. "Normal depletion" corresponds to depletion down to 0.1–0.5, while "excess depletion" refers to reduction to 0.001–0.1 of abundances in these chondrites. (Reprinted with permission from Larimer (La67).)

The chondrites can be divided into three groups. *Carbonaceous* chondrites, designated by a letter C, are highly oxidized. In particular, their iron content is always strongly oxidized, meaning that each iron atom donates two or three of its outer shell electrons to other elements in the chondritic mineral. Carbonaceous chondrites derive their name from the carbon-rich compounds they contain. In contrast, *enstatites*, or E *chondrites*, are highly reduced, containing iron only in the metallic form or as troilite. Between these two extremes we find *ordinary*, or *O chondrites*.

The carbonaceous chondrites are subdivided into three classes. Type I is virtually free of chondrules and consists largely of a mineral matrix. Type III consists of 70 to 80% chondrules with very little matrix in between. Type II is intermediate to these. Chemical analyses show that most chondritic meteorites are quite strongly depleted in the more volatile elements when compared to type I. This is because the matrix is richer in volatile elements than are the chondrules.

We now ask why these two chondritic constituents, the chondrules and the matrix, should differ so greatly in their content of volatiles.

The *chondrules* are thought to have been the first solids to condense out of a Solar Nebula which began as a high-temperature, gaseous mass. Were these spherical inclusions already present in the early Solar Nebula and do they therefore contain information that could be used to infer primitive conditions? Studies on the X-ray flaring of T Tauri stars suggest that chondrules may form in T Tauri disks through flash-heating. Magnetic fields connecting the rotating central star and the surrounding disk can suddenly reconnect to release magnetic pressure and energy, and accelerate charged particles to high energies. These might heat and melt dust aggregates, which could then cool and condense as chondrules (Gr97), (Sh97). The most refractory materials would have condensed out first, forming the chondrules. As Table 10.1 suggests, highly volatile material such as bismuth, Bi, lead, Pb, and indium, In, would have remained in vapor form and would only have condensed at much lower temperatures. Those chondrites rich in chondrules, therefore, contain largely those refractory constituents which condensed out first.

Actual depletion for ordinary chondrites relative to a cosmic abundance distribution of elements is shown in Fig. 10.8. We see that bismuth and indium are depleted

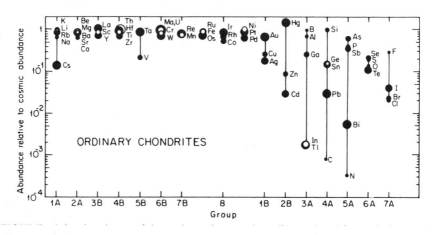

Fig. 10.8. Abundance of the various elements in ordinary chondrites, relative to the cosmic abundance given in Table 1.1 (after Anders (An72)).

by two to three orders of magnitude.

Two possible condensation sequences showing the nebular temperatures at which different compounds condensed are shown in Fig. 10.9. The actual condensation temperatures depend to some extent on whether the nebula cools slowly or rapidly. The slower cooling assumes that complete diffusion can take place with the formation of alloys and solutions to the limit of solubility of all the substances involved. This permits the condensation temperatures for minor elements to be higher and also widens their condensation ranges.

A fairly detailed picture of the evolution of the early Solar Nebula is emerging. The nebula started at temperatures in excess of perhaps 1800 K, and cooled to below 400 K before the planets formed. Evidence for this lower temperature comes from the long chain hydrocarbons in the carbonaceous chondrites which can only form at low temperatures in reactions of the type

$$20CO + 41H_2 \Leftrightarrow C_{20}H_{42} + 20H_2O \qquad (10\text{-}55)$$

through the interaction of residual carbon monoxide, CO, with hydrogen. These reactions, however, only take place in the temperature range between 300 K and

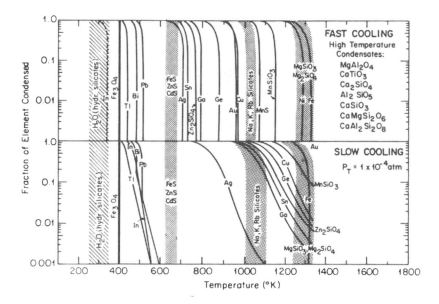

Fig. 10.9. Condensation sequence of gas whose initial composition corresponds to the cosmic abundance of elements. For grains of radius 10^{-5} cm the upper sequence applies for times of minutes or hours, while the lower sequence corresponds to cooling times of years or centuries. The shaded areas show condensation or chemical transformation of major constituents (after Anders (An72)).

400 K for pressures likely to prevail. At temperatures above 450 K, these gases react differently, giving

$$CO + 3H_2 \Leftrightarrow CH_4 + H_2O. \tag{10-56}$$

The double arrows in these two reactions show that they can go in either direction, but the hydrocarbons at the right are favored toward lower temperatures and the higher CO content seen on the left side of the equations is favored at higher temperatures (Fig. 10.10). CO, of course, is a constituent of interstellar clouds and is expected to be present in the early nebula. The production of hydrocarbons from CO and H_2 proceeds in the laboratory through the Fischer–Tropsch reaction, which is just the reaction (10–55) catalyzed by iron or cobalt in industrial syntheses. That this reaction is responsible for the carbon compounds found in meteorites is indicated by Fig. 10.11, which shows that, among about 10^4 possible hydrocarbon molecules that can be formed using 16 carbon atoms and an arbitrary amount of hydrogen, only six are present in appreciable abundance in the meteorite analyzed. Five of these, all underlined in the figure, are common to both samples. Acenaphthene, not detectable in the synthetic example, has been seen in other products prepared by the Fischer–Tropsch method at higher temperatures.

All this still leaves uncertain the cooling rate of the Solar Nebula. This depends on whether the nebula was self-shielding, in that the inner parts prevented solar

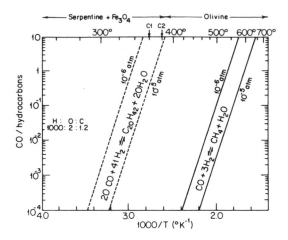

Fig. 10.10. Equilibrium between CO and H_2 at different temperatures and pressures. If equilibrium is maintained on cooling, CO is largely converted into CH_4 before more complex molecules can be formed at lower temperatures. However, this reaction is slow, and some CO may survive to lower temperatures where complex organic molecules can form, particularly if Fe_3O_4 (iron rust) and hydrated silicates such as serpentine, $Mg_3Si_2O_5(OH)_4$, form at 380 to 400 K. Both of these are effective catalysts for the Fischer–Tropsch reaction (see text) (after Anders (An72)).

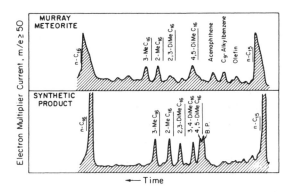

Fig. 10.11. Gas chromatogram of hydrocarbons in the range containing 15–16 carbon atoms. The synthetic product was made by a Fischer–Tropsch reaction (see text) (after Anders (An72)).

heating of the outer parts, or whether solar heating played an important role. The cooling rates under these two conditions are shown in Fig. 10.12. Either way, the

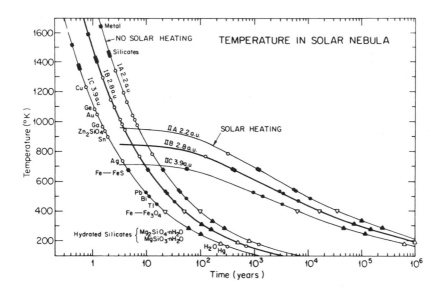

Fig. 10.12. Cooling curves for the Solar Nebula with and without heating by the protosun while still on the highly luminous Hayashi track. Presumably the initial cooling follows curves IA, IB, and IC at 2.2, 2.8, and 3.9 AU, respectively, from the Sun. Later on, curves IIA, IIB, and IIC are more likely to be germane. Accretion temperatures of different types of chondrites are indicated by symbols and are inferred from the condensation temperatures of Pb, Bi, In, Tl, and H_2O, as well as the threshold of the $Fe \rightarrow Fe_3O_4$ reaction. A time scale for accretion of chondritic material of $10^4 - 10^5$ yr is implied. The precise distances at which the various chondrites form is uncertain. (Reprinted with permission from Larimer and Anders (La67a).)

early Solar Nebula cooled in a remarkably short time, perhaps 10^4 to 10^5 yr, a mere instant when compared to the Sun's lifetime as a main sequence star $\sim 5 \times 10^9$ yr.

10:12 Nascent Planetary Disks

The gas in a nascent planetary disk is bound by gravity. Consider two parallel disks with identical surface mass density, respectively located in the planes $z = -z_i$ and $z = z_i$. Then, for disks with radii $R \gg 2z_0$ the results of Problem 9–11 tell us that a mass placed anywhere on axis between the planes will experience no net force. Accordingly, if the gas in a nascent disk has a vertical number-density distribution $n(z)$ symmetrically placed above and below a central plane, then a particle of mass M at some height z above the plane will feel a net force solely due to matter distributed between distances $-z_i$ and z_i from the central plane, and this force will be proportional to the surface density $\sigma(z_i)$ of material between these two planes, $F = -2\pi G \sigma(z_i)$. For a predominantly gaseous disk with atoms or molecules of mass m,

$$\sigma(z) = mN(z) \equiv m \int_{-z}^{+z} n(z)dz , \tag{10-57}$$

where $N(z)$ is the column density viewed perpendicular to the symmetry plane. The increment of pressure dP contributed by an element of thickness dz at height z above the plane is

$$dP = -2\pi G n(z) m \sigma(z)dz . \tag{10-58}$$

For an isothermal ideal gas at temperature T, this pressure gradient amounts to a density gradient

$$\frac{1}{n(z)} \frac{dn(z)}{dz} = \frac{d\ln n(z)}{dz} = - \left(2\pi G m^2/kT\right) N(z) . \tag{10-59}$$

Differentiating this, and restricting ourselves to the space $z \geq 0$ above the plane, we obtain

$$\frac{d^2 \ln n}{dz^2} = \frac{\ddot{n}(z)}{n(z)} - \left[\frac{\dot{n}(z)}{n(z)}\right]^2 = -\frac{4\pi G m^2}{kT} n(z) , \tag{10-60}$$

where dots indicate differentiation with respect to z. For a number density n_0 in the central plane, this yields, respectively, a density $n(z)$ and column density $N(z)$,

$$n(z) = \frac{n_0}{\left[1 + (2\pi n_0 G m^2/kT)^{1/2} |z|\right]^2} , \tag{10-61}$$

$$N(z) = \frac{2n_0 \left(kT/2\pi n_0 G m^2\right)^{1/2}}{\left[1 + (2\pi n_0 G m^2/kT)^{1/2} |z|\right]} . \tag{10-62}$$

The scale height $|H|$, above and below the symmetry plane, where the number density drops to $n_0/2$ is

$$|H| = \left(kT/2\pi n_0 G m^2\right)^{1/2} . \tag{10-63}$$

The surface density of the disk is

$$\sigma = 2 \int_0^\infty mn(z)dz = (n_0 kT/2\pi G)^{1/2} . \tag{10-64}$$

Note that this appears to depend only on n_0 and T, but does indirectly depend on the atomic or molecular mass m of the ideal gas constituents, by virtue of the assumption that the disk is gravitationally bound.

An important limitation of this model is that it does not accurately depict a protosolar disk rotating about a central Sun. The gravitational field of that configuration is dominated by the Sun, and the rotational motion provides a measure of stability.

In contrast, thin disks supported solely by their own gravitational fields invariably are unstable.

PROBLEM 10–4. Consider an early Solar Nebula consisting largely of H_2 at 2000 K orbiting a central Sun. If the nebular mass is $1M_\odot$, and the projected density of matter onto a central plane of this rotating disk is constant out to a distance of about 10^{14} cm, roughly 7 AU, derive the scale height H and show that the expected nebular pressure is $\sim 2 \times 10^{-3}$ atm. If the surface density increases nearer the Sun, the gas pressures found in the realm of Mercury, Venus, Earth, and Mars, should correspondingly increase, falling within the range covered in Table 10.1.

10:13 Formation of Primitive Condensates in the Early Solar Nebula

Given that the early Solar Nebula had a very high temperature, and that much of the interstellar dust would have evaporated, we can ask how rapidly new grains would grow as the nebula cooled.

Once condensation temperatures are reached, grain growth is quite rapid. We imagine a seed grain, perhaps an interstellar grain that has survived, a grain that might be entering the cooling nebula from a surrounding region which had not participated in the collapse, or perhaps a seed spontaneously formed from the vapors. Consider that the seed has radius a and is located in the plane of the nebular disk. We know that all the freely orbiting material will pass through the disk twice per orbital period P. At the Earth's distance from the Sun, this means that material passes through this plane twice yearly. In the answer to Problem 10–4 we saw that the areal density of this matter was $\sigma = 6.4 \times 10^4$ g cm^{-2}. If a fraction, say $f = 10^{-4}$, of this mass condenses at a particular temperature, the seed will grow at a rate

$$\frac{da}{dt} = \frac{\pi a^2 \sigma f}{4\pi a^2 \rho_{\rm gr}} \left(\frac{2}{P} \right) = \frac{\sigma f}{2\rho_{\rm gr} P} . \tag{10-65}$$

For a grain density $\rho_{\rm gr} = 3$ g cm^{-3}, we then obtain a growth rate at the Earth's distance from the Sun of roughly 1 cm yr^{-1}. However, since there is only $f\sigma \sim 6$ g cm^{-2} of condensable material in the disk, we see that the sweep-up will only take a few years, even if relatively few seed particles are initially present. Nucleation of seeds is almost inevitable if only because Galactic cosmic rays which always abound can produce nucleation in any supersaturated vapor — as they also do in a Wilson cloud chamber.

10:14 Formation of Planetesimals

Once centimeter-sized particles or at least the observed millimeter-sized chondrules form in the protoplanetary nebula, their scale height is quite small. The thermal velocity of a millimeter-sized particle with mass $\sim 10^{-3}$ g and temperature 10^3 K is $(3kT/m)^{1/2} \sim 2 \times 10^{-5}$ cm s^{-1}, so that the scale height, obtained as in Problem 9–11, is effectively less than the chondrule size even if we use an areal density due to condensed matter alone. The layer of condensed particles is therefore exceedingly thin, and the velocity of sound in the disk — the speed with which sound propagates through collisions by grains — is extremely low. The one factor that could make the speed of sound in the disk higher would be if the grains were touching, so that the bulk speed of sound in the material would become important. That the scale heights in disks can be very thin is indicated by the rings of Saturn, which are only ≤ 1.5 km thick, a spread that may be due to perturbations by Saturn's satellites (Ni96).

Goldreich and Ward (Go73) have considered the stability of such a disk. They start with a dispersion relation, similar to the Jeans criterion but applicable to rotating axisymmetric disks in which matter is orbiting along Keplerian trajectories:

$$\omega^2 = k^2 c_s^2 + \Omega^2 - 2\pi G \sigma k f . \tag{10-66}$$

Here Ω is the frequency of the orbital rotation in radians per second, k is the wave number, $k = 2\pi/\lambda$, and λ is the wavelength of the disturbance, while ω is its angular frequency. When that frequency goes from a real to an imaginary value, instability sets in. Because the speed of sound c_s is essentially negligible, the criterion for instability is

$$2\pi G \sigma k f > \Omega^2 . \tag{10-67}$$

The minimum wavelength that is unstable then becomes

$$\lambda_{\min} = 4\pi^2 G \sigma \Omega^{-2} f . \tag{10-68}$$

The mass contained in such a fragment is $f\sigma\lambda_{\min}^2$, so that the minimum condensation mass would be

$$M_{\min} = 16\pi^4 G^2 \sigma^3 \Omega^{-4} f^3 , \tag{10-69}$$

which has a value of order 10^{18} g, while $\lambda_{\min} \sim 4 \times 10^8$ cm. For unit density we obtain *planetesimals* roughly 10 km in size.

We note a number of points. First, if equation (10–66) is multiplied by the square of some wavelength λ and by a local density ρ, we see that the first term on the right is comparable to the internal energy of the gas; the second term depicts a rotational kinetic energy; and the third term corresponds to self-gravitational attraction. When the wavelength is sufficiently large so that the self-gravity exceeds the rotational and internal energies combined, instability and contraction can set in.

Second, the maximum wave number $k = 2\pi/\lambda_{\min} \sim 10^{-8}$ cm^{-1} when multiplied by a speed of sound derived from grain velocities, is far smaller than Ω which, at the distance of the Earth from the Sun, has a value $\sim 2 \times 10^{-7}$ rad s^{-1}. If the speed of sound were higher in the disk, and the first term on the right became dominant over the second, the instability criterion would become

$$\frac{2\pi}{\lambda_{\min}} = k_{\max} < \frac{2\pi G\sigma f}{c_s^2}, \qquad \lambda_{\min} > \frac{c_s^2}{G\sigma f}, \qquad (10\text{-}70)$$

and the minimum wavelength would be $\sim 2.5 \times 10^6 c_s^2$ cm. The argument for the formation of planetesimals through gravitational instabilities arising in a disk, therefore, depends quite crucially on the speed of sound being low.

Third, the speed with which such bodies form is going to be of order $\omega^{-1} \sim \lambda_{\min}/2\pi c_s$, which even for $c_s \sim 2 \times 10^{-5}$ cm s^{-1} and $\lambda_{\min} \sim 4 \times 10^8$ cm is only about a hundred thousand years, quite brief compared to the age of the Solar System. Again, the speed of sound is quite critical in this estimate.

Once planetesimals have formed, further growth of planets may occur through their gravitational accretion into large bodies. Just how this takes place is not satisfactorily understood. We do not know what determined the distribution of masses of the planets in our Solar System. Nor do we know why some massive planets seen in distant stellar systems orbit their parent stars at radii much smaller than Jupiter's distance from the Sun.

10:15 Condensation in the Primeval Solar Nebula

In Section 1:10 we presented some current views on the origin of the Solar System.

PROBLEM 10–5. Just after the Sun first formed, it was surrounded by a dense cloud of gas from which the planets eventually condensed. As a first step, small grains probably formed. Suppose the mass was evenly distributed throughout the nebular disk, with a scale height $H \sim 1$ AU, that its total mass was twice that of all planets combined (see Table 1.4), that the radius was $R = 10$ AU, that the temperature at each point was in thermal equilibrium with solar irradiation, and that the initial abundance was similar to the Solar System abundance (Table 1.1). Making use of Table 10.2, calculate an approximate distance from the Sun at which iron would have condensed. Do the same for carbon. Would water or ice have been able to condense within the nebula? Note that the Sun may have been on the Hayashi track (Fig. 1.5) at that time. Assume its luminosity was ten times greater than now.

PROBLEM 10–6. The action of light pressure may have tended to produce homogeneity in the Solar Nebula. Consider the outward-directed flux amounting to $\alpha L_\odot/4\pi r^2$ at a distance r from the Sun, where α is a factor of order unity. The orbital velocities of two grains, both orbiting the Sun at the distance of Jupiter would then depend on their densities, as well as their interaction with light. Let both grains have radii $s \sim 10^{-3}$ cm, but let one particle have a density 2 g cm^{-3}, while the other's is 4 g cm^{-3}. Assume both grains to be spherical and black, absorbing light with a cross-section πs^2. Show how the orbital velocity differs as a function of s, the difference being greatest for small s. Note that grains with large density and/or

Table 10.2. Relation Between Temperature and Vapor Pressure Compiled from (Ro65), (Du62), and (Le72).[a]

Vapor Pressure	10^{-11}	10^{-10}	10^{-9}	10^{-8}	10^{-7}		torr[b]
Atomic carbon at	1695	1765	1845	1930	2030		K
Atomic iron at	1000	1050	1105	1165	1230		K

As seen from equation (10–54), most solid substances obey a vapor-pressure–temperature relationship of the form $\log_{10} P = A - B/T$.

At low pressure:
P (in torr):
$$\begin{cases} \text{carbon} & A = 12.73 \\ \text{iron} & A = 9.44 \\ \text{NaCl} & A = 7.9 \end{cases} \quad \text{and} \quad \begin{array}{l} B = 4.0 \times 10^4 \text{ K} \\ B = 2.0 \times 10^4 \\ B = 8.5 \times 10^3 \end{array}$$

For water the following data are available:

$$H_2O \quad \begin{cases} 7 \times 10^{-9} & 3 \times 10^{-10} & 7.4 \times 10^{-15} & 14 \times 10^{-22} & \text{torr} \\ 143.2 & 133.2 & 123.2 & 90.2 & \text{K} \end{cases}$$

For hydrogen:

$$H_2 \quad \begin{cases} 3.1 \times 10^{-7} & 8.8 \times 10^{-9} & 7.5 \times 10^{-11} & 4.5 \times 10^{-13} & \text{torr} \\ 4.0 & 3.5 & 3.0 & 2.6 & \text{K} \end{cases}$$

[a] Parts reprinted by special permission from Rosebury, *Handbook of Electron Tube and Vacuum Techniques*, 1965, Addison-Wesley, Reading MA. Other parts reprinted from (Du62) ©1962, John Wiley and Sons. We note that the inner planets consist primarily of low vapor-pressure material and that, by and large, the outer planets contain more volatile substances.
[b] 1 torr $= 133 \times 10^3$ dyn cm^{-2} $= 1.32 \times 10^{-3}$ atm.

size differences would therefore collide more frequently and be destroyed, whereas grains with nearly identical properties would tend to survive longer.

PROBLEM 10–7. After small particles and chunks were formed through condensation, a second stage of condensation seems to have taken place in which gravitational attraction played a dominant role. Before this time, particles presumably had acquired almost identical low eccentricity, low inclination orbits at any given distance from the Sun, and the relative velocities of these grains must have been small. This would have come about because high- or low-velocity grains would be eliminated preferentially through more frequent destructive collisions with other bodies.

(a) At what size would a body whose density ρ is 3 g cm^{-3} have a gravitational capture cross-section that is twice as large as its geometric cross-section? Assume a relative velocity V_0 for particles to be captured. The result of Problem 3–11 may be useful.

(b) Derive the growth rate of a body with $\rho = 3$ g cm^{-3} moving through a nebula whose density is $\rho_0 = 3 \times 10^{-12}$ g cm^{-3}. Let its relative velocity be $V_0 = 1$ km s^{-1} and start at a time when its gravitational capture cross-section is twice its geometric cross-section.

(c) Show that the mass growth for a spherical gravitating body, whose capture cross-section is much greater than its geometric cross-section and whose density $\rho = 3M/4\pi R^3$ has a fixed value, is proportional to $M^{4/3}$ or R^4. More massive bodies therefore have a higher mass capture rate than lower mass bodies whose geometric capture cross-section only allows them to capture mass at a rate proportional to R^2.

PROBLEM 10–8. Suppose that a grain stays spherical as it grows through capture of matter. It moves through the Solar Nebula at $V_0 = 1\,\mathrm{km\,s^{-1}}$, escapes destructive collisions by chance, and grows from a radius of $\sim 10^{-8}$ cm — one molecule — up to 10 km. If the nebular density is $3 \times 10^{-12}\,\mathrm{g\,cm^{-3}}$, of which a 1% nonvolatile fraction can be captured, and the particle's density is $2.5\,\mathrm{g\,cm^{-3}}$, show that the growth time is roughly 10^8 yr.

Answers to Problems

10–1. Substitute a new variable, u, given by $r = r_0 \sin^2 u$, so that $dr = 2r_0 \sin u \cos u\, du$, which gives the desired result through the integral

$$\int_{\pi/2}^{0} \sin^2 u\, du = -\frac{\pi}{4} .$$

10–2. Set the surface mass density on the sphere equal to $\sigma = M/4\pi R^2$ and consider a point at some off-center distance a. The potential at that point is

$$V = -\int_0^{\pi} \frac{2\pi\sigma G R^2 \sin\Theta\, d\Theta}{\sqrt{(R\cos\Theta - a)^2 + R^2 \sin^2\Theta}} ,$$

where normal polar coordinates have been used. Integration leads to

$$V = -\frac{2\pi\sigma G R}{a}(R^2 - 2aR\cos\Theta + a^2)^{1/2}\Big]_0^{\pi} = -4\pi\sigma G R .$$

Furthermore, because any spherical distribution can be built up from a continuous distribution of spherical shells, the potential anywhere within an empty central sphere will be constant throughout.

10–3. To work out this problem, it may be best to derive equation (10–29) first, since equation (10–28) then follows from setting $Q = 0$. To derive (10–29) we may start with equation (9–31) and substitute for the outflow parameters v_o, ρ_o, and P_o in terms of the corresponding inflow parameters by means of equations (9–26), (9–34), and (9–35). Then, for extremely high velocities, assume that $\gamma_i P_i/\rho_i = c_i^2 \ll v_i^2$ is negligibly small.

10–4. For a uniform disk of mass $M_\odot$ and radius $R = 10^{14}$ cm, the surface density $\sigma = M_\odot/\pi R^2 = 6.4 \times 10^4$ g cm^{-2}, which by (10–64) equals $(n_0 kT/2piG)^{1/2}$. This gives $n_0 \sim 6 \times 10^{15}$ cm^{-3}. The scale height then is obtained from (10–63), $H \sim 3 \times 10^{12}$ cm. We also obtain the pressure $P_0 = n_0 kT \sim 2 \times 10^3$ dyn cm$^{-2} \sim 2 \times 10^{-3}$ atm.

10–5. (a) The volume of the nebula is $V = \pi R^2 H \sim 10^{42}$ cm^3. Its mass is $M \sim 5 \times 10^{30}$ g. The density at each point then is $\rho \sim 5 \times 10^{-12}$ g cm^{-3}. From Table 1.2, the abundance of iron is $\rho_{Fe}/\rho_{\text{tot}} \sim 1.3 \times 10^{-3}$; this makes the density of iron $\rho_{Fe} \sim 6.5 \times 10^{-15}$ g cm^{-3}, and the pressure, if in the form of vapor, $P_{Fe} = (\rho_{Fe}/m_{Fe})kT \sim 10^{-8}T$ dyn cm$^{-2} = 7 \times 10^{-12}T$ torr.

According to the vapor pressure formula given in Table 10.2, the vapor pressure of iron at $T = 1140$ is $\sim 8 \times 10^{-9}$ torr, which corresponds to the vapor pressure of iron in the nebula at this temperature. If the early Sun had a luminosity ten times higher than the solar luminosity today, equation (4–81) tells us that iron could have condensed at distances greater than

$$R \sim \left(\frac{10L_\odot}{16\pi\sigma T^4} \right)^{1/2} \sim 3 \times 10^{12}\,\text{cm} \sim 0.2\,\text{AU} \ .$$

(b) An identical calculation for carbon gives a vapor pressure of $\sim 1.7 \times 10^{-7}$ at a temperature of 2050 K, indicating that carbon will condense approximately at $R = 10^{12}$ cm.

(c) Under the circumstances described the temperature out to 10 AU everywhere exceeds $T \sim 158$ K, and $P_{H_2O} \sim 2 \times 10^{-10}T$ torr exceeds $\sim 3 \times 10^{-8}$ torr everywhere. Extrapolating the H_2O data in Table 10.2, we see that at temperatures ~ 150 K, the equilibrium vapor pressure of water ice substantially exceeds 10^{-8} torr. Pure water is, therefore, precluded from condensing in this nebula. However, the presence of other substances such as ammonia, NH_3, can lower the vapor pressure of H_2, which might permit water to condense out in solution, at the colder edges of the nebula.

10–6. For grains in circular orbits:

$$r\omega^2 = \frac{3}{4\pi\rho s^3} \left(\frac{4\pi M\rho s^3 G}{3r^2} - \frac{\alpha L_\odot \pi s^2}{4\pi c r^2} \right),$$

$$\text{velocity} = \omega r = \left[\frac{1}{r} \left(MG - \frac{3\alpha L_\odot}{16\pi\rho s c} \right) \right]^{1/2}.$$

The velocity difference for the grains with respective densities $\rho = 2$ and 4 g cm^{-3}, and $s = 10^{-3}$ cm is $\Delta v \sim 10^4 \alpha$ cm s^{-1}.

10–7. (a) From Problem 3–11 we see that a body with radius R and density ρ has a total capture cross-section equal to twice its geometric cross-section if

$$\frac{4\pi}{3} \frac{2G}{V_o^2} \rho R^2 = 1 \ .$$

(b) For an impact parameter s,

$$\frac{dM}{dt} = 4\pi\rho R^2 \frac{dR}{dt} = \rho_o V_o \pi s^2 = \rho_o V_o \pi \left[R^2 + \frac{2MGR}{V_o^2} \right] ,$$

$$\int \frac{4\rho}{\rho_o V_o} \left[1 + \frac{8\pi}{3} \frac{GR^2\rho}{V_o^2} \right]^{-1} dR = \int dt = \tau .$$

Initially, $dM/dt = 2\rho_o V_o \pi R^2 = (3\rho_o/4\rho)(V_o^3/G) \sim 10^{10} \, \mathrm{g \, s^{-1}}$.

(c) For large bodies $dM/dt \propto R^4$ or $M^{4/3}$, because $M \propto R^3$.

10–8. Initially, for a particle radius s, $\dfrac{dM}{dt} \sim 4\pi s^2 \dfrac{ds}{dt}\rho = 10^{-2}\pi\rho_o V_o s^2$,

$$s = 10^{-2}\frac{\rho_o V_o}{4\rho} t .$$

From Problem 10–7(a) we see that gravitation takes over when $s \sim [3V_o^2/8\pi\rho G]^{1/2} \sim 10^8$ cm, so that gravitation can be neglected for a body with $s \lesssim 10^6$ cm $= 10$ km. The growth time then is

$$t \sim \frac{10^6}{3 \times 10^{-10}} \sim 3 \times 10^{15} \, \mathrm{s} \sim 10^8 \, \mathrm{yr} .$$

11 The Universe We Inhabit

11:1 Questions About the Universe

In preceding chapters we discussed the appearance of stars and stellar systems and we looked in some detail at the immediate surroundings of the Sun, the one star to which we have easy access. Now, we want to examine the environment in which the Sun, the stars, and the stellar systems are embedded. We want to learn about the properties of the Universe.

The first questions we would impulsively ask are:

(1) What is the shape of the Universe?
(2) How big is it?
(3) How long has it existed?
(4) What are its contents?

While these are some of the most basic questions to ask, obtaining the requisite data remains difficult. As a result, we still have no more than partial answers.

We can take two approaches to make headway. The first is the observational approach. We attempt to observe what the Universe is "really" like. The other approach is synthetic. We construct hypothetical models of the Universe and see how the observations fit them. This second procedure might at first glance seem superfluous. It might seem that all we need are observations; but this is not so.

Any observation has to be interpreted, meaning that it needs to be understood within the framework of theory, even if that theory consists of nothing more than the prejudices that constitute common sense. Common sense itself implies a model. It is three-dimensional; time measurements can be completely divorced from distance measurements; bodies obey the Newtonian laws of motion; there are laws of conservation of energy and of momentum. Though common sense is quite useful at times, it can lead to great misconceptions if uncritically applied.

11:2 Isotropy and Homogeneity of the Universe

If we look out into the Universe as far as the best available telescopes allow, we find that no matter which direction we look, essentially the same picture presents itself. We find roughly the same kind and number of galaxies at a given distance in all directions. There may be statistical variations but they appear to be random. The general coloration of galaxies is also the same independent of direction. The only systematic color differences we detect are those associated with distance, but the universal red shift of the spectra of distant galaxies does not appear to change from one part of the sky to another.

Strictly speaking, all this is true only when we look at fields of view outside the plane of our own Galaxy. The Milky Way absorbs light so strongly that we always have to make allowances for its presence.

Independence of direction is called *isotropy*. As far as we can tell, the Universe is isotropic. There are no indications of any preferred directions, except for the flow of time to be discussed in Section 11:18.

Next we take into account all those effects associated with distance. We ask ourselves whether conditions at large distances from us appear to be different from those nearby. Is the universal red shift the only effect we see, or are there other distance-dependent factors? If the red shift indeed were the only effect, then we could postulate an expanding model to explain all observations. The red shift would be taken to be a Doppler shift caused by the recession of distant galaxies. We would imply that if it were not for this cosmic expansion, distant parts of the Universe would appear identical to our local environment. In such a model no structural differences would exist in different parts of the Universe and the Cosmos would appear to be *homogeneous*.

But this is not what we observe. Galaxies detected at large distances appear quite different from those nearby because the Universe is very large and the information conveyed by means of light signals sometimes takes billions of years to reach us. A distant galaxy we view today appears not as it would to a local observer stationed near that galaxy, but rather as it would have looked to such an observer many æons ago when the galaxy was younger. So, we must expect that distant galaxies will appear progressively younger the farther away we look. Quasars also are found to be most prevalent at a distance of 10 – 12 billion light years, as Fig. 11.1 shows.

Does this mean that the Universe is inhomogeneous?

Not necessarily! And this is the place where theoretical models begin to become important. We have to consider the possible existence of two entirely different models, an *evolving* model and a *self-regenerating* model. These models will have different histories.

In most evolving models, matter initially is quite evenly distributed. At some more or less narrowly defined stage, however, matter aggregates to form galaxies. These recede from each other in a cosmic expansion observed as a red shift. In this model more distant galaxies should consistently appear younger. We conclude that the distant galaxies should at least appear "different" from those nearby. If some such difference could be firmly established we would have strong evidence in support of an evolving Universe.

A self-regenerating model takes a different view. Such a universe maintains exactly the same appearance at all epochs and for all time. Such models take two quite different forms. Historically, the first of these was the *steady state* universe. It pictured distant galaxies streaming away from us. But any depletion due to the cosmic

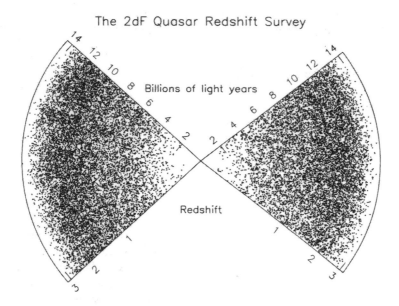

Fig. 11.1. The distribution of quasars across the sky appears to be independent of direction, indicating that the Universe is isotropic. But we observe few quasars close by. At increasing red shifts, their numbers first rise and then decline. Although this might seem an indication of cosmic inhomogeneity, a more detailed analysis reveals that the Universe indeed is homogeneous but that it is evolving: The population of quasars peaked at an epoch corresponding to red shift $z \sim 2$, and thereafter declined. Courtesy of Scott Croom, Brian Boyle and the 2dF QSO Redshift Survey (Cr05).

expansion was replenished through the creation of new matter to form a replacing generation of galaxies. The density of the Universe was thus kept constant. The assortment of galaxies in a given volume remained statistically identical at epochs separated by many æons. There would always be a mixture of young and old galaxies occupying any given volume and the ratio of these galaxies would also remain constant. It did not matter whether we viewed a distant region today, or several æons from now. The Universe would always look roughly identical even though the individual galaxies occupying a given region would no longer be the same.

In a more recent version of such a theory, the currently observed evolutionary phase of the Universe is preceded by a period of rapid *inflation* in which the Universe expands at an exponential rate, regenerating itself until its radius has expanded by a factor of order 10^{25}, all the time maintaining a steady temperature and density.

The remote future of the Universe also appears to be self-regenerating. The discovery that the energy density of the Cosmos is now dominated by a *dark energy* of unknown origin suggests that the Universe will forever continue to expand at an exponential pace, regenerating its dark energy all the time. While the nature and consequences of inflation are still being studied, and the sources of dark energy are a topic of strong current interest, finding new tests for distinguishing different cosmological models remains of the greatest importance. Let us see what these entail.

11:3 Cosmological Principle

Some postulates about the Universe have to be granted before any theory can be developed. These postulates or axioms must then be shown not only to be internally consistent, but also to be borne out by observational consequences.

A particularly demanding postulate is the *perfect cosmological principle*. This principle predicted that for any observer located at an arbitrary position, at an arbitrary time in the history of the Universe, the Cosmos will present exactly the same aspect as that observed by an observer at some other location at the same or even at some completely different epoch. This principle led to the development of the steady state model. This has since been discarded because it failed to account for several observations. The first is the cosmic microwave background radiation exhibiting a blackbody spectrum; no mechanism has been found by which a universe, with as low a density for all time as that observed today, can maintain and replenish a ubiquitous blackbody radiation bath. The second is a high population of small galaxies and a low fraction of large galaxies observed at high red shifts. A third is the prevalence of quasars at red shifts $z \sim 2$, seen in Figure 11.1.

A more modest postulate than the perfect cosmological principle is the simpler *cosmological principle* (Bo52). Its main hypothesis is that our position in space and time is not unusual. Hence our local physics, and our locally made observations of the Universe should not markedly differ from those made by other observers located in different regions of the Universe. This means that the Universe is homogeneous at any given epoch, but may evolve from one epoch to another. In contrast to the perfect cosmological principle, the cosmological principle does appear to hold true.

Both of these principles are extensions of the Copernican hypothesis that we should in no way consider ourselves favored observers. Though it applies only in a statistical sense, since one galaxy obviously looks different from its neighbors, the cosmological principle is useful when used in conjunction with a number of simple abstract concepts.

The first of these is that of a substratum. The *substratum* in any cosmic model is a matrix of geometrical points all of which move in the idealized way required by the model. Real galaxies have random velocities, but we would expect their mean motions to be zero with respect to the substratum. We might also expect that the 2.73 K microwave background radiation would appear isotropic to an observer at rest in the substratum. A state of rest relative to the substratum can therefore be determined in a number of practical ways and plays a fundamental role in cosmology. We call a particle at rest in the substratum a *fundamental particle* and an observer who is similarly stationary a *fundamental observer*.

The watch carried by a fundamental observer measures *proper time*, which generally differs from time registered by clocks in motion relative to the substratum.

The proper time of a fundamental observer can be considered to define a *world time* scale that could be used by all fundamental observers comparing measurements. For example, in describing the evolution of a cosmic model we normally think of a *world map* that describes the appearance of the Universe at one particular world time. In contrast, we can also think of a *world picture* that is just the aspect the Universe presents to a particular fundamental observer at any given time. To see the difference between these concepts, we note that all galaxies are at rest in a world map, but the map may be expanding. On the other hand, in a world picture, distant galaxies would appear to recede from the observer — at least at the present epoch.

11:4 Homogeneous Isotropic Models of the Universe

Observations made to date do not indicate that there are any preferred directions or unusually dense regions in the Universe. The data are compatible with a homogeneous isotropic model of the Universe, that is, a universe in which there are no select locations or directions. An observer placed at any location in the Universe would see distant galaxies red-shifted, in apparent recession, no matter what direction he chose to observe.

In order to construct a model of such a universe, we assume that the red shift indicates a genuine expansion. This assumption has become entrenched in cosmology, primarily through default. When the red shift was first discovered, a number of explanations were advanced. One by one the competing hypotheses have been eliminated — found incompatible with observations, or unlikely on other grounds. The expansion of the Universe is the only hypothesis that has survived all tests and now is accepted as the actual cause of the red shift.

We can visualize a universe in which an observer O' at any point sees all other observers in distant galaxies receding from her. A simple model in two dimensions consists of a rubber sheet (Fig. 11.2). Let spots be painted on this sheet in some random manner. If the sheet is now stretched in length L and width W, by fixed amounts of αL and αW, respectively, then all distances are increased by a fractional amount α. If the spots on the sheet represent galaxies, then a galaxy that initially was at some distance $r = (x^2 + y^2)^{1/2}$ from a given galaxy, will later be at distance $(1+\alpha)r = \{[(1+\alpha)x]^2 + [(1+\alpha)y]^2\}^{1/2}$, where x and y are Cartesian coordinates along the directions L and W.

A flat rubber sheet is not the only two-dimensional model for an expanding homogeneous isotropic universe. Take a rubber balloon and paint spots on it to represent galaxies. At a given instant let the radius of the balloon be a. Let the angle subtended by two galaxies at the center of the balloon be χ. The distance between galaxies measured along the surface is the arc length $a\chi$. If the balloon is expanded the angle χ remains constant, but the radius increases to some new value, say $a' = (1 + \beta)a$, where β is the fractional increase in the radius. The distance between galaxies is now $(1 + \beta)a\chi$, and the fractional increase is independent of χ. This means that if the Universe is homogeneous and isotropic at a given instant an isotropic expansion will keep it that way.

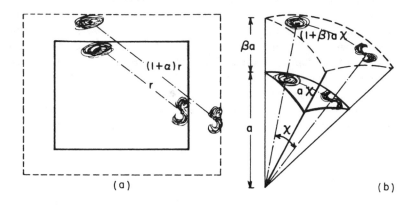

Fig. 11.2. (a) Expansion in a flat universe; (b) Expansion in a curved universe.

If the time rate of change of β is $\dot{\beta}$, the recession velocity between the two galaxies is $a\dot{\beta}\chi$, which increases linearly with angle χ. Dividing the recession velocity by the distance, we obtain the ratio $a\dot{\beta}\chi/a\chi = \dot{\beta}$. We talk about a linear *distance–velocity relation* because, for increasing separation, the recession velocity increases in proportion to the separating distance.

A sphere of radius a is described by the equation

$$x_1^2 + x_2^2 + x_3^2 = a^2, \tag{11-1}$$

where x_1, x_2, x_3 are three mutually orthogonal Cartesian coordinates. An element of length dl on the sphere is given by

$$dl^2 = dx_1^2 + dx_2^2 + dx_3^2. \tag{11-2}$$

Eliminating the coordinate x_3 by means of equation (11–1), we find

$$dl^2 = dx_1^2 + dx_2^2 + \frac{(x_1\,dx_1 + x_2\,dx_2)^2}{a^2 - x_1^2 - x_2^2}\,. \tag{11-3}$$

In terms of spherical polar coordinates, we can write dl^2 as

$$dl^2 = a^2(d\theta^2 + \sin^2\theta\,d\phi^2)\,. \tag{11-4}$$

We can repeat this procedure for a four-dimensional sphere in an exactly analogous way. Here we do not deal with a two-dimensional surface or a three-dimensional space. Rather, we work with a space showing isotropy and homogeneity in three dimensions; and analogously to the three-dimensional approach of equations (11–1) to (11–4), we want to investigate the properties of a three-dimensional *hypersurface* on a four-dimensional *hypersphere*. Problem 11–1 will show that the relation corresponding to equation (11–4) then has the form

$$dl^2 = a^2[d\chi^2 + \sin^2\chi(\sin^2\theta\,d\phi^2 + d\theta^2)]. \tag{11-5}$$

PROBLEM 11–1. Show how relation (11–5) is obtained by starting with an equation for a hypersphere

$$x_1^2 + x_2^2 + x_3^2 + x_4^2 = a^2. \tag{11-6}$$

Continue by showing that in terms of three-dimensional polar coordinates, we have

$$dl^2 = dr^2 + r^2\,d\theta^2 + r^2\sin^2\theta\,d\phi^2 + \frac{(r\,dr)^2}{a^2 - r^2}. \tag{11-7}$$

Then substitute the new variable

$$r = a\sin\chi. \tag{11-8}$$

Consider a sphere of radius R in a conventional three-dimensional space. On a two-dimensional spherical surface the distance along the sphere is given by $R\theta$. A circle about $\theta = 0$ on this surface has length $2\pi R\sin\theta$. At increasing distance from the origin the size of the circle increases to a maximum value $2\pi R$ at a distance $\pi R/2$. After that it decreases and shrinks to a geometric point at the *antipodal position* — at distance πR.

PROBLEM 11–2. Show that on a four-dimensional hypersphere:
  (i) The ratio of the circumference of a circle to its radius is less than 2π.
  (ii) The surface area of a sphere is

$$S = 4\pi a^2\sin^2\chi. \tag{11-9}$$

(iii) As the angle χ increases the sphere grows and the surface of the sphere increases reaching a maximum value $4\pi a^2$ at distance $\pi a/2$ before shrinking to a point at distance πa. Show that the element (11–5) defines the total volume

$$V = \int_0^{2\pi}\int_0^\pi\int_0^\pi a^3\sin^2\chi\sin\theta\,d\chi\,d\theta\,d\phi. \tag{11-10}$$

so that

$$V = 2\pi^2 a^3. \tag{11-11}$$

We can denote a parameter

$$\lambda \equiv \frac{k}{a^2} \quad \text{where } k = 0, \pm 1 \tag{11-12}$$

that defines the *curvature* of a space. The curvature of cosmological models is defined by the *Riemann curvature constant* k that can only assume values $+1$, 0, or

−1, respectively, describing universes of positive, zero, and negative curvature. The constant k denotes the algebraic sign of the parameter λ of equation (11–12).

When the *radius of curvature* is infinite, $k = 0$, $\lambda = 0$, and the space has zero curvature. Such a space is said to be *flat* or *Euclidean*. When $\lambda > 0$, the space has positive curvature. We can also define spaces of negative curvature for which $k = -1$, $\lambda < 0$ if, as in (11–13) below, we replace the right side of (11–6) by $-a^2$. Note that the two two-dimensional universes described earlier have different curvature constants. The sheet model is Euclidean; the balloon has positive curvature.

PROBLEM 11–3. In a space of negative curvature, a *hyperbolic space*, sometimes called a *pseudospherical space*:

$$x_1^2 + x_2^2 + x_3^2 + x_4^2 = -a^2, \tag{11-13}$$

where a is real.

(i) Show that

$$dl^2 = r^2(\sin^2\theta\, d\phi^2 + d\theta^2) + (1 + r^2/a^2)^{-1}\, dr^2, \tag{11-14}$$

where r can have values from 0 to ∞.

(ii) Defining $r = a \sinh\chi$ (where χ goes from 0 to ∞)

$$dl^2 = a^2\{d\chi^2 + \sinh^2\chi(\sin^2\theta\, d\phi^2 + d\theta^2)\}. \tag{11-15}$$

Show that the ratio of the circumference of a circle to its radius is greater than 2π.

(iii) Show that the surface of a sphere is

$$S = 4\pi a^2 \sinh^2\chi \tag{11-16}$$

which increases without limit.

(iv) The volume of the space is

$$V = \int_0^{2\pi} \int_0^\pi \int_0^\infty a^3 \sinh^2\chi \sin\theta\, d\chi\, d\theta\, d\phi \tag{11-17}$$

which is infinite.

To summarize, we note that a space of positive curvature has a finite volume and is closed. Increasing χ beyond a value π returns us to a region already defined by χ values between 0 and π. The volume of a closed space is finite and given by equation (11–11). The space of negative curvature is *open*. The volume of an open space is infinite.

A self-replicating inflationary universe can exist only in a flat space. In a curved expanding space the radius of curvature progressively increases, so that the universe is not precisely replicating itself. A rapid inflationary phase is still possible in a curved universe, but the cosmos then cannot be strictly self-replicating.

In our balloon model of a universe, a galaxy close to an observer subtends a large angular diameter. Were this galaxy moved to increasing distances it would subtend progressively smaller angular diameters until a minimum value was reached at a distance $\pi a/2$, where a is the radius of curvature of the balloon. Beyond this distance the angular diameter of the galaxy once again would increase until it reached a maximum value of 2π when seen at the antipodal point of the balloon — that is, at a distance πa. An observer could then look in any direction he pleased and see the galaxy at one and the same distance from him. In exact analogy similar features characterize three-dimensional hypersurfaces. Figure 11.3 illustrates these effects.

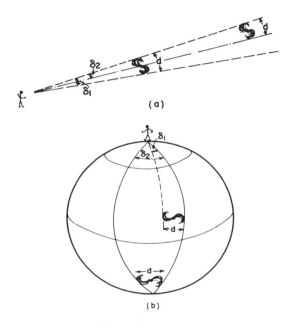

Fig. 11.3. (a) Distance–angular–diameter relation in a flat space. (b) Distance–angular–diameter relation on the surface of a three-sphere.

Attempts to detect a positive curvature, by searching for a minimum angular diameter of distant galaxies, are made difficult by the finding that galaxies evolve, and that small galaxies, in the course of time, merge to form larger galaxies.

11:5 Olbers's Paradox

Consider a Euclidean space uniformly filled with stars. The amount of starlight emitted in a shell at distance r to $r + dr$ from an observer is proportional to the volume, $4\pi r^2\, dr$. A fraction proportional to $1/r^2$ is incident on the observer's telescope. From each spherical shell of thickness dr the observer therefore receives an amount of light proportional to dr alone. On integrating out to infinite distance, the light received by the observer should have infinite brightness. This infinity arises only because we have not taken into account the self-shadowing of stars. A foreground star will prevent an observer from seeing a star in a more distant shell, provided both stars lie along the same line of sight. When shadowing is taken into account we find that the sky should only be as bright as the surface of a typical star, not infinitely bright. Of course, that still is much brighter than the night sky.

To someone who strongly believed in Euclidean space and in the infinite size and age of the Universe this would appear paradoxical. Wilhelm Olbers, who advanced this argument in 1826, saw that such a cosmological view could not be held.

If we try to circumvent the argument by introducing curved space, no advance can be made. In such a space the area of a sphere drawn about an observer is of the form of equations (11–9) or (11–16) — the surface area $S = 4\pi a^2 \sigma^2(\chi)$ is a function of distance χ alone. The number of stars in a spherical shell is proportional to $S(\chi)\, d\chi$. But the amount of light reaching the observer from that shell is also reduced by a factor $S(\chi)$. These two factors cancel to give the same distance independence obtained for a flat space.

We could next argue that interstellar dust might absorb the light. But in an infinitely old universe dust would come into radiative equilibrium with stars and would emit as much light as was absorbed. The dust would then either emit as brightly as the stars, or else it would evaporate into a gas that either transmitted light or else again emitted as brightly as the stars.

The paradox can only be resolved if at least one of three possibilities applies (Bo52), (Ha65):

(a) The Universe is very young. Stars have not been shining for very long; light from great distances could not yet have reached us.

(b) The constants of physics vary with time. Because these constants affect the rate at which stars emit light, it could be that stars only started shining brightly in recent times.

(c) There are large recessional velocities of stars at great distances. Their spectral shifts and apparently diminished luminosity lead to a lower sky brightness.

The correct resolution to Olbers's paradox appears to be that stars are too sparse in the Universe, with a stellar density of only 10^{10} Mpc^{-3}, and that they shine for too short a time, only about $t = 10^{10}$ yr. Each star can then be thought of as occupying a volume $V_0 = 3 \times 10^{63}$ cm^3. If the star's surface area is $\sigma = 10^{23}$ cm^2 and it radiates for time $t \sim 15$ Gyr, roughly the age of the Universe, it could only fill a volume $V_s = c\sigma t = 10^{51}$ cm^3 without diluting the radiation leaving its surface. Since $V_s/V_0 = 3 \times 10^{-13}$, we see that when the radiation of all stars is diluted to fill the entire Universe, the radiation density and hence the surface brightness of the sky is diminished by a factor of 3×10^{12} when compared to the surface brightness of a typical star. Doppler shifts and diminished intensity due to the expansion of the Universe are relatively minor factors (We87). The darkness of the sky is largely due to the low density of stars, their youth, and their finite energy resources.

11:6 Measuring the Geometric Properties of the Universe

It is possible, at least in principle, to determine the size and curvature of the Universe on the basis of astronomical observations (Ro55)*, (Ro68)*. Let us examine the simplest relationships between directly observed quantities and the more abstract geometrical properties of a homogeneous, isotropic universe. The greatest asset of such relations is an independence of the specific dynamics — potentially governed by general relativity — used to describe cosmic expansion and evolution. In effect, we can obtain a geometric description of the Universe as it appears at the present world time, without needing to make assumptions about how the Universe evolved before reaching its present state or how it will evolve in the future. This approach, though greatly simplified, can yield substantial information.

Consistent with the spaces discussed in Section 11:4, one can show, on group-theoretical grounds (Ro33, Wa34), that the most general metric describing homogeneous isotropic spaces is the Robertson–Walker metric — often called the *Friedmann–Robertson–Walker, FRW* metric, after three early pioneers of cosmology. Here

$$ds^2 = c^2\,dt^2 - dl^2 \quad \text{with} \quad dl^2 = a^2(t)\{d\chi^2 + \sigma^2(\chi)[d\theta^2 + \sin^2\theta\,d\phi^2]\}\,, \quad (11\text{-}18)$$

where dl^2 is the metric of a three-dimensional homogeneous, isotropic space. The function $\sigma(\chi)$ has the form $\sin\chi$, χ, or $\sinh\chi$, depending on whether the Riemann curvature constant of the three-space is $k = 1$, 0, or -1. The *comoving coordinate interval* $d\chi$ remains unchanged in a pure expansion or contraction of the *scale factor* $a(t)$. For positively curved universes, $k = +1$ and the scale factor can be taken to be the radius of curvature of the universe. For $k = 0$ or -1 universes, the scale factor has no absolute length and serves mainly to compare the dimensions of the universe at different epochs.

In this notation:

(a) The *world line* of a stationary galaxy is a curve, with χ, θ, and ϕ constant. Along this curve ds measures the world time interval dt (Section 11:3 and Fig. 11.4).

(b) The world line of any light signal is a *null geodesic*, meaning that it is characterized by $ds = 0$.

(c) If we choose a specific world time — $t = $ constant, $dt = 0$ — we can measure spatial distances within the Universe with the aid of the metric $-ds^2$. The curvature of the Universe k/a^2 will then be fully determined if we can ascertain the value of k and of $a(t)$. To this end, consider a world diagram representing an observer O located at $(\chi, \theta, \phi) = (0, 0, 0)$ and a galaxy at $(\chi_0, \theta_0, \phi_0) = $ constant. As $a(t)$ changes with time, a constant interval $d\chi$, in the three-space (11–18) will lead to a changing value of $ds^2 - c^2\,dt^2$. In particular, for a light beam traveling from an event occurring at world time t_1 to reach an observer at world time t_0, (Fig. 11.4) we can set $ds = 0$, and then the equations (11–18) reduce to $c\,dt = a(t)d\chi$. Integration along a fixed line of sight (θ, ϕ) then leads to

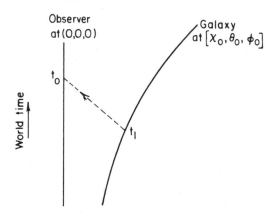

Fig. 11.4. Relation between galaxy and observer (after H. P. Robertson (Ro55)). Courtesy of the Publications of the Astronomical Society of the Pacific.

$$\int_{t_1}^{t_0} \frac{c\, dt}{a(t)} = \chi(t_1, t_0) \, . \tag{11-19}$$

This is the relation between *distance parameter* χ, often also called the *comoving distance*, χ, and the span of world time $(t_1 - t_0)$. We can also define a proper distance which, in an expanding universe at time t, is defined as the distance between the observer and the emitting source along a surface of constant proper time $dt = 0$. Then $ds = a(t) d\chi$, which integrates to

$$\text{proper distance } D(t) = a(t)\chi \, . \tag{11-20}$$

In the limit of infinitesimal time intervals, we also define a

$$\text{conformal time increment } d\tau \equiv dt/a(t) = d\chi/c \, , \tag{11-21}$$

where dt is a proper time interval.

The use of comoving and conformal coordinates places space and time on the same footing, rendering them unaffected by a change solely in scale factor $a(t)$.

We will keep referring to a comoving distance or a *distance parameter* here instead of a *distance* because it is not quite clear just what we would like to call "distance." We will see further on, in equation (11–34), that the apparent luminosity of distant objects makes the quantity $a(t_0)\sigma(\chi)(1 + z)$ another useful measure of distance. Here z is a measure of the red shift defined by equation (11–23), below. We discuss this here only because it is annoying not to have an exact analogue to all the concepts we normally like to attribute to distance. However, in the more general mathematical spaces that are useful in cosmology, we do not have all these properties embodied in a single parameter.

If the light emitted in a time interval t_1 to $t_1 + dt_1$ is received between times t_0 and $t_0 + dt_0$, and the frequency of the emitted signal is ν_1 while that of the received signal is ν_0, then

$$\nu_0 \, dt_0 = \nu_1 \, dt_1 \tag{11-22}$$

because the total number of oscillations in the wave is conserved during propagation. In terms of wavelength $\lambda = c/\nu$, we have

$$z \equiv \frac{\lambda_0 - \lambda_1}{\lambda_1} = \frac{dt_0}{dt_1} - 1 = \frac{a(t_0)}{a(t_1)} - 1. \tag{11-23}$$

Equation (11–23) defines the measured *red-shift parameter* z for radiation emitted at time t_1. Note that λ, along with all other physical measures of length, stays proportional to the scale factor a.

PROBLEM 11–4. Equation (11–23) is not yet in a useful form because we do not know how $a(t)$ varies with time. However, if we make the assumption that $a(t)$ varies regularly, so that a Taylor expansion may be used to determine $a(t_1)$ in terms of derivatives of $a(t_0)$, show that

$$z = \left(\frac{\dot{a}_0}{a_0}\right)(t_0 - t_1) + \frac{1}{2}\left[2 - \frac{\ddot{a}_0 a_0}{\dot{a}_0^2}\right]\left(\frac{\dot{a}_0}{a_0}\right)^2 (t_0 - t_1)^2$$
$$+ \frac{1}{6}\left[6 - 6\frac{a_0 \ddot{a}_0}{\dot{a}_0^2} + \frac{\dddot{a}_0}{a_0}\right]\left(\frac{\dot{a}_0}{a_0}\right)^3 (t_0 - t_1)^3 + \cdots, \tag{11-24}$$

where $\dot{a}_0$, $\ddot{a}_0$, and $\dddot{a}_0$ are the first, second, and third time derivatives of $a(t)$ evaluated at t_0, the time of observation.

Similarly expanding $a(t)$, the scale factor at some arbitrary time, as a Taylor series around t_0, we obtain

$$a(t) = a_0[1 + \dot{a}_0(t - t_0) + \frac{1}{2}\ddot{a}_0(t - t_0)^2 + \frac{1}{3!}\dddot{a}_0(t - t_0)^3 + \cdots]. \tag{11-25}$$

This can be rewritten in terms of three quantities, the *Hubble constant*, which defines the rate of expansion,

$$H(t) \equiv \frac{\dot{a}(t)}{a(t)}, \qquad H_0 \equiv \frac{\dot{a}_0}{a_0}, \tag{11-26}$$

the *deceleration parameter*, a negative acceleration rate,

$$q(t) \equiv -\frac{a\ddot{a}}{\dot{a}^2}, \qquad q_0 \equiv -\frac{a_0 \ddot{a}_0}{\dot{a}_0^2} = -\frac{\ddot{a}_0}{a_0}\frac{1}{H_0^2}, \tag{11-27}$$

and *jerk*, the rate at which the acceleration changes,

$$j(t) \equiv \frac{\dddot{a}}{a}\left(\frac{a}{\dot{a}}\right)^3, \qquad j_0 = \frac{\dddot{a}_0}{a_0}\left(\frac{1}{H_0}\right)^3. \qquad (11\text{-}28)$$

In each of the expressions, parameters with subscript zero refer to the present epoch. With these parameters, we can rewrite (11–24) as

$$z(t) = H_0(t_0 - t_1) + \frac{1}{2}H_0^2[1 + q_0](t_0 - t_1)^2 + \frac{1}{6}j_0(t_0 - t_1)^3. \qquad (11\text{-}29)$$

11:7 Angular Diameters and Number Counts

We can next ask about the angular diameter δ subtended by an observed galaxy. Let the intrinsic, locally measured diameter of the galaxy be D. Equation (11–18) shows that, at world time t_1, the circumference of a major circle centered on the observer and drawn transverse to the line of sight through the galaxy is $2\pi a_1\sigma(\chi)$. This corresponds to the full range of azimuthal angles $0 \lesssim \phi \lesssim 2\pi$ in a plane that we can arbitrarily designate $\theta = \pi/2$. The linear diameter of the galaxy therefore subtends a segment $D/2\pi\sigma(\chi)a_1$ of a full circle, and an angular diameter

$$\delta = \frac{D}{a_1\sigma(\chi)} \qquad (11\text{-}30)$$

radians. To convert this into values of $a(t)$ measured by the observer at epoch t_0, we invoke equation (11–23) to obtain

$$\delta = \frac{(z+1)D}{a_0\sigma(\chi)}. \qquad (11\text{-}31)$$

Note that this is a factor of $(z+1)$ larger than the fraction of a radian that D actually subtends at time t_0. This is because the galaxy was nearer to the observer when it emitted light at epoch t_1 than when its light arrives, at t_0.

The second relation of interest to observational cosmology is the dependence on χ of the number of galaxies $N(\chi)$ whose comoving distance is less than or equal to a given value χ. The comoving number density of galaxies n in the three-space defined by the metric dl^2 in (11–18), is independent of t if evolutionary effects are neglected. In a homogeneous model it is also independent of χ, so that

$$N(\chi) = 4\pi n \int_0^\chi \sigma^2(\chi)\,d\chi. \qquad (11\text{-}32)$$

Figure 11.5 illustrates these ideas.

PROBLEM 11–5. Show that (11–32) can be expanded into the series relation

$$N(\chi) = \frac{4\pi n}{3}\chi^3\left(1 - \frac{k}{5}\chi^2 + \cdots\right). \qquad (11\text{-}33)$$

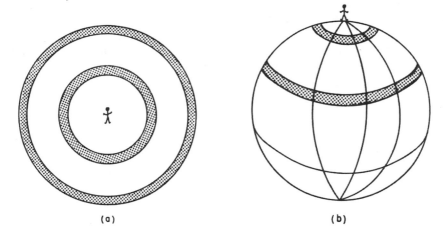

Fig. 11.5. (a) Distance–number–count relation in (a) a flat space, (b) on a spherical surface. Because the circle or surface drawn about the observer at any given distance is always smaller than the corresponding circle or surface in a flat space, the number of galaxies counted at any distance in a spherical universe will also be lower than the number counted in a flat — Euclidean — universe.

Both relations (11–31) and (11–33) depend on a knowledge of z, if distance is derived from red shift. However, z is often hard to measure at great distances because galaxies there are quite faint. We might therefore prefer to deal with the total observed flux, a readily determined quantity, rather than the red-shift parameter. To do this, we need to know more about the apparent luminosity of distant galaxies as seen by an observer today. To determine this, we make the further assumption that photons are conserved and that their energy is related to frequency by the Planck expression $\mathcal{E} = h\nu$ with h a universal constant independent of world time.

If L_1 is the bolometric luminosity of the galaxy at the time of emission, then the bolometric flux $\mathcal{F}_0$ reaching an observer O is

$$\mathcal{F}_0 = \left(\frac{L_1}{4\pi a_0^2 \sigma^2(\chi)} \right) \left(\frac{1}{(1+z)^2} \right) . \tag{11-34}$$

Here the first term represents the geometrical dilution of radiation, since $4\pi a_0^2 \sigma^2$ is the surface area of the three-space of (11–18) drawn about the emitting galaxy and given by (11–9) or (11–16). The second term represents the reddening. The term $(1 + z)$ appears squared. One reduction by $(z + 1)$ is due to the decrease in spectral frequency and, hence, the decrease in energy per arriving photon. A second reduction by $(z + 1)$ enters because all conceivable frequencies are lowered, including the rate at which photons emitted by the galaxy arrive at the observer. In unit time interval the observer sees $(1 + z)$ fewer photons than were emitted at the galaxy in unit time. This corresponds to an apparent slowing down of clocks and again means that less energy arrives at O per unit time interval.

If we deal with a nearly monochromatic source whose luminosity ΔL_1 in frequency interval $d\nu_1$ is $\Delta L_1 = L(\nu_1)\,d\nu_1$, then the flux received is

$$\mathcal{F}(\nu_0)\,d\nu_0 = \frac{L(\nu_1)\,d\nu_1}{4\pi a_0^2 \sigma^2(\chi)(1+z)^2}\,, \qquad (11\text{-}35)$$

where subscripts 1 and 0, respectively, refer to emission and reception times. Through (11–22) and (11–23) we obtain $d(\nu_1) = (1+z)d(\nu_0)$ and the expression

$$\mathcal{F}(\nu_0) = \frac{L(\nu_1)}{4\pi a_0^2 \sigma^2(\chi)(1+z)}\,. \qquad (11\text{-}36)$$

This is the spectral line flux to be expected from individual distant quasars and galaxies.

Returning to equation (11–34) we can make use of the ratio of bolometric luminosity of a source and its apparent bolometric magnitude at the observer as a measure of the *luminosity distance* of the source. This is defined as

$$\mathcal{D}_L \equiv \sqrt{\frac{L_1}{4\pi(1+z)^2 \mathcal{F}_0}} = a_0 \sigma(\chi)\,. \qquad (11\text{-}37)$$

In principle, it should be possible to determine the history of cosmic expansion by plotting the luminosity distance versus the red shift. If the expansion rate of the Universe had once been high and then had suddenly stopped, we would see distant sources highly redshifted while nearby sources were not redshifted at all. As we will see in the next section, increasing sensitivities have now made such observations possible.

The flux density $\mathcal{F}(\nu_0)$ in equation (11–36) is often written as $S(\nu)$. The relations $N(\chi) \propto \chi^3$ and $\mathcal{F}(\nu_0) \propto \chi^{-2}$, respectively, in (11–33) and (11–36), confirm that in a Euclidean universe Seeliger's theorem — the relation between N and S given in equations (2–4) and (2–5) — holds universally.

11:8 The Flux from Distant Supernovae

As mentioned in Section 2:8, supernovae of type Ia are found to have uniform luminosities in the local universe. Once the *light curve* — given by the rest-frame color, the rise time and rate of decline of luminosity — is determined, the luminosity is found to be essentially identical for all SNe Ia. This has permitted their use as distance indicators, as in Fig. 11.6.

Equation (11–23) shows that the red shift $z = \Delta\lambda/\lambda$ is just the fractional expansion the Universe has undergone over some time interval $(t_0 - t_1)$ referred

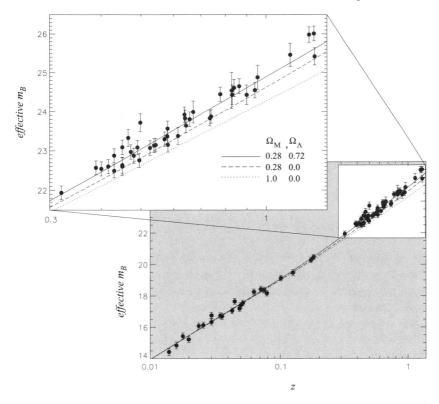

Fig. 11.6. The magnitudes m_B of supernovae of type Ia selected for exhibiting low extinction within their parent galaxies, and plotted against their red shifts z. The magnitudes m_B take red shift into account and correspond to the blue light emitted in the rest-frame of the supernova. The implications of the three curves marked by different values of the symbols Ω_M and Ω_Λ are discussed below, in Section 11:12. The data were culled from surveys of both nearby and distant supernovae published by different groups in five separate studies. Courtesy of Saul Perlmutter (Pe05).

to the present epoch t_0. More generally dz/dt refers to the expansion rate between any two epochs t and $t + dt$. Differentiating with respect to time gives $dz/a_0 H(z) = dt/a(t)$, where we have written the Hubble constant as a function of red shift: $H(z) = (1/a)(da/dz)(dz/dt)$. When we see light from a distant supernova redshifted by an amount z, $H(z)$ gives the Hubble constant of the Universe at the epoch when the supernova emitted the light. If the rate of expansion is not constant, then the Hubble constant changes along a photon's light path as it travels toward us. The comoving coordinate of the emitting supernova is then obtained by integration along the light path derived from (11–19)

$$\chi(z) = \frac{c}{a_0} \int_0^{z_1} \frac{dz}{H(z)} , \qquad (11\text{-}38)$$

where z_1 is the red shift at the epoch of emission. By inverting this relation and comparing the comoving distances of supernovae at different red shifts, we can recover the history of the cosmic expansion, i.e., the history of the Hubble constant over past epochs.

Returning to (11–29) we might make use of the proper distance $\mathcal{D}$ from (11–20) to substitute $\mathcal{D}/c$ for $(t_0 - t_1)$ and obtain a relation between distance and red shift, independent of the spatial curvature k. However, $\mathcal{D}$ is not a directly measurable quantity. On the other hand, while the luminosity distance $\mathcal{D}_L$ can be directly assessed, it does depend on curvature through its dependence on $\sigma(\chi)$. Considerations that will arise in Chapter 12 tell us that space is at least very close to flat, and for a flat space, $\mathcal{D}_L = \mathcal{D}$. Substituting $\mathcal{D}_L/c$ for $(t_0 - t_1)$ in equation (11–29) therefore gives us a direct link between the luminosity distance and red shift.

$$z(t) = H_0(\mathcal{D}_L/c) + \frac{1}{2}H_0^2[1 + q_0](\mathcal{D}_L/c)^2 + \frac{1}{6}j(t)(\mathcal{D}_L/c)^3 . \tag{11-39}$$

Since equation (11–37) relates $\mathcal{D}_L$ to the known bolometric luminosity of SNe Ia and the observed bolometric flux $\mathcal{F}$, this leads by way of equation (A–3) of the Appendix to a relation between the observed magnitudes of SNe Ia and their red shifts.

$$m_0 = M_1 + 5\log\left[\sigma(\chi)(1 + z)\frac{a_0}{10\,\mathrm{pc}}\right]$$

$$= M_1 + \log\left[\frac{\mathcal{D}_L}{10\,\mathrm{pc}}\right] = M_1 + 5\log\left[\frac{L_1/4\pi\mathcal{F})^{1/2}}{10\,\mathrm{pc}}\right] , \tag{11-40}$$

where a_0 is now measured in parsecs and the division by $10\,\mathrm{pc}$ reflects that absolute magnitude M_1 always refers to the apparent magnitude of an object at a distance of $10\,\mathrm{pc}$ (as explained in Section A:7(e)).

Equations (11–39) and (11–40) now provide direct links between the observed bolometric magnitudes of SNe Ia, their red shifts, and their luminosity distances. Equation (11–39) shows the further relation to the deceleration parameter.

Figure 11.7 shows the apparent magnitudes of distant SNe Ia as a function of red shift, referred to their expected magnitudes if the deceleration parameter had been zero, $q(t) = 0$ throughout.

To date, the most distant SNe Ia reliably observed extend no further than red shifts $z \lesssim 1.7$. But the data of Figs. 11.6 and 11.7 suggest that the cosmic expansion was slowing down before the epoch $z \sim 0.8$ and has been accelerating ever since.

The observational value of q_0 is still somewhat uncertain and, as we just saw, appears to be changing. As Table 11.1 shows, we can relate potential q_0 values to the curvature k of space.

PROBLEM 11–6. From the definition of q_0, and the exponential expansion of an inflationary phase, $a(t) \propto e^{tH}$, show that the cosmic scale factor expands exponentially. Show also that for these models, $q_0 = -1$.

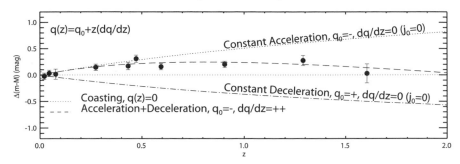

Fig. 11.7. The apparent magnitudes of supernovae of type Ia plotted as a function of red shift. The data points shown are weighted averages for supernovae found in fixed redshift bins. The data and the curves depicting different models of the expansion history of the Universe are shown relative to an eternally coasting model, for which the deceleration parameter $q(z) = 0$ for all time. The best-fit curve through the data points is indicated by the dashed curve, which assumes $q(t) = q_0 + z(dq/dz)$ with $q_0 < 0$ and $dq/dz > 0$. It indicates that the expansion of the Universe was decelerating at a red shift $z \sim 1.7$, where the slope of the curve is parallel to the constant deceleration curve, then coasted for a while, and now is accelerating. The point of inflection, where the acceleration drops to zero, appears to be around $z = 0.8$ though the data are too sparse to define the epoch precisely. The fiducial lines are drawn for $\Omega_\Lambda = 0.73$, $\Omega_M = 0.27$, parameters discussed below, in Section 11:12. (Courtesy of Adam Riess (Ri04).)

Table 11.1. Values of the Deceleration Parameter q_0 in Different Cosmological Models.

	k	q_0
For evolving models with zero	$+1$	$> 1/2$
cosmological constant and	0	$= 1/2$
pressure (see Section 11:12 below)	-1	$0 \le q_0 < 1/2$
Inflationary and steady state models	0	$q_0 = -1$

11:9 Magnitudes and Angular Diameters of Galaxies

In principle, the arguments just presented apply not only to supernovae but also to galaxies. However, galaxies significantly evolve over cosmic time. In order to take this into account, one needs to replace M_1 in (11–40) by its expanded form

$$M_1 = M_0 - \dot{M}_0(t_0 - t_1) + \cdots . \tag{11-41}$$

To the extent that this may prove successful one can then obtain a relation between a galaxy's magnitude and its angular diameter. If the flux $\mathcal{F}_0$ in equation (11–34) is integrated over the received spectral frequency range $d\nu_0$ and divided by the square of the angular diameter δ in (11–31), we obtain a relation between magnitude and angular diameter that could be applied to distant galaxies if they were all alike.

PROBLEM 11–7. Show that to lowest order

$$m_0 = 2.5 \log \Delta \mathcal{F}_0 = 2.5 \left[2 \log \delta - 3 \log(1 + z) + \log \frac{\Delta L_1}{4\pi D^2} \right] . \qquad (11\text{-}42)$$

A useful lowest-order approximation is

$$(t_0 - t_1) = H_0 z \qquad (11\text{-}43)$$

so that

$$m_0 = M_0 - \dot{M}_0 H_0 z + 5 \log \left[\sigma(\chi)(1 + z) \frac{a_0}{10 \,\text{pc}} \right] . \qquad (11\text{-}44)$$

In most of the cosmological models considered today, curvature effects such as those illustrated in Figs. 11.6 and 11.7, are observable, if at all, only at such large distances that galaxies and the stars inside them presumably evolve significantly during the time their signals take to reach us. How well we can then define the time derivatives $\dot{M}_0$ or $\ddot{M}_0$ is not at all clear. Galaxies can suffer catastrophic structural changes as evidenced, say, by the explosion of material from the nucleus of the galaxy M82 or the extremely powerful radio "jet" of the giant spherical galaxy M87. For many quasars and violently active galactic nuclei, AGNs, even greater short-term variations in luminosity and spectral energy distribution may be expected. These traits make galaxies, quasars, and AGNs rather unreliable distance indicators.

11:10 Dynamics on a Cosmic Scale

As we saw in our discussion of black holes in Chapter 5, Einstein's general theory of relativity makes gravitation inseparable from the geometry of space. General relativistic cosmology posits this same strong interdependence (Ei17). In the remainder of our discussions, we will assume that general relativity provides a good description of the Universe and its evolution, even though we recognize two limitations:

(a) General relativity has been tested out to a scale no larger than the Solar System, $o(10^{13}\,\text{cm})$, and it is not clear that the same laws hold on the scale of the Universe, $o(10^{28})\,\text{cm}$. Few laws of physics span such large ranges.

(b) We have found no way to date of incorporating quantum physics in general relativity, so that our description of the early Universe must be uncertain for epochs when the Cosmos may have been no older than $\sim 10^{-43}$ s and, given the limited speed of light c, regions separated by more than $\sim 10^{-33}$ cm could never have been in causal contact.

11:11 Einstein's Field Equations

In the early 1930s, Georges Lemaître, H. P. Robertson, and A. G. Walker found that in an isotropic homogeneous space the general relativistic field equations of Einstein reduce to two simple differential equations in the *scale factor*, a (Le31), (Ro33), (Wa34). The first of these equations gives the mass density ρ, including the equivalent mass of all radiation and matter, in terms of a *cosmological constant* Λ, the curvature of space k, and the expansion or contraction speed $\dot{a}/a$, relative to a scale factor a whose dimensions are length.

$$\frac{8\pi G\rho}{3} = \frac{-\Lambda c^2}{3} + \left(\frac{\dot{a}^2 + kc^2}{a^2}\right), \tag{11-45}$$

where G is the gravitational constant, and dots represent differentiation with respect to world time. The constant Λ corresponds to a tension in the cosmic substrate so that work has to be done on the Universe in order to expand it; alternatively work can be derived during an expansion, depending only on whether Λ is taken to be negative or positive. The ratio $\dot{a}/a \equiv H$ is the Hubble constant at any given epoch; despite its name, H generally evolves with world time.

The second of the equations gives the change in the mass density of the Universe on expansion or contraction, including work done against the pressure of the cosmic fluid, P. This is the energy conservation equation

$$\frac{d\rho}{dt} + 3\left(\rho + \frac{P}{c^2}\right)\frac{\dot{a}}{a} = 0, \quad \text{or} \quad \frac{8\pi GP}{c^4} = \Lambda - \left(\frac{2a\ddot{a} + \dot{a}^2 + kc^2}{c^2 a^2}\right). \tag{11-46}$$

The equation on the right is obtained by inserting the time derivative of (11–45) into the equation on the left.

11:12 The Density Parameter Ω

If we set Λ and $k = 0$ in equation (11–45), we can define a *critical density* ρ_{crit}

$$\rho_{\text{crit}} \equiv \frac{3H^2}{8\pi G}. \tag{11-47}$$

For today's Hubble constant, $H_0 \sim 70\,\text{km s}^{-1}\,\text{Mpc}^{-1}$, $\rho_{crit} = 9.7 \times 10^{-30}\,\text{g cm}^{-3}$. A look at equation (11–46) tells us that this density is critical in the sense that the acceleration $\ddot{a}$ will asymptotically approach a value of zero as the density progressively declines through expansion, so that $\dot{a}$ approaches zero provided the pressure P also becomes negligible.

$$\ddot{a} = -\frac{\dot{a}^2}{2a} \quad \text{and} \quad q_0 = \frac{1}{2}, \tag{11-48}$$

where we have made use of equation (11–27). The Universe then continues to expand indefinitely, but the velocity of expansion decelerates monotonically. On integrating with respect to world time,

$$\ln[\dot{a}a^{1/2}] = \text{constant} \tag{11-49}$$

or

$$Ha^{3/2} = \text{constant}, \qquad H(z+1)^{-3/2} = \text{constant}. \tag{11-50}$$

We may then talk about an apparent age τ_A for the Universe given by

$$\tau_A \equiv \frac{1}{H} = \left[\frac{3}{8\pi G\rho}\right]^{1/2}. \tag{11-51}$$

This age is proportional to $a^{3/2}$. For late stages in the evolution of the universe, when it is filled with nonrelativistic matter, but before domination by dark energy Λ sets in,

$$\tau_A \propto a^{3/2}, \quad \rho \propto a^{-3} \quad \text{[matter-dominated era]}. \tag{11-52}$$

In contrast, for $\Lambda = k = 0$ and early epochs, when the universe is intensely hot and the mass–energy density is dominated by radiation,

$$P = \frac{\rho c^2}{3} \quad \text{[radiation-dominated era]} . \tag{11-53}$$

Equation (11–51) still holds, though we now have

$$\rho \propto a^{-4} \quad \text{[radiation-dominated era]} , \tag{11-54}$$

as is readily understood if we consider the number of quanta of radiation to be constant during this era. In that case, the number density of quanta is proportional to a^{-3}, but because of the red shift the energy per quantum decreases as a^{-1}. The product of number density and energy per quantum of radiation is therefore proportional to a^{-4}. This holds equally well for photons and other quanta lacking rest–mass. To a good approximation, it also holds for massive particles at very high energies, where the particles travel at speeds close to the speed of light and their rest–masses are small compared to their total mass energies. From equation (11–54) we then have

$$H - \frac{1}{\tau_A} \propto a^{-2} \propto (z+1)^2 \quad \text{[radiation-dominated era]}. \tag{11-55}$$

In summary:

$$\left.\begin{array}{l} \tau_A = H^{-1} \propto [a(t)]^{n/2} \\ \propto [z+1]^{-n/2} \end{array}\right\} \qquad \left\{\begin{array}{l} n = 3 \text{ [matter-dominated era]}, \\ n = 4 \text{ [radiation-dominated era]}. \end{array}\right. \tag{11-56}$$

PROBLEM 11–8. (a) Show that the actual age or world time, t_a, of the Universe, when $k = \Lambda = 0$, is just

$$t_a = (2H)^{-1} \tag{11-57}$$

for a relativistic equation of state, $P = \rho c^2/3$,

(b) At a later epoch, when the pressure has dropped to $P = 0$ and the Universe is matter-dominated, show that the world time is related to the Hubble constant by $t_a = 2/3H$, provided $t_a \gg t_0'$, where t_0' marks the epoch when the pressure drops well below the relativistic value. Because the epoch during which the Universe has been matter-dominated is far longer than the earlier relativistic epoch, the age of the Universe today is

$$t_a \sim \frac{2\tau}{3} = \frac{2}{3H} . \tag{11-58}$$

(c) For the very latest stages of a universe with cosmological constant but a flat space, $k = 0$, the cosmological constant Λ must eventually dominate the mass–energy density. Show that the asymptotic age of the universe then approaches

$$t_a \sim (3/\Lambda c^2)^{1/2} \ln(a/a_x) = \frac{\ln(a/a_x)}{H} = \frac{\ln(1+z_x)}{H} , \tag{11-59}$$

where $a_x \ll a$ is the scale factor and $z_x \gg 1$ is the red shift at the crossover from the matter-dominated to the Λ-dominated era.

Right now, the density of atoms and ions, usually referred to as the *baryonic density* ρ_B of the Universe is well below the critical density, as is the density of dark matter discussed in Section 9:17. But the value of the cosmological constant appears sufficiently high to make the overall mass–energy density equal to ρ_{crit}.

In order to describe the evolution of the Universe for different possible density values, we frequently make use of a *density parameter* Ω, defined by

$$\Omega \equiv \frac{\rho}{\rho_{\text{crit}}} . \tag{11-60}$$

If we assume that the pressure in the Universe has been negligible in recent epochs, $P = 0$, we obtain from (11–46) that

$$\frac{\ddot{a}}{a} = -\frac{\Omega_M H^2}{2} + \frac{\Lambda c^2}{3} \quad \text{and} \quad q = \frac{\Omega_M}{2} - \Omega_\Lambda , \quad \text{with} \quad \Omega_\Lambda \equiv \frac{\Lambda c^2}{3H^2} . \tag{11-61}$$

Here Ω_M is the density parameter for matter, including dark matter and baryonic matter, and the *deceleration parameter* q is given by (11-27). The curvature $\lambda = k/a^2$ expressed in (11–12) can be used to define a *curvature density parameter* Ω_k as

$$\Omega_k \equiv -\left(\frac{c}{H_0}\right)^2 \frac{k}{a_0^2}, \quad \text{so that} \quad \Omega_M + \Omega_\Lambda + \Omega_k = 1 . \tag{11-62}$$

This emphasizes that the curvature of space dynamically acts like a mass density. The expression on the right of (11–62) is just a different way of writing the Einstein equation (11–45). Figures 11.6 and 11.7 indicate that the present expansion rate of the Universe may be dominated by *dark energy* due to either a cosmological constant Λ, or to some variable form of energy often referred to as *quintessence*, which may not have remained constant throughout the evolution of the Universe. Its density parameter today is $\Omega_\Lambda \sim 0.7$, while $\Omega_{DM} \sim 0.23$ for dark matter and $\Omega_B \sim 0.04$ for baryonic mass, with uncertainties of order 0.04 in both Ω_Λ and Ω_{DM}. As seen in Table 13.1, the Universe appears to be flat, i.e., $k = 0$; today's total density parameter $\Omega_T(t_0) \equiv \Omega_0 = \Omega_\Lambda + \Omega_{DM} + \Omega_B \equiv \Omega_\Lambda + \Omega_M = 1.02 \pm 0.02$ (Be03, Sp03). For a constant dark energy density Λ, the cross-over from matter- to dark-energy-domination occurs when $\Omega_M(1 + z)^3 = \Omega_\Lambda$. For $\Omega_M = 0.27$ and $\Omega_\Lambda = 0.7$, this occurs for $z = 0.37$ when, judging from (11–59), the Universe was very roughly $\sim 4 \times 10^9$ yr younger, or $\sim 10^{10}$ yr old. For the same current values of Ω_M and Ω_Λ, (11–61) shows that the Universe transitioned from decelerating to accelerating expansion, $q = 0$, at $z = 0.73$, in accord with the turnover region in Figs. 11.6 and 11.7. The SNe Ia data, in this way, are useful for setting bounds on the ratio Ω_M/Ω_Λ.

11:13 Some Simple Models of the Universe

(a) Static Universe of Einstein

Before the expansion of the Universe had been discovered, Einstein (Ei17), quite understandably, proposed a static cosmic model of the Universe. His general relativistic field equations, (11–45) and (11–46) allowed Einstein to calculate the density of this universe in terms of its radius, on the assumption that the pressure was negligibly low, $P = 0$. The Einstein universe is spherical ($k = 1$) and has constant radius of curvature a (Fig. 11.8).

PROBLEM 11–9. For *Einstein's universe* $a =$ constant and $k = 1$.

(i) Show that these two assumptions lead to

$$\Lambda = \frac{1}{a^2} + \frac{8\pi G P}{c^4} \qquad (11\text{-}63)$$

and that the density of the universe is fixed at

$$\rho = \frac{c^2}{4\pi G a^2} - \frac{P}{c^2}. \qquad (11\text{-}64)$$

We know that $P/c^2 \ll \rho$, at present. If we lived in an Einstein universe, Λ would have to have a value $\sim a^{-2}$, and $\rho \sim c^2/4\pi G a^2$.

(ii) Show that if $k = 0$ and $P = 0$ a static universe would require $\Lambda = 0$ and $\rho = 0$, leaving a undefined.

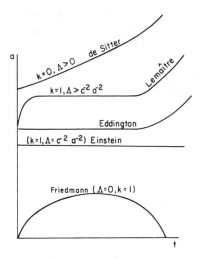

Fig. 11.8. Some cosmological models. The scale of $a(t)$ and of t is not the same for different curves. The only important consideration in this figure is the shape of each curve, rather than its displacement along the ordinate or exact dimensions.

(b) de Sitter Model

Shortly after Einstein proposed the static model in 1917, Willem de Sitter pointed out that the general relativistic field equations permit the description of a second model, one that exists in a flat space, $k = 0$. It is an expanding model. At first it had no more than academic interest (de17). But the finding by Edwin P. Hubble of progressively increasing red shifts in distant galaxies and the implied *cosmic expansion* raised its significance (Hu29). Its main drawback is that the density of such a universe must be zero. However, cosmic densities are low anyway, and this was not considered an overriding difficulty.

PROBLEM 11–10. de Sitter's universe is flat and empty, $k = \rho = P = 0$. Show that the scale factor a of the expanding universe obeys

$$a = a_0 e^{(\Lambda c^2/3)^{1/2}t} \tag{11-65}$$

and the age of the universe is given by (11–59). The actual Universe appears to be headed toward an exponential expansion, but the pressure P appears to be proportional to Λ and negative.

(c) Eddington Model

In 1930, Georges Lemaître and Arthur Stanley Eddington discovered that the Einstein universe is unstable (Ed30). A small deviation from the perfect conditions postulated by Einstein would result in either a continuing expansion or else an accelerating collapse. They set up a model invoking this instability, thinking that galaxies might be able to form at the unstable stage.

Eddington then proposed a model of the Universe that starts out in an Einstein state, becomes perturbed by processes involving the formation of the galaxies from an initially uniform distribution of gas, and goes over into a uniform expansion. One difficulty with this model is that the formation of galaxies is more likely to result from an instability that would lead to contraction rather than expansion of the Universe. The model was interesting because it focused attention on cosmology as not just a matter of geometry. A model must also be able to account for the physical state of matter found in the Universe. Galaxies are likely to have condensed out of an initially uniform gas. If this gas was in rapid expansion, how was it possible to counteract the expansion in order to force the gas to contract into galaxies? Eddington and Lemaître tried to make a plausible guess.

PROBLEM 11–11. Prove the instability of the Einstein universe for $P = 0$. Note that an infinitesimal expansion makes $\rho < c^2/4\pi G a^2 - P/c^2$, so that even with $\dot{a} = 0$, we have $\ddot{a} > 0$ and the expansion must continue. The proof for an initial contraction is similar.

(d) Lemaître Model

Lemaître (Le50) also proposed another model. The universe starts out in a highly contracted state and initially expands at a rapid rate. The expansion is slowed down and brought to a halt in a state that is nearly identical with the Einstein state. Galaxies form at this stage and give rise to a new expanding phase that continues indefinitely (Fig. 11.8). As we saw, in Figs. 11.6 and 11.7, recent measurements on the magnitudes of supernovae at high red shifts show them to be somewhat fainter than would be expected if the Universe was expanding at constant speed. They indicate that the Universe was decelerating before red shift $z \sim 1$, was freely coasting around $z \sim 0.5$ and now is accelerating (Ri01). This has been interpreted as a speeding up of the expansion and a positive constant Λ or a quintessence with an as yet unknown equation of state. Although the Universe never entirely stopped expanding, the resemblance to the Lemaître model is evident.

PROBLEM 11-12. For a universe with quintessence Λ, that makes no contribution to the pressure P but does add to the density in the same way as a cosmological constant, show that the acceleration $\ddot{a}$ goes to zero for a density ρ_o for which

$$\Lambda = \frac{4\pi G \rho_o}{c^2} \quad \text{for nonrelativistic matter,}$$

$$= \frac{8\pi G \rho_o}{c^2} \quad \text{for relativistic matter.} \tag{11-66}$$

If an inflection is to occur for vanishing expansion, $\dot{a}/a \Rightarrow 0$, show that the curvature of the universe at this epoch has to be positive $k = +1$, and the radius of curvature a_o is

$$
\begin{aligned}
a_o &= \left(\frac{c^2}{4\pi G\rho_o}\right)^{1/2} = \Lambda^{-1/2} \quad \text{nonrelativistically,} \\
&= \left(\frac{3c^2}{16\pi G\rho_o}\right)^{1/2} = \left(\frac{2\Lambda}{3}\right)^{-1/2} \quad \text{relativistically.} \qquad (11\text{-}67)
\end{aligned}
$$

A variant of the model of Problem 11–12, which goes through an inflection but never ceases to expand, may also be considered.

PROBLEM 11–13. We can ask for the value of a cosmological constant Λ and density ρ at which a universe with $k = 0$ would undergo inflection for different equations of state $P(\rho)$:

(a) Show that for a matter-dominated universe with $P = 0$, $\ddot{a} = 0$ for $\Omega_\Lambda \equiv \Lambda c^2/3H^2 = 1/3$, $\Omega_M \equiv 8\pi G\rho_M/3H^2 = 2/3$.

(b) Show that for a radiation-dominated universe with $P/c^2 = \rho_r/3$, $\ddot{a} = 0$ occurs for $\Omega_\Lambda = 1/2$, $\Omega_{\rho_r} = 1/2$.

(c) Current observations suggest that the equation of state for dark energy, written as the ratio of pressure to energy density, is

$$
w = \frac{8\pi G P_\Lambda}{c^4 \Lambda} \sim -1 \qquad (11\text{-}68)
$$

and that the pressure due to matter P_M is negligible even when the density of matter ρ_M is appreciable (Sp03). Show that the epoch of inflection, $\ddot{a} = 0$ occurred when $\Lambda = 8\pi G\rho/5c^2$. If Λ is constant throughout time, and the best estimate for conditions today is $\Omega_\Lambda = \Lambda c^2/3H^2 \sim (0.7/0.27)(8\pi G/3H^2 c^2)\rho_0 = (0.7/0.27)\Omega_M$, this means that inflection occurred when $\rho = 5(0.7/0.27)\rho_0$, i.e., at $z = 1.35$. See Fig. 11.7.

(e) Friedmann Models

The relativistic cosmological models described so far have had one feature in common. They all involve a nonzero *cosmological constant* Λ in the relativistic field equations. Alexander A. Friedmann set this constant equal to zero, essentially denying its existence (Fr22). The Friedmann models can have Riemann curvature $k = -1, 0,$ or $+1$. Some of the models start in an extremely dense state and continue to expand. Others start out in a dense state, expand, eventually start contracting, and collapse back into the initial dense state. This cycle may repeat itself so that such models oscillate.

An important feature of both the Lemaître and Friedmann models is the high initial density of the universe. This was seized on by George Gamow, Ralph Alpher, and Robert Herman who set out to compute the temperature and pressure that must have existed at early times and thus establish the nuclear reactions that should have taken place (Aℓ01a). The resulting constituents should have the chemical composition of matter observed in the early phases of the Universe before nucleosynthesis in stars followed by stellar explosions would have significantly altered the chemical makeup of stars and interstellar matter. This early chemical composition should be close to that found on the surfaces of the oldest stars observed in the galaxy. The models developed by Gamow, Alpher, and Herman had the strength of making predictions that could be tested. In Section 12:13 we will see how well these predictions were borne out!

(f) Einstein–de Sitter Model

A special case among the Friedmann models is one in which not only the cosmological constant, but also the curvature k is zero, and the pressure term in (11–46) is negligibly small compared to the mass density $P/c^2 \ll \rho$. This model is often referred to as the Einstein–de Sitter universe.

PROBLEM 11–14. Show that (11–45) and (11–46) applied to the Einstein–de Sitter model yield the relations

$$\frac{\ddot{a}}{a} = -\frac{4\pi G\rho}{3} \quad \text{and} \quad H^2 q_0 = \frac{4\pi G\rho}{3} . \tag{11-69}$$

More generally, for $P = \Lambda = 0$ obtain also

$$(2q_0 - 1) = \frac{kc^2}{H^2 a^2} . \tag{11-70}$$

PROBLEM 11–15. For a Friedmann universe ($\Lambda = 0$) prove that:
   (a) With $k = +1$ and an initially dense universe for which $P/c^2 = \rho/3$, the solution to equations (11–45) and (11–46) has the parametric form

$$a_+ = b_{(+o)} \sin x, \qquad t_+ = b_{(+o)}(1 - \cos x)/c, \tag{11-71}$$

where x is a parameter and $b_{(+o)}$ is a constant.
   (b) For late stages of a universe with $k = +1$ and $P = 0$, show similarly that

$$a_+ = a_{(+o)}(1 - \cos x), \qquad t_+ = a_{(+o)}(x - \sin x)/c. \tag{11-72}$$

Note that x grows monotonically with t_+, so that $(1 - \cos x)$ eventually must approach zero. The universe first grows, but later on collapses.
   (c) For the dense stage, of a *hyperbolic universe* ($k = -1$)

$$a_- = b_{(-o)} \sinh y, \qquad t_- = b_{(-o)}(\cosh y - 1)/c. \qquad (11\text{-}73)$$

(d) For the late stage of a hyperbolic universe

$$a_- = a_{(-o)}(\cosh y - 1), \qquad t_- = a_{(-o)}(\sinh y - y)/c. \qquad (11\text{-}74)$$

We have looked at so many different solutions of equations (11–45) and (11–46) because different portions of the Universe may at early times have had different densities and curvatures. As we will see in Chapter 13, such curvature fluctuations could explain the structure on small scales that we see in the Universe today. The Universe we survey today consists of many regions that, as we will see in Section 11:15, once were out of causal contact. We will need to understand how all these different segments may have evolved separately before eventually merging.

11:14 Self-Regenerating Universes

First described by Hermann Bondi and Thomas Gold (Bo48) and by Fred Hoyle (Ho48), such a universe is flat and provides the same aspect at all times and in all places. The expansion rate is uniform in space and time. Old galaxies and young ones are statistically distributed in some fixed ratio at all distances from an observer.

Inflationary cosmological models, which exhibit a somewhat similar regeneration as the universe expands by some twenty-five orders of magnitude, were first proposed by Alan Guth (Gu81). Such universes, however, need not initially have a flat space. The only requirement is that the universe expands to such an extent that its curvature today be very small and difficult to measure — just as the curvature of a large sphere is difficult to measure when we examine only a small portion of its surface. We will investigate inflation in more detail in Chapter 12.

The discovery that the expansion of the Universe today is largely driven by *dark energy*, suggests that the remote future of the Universe will also be self-regenerating. The part of the Universe within any given observer's event horizon will gradually empty itself of stars and galaxies as these expand away and disappear across the horizon, so that most regions of the Universe will eventually contain nothing except dark energy that will forever continue to replenish itself.

The most surprising aspect of all these models is the suggestion that matter or energy is continually created. It is created from empty space — from nothing!

Or is it? Some theories postulate new fields from which matter or energy would be created. But thus far this has been done mainly to keep the conservation laws of physics intact. The new fields — called the C-field in the traditional steady state theory; referred to as the *Higgs scalar field* ϕ in inflation; and considered to be some form of *vacuum energy*, usually called *dark energy* to account for today's expansion of the Cosmos, all serve this purpose.

An observer in a steady state universe would expect to see matter created locally and we might wonder whether the rate of creation might be observed directly. We can readily compute what that rate should be. Consider a spherical volume with radius r expanding at some rate directly proportional to r. Call this rate Hr,

$$\frac{dr}{dt} = Hr \ .$$ (11-75)

The rate at which the volume expands is

$$\frac{d(4\pi r^3/3)}{dt} = 4\pi r^2 \left(\frac{dr}{dt}\right) = 4\pi r^3 H \ .$$ (11-76)

If the density of the sphere is to be maintained constant at some value ρ_0 during this expansion, the increased volume must be filled with matter at density ρ_0 so that the rate of matter creation is $4\pi r^3 H \rho_0$ in a sphere of radius r. Dividing by the volume of the sphere, we find the rate of matter creation in unit volume to be $3H\rho_0$

$$\dot{\rho}_0 = 3H\rho_0.$$ (11-77)

An estimate of the value of ρ_0 can be obtained by taking the number density of galaxies and multiplying by a typical galactic mass. The matter density so obtained is of order $\sim 10^{-30}$ g cm^{-3}. With a Hubble constant of 70 km s^{-1} Mpc^{-1} we obtain a creation rate $3H\rho_0 \sim 10^{-47}$ g cm^{-3} s^{-1}. If the baryonic fraction of matter were created in the form of hydrogen, this would imply a creation rate of the order of one atom per cubic meter in ~ 40 Gyr.

As we will see in Chapter 12, the creation rate in the self-regenerating phase of the inflationary universe, in contrast, was fast and furious. But we don't yet know how to directly test for this, so many æons later.

11:15 Horizon of a Universe

When a man on a cruising ocean liner wants to determine the distance of the horizon, he only needs to drop a buoy overboard and determine the buoy's range at the last instant before it disappears over the horizon. If the man is quick enough, he may then be able to shin high up on the ship's mast and briefly see the buoy again before it finally disappears over the horizon a second time. Two points are worth noting.

First, the distance of the horizon depends on the position of the observer. If a preferred horizon distance is to be defined, it should be selected in terms of some fundamental observer placed at some specific height above the ocean surface.

Second, no matter how high above the surface the observer climbs, there is an absolute horizon beyond which he can never see. He cannot see further than halfway to the antipodal point. His absolute horizon divides the surface of the Earth into two hemispheres.

An observer placed at a given location in a universe will also be able to define a horizon beyond which he cannot see. This horizon can be specified in a number of different ways. The distance to the horizon may depend on the speed at which the observer is moving, so that a horizon is best defined in terms of a fundamental observer O who is at rest with respect to the mean motion of galaxies in his local environment.

To define concepts, let us return to the FRW metric given by equations (11–18).

(a) The *proper distance* between a fundamental observer and some fundamental particle is defined for a given world time, t_1 by $dt = 0$,

$$\mathcal{D}(t_1) = \int ds = a(t_1) \int d\chi = a(t_1)\chi \,. \tag{11-78}$$

The velocity of particle then has two components: the first is the recession due to the expansion of the substratum, the second is a peculiar velocity v_p relative to the substratum.

$$v(t_1) = \dot{a}(t_1)\chi + a(t_1)\dot{\chi} = H(t_1)\mathcal{D}(t_1) + v_p \,, \tag{11-79}$$

where the first term on the extreme right is the *Hubble expansion velocity*, frequently called the *Hubble flow* or, more simply, the *recession velocity*. When the peculiar velocity is zero, the red shift tells us the distance to the emitter, rather than its velocity. The red shift thus plays a strikingly different role in general relativistic cosmologies from what might be expected if we tried to view the cosmic expansion in terms of special relativistic Doppler velocities. Distant galaxies move away from us because space expands, *not* because they move away from us relative to a stationary substratum.

As already noted in the derivation of equations (11–19), for a photon the peculiar velocity is the speed of light, and the comoving distance to the point from which light is reaching us was obtained by integrating over its past light cone. This is illustrated in the bottom panel of Fig. 11.9.

(b) The *Hubble radius R_H*, which defines a volume known as the *Hubble sphere*, is the distance at world time t_1 of a fundamental particle having a recession velocity equal to the speed of light,

$$R_H(t_1) = c/H(t_1) \,. \tag{11-80}$$

Returning now to the discussion of cosmic horizons, a classification due to W. Rindler (Ri56)* and further elucidated by Davis and Lineweaver (Da04)*, defines three kinds of horizons: an *event horizon*, a *particle horizon*, and finally an *absolute horizon*:

(c) In some cosmological models, remote parts of the universe recede from an observer at ever-increasing speeds, for instance, during the exponentially expanding phase of an inflationary universe or at all epochs in a steady state model. In such a universe there will exist a world time t_1 (Fig. 11.9) at which the distance of a fundamental particle P from a fundamental observer at A increases at precisely the speed of light. Before t_1 the particle P, which may be a galaxy, can emit radiation that eventually reaches A; but after t_1 radiation emitted by the galaxy cannot reach

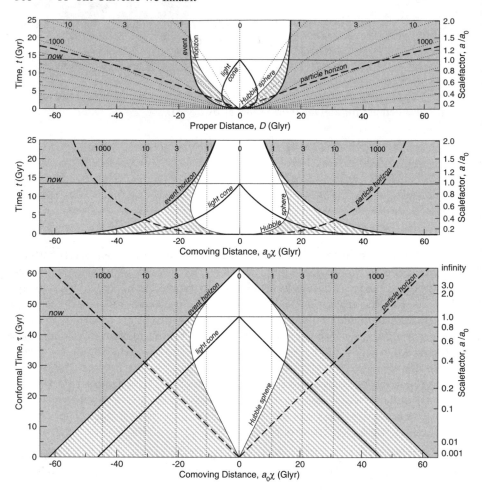

Fig. 11.9. Space–time diagrams for a general relativistic universe with $\Omega_M = 0.3$, $\Omega_\Lambda = 0.7$ and $H_0 = 70$ km s^{-1} Mpc^{-1}. In all three plots, vertical dotted lines show the world lines of comoving objects; each is marked with the red shift at which a comoving galaxy would appear. Ours is the central vertical world line $z = 0$. Time and conformal time, and their relative scale factors $a \equiv a/a_0 = (1 + z)^{-1}$ are plotted against comoving distance $a_0\chi$ measured in units of 10^9 light years (300 Mpc). The present epoch is shown by the horizontal line marked *now*, whose intersection with our world line marks the apex of our past light cone. In the top two panels the present age of the Universe is marked as $\sim$13.5 Gyr. In the bottom panel the conformal time τ going back to a red shift $z \sim 1100$ is given. This is the red shift of the microwave background; as discussed in Chapters 12 and 13, this is the greatest distance from which light can reach us in our Universe. 1 Glyr $= 10^9$ light years. The comoving distance to the surface from which the microwave background has reached us currently lies at 46 Glyr as explained in Problem 11–17, below. The figure does not attempt to reach farther back. (For additional explanations see the text.) (Courtesy of Tamara M. Davis and Charles H. Lineweaver (Da04).)

him because the intervening distance is increasing at a rate greater than the speed of light. In terms of (11–19), the event horizon lies at comoving distance

$$\chi_{eh}(t_1) = \int_{t_1}^{\infty} \frac{c\,dt}{a(t)}. \qquad (11\text{-}81)$$

The cosmic event horizon is somewhat analogous to the event horizon for matter falling into a black hole (Section 5:13). Events prior to t_1 can be transmitted to the observer whereas events subsequent to t_1 may forever remain hidden from him – the difference being that, if the expansion of the Universe sufficiently slows down, a galaxy that was previously receding faster than the speed of light may again come into view. This is apparent in the top panel of Fig. 11.9, where the teardrop-shaped *past light cone* traces the distances from which light was emitted at different epochs t_1. Particles that were initially receding faster than the speed of light become visible as the original expansion of the Universe decelerates. Later, as dark energy begins to play a dominant role $\Omega \to 1$ and the Universe accelerates, such regions will once again cross the event horizon never to be seen again. As shown in the bottom panel, the past light cone given by (11–19) in comoving coordinates approaches the event horizon (11–81) at $t = t_{\infty}$.

(d) In other cosmological models a different type of horizon is important. Take an inflationary model in which matter initially is highly compact. At zero time the universe begins an exponential expansion. Two particles, P and A, initially quite far apart, rapidly recede from each other. Because of the large separation, light emitted by particle P at time $t = 0$ will not reach A until well after the cosmic expansion has slowed down, at t_0. Before t_0 the observer is quite unaware of the existence of P. After t_0 he can receive messages emitted at P. Particle P is said to enter the observer's *particle horizon* at time t_0 (Fig. 11.8). In terms of (11–19), the distance to the particle horizon is

$$\chi_{ph}(t_1) = c \int_{0}^{t_0} \frac{dt}{a(t)}. \qquad (11\text{-}82)$$

We can then define a particle horizon for any fundamental observer and world time t_0. It is a surface that divides all fundamental particles into two classes: those that have already been observable and those that have not.

For electromagnetic radiation, the particle horizon currently does not have a great deal of significance, because we are unable to see farther than the surface of last scatter. However, neutrinos and gravitational radiation could be reaching us from greater distances and be detected by sufficiently sensitive detectors. For these two types of radiation the particle horizon could then impose a more distant surface from beyond which no signals would reach us.

PROBLEM 11–16. Show that during early epochs, when curvature and dark energy may be neglected, the particle horizon lies at about the Hubble radius in the radiation-dominated phase and at about twice the Hubble radius in the matter-dominated phase of the Universe.

PROBLEM 11–17. The past light cone shown in the bottom panel of Fig. 11.9 stretches back to the surface from which the microwave background radiation reaches us. At this *surface of last scatter*, lying at red shift $z \sim 1100$, the radiation is scattered by electrons one last time before the cosmic expansion makes the Universe essentially transparent. As implied by equation (11–19), the past light cone just traces out the distances $\chi(t_1, t_0)$ to points that emitted radiation at different times t_1. The comoving distance to the emitting surface can be evaluated by means of the following steps:

(a) Show that equation (11–45) can be rewritten as

$$\dot{a} = a_0 H_0 \left[\Omega_M \left(\frac{a_0}{a} \right) + \Omega_\Lambda \left(\frac{a}{a_0} \right)^2 \right]^{1/2} . \tag{11-83}$$

(b) Using this expression, equation (11–19) and the identity $dt/a(t) = da/(\dot{a}a)$, evaluate the distance $a_0\chi(t_1, t_0)$ back to the surface of last scatter, and verify that it corresponds to the 46 Glyr shown in Fig. 11.9.

In some models both particle and event horizons exist. The inflationary model with dark energy that currently best describes our Universe is of this kind. Because there is an initial explosion from a compact state, a particle horizon will exist; because there is a subsequent acceleration governed by dark energy an event horizon will also come into play.

(e) We may wonder about the distance of the horizon from a moving observer. Clearly, if the observer accelerates himself toward a fast receding galaxy, his event horizon can be extended as indicated in Fig. 11.10.

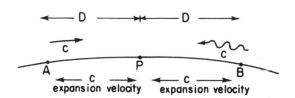

Fig. 11.10. The absolute horizon for an observer at A (see text).

A number of results can be proven (Ri56):

(i) In a model without an event horizon, a fundamental observer can sooner or later observe any event.

(ii) In a model having an event horizon, but no particle horizon, an observer can be present at any one specified event, provided he is willing to travel and provided he starts out early enough.

Statement (i) depends on the absence of particles receding faster than the speed of light. Statement (ii) hinges on the fact that any given particle must have been within an observer's event horizon at some time in the distant past.

(iii) In a model with both event and particle horizons an observer originally attached to a fundamental particle finds that there exists a class of events absolutely inaccessible to him, no matter how he travels through space. This class of events defines an absolute horizon as shown by the following argument.

Suppose a fundamental observer were placed at some position A in the universe. There can then exist a critical particle P that initially recedes at exactly the speed of light and that enters A's particle horizon at time $t = \infty$. Let the initial distance (11–20) between P and A be D. Next, consider a fundamental observer at a point B situated along the line of sight AP but at a distance D beyond P. Again, P will enter B's particle horizon at $t = \infty$. By moving at the speed of light toward P, observer A would reach P at $t = \infty$. B would be receding at the speed of light relative to P and would, therefore, enter A's particle horizon on A's arrival at P at time $t = \infty$; but all particles beyond B would forever remain inaccessible to A. Position B defines an absolute horizon for a fundamental observer at an initial position A (Fig. 11.10).

11:16 Topology of the Universe

Thus far we have assumed that the Universe is *simply connected* — that it has the simplest *topological* structure. In the two-dimensional models, we have talked about spherical surfaces, or planes, or hyperbolical surfaces of negative curvature.

There exist more complicated surfaces some of which can be easily constructed. If we take a rectangular sheet and label the four edges a, b, c, d, as shown in Fig. 11.11(a) we can obtain a cylindrical surface by joining edges a and b. But there are several ways of joining a and b. We can give the sheet a twist, as indicated by the arrows, and obtain a Møbius strip as in Fig. 11.11(b).

If edges a and b are joined and edges c and d are joined also, we can obtain a torus or a Klein bottle depending on whether the sheet is given no twist or one twist — Fig. 11.11(c).

The Møbius strip and Klein bottle are *re-entrant*. Starting on the exterior side of a surface, we are able to reappear at the starting point but on the interior of the surface without ever crossing an edge or perforating the surface. There is, however, a change in directions. An arrow pointing in a particular direction appears reversed when it returns to its starting point (on the opposite surface of the strip).

In some re-entrant models an observer O' may return to her own past. In still others she might return to her starting point with her arrow of time reversed with respect to her surroundings. Some spaces of negative curvature are not open when complex topological forms are allowed (He62). Although general relativity determines the

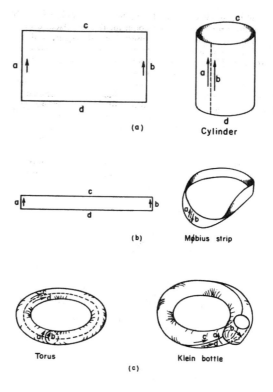

Fig. 11.11. Simply connected and re-entrant topologies.

geometry — i.e. the curvature of space-time — it only constrains *topology*; it does not dictate it.

How could we tell if we were living, say, in a universe with the topology of the torus in Fig. 11.11(c). If the circumference of the torus defined by the length of edges $a = b$ is shorter than the distance to where the cosmic microwave background radiation originated, we could look along opposite directions parallel to these two edges on the surface of the torus to see whether we can identify matching features in the background surface brightness distribution, viewed along these two directions. The same could be done looking along opposite directions on the surface parallel to edges c and d. In a universe described by a three-dimensional, rather than a two-dimensional hypersurface, the topology can be more complex. There could then be numerous pairs of directions along which we could look to distinguish whether the universe is simply or multiply connected. Light reaching the observer from different directions would be traversing the universe along quite distinct trajectories, and there would be little correlation in the brightness distribution pattern observed at widely spaced angular separations. The actually observed microwave background does lack such correlations at large angular separations, and this has led to the suggestion that space might be closed — positively curved — and have a complex topology, specifically the topology of a twelve-sided, so-called *Poincaré dodeca-hedral space* (Lu03). A search for matching pairs of patterns in the observed microwave background distribution, however has failed to confirm this (Co04).

11:17 Do the Fundamental Constants of Nature Change with Time?

Has the speed of light been the same throughout the history of the Universe? Do Planck's constant or the gravitational constant, or the charge of the electron change very slowly but yet significantly on a cosmic scale?

The first person to worry about this problem and to come up with some quantitative indicators was P. A. M. Dirac (Di38). He noted that the fundamental constants of Nature could be arranged in groups that were dimensionless numbers of order 10^{39} or $(10^{39})^2 = 10^{78}$. The ratio of the gravitational to the electromagnetic force of attraction between proton and electron, or alternatively the ratio of the gravitational to the electromagnetic Bohr orbit, is one such example

$$\frac{e^2}{m_e m_p G} = 10^{39}. \tag{11-84}$$

If this ratio is to be constant, then we would expect the mass and charge of the electrons or protons to change if cosmic evolution affected the value of the gravitational constant.

The radius of the Universe out to the cosmic horizon R_H is of the order of 10^{28} cm, while the Compton wavelength of the electron is

$$\lambdabar_c = \frac{\lambda_c}{2\pi} = \frac{\hbar}{m_e c} = 4 \times 10^{-11} \text{ cm}, \tag{6-168}$$

so that $R_H / \lambdabar_c \sim 10^{39}$ as well. In general, such numbers can be constructed by taking the ratio of a cosmic and a microscopic quantity. We may not expect to get ratios of precisely 10^{39} in each case, but the exponents cluster remarkably closely around the numbers 39 and 78. The number of atoms in the Universe obtained by dividing the mass of the Universe M by the mass of the proton m_p is

$$N = \frac{M}{m_p} = 10^{78}. \tag{11-85}$$

The puzzling thought about this ratio is that the flow of particles across the cosmic horizon would clearly destroy its constancy on a time scale of the order of an inverse Hubble constant $\tau_A = H^{-1} \sim 4 \times 10^{17}$ s. Over a period of 10^{10} yr, N would change appreciably, in violation of Dirac's argument that dimensionless constants should not change.

Dirac argued that if the clustering of such numbers around the values of 10^{39} or its multiples was no coincidence, then it indicated that microscopic and macroscopic

— atomic or subatomic and cosmic — quantities were related. Since the Universe is expanding, then because of the changes in the size of the Cosmos, there should be corresponding changes on an atomic scale to keep the dimensionless expressions constant. How this might happen he did not explain.

Gravitational fields have been considered the only suitable candidates for inducing change, ever since Einstein tried to incorporate Mach's principle into his general relativistic theory of gravitation. *Mach's principle*, named after nineteenth and early twentieth century scientist Ernst Mach, takes on various forms, but states that the local properties of the Universe should somehow be dictated by its grand structure. Dirac's hypothesis of constant dimensionless numbers is another version of Mach's principle. Its quest is to unite the very large-scale behavior of matter with physics on an atomic and nuclear scale.

The question of ultimate interest is whether the fundamental constants of Nature remain constant or change as the Universe evolves? A natural time scale over which we might expect the constants to change in response to an expanding universe is the age of the Cosmos, of order H^{-1}. Let us first examine what we know about the gravitational constant G.

Between 1969 and 1971 astronauts placed laser-ranging retroreflectors on the Moon. Laser signals from Earth bounced off these reflectors show no changes in the Moon's orbit in over three decades. This places an upper limit of $|\dot{G}/G| < 10^{-11}$ per year (No96). More involved and somewhat less direct arguments restrict even this low rate by another order of magnitude (Th96).

An interesting study on the possible variability of Planck's constant with world time was done by Wilkinson (Wi58). He was interested in the integrated effect of changes in Planck's constant over a period of the order of the age of the Earth. The age of the Earth and of meteorites can be determined separately from a number of different radioactive decay schemes, some of which involve alpha-particle emission and others beta-decay. These two processes have quite different physical bases, and we would not expect the ages given by beta- and alpha-decay schemes to be the same if the fundamental constants of Nature varied appreciably.

The evidence cited by Wilkinson comes from a study of *paleochroic haloes*. These haloes are spherical shells observed in rocks that have small inclusions of radioactive material. As the material decays, any alpha particles will give rise to a thin visible shell at the end of the particle's path through the rock, where most of the energy is dissipated. Corresponding to individual alpha-particle energies, we then obtain individual shells. These shells are easily identified with given alpha-decay schemes. Two statements can then be made.

(i) The physics of charged particle transit through material, a process that is purely electromagnetic, is invariant over a period of order 2×10^9 yr or perhaps slightly more. Otherwise the shells would be diffuse, not thin. This is instructive because the alpha-decay scheme discussed by Wilkinson involves both electromagnetic and nuclear forces.

(ii) Some alpha-emitting nuclei also have the possibility of undergoing beta decay. The ratio of these two decay rates is called the *branching ratio*. Wilkinson was able to state that if the branching ratio had increased or decreased by an amount of order 10 over the past 2×10^9 yr, he would have found that certain haloes produced by alpha particles should have been absent and others much stronger than predicted. Because no such anomalies were found, any changes taking place over the past few æons in the many fundamental physical constants involved were probably small.

Observations on a variety of different spectral lines, all similarly red-shifted in distant quasars, permit us to also conclude that the relative masses of the electron and proton cannot have changed by more than a few parts in 10^{14} per year, and that the fine-structure constant must have remained comparably constant (Co95a). Similarly, observations of distant quasars have shown the fine-structure constant α to be constant to a few parts in 10^{16} per year, and laboratory experiments conducted over a five-year period show the current rate of change of the fine-structure constant to have upper limits of the order of a few parts in 10^{15} per year (Fi04).

If there are changes in the fundamental physical constants they are so small that they are unlikely to have played a role in the evolution of the Universe – or vice versa. The expansion of the Universe apparently has no significant influence on the fundamental constants of Nature. Mach's principle appears to be invalid.

11:18 The Flow of Time

We tend to think that time always increases. But with respect to what? And what do we mean by "increase"?

The simplest answer would be to say that time is that which is measured by a clock. But as we saw in Section 3:10 there are different types of clocks and we might wish to compare them to see whether they all are running at the same rate or whether there might be, say, a systematic slowing down of one type of clock relative to the others.

Here we have assumed that all possible clocks will always run in one direction only. In that case, however, we would never be able to decide whether time is running "forward" or "backward" because these two directions would be indistinguishable.

For gravitational and electromagnetic processes we do not know how to define a direction of time's flow. The physics that describes the orbiting of the Earth around the Sun holds equally whether the Earth moves in a direct or retrograde orbit. Under a time reversal, the Earth and all the other planets would return along the same orbits that had led them to their current positions in the Solar System. Such orbits would be no different from a set of future orbits that could have been predicted from a simple reversal of all velocities involved. If we recorded the motion of the planets on film, we would detect no violations of the laws of Nature whether we ran the film forward or backward.

Similarly we could use the orbital motion of an electron in a magnetic field to define time. Here the orbit in which the electron travels is identical to one that a positron would travel if it were going along the same path backward in time.

Both these examples exhibit a basic symmetry that seems to pervade all natural physical processes. If we reverse the flow of time T, reverse the sign of the electric charge of matter C, and reverse the sign of all positions and motions, in what is called a *parity operation* P, then the observed results are indistinguishable from an original process in which none of these reversals or reflections took place. The operation C is called *charge conjugation*; and T is called the *time-reversal operation*. The currently favored physical theories, the so-called *local field theories*, require that under the combined operations CPT all physical processes remain invariant.

Because of these symmetries, it would appear impossible for us to know whether we are living in a world in which time is running forward and the Universe is expanding, or whether time is running backward and the Universe is contracting. These cosmic motions are independent of electric charge so that a charge conjugation would not be noticeable either. We would just assume we were made up of matter, but actually it might be what we currently call antimatter.

How, then, do we determine the direction in which time is flowing?

The second law of thermodynamics was long believed to define the direction of time in a unique way. The law states that as time increases, any isolated system tends toward increasing disorder. Light initially concentrated near the surface of a star flows out to fill all space. The reverse never happens. Light that fills space does not converge and flow into a single compact object. Such ordered motions, although strictly permitted by a simple time-reversal argument, are possible but highly improbable. The second law of thermodynamics states that as time increases, greater randomness comes about because there are many states of a system in which the system is disordered and only few in which it has a high degree of order. If any given state is as likely to occur as any other state, then the chances are that the evolved system will be found in one of the many disordered states rather than in one of the very few ordered configurations.

If tendency toward disorder depends on the cosmic expansion, as suggested by T. Gold (Go62), then the flow of time should be well correlated with the flow toward disorder. But what would happen if the Universe were contracting rather than expanding? Would the second law still hold and time run in the accustomed manner? We do not know.

We wish that there might be more straightforward ways of determining the direction of the flow of time. A possibility of this sort has come into sight in recent decades.

In 1956 Lee and Yang (Le56) pointed out that parity might be violated in weak interactions. This was swiftly verified in a variety of experiments. *Parity, P,* is the ability to differentiate mirror-symmetric objects, such as a right hand from a left hand. Experiments on beta decays, which involve solely the weak force of nature, showed that all neutrinos have a left-handed spin, never a right-handed, whereas antineutrinos have a right-handed spin. Nature evidently can tell left from right.

However, it then seemed that invariance under a combination of operations CP should hold universally true. A positron moving in a negative direction should invariably trace out the same path as an electron moving in a positive direction at an identical velocity. If CP invariance held, together with the above-mentioned CPT invariance, the laws of physics still would automatically remain invariant also under time reversal, T.

In the mid-1960s, however, a group at Princeton University (Ch64) discovered that violations of CP symmetry occasionally occurred in the decay of neutral K-mesons. Since then this has been fully confirmed in measurements on both neutral K and neutral B mesons.

In the high-energy collision of protons and antiprotons one produces both neutral K mesons K^0 and their antiparticles $\bar{K}^0$. A K^0 is always accompanied by production of a π^+ and a K^-; a $\bar{K}^0$ is always accompanied by a K^+ and a π^-. This makes it easy to know whether a K^0, a *kaon*, or its antiparticle, an *antikaon* has been produced. This is important, because the K^0 and $\bar{K}^0$ spontaneously interconvert; each *oscillates* between being a K^0 and a $\bar{K}^0$ a part of the time. Both these particles are unstable. The K^0 decays into a π^+ and an e^-, and a $\bar{K}^0$ decays into a π^- and an e^+. By observing the nature of the particle initially produced, and the decay products when it decays, we can determine what fraction of the time the K^0 remains a K^0 and what fraction of the time it spends as its antiparticle $\bar{K}^0$, and vice versa. Measurements show that there is a slightly greater probability, $\sim 6 \times 10^{-3}$, that the antikaon will turn into a kaon than the other way around. We may think of this as the time required for the kaon–antikaon transformation being longer than the antikaon–kaon transformation.

Now, the time taken for a kaon–antikaon transformation equals the time-reversed rate for the antikaon-kaon transformation. If we recorded the sequence on film, the time taken would be the same running the film forward as backward. But because the observed kaon–antikaon transformation is slower than the observed antikaon–kaon transformation, this is a violation of time-reversal invariance (Pe98, Sc99).

Nature does seem to know which way time flows. However, whether this ability to distinguish the direction of time also determines its direction, is not yet resolved.

In Section 12:12 we will discuss the preponderance of matter over antimatter in the Universe, an asymmetry that may be related to a violation of CP or time-reversal T symmetry. If so, however, the specific violation accounting for this preponderance of matter still eludes us (Qu03).

11:19 Branes and Compact Dimensions

The three-dimensional hypersurfaces we discussed in Section 11:4 can be thought of as three-dimensional membranes — *3–branes* — in a four-dimensional space. A recent set of world models suggests that our Universe is just one of many branes, parallel universes, that are stacked in layers in a higher-dimensional space having n extra spatial dimensions beyond the three infinite spatial dimensions that constitute our brane. Figure 11.12 shows such a brane that happens to fold back on itself. But separate branes stacked on top of each other also have been postulated. The spacing between these surfaces could be as small as a fraction of a millimeter. A

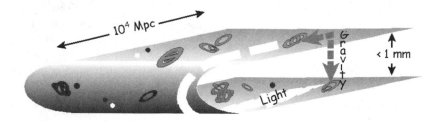

Fig. 11.12. A brane model of the Universe. Branes are three-dimensional worlds that may be separated from each other by a gap less than 1 mm across. Branes may fold over, as shown in this illustration, or they may run parallel to each other indefinitely. Light, neutrinos, and relativistic particles can propagate solely within each brane, but gravitational fields set up in one brane may propagate across the gap to another.

currently favored model is one in which $n = 2$ additional *noncompact* dimensions exist. Noncompact dimensions are dimensions that could be infinite, in contrast to *compact dimensions* that might, e.g., represent particle spins, which are restricted to a finite domain (Ra99, Ar02).

Vibrations of branes could occasionally cause them to collide and produce all the energy and matter in the universe. Such collisions might repeatedly occur, giving rise to an endless series of universes. One consequence could be that the cosmic expansion we are currently witnessing is not the result of a single explosive origin of our universe, but rather is due to a collision between an adjacent brane and ours.

Brane worlds have been invoked to explain why gravity is so weak compared to electromagnetic and nuclear forces — i.e., why the gravitational attraction between two electrons is 42 orders of magnitude weaker than their electrostatic attraction. The argument is that although photons, neutrinos, and all particles are restricted to adhere to a single brane, gravity propagates in all the added dimensions as well, because gravity is what determines the structure of space–time. On scales smaller than the separation between branes, gravitational forces might then be much stronger, becoming comparable to electromagnetic forces on very small scales. Experiments designed to search for changes in the gravitational constant G on scales below 100 μm, to date have failed to uncover dramatic changes (Ch03). Predictions made by various versions of the theory can also be tested with accelerators that slam elementary particles into each other at energies of order 10^{12} eV. It is heartening that these theories can be experimentally tested — and potentially accepted or rejected.

Additional Problem

11–18. A supernova exploded at $(z + 1) \sim 20$. Using (11–38), and (11–56) with $n = \frac{3}{2}$, show that its distance at the epoch of explosion t_1 was

$$a_1\chi(z) = \frac{a_0\chi(z)}{(z+1)} = \frac{1}{(z+1)} \frac{c}{H_)} \int_0^z \frac{dz}{(z+1)^{3/2}} \sim 10^{27} \text{ cm}.$$

If its luminosity was 10^{46} erg s^{-1} show, with the help of (11–34), that its bolometric flux observed today is $\mathcal{F} \propto (z+1)^{-4} \sim 5 \times 10^{-15}$ erg cm^{-2} s^{-1}.

Answers to Selected Problems

11–1. Writing $x_4^2 = a^2 - r^2$, with $r^2 = x_1^2 + x_2^2 + x_3^2$, we have

$$dx_4 = \frac{(r\,dr)}{a^2 - r^2}^{11/2}, \text{ and } dx_1^2 + dx_2^2 + dx_3^2 = dr^2 + r^2\,d\theta^2 + r^2\sin^2\theta\,d\phi^2.$$

This leads to (11–7) which gives (11–5) on substituting (11–8).

11–2. The radius of the circle is $a\chi =$ constant. From (11–5) we see that an element at distance $a\chi$ has an increment of length squared

$$dl^2 = a^2 \sin^2\chi(\sin^2\theta\,d\phi^2 + d\theta^2).$$

(i) We can always choose $\theta = \pi/2$ as the plane of the circle; then the circumference becomes

$$\oint dl = \int_0^{2\pi} a \sin\chi\,d\phi = 2\pi a \sin\chi.$$

The ratio of circumference to length therefore is $\chi^{-1} 2\pi \sin\chi \le 2\pi$.
(ii) The area of an element on the sphere is

$$d\sigma = (a \sin\chi \sin\theta\,d\phi)(a \sin\chi\,d\theta).$$

The area of the whole sphere therefore is

$$\int\int a \sin\chi \sin\theta\,d\phi\, a \sin\chi\,d\theta = 4\pi a^2 \sin^2\chi.$$

(iii) The element of three-dimensional volume suggested by (11–5) is

$$dV = (a\,d\chi)(a \sin\chi\,d\theta)(a \sin\chi \sin\theta\,d\phi) = a^3 \sin^2\chi \sin\theta\,d\theta\,d\phi\,d\chi$$

from which (11–10) and (11–11) follow.

11–3. (i)

$$x_4^2 = -a^2 - r^2, \qquad dx_4^2 = -\frac{r^2\,dr^2}{(a^2+r^2)},$$

$$dl^2 = dr^2 + r^2(\sin^2\theta\,d\phi^2 + d\theta^2) - \frac{r^2\,dr^2}{a^2+r^2}.$$

This is equivalent to (11–14).

(ii) With $r \equiv a \sin h\chi$

$$dl^2 = a^2 \sinh^2 \chi (\sin^2 \theta \, d\phi^2 + d\theta^2) + \frac{a^2 (d \sinh \chi)^2}{1 + \sinh^2 \chi} \, ,$$

but $1 + \sinh^2 \chi \equiv \cosh^2 \chi$ and $d \sinh \chi = \cosh \chi \, d\chi$, so that we obtain (11–15). For $a\chi = $ constant, $\theta = \pi/2$, $dl = a \sinh \chi \, d\phi$, and a circle has circumference $2\pi a \sinh \chi \geq 2\pi a \chi$.

(iii) The element of surface is then seen to be

$$d\sigma = (a \sinh \chi \sin \theta \, d\phi)(a \sinh \chi \, d\theta) \, .$$

Integration over all values of θ and ϕ for $a\chi = $ constant then gives (11–16).

(iv) Similarly, the volume element is

$$dV = (a \sinh \chi \, d\theta)(a \sinh \chi \sin \theta \, d\phi)(a \, d\chi) \, ,$$

which leads to (11–17).

11–4. Equation (11–23) states $1 + z = a(t_0)/a(t_1)$. A Taylor expansion in $\Delta \equiv (t_1 - t_0)$ gives

$$1 + z = \frac{a(t_0)}{a(t_0 + \Delta)} = 1 - \frac{\dot{a}_0}{a_0} \Delta + \frac{1}{2} \left(\frac{\dot{a}_0^2}{a_0^2} - \frac{\ddot{a}}{a_0} \right) \Delta^2 + \cdots ,$$

where $a_0 \equiv a(t_0)$. This yields (11–24).

11–5.

$$\sinh^2 \chi = \left(\chi + \frac{\chi^3}{6} + \cdots \right)^2 = \chi^2 + \frac{\chi^4}{3} + \cdots = \chi^2 - \frac{k\chi^4}{3} + \cdots , k = -1.$$

$$\sin^2 \chi = \left(\chi - \frac{\chi^3}{6} + \cdots \right)^2 = \chi^2 - \frac{\chi^4}{3} + \cdots = \chi^2 - \frac{k\chi^4}{3} + \cdots , k = 1.$$

$$N(\chi) = \frac{4\pi}{3} n\chi^3 \left(1 - \frac{k}{5} \chi^2 + \cdots \right) .$$

11–6.

$$q_0 = -\frac{a_0 \ddot{a}_0}{\dot{a}_0^2} \, . \quad \text{For exponential expansion :}$$

$$\dot{a}_0 = a_0 H, \quad \ddot{a}_0 = a_0 H^2, \quad \therefore \ q_0 = -1.$$

11–8. (a) Because $k = \Lambda = 0$, and $P = \rho c^2/3$, $\ddot{a} = 0$, in equation (11–46). We know that for a relativistic gas $\rho \propto a^{-4}$, so that (11–45) can be written as

$$H = \frac{\dot{a}}{a} = \left(\frac{8\pi G \rho_0}{3} \right)^{1/2} \frac{a_0^2}{a^2} \, ,$$

where a_0 and ρ_0 are values chosen for some initial time $t_0 = 0$. On integrating, we obtain a world time t_a:

$$a_0^{-2} \left(\frac{3}{8\pi G \rho_0} \right)^{1/2} \int_0^a a \, da = \int_0^{t_a} dt = t_a \sim \frac{1}{2H} \quad \text{for } a_0 \ll a.$$

(b) When the pressure can be taken as negligible, $P = 0$ and

$$\frac{\dot{a}}{2a} = -\frac{\ddot{a}}{\dot{a}},$$

which integrates to $\dot{a} = Aa^{-1/2}$, where

$$H = \frac{\dot{a}}{a} = \left(\frac{8\pi G \rho_0' a_0'^3}{a^3} \right)^{1/2} \equiv \frac{A}{a^{3/2}}$$

defines the value of the constant of integration A. A second integration then yields, for $t_a \gg t_0'$,

$$t_a \sim t - t_0' = \frac{2}{3} \left(\frac{1}{8\pi G \rho_0' a_0'^3} \right)^{1/2} (a^{3/2} - a_0'^{3/2}) \sim \frac{2}{3H} \quad \text{for } a_0' \ll a.$$

(c) This follows from integration of (11–45) for $\rho = k = 0$.

11–9. (i)

$$\Lambda = (a^2)^{-1} + \frac{8\pi GP}{c^4} \quad \text{because} \quad \dot{a} = \ddot{a} = 0 \text{ in (11–46)}.$$

(11–64) then follows from (11–45) and (11–46).

(ii) The result follows from substitution of the given values in (11–46) and (11–45).

11–10. (11–45) and (11–46) yield

$$\Lambda = \frac{3\dot{a}^2}{c^2 a^2} = \frac{2a\ddot{a} + \dot{a}^2}{c^2 a^2},$$

which has the solution (11–65). The age is $a/\dot{a} = H^{-1} = \sqrt{3/\Lambda c^2}$.

11–11. Initially, by (11–45), (11–46), $6\ddot{a} = ac^2[2\Lambda - 8\pi G\rho/c^2]$. If ρ decreases because of a disturbance, $\ddot{a} > 0$, the universe expands, which decreases ρ further, and so on (Ed30).

11–12. The first set of expressions is obtained from equations (11–45) and (11–46), on setting $\ddot{a} = 0$, with $P = 0$ for nonrelativistic matter and $P = \rho c^2/3$ for relativistic matter. The second set of expressions is then obtained by setting $\dot{a}/a = 0$ and substituting the respective values for Λ into equation (11–45).

11–13. (a) From (11–46) and the definition of Ω_Λ in (11–61), $\Omega_\Lambda = \frac{1}{3}$ follows at once in a matter-dominated universe. Equations (11–45) and (11–60) then lead to $\Omega_M = \frac{2}{3}$, a result that also is apparent because $\Omega_\Lambda + \Omega_M = 1$ in a flat universe, $k = 0$.

(b) In the radiation-dominated case, (11–45) and (11–46) lead to $\Omega_\Lambda = \Omega_M$, and for $k = 0$ each must be $\frac{1}{2}$.

(c) For $8\pi G P_\Lambda / c^4 = -\Lambda$, (11–46) with (11–45) lead to $\Lambda = 8\pi G \rho_M / 5c^2$, or $\Omega_\Lambda = \Omega_M / 5$, for $\ddot{a} = 0$. Because today's value of $\Omega_\Lambda \sim (0.7/0.27)\Omega_M$, the density ρ_M at inflection must have been $5 \times 0.7/0.27 \sim 13$ times higher, so that $(z + 1)$ at inflection would have been 2.35 and $z = 1.35$.

11–14. (11–45) and (11–46) give

$$-2a\ddot{a} = \dot{a}^2 + kc^2 = \frac{8\pi G \rho a^2}{3} \, .$$

For $\Lambda = P = 0$,

$$-\left(\frac{\ddot{a}a}{\dot{a}^2}\right) = q_0 \quad \text{and} \quad 2q_0 - 1 = \frac{-2a\ddot{a}}{\dot{a}^2} - 1 = \frac{kc^2}{\dot{a}^2} = \frac{kc^2}{H^2 a^2} \, .$$

11–15. (a)

$$\frac{8\pi G P}{c^4} = -\left[\frac{2a\ddot{a} + \dot{a}^2 + c^2}{c^2 a^2}\right] = \frac{8\pi G \rho}{3c^2} = \frac{c^2 + \dot{a}^2}{c^2 a^2}$$

so that initially $\ddot{a}a = -(c^2 + \dot{a}^2)$. We try the solution (11–71) that gives

$$\dot{a}_+ = c \cot x, \quad \ddot{a}_+ = \frac{-c^2}{b_{(+o)} \sin^3 x} \, .$$

These satisfy the differential equation above.

(b) The trial solution (11–71) satisfies the equation

$$2\ddot{a}a = -(\dot{a}^2 + c^2)$$

which follows from (11–46) at this stage of evolution. We see this since

$$\dot{a}_+ = c \sin x [1 - \cos x]^{-1}, \quad \ddot{a}_+ = -c^2 [a_{(+o)}(1 - \cos x)^2]^{-1}.$$

(c) The equation to be satisfied now is $\ddot{a}a = c^2 - \dot{a}^2$ which, following the above procedures, is fulfilled by (11–73).

(d) Similarly (11–74) satisfies $2\ddot{a}a = -\dot{a}^2 + c^2$ for the late stages of a hyperbolic universe.

11–16. For $\Lambda = k = 0$, and $n = 3$ or 4, respectively, for a matter- or radiation-dominated universe, we have

$$\left(\frac{\dot{a}}{a}\right)^2 = \frac{8\pi G \rho_0}{3}\left(\frac{a_0}{a_1}\right)^n$$

from (11–45), so that (11–82) yields

$$\chi_{ph} = c \int_0^{t_0} \frac{dt}{a} = c \left(\frac{3}{8\pi G \rho_0 a_0^n} \right)^{1/2} \int_{a_1}^{a_0} a^{(n-4)/2} da .$$

For $a_1 \ll a_0$, this yields $a_0 \chi_{ph} = R_H$ for $n = 4$ and $a_0 \chi_{ph} = 2R_H$ for $n = 3$.

11–17. (a) We can rewrite equation (11–45) as

$$\left(\frac{\dot{a}}{a} \right)^2 = \left(\frac{\dot{a}_0}{a_0} \right)^2 \left[\Omega_M (z+1) + \frac{\Omega_\Lambda}{(z+1)^2} \right] (z+1)^2 .$$

On setting $(z+1) = a_0/a_1$ we obtain the desired result.

(b) Noting that $dt/a(t) = da/(\dot{a}a)$, we obtain the distance to the surface emission at $z \sim 1100$ from equation (11–19) as

$$\chi(t_1, t_0) = \frac{c}{\dot{a}_0} \int_{a_1}^{a_0} \frac{da}{[\Omega_m a_0 a + \Omega_\Lambda (a^4/a_0^2)]^{1/2}} .$$

This may be integrated to obtain the distance to the surface of last scatter, 46 Glyr, shown in the bottom panel of Fig. 11.9. A rough estimate is obtained by neglecting the second term in the denominator, which remains relatively small until $z \sim 0.5$.

12 An Astrophysical History of the Universe

Having gained some insight into the present appearance of the Universe and the factors that determine its evolution, it is natural to ask whether the Cosmos ever had a beginning and what, if anything, came before it. In Section 11:3 we saw why the Universe cannot be in a steady state but must be evolving. It either had a beginning, or it has oscillated, alternately expanding and contracting. An evolving universe, however, raises two new problems of its own, which we now take up in turn.

12:1 The Isotropy Problem

The first of these concerns the near-perfect isotropy of the microwave background radiation. Admittedly, as we saw in Section 5:9, the Sun and nearby galaxies ranging out to ~ 100 Mpc and moving as a group, exhibit peculiar velocities of the order of several hundred kilometers per second with respect to the microwave background radiation. Figure 12.1 summarizes some of this information. But, averaged over the largest scales, $\gtrsim 10^3$ Mpc, the distribution of galaxies appears to be at rest with respect to the background. This raises a puzzling question, "How could the cosmic microwave background radiation be as homogeneous and isotropic as it is observed to be?" This radiation is reaching us from great distances. Yet it must at some time have been in local thermal equilibrium to have a blackbody spectrum. But how can we account for thermal equilibrium in radiation arriving from portions of the sky in diametrically opposite directions? The radiation from each of these directions is only just now reaching us after a long journey through space. How could two such widely separated regions ever have been in causal contact, let alone thermal equilibrium?

Equation (11–19) tells us that the rate at which light can traverse an increment of comoving distance $d\chi$ in time dt is

$$\frac{d\chi}{dt} = \frac{c}{a(t)} \ . \tag{12-1}$$

PROBLEM 12–1. Using equations (12–1) and (11–56) to (11–58), show that the time taken for light to cross a parameter distance 2χ is always more than twice the time it took to traverse χ in the first place, whether a universe is in a relativistic

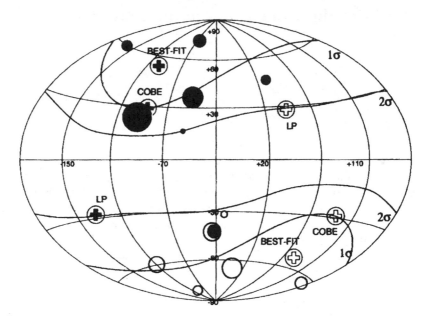

Fig. 12.1. Estimates of the motion of the Local Group of galaxies. The direction of motion relative to the microwave background radiation is designated COBE, signifying that it was derived from data obtained with the *Cosmic Background Explorer, COBE* satellite. More recent observations with the *Wilkinson Microwave Anisotropy Probe, WMAP* provide an even more precise determination in good agreement with the position shown. Residual velocity measurements — deviations of red-shift distances from other distance indicators — are shown as follows: Filled and open points represent supernovae, respectively, with negative and positive residual velocities. The areas of these points correspond to the magnitude of the velocity residual. Filled and open crosses show the direction the Local Group is approaching or leaving, according to the best fit to available data, respectively, on supernovae in distant galaxies, and data on the brightest galaxies in clusters reported by Lauer and Postman (La94). Galactic coordinates are used with the Milky Way plane at zero latitude and the central meridian at 335°, i.e., removed 25° from the Galactic center (Ri95).

phase or is filled with nonrelativistic matter. Show also that the comoving distance χ that can be traversed in time t increases monotonically in proportion to $t^{1/2}$ in a radiation dominated era and $t^{1/3}$ in a matter-dominated era.

Since the microwave radiation from two opposite directions in the sky is only just now reaching us, it would appear that the sole way these regions could have reached thermal equilibrium, in either a radiation- or a matter-dominated era, is to postulate an epoch, early in the evolution of the universe, during which the Universe was not expanding or only expanding slowly. Though a cosmological constant can bring about a pause in expansion, the current value of Λ is so minuscule that it cannot have sufficiently reduced the expansion at early epochs when radiation and matter densities were the dominant terms in equations (11–45), (11–46). These two equations are rewritten here for convenience,

$$\frac{8\pi G\rho}{3} = \frac{-\Lambda c^2}{3} + \left(\frac{\dot{a}^2 + kc^2}{a^2}\right).$$ (12-2)

$$\frac{d\rho}{dt} + 3\left(\rho + \frac{P}{c^2}\right)\frac{\dot{a}}{a} = 0, \quad \text{or} \quad \frac{8\pi GP}{c^4} = \Lambda - \left(\frac{2a\ddot{a} + \dot{a}^2 + kc^2}{c^2 a^2}\right).$$ (12-3)

12:2 The Flatness Problem

A second problem that appeared puzzling for some time was the apparent absence of any curvature in the Universe. This came to be known as the *flatness problem*. Let us rewrite equation (12–2) in the form

$$\frac{4\pi G\rho a^3}{3a} = \frac{kc^2}{2} + \frac{\dot{a}^2}{2} - \frac{\Lambda c^2 a^2}{6}.$$ (12-4)

We then note that the term on the left has the form of a potential, whereas the second term on the right has the form of a kinetic energy per unit mass. If we consider conditions in the remote past, when the Universe was very young and more compact, we see from (12–4) that the cosmological constant term would have declined in proportion to a^2 and become negligibly small as a declined. In contrast, equation (11–54) shows that the potential term was proportional to a^{-2}, and therefore very large. $\dot{a}^2$ must therefore also have been very large, and the difference between these two quantities, namely kc^2, would have remained constant, independent of a. In order for this term to be small today, it must have been a fantastically small fraction of the potential or the kinetic energy per unit mass when the Universe was very young. For all practical purposes the Universe must have been very close to Euclidean at all early times. The most sensible conclusion to draw was that the Universe had always been flat, and that the Riemann curvature constant had always been $k = 0$, and still was. In general relativistic cosmologies, the curvature constant is a property of the space that remains invariant, unaffected by the dynamics of evolution.

The problem with this was that today's Hubble constant and hence $\dot{a}^2$ were too high for the present-day mass density on the left of (12–4) to balance. It appeared that the Universe had to be open and $k = -1$. This was before the recognition that the dark matter density of the Universe appreciably exceeds the baryonic density, and before the existence of dark energy had been discovered and found to readily balance the equation with $k = 0$.

With this realization, the flatness problem went away. However, both the isotropy and the flatness problems have had an historically important impact on cosmological thought and were two of the original reasons for adopting the inflationary cosmological model, which we will discuss below, in Section 12:6. Before that, however, we need to delve into the origins of the microwave background radiation.

12:3 Where Did the Microwave Background Radiation Originate

Let us ask ourselves first whether we can identify any plausible sources of energy giving rise to the microwave background radiation. Could this energy, for example, have been generated in stars? Could a large number of discrete sources have produced this uniform background?

If the Universe is isotropic and homogeneous, we can imagine dividing it into individual cubicles separated by totally reflecting walls that expand along with the Cosmos. The radiation accumulated due to past emission by sources within each cubicle, as measured today, would then be precisely the same as though no walls had existed at all. Successive reflections off the receding walls would systematically redshift the radiation.

Consider such a cubicle that currently has unit volume. At some past epoch characterized by red shift z, the sources within the cubicle emitted an energy increment $\Delta\varepsilon$. The red shift will have reduced this to $\Delta\varepsilon(1+z)^{-1}$, by today. Because the volume of the cubicle will, in the meantime, have increased by $(1+z)^3$, the radiation density in the cubicle will have declined by $(1+z)^4$, but the volume would have increased by $(1+z)^3$.

Let us next ask whether the energy now found in the microwave background could have been produced through the conversion of hydrogen into heavier elements in stars? We currently observe that a fraction $f < 0.30$ of the mass of atomic matter is in the form of elements heavier than hydrogen (Lo03). Most of this is in the form of helium. As discussed in Chapter 8, the liberated energy is 0.029 hydrogen masses m_H per helium nucleus formed. For a contemporary baryonic density, ρ_B, we therefore would expect the radiation density in the Universe to be

$$\rho_{\rm rad} = \frac{\Delta\varepsilon}{(1+z)^4} \leq \frac{0.029 f \rho_B c^2}{4(1+z)} \leq \frac{8 \times 10^{-13}}{(1+z)} \;\; {\rm erg\ cm}^{-3}, \qquad (12\text{-}5)$$

where the baryonic density, today, is $\rho_B(t_0) = \Omega_B \rho_{crit} \sim 4 \times 10^{-31}\,{\rm g\ cm}^{-3}$ and an energy injection epoch, z, is assumed. The dependence on $(1+z)^{-1}$ is readily understood if we consider the ratio of photon-to-nucleon number densities to remain constant during cosmic expansion. Then the energy per photon decreases with increasing red shift as $(1+z)^{-1}$, while the mass–energy for nucleons remains constant.

Today's microwave background corresponding to a temperature of 2.725 K provides a radiation density of 4×10^{-13} erg cm^{-3}. If the hydrogen-to-helium conversion energy had been recently injected into the Universe and had then been rapidly thermalized, it would be possible to account for the observed radiation density, $\rho_{\rm rad}$, only if the flux had been generated at epochs $z \leq 2$. Otherwise the red-shifted energy density would fall short of the currently observed background radiation.

Current observations of the most distant quasars and galaxies indicate that the Universe was largely transparent by the epoch $z = 6$. Within galaxies not quite that distant, for which infrared observations are available, about half the radiation produced in stars appears to have escaped as ultraviolet, visible, or near-infrared radiation, while another half was locally absorbed by dust and re-emitted in the far infrared. But dust re-emission would have left its imprint on the background radiation in marked deviations from its observed spectrum now known to be blackbody to within <1% of peak intensity. Moreover, the high transparency of the Universe by $z = 2$ would have prevented this infrared flux from becoming isotropically redistributed or thermalized.

The generation of sufficient energy through collapse of massive objects to form black holes can also be ruled out. Table 1.6 shows that the total mass in black holes today is of order 0.1% of the total baryonic mass, and it appears that far less than 1% of the collapse energy may be emitted as electromagnetic radiation. Nor would this radiation have a blackbody spectrum.

Arguments along such lines rule out the possibility that the microwave background could have been recently generated. We are driven, instead, to consider a primordially hot Universe, whose thermal history is indicated in Fig. 12.2.

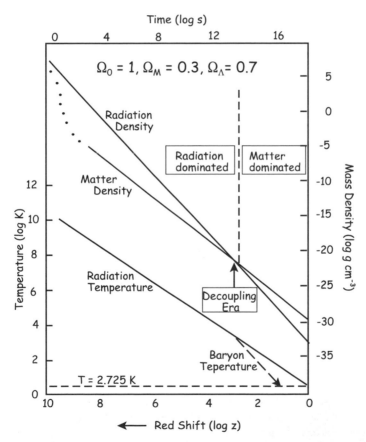

Fig. 12.2. Density/temperature history of the Universe since the Cosmos was one second old. Based on a drawing by M. Longair (Lo78).

12:4 Looking Back in Time

We now ask ourselves how, using only data available today, we might determine conditions that existed in the Universe during earlier epochs. Some quantities we will need to know depend only on the red shift z. These parameters are tabulated in Table 12.1 and represent an adiabatic extrapolation back in time based solely on volumetric arguments. Because the microwave radiation is thermal, the density is related to temperature by the blackbody expression (4–74), $\rho = aT^4 \propto (z+1)^4$. The rest-mass density of matter, in contrast, is proportional to $(z+1)^3$, throughout, and a cosmological constant would have remained invariant.

Table 12.1. Evolution of Different Quantities with Red Shift.[a,b]

Red Shift $(z+1)$	T_{photons} K	n_{photons} cm^{-3}	ρ_{photons} g cm^{-3}	ρ_M g cm^{-3}	Remarks
1.1×10^9	3×10^9	5.4×10^{29}	6.7×10^2	3.5×10^{-3}	
3×10^8	8×10^8	1.11×10^{28}	3.7	$\sim 7\times10^{-5}$	Element formation era[c]
10^4	2.73×10^4	4.1×10^{14}	4.6×10^{-18}	2.6×10^{-18}	
					⇑ Radiation dominated[d]
					⇓ Matter dominated[d]
1100	3.00×10^3	5.5×10^{11}	6.7×10^{-22}	3.5×10^{-21}	Baryon decoupling era[e]
1000	2.73×10^3	4.1×10^{11}	4.6×10^{-22}	2.6×10^{-21}	
100	272.5	4.1×10^8	4.6×10^{-26}	2.6×10^{-24}	
10	27.25	4.1×10^5	4.6×10^{-30}	2.6×10^{-27}	
1	2.725	4.1×10^2	4.6×10^{-34}	2.6×10^{-30}	Present

[a] Assumes a current value of $\Omega_M = \Omega_B + \Omega_{DM} \sim 0.27$, with baryonic matter alone, $\Omega_B \sim 0.04$, $H_0 = 70\,\text{km s}^{-1}\,\text{Mpc}^{-1}$ and $\rho_{crit} = 9.7 \times 10^{-30}\,\text{g cm}^{-3}$. ρ_M is the rest-mass density. The mass density of photons is $\rho_{\text{photons}} = aT^4_{\text{photons}}/c^2$.
[b] Entries in this table are derived from current estimates, which appear in the last row. These are extrapolated backward in time assuming adiabatic evolution.
[c] As we will see in Section 12:13, helium and trace quantities of other light elements form from protons and neutrons at this epoch.
[d] The radiation- and matter-domination epochs are discussed in Section 12:16.
[e] At high temperatures matter and radiation are coupled and in thermal equilibrium. As the Universe expands and cools to sufficiently low temperatures and densities, baryons and radiation decouple and cool to different temperatures. This is discussed quantitatively in Section 13:5.

This extrapolation tells us that the Universe could at one time have been extremely hot and compact, quite possibly at temperatures and densities far higher than those envisaged in Table 12.1. Let us, therefore, tackle the question from a different perspective and ask, "What is the highest temperature, highest density universe we could image?"

12:5 The Planck Era

Let us first ask about the earliest, shortest time span over which the laws of physics, as we currently understand them, might have been valid. This is determined by the range of validity of general relativity. At extreme densities and over extremely short time intervals, this theory will need to be replaced by a quantized theory of gravitation that still awaits discovery. However, as we saw in Chapter 5, it is possible to list conditions under which general relativity can be trusted.

In Section 5:20 we found that the most compact mass we could envision was the *Planck mass*

$$m_p \equiv \left(\frac{\hbar c}{G} \right)^{1/2} = 2.18 \times 10^{-5} \, \text{g}. \tag{12-6}$$

The diameter of this mass is the *Planck length*

$$l_p \equiv \left(\frac{\hbar G}{c^3} \right)^{1/2} = 1.61 \times 10^{-33} \, \text{cm}. \tag{12-7}$$

The shortest time during which it makes sense to talk about such a mass is the length of time light would take to traverse the Planck length. This is the *Planck time*,

$$t_p = \left(\frac{\hbar G}{c^5} \right)^{1/2} = 5.38 \times 10^{-44} \, \text{s}. \tag{12-8}$$

Over shorter intervals than this, one end of the Planck mass distribution would cease to be aware of the presence of the other end, so that the laws of causality would no longer apply. The Planck time must therefore be considered the earliest time in the existence of the Universe for which equations (12–2) and (12–3) could apply. There is no guarantee that they apply that early, but there is reason to expect that they cannot apply any earlier.

Finally, dividing the Planck mass by the cube of the Planck length gives us a measure of the density the early Universe could have attained. The laws of relativity could not be expected to apply at higher densities than this *Planck density*

$$\rho_p \equiv \frac{c^5}{\hbar G^2} = 5.18 \times 10^{93} \, \text{g cm}^{-3}. \tag{12-9}$$

It makes sense to also ask about the temperature that might have existed at the Planck time. We can proceed by considering two possibilities: only bosons being present or only fermions being present. For bosons, having two possible spin values, the density–temperature relation has the form of equation (4–74). For fermions restricted to a single spin value, the energy density in the relativistic extreme is

$$\rho = \frac{4\pi c}{h^3} \int_0^\infty \frac{p^3 \, dp}{e^{pc/kT} + 1} = \frac{7\pi^5}{30 h^3 c^3} (kT)^4 = \frac{7}{16} a T^4, \tag{12-10}$$

where a is the radiation constant 7.57×10^{-15} erg cm^{-3} K^{-4} — not to be confused with the scale factor for which we have used the same symbol a.

For a mixture of bosons with statistical weight g_{bi} and fermions with statistical weight g_{fj}, we then find

$$T = a^{-1/4} \rho^{1/4} \left(\frac{1}{2} \sum_i g_{bi} + \frac{7}{16} \sum_j g_{fj} \right)^{-1/4} c^{1/2}. \qquad (12\text{-}11)$$

The factor $c^{1/2}$ enters because equation (12–9) refers to energy density, rather than mass density. If neutrinos and antineutrinos were present at such early epochs, we would have to set $g_{fj} = 1$ for each neutrino species because each has only a single spin state. For electrons or positrons, $g_{fj} = 2$, corresponding to the two possible spin states for these particles. For photons, $g_{bi} = 2$, again corresponding to two possible spin states. If the number of available species was of order unity, the temperature would have been

$$T_p \sim \left(\frac{c^2 \rho_p}{a} \right)^{1/4} \sim 10^{32} \, \text{K}. \qquad (12\text{-}12)$$

The mass contained within the present-day horizon of our Universe is about 10^{55} g. At the Planck time this mass would have been contained in a volume of order 10^{-39} cm^3, having diameter $\sim 10^{-13}$ cm. This span, however, could only be traversed at the speed of light, in 3×10^{-24} s, a period which is long compared to the Planck time. We again encounter the difficulty here of causally connecting one end of the Universe to the other during these initial states. This is not surprising because we had set up equation (12–7) with the idea that the Planck time simply connects different portions of a Planck mass causally — not a mass 60 orders of magnitude larger and hence 20 orders of magnitude more extended.

12:6 Inflationary Cosmological Models

To overcome this difficulty with causality, while also resolving the isotropy and the flatness problems, Alan Guth in 1981 introduced an inflationary cosmological model that evolves through several successive stages (Gu81).

At first the cosmos is compact and expanding sufficiently slowly to permit widely separated regions to come into causal physical contact and reach equilibrium. This gives way to vastly more rapid expansion of such enormous proportions that regions which had previously attained equilibrium grow to a size far larger than the currently observed Universe. This inflationary phase, in turn, is succeeded by a stage in which the Universe is filled with radiation and particles at high temperature. The expansion continues, though at a decreasing rate; the density and temperature progressively decline, much in the fashion we had deduced in Table 12.1, ultimately permitting the formation of clusters of galaxies and voids that we observe all around us today. The expansion during the inflationary phase is so extensive that the Universe we observe out to the most distant galaxies is only a tiny portion of a much larger Cosmos beyond our event horizon, whose actual size may be incalculably large.

Let us now examine how such a cosmos could take shape.

Starting in the Planck era, $t \sim 10^{-43}$ s, an interval of rapid relativistic expansion begins, during which the density and hence the temperature both drop by many orders of magnitude. During this time the scale factor a, the temperature T, the world time t, and the density obey the relations (11–57) and (12–2)

$$a \propto t^{1/2} \propto T^{-1} \propto \rho^{-1/4}. \tag{12-13}$$

This continues until $t \sim 10^{-35}$ s by which time the temperature has dropped from 10^{32} K to 10^{28} K, where particles and/or radiation have energies of 10^{15} GeV. The energy density has correspondingly dropped by a factor of 10^{16}, so that the expansion no longer is dominated by the matter and radiation density of the Universe but rather by a vacuum energy density, assumed to have an energy density corresponding to a temperature of $\sim 10^{15}$ GeV or $\sim 10^{28}$ K.

This vacuum energy density — a potential energy density — is dominated by the so-called *Higgs scalar field*. Its form is not known but it is often written as

$$V(\phi) = \frac{\lambda}{4}(\phi^2 - \eta^2)^2, \tag{12-14}$$

sketched in Fig. 12.3.

Triggered by any small inhomogeneity, the vacuum energy density can slowly relax to the state of lowest energy density, $\phi^2 = \eta^2$, through a gradual expansion resembling the expansion of a Friedmann–Lemaître–Robertson–Walker (FLRW) model, in which the vacuum energy density acts somewhat like a cosmological constant Λ.

As the expansion proceeds, the Higgs field drives the expansion of the Universe. Inserted into equations (12–2) and (12–3), the vacuum energy density acts very much like a density due to a slowly varying cosmological constant Λ. The slow variation reflects the gradual change in $V(\phi)$ as ϕ^2 approaches η^2. The word "slow" needs to be explained. The entire inflationary expansion lasts only $\sim 10^{-33}$ s. "Slow" means that the vacuum energy is gradually relaxing toward η during this period. As ϕ approaches closer to η, the potential drops appreciably more rapidly.

We can describe this sequence of events quantitatively. Substituting equation (12–2) into (12–3) we obtain

$$\frac{8\pi G P}{c^2} = \frac{2\Lambda c^2}{3} - 2\frac{\ddot{a}}{a} - \frac{8\pi G \rho}{3}. \tag{12-15}$$

To find the expansion rate we need to postulate an *equation of state* during this phase — a relation between pressure and density. If we adopt a negative pressure, and consider the density to be dominated by the Higgs field,

$$\frac{P}{c^2} = -\rho = -\frac{\Lambda c^2}{8\pi G}, \tag{12-16}$$

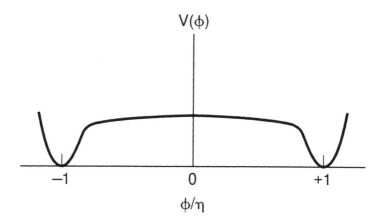

Fig. 12.3. Sketch of the inflationary potential. Note the symmetry about the vertical axis, $\phi/\eta = 0$, the initially shallow decline in $V(\phi)$ as ϕ approaches $\pm\eta$, and the much steeper decline in $V(\phi)$ near $\phi = \pm\eta$.

we find from (12–3) that the density remains constant. If we set the expansion to be exponential,

$$a = a_{oi}e^{Ht} = a_{oi}\exp\left[\left(\frac{2\Lambda c^2}{3}\right)^{1/2}t\right],\qquad (12\text{-}17)$$

where a_{oi} is the scale factor at onset of inflation, equations (12–15) to (12–17) are seen to be consistent with (12–2) and (12–3) provided the Riemann curvature term kc^2/a^2 in equation (12–2) can be considered negligibly small. This is certainly close to correct because the exponential expansion of the scale factor, a, rapidly reduces any curvature that may have been present initially. Equation (12–17) reflects a re-generation rate (11–77) and, once again, recalls that the inflationary stage mimics the steady-state universe in apparently regenerating itself out of nothing.

The mass–energy density $\Lambda c^2/8\pi G$ at $t = 10^{-35}$ s, is of the order that the radiation density would have at a temperature of 10^{28} K, which is $\sim 10^{77}$ g cm^{-3}. The exponential expansion rate at this epoch must then be of order

$$\left(\frac{2\Lambda c^2}{3}\right)^{1/2} \sim 4 \times 10^{35}\,\text{s}^{-1}. \qquad (12\text{-}18)$$

In the course of 1.5×10^{-34} s, the Universe is able to expand by a factor of $\sim e^{60}$ or $\sim 10^{26}$. All this time the Higgs scalar field, which we have represented by a cosmological constant, remains roughly constant; the energy required to keep up the expansion is provided by the release of tension through the equation of state in which the pressure term is just the negative of the energy density.

At the end of this inflationary expansion, the dimensions of the region encom-passing the Universe that lies within our current horizon are of the order of 10 cm.

Moreover, it has only taken the Universe $\sim 10^{-34}$ s to expand to this size. At the speed of light the same expansion would have taken 10^{-10} s. This emphasizes that space itself can expand at an arbitrary speed, limited only by equations (12–2) and (12–3). Matter and radiation, in contrast, cannot traverse space any faster than the speed of light, c.

Piecing together this series of events, we see that the Universe we now observe around us has come from a region only about 10^{-25} cm in diameter at the start of the inflationary phase. During the preceding 10^{-35} s this region could have been traversed by radiation, come to pressure equilibrium, and attained homogeneity. There is no guarantee that this actually happened, but the model makes a good case for it. Note also that the region $\sim 10^{-25}$ cm in diameter is more than ten orders of magnitude smaller than the estimate of 10^{-13} cm we had reached in Section 12:5 without considering inflation. The difference arises from the hypothesis that most of the mass–energy density observed in the Universe today originates not from the mass–energy present during the Planck era, but rather from the vacuum energy density that is released following inflation.

A feature of the inflationary model is that it is also consistent with the flatness of the Universe, the lack of observed curvature, and hence the value $k = 0$ of the Riemann curvature constant. This is because any curved surface when sufficiently expanded will appear flat in a small region around any chosen point. The inflation increases the scale of the Universe so enormously and balloons the Universe out to such a large extent, that locally, within any horizon defined by the speed of light, the curvature becomes negligible.

Of importance in considering the inflationary model is the recognition that the portion of the Universe we now see is only an unimaginably small fraction of a larger Universe which will forever remain unknown to us — out of touch, beyond physical reach, beyond study by physical means. Because physics normally confines itself to statements about systems that can be examined observationally or through experiment, the proposition that such remote realms of the Universe exist, though they can never be observed, breaks with traditional ideas about the range of permissible scientific inference!

PROBLEM 12–2. Fundamental particle physics suggests that the Universe should have been filled with magnetic monopoles in the earliest moments of existence of the Universe. *Magnetic monopoles* are carriers of magnetic charge, similar in fashion to electrons that carry a unit of electric charge. These particles were considered by Polyakhov and independently by t'Hooft in 1974 (Po74), (Ho74) and are likely to have energies of order 10^{16} GeV apiece. Show that after inflation, the Universe could have contained fewer magnetic monopoles than one for each galaxy that would eventually form.

12:7 The Post-Inflationary Stage

Toward the end of inflation, a phase transition sets in. The extreme expansion during the inflationary phase will have drastically lowered temperatures as ϕ slowly approached η. However, as the vacuum energy is released, rapid reheating occurs. The energy of the false vacuum becomes available, much as energy is liberated when supercooled liquid water freezes to form ice.

Conservation of energy and the establishment of thermal equilibrium at this stage require that the enormous reservoir of vacuum energy give rise to large numbers of photons and particles. If only electromagnetic radiation were produced — and produced suddenly — its temperature would correspond to $T \sim 10^{14}$ GeV. But the vacuum energy does not decay all at once and, besides photons, an appreciable array of other particles and antiparticles arises making for a high entropy. The reheating at the end of inflation is an essential feature of the theory. Just how it comes about, however, is still being debated (Ko96). A relevant factor may be that the temperature at this stage roughly corresponds to the temperature $T_{GUT} \sim 10^{15}$ GeV, at which *grand unified theories of particle interactions, GUTs,* expect the strong and weak nuclear forces, as well as electromagnetic and gravitational forces, to be roughly equal to each other. Below this energy, the strong and the electroweak forces no longer remain comparable, symmetry is broken, and a phase transition can set in.

During reheating, the equation of state becomes that for a relativistic gas,

$$P = \frac{\rho c^2}{3} . \tag{12-19}$$

Once this new phase of the Universe begins, the continuing expansion leads to an adiabatic temperature and density decline, during which unstable particles decay while some others, like neutrinos, decouple from the electromagnetic radiation field. The Universe is now approximated by a FLRW model that continues to expand and cool right down to the present era.

Not all of the vacuum-energy density from the inflationary state needs to be transformed into particles and radiation. Some of it may stay on as a low-level residual cosmological constant, Λ, whose energy density $c^2\Lambda/8\pi G$ is negligible at the time

$$\frac{c^2 \, |\, \Lambda \, |}{8\pi G} \ll \frac{3H_o^2}{8\pi G} \tag{12-20}$$

in order to be consistent with today's observed expansion rate.

At the world time t_{ei} marking the end of inflation, the total energy density is almost unaltered from its original value at the onset of inflation T_{oi}, a time when strong, electromagnetic, and weak interactions all have a universal strength of about 10^{15} GeV, corresponding to a temperature of roughly 10^{28} K. The red shift at which inflation ends, therefore, is $z_{ei} \sim 10^{28}/3 \sim 10^{27.5}$, where the factor 3 roughly corresponds to today's microwave background radiation temperature. Because the Universe expands by a factor of $\sim 10^{26}$ during inflation, we see that the red shift at onset of inflation would have been $z_{oi} \sim 10^{53.5}$.

12:8 The Riemann Curvature Constant

An important consequence of inflation is this: following the onset of the immensely rapid inflationary stage, there once existed, at world time t_1, a volume V_1 that ultimately evolved into a volume of the Universe V_0 stretching out to some red shift, say $z \sim 1$, at the present world time t_0 (Pe93)*.

To show this, let us consider the comoving distance $\chi(t_1, t_0)$ that a light ray is able to traverse in time $(t_0 - t_1)$. From (11–19) or (11-82) we see that this is

$$\chi(t_1, t) = \int_{t_1}^{t_0} \frac{c\, dt}{a(t)} = \int_{a_1}^{a_0} \frac{c\, da}{a \dot{a}} . \tag{12-21}$$

This finite comoving distance tells us that the rapid inflationary expansion physically isolated volume V_1 from the rest of the Universe and kept it isolated until recent times. The region within V_1 evolved to its present appearance through a purely local set of physical processes, unaffected before epoch t_0 by anything that would have gone on elsewhere in the Universe, beyond comoving distance $\chi(t_1, t_0)$.

This will be an important consideration when we look at the formation of cosmic structures in Chapter 13. There, we will see that the size and distribution of clusters of galaxies and voids can be related to density fluctuations and, more generally, the curvature of space k in some volume V_1 during the Planck era, possibly well before the inflationary phase set in.

Let us examine this curvature quantitatively to show, as we previously assumed, that its effect was essentially negligible on the evolution of the Universe, at least until rather recent epochs. Galaxies we observe at moderate red shift $z \sim 1$ today, lie at a comoving distance χ_0, given to rough approximation by

$$\chi_0 a_0 \sim cH_0^{-1}. \tag{12-22}$$

This approximation involves a scale factor that increases by no more than a factor of $(z + 1) \sim 2$, and a Hubble constant that decreases by only a factor of order $(z + 1)^{3/2} \sim 2.8$ during the traversal of χ_0. We now assume that, at some initial epoch during the inflationary era, t_1, we would have had a scale factor a_1 and a corresponding Hubble constant H_1 that would have again permitted radiation to cross a comparably large comoving distance χ_0 in roughly its own Hubble time

$$\chi_0 a_1 \sim cH_1^{-1}. \tag{12-23}$$

Expressions (12–22) and (12–23) lead to the conclusion that the red shift at t_1 is, to lowest approximation,

$$z_1 \sim \frac{a_0}{a_1} = \frac{H_1}{H_0} . \tag{12-24}$$

Equations (12–22) and (12–23) also show that the product $\chi a H$ is roughly the same at the present epoch t_0, and at the early world time t_1. And (12–2) tells us that

$$(Ha\chi)^2 = \frac{8\pi G\rho(a\chi)^2}{3} + \frac{\Lambda(a\chi c)^2}{3} - k\chi^2 c^2. \qquad (12\text{-}25)$$

Equations (12–22) and (12–23) show that the left side of (12–25) has the same value $\sim c^2$ at epochs t_0 and t_1, and for an identical value of χ the last term on the right of (12–25) remains identical as well. Hence, the curvature term has the same fractional significance relative to $(Ha\chi)^2$ at epoch t_1 as it does at the present epoch t_0.

Let us now examine the evolution of the different terms in expression (12–25) as we successively look back in time to epochs t_{ei} marking the end of inflation, and even farther back in time to the onset of inflation at t_{oi} to determine the epoch at which (12–23) would have held. We can see from (11–56) that $H \propto a^{-3/2}$ in the matter-dominated era which, as Table 12.1 indicates, lasted from the epoch of red shift $(z + 1) \sim 10^{3.5}$ to the present, $(z + 1) \sim 1$. During the preceding radiation-dominated era going back to the end of inflation $(1 + z_{ei}) \sim 10^{27.5}$, $H \propto a^{-2}$. During inflation — from $(1 + z_{ei}) \sim 10^{27.5}$ to $(1 + z_{oi}) \sim 10^{53.5}$ H remains constant.

This means that the left side of (12–25) first increases by a factor of $10^{3.5}$ as we go back in time from $(1 + z) = 1$ to $(1 + z) = 10^{3.5}$. It then increases by a further factor of 10^{48}, going back to red shift $z_{ei} = 10^{27.5}$ for a total increase of $10^{51.5}$. During inflation, the left side then drops in proportion to $a^2 \propto z^{-2}$, in going from z_{ei} back to z_{oi}, and diminishes by a factor of $\sim 10^{51}$.

This explains how the left side of (12–25) can have roughly the same value at t_0 and some arbitrary epoch t_1 following the onset of inflation at t_{oi}. Figure 12.4 plots the evolution of the product Ha to show that it had the same value following the onset of inflation as it does today, so that (12–23) indeed would have correctly described the Universe at epoch t_1.

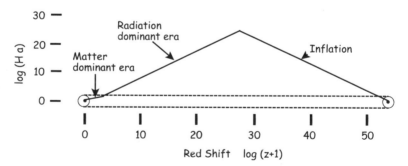

Fig. 12.4. Mapping of the Universe at inflationary times onto today's observed Universe. The diameter of the Hubble sphere at all epochs corresponds to the separation between the two horizontal dotted lines. The black circular spot on the left, encloses today's observed Universe out to $z \sim 1$. The black circular spot on the right encloses the corresponding region following the onset of inflation that eventually came to be mapped onto the black circular spot on the left, i.e., onto the portion of the Universe we see stretching out to $z \sim 1$ today.

PROBLEM 12–3. Show that the sum of the first two terms on the right of (12–25) follows the same scaling as the left side of the equation in each of the three intervals ranging from $(z + 1) = 1$ back in time, respectively, to $10^{3.5}$, $10^{27.5}$, and $10^{53.5}$, by assuming that the term containing Λ is negligible compared to the term containing ρ from z_0 to the end of inflation, $z_{ei} \sim 10^{27.5}$, and that the *false vacuum* of the *Higgs field*, designated by the term involving Λ during inflation, then dominates back to $z_{oi} = 10^{53.5}$ and the onset of inflation.

PROBLEM 12–4. Convince yourself that, if the end of inflation occurred at red shift $z_{ei} = 10^{27.5}$, the onset of inflation must have taken place at $z_{oi} \gtrsim 10^{53.5}$, so that the inflationary expansion of the Universe must have increased the scale factor by at least 10^{26}, for the homogeneity of the Universe to be explained by inflation.

Suppose now that the curvature term at the present epoch was significant though difficult to discern. Then at world time t_1, it would have been equally significant relative to the other three terms in equation (12–25). At even earlier times during the inflationary period, $k\chi^2 c^2$ would have grown in importance, as $\Lambda(a\chi c)^2/3$, the other dominant term in (12–25), would have declined in proportion to a^2. The effects of this increased curvature would be apparent today in observations of regions lying beyond $z \sim 1$ — i.e., outside the volume V_0 we had considered. This is because we observe these more remote regions as they appeared at a world time $t \ll t_0$. At that epoch, regions from an era much earlier than t_1 would have been mapped onto our universe, and those regions would correspondingly have exhibited larger curvature. Because regions lying beyond $z \sim 1$ show no such enhanced curvature, we conclude that curvature was a minor effect even in the early phases of inflation. The global value of the Riemann curvature constant appears to have been $k = 0$ throughout time. This partly justifies our assumption in Sections 12:6 and 12:7 that curvature effects could be neglected.

Local curvature fluctuations on small scales are not ruled out by this argument. In Chapter 13 we will see their importance for seeding the formation of cosmic structures.

12:9 Quark–Gluon Plasma

Let us return to the post-inflationary era when the phase transition from the false vacuum to the present-day vacuum sets in. What kinds of particles are created at that epoch cannot be ascertained right now because neither the particles created in high-energy accelerators, nor the most energetic observed cosmic-ray events reach energies of 10^{14} GeV. The highest-energy cosmic-ray particles observed have energies just above 10^{11} GeV. More important, perhaps, the energetic cosmic-ray particles we do observe are isolated, whereas they would have been densely packed at early epochs.

We know that protons, neutrons, and the atomic nuclei in which they are assembled consist of quarks and gluons. Accelerator experiments in which high-energy gold nuclei are smashed into gold targets, fleetingly produce a dense aggregate of subnuclear particles as two gold nuclei interpenetrate. These experiments show that quarks and gluons, instead of being confined to well-identified particles then behave more like free particles. At densities higher than those encountered in ordinary nuclei, quarks and gluons are expected to produce a *quark–gluon plasma*, QGP. Such high densities would have existed in the post-inflationary era at temperatures above 10^{13} K corresponding to times earlier than $\sim 10^{-6}$ s.

Quarks are subnuclear particles characterized not only by an electric charge, but also by what is called a *color*, though it has nothing to do with colors the human eye can see. The quarks interact with each other through eight massless particles — the gluons — which are massless but differ from electromagnetic radiation. Collectively, the gluons and quarks are referred to as *partons*, since they occur only as parts of other more complex particles — never as individual, isolated quarks or gluons.

Six types or *flavors* of quarks have been found. They are designated as follows:

Table 12.2. Quarks, their Masses, and Electric Charges.

Quark	$|$Mass (Energy/c^2)$|$	Electric Charge
up (u)	~ 5 MeV/c^2	$+2/3$
down (d)	~ 10 MeV/c^2	$-1/3$
strange (s)	~ 150 MeV/c^2	$-1/3$
charm (c)	~ 1.3 GeV/c^2	$+2/3$
bottom(b)	~ 4.2 GeV/c^2	$-1/3$
top(t)	~ 173.5 GeV/c^2	$+2/3$

The behavior of quarks and gluons is described by *quantum chromodynamics*, QCD, the theory that deals with the structure of baryons, of which the protons and neutrons are the most stable, and mesons, among which the pions and kaons are the least massive. Collectively, the baryons and mesons are called *hadrons*. The baryons, the class of particles that includes protons and neutrons contain sets of three characterizing quarks, whereas the mesons, including pions, kaons, and others, are characterized by a quark and an antiquark. The charged mesons necessarily have an antiquark of a different flavor than the quark. Even though the electrical charges of the quarks are multiples of one-third the electron charge, as shown in Table 12.2, the total charge of every hadron is either zero or a positive or negative integer multiple of the electron charge. The characterizing set of quarks for the proton are (uud). For the neutron they are (ddu), respectively yielding a net charge of 1 and 0. Among quarks, the d and u quarks have the lowest masses, making them the most stable. Among hadrons, the protons and neutrons are the most stable.

At early epochs when the Universe was still extremely hot, a quark–gluon plasma consisting of all six types of quarks is thought to have prevailed. As the temperature dropped below 2×10^{15} K, roughly 175 GeV, at $\sim 10^{-10}$ s, the t quark and its antiparticle $\bar{t}$ would begin to annihilate and decay into lower-mass quarks, leaving only the five lower-mass quarks and their antiquarks in the quark–gluon plasma. The other massive quarks would successively drop out in the same way, ultimately leaving only the very low mass quarks, u and d, the strange quark s, and possibly their antiparticles in the quark–gluon plasma as the temperature dropped to ~ 250 MeV.

12:10 The Origin of Baryonic Mass

Let us now set up a cosmic chronology from as far back as experimental evidence today permits us to reach. At a temperature of ~ 250 MeV $\sim 3 \times 10^{12}$ K, the Universe is 10 μs old, and u and d quarks are fully relativistic, as indicated by their low masses listed in Table 12.2.

PROBLEM 12–5. Use equation (12–10) to estimate the mass density per quark species as $\sim 3 \times 10^{14}$ g cm^{-3} at a temperature $T \sim 3 \times 10^{12}$ K. The combined energy density of u and d quarks alone should, therefore, be $\sim 6 \times 10^{14}$ g cm^{-3} at this temperature.

PROBLEM 12–6. The radii of protons and neutrons are $r \sim 10^{-13}$ cm. Their masses, given in Appendix B:1, are $m_P \sim m_N \sim 1.67 \times 10^{-24}$ g. Show that the mass density of the matter contained in either particle is $\sim 4 \times 10^{14}$ g cm^{-3}.

From this we see that, as the temperature of the Universe drops through $T \sim 3 \times 10^{12}$ K, the mass density is rapidly dropping to the density of nucleons. Before the temperature drops this low, the quarks are able to roam widely, because quarks are free to move around as long as the quark plasma is sufficiently dense. This property is called *asymptotic freedom*. But the binding force on a quark becomes essentially infinite if it attempts to stray beyond a distance of $\sim 10^{-13}$ cm from its nearest neighboring quark. As the Universe expands and the density drops, the quark–gluon plasma reaches the *freeze-out density*, at which it breaks up into hadrons. By assembling themselves into hadrons, the quarks are able to maintain a close distance to nearest-neighboring quarks. Removing them farther from a nearest neighbor would require enormous energies. This explains why no isolated quark has ever been observed.

By smashing two gold nuclei into each other at energies of 200 GeV, at the Relativistic Heavy Ion Collider, at the Brookhaven National Laboratory, a quark–gluon plasma may have been fleetingly produced. A large fraction of the kinetic energy of interacting *nucleons* — protons and neutrons making up the gold nuclei — is converted into quarks, antiquarks, and gluons at high temperatures. As this plasma expands and cools, the freeze-out temperature is reached at which hadrons form. These particles no longer interact with each other but stream out of the collision target into an ambient set of detectors that measure their properties. Although the aggregate produced in these collisions has some of the attributes theoretically expected of a quark–gluon plasma, further experiments are required to establish an equation of state and the nature of the phase transition from the plasma to ordinary hadronic matter (Lu03a).

It is interesting to note that the expected phase transition at early cosmic times occurs when the quark–gluon plasma is still relativistic. To understand why this is, we need to note that the mass–energy of two u quarks and one d quark adds up to only $\sim$20 MeV/c^2, whereas the rest-mass of the proton is $\sim$938 MeV. Why then are the protons and neutrons so massive (Wi99)?

PROBLEM 12–7. To explain this mass difference, note that the radius of the proton is $r_p \sim 10^{-13}$ cm.

(a) Equation (7–1) then suggests that the momentum of each quark has to be $p \sim \hbar/r_p \sim 10^{-14}$ g cm s^{-1}. Show that the quarks are moving relativistically.

(b) Show that the individual quark masses then are $\sim p/c \sim 3.5 \times 10^{-25}$ g or $\sim$200 MeV/c^2 apiece, and that the three quarks alone can account for nearly $\frac{2}{3}$ of the rest–mass of the proton and similarly for the neutron. But this approach does not take the mass of the gluons into account.

(c) The gluons binding the quarks set up a potential that allows the quarks to move freely when they are close to each other, but sharply prevents them from venturing beyond r_p. We can then think of the gluons as setting up a potential well with close-to-vertical walls in which the three quarks are trapped. The lowest energy of each quark will then be a standing wave state with wavelength twice the potential well diameter, amplitude zero at the well walls and maximum amplitude at the well center. Show that the total mass of the three quarks in the potential set up by the gluons should then be $\sim$930 MeV/c^2.

Although Problem 12–7 is a drastic oversimplification — because gluons and quarks are inseparably connected — the masses of protons and other hadrons can indeed be calculated in computations that seek out stable configurations of partons mutually interacting to produce standing waves (Wi99). Such calculations show that most of the baryonic mass is due to the dynamics inside the hadrons. As already mentioned in Chapter 5, inertial mass should be considered to be a measure of energy content, so that we should be writing $m = \mathcal{E}/c^2$, as Einstein originally did (Ei05b), rather than the more conventionally quoted expression $\mathcal{E} = mc^2$.

In summary, most of the baryonic mass we find in the Universe is not due to the rest-mass of quarks, or any other particles, but rather a consequence of the binding forces that confine partons to small spaces.

The origin of the electron mass, however, appears to be different and is not yet understood.

12:11 Leptons and Antileptons

The electron is a particle that belongs to the family of *leptons*. This family comprises the electron e and its neutrino ν_e, the muon μ and its neutrino ν_μ, and the τ, pronounced "tau," and its neutrino ν_τ. All six are spin-$\frac{1}{2}$ fermions. The mass of the μ is $207\, m_e$ or $\sim 106\,\text{MeV}$, and the mass of the τ is $1.78\,\text{GeV}$. These six particles and their corresponding antiparticles, would all have been mixed in with the quark–gluon plasma at early epochs. As the temperature of the Universe dropped, the τ particles would have annihilated against their antiparticles $\bar{\tau}$ as the temperature dropped through $T \sim 1.8\,\text{GeV} \sim 2 \times 10^{13}\,\text{K}$.

12:12 The Matter–Antimatter Asymmetry

Today, the Universe consists almost entirely of matter. Trace quantities of antiprotons have been detected at levels of a few parts per million in the cosmic-ray flux at energies in the 200 to 600 MeV range, and at a level of one part in 10^4 at energies around 10 GeV. These antiparticles are probably formed in collisions of higher-energy cosmic rays with the interstellar medium (Ho96), (Mo97a). Aside from such traces, the Universe appears to be devoid of antimatter.

Antimatter is hard to detect at a distance. Antihydrogen, which has been created in small quantities at accelerators, is expected to give off a spectrum identical to hydrogen. Hence, a distant galaxy would look the same whether it was composed of matter or antimatter. However, because galaxies and clusters of galaxies interact, annihilation of matter and antimatter at cluster boundaries would be readily detected through the emission of annihilation radiation — $\lesssim 100\,\text{MeV}$ gamma rays. Because significant annihilation radiation is not evident, we conclude that the Universe consists overwhelmingly of matter.

We do not know how to account for this predominance. The most likely cause of the asymmetry is a weak interaction that involved *CP*-violation (see Section 11:18) during an early era when the Universe was still very hot. One possibility is that a species of very massive neutrinos existed at these early epochs and that *CP*-violation in the subsequent decay of these particles generated a matter–antimatter asymmetry in its decay products. This would express itself as a predominance of leptons over antileptons, and a precisely compensating predominance of baryons over antibaryons. The difference between the number of leptons and antileptons is known as the *lepton number*, and the difference between the number of baryons and antibaryons is called the *baryon number*. Although the decay of such a massive neutrino would violate both lepton and baryon number, the difference between lepton and baryon number would be conserved (Qu03).

The matter/antimatter symmetry could also have been broken at a later epoch, $t \sim 10^{-11}\,\text{s}$, when energies would have dropped to the weak interaction scale $T \sim 300\,\text{GeV}$, and similar symmetry violations could have taken place. Either way, the violations need to be only of the order of one part in $\sim 10^9$ to produce today's observed baryon-to-photon ratio (Tu97). A vast amount of matter and antimatter may then have annihilated early in the evolution of the Universe to produce an equivalent amount of radiation and leave behind just one particle for every original set of $\sim 10^9$ particle–antiparticle pairs that annihilated, melding into the cosmic background radiation.

A number of *CP*-violating proposals have been advanced. The solution of the puzzle, however, must be found experimentally. Experiments at high-energy accelerators show that interactions of K^0 and $\bar{K}^0$ mesons, and similarly of B meson systems, violate both *C* and *CP* conservation. So we know that asymmetries of this kind do arise in Nature though, so far, not on a scale sufficient to account for the observed preponderance of cosmic matter over antimatter.

12:13 Early Element Formation

Let us now return to our chronology.

(i) We briefly interrupted this survey with the Universe $10\,\mu$s old. The post-inflationary phase is by now long established. Temperatures and densities have been evolving relativistically, all along, in accordance with equation (12–13), and the temperature has dropped to $T \sim 3 \times 10^{12}$ K. The densities now are so low that the quark–gluon plasma cannot be maintained, and the plasma breaks up into individual particles in which the quarks can remain closely confined — mainly pions that contain quark–antiquark pairs of u and d. The pions and antipions mutually annihilate almost as soon as they are formed, releasing neutrinos, photons, and energy to heat all the remaining cosmic constituents, mainly photons, electrons, positrons, neutrinos, and antineutrinos. Also emerging at this epoch are minuscule numbers of protons and neutrons, roughly one for every $\sim 10^9$ ambient particles and photons.

(ii) The Universe is a little more than 10 ms old (Bo85). The temperature has dropped to below 10^{11} K, and neutrons and protons have number densities $n(\mathcal{N})$ and $n(P)$ that reflect thermal equilibrium through a Boltzmann factor corresponding to the difference, $c^2\Delta m$, in their mass–energies (Aℓ53).

$$\frac{n(\mathcal{N})}{n(P)} = \exp\left(-\frac{c^2\Delta m}{kT}\right) . \tag{12-26}$$

This equilibrium is maintained by interactions with ambient electrons, positrons, and electron neutrinos as well as antineutrinos, through reactions of the type

$$\nu + \mathcal{N} \Leftrightarrow P + e^-, \tag{12-27}$$

$$e^+ + \mathcal{N} \Leftrightarrow P + \bar{\nu}, \tag{12-28}$$

$$\mathcal{N} \Leftrightarrow P + e^- + \bar{\nu}. \tag{12-29}$$

Just as in equation (12–10), the energy density in neutrinos plus antineutrinos is related to the energy density of photons by the ratio $\frac{7}{8}$. The neutrino/antineutrino energies are still sufficiently high at this temperature to keep electron–positron pairs in equilibrium abundance through the reactions

$$e^+ + e^- \Leftrightarrow \nu + \tilde{\nu}. \tag{12-30}$$

(iii) At $t \sim 0.1$ s, the temperature is $T \sim 3 \times 10^{10}$ K. The weak interactions now are too slow to compete with the expansion rate, and the neutrinos decouple from matter. The energy at this stage is still of the order of a few MeV per particle, and hence electron–positron pairs persist in equilibrium with photons.

$$e^+ + e^- \Leftrightarrow \gamma + \gamma. \tag{12-31}$$

(iv) At $t \sim 1$ s, $T \sim 10^{10}$ K, particle energies are of the order 1 MeV. Reactions of the type (12–28) and (12–29) can no longer maintain the equilibrium between neutron and proton number densities required by equation (12–26), and the ratio of neutrons to protons freezes out at a value characteristic of $\sim 10^{10}$ K. The neutron–proton mass difference is $c^2 \Delta m = 1.293$ MeV, and detailed calculations give a neutron-to-proton ratio of $\sim \frac{1}{6}$ at freeze-out. Thereafter, the ratio changes primarily as the neutrons decay into protons with a half-life of order 10.23 ± 0.02 min — often referred to in terms of the mean life 885.4 ± 1.3 s (Ar00).

PROBLEM 12–8. Show that the neutron half-life is 0.693 times its mean life.

(v) At $t \sim 10$ s, $T \sim 3 \times 10^9$ K, energies drop below the electron–positron rest–mass and these particles annihilate in pairs, heating the radiation and matter remaining behind. No further pairs are produced, because the photon energies now are too low. But the heat generated in annihilation causes a temperature difference between the photon bath and the bath of neutrinos that earlier decoupled from matter and radiation. Henceforth, the neutrino temperature uniformly remains below the photon temperature. If neutrinos have a rest–mass very close to zero, their contemporary background temperature — as we will see in Problem 12–9 below — should be found to be 1.95 K, whereas the radiation temperature is known to be just below 2.73 K. This neutrino bath has not yet been detected; we do not currently have sufficiently sensitive apparatus.

At this stage, deuterons can already be formed through the reaction

$$\mathcal{N} + P = D + \gamma \tag{12-32}$$

but they are quickly destroyed by photodissociation.

(vi) At $t > 10^2$ s and $T \sim 10^9$ K this photodissociation becomes less frequent as photon energies decline. Now, the following reactions all can set in (Fig. 13.13) to produce the stable isotopes, D, ^{3}He, and ^{4}He.

$$\left. \begin{array}{l} D + \mathcal{N} \Leftrightarrow {}^3H + \gamma, \\ D + D \Leftrightarrow {}^3H + P, \end{array} \right\} \quad \left\{ \begin{array}{l} {}^3H + P \Leftrightarrow {}^4He + \gamma, \\ {}^3H + D \Leftrightarrow {}^4He + \mathcal{N}, \end{array} \right.$$

(12–33)

$$D + P \Leftrightarrow {}^3\text{He} + \gamma, \Big\}$$
$$D + D \Leftrightarrow {}^3\text{He} + \mathcal{N}, \Big\}$$

$$\begin{cases} {}^3\text{He} + \mathcal{N} \Leftrightarrow {}^4\text{He} + \gamma, \\ {}^3\text{He} + D \Leftrightarrow {}^4\text{He} + P, \\ {}^3\text{He} + {}^3\text{He} \Leftrightarrow {}^4\text{He} + 2P. \end{cases}$$

It is not possible to form stable elements of mass 5 or 8 at these temperatures and densities, though some traces of beryllium and lithium isotopes of mass 7 can be formed, the latter in quantities that should be measurable,

$${}^4\text{He} + {}^3\text{He} \rightarrow {}^7\text{Be} + \gamma,$$

(12-34)

$${}^4\text{He} + {}^3\text{H} \rightarrow {}^7\text{Li} + \gamma.$$

Figure 12.5 shows the final abundances for several primordially produced elements expected for different baryon-to-photon ratios η. The figure also shows the current uncertainties in the observed values of the primordial densities. These uncertainties arise from alternative mechanisms for light-element production, through nucleosynthetic processes in stars or spallation of heavy elements in circumstellar or interstellar spaces by cosmic-ray particles. Measurements on the D/H ratio in a particularly metal-poor damped Ly-α system at red shift $z = 2.076$ yields a value $\sim 2 \times 10^{-5}$, consistent with $\Omega_B h^2 \sim 0.025$, somewhat outside the range of the best helium abundance values (Pe01).

12:14 The Entropy of the Universe

Equation (4–134) tells us that the entropy S of blackbody radiation at temperature T in a volume V is $S = 4aT^3V/3$. For an adiabatic expansion, $dS = 0$, and the product

$$VT^3 = \text{constant}   \text{(adiabatic process)}.$$ (12-35)

The entropy is strongly dominated by the radiation. The baryons — because of their scarcity — contribute only negligibly. The entropy per baryon, then, is just the entropy per unit volume divided by the number of baryons in the same volume

$$\frac{S}{Vn_B} = \frac{4aT^3}{3n_B} = 10^{-14} \frac{T^3}{n_B} \text{ erg K}^{-1}$$ (12-36)

or, because the photon number density (equation (4–74)) is $20T^3 \text{ cm}^{-3}$,

$$\frac{S}{Vn_B} = 5 \times 10^{-16} \frac{n_\gamma}{n_B},$$ (12-37)

where n_γ is the photon density. This does not take neutrino entropy into account, which is comparable to the photon entropy for each type of neutrino (electron-, muon-, or tau-neutrino) present though somewhat smaller: first, because the statistical weights of neutrinos and antineutrinos are lower, as discussed in connection with equation (12–11); and second, because, as we will see in Problem 12–9, the present-day neutrino temperature is lower than the photon temperature by an added factor of $(11/4)^{1/3}$.

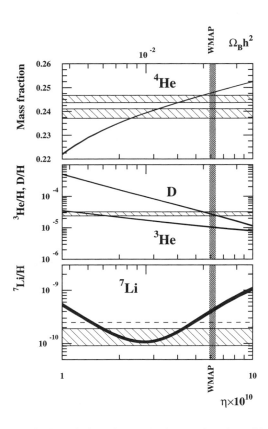

Fig. 12.5. Predictions of primordial nucleosynthesis as a function of baryon-to-photon ratios, η, in the Universe. The continuous curves represent theoretical predictions. The hatched bands indicate the range of primordial light element abundances consistent with observations of ^{4}He and D in metal-poor extragalactic clouds and with ^{7}Li observations in metal-poor stars in the globular cluster NGC 6397. ^{3}He values were obtained from measurements on HII regions, but may not accurately reflect primordial abundances because of contamination by cosmic-ray spallation products and other sources. The dashed line in the box labeled ^{7}Li indicates a 95% confidence level observational upper limit for the abundance of this isotope. The range of $\Omega_B h^2$ shown in this figure is consistent with cosmic microwave background observations (Sp03). h is the Hubble constant in units of 100 km s^{-1} Mpc^{-1}. A Hubble constant of 70 km s^{-1} Mpc^{-1} corresponds to $h = 0.7$. Courtesy of Coc et al. (Co04a).

Equation (12–10) gives the total energy density per fermion of a given spin. Neutrinos and antineutrinos have only one possible spin each. Electrons and positrons have two spin modes each. For the era when all these species were in thermal equilibrium with photons, an epoch characterized by temperatures in the range $10^{12} \geq T \geq 10^{10}$ K, the total energy density in photons and electrons and electron-, muon-, and tau-neutrinos — plus their respective antiparticles — was

$$\rho = aT^4 + 4\left(\frac{7}{16}\right)aT^4 + 3 \times 2\left(\frac{7}{16}\right)aT^4 = \frac{86}{16}aT^4, \qquad (12\text{-}38)$$

where it is important to note that photons are their own antiparticles. Equation linebreak(4–134), which applies equally to photons and relativistic particles, then leads to an entropy per unit volume, $s = S/V$,

$$s = \frac{43}{6}aT^3 \qquad (12\text{-}39)$$

at these high temperatures. Because the subsequent evolution of the Cosmos can be considered to be adiabatic, the total entropy remains constant, and the entropy per unit volume is inversely proportional to the scale factor cubed,

$$sa^3 = \text{constant.} \qquad (12\text{-}40)$$

Here again it is important not to confuse the scale factor a with the radiation constant for which we have used the same symbol in equations (12–36) to (12–39).

PROBLEM 12–9. Show that the annihilation of electron–positron pairs raises the photon temperature by a factor of $(11/4)^{1/3}$ compared to that of the decoupled neutrino background bath. Note that entropy has to be conserved in the annihilation. If the neutrino rest–mass is close to zero, show that the temperature of primordial neutrinos, today, should be ~ 1.95 K.

12:15 A More Precise Extrapolation Back in Time

The rest–mass of neutrinos has not yet been measured, but appears to be well below 1 eV or $<10^{-33}$ g. The neutrinos must, therefore, be relativistic throughout the radiation-dominated era, and contribute appreciably to the overall radiation density. This follows from equations (12–10) and (12–38), where the photon and neutrino terms are seen to be of comparable magnitude. However, care has to be taken in the application of (12–38) because, as we saw in Problem 12–9, the neutrino temperature is lower than the photon temperature T_{ph} by a factor of $\sim(4/11)^{1/3}$. Equation (12–38) then tells us that the relative contributions of photons and the three neutrino species to the total radiation density are, respectively, aT_{ph}^4 and $(42/16)a[(4/11)^{1/3}T_{ph}]^4 \sim 0.681aT_{ph}^4$. A more careful calculation which takes into account that the neutrinos still are able to absorb part of the energy liberated in the electron–positron annihilation era, raises the neutrino contribution to the density to $0.685aT_{ph}^4$. This makes the total radiation mass density $\rho_{radiation} \sim 1.685aT_{ph}^4c^{-2}$.

With this value for the radiation mass density, we can now return to the extrapolation back in time that Table 12.1 provided, and fill in rather more information on the assumption that the Universe is flat — the Riemann curvature constant $k = 0$. At each epoch $(z + 1)$, we can now calculate a Hubble constant based on the mass–energy density of the Universe at the corresponding red shift. The age of the Universe is then given by (11–57) and (11–58), respectively, for the radiation-dominated and mass-dominated epochs. For the most recent dark-energy-dominated era, we take today's directly observed Hubble constant, here taken to be $H_0 = 70$ km s^{-1} Mpc^{-1}, and extrapolate back in time, using (11–59) and assuming a dark energy density parameter $\Omega_\Lambda = 0.7$ and a dark-plus-baryon mass-density parameter $\Omega_M = 0.27$. Equality of radiation and matter densities then comes at $(z + 1) = 3230$, whereas neglect of the neutrino contributions would have led to equality at $(z + 1) \sim 5400$ — a significant difference. These considerations are now factored into Table 12.3, which provides a more detailed extrapolation back in time than was possible in Table 12.1, solely on the assumption of an adiabatic expansion of a gas consisting of particles with rest–mass and photons. Like Table 12.1, Table 12.3 assumes that today's Hubble constant is $H_0 \sim 70$ km s^{-1} Mpc^{-1}. The assumption of $k = 0$ implies that the mass–energy density is critical throughout, so that it scales as $\rho_{crit} = \rho_{rad}(z + 1)^4 + \rho_M(z + 1)^3 + \rho_\Lambda$, relative to current values of ρ_{rad}, ρ_M, and ρ_Λ. Equation (11–47) then yields the dependence of the Hubble constant on critical density and thereby on red shift.

12:16 The First 400,000 Years

The eventful history we have sketched of the Universe, starting with the Planck era and continuing through formation of the light elements, has all taken place within the first few minutes of existence. The next 400,000 years, in contrast, pass rather quietly, as Tables 12.1 and 12.3 indicate. The Universe keeps expanding adiabatically, with a diminishing Hubble constant in full accord with equation (11–57). Since the Hubble constant determines the distance to the particle horizon, the volume of the Universe within which physical contact and potentially pressure equilibrium can be established keeps growing, as Table 12.3 shows. Because the mass–energy of radiation declines as a^{-4}, while the matter density declines only as a^{-3}, a cross-over point is reached at which the mass density of the Universe switches from radiation domination to matter domination. Tables 12.1 and 12.3, which we drew up by an extrapolation backward in time, show that this happens at $z \sim 3230$ when the Universe is roughly 70,000 years old.

Throughout this time, and for another 300,000 years, the Universe remains opaque. Light cannot penetrate any appreciable distance without being Thomson-scattered by free electrons. Throughout, most of the matter density in the Universe is due to dark matter. We see from Table 12.1 that the baryonic matter density at

Table 12.3. Evolution with Riemann Curvature Constant $k = 0^{a,b}$.

Red Shift $(z+1)$	Hubble Constant $(\mathrm{km\,s^{-1}\,Mpc^{-1}})$	t_{age}	Hubble Radius $R_H = (c/H)$	Rest-Mass within R_H (g)
1.1×10^9	7.5×10^{17}	20 s	0.08 AU	2.5×10^{34}
3×10^8	5.6×10^{16}	270 s	1.1 AU	1.2×10^{36}
10^4	7.2×10^7	6,600 yr	4 kpc	2×10^{49}
				⇑ Radiation dominated
3200^c	9.2×10^6	6.9×10^4 yr	45 kpc	9×10^{50}
				⇓ Matter dominated
1100^d	1.5×10^6	4.8×10^5 yr	220 kpc	3.0×10^{51}
1000	1.3×10^6	5.5×10^5 yr	230 kpc	3.6×10^{51}
100	3.6×10^4	1.75×10^7 yr	8.3 Mpc	1.7×10^{53}
10	1,140	5.5×10^8 yr	260 Mpc	5.3×10^{54}
1	70	1.37×10^{10} yr	4,300 Mpc	2.3×10^{55}

^{a} Assumes $\Omega_0 = 1$, throughout, as implied by $k = 0$, and current values $\Omega_\Lambda = 0.7$, $\Omega_M \sim 0.27$, with a Hubble constant $H_0 = 70$ $\mathrm{km\,s^{-1}\,Mpc^{-1}}$. The Hubble constant at earlier epochs is calculated from the rest-mass density of matter plus a radiation density taken as the mass density of photons plus three species of neutrinos and their antineutrinos — $1.685aT_{ph}^4/c^2$.

^{b} Values for early epochs are derived from radiation densities $1.685aT_{ph}^4$. Matter densities are obtained from Table 12.1. Ages t_{age} are derived from equations (11–51) and (11–57) to (11–59). The Universe is radiation-dominated before $z \sim 3200$ and matter-dominated thereafter until Λ-domination sets in at $z \sim 0.37$.

c Equality of radiation and matter densities.

^{d} Decoupling era.

$T = 3000\,\mathrm{K}$, for $\Omega_B = 0.04$ and $\Omega_M = 0.23$, is about 300 $\mathrm{cm^{-3}}$. This is the number density of electrons at the onset of the *recombination era*. [1] The Thomson scattering cross-section $\sigma_e = 6.652 \times 10^{-25}\,\mathrm{cm^2}$ thus restricts the mean free path of radiation to ~ 1 kpc, less than 1% of the distance to the cosmic horizon at this epoch.

Electrons and protons, however, rapidly combine as the temperature drops through 3000 K at $z \sim 1100$ and the Universe, for the first time in its history, becomes transparent. Radiation can now freely travel across the entire Cosmos — a matter of the greatest importance to us, as we try to unravel the history of the Universe. It permits us to look back in time and directly view conditions when the Universe was only 400,000 years old. In Chapter 13, we will see how this look back in time helps us to deduce how galaxies and clusters of galaxies came to be formed.

---

[1] "Recombination era" is an unfortunate but generally accepted misnomer. It implies that electrons and protons had at some previous era already existed in a combined state as hydrogen atoms. However, as the history portrayed throughout this chapter shows, until $\sim$400,000 years had passed, the Universe was consistently too hot to permit electrons to stay attached to protons.

12:17 Last Impact and Decoupling of Matter from Radiation

Because the photon number density exceeds the number density of atoms by a factor of $\sim 10^9$, the photons as a whole decouple from matter rather quickly, whereas matter takes a long time to decouple from the radiation. During this longer interval, the radiation continues to Thomson scatter off electrons.

On average, last impact for a photon on an electron occurs when

$$[x n_H \sigma_e c] t_{\text{age}} \sim 1 \, . \tag{12-41}$$

Since the ionization fraction x rapidly drops through 10^{-3} at ~ 3200 K and $n_H \sim 300$ cm^{-3}, we see that $x n_H \sigma_e c$ drops to below 10^{-14} s^{-1}, while the age of the Universe at this epoch, as given in Tables 12.1 and 12.3, is $<10^{14}$ s. For most photons, scattering at the *surface of last impact*, therefore, comes about rapidly at decoupling around 3200 K.

12:18 Observational Evidence

We have sketched a rather detailed history of the Universe in this chapter. We should still ask,"How far back in time can we actually trust the sequence of events described? What is the solid evidence?"

The observational evidence available today does not actually require temperatures and densities ever to have been higher than those needed to bring the abundance of neutrons and protons into rapid thermal equilibrium on a time scale short compared to the neutron half-life. This temperature is 10^{10} K, prevailing in the Universe at epoch $t = 1$ s. As the Universe cooled below this temperature, the neutron-to-proton ratio froze out at a value characteristic of 10^{10} K, but with a subsequent decay of neutrons in the following minutes leading to the formation of ^{4}He and minor traces of other light elements when the Universe had reached an age of a few hundred seconds, by which time the temperature had dropped below 10^9 K.

With improved observational determinations for the initial abundance of helium and the current matter density, evidence for early temperatures as high as 10^{11} K would be strengthened. At these temperatures all light or massless neutrino species are kept in thermal equilibrium through neutral current weak interactions with electrons and positrons

$$e^+ + e^- \Longleftrightarrow \nu_i + \bar{\nu}_i \quad \text{where} \quad i = e, \mu, \tau \, . \tag{12-42}$$

As the temperature drops through 3×10^{10} K, at $t \sim 0.1$ s, the weak interactions become too slow to keep up with the expansion rate, and the neutrinos and antineutrinos decouple from the electrons and positrons. The number of neutrino species existing in the Universe, nevertheless, still can affect the total helium abundance produced later, at $t \geq 10^2$ s and $T \leq 10^9$ K, through their contribution to the density in equation (12–2) and, thereby, to a higher expansion rate. The amount of helium ^{4}He produced is increased if the expansion is faster, because the neutrons present in the early Universe have less time to decay. This is a quite significant effect, increasing the total amount of helium in the Universe by 4 to 6% for each added neutrino species. Current observations of early helium abundances agree best with ^{4}He production by three neutrino species. Recent accelerator experiments on Z° boson decay show that only three kinds of neutrinos exist in Nature. If improved measurements on the early helium abundance and the total matter density in the Universe confirm that a mass density requiring three species of ambient neutrino radiation baths existed at helium formation, we would have to conclude that initial temperatures at one time had been sufficiently high to produce all three neutrino species in thermal equilibrium with radiation and matter. This would have required temperatures $\sim 10^{11}$ K. To date, these appear to be the strongest observational demands for high temperatures in the early Universe.

To reach much farther back in time, we would need to better understand physical processes at higher temperatures and densities. Current accelerator experiments on the decay of Z-bosons are beginning to yield insights into the state of quarks and gluons that may form a quark–gluon plasma at temperatures of 10^{12} K — the temperature of the Universe at age 10^{-2} s (Wi98). To produce the Z-bosons, however, accelerators capable of producing particles at energies in excess of 1 TeV $\equiv 10^3$ GeV $\sim 10^{16}$ K have been required. To go to even higher temperatures, we may have to turn to Nature. Gamma rays at energies up to 50 TeV have now been detected coming from the Crab Nebula (Ta98). Far higher-energy, 3×10^{20} eV $\equiv 3 \times 10^{11}$ GeV, cosmic-ray particles have also been observed; their energies are only a factor of $\sim 10^3$ lower than the energy of matter at the postulated phase transition that ends inflation. The study of such naturally occurring high-energy particles may help us elucidate events at epochs $\lesssim 10^{-27}$ s. By understanding the physics of matter in this early era, we may find observational tests to probe the defining predictions of inflation.

Answers to Selected Problems

12–1. From equation (11–56) we see that

$$a(t) \propto t^{1/2} \quad \text{for a relativistic gas,}$$
$$\propto t^{2/3} \quad \text{for nonrelativistic matter.}$$

If we insert these relations in equation (12–1) and integrate, we see that the time $t_1 - t_0$ required to cross distance parameter χ can be derived, respectively, for these two cases, from

$$\chi \propto (t_1^{1/2} - t_0^{1/2}) \quad \text{relativistic}$$

and

$$\chi \propto (t_1^{1/3} - t_0'^{1/3}) \quad \text{nonrelativistic.}$$

For the relativistic case doubling χ requires an added time interval that can be derived from $t_2^{1/2} - t_1^{1/2} = t_1^{1/2} - t_0^{1/2}$. Setting $t_0 = 0$, this yields $t_2 = 4t_1$. For the nonrelativistic case, a similar argument leads to $t_2 = 8t_1$.

12–2. During inflation, the number density of all particles decreases by $\sim (10^{26})^3 = 10^{78}$. We saw that the initial mass density before inflation corresponded to a temperature $T \sim 10^{28}$ K, or 10^{77} g cm^{-3}, so that a volume 10 cm in diameter, which defines all matter ultimately to become part of the Universe within present horizons, would have originated in a volume element 10^{-75} cm^3 before inflation. The mass content of this region would have been 100 g. Even if filled entirely with monopoles of energy 10^{16} GeV $= 1.8 \times 10^{-8}$ g, it could at most have contained 10^{10} monopoles. The number of galaxies within the current cosmic horizon is of order 10^{11}.

12–3. Except for a brief recent interval starting at $z \leq 2$, matter or radiation have dominated the right side of (12–25) all the way back to z_{ei}, with $\rho \propto H^2$, as follows from (11–52), (11–54) and (11–56). From z_{ei} back to z_{oi}, the second term dominates the right side of (12–25), and both Λ and H remain constant.

12–4. A short inflationary period would mean that regions which had not been in causal contact before world time t_{oi} would be included in the volume V_1 that maps onto the observed volume V_0 today. The volume V_0 would then include regions that had never been in causal contact and would appear, in violation of observations, to be chaotic rather than homogeneous. The inflationary scenario was designed to solve the homogeneity problem, but can do this only if the inflationary era is sufficiently long to permit small regions, which could have established firm causal contact by $t_{oi} \sim 10^{-35}$ s, to be mapped onto the observed universe.

12–7 (a) $p \sim \hbar/r_p = 1.05 \times 10^{-14}$ g cm s^{-1}. Because the quark rest-mass is ~ 10 MeV $\sim 2 \times 10^{-26}$ g, its motions must be highly relativistic. (b) The individual quark masses are $p/c \sim 3.5 \times 10^{-25}$ g ~ 200 MeV/c^2, while the proton mass is 938 MeV/c^2. (c) The wavelength of the standing wave is $\lambda \sim 4r_p$. Its frequency is $\nu = c/\lambda$, and the associated mass per quark is $h\nu/c^2 = h/(4r_p c) \sim 5.5 \times 10^{-25}$ g ~ 310 MeV/c^2.

12–8. For exponential decay, half the number of initially present neutrons n_0 is left when $n_0 e^{-\alpha t} = n_0/2$, which occurs for $t_{1/2} = \ln 2/\alpha = 0.693/\alpha$. The mean life is given by

$$t_{\text{mean}} = -\frac{1}{n_0} \int_0^\infty t \left(\frac{dn}{dt} \right) dt = \frac{1}{\alpha} \ .$$

12–9. Equation (12–38) tells us that the energy density for electrons plus positrons, is $7aT^4/4$ in the relativistic limit. Accordingly, the entropy (4–134) is also 7/4 as high as that of photons. Since entropy is conserved in adiabatic cooling, and the photon entropy is dependent on the temperature alone, regardless of the presence of any other particles, the electron and positron entropies also must depend only on the instantaneous temperature and must continue to scale in proportion to the photon entropy. During annihilation, which can be considered a phase transition, entropy is also conserved and temperature remains essentially constant so that the volume must expand by a factor $(11/4)$ to have an identical photon energy density and entropy density at the beginning and end of the annihilation period. During this expansion, the neutrinos are cooled in proportion to the linear expansion and hence a photon-to-neutrino temperature ratio $(11/4)^{1/3} = 1.40$ becomes established at annihilation and is maintained forever after, making the neutrino temperature $T_\nu \sim 1.40 \times 2.725 = 1.95\,\mathrm{K}$ today. If the neutrino rest–mass is significant, cooling on expansion makes the neutrinos nonrelativistic. They then cool to appreciably lower temperatures because their adiabatic constant is $\gamma = 5/3$, whereas photons have $\gamma = 4/3$.

13 The Formation of Cosmic Structures

In this chapter we will consider the origins of cosmic structure on the scale of galaxies and clusters of galaxies. The problem of galaxy formation is one of the most complex in all of astrophysics. Considering that the observed portion of the Universe contains a hundred billion galaxies, they are certainly no accident. Yet we have only vague theories of how they came into being.

We can only guess that substantial primordial fluctuations existed at the dawn of time when the Universe was intensely hot. Surviving through enormous expansion until the Cosmos had appreciably cooled, the fluctuations seeded the formation of small protogalaxies. The protogalaxies then collided to form the larger galaxies now observed in clusters throughout the ambient Universe. The earliest galaxies appear to have formed within no more than a billion years after the birth of the Cosmos.

We will assemble the individual concepts that have been offered in partial explanation of galaxy formation. These do not yet fit neatly together; nor do they provide clear answers to many questions. But they provide a plausible path through the maze and may serve as a guide until a more complete theory of galaxy formation is developed.

13:1 The Inhomogeneous Universe

Averaged over sufficiently large scales, the Cosmos appears homogeneous and isotropic. Its mean density is everywhere the same; the kinds of structures we see differ little from each other. Yet, on small scales the Universe is far from homogeneous. Structure is apparent everywhere (Fig 1.12).

The range of velocities we observe reflects similar inhomogeneity. On the largest scales the Universe is expanding. On smaller scales, individual galaxies, small groups, and large clusters of galaxies remain gravitationally bound rather than taking part in the expansion.

What determines these differences in scale? How did the clusters, stretching across vast regions ever manage to aggregate in an explosively expanding Universe? How could this global expansion have been locally reversed so that matter would assemble itself into the clumpy structures ubiquitous today?

13:2 Primordial Seeds

In a classical paper written in 1946, E. M. Lifshitz showed that galaxy formation in homogeneous, isotropic expanding universes appears impossible unless we invoke substantial primordial fluctuations (Li46). Only if there were pre-existing inhomogeneities could we explain a propensity for collapse. Galaxy formation might be particularly simple if primordial seeds were present.

A quarter century later, E. R. Harrison (Ha70a) and Ya. B. Zel'dovich (Ze72) proposed that primordial fluctuations, present in the Cosmos since the earliest epochs, would survive the drastic expansion of the Universe to provide the seeds around which galaxies would later form once the expansion had sufficiently slowed. During the early era of rapid expansion, these fluctuations persist because pressures cannot propagate across cosmic horizons.

What Harrison had in mind when he made his proposal, well before Guth's (Gu81) inflationary theory had been introduced, were density fluctuations that were beyond each others' horizons and would grow as if they were separate universes. He argued that the amplitude of these fluctuations composed of a superposition of components having different wavelengths λ should decrease as λ^{-1}. This would prevent the formation of structures larger than any we observe in the Universe today (Ha70a). Fluctuations at small wavelengths could initially have been large, but were no great concern because they would eventually be damped out by diffusion and drag. Zel'dovich further postulated that the magnitude of the density fluctuations $\delta\rho/\rho$, which throughout the radiation-dominated era would have equaled $4\delta T/T$, should be of order 10^{-5} to 10^{-4}. These were sufficiently large to promote the aggregation of matter and formation of clusters of galaxies despite the continuing cosmic expansion following the decoupling of matter from radiation. They also were small enough to avoid widespread formation of massive black holes, which are observed to be relatively rare (Ze72). Observations of the microwave background largely bear out the order of magnitude of the fluctuations that Zel'dovich postulated but leave us no closer to explaining their origins.

13:3 The Seeds of Structure

In the inflationary model, the Universe goes through a rapid expansion during which neighboring portions of space disappear over each other's event horizons, carrying their locally imprinted density and velocity fluctuations with them. Once inflation comes to a halt, and all the energy in the inflationary fields is converted into matter and radiation, these fluctuations persist until previously isolated regions start to enter each other's particle horizons to begin locally interacting again. During both the radiation- and matter-dominated eras, the Hubble radius $R_H = c/H$, which we encountered in equation (11–80), remains directly proportional to the age of the Universe, in conformance with (11–56). It continues to grow as the cosmic expansion slows down. Fluctuations are smoothed out through the establishment of pressure equilibrium and dissipative effects, first in small regions and progressively over larger volumes.

Let us now see whether the inflationary sequence can help us to understand the formation of the clusters of galaxies and voids that constitute the small-scale inhomogeneities of the Universe. One possibility worth examining is whether thermal fluctuations at the extreme temperatures of the Planck era could have provided seeds of a magnitude $(\delta\rho)/\rho \sim 10^{-5}$ that Zel'dovich postulated.

Energy fluctuations within a volume V just before onset of inflation at epoch t_{oi}, i.e., just before the region becomes causally isolated from the rest of the Universe, can be derived from equation (4–118). If the temperature dependence of the enclosed energy density is proportional to $a_{oi}T^4$, as (12–38) implies, then

$$\frac{\langle(\Delta\mathcal{E})^2\rangle^{1/2}}{\langle\mathcal{E}\rangle} = \frac{(kT^2)^{1/2}}{\langle\mathcal{E}\rangle}\left(\frac{\partial\langle\mathcal{E}\rangle}{\partial T}\right)^{1/2} = \frac{(4ka_{oi}T^5V)^{1/2}}{(a_{oi}T^4V)} = \left(\frac{4kT}{a_{oi}T^4V}\right)^{1/2},$$

(13-1)

where a_{oi} is a constant whose magnitude judging by (12–38) is of order $N^{1/2}a/2$, where N is the number of particle species whose fluctuations we assume here to be uncoupled and to add in quadrature, and a is the radiation constant.

We saw in Section 12:6 that the temperature at onset of inflation was $T_{oi} \sim 10^{28}$ K, and that the diameter of a region in causal contact at the time was $\sim 10^{-25}$ cm, corresponding to a spherical volume defined by the Hubble radius R_{oi} at t_{oi}, $V_{oi} \sim 5 \times 10^{-76}$ cm^3. Substituting these values into (13–1) indicates that the root mean square mass–energy fluctuations for radiation over this volume would be of the order of

$$(\delta\rho)/\rho \equiv \langle(\Delta\rho)^2\rangle^{1/2}/\langle\rho\rangle = \langle(\Delta\mathcal{E})^2\rangle^{1/2}/\langle\mathcal{E}\rangle \sim 10^{-5}$$

(13-2)

provided $N^{1/4}$ is of order unity.

Although fluctuations roughly of this magnitude are observed in the cosmic microwave background radiation, equation (13–1) also requires the fluctuations to be proportional to $(T^3V)^{-1/2}$. We therefore need to follow the development of the fluctuations through the inflationary phase, reheating, and the initially relativistic and subsequently radiation-dominated eras that follow. The end of inflation releases vacuum energy that reheats the Universe to roughly the same temperature as at onset, T_{oi}. So, a volume V of the same size as V_{oi} at onset of-inflation, would have fluctuations of similar amplitude right after reheating. The subsequent evolution of the fluctuations, however, requires additional analysis, for which we first need to define a few concepts.

In the radiation-dominated era, radiation and particles are strongly coupled and behave as a fluid. Fluctuations in this fluid can be described by a superposition of standing-wave Fourier components of wavelength λ or spatial wave number $\mathbf{k} = (2\pi/\lambda)\varepsilon_{\mathbf{k}}$ in comoving coordinates. We assume that the phases of the different frequency components as well as their orientations in space are uncorrelated. Such fluctuations are said to be *Gaussian*. The volume element in wave number space is $d^3k = 4\pi k^2 dk$, reminiscent of the partition function (4–65) but — because these waves are acoustic, i.e., longitudinal — lacking the factor 2 that (4–65) included for the two states of polarization.

For a fluctuation with wave number $\mathbf{k}$ the amplitude of the density deviation, $\Delta\rho_{\mathbf{k}}$, from the mean value $\langle\rho\rangle$, can be positive or negative. The probability $P(|\Delta\rho_{\mathbf{k}}|^2)$ for the square of this amplitude to have the value $|\Delta_{\mathbf{k}}|^2$ is called the *power spectrum* of the fluctuations. The normalized standard deviation, i.e., the square of the *mass density contrast* $\delta\rho/\rho$, then is obtained by integrating over the probability of occurrence $|\Delta_k|^2$ of a fluctuation with wave number k,

$$\left(\frac{\delta\rho}{\rho}\right)^2 \equiv \frac{\langle(\Delta\rho)^2\rangle}{\langle\rho\rangle^2} \propto \int_0^{k_{max}} P(|\Delta_{\mathbf{k}}|^2)k^2 dk \ . \tag{13-3}$$

If the most probable value of $|\Delta_{\mathbf{k}}|^2$ is proportional to k^n, or λ^{-n}, that is,

$$\text{if } P(|\Delta_{\mathbf{k}}|^2) \propto k^n, \text{ then } (\delta\rho)/\rho)^2 \propto k_{max}^{n+3} \propto \lambda_{min}^{-(n+3)} \quad \text{for } n > -3 \ . \tag{13-4}$$

Here k_{max} is the maximum wave number and λ_{min} is the corresponding minimum wavelength of the fluctuations. For a standing wave with wavelength λ, regions of maximum density are separated by a distance λ, the space between them having lower density. A density enhancement, therefore, has a volume $V_{\lambda_{min}}$ of order λ_{min}^3, and for the fixed mean density $\langle\rho\rangle$ its mass M is proportional to $V_{\lambda_{min}}$.

$$\frac{\delta\rho}{\rho} \propto \lambda_{min}^{-(n+3)/2} \quad \propto V_{\lambda_{min}}^{-(n+3)/6} \quad \propto M^{-(n+3)/6} \ . \tag{13-5}$$

The spectrum postulated by Harrison and Zel'dovich has $n = 1$.

13:4 Evolution of Inhomogeneities

Given the existence of fluctuations, we need to ask how they evolve as the Universe expands. For a homogeneous isotropic space the Einstein field equations of general relativity reduce to (11–45), (11–46):

$$\frac{8\pi G\rho}{c^2} = -\Lambda + 3\left(\frac{\dot{a}^2 + kc^2}{c^2 a^2}\right) \tag{13-6}$$

and

$$\frac{8\pi GP}{c^4} = \Lambda - \left(\frac{2a\ddot{a} + \dot{a}^2 + kc^2}{c^2 a^2}\right), \quad \text{or} \quad \frac{d\rho}{dt} + 3\left(\frac{P}{c^2} + \rho\right)\frac{\dot{a}}{a} = 0, \tag{13-7}$$

where, as before, G and Λ are the gravitational and cosmological constants, k is the Riemann curvature constant, and ρ and P are the mass density and pressure of the Universe. Dots indicate differentiation with respect to world time.

We can rewrite equation (11–62), which is equivalent to (13–6) in terms of the deviations from a flat universe that fluctuations induce,

$$(1 - \Omega_T) \equiv 1 - (\Omega_\Lambda + \Omega_M + \Omega_r) = \Omega_k \equiv -\frac{kc^2}{H^2 a^2} . \qquad (13\text{-}8)$$

Here, Ω_T is the total density parameter summed over the mass of radiation, baryonic and dark matter, and dark energy, and Ω_r is the density parameter due to radiation, which was neglected in (11–62), where we had assumed a pressure $P = 0$.

$$\Omega_T = \left(\frac{(\rho_r + \rho_{DM} + \rho_B + \rho_\Lambda)}{\rho_{crit}} \right) \equiv \left(\frac{8\pi G}{3H^2} \right) \rho_T . \qquad (13\text{-}9)$$

While Ω_Λ appears to dominate this equation today, it was negligible at early times when the radiation and mass densities were far higher whereas a cosmological constant would have remained unchanged. Multiplying (13–8) by $a^2 \rho_{crit}$, we obtain

$$(\Omega_T - 1)\rho_{crit} a^2 \equiv a^2 \delta\rho_T = \frac{3kc^2}{8\pi G} = \text{constant} , \qquad (13\text{-}10)$$

where $\delta\rho_T$ is the deviation from the density of a flat universe. Fluctuations cannot have been damped on scales large compared to the distance that pressures propagated at the speed of sound c_s in a Hubble time $1/H$. Rather, on superhorizon scales, density fluctuations would have continued to expand precisely as in a homogeneous universe of density $\rho_T + \delta\rho_T$.

While equation (13–10) tells us that $(\delta\rho_T)a^2$ is constant, (11–54) shows that the overall density ρ_T of the Universe in the radiation-dominated era, also kept the product $\rho_T a^4$ very nearly constant. Accordingly, as long as the Universe was radiation-dominated,

$$\frac{\delta\rho}{\rho} = \frac{a^2}{a_i^2} \left(\frac{\delta\rho_i}{\rho_i} \right) \propto \frac{a^2}{a_i^2} M_i^{-(n+3)/6} , \qquad (13\text{-}11)$$

where we have made use of (13–5) and dropped the subscript T since the radiation density is tantamount to the total density. The subscript i denotes initial values at the onset of the radiation-dominated era; values without subscript refer to any later epoch during this era. The mass M_i in this equation refers to the mass scale, dominated by radiation, for which the initial fluctuation has a density contrast $(\delta\rho_i)/\rho_i$.

We should still note that the Universe may exhibit density fluctuations of different kinds. As noted in (13–11) the density contrast $\delta\rho/\rho$, which is always a positive quantity, increases in proportion to a^2 during the radiation-dominated era, for both positive and negative values of $\delta\rho$. For a globally flat universe, local values of Ω_T then deviate from unity, local values of Ω_k deviate from zero, and local values of the Riemann curvature constant fluctuate between $k = 1$ or -1, in passing from one locale to another.

Such fluctuations are called *curvature fluctuations*. A four-dimensional universe with curvature fluctuations has the appearance of a three-dimensional spherical hypersurface, with wrinkles protruding into a fourth dimension. It is as though we were viewing a four-dimensional orange, whose surface irregularities extend into the fourth dimension.

A different type of universe, which originally has the same value of Ω_T throughout, but with different ratios of Ω_r, Ω_{DM}, Ω_B, and Ω_Λ in different regions, is said to initially have *isocurvature fluctuations*, because the initial curvature is identical throughout. Only later, when the density of radiation has sharply decreased relative to dark or baryonic matter, will overall curvature fluctuations show up, and then only if the radiation and matter densities have insufficient time to redistribute themselves homogeneously.

During the radiation-dominated era, the curvature k and cosmological constant Λ in (13–6) can be neglected. Problem 11–16 then tells us that the particle horizon roughly equals the Hubble radius.

The particle horizon monotonically grows during this era, revealing progressively increasing portions of the Universe that were beyond view and beyond causal contact ever since onset of inflation. It is as though we were initially viewing just a tiny portion of a vast, enormously detailed map, remote regions of which remained hidden from us. As time went on, small ponds, then lakes, then seas, and finally entire oceans on this map would come into view to be discerned. Similarly, as the cosmic particle horizon grows, perturbations of ever-increasing size are said to "enter" the horizon — and thus come into view.

The rest-mass of matter the horizon already encloses when a perturbation having a scale equal to the Hubble radius $R_H = c/H$ comes into view, is predominantly that of dark matter, $M_{DM} \sim (4\pi/3)\rho_{DM}(c/H)^3$. While $\rho_{DM} \propto a^{-3}$ use of equation (11–55) tells us that $(c/H)^3 \propto a^6$, so that $M_{DM} \propto a^3$ and, from (13–11), the fluctuations of matter and matter density have contrast

$$\left(\frac{\delta M}{M}\right) = \left(\frac{\delta\rho}{\rho}\right) \propto M_{DM}^{2/3} M_i^{-(n+3)/6} . \qquad (13\text{-}12)$$

If dark matter, baryonic matter, and radiation are thoroughly mixed throughout the radiation-dominated era, the initial mass spectrum expressed in terms of M_i is also a good representation of the mass spectrum of its different components. As the perturbation enters the particle horizon, the dark matter has not yet begun to flow and aggregate relative to its surroundings. Hence, it does not yet exert a gravitational tug that will eventually redistribute the ambient plasma and its coupled radiation field. The left side of (13–12) accordingly remains a good approximation to the density contrast in both the dark and baryonic matter. This simplification permits us to rewrite the mass spectrum for fluctuations. Dropping all subscripts,

$$\left(\frac{\delta M}{M}\right) \propto M^{2/3} M^{-(n+3)/6} = M^{(1-n)/6} . \qquad (13\text{-}13)$$

By concentrating on increasingly large fluctuations, matched in size to the growing particle horizon at the time they enter it, we have shown that equation (13–13) holds for fluctuations on all scales. For the Harrison–Zel'dovich spectrum, which has $n = 1$, (13–13) tells us that the amplitudes of density and matter fluctuations are independent of mass. This independence is called *scale invariance*.

13:5 The Coupling of Radiation and Matter

Let us return to Tables 12.1 and 12.3 where we displayed the evolution of the number densities of photons, the radiation temperatures, and the densities of baryons as functions of red shift z to show how they attained their currently observed values. The temperature at decoupling is also displayed. As the Cosmos cools through the temperature range from 10^4 and 10^3 K, the density of photons capable of ionizing hydrogen drops by more than 50 orders of magnitude. As Fig. 13.1 shows, around

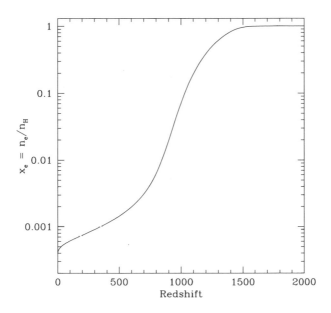

Fig. 13.1. The ratio of electrons to hydrogen atoms as a function of red shift for the assumed parameters $\Omega_T = 1.0, T_0 = 2.725 K, \Omega_B = 0.04, \Omega_A = 0.73$, a helium abundance by mass, $Y = 0.24$, and a Hubble constant $H_0 = 70$ km s^{-1} Mpc^{-1}. Courtesy of Sara Seager, who provided this updated version of an earlier plot by Seager, Sasselov, and Scott (Se99).

4000 K or $z \sim 1465$ the ionization rate per atom drops precipitously, while the recombination rate changes much more slowly. Electron and ion densities progressively decline, although a low level of ionization persists forever.

We can understand this in terms of equation (4–72). At temperature $T \ll h\nu/k$ the number density of photons n_i with energies above the ionization energy for hydrogen, $h\nu_i$, is

$$n_i \sim \int_{\nu_i}^{\infty} \frac{8\pi\nu^2}{c^3} e^{-h\nu/kT} d\nu. \qquad (13\text{-}14)$$

The ionization rate per atom then is

$$-\frac{1}{n_H}\frac{dn_H}{dt} = n_i c \sigma_i, \qquad (13\text{-}15)$$

where σ_i is the ionization cross-section at the *ionization edge* $\sim 10^{-17}\,\text{cm}^{-2}$. Since both the number of photons in this temperature range and the ionization cross-section rapidly drop above the ionization edge, use of this cross-section suffices. At $10^4\,\text{K}$, $n_i c\sigma_i \sim 82\,\text{s}^{-1}$. By $4000\,\text{K}$, the rate has dropped to $1.7 \times 10^{-9}\,\text{s}^{-1}$, and at $3000\,\text{K}$ it is $2.4 \times 10^{-15}\,\text{s}^{-1}$, meaning that the ionization rate drops below the Hubble expansion rate which dominates the photon cooling.

As Table 9.2 and equation (9–5) show, the compensating recombination rate is

$$n_e \alpha_c \sim 3 \times 10^{-13}\left(\frac{10^4}{T}\right)^{1/2} x n_H \quad \text{s}^{-1}, \qquad (13\text{-}16)$$

where x is the *ionization fraction* $x \equiv n_e/n_H$, and α_c is the sum of recombination coefficients to all excited states except the ground state, which produces an ionizing photon and, therefore, does not lead to permanent recombination. Below $4000\,\text{K}$, x rapidly declines. At red shift $z \sim 10^3$, it is $\sim 10\%$. By $z \sim 900$, as Fig. 13.1 indicates, x has dropped to below 2%. From Table 12.1, we can obtain the hydrogen density at this epoch as $n_H \sim 200\,\text{cm}^{-3}$, so that $n_e \sim 4\,\text{cm}^{-3}$. Then $n_e \alpha_c \sim 2.4 \times 10^{-12}\,\text{s}^{-1}$, and the age of the Universe is $\sim 10^{13}\,\text{s}$. Electrons and protons thus continue to combine for a little while longer, with essentially no competing ionization until the ionization fraction drops to $\leq 10^{-4}$. There the recombination rate also drops below the Hubble expansion rate and x remains at a value between 10^{-4} and 10^{-5}.

13:6 Cooling of Gas After Decoupling

Given that the adiabatic constants for radiation and atomic matter, respectively, are $\gamma = \frac{4}{3}$ and $\frac{5}{3}$, we might expect from equations (4–129) that, after decoupling, the radiation temperature would drop in proportion to $(1+z)$, whereas the gas temperature would drop faster, in proportion to $(1 + z)^2$. However, even when decoupling is almost complete, the gas can still be heated by photons, even as it cools adiabatically, because the photon number exceeds the number of atoms by a factor of $\sim 10^9$. The heating is indirect. The photons can only share their energy through Thomson scattering off the very small fraction x of electrons that have not combined with protons to form atoms. But the Thomson scattering cross-section is sufficiently high, and the photon energy density $\rho_{\text{rad}} = aT^4$ is large enough, so that the energy gained by the electrons through scattering can be shared in subsequent collisions with atoms to keep the gas heated despite appreciable cooling through cosmic expansion.

For a nonrelativistically moving electron, the apparent temperature of the microwave background radiation at an angle θ relative to the electron's direction of motion is $T(\theta) = T(1 + (V/c)\cos\theta)$, where the low velocities permit us to neglect the factor $\gamma(V)$ in equation (5–49). Here T is the rest-frame temperature at any particular red shift z. Referring to the coordinate system of Fig. 5.8, except that we designate the polar angle by θ rather than θ', we can see that the net momentum transferred to an electron per unit time by radiation incident from an annular element of solid angle $2\pi \sin\theta d\theta$ is

$$\frac{dp}{dt} = \frac{aT(\theta)^4}{4\pi}\sigma_e \frac{V}{c}(\cos\theta)2\pi \sin\theta d\theta . \tag{13-17}$$

Here σ_e is the Thomson scattering cross-section, and the term $(\cos\theta)$ gives the non-cancelling momentum component directed opposite to the motion. This produces a drag force $\mathcal{F}$ obtained by integrating over all angles

$$\mathcal{F} = \int \frac{aT^4\sigma_e}{2}\cos\theta \left(1 + 4\frac{V}{c}\cos\theta\right)\sin\theta d\theta = \frac{4aT^4\sigma_e V}{3c} . \tag{13-18}$$

The work done on the radiation bath by the motion of the electron per unit time is $\mathcal{F}V = 4aT^4\sigma_e V^2/3c$. If the temperature of the electrons is T_e, the mean square velocity is $\langle V^2\rangle = 3kT_e/m_e$, where m_e is the electron mass and k is the Boltzmann constant. The work done on the radiation bath then becomes $\mathcal{F}V = 4aT^4\sigma_e kT_e/m_e c$, and represents the energy loss of the electrons. If the electrons and radiation were in thermal equilibrium, the heating of the electrons by the radiation would just equal the loss of energy from the electrons to the radiation. So, when the gas and radiation temperatures differ, the energy exchange between the two is

$$\frac{d\mathcal{E}}{dt} = \frac{4akT^4\sigma_e}{m_e c}(T - T_e) . \tag{13-19}$$

For an ionization fraction $x \ll 1$, the electrons share their energy with $\sim 1/x$ atoms, so that the temperature of the atoms changes at a rate

$$\frac{dT_a}{dt} = \left(\frac{x}{1+x}\right)\frac{2}{3k}\frac{d\mathcal{E}}{dt} = \left(\frac{x}{1+x}\right)\frac{8aT^4\sigma_e}{3m_e c}(T - T_e). \tag{13-20}$$

The heating rate is a sensitive function of the radiation density and the relative fraction of atoms and electrons but remains substantial until the epoch $z \sim 200$ when the radiation temperature is still several hundred degrees. At lower red shifts, however, radiative heating becomes negligible and the gas does adiabatically cool in proportion to $(1+z)^2$.

13:7 Photon Drag

Before decoupling, radiation and matter form a tightly bound fluid. The radiation scattered by an electron per unit time has energy

$$\frac{d\mathcal{E}}{dt} = aT^4 c\sigma_e. \tag{13-21}$$

As (13–18) shows, the drag on electrons then reduces any systematic velocity v of atomic matter with respect to the ambient radiation bath, at a rate

$$\left(\frac{1}{v}\right)\frac{dv}{dt} = \frac{d\ln v}{dt} = \frac{4aT^4x\sigma_e}{3m_H c} , \tag{13-22}$$

where, for simplicity, we have acted as though hydrogen were the only major constituent and have neglected corrections to account for the admixture of helium.

For largely neutral gas, the main mode of interaction between electrons and atoms is *charge exchange*, in which an incident unbound electron exchanges places with an electron originally bound to its hydrogen nucleus. If the impinging electron has substantial momentum, it transfers this to the nucleus to which it now becomes bound. The charge exchange cross-section is $\sim 10^{-15}$ cm^2, large compared to most other atomic cross-sections. In Section 13:5, we saw that at $z \sim 900$ the ionization fraction is $x \sim 2 \times 10^{-2}$ and the density of neutral atoms is $\sim 2 \times 10^2$ cm^{-3}. An electron moving with speed $(3kT/m)^{1/2} \sim 2 \times 10^7$ cm s^{-1} then interacts with an atom roughly every 2.5×10^5 s, which is rapid compared to the time scale of cosmic evolution.

PROBLEM 13–1. Using Tables 12.1 and 12.3, Fig. 13.1, and the work of Section 11:12, convince yourself that the time constant for damping motion is only of order 2×10^7 s at red shift $z = 10^4$, much shorter than the age of the Universe at that time, roughly 2×10^{11} s. By comparison, at $z \sim 800$, the damping time constant has increased to $\sim 7 \times 10^{13}$ s because both the radiation density and the ionization fraction have dropped. The damping time is then considerably greater than the instantaneous age of the Universe $\sim 2 \times 10^{13}$ s, and matter can move almost freely through the radiation background.

The radiation drag experienced by matter before decoupling is referred to as *Silk drag* or *Silk damping*. It efficiently damps out small-scale turbulence and oscillations (Si68).

PROBLEM 13–2. (a) Using Tables 12.1 and 12.3 convince yourself that at $z \sim 10^4$ a photon can travel though the plasma no more than a distance of $\sim 10^{18}$ cm, or for a time ~ 1 yr, before being Thomson scattered. Compare this distance to the scale of the cosmic horizon at this epoch to see that matter and radiation act as a tightly coupled fluid. (b) Show that the speed of sound in this photon/baryon plasma can be obtained from (9–25), (4–43), (4–125) and (4–129), and is

$$c_s = \left(\frac{\partial P_r/\partial T + \partial P_B/\partial T}{\partial \rho_r/\partial T + \partial \rho_B/\partial T}\right)^{1/2} = \left(\frac{(4/3)\rho_r c^2 + (5/3)(1+x)\rho_B kT/m_H}{(4\rho_r + 3\rho_B)}\right)^{1/2}. \tag{13-23}$$

The mass density of dark matter does not enter this equation because it does not affect the dependence of pressure on density. (c) Show that well before decoupling, when $\rho_r = aT_0^4(1+z)^4/c^2 \gg (3H_0^2/8\pi G)\Omega_B(1+z)^3 = \rho_B$, the speed of sound remains essentially constant at $c_s = c/\sqrt{3} \sim 1.7 \times 10^{10}$ cm s^{-1}. As Table 12.1 indicates, this is the situation at $z \gtrsim 10^4$. Show also that at decoupling, where $(z + 1) \sim 1100$, the radiation pressure still dominates and, although the Universe is now matter dominated, the baryonic term ρ_B in the denominator of (13–23) is still somewhat smaller than ρ_r. Convince yourself that the speed of sound just before decoupling is $\sim [(4/3)(aT_0^4(1+z)]^{1/2}/[(3H_0^2/8\pi G)\Omega_B+4aT_0^4(1+z)/c^2]^{1/2}$ $\sim 1.4 \times 10^{10}$ cm s^{-1} , while after decoupling it is $(5(1 + x)kT/9m_H)^{1/2}$ $\sim 4 \times 10^5$ cm s^{-1}. Figure 13.1 can be used to obtain $x \sim 0.2$.

13:8 Oscillations Around the Decoupling Era

Before decoupling, Silk damping firmly couples matter to radiation, giving rise to a plasma that can oscillate — contracting in response to gravitational self-attraction, rebounding as the radiation pressure rises. Such oscillations are acoustic waves. To see this we need to look more carefully at the gas dynamics. We may use the same approach and notation as in Sections 9:3 and 10:3 but will have to keep in mind that we now are dealing with a perturbed fluid comoving with an expanding universe. We begin with the *continuity equation*

$$\frac{\partial \rho}{\partial t} + \nabla \cdot (\rho \mathbf{v}) = \frac{\partial \rho}{\partial t} + \rho \nabla \cdot \mathbf{v} + \mathbf{v} \cdot \nabla \rho = 0 , \tag{13-24}$$

where ρ is the mass density. We also need the *Euler equation*

$$\frac{\partial \mathbf{v}}{\partial t} + (\mathbf{v} \cdot \nabla) \mathbf{v} + \frac{\nabla P}{\rho} + \nabla \phi \equiv \frac{d}{dt}\mathbf{v} + \frac{\nabla P}{\rho} + \nabla \phi = 0 , \tag{13-25}$$

and the *Poisson equation*

$$\nabla^2 \phi = 4\pi G \rho . \tag{13-26}$$

As in Chapters 9 and 10, we again consider a perturbed fluid with density $\rho = \rho_0 + \rho_1$, pressure $P = P_0 + P_1$, velocity $\mathbf{v} = \mathbf{v_0} + \mathbf{v_1}$, and gravitational potential $\phi = \phi_0 + \phi_1$, where subscripts 0 and 1, respectively, indicate the unperturbed and perturbed states.

We are interested in conditions around the decoupling era, where the Universe is already mass dominated. We take the unperturbed velocity to be given by the Hubble flow. Then, at any epoch t, the unperturbed conditions in the adiabatic flow are

$$\rho_0(t) = \rho_0(t_0)\left(\frac{a(t_0)}{a(t)}\right)^3 , \quad \mathbf{v_0} = \frac{\dot{a}}{a}\mathbf{r}, \quad \frac{\partial P}{\partial \rho}\bigg]_S = c_s^2, \quad \nabla \phi_0 = \frac{4\pi G\rho_0(t)}{3}\mathbf{r} ,$$
$$\tag{13-27}$$

where t_0 refers to the present era, and the scale factor a satisfies (13–6) and (13–7). The third equality corresponds to (9–25), with c_s the speed of sound, and the fourth represents the force field in the Hubble flow.

Provided we restrict ourselves to regions well within the Hubble radius, where wavelengths are short with respect to R_H, we can carry out a perturbation analysis using a Newtonian approximation. This restriction implies that we do not have to worry about fluctuations within which expansion velocities approach or exceed the speed of light. Nor do retardation effects and their accompanying causal uncertainties arise for regions small compared to the Hubble radius, where signals can propagate in time intervals short compared to the age of the Universe.

In carrying out the Newtonian approximation, it is useful to first identify the unperturbed solutions of equations (13–24) to (13–26). These, respectively, reduce to

$$d\rho_0/dt + (\dot{a}/a)\rho_0 \nabla.\mathbf{r} = 0 ,$$

$$(\ddot{a}/a) + 8\pi G\rho_0/3 = 0 ,$$

both consistent with (13–6) and (13–7) under Newtonian conditions, where pressure terms are small and the cosmological constant can be ignored. The unperturbed Poisson equation is

$$\nabla^2\phi_0 = 4\pi G\rho_0 .$$

With these primary expressions identified, we can proceed to examine the smaller terms in the first-order perturbed equations (Ko90*). The perturbed continuity equation becomes

$$\frac{\partial\rho_1}{\partial t} + 3\frac{\dot{a}}{a}\rho_1 + \frac{\dot{a}}{a}\mathbf{r} \cdot \nabla\rho_1 + \rho_0\nabla\mathbf{v_1} = 0 . \tag{13-28}$$

where $\dot{a}/a$, ρ_0, ρ_1, and $\mathbf{v_1}$ all refer to time t. The perturbed first-order Euler equation is

$$\frac{\partial\mathbf{v_1}}{\partial t} + \frac{\dot{a}}{a}\left(\mathbf{v_1} + \mathbf{r} \cdot \nabla\mathbf{v_1}\right) + \frac{c_s^2}{\rho_0}\nabla\rho_1 + \nabla\phi_1 = 0 , \tag{13-29}$$

The first-order perturbations of the potential obey the Poisson relation

$$\nabla^2\phi_1 = 4\pi G\rho_1 . \tag{13-30}$$

These three equations can be simplified by postulating that the perturbations in the density contrast $\delta \equiv \rho_1/\rho_0$, velocity $\mathbf{v_1}$, and potential ϕ_1, all exhibit an identical form, a superposition of sinusoidal components with wave number $\mathbf{k}$ in the volume of interest V.

$$\varphi(\mathbf{r}, t) = (2\pi)^{-3} \int \varphi_k e^{-ia(t_0)\mathbf{k}\cdot\mathbf{r}/a(t)} dV . \tag{13-31}$$

Here φ stands for δ, $\mathbf{v_1}$, or ϕ_1. Since $\mathbf{r}[a(t_0)/a(t)]$ is seen to be a comoving coordinate, $\mathbf{k}$ is a corresponding comoving wave number, and the wavelength of the oscillations expands with the Universe. Alternatively, if $\mathbf{r}$ is taken to be a vector whose physical length is measured in megaparsecs, $[a(t_0)/a(t)]\mathbf{k}$ has to be considered a physical wave number with dimensions given in reciprocal megaparsecs, Mpc^{-1}. The corresponding physical wavelength $\lambda = 2\pi(a(t)/a(t_0)k)$ then has the same units of length as $\mathbf{r}$.

PROBLEM 13–3. Show that, with (13–31), equations (13–28) to (13–30) reduce, respectively, to (13–32) to (13–34). Using the notation $\dot{\delta} \equiv d\delta/dt = (\partial\delta/\partial t) + v.\nabla\delta$ throughout, these read:

$$\dot{\delta}_k - \frac{ia(t_0)\mathbf{k}}{a(t)} \cdot \mathbf{v_k} = 0, \qquad \ddot{\delta}_k - i\left(\frac{a_0}{a}\right)\mathbf{k} \cdot \dot{\mathbf{v}}_k + \left(\frac{\dot{a}}{a}\right)\dot{\delta}_k = 0, \qquad (13\text{-}32)$$

where $\mathbf{v_k}$ is the perturbed velocity component having wave number k,

$$\frac{d\mathbf{v_k}}{dt} + \left(\frac{\dot{a}}{a}\right)\mathbf{v_k} - i\mathbf{k}\left(\frac{a(t_0)}{a(t)}\right)\left(c_s^2 + \phi_k\right)\delta_k = 0, \qquad (13\text{-}33)$$

and

$$\phi_k = -\frac{4\pi G\rho_0}{k^2}\left(\frac{a(t)}{a(t_0)}\right)^2 \delta_k. \qquad (13\text{-}34)$$

Further show that these three equations can be combined to yield a differential equation for the evolution in density contrast, δ_k:

$$\ddot{\delta}_k + 2\left(\frac{\dot{a}}{a}\right)\dot{\delta}_k + \left[(c_s k)^2\left(\frac{a(t_0)}{a(t)}\right)^2 - 4\pi G\rho_0\right]\delta_k = 0. \qquad (13\text{-}35)$$

In deriving these results, it is useful to remember that we are dealing with acoustic, i.e., longitudinal, waves and that this affects the calculation of divergences.

We can construct any arbitrary perturbation by superposing Fourier components of the form (13–31). Accordingly, equations (13–32) to (13–35) will hold for any perturbation if the individual terms are summed over wave number components $\mathbf{k}$. The velocity components included in these equations can be either transverse to or aligned with $\mathbf{r}$ and $\mathbf{k}$ and represent, respectively, rotational or irrotational motions. However, only the irrotational modes affect the evolution of the density contrast, as equation (13–32) indicates.

It is important to note that the mass density ρ in equations (13–27) to (13–35) refers to the sum of the densities of dark matter ρ_{DM} and baryonic matter ρ_B, and that the dark matter might have several different components. When both baryonic and dark matter are present, and the dark matter may comprise several different weakly interacting constituents, we can more generally write separate equations of the form (13–35) for each component i,

$$\ddot{\delta}_{ik} + 2\left(\frac{\dot{a}}{a}\right)\dot{\delta}_{ik} + \left[(c_{si}k)^2\left(\frac{a(t_0)}{a(t)}\right)^2 \delta_{ik} - 4\pi G\rho_0 \sum_j \frac{\rho_j}{\rho_0}\delta_{jk}\right] = 0, \qquad (13.35\text{a})$$

where c_{si} is the speed at which sound propagates through the ith component. ρ_j/ρ_0 is the fractional mass density of the jth component. Not knowing the nature of the dark matter we can only make assumptions about the rate at which sound propagates through it or — if it has several constituents — the speed of sound in each.

Notice that when there is no cosmic expansion, $\dot{a}/a = 0$, we recover equation (10–9) and consequently also the Jeans length (10–13) and Jeans mass (10–17).

In the matter-dominated cosmological model with $\Omega_0 = 1$, (11–58) tells us that $\dot{a}/a = 2/3t$, so that (13–6) yields $4\pi G\rho_0 = 2/3t^2$. We can then identify two different regimes. For large values of $\mathbf{k}$ (i.e., small wavelengths) gravitational effects can be neglected in (13–35) and (13–35a), and baryonic density perturbations will oscillate. For $\Omega_0 = 1$, $a(t)/a(t_0) = (t/t_0)^{2/3}$, the oscillation takes on the form

$$\delta(t) \propto e^{\pm i\omega t} \quad \text{with} \quad \omega = 3c_s k[a(t_0)/a(t)] . \tag{13-36}$$

Equation (10–12) shows that oscillating waves propagate at a fraction of the speed of sound. The maximum distance they can travel in a Hubble time is $\sim c_s/H$. If we consider this to be the time required to propagate half a wavelength, to effect maximum condensation or maximum rarefaction, the longest observable wavelength has twice this dimension $\sim 2c_s/H$. At $z = 10^4$, where $c_s \sim c/\sqrt{3}$, this maximum wavelength is $2c/[H\sqrt{3}] \sim 1.6 \times 10^{22}$ cm, or just over 5 kpc for the Hubble constant listed in Table 12.3. The baryonic Jeans length at this epoch is $c^2[\pi/(3aT^4G)]^{1/2} \sim 20$ kpc, and thus appreciably larger than the Hubble radius. This tells us that none of the fluctuations entering the particle horizon at this or earlier epochs steadily contract, but rather will oscillate. Even just before decoupling, at $z \sim 1100$, the Jeans length still is roughly comparable to the Hubble radius.

Regions of excess density tend to expand, setting up pressure gradients in the radiation-coupled fluid that provide a restoring force to imposed oscillations. Because of the tight coupling between radiation and matter, the density oscillations are also temperature oscillations with heating and cooling, respectively, marking the compressed and rarefied phases of a standing sound wave. In the absence of strong viscous damping the oscillations continue until decoupling of matter from radiation at the epoch of recombination.

As already mentioned in Chapter 12, the term *recombination epoch* is unfortunately misleading. It implies that there was a prior epoch when electrons and protons had been combined as neutral hydrogen. In modern cosmology no such epoch ever existed. Electrons and protons never combined into atoms until after $\sim$400,000 years of continuing expansion, when the temperature finally fell below 4000 K.

Oscillating modes that happen to be at maximum or minimum density contrast at decoupling, when radiation becomes free to escape a density inhomogeneity, correspond to peaks in the spatial power spectrum observed in the temperature distribution across the sky. A wave that completes half an oscillation by decoupling, and is thus caught at an extremum, sets the physical scale of the first peak in the power spectrum. Both maxima and minima contribute to the peaks in the power spectrum, and successive peaks thus correspond to integral multiples of this oscillation period. Because the speed of sound is independent of the wavelength, the period of the oscillation is proportional to the wavelength, and successive peaks in the power spectrum correspond to acoustic waves with wavelengths that are integer fractions of the wavelength — i.e., integer multiples of the wavenumber — at the first peak.

At decoupling, $(z + 1) \sim 1100$, the speed of sound obtained in Problem 13–2 is $\sim 1.4 \times 10^{10}$ cm s^{-1}, so that the maximum wavelength that can appreciably grow in a Hubble time is $\lambda_{max} \sim 2c_s/[H_0(z + 1)^{3/2}] \sim 3.2 \times 10^{23}$ cm ~ 110 kpc. The dimension of either the rarefaction or the condensation part of this wave, however, is just half the wavelength, $\lambda_{max}/2$. From equation (11–31) we see that the angle δ_{max} subtended by this half wavelength at decoupling, viewed from our distance today, is

$$\delta_{max} \sim \frac{\lambda_{max}}{2[a_0\sigma(\chi)/(z + 1)]} \sim \frac{c_s}{(z + 1)^{1/2}c} \sim 0.015 \text{ radians} \sim 0.85° . \quad (13\text{-}37)$$

The first peak in the power spectrum should correspond to roughly this angular extent.

In terms of the spherical harmonics defined in Section 6:25, we can associate a feature of angular extent δ_{max} on the sky with the first harmonic $\ell_1 \sim \pi/\delta_{max}$. Higher harmonics should then be given by

$$\ell_n \sim n\ell_1, \text{ where } \ell_1 = \frac{\pi}{\delta_{max}} \sim 210 . \quad (13\text{-}38)$$

The actually observed value of the microwave background fluctuation peak lies close to $\ell_1 \sim 220$, in reasonable agreement with our rough estimate. The peak is broad $100 \leq \ell \leq 350$, with a peak temperature amplitude $\Delta T \sim 75\mu$K (Pa03). On larger scales than this (i.e., $\ell < 90$ or equivalently $\theta > 2°$) patches of the sky will have been out of causal contact at decoupling. Fluctuations on those scales should, therefore, reflect the unprocessed primordial spectrum. In conformance with the hypothesis of Zel'dovich in Section 13:1, we should expect to see fluctuations of order 10^{-5} in the surface brightness and temperature of the microwave background radiation on these larger scales. We will discuss the fluctuation spectrum in more detail in Section 13:16. In the meantime, however, a glance at Fig. 13.8 may elucidate the extent to which observations support the theoretical structure we just established.

13:9 The Jeans Criterion

Thus far, we have focused our attention on small wavelength oscillations — oscillations with large wave numbers. Returning to (13–35) we see that, for small physical wave numbers, $[a(t_0)/a(t)]k^2 \ll [a(t_0)/a(t)]k_J^2 = (2\pi/\lambda_J)^2 \equiv 4\pi G\rho_0/c_s^2$, wavelengths become longer than the Jeans length,

$$\lambda_J = c_s \left(\frac{\pi}{G\rho_0(t)}\right)^{1/2} = c_s \left[\frac{\pi}{G\rho_0(t_0)} \left(\frac{a(t)}{a(t_0)}\right)^3\right]^{1/2} , \quad (13\text{-}39)$$

which we note is the same as the Jeans length in a nonexpanding medium. In this regime, terms in k^2 can be neglected and, for a flat space, where $\dot{a}/a = 2/3t$ and $\rho_0 = 3H_0^2/8\pi G$, (13–35) yields

$$\ddot{\delta}_k + \left(\frac{4}{3t}\right)\dot{\delta}_k - \left(\frac{2}{3t^2}\right)\delta_k = 0 . \qquad (13\text{-}40)$$

PROBLEM 13–4. Show that this has a *power-law* solution of form $\delta = At^n$ for the density contrast

$$n(n-1) + \frac{4}{3}n - \frac{2}{3} = 0 \qquad (13\text{-}41)$$

meaning that both $n = -1$ and $n = 2/3$ are solutions.

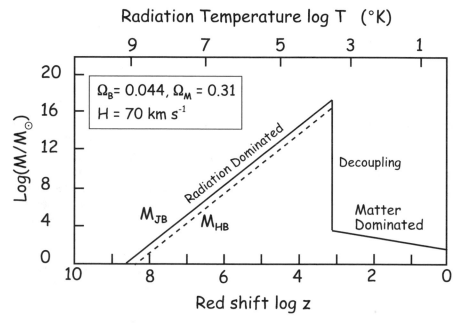

Fig. 13.2. The baryonic Jeans mass M_{JB} and the baryonic mass within the Hubble radius M_{HB} as a function of red shift. Based on a drawing by E. W. Kolb and M. S. Turner (Ko90).

The first of these is a decaying mode that dies out, whereas the second continues to grow. However, in contrast to the exponential growth of a Jeans mass in a non-expanding interstellar cloud, the growth of a cosmologically expanding excess-density region proceeds at merely the $\frac{2}{3}$ power of time. This is slow because the cosmic expansion works against gravitational contraction.

PROBLEM 13–5. (a) Knowing the Jeans length for the plasma from (13–39) obtain the baryonic Jeans mass given by (10–17) as $M_{JB} = (4\pi/3)\rho_B(t)(\lambda_J/2)^3$ and show that it increases as $\sim(z+1)^{3/2}$ from low red shifts to the epoch of decoupling, by assuming that the temperature of the baryons rises roughly as $(z + 1)^2$ over this epoch. (b) Show also that, for $H_0 = 70 \, \text{km s}^{-1} \, \text{Mpc}^{-1}$, $\Omega_T = 1$, $\Omega_B = 0.044$ and $\Omega_M = 0.31$, M_{JB} rises dramatically, by many orders of magnitude, as one proceeds to higher red shifts. The era of full coupling of radiation to matter is reached over a narrow redshift range about $(z + 1) \sim 1,100$, where the speed of sound rather abruptly rises from $\sim 2.5 \times 10^5$ to $\sim 1.4 \times 10^{10} \, \text{cm s}^{-1}$, as seen from Problem 13–2. (c) Show that the baryonic Jeans mass after decoupling is $\sim 5 \times 10^4 M_\odot$ and that this is well within the Hubble radius $c/H \sim 3.5 \times 10^{23} \, \text{cm}$ at this epoch. (d) In contrast, show that the M_{JB} just before decoupling is higher by the rise in c_s^3, a factor of $\sim 1.8 \times 10^{14}$, for a total baryonic mass of $\sim 9 \times 10^{18} M_\odot$, and that the Jeans length abruptly rises to $\sim 1.5 \, \text{Mpc}$, which exceeds the Hubble radius $R_H = c/H$ at that epoch by a factor of ~ 13. Figure 13.2 illustrates this evolutionary sequence. (e) Finally, show that, before decoupling, M_{JB} diminishes in proportion to $(1 + z)^{-3}$ as one goes farther back in time, and that the Hubble radius declines in the same proportion, so that the baryonic mass within the horizon is always three orders of magnitude below the Jeans mass.

Because the Jeans length diminishes for low ratios of c_s^2/ρ, contraction of condensations is most likely to be induced by *dark matter* (Ki05). Though collisionally decoupled from ordinary matter and radiation, and contributing neither to the pressure nor the speed of sound, dark matter dominates density ρ_0 in (13–35) and hence may be expected to lower the Jean length and Jeans mass. However, very little is known about the physics of dark matter.

13:10 Condensation on Superhorizon Scales

At first glance, large-scale initial fluctuations imprinted on the Universe before inflation might appear incapable of growing, since they would extend beyond the particle horizon and apparently be out of physical contact and unable to contract. This view, however, is false and we need to understand why.

Let us consider fluctuations present in the earliest phases of the Universe, when the expansion of the Universe may have been slow just before onset of inflation. At this early epoch, the gravitational potential from any mass condensation reaches out to a particle horizon that may be at much greater distances than the horizon encountered later, during the decoupling era. The gradient of this initially established gravitational potential can be envisaged in terms of lines of force marking the gravitational force field. As inflation progresses and the Universe expands at exponentially increasing rates, these lines of force remain frozen in comoving space. No dissipative mechanism exists to wipe them out. Thus, throughout inflation and the post-inflationary period, matter in widely separate regions of the Universe remains gravitationally coupled even though separated by many times the distance to the instantaneous Hubble radius. Matter and radiation will keep flowing, albeit slowly, along the primordially imprinted gravitational lines of force regardless of the circumstance that the attracting center lies outside the particle horizon. Perturbations on such *superhorizon scales*, therefore, continue to grow. At a speed of sound c_s, which we saw to be a significant fraction of the speed of light c, fluctuations can grow appreciably during the Hubble time just before decoupling. Fluctuations at substantially longer wavelengths than R_H will not significantly grow. The fraction of a wavelength the fluid can move in unit time c_s/λ diminishes with increasing wavelength. Because fundamental oscillations quite generally are accompanied by higher harmonics, these harmonics more readily grow and leave a pronounced mark on the spatial fluctuation spectrum in the microwave background radiation seen in Fig. 13.8. We may, therefore, expect to see significant spatial structure in the microwave background spectrum that peaks on a scale comparable to the particle horizon at decoupling, which currently subtends an angle of order $\sim 1°$ on the sky. (Hu97)*.

Let us still return to the discussion on the potential seeds of structure that we left unsettled in Section 13:3. Extrapolating back in time, we note that the comoving diameter of a region subtending $1°$ on the sky is $\sim 10^{26}$ cm today. In Section 12:7, we estimated that the end of inflation occurred around $z \sim 10^{27.5}$, so that this region would have spanned a diameter of $1/30$ cm at that epoch and a volume $V_{ei} \sim (1/30)^3$ cm^3. Judging from (13–1) and (13–2) a region of this size reheated to $T_{oi} \sim 10^{28}$ K (i.e., 10^{15} GeV) would have had thermal fluctuations of order $(\delta\rho)/\rho \sim 10^{-5}(V_{oi}/V_{ei})^{1/2} \sim 10^{-40.5}$ at that epoch. But (13–11) suggests that these fluctuations should have grown in proportion to a^2 between the end of inflation and the end of radiation domination (i.e., by a factor of 10^{48}) making $(\delta\rho)/\rho \sim 10^{7.5}$ — many orders of magnitude greater than observed in the microwave background. Although this is disconcerting, we are dealing with several parameters that still are uncertain by many orders of magnitude, the temperature and number of particle species at reheating, the red shift at end of inflation, etc. However, the indications at least are that readily explained fluctuations at primordial times could account for the seeds around which cosmic structure eventually evolved.

13:11 A Swiss-Cheese Model

Let us now turn to the question of how curvature fluctuations of the type that lead to acoustic oscillations or eventual collapse may be incorporated into a general relativistic model of the Universe. We will look into this question in some detail with a model that is simple but that, nevertheless, permits us to gain quantitative insight into the effects of Riemann curvature k, a cosmological constant Λ, and other factors. Given the rate at which current models of the Universe are changing in response to observations, it is worthwhile having such a simple model with which the effects of different factors, such as the influence of quintessence, might be estimated analytically, without the need for a full simulation with more powerful, but possibly less transparent, computer models.

In Chapter 5, we saw that empty space surrounding an isolated mass M is characterized by the *Schwarzschild* metric, equation (5–62),

$$ds^2 = \left(c^2 - \frac{2MG}{r}\right) dt^2 - \left(1 - \frac{2MG}{rc^2}\right)^{-1} dr^2 - r^2(\sin^2 \theta \, d\phi^2 + d\theta^2). \quad (13\text{-}42)$$

This metric defines the trajectories of particles and the paths along which light beams propagate in an empty Universe surrounding a point mass. The significance of this metric, however, is far greater, as demonstrated in a powerful theorem derived by the mathematician George D. Birkhoff in 1923 (Bi23). Birkhoff showed that a metric of precisely the Schwarzschild form must hold in empty space surrounding any spherically symmetric mass distribution M, even when this empty space itself is embedded in a larger, spherically symmetric distribution of matter. Moreover, he showed that this metric must be *static*, invariant in time.

In Chapter 11, on the other hand, we saw that the evolution of a simply connected, homogeneous, and isotropic three-dimensional space is described by the quite different Friedmann–Robertson–Walker (FRW) metric (11–18). This metric is spherically symmetric around arbitrarily selected points

$$ds^2 = c^2 \, dt^2 - a^2(t)\{d\chi^2 + \sigma^2(\chi)[d\theta^2 + \sin^2 \theta \, d\phi^2]\}, \quad (13\text{-}43)$$

where $d\chi$ is the comoving element of space and $\sigma(\chi)$ takes on the form $\sin \chi$, χ, or $\sinh \chi$, depending on whether the *Riemann curvature constant* of the three-space is $k = 1, 0$, or -1.

Equations (13–42) and (13–43) taken together permit us to describe the gravitational field of a spherically symmetric galaxy in an inhomogeneous Universe filled with similar galaxies. Consider such a galaxy inside a concentric spherical vacuole — a bubble of empty space embedded in an ambient isotropic, homogeneous, expanding universe (Fig. 13.3). In two ground-breaking papers Einstein and Straus (Ei45) showed how such galaxies fit into a Universe with zero pressure, $P = 0$. Since the pressure in empty space surrounding galaxies and clusters of galaxies is currently quite low, this model helps to describe the Universe observed today.

Within a well-defined radius, r_{ve}, the central mass M is surrounded by vacuum and, as Birkhoff's theorem requires, the Schwarzschild metric applies. Outside this empty region, the FRW metric describes a homogeneous, isotropic, expanding space. The spherical boundary between these two regions expands at a rate commensurate with the overall cosmic expansion. As the external universe expands, its density, ρ_e, drops. Correspondingly, the vacuole radius expands in a way that keeps the gravitational mass of the central condensation M, divided by the vacuole volume V, at a density $\rho_v = M/V$ which at all times stays exactly equal to ρ_e. In effect, the ambient universe becomes gravitationally unaware of the existence of the embedded mass, and equations (13–6) and (13–7) ensure a uniform expansion of the ambient universe.

Einstein and Straus noted that, in this fashion, any number of spherical mass concentrations, each enveloped in its own nonoverlapping spherical vacuole could be embedded in an otherwise homogeneous, isotropic substratum, as sketched in Figs. 13.3 and 13.4. They proposed this as a suitable description of structure observed in the Universe. Because of the vacuoles surrounding the mass distributions, their model is often referred to as a *Swiss-cheese* model.

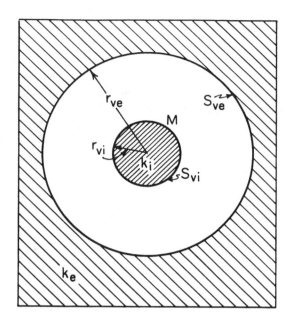

Fig. 13.3. Spherical mass distribution in an ambient universe. A spherical mass M is enveloped in a concentric sphere of empty space — a vacuole embedded in a homogeneous, isotropic, external space of Riemann curvature constant k_e. The interior of M may also be homogeneous and isotropic, but with a different curvature constant, k_i. The surface S_{vi} separates the interior of the mass from the surrounding vacuum; the surface S_{ve} separates the vacuum from the external universe. The respective radii of these two surfaces are r_{vi} and r_{ve}.

Let us see how the Einstein–Straus fitting of the two metrics works. At a radial position r_v in the vacuole surrounding a mass M, the Schwarzschild metric (13–42) can be rewritten as

$$ds^2 = c^2(1 - 2MG/r_v c^2)\, dt_v^2 - \frac{dr_v^2}{(1 - 2MG/r_v c^2)} - r_v^2\, d\Omega^2, \qquad (13\text{-}44)$$

where $d\Omega^2$ is the increment of solid angle, r_v is a radial coordinate, θ and ϕ are angular coordinates, and t_v is *proper time*, measured by a freely falling observer, instantaneously at rest at a point $(r_v, \theta, \phi) = $ constant.

In a universe with cosmological constant Λ, Pirani correspondingly showed that the Schwarzschild metric in the vacuole reads (Pi54):

$$ds^2 = -\left(1 - \frac{2MG}{r_v c^2} - \frac{\Lambda r_v^2}{3}\right) c^2\, dt_v^2 + \left(1 - \frac{2MG}{r_v c^2} - \frac{\Lambda r_v^2}{3}\right)^{-1} dr_v^2 + r_v^2\, d\Omega^2. \quad (13\text{-}45)$$

The metric in the external universe retains the form (13–43) even with the inclusion of the cosmological constant, but the expansion rate of the universe given by $a(t)$ is

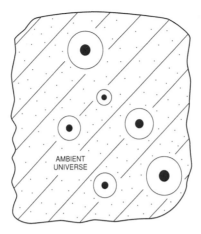

Fig. 13.4. Swiss-cheese model of the Universe. Spherical mass distributions M, potentially representing spherical clusters of galaxies each encapsulated in its own vacuole, are embedded in an external matrix of homogeneous, isotropically distributed matter of density ρ_e. As the universe expands the volume V of each vacuole grows to keep the density $M/V = \rho_e$. In the limit, the Universe can be entirely filled with individual spherical clusters of various sizes, with virtually no homogeneous matrix in between.

then given by (13–6) and (13–7) with the cosmological constant included. For now, we will disregard the cosmological constant, but will return to it later.

A convenient feature of proper time t_v is that it readily permits observers at different radial positions within the vacuole to compare clock rates as shown by equation (5–60). Setting $d\theta = 0$, $d\phi = 0$, we find that the ratio of proper times at an arbitrary radial position r_v within the vacuole, and at the vacuole's boundary to the external ambient Universe r_{ve} is

$$\frac{dt_v}{dt_{ve}} = \frac{(1 - 2MG/c^2 r_v)^{1/2}}{(1 - 2MG/c^2 r_{ve})^{1/2}} . \tag{13-46}$$

In the limit $r_{ve} = \infty$, this reduces to the first of the two expressions (5–60).

PROBLEM 13–6. Substitute the results of Problems 11–1 to 11–3 into equation (11–18), to show that the metric (13–43) can be expressed as

$$ds^2 = c^2 dt_e^2 - a^2(t_e)\left[\frac{d\chi^2}{(1 - k_e\chi^2)} + \chi^2 d\Omega^2\right], \tag{13-47}$$

where $a(t_e)$ is the ambient scale factor, t_e is the time measured in the external universe, and k_e is the Riemann curvature constant in the external universe. (Note that the parameter χ, as used here and in the remainder of this chapter, plays the role of $r = a(t_e)\sigma(\chi)$ in equations (11–18) and (13–43).)

In Section 5:13 we saw that the radius r of a sphere is defined in terms of the square root of the surface area of the sphere, $r \equiv (S/4\pi)^{1/2}$. To obtain the radial distance between spheres with radii r and $r + \Delta r$, for $\Delta r \ll r$, in the vacuole where space is defined by the Schwarzschild metric, we can set the intervals $d\Omega$ and dt_v equal to zero. The radial separation then is the interval

$$|\Delta s| = \frac{\Delta r}{(1 - 2MG/r_v c^2)^{1/2}}. \tag{13-48}$$

Similarly, in the ambient Universe defined by the Friedmann–Robertson–Walker metric the radial separation becomes

$$|\Delta s| = \frac{a(t_e)\Delta\chi}{(1 - k_e\chi_o^2)^{1/2}}, \tag{13-49}$$

where χ_o is the comoving radius of S_{ve}. We now note that points lying within the surface S_{ve}, which separates the spherically symmetric vacuole from an ambient FRW Universe, have to be identified, respectively, in terms of the Schwarzschild and FRW coordinates; and the metric at each point must be continuous across S_{ve}. But because r_{ve} and r_e, at S_{ve} are defined as $(S_{ve}/4\pi)^{1/2}$, this will happen when $r_{ve} = r_e$ at a time $t_e = t_{ve}$, where t_{ve} is the proper time measured within the vacuole at the interface S_{ve} and t_e is *world time*, as defined in Section 11:3.

As the Universe expands, the surface separating the vacuole from the ambient Universe increases to $S_{ve} + \Delta S_{ve} = 4\pi(r_{ve} + \Delta r_v)^2 = 4\pi(r_{ve} + \Delta r_e)^2$.

We now see what happens when S_{ve} grows by a radial increment $\Delta r_v = \Delta r_e$. Crossing the intervals Δs of equations (13–48) and (13–49) at the speed of light requires respective time increments

$$\Delta t_v = \frac{\Delta r_v}{c(1 - 2MG/r_v c^2)^{1/2}} \tag{13-50}$$

and

$$\Delta t_e = \frac{a_e(t)\Delta\chi}{c(1 - k_e\chi_o^2)^{1/2}} \equiv \frac{\Delta r_e}{c(1 - k_e\chi_o^2)^{1/2}}. \tag{13-51}$$

Because the unit of time, in either domain, is the time required by light to traverse a distance c, we can arbitrarily set $\Delta r_v = \Delta r_e = c$ to compare unit time intervals in the two domains

$$\frac{dt_v}{dt_e}\bigg]_{ve} = \frac{(1 - k_e\chi_o^2)^{1/2}}{(1 - 2MG/r_{ve}c^2)^{1/2}}. \tag{13-52}$$

As Schücking (Sc54) first remarked, the clock rates across S_{ve} change as a function of time and differ from each other. Combining expressions (13–46) and (13–52) we obtain the general expression contrasting clock rates at arbitrary points r_v and r_e,

$$\frac{dt_v}{dt_e} = \left(\frac{dt_v}{dt_{ve}}\right)\left(\frac{dt_{ve}}{dt_e}\right) = \frac{[1 - 2MG/(r_v c^2)]^{1/2}[(1 - k_e\chi_o^2)^{1/2}]}{[1 - 2MG/(r_{ve}c^2)]}. \tag{13-53}$$

These comparisons between *proper time* t_v and *world time* t_e hold for freely falling observers at rest at locations r_v and comoving, *fundamental observers* in the ambient universe. We should however note that, in comparing clock rates with observers at greater distances in the universe beyond the interface S_{ve}, the additional time dilatation arising from cosmic expansion must also be considered.

Substitution for dt_v in equation (13–44) permits us to write the metric in the vacuole as

$$ds^2 = c^2 dt_e^2 (1 - 2MG/r_v c^2) \left[\frac{(1 - k_e \chi_o^2)}{(1 - 2MG/r_{ve} c^2)^2} \right] - \frac{dr_v^2}{(1 - 2MG/r_v c^2)} - r_v^2 d\Omega^2 \,.$$

(13-54)

As Schücking (Sc54) noted, somewhat modified metrics of this kind exist also for models of the Universe characterized by a cosmological constant. We can see from the form of (13–45), that we can obtain the expressions characterizing a universe with cosmological constant, by replacing terms of the form $2MG/(rc^2)$ by $[2MG/(rc^2) - \Lambda r^2/3]$, in equations (13–46), (13–48), (13–50), and (13–52) to (13–54).

We may still note that no cylindrically symmetric Swiss-cheese models appear to exist. An expanding or contracting FRW universe cannot be mated to any cylindrically symmetric static metric across a surface without disturbing this symmetry (Se97). This is in contrast to the spherically symmetric cases we have so far discussed, where an important feature is the static character of the Schwarzschild metric.

13:12 Birkhoff's Theorem and "Why Galaxies Don't Expand"

Let us now look at an important consequence of Birkhoff's theorem and its requirement that the metric be static in the empty space surrounding a central spherical mass aggregate M, which could be an isolated star, spherical galaxy, or spherical cluster. Being static, the space does not expand. Even when the vacuole surrounding a galaxy grows as the surrounding Universe expands, the space within the vacuole remains static. The metric (13–54) has no time-dependent terms corresponding, for example, to the scale factor $a(t_e)$ that signifies the cosmic expansion in (13–47). As the universe expands, the surface S_{ve} is pulled along with it, revealing in its wake a static Schwarzschild space of progressively larger radius r_{ve}. It is as though increasing portions of an underlying Schwarzschild space were being uncovered as the expanding Universe peels away from the mass M.

The static character of the Schwarzschild metric which, as (13–45) shows, also holds when a cosmological constant is included, tells us at once that galaxies like the Magellanic Clouds that orbit larger galaxies like the Milky Way do not participate in the cosmic expansion. They travel along trajectories that are Newtonian in the limit $MG/r_v c^2$ and $\Lambda r_v^2 \ll 1$. Even when $MG/r_v \sim c^2$, close to the surface of a neutron star or a black hole, where the curvature of space becomes appreciable and the trajectories followed by particles and photons no longer approximate Newtonian trajectories, the space is still static and isolated from the cosmic expansion.

Birkhoff's theorem answers the question of why the Universe can expand globally, while stars and galaxies orbit each other along Newtonian trajectories, apparently heedless of the cosmic expansion.

13:13 Curvature Fluctuations

The Universe we observe appears to be rather close to being *Euclidean*, or flat, having a value $k_e = 0$. As equation (13–8) tells us, regions of somewhat higher density than the mean may, therefore, be represented by domains with positive Riemann curvature constant, $k = +1$, and regions of lower density may be thought of as domains with negative curvature, $k = -1$. The Swiss-cheese model permits us to embed spherical, dense, homogeneous, isotropic domains with curvature k_i inside less dense, homogeneous, isotropic, external regions with curvature k_e, provided they are separated by a concentric spherical vacuole. The mass of the internal domain, again, has to equal the mass the vacuole would have contained if filled at the density of the external universe. The boundary conditions at the surface S_{vi} separating the internal domain from the vacuole are of the same form as those prevailing at S_{ve}. In full analogy to equation (13–54) we can therefore write the metric in the vacuole as

$$ds^2 = c^2\, dt_i^2(1-2MG/r_vc^2)\left[\frac{(1 - k_i\chi_o^2)}{(1 - 2MG/r_{vi}c^2)^2}\right] - \frac{dr_v^2}{(1 - 2MG/r_vc^2)} - r_v^2\, d\Omega^2.$$
(13-55)

Similarly all the other relations, (13–46) to (13–53), still hold with subscripts i replacing subscripts e.

We can now consider a sphere with radius $a(t)\chi$ and curvature constant k_i embedded in an external universe with curvature constant $k_e < k_i$. At very early times in the expansion of the universe the densities (13–9) in the two regions are the same, $\rho_i = \rho_e$. The two domains interface at surfaces $S_{vi} \sim S_{ve}$ separated by a negligibly thin vacuum shell. The radii of these shells can be taken to be $a(t_i)\chi_o$ and $a(t_e)\chi_o$, respectively. Initially, the expansion rate of the external domain will be only slightly faster than that of the internal sphere because the curvature terms in equations (13–6) and (13–7) are also all but negligible at these epochs. As the expansion proceeds, the curvature terms do become significant, the expansion of the external domain becomes progressively faster than that of the internal sphere, and the thin vacuum shell develops into a sizeable vacuole. The clock rates at one and the same comoving radial coordinate χ_o, respectively, interior and exterior to the two vacuole surfaces, are related by (Ha92), (Ha95)

$$dt_i = \left(\frac{1 - k_i\chi_o^2}{1 - (2MG/r_{vi}c^2)}\right)^{-1/2} dt_v\bigg]_{vi}$$
$$= \left(\frac{1 - k_e\chi_o^2}{1 - k_i\chi_o^2}\right)^{1/2}\left(\frac{1 - (2MG/c^2\chi_o a_i)}{1 - (2MG/c^2\chi_o a_e)}\right) dt_e.$$
(13-56)

This makes the rate of evolution of the inner and outer homogeneous, isotropic domains dependent upon the time evolution of a_i and a_e specified by (13–6) and (13–7).

PROBLEM 13–7. Show, by differentiating the following three expressions with respect to time, that they are solutions of equations (13–6) and (13–7) for matter-dominated, zero-pressure, $P = 0$, $\Lambda = 0$ universes:

$$t = \frac{a_m}{c}\left[\frac{a}{a_m}\left(\frac{a_m}{a}+1\right)^{1/2} - \coth^{-1}\left(\frac{a_m}{a}+1\right)^{1/2}\right] \quad \text{for} \quad k = -1, \quad (13\text{-}57)$$

$$t = \frac{2}{3}\frac{a_m}{c}\left(\frac{a}{a_m}\right)^{3/2} \quad \text{for} \quad k = 0, \quad (13\text{-}58)$$

$$t = \frac{a_m}{c}\left[\frac{(2n+1)\pi}{2} - \tan^{-1}\left(\frac{a_m}{a}-1\right)^{1/2} - \frac{a}{a_m}\left(\frac{a_m}{a}-1\right)^{1/2}\right],$$
$$\text{for } k = +1 \text{ and } n = 0, 1, 2, \ldots . (13\text{-}59)$$

Here,

$$a_m = \frac{8\pi G\rho a^3}{3c^2} \quad (13\text{-}60)$$

is the scale factor at maximum expansion of the $k = +1$ domain, which can be seen, with the help of equations (11–11) and (11–17), to be a constant since it is proportional to the total mass enclosed by a comoving region in a matter-dominated universe.

Because a_m is constant during the matter-dominated, zero-pressure era, it is a suitable standard of length for epochs during which the primary contributor to the mass density is matter, rather than radiation or a cosmological constant. Table 12.3 shows that the Universe was matter dominated after $z \sim 10^3$. Even if $\Omega_\Lambda \sim 0.7$ today, the Universe must have been matter dominated before $z \sim 2$ when the matter density would have been $(z + 1)^3 > 27$ times higher than today. So, equations (13–57) to (13–59) should be instructive for studying the era characterized by red shifts $2 \leq z \leq 10^3$, during which the formation of galaxies and clusters took place.

The solutions to equations (13–57) to (13–60) are plotted in Fig. 13.5 and provide a direct comparison of the evolutionary rates in models with different Riemann curvature constants k, but with identical initial matter density ρ for the early stages of matter-dominated evolution. For $k = +1$, Fig. 13.5 shows only the first oscillatory period, for which $n = 0$ in (13–59). For $t \gtrsim a_m/c$, the physics of collapse for regions with $k = 1$ becomes more complex.

Figure 13.5 also shows that at epochs $\gtrsim a_m/c$, the Universe is *curvature dominated*, meaning that the expansion rate in (13–6) is largely determined by the curvature term, rather than by the density of matter or radiation, which initially were almost homogeneously distributed throughout the Universe.

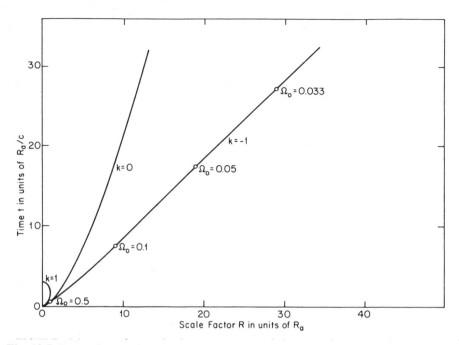

Fig. 13.5. Rates of expansion in zero-pressure, Friedmann–Robertson–Walker (FRW) universes with $\Lambda = 0$ all of which have the same density at some early time, t. The scale factor is measured in units of $R_a \equiv a_m$ the maximum scale factor for a universe with Riemann curvature constant $k = +1$. The unit of time is $R_a/c \equiv a_m/c$. For the open universe, $k = -1$, the decreasing density parameter Ω_o is indicated at various epochs. Beyond epoch $R_a/c = a_m/c \lesssim 1$ the universe is *curvature dominated*. Its expansion rate is determined by the Riemann curvature constant k rather than by the density of matter or radiation.

PROBLEM 13–8. (a) Convince yourself that equations (13–57) and (13–59), respectively, are just a different way of writing equations (11–74) and (11–72). Notice that the maximum value of a in (11–72) occurs for $x = \pi$, where $a = 2a_{(+o)} \equiv a_m$, and world time $t = \pi a_{(+o)}/c = \pi a_m/2c$. Similarly the epoch of minimum extension after collapse is $t \sim 2\pi a_{(+o)}/c = \pi a_m/c$.

(b) At early epochs, the Universe was very dense and the Riemann curvature term in equations (13–6) and (13–7) was negligibly small. Equations (11–72) and (11–74) should therefore give identical growth rates in the limit $t \to 0$. By expanding the trigonometric and hyperbolic functions in these two equations, convince yourself that by setting parameters $y = x$ and $a_{(+o)} = a_{(-o)}$ we obtain identical expansion rates at these early epochs.

PROBLEM 13–9. Let the epoch of primordial collapse and the formation of the earliest stars have occurred in a $k = +1$ domain when the expansion of the universe had sufficiently slowed, at a scale factor $a \sim a_m$. Equation (11–72) places this time of collapse at $t_+ = \pi a_{(+o)}/c = (\pi/2)a_m/c$. The first stars are believed to have formed at red shift $(z + 1) \sim 20$. If this collapse occurred in a $k = +1$ domain embedded in either a $k = 0$ or a $k = -1$ universe, estimate either from equations (11–71) to (11–74) or from (13–57) to (13–60), or a combination of these, the scale factor in the ambient universe at the initial epoch of collapse and also today, at $z = 0$. How would the required sizes of the vacuoles surrounding the $k = +1$ regions today compare to those of the voids seen in Figs. 1.12 and 13.6?

13:14 Primordial Collapse and the Density Parameter Ω_0

Problem 13–9 and Fig. 13.5 show that a universe which initially has the same density everywhere, but has embedded domains of Riemann curvature $k = +1, 0$, or -1, respectively, would fragment into regions with scale factors roughly in the ratio of $1 : 1.8 : 2.3$ at the epoch of initial collapse $t_+ = \pi a_m/2c$ of the $k = +1$ domain.

If this collapse occurred at red shift $(z + 1) \sim 20$ the scale factor in $k = 0$ domains today would be at least a factor of $\sim 1.8 \times 20 = 36$ larger than in the $k = +1$ domains, which could not have grown after onset of collapse. The scale factor in $k = -1$ domains would be a factor of 46 larger. In a $k = 0$ universe, the voids in Fig. 13.6 would then have to be at least a factor of ~ 36 times larger than the populated regions and the mean density of the universe would have to be of order 10^4 to 10^5 times larger in the collapsed regions than in the ambient universe. This ratio is in rough agreement with observations. Highly massive clusters of galaxies, more massive even than the Pisces cluster of Fig. 13.7, can have masses $\sim 10^{49}$ g and radii of order 2 Mpc, i.e., mean densities of $\sim 4 \times 10^{-26}$g. The matter density of a galaxy is of the same order $\rho_M \sim 10^{-25}$g cm^{-3}. In contrast the cosmic matter density is $\sim 3 \times 10^{-30}$ g cm^{-3}.

These densities might be compatible with a $k = -1$ universe, but the age of that model would then be far too short. The ages of the oldest stars in the Galaxy suggest that the Universe is well over 10^{10} yr old today. This age is compatible with a curvature constant $k = 0$ traced back to early times, as in Tables 12.1 and 12.3, which were based on extrapolations with equations (11–55) and (11–56) for such a Euclidean universe. But, as Fig. 13.5 and equations (13–57) and (13–58) show, a scale factor of order $a_e \sim 40a_m$ is reached far more quickly in a $k = -1$ universe than in a $k = 0$ model. This would make a $k = -1$ model younger than the age of the oldest stars.

In a universe that maintains $\Omega_T = 1$ throughout its evolution, regions with $k = +1$ have to be balanced by regions $k = -1$ at early times, when the cosmological constant Λ can be neglected. Later, however, Λ can dominate the expansion, as it does today. As (13–6) shows, the effect of a positive cosmological constant on the expansion rate has the same sign as curvature $k = -1$ except that its influence on the expansion persists, whereas that of the $k = -1$ domain diminishes as $1/a^2$.

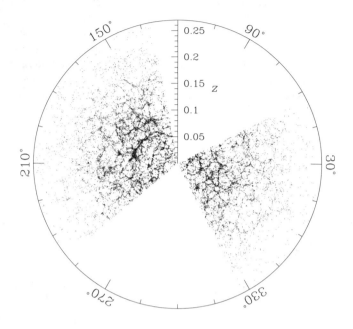

Fig. 13.6. The distributions of galaxies and voids in a $\pm 1.5°$ wide declination strip about the celestial equator, out to red shifts of roughly $z = 0.2$. The red shift of the galaxy concentrations is given by the radial distance, on a scale indicated along the vertical axis. The right ascension is given in degrees; zero indicates a right ascension of zero hours, and the hour angle increases by an hour every $15°$. The largest filamentary structures definitively verified in such surveys stretch over distances of order 80 Mpc. Filaments longer than this may constitute chance interconnections of smaller features (So04). Compare the structures seen here to those on scales a factor of 6 smaller, in Fig. 1.12, and note both the similarities and differences. Courtesy of Dr. Michael Blanton and the Sloan Digital Sky Survey team (Bℓ03).

At first glance it would appear that the $k = 0$ universe also runs into difficulties because the matter density of the Universe is well below the critical density; $\Omega_M < 1$, whereas $k = 0$ requires $\Omega_0 = 1$. This originally led to the postulate, now borne out by observations, that there is an additional source of mass in the Universe provided by either a cosmological constant Λ or a less-well-defined, variable mass density termed *quintessence*, either of which could contribute a density parameter Ω_Λ making $\Omega_0 = \Omega_M + \Omega_\Lambda = 1$ (Pr95).

In summary, we see that the Swiss-cheese model leads to a number of insights. It provides a formalism in which stars, galaxies, and clusters can remain stable, while the external universe expands. It shows how regions of minute initial overdensity, implying a local Riemann curvature $k = +1$, can collapse on time scales that are compatible with observations in an ambient universe with $k = 0$. And it leads to density contrasts compatible with those observed for galaxies and clusters, today. We saw in Problem 13–5 that the Jeans mass at decoupling — the smallest mass that can collapse at that epoch — is $\sim 5 \times 10^4 M_\odot$, roughly the mass of a globular cluster. But, as Problem 13–5 also shows, the total mass within the Hubble radius at that epoch is $\sim 10^{12}$ times higher, indicating that larger aggregates, such as galaxies and clusters of galaxies, can also collapse at that time. We will return to the question of initial collapse in Section 13:22, where we will attempt to establish a relation between the primordial fluctuation spectrum of Section 13:3 and the mass spectrum of observed astronomical condensations.

13:15 Inhomogeneities in the Microwave Background Radiation

Equations (11–72) and (11–74) allow us to estimate the density contrast between $k = -1, 0$ and $+1$ domains at $z \sim 10^3$, the era of decoupling, if we assume that a_+ reached maximum value at the epoch of first star formation $(z + 1) \sim 20$.

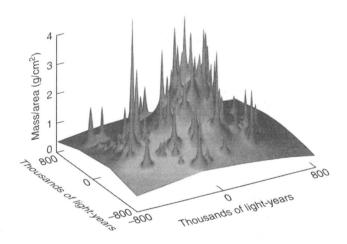

Fig. 13.7. Mass distribution in the Pisces cluster of galaxies shown in Fig. 1.11. The concentration of mass was derived from gravitational lensing of light from background galaxies. Note that most of the mass resides not in individual galaxies shown as spikes, but rather in a more broadly distributed hump of dark matter emitting no light. The mass enclosed within a radius of ~ 150 kpc is $\sim 2 \times 10^{14} M_\odot$ (Ty98) (courtesy of J. A. Tyson, 1997).

PROBLEM 13–10. Clusters of galaxies have by now been detected out to distances $z \gtrsim 3$ (St98). Consider a space with curvature $k_i = +1$ with mass equivalent to that of a large cluster of galaxies, such as the highly massive cluster A3627, with $M \sim 10^{49}$ g (Kr96). Its radius today is of order 2 Mpc. Let this cluster be part of a Swiss-cheese universe in which the external space has curvature $k_e = 0$. Take the size of the cluster today to reflect the size it would have reached at maximum expansion to a scale factor a_m at $(z + 1) = 20$, as in Problem 13–9. (a) Calculate the density contrast for matter at $z \sim 10^3$, right after decoupling, by calculating the relative values of the scale factors a_-, a_0, and a_+, and (b) show that radiation escaping the $k = +1$ domain at that time would have suffered a red shift, or equivalently a temperature shift, of order $\sim 10^{-5}$ in climbing out of the potential well and into the ambient universe during that era. This is of the order of the temperature fluctuations observed in the microwave background radiation. This redshifting of radiation, which we will discuss further in Section 13:18 is called the Sachs–Wolfe effect (Sa67a).

13:16 The Microwave Background Temperature Fluctuations

The microwave background radiation is virtually isotropic. In Section 5:9, we discussed the dipole anisotropy arising from our own motion relative to the background. But once this dipole component is subtracted out, the observed deviations from anisotropy are remarkably small and are generally expressed as deviations ΔT from the mean temperature $T \sim 2.725$ K. Along any given direction $\mathbf{n} = (\theta, \phi)$ the normalized temperature fluctuations are of order $\Delta T(\mathbf{n})/T \sim 10^{-5}$. Since the radiation appears to emanate from a sphere that envelops us, we can best express the fluctuations as superpositions of *spherical harmonics* with axes oriented along different directions, $\mathbf{n}$. From Section 6:25, we then have (Hu02)

$$\frac{\Delta T(\mathbf{n})}{T} = \sum_{\ell,m} a_{\ell m} Y_{\ell m}(\mathbf{n}) \quad ; \quad a_{\ell m} = \int \frac{\Delta T(\mathbf{n})}{T} Y_{\ell m}^*(\mathbf{n}) d\Omega \right) , \quad (13\text{-}61)$$

where $Y_{\ell m}(\mathbf{n})$ is given by (6–192).

As in Section 13:3, we will assume that the standing waves giving rise to the temperature fluctuations are *Gaussian*; the phases of the different harmonics contributing to a fluctuation are uncorrelated; their polar directions $\theta = 0$ and azimuthal angles $\phi = 0$ are randomly distributed across the sky. Current observations indicate that the cosmic microwave background radiation (CMBR) fluctuations indeed are Gaussian to the accuracy that measurements permit.

We saw in Section 13:3 that the squares of the amplitudes are proportional to the wave number k, for a Harrison – Zel'dovich spectrum $n = 1$. But k is directly proportional to the wavenumber m in the function $e^{im\phi}$. This can be seen from Fig. 6.18, where the number of azimuthal segments on a sphere is given by m. A

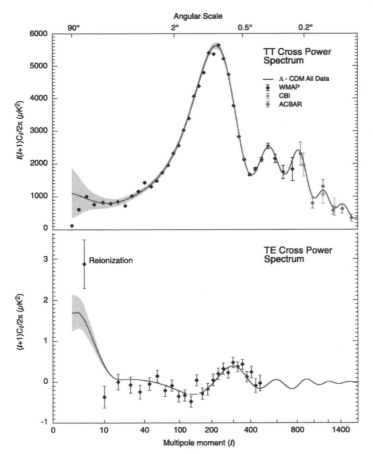

Fig. 13.8. The cosmic microwave background angular power spectrum obtained during the first year of operations of the *Wilkinson Microwave Anisotropy Probe* mission, *WMAP*. (a) Upper panel: The designation *TT* refers to the temperature square of the temperature fluctuations $|\Delta T_\ell|^2$ defined by equation (13–63) and plotted here as a function of ℓ. The peaks in the plot reflect the acoustic oscillations of matter just before its decoupling from radiation. The units on the abscissa are in angular wave number $\ell \sim 180°/\theta_\ell$, where θ_ℓ is the angular extent of a detected oscillation on the sky. The prime peak of this plot ranges around angular scales of $1°$. The solid line tracing the data points is consistent with a cosmological model that assumes the existence of cold dark matter and a cosmological constant — a ΛCDM model. The shaded band represents the *cosmic variance* for this model, an uncertainty that arises from the limited number of independent large-scale statistical samples one can obtain on the sky, given that the entire celestial sphere subtends only 4π steradians. (b) Lower panel: The designation *TE* refers to the degree of polarization of the electric vector in the background radiation cross correlated with temperature. Note that this is a plot of $(\ell + 1)C_\ell/2\pi$, lacking the additional factor of ℓ of the cross-correlation in the upper panel. As explained in Section 13:27, anisotropic Thomson scattering of the background radiation by free electrons produces this linear polarization at optical depths $\tau \lesssim 1$. The highest peak in this plot shows correlation on much larger angular scales than in the *TT* plot. This tells us that the scattering takes place at red shifts $z \sim 20 \pm 10$ — much lower than the red shift $z \sim 1,100$ at which the *TT* correlations are observed. After Bennett et al. (Be03).

measure of amplitude $a_{\ell m}$ may therefore be defined as $|a_{\ell m}|^2 = |a_\ell|^2|m|$. The *angular power spectrum* C_ℓ is defined as the ensemble average over the entire sky of the mean square amplitudes $|a_{\ell m}|^2$,

$$C_\ell \equiv \langle |a_{\ell m}|^2 \rangle = \frac{1}{2\ell+1} \sum_{m=-\ell}^{\ell} |a_{\ell m}|^2 \sim \frac{1}{2\ell+1}|a_\ell|^2 \sum_{m=-\ell}^{\ell} |m|. \qquad (13\text{-}62)$$

From equation (6–194) we see that the different values of ℓ and m modulate the functions $Y_{\ell m}$ without changing their amplitudes. On small sections of the sky and for high spatial frequencies, the spherical harmonics can be represented as a superposition of sinusoidal components. The amplitudes of these components can then be deduced by autocorrelating the modulation of the CMBR surface brightness observed along circular sweeps across the sky. Multiplying a sine wave $a_\ell \sin \ell\theta$ by itself displaced by a variable phase α, i.e., by $a_\ell \sin(\ell\theta + \alpha)$, leads to a sinusoidally modulated autocorrelation product of amplitude a_ℓ^2, with peak power for α equaling integer multiples of π/ℓ. In terms of the equivalent spherical harmonics the temperature fluctuations are given by

$$|\Delta T_\ell|^2 = \langle \Delta T^* \Delta T \rangle$$

$$|\Delta T_\ell|^2 \sim \frac{|a_\ell|^2}{(2\ell+1)} \sum_{m=-\ell}^{\ell} |m| \int_0^\pi \sin\theta d\theta \sum_{m=-\ell}^{\ell} Y_{\ell m}(\theta,\phi) \sum_{m'=-\ell}^{\ell} Y_{\ell m'}^*(\theta,\phi)$$

$$= \left(\frac{\ell(\ell+1)C_\ell}{2\pi} \right). \qquad (13\text{-}63)$$

Here, the top expression indicates that the amplitudes are derived from the autocorrelation function. Asterisks denote the complex conjugate. For the last equality on the right, we have made use of equation (6–194) to obtain the products $\int \sin\theta d\theta Y_{\ell m} Y_{\ell' m'}^*$ as $\delta_{\ell\ell'}/2\pi$, the sweeps across the sky being represented as changes in coordinate θ alone. The final term shown in (13–63) is the quantity plotted in the upper panel of Fig. 13.8. A more rigorous derivation of this important expression is due to Abbott and Wise (Ab84).

The *Wilkinson Microwave Anisotropy Probe, WMAP* mission, an all-sky survey at wavelengths of 3.2, 4.9, 7.3, 9.1, and 13 mm, conducted observations from which the surface brightness or temperature power spectrum across the sky were deduced, as shown in Fig. 13.8.

Knowing the distance to the *surface of last scatter* — the surface from which the microwave background radiation originates — this angular power spectrum can be compared to the power spectrum (13–3) for the mass density contrast in the radiating surface.

The prediction by Harrison (Ha70a) and Zel'dovich (Ze72) that the spectrum of fluctuations should be scale invariant seems to be borne out by the observations. The spectral index n derived from measurements with the *Wilkinson Microwave Anisotropy Probe* is $n \sim 0.93 \pm 0.3$ for the scalar part of the density fluctuations. This appears to vary somewhat when measured on different scales, and may also be augmented by a smaller component due to gravitational waves, often referred to as a *tensor component* (Be03). Whether the data are strictly compatible with the Harrison–Zel'dovich spectrum will have to await more precise observations.

The actually observed microwave background spectrum is highly complex and the derived value of n has to be culled out from a fluctuation spectrum that records the entire history of the decoupling of matter from radiation as well as the subsequent aggregation of matter into gravitationally bound clumps. In the next few sections we will see how these processes become imprinted on the microwave background fluctuation spectrum.

13:17 The Three-Dimensional Power Spectrum of Galaxies and Clusters

All structure in the Universe may originate from primordial fluctuations antedating inflation. If so, fluctuations that give rise to microwave background inhomogeneities should also leave their mark on the distribution of galaxies and clusters. This distribution must reflect not only the relative amplitudes of the primordial fluctuations, but also their decline at high spatial frequencies through Silk damping. Figure 13.9 shows the *power spectrum* $P(k)$ of the microwave background radiation, clusters of galaxies, and other cosmic structures, indicating a seamless evolution from seeding by primordial fluctuations to smaller-scale structures shaped by subsequent evolution. $P(k)$ is the same function we had previously called $P(|\Delta_\mathbf{k}|^2)$ to emphasize that it referred to fluctuations.

13:18 The Observed Imprint of Oscillations

Where an acoustic wave produces a compression before decoupling, the density of both the baryons and photons is increased, producing a gravitational potential well. During compression the radiation temperature rises. If decoupling occurs just as the compression reaches its maximum, the radiation flows out of the region and cools back to its original temperature on expanding to its original volume. But now it also experiences further cooling because it has to climb out of the gravitational well created by the dark matter and baryons. This redshifts, or equivalently cools the radiation. A region in which matter is more concentrated, therefore, appears cooler on a temperature map of the microwave radiation. Likewise a rarefied region appears warmer. This effect is called the Sachs–Wolfe effect (Sa67a).

The Sachs–Wolfe effect imprints its own signature on the power spectrum. Acoustic waves that have gone through half a compression cycle contribute to a deviation from temperature uniformity, as do waves that have undergone half of a rarefaction cycle. But waves that have gone through a complete cycle are back where they started and are neither at peak compression nor peak rarefaction. They, therefore, produce less of an imprint on the power spectrum than waves caught at their extrema. For this reason, only the odd half-integer frequencies, with $\ell = \ell_1, \ell_3, \ell_5 \ldots$

are expected to exhibit strong peaks in the power spectrum. This is roughly borne out by Fig. 13.8, where the second, third, and fourth peaks fall at wave numbers ℓ roughly at multiples $3 : 5 : 7$ of $\ell \sim 160$. Only the first peak lies at a rather higher wave number $\ell \sim 220$.

Shorter wavelength fluctuations are more readily damped by the Silk drag and through diffusion than are longer-scale wavelengths, because it takes longer both for radiation and sound waves to propagate across larger fluctuations. Because of this, the peaks in the power spectrum are progressively lower at high values of ℓ, as seen in Fig. 13.8.

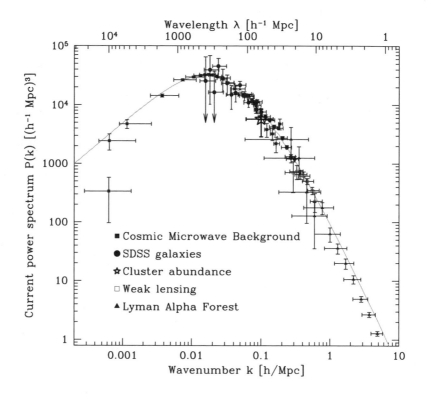

Fig. 13.9. Three-dimensional power spectrum $P(k)$ of the microwave background spectrum, galaxies from the Sloan Digital Sky Survey with mean red shifts $z \sim 0.1$, clusters of galaxies, Ly-α forest absorbers, and weak-lensing sources. Note the slope $n \sim 1$ of the solid line fit at low wave numbers, corresponding to the Harrison–Zel'dovich postulate and the correspondingly anticipated power spectrum of the microwave background fluctuations. At higher wave numbers, where Silk damping plays an increasingly prominent role, the spectrum turns over and the slope progressively declines through $n = -1$ to -2 at the highest wave numbers (Te04).

The astrophysics literature, unfortunately, sometimes refers to the peaks in the power spectrum as *Doppler peaks*. This is misleading, since the acoustic waves are neither expanding nor contracting at their extrema where they make their major imprint on the power spectrum. Nonetheless, the misnomer has stuck.

A number of effects other than acoustic oscillations also leave an imprint on the microwave background power spectrum, increasing the complexity of correctly interpreting observations. In one way or another each provides information on the distribution of matter along the line of sight to the surface of last scatter. One such effect is the Sunyaev–Zel'dovich effect described in Section 6:23. Another is grav-itational deflection of light by clumpy mass distributions along the line of sight, as expected from the discussion of Section 5:14. A third effect is large-angle Thomson scattering off ionized intergalactic clouds along the line of sight, leading to a polar-ization of radiation as explained in Sections 6:15 to 6:17 and called for by equation (6–114). A fourth effect we will encounter later is the Rees–Sciama effect — a red-shifting of radiation passing through massive expanding clusters on its way from the surface of last scatter.

An immediate result of the constancy of the slope $n \sim 1$ of the power spec-trum $P(k)$ in Fig. 13.9, at low wave numbers, is the derivation of an upper limit to neutrino masses. Equation (9–95) specifies a minimum mass that neutrinos would require if they were to account for the dark matter in galaxies. Similar considera-tions hold for mass aggregates on all scales. The good agreement of the slope of the power spectrum with that predicted by Harrison and Zel'dovich tells us that neutri-nos are not diffusing out of low-mass galaxies while staying trapped in large clusters of galaxies. Now, if the masses of neutrinos were higher than $\sim$0.23 eV they would, indeed, find themselves trapped in the largest mass aggregates, and would then also contribute significantly to their masses. If so, $P(k)$ would not have the observed constant slope at low wave numbers. Hence $m_\nu < 0.23$ eV appears to be an upper limit to neutrino masses. It is then also possible to sum over all the relic neutrino and antineutrino species, by using (4–74) and (12–38) to scale the number densities of neutrinos to those of photons in the microwave background. Such arguments lead to the upper limit to the neutrino density parameter $\Omega_\nu < 0.015$ cited in Table 13.1.

13:19 Oscillations and Fundamental Cosmological Parameters

As we have seen, the physics determining the microwave background oscillations is rich. Equation (13–23) for the speed of sound, which also determines the size of the longest wavelength oscillations, is a strong indication of the value of Ω_B, given that Ω_r is already known from the temperature of the microwave background radiation.

A good part of the celestial sphere can be examined for telltale signs of phys-ical conditions at the time of decoupling. The imprint on radiation left by its pas-sage across the Universe on its way to reaching us, also provides valuable data. Microwave background observations, therefore, are a gold mine of information on a variety of fundamental properties of the Universe. A tabulation of best-fit cos-mological parameters obtained through the *Wilkinson Microwave Anisotropy Probe*

Table 13.1. A Compilation of "Best Fit" Cosmological Parameters.[a]

Description	Symbol	Value	Uncertainty
Total density parameter	Ω_T	1.02	± 0.02
Dark energy density parameter	Ω_Λ	0.73	± 0.04
Baryon density parameter	Ω_B	0.044	± 0.004
Baryon density (cm^{-3})	n_B	2.5×10^{-7}	$\pm 1 \times 10^{-8}$
Matter density parameter	Ω_M	0.27	± 0.04
Light neutrino density parameter	Ω_ν	< 0.015	95%CL[b]
CMB temperature[c]	T_{CMB}	2.725	± 0.002
CMB photon density (cm^{-3})	n_ν	410.4	± 0.9
Baryon-to-photon ratio	η	6×10^{-10}	$\pm 3 \times 10^{-11}$
Scalar spectral index (on a $\sim$ 20 Mpc scale)	n_s	0.93	± 0.03
Tensor-to-scalar ratio (on a $\sim$500 Mpc scale)	r	< 0.9	95% CL[b]
Thickness of decoupling zone (FWHM)	Δz_{dec}	195	± 2
Hubble constant km s^{-1} Mpc^{-1}	H	71	+4/-3
Age of the Universe (Gyr)	t_0	13.7	± 0.2
Red shift at matter-radiation equality	z_{eq}	3233	+194/-210
Age at decoupling (kyr)	t_{dec}	379	$+8/-7$
Sound horizon at decoupling (deg)	θ_s	0.598	± 0.002
Sound horizon at decoupling (Mpc)	r_s	147	± 2
Red shift at decoupling	z_{dec}	1089	± 1
Age at reionization (Myr 95%CL[b])	t_r	180	+220/-80
Red shift at reionization	z_r	20	+10/-9
Reionization optical depth	τ	0.17	± 0.04

[a] Based on a larger compilation by Bennett et al. (Be03) derived from the WMAP, COBE, and other microwave background surveys, as well as statistical surveys of galaxies.
[b] CL = confidence level.
[c] Cosmic Microwave Background Explorer (COBE) data.

mission, other cosmic microwave background (CMB) surveys, and statistical surveys of galaxy distributions, is provided in Table 13.1 (Be03).

One important conclusion that jumps out from the WMAP data is that the curvature of the Universe is vanishingly small. The Universe appears to be flat with Riemann curvature constant $k = 0$. This assertion is based on the finding that the first peak in the microwave background power spectrum lies at $\ell_1 \sim 200$.

Let us recall that equations (13–37) and (13–38) derived the location of this peak by attributing it to acoustic oscillations at the speed of sound expected in a decoupling gas at a temperature ~ 3000 K and red shift $z \sim 1100$. We then derived the angular scale of the largest oscillations expected by dividing half this wavelength $\lambda_{max}/2$ by the comoving distance $\mathcal{R}$ at which its radiation was emitted. Determining the angle in this way is correct only if the universe is flat. In a curved universe, equation (11–30) tells us — in somewhat different notation — that the observed angular diameter would not have been $\lambda_{max}/2\mathcal{R}$, but rather $\lambda_{max}/(2\mathcal{R}\sigma(\chi))$. For a positively curved universe, χ is the angle subtended by a segment of length $\mathcal{R}$ on the three-dimensional surface of a four-dimensional sphere, as seen from the center of that sphere; and $\sigma(\chi) = \sin\chi$ is always less than unity. The observed angle subtended by an acoustic oscillation of wavelength λ_{max} would, therefore, appear larger in a positively curved universe than in flat space. Similarly, if the Universe had negative curvature, $k = -1$, $\sigma(\chi) = \sinh\chi \geq 1$, and the observed angle would appear smaller. The agreement between the calculated and observed angles subtended by acoustic oscillations, therefore, indicates that the Universe cannot be appreciably curved over distances comparable to those at which we observe the surface of last scatter. The Universe appears to be close to flat, if not actually flat with $k \equiv 0$.

The Hubble radius at decoupling, given in Table 12.3, is $\sim$220 kpc, a few times larger than the radius of the cluster of Problem 13–10 at that epoch. Such a comoving region's diameter would subtend an angle comparable to the angular range of $1°$ on the sky in which the cosmic microwave background radiation exhibits its strongest fluctuations. This suggests that primordial curvature fluctuations evolved to give rise to mass concentrations comparable to those of clusters of galaxies observed today, and also are responsible for the microwave anisotropy peaks in Fig. 13.8.

13:20 The Rees–Sciama Effect

Rees and Sciama first considered the wavelength shifts experienced by the cosmic microwave background radiation on traversing a spherical density inhomogeneity on its way to reaching Earth (Re68a). This is a complex problem involving a variety of competing effects that make contributions of comparable magnitudes. Both blue and red shifts are possible. Viewed in terms of the Swiss-cheese model, radiation that passes solely through the empty regions of a vacuole, bypassing the massive inner domains, undergoes a less complex history than radiation that passes through both the vacuole and the inner k_+ region. Passage of radiation through one or more vacuoles on its way to reaching Earth, can lead to CMBR wavelength shifts comparable to those produced by density inhomogeneities in the surface of last scatter. This particular type of perturbation is called the Rees–Sciama effect (Re68a).

PROBLEM 13–11. Consider the cluster of Problem 13–10. By $(z + 1) \sim 20$ the radius of the central homogeneous, overdense sphere has expanded to its full extent $\sim$2 Mpc, whereas the radius of the vacuole continues to expand. The expansion rate at the vacuole surface is governed by the Hubble constant and is $\sim$12,500 km s^{-1}. Its expansion during the interval taken by light to traverse it can be significant. Consider the microwave background radiation that passes through the inhomogeneity, and qualitatively estimate the factors that contribute to the induced wavelength shifts.

13:21 Formation of the Largest Structures

Let us look now at the geometric shape of condensations that gravitational collapse may induce. A commonly accepted starting point in the discussion of structure formation is the assumption that condensation has taken place around primordial fluctuations that *bias* where a new galaxy or cluster of galaxies forms. The concept of bias, b, can be quantified and is particularly important. A hint at how the largest scale structures formed comes from observations. As we survey the disposition of galaxies on the very largest scales, we see them aggregated into *sheets*— sometimes referred to as *walls* or *filaments* — that intersect and appear to enclose empty regions called *voids*. Where the walls surrounding the voids intersect, galaxies have concentrated to form major clusters, as in Fig. 13.6. The origins of both the walls and the clusters can be explained by the dynamics of free fall. The epoch of collapse comes sufficiently early for the density parameter Ω_Λ to have been negligible compared to Ω_M.

PROBLEM 13–12. Neglecting cosmic expansion, show that the free-fall collapse time for a thin sheet of uniform initial density ρ_o is only

$$t_{\rm ff} = (2\pi G\rho_o)^{-1/2} \,, \tag{13-64}$$

in contrast to the free-fall time for a sphere of the same initial density, for which we had previously found equation (10–4), $t_{\rm ff} = (3\pi/32G\rho_o)^{1/2}$ (Dr96).

The tendency for an arbitrary, irregularly shaped region, therefore, is to collapse more readily along one dimension, into a sheet rather than into a spherical aggregate. This will be true even in the presence of cosmic expansion. Once such sheets are formed, mass concentrations on the lines where these sheets intersect would again collapse one-dimensionally along the intersecting lines to form clusters of galaxies. It is worth noting that no particular scale is favored by the free fall. Collapse on all scales takes equally long, and is solely determined by the density.

The initial mass density ρ_o may be expressed in units of the critical density $3\Omega_M H^2/8\pi G$ to yield the free-fall collapse time

$$t_{\rm ff} = (4/3\Omega_M H^2)^{1/2} = (3/\Omega_M)^{1/2} t_a, \tag{13-65}$$

where t_a is the age of the matter-dominated Universe at the epoch of collapse (11–58). A high matter-density accelerates collapse. Isolated regions with appreciably higher density than the ambient universe collapse faster than regions of lower density.

When gravitational collapse does occur, flattened regions, often called *pancakes*, are favored. The roughly spherical galaxies and clusters of galaxies that we observe today are the result of smaller-scale fragmentation and relaxation of higher-density domains that may once have resided within lower-density sheets or walls. As those low-density walls of galaxies continue to slowly collapse, they will most probably break up to form new clusters of galaxies or be gravitationally pulled into the nearest previously formed cluster, making it even larger. Thus, the clusters we see today are not a final product. Galaxies are continually merging into them, adding both mass and angular momentum to keep the cluster stirred up. Æons from now, the walls and voids are unlikely to survive. Clusters of galaxies will remain, but will exhibit increasingly compact cores as the galaxies within them collide, merge, and become progressively more massive.

13:22 Press–Schechter Condensation

In depicting the Swiss-cheese models of galaxies and clusters of galaxies, we refrained from discussing how the central condensations contract once they have reached their maximum dimensions. Let us now look at the actual process of collapse.

Observations out to great distances permit us to determine the sizes of galaxies and clusters at early times. They generally are smaller and less massive than the galaxies and clusters observed today, which appear to have been accreting matter continuously over the æons since the first stars formed.

In 1974 Press and Schechter suggested a way to understand the successive formation of increasingly extended mass condensations in an expanding FRW universe with primordial inhomogeneities (Pr74). They postulated that a locally overdense region of gravitationally interacting particles initially condenses to produce relatively small aggregates. These aggregates then would act like a new generation of individual particles of larger mass that would condense into a next generation of larger aggregates remaining embedded in a much more extended, but less overdense region collapsing more gradually. Press and Schechter proposed that, in this fashion, a succession of generations of aggregates of ever-increasing size and mass would form. They proposed that this process would generate a *self-similar* mass spectrum of aggregates, independent of the original mass spectrum within the overdense region. Rather than considering condensations to be in equilibrium with their surroundings as the Einstein and Straus Swiss-cheese models did, Press and Schechter envisaged a system continually out of equilibrium, evolving to form condensations of progressively larger masses.

This approach is consonant with the considerations of Section 13:4, where we found that small fluctuations enter the particle horizon at early times, while larger ones, within which the small fluctuations may be embedded, only appear as the particle horizon further expands. We also found that the mass spectrum should be scale invariant, i.e., self-similar.

The distribution of aggregates with different masses in the Press–Schechter depiction is related to the total number n_* of aggregates by

$$n_* \equiv \int_0^\infty n(m) dm \ . \tag{13-66}$$

The mean mass is

$$m_* \equiv n_*^{-1} \int_0^\infty n(m)m\,dm \equiv \rho/n_* , \qquad (13\text{-}67)$$

where ρ is the mean density. The gas of aggregates, however, can be further characterized by a lumpiness specified by a probability distribution function. Let us assume, as we did in Sections 13:3 and 13:4, that the primordial cosmic fluctuations are Gaussian. The density contrast $\delta \equiv (\delta\rho)/\rho$ associated with a volume element V then has a probability distribution,

$$p(\delta, V) = \frac{1}{\sqrt{2\pi}\sigma} \exp - \left(\frac{\delta^2}{2\sigma^2} \right) , \qquad (13\text{-}68)$$

where $\sigma^2 = \langle \delta^2 \rangle$ is the standard deviation of the density contrast. For a randomly distributed set of particles, the variance of mass in a given volume is proportional to the volume, and the standard deviation per unit volume, written as σ, is

$$\sigma \equiv \langle \delta^2 \rangle^{1/2} = \frac{\left[\langle M^2 \rangle - \langle M \rangle^2 \right]^{1/2}}{M} . \qquad (13\text{-}69)$$

Because the early Universe is close to flat, and is matter dominated at decoupling, an overdense volume, with density contrast δ, expands only until it reaches its maximum size at a scale factor a_m, given by equation (13–60). As (13–6) shows, the decelerating cosmic expansion $\dot{a}^2/a^2$ drops to zero at a smaller value of a when ρ is high. Accordingly, when the density within the overdense region varies from one location to another, the most overdense regions reach their maximum size early in the expansion, while less overdense domains reach maximum size later. Since most of the mass in the Universe is in the form of dark matter, the overdense domains correspond to *dark matter haloes* to which baryonic matter becomes attracted.

Once a halo reaches maximum size it begins to contract and, being lumpy rather than perfectly homogeneous, it relaxes to eventually give rise to a bound virialized aggregate. Thereafter, this aggregate acts as a point mass, gravitationally attracting other haloes that have also formed within an even larger ambient overdense region — which, in turn, eventually also collapses and fragments.

The probability P that a halo will already have reached maximum size and become bound by a time t_1 depends on whether its original density contrast δ was sufficiently high, $\delta > \delta_1$.

$$P = \int_{\delta_1}^\infty p(\delta, V)\,d\delta = \frac{1}{2} \left(1 - \frac{2}{\sqrt{\pi}} \int_0^{\delta_1/\sqrt{2}\sigma} e^{-t^2}\,dt \right) \equiv \frac{1}{2} \left(1 - \mathrm{erf}\, \frac{\delta_1}{\sqrt{2}\sigma} \right) , \qquad (13\text{-}70)$$

where the integral in the middle expression is the *error function* erf. Because we are assuming the fluctuations to be Gaussian, we can take advantage of equation (13–5) and set $\sigma^2 = AM^{-(3+n)/3}$, where A is some constant. Then

$$\frac{\delta_1}{\sqrt{2}\sigma} = \frac{\delta_1}{\sqrt{2A}}M^{(3+n)/6} \equiv \frac{1}{\sqrt{2}} \left(\frac{M}{M_1} \right)^{(3+n)/6} . \qquad (13\text{-}71)$$

Here, we have replaced the density contrast δ_1 by the mass scale M_1 to which it refers at time t_1,

$$M_1 \equiv (A/\delta_1^2)^{3/(3+n)} . \tag{13-72}$$

From (13–10) we know that $a^2 \delta\rho$ is constant, and during the matter-dominated era, $a^3 \rho$ is constant as well. This makes $\delta \propto a \propto (1+z)^{-1}$, where we have invoked (11–56). Hence A must be proportional to $(1+z)^{-2}$ and $M_1 \propto (1+z)^{-6/(n+3)}$. In terms of its value today, M_0, we can set

$$M_1 = M_0/(z+1)^{6/(n+3)} \tag{13-73}$$

at red shift z.

That a volume with density contrast δ_1 has already condensed does not prevent it from becoming part of a larger, more massive volume of lower density contrast δ_2 that condenses out at time t_2. So V, the fraction of the volume that becomes self-bound with masses above M, by some maximum time or cosmic age, t_{max}, is the time-derivative of P.

$$\frac{dP}{dt_{max}} = \frac{-1}{\sqrt{\pi}} e^{-t_{max}^2} \tag{13-74}$$

Let us denote the number density of haloes less massive than M at red shift z by $n(M, z)$. Then the number density of haloes of mass M between M and $M + dM$ at epoch z is

$$\frac{dn(M, z)}{dM} = -\frac{\rho}{M} \frac{dP}{dM} . \tag{13-75}$$

$$\frac{dn(M, z)}{dM} = \frac{1}{\sqrt{8\pi}} \left(1 + \frac{n}{3}\right) \frac{\rho}{M^2} \left(\frac{M}{M_1}\right)^{(n+3)/6} \exp\left[-\frac{1}{2}\left(\frac{M}{M_1}\right)^{(n+3)/3}\right], \tag{13-76}$$

where we have substituted (13–71) in (13–70). Figure 13.9 suggests that $n = -1$ for the largest cosmic structures, and that this gradually diminishes further to $n = -2$. If we choose $n = -1.5$ to represent the scale of clusters of galaxies

$$\frac{M^2}{\rho} \frac{dn(M, z)}{dM} = \frac{3(z+1)}{4\sqrt{2\pi}} \left(\frac{M}{M_0}\right)^{(1/4)} \exp\left[-\frac{(z+1)^2}{2}\left(\frac{M}{M_0}\right)^{1/2}\right] . \tag{13-77}$$

To represent haloes of lower mass, a value of n closer to -2 needs to be invoked. Figure 13.10 shows the computed evolution of these halo number densities in terms of red shift.

As Press and Schechter recognized, the process they described results in the capture of only half of all the mass in the Universe, because the underdense regions in a nearly flat universe carry another half of the mass and do not spontaneously condense. However, the two authors proposed that this mass would eventually accrete onto previously formed aggregates, essentially doubling the masses of all those condensations. This seems to be borne out by computer simulations, and the right side of (13–76) should therefore be multiplied by a factor of 2.

Although (13–76) includes a number of simplifying assumptions its overall validity appears to be borne out by observations. Structure formation is a continuing process. Clusters of galaxies are still growing as galaxies from their surroundings are gravitationally attracted and fall in. Within the Galaxy, we are also witnessing the capture of matter tidally stripped from smaller companions. And, as we will see later, in Fig. 13.16 and Section 13:33, in densely populated regions of clusters, we can witness the merger of fully developed galaxies triggering the widespread formation of massive young stars and a spectacular rise in luminosity.

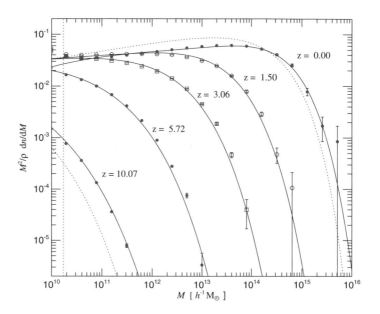

Fig. 13.10. The comoving number density $n(M, z)$ of haloes less massive than M, plotted in terms of the product $M^2 \rho^{-1}(dn/dM)$ as a function of M. Here ρ is the mean density of the Universe. The flattening of the curves, out to masses of order $\gtrsim 10^{14} M_\odot$ at low red shifts, indicates that $n(M)$ is approximately proportional to $1/M$, for masses in this range, in today's Universe. The two dotted curves designated $z = 10.07$ and $z = 0$ depict the fractional amounts of mass in each mass range that the Press–Schechter formalism condenses out by the corresponding red shifts. Solid lines represent an analytical fit to extensive numerical simulations indicated by the individual symbols. The vertical dotted line is drawn at a mass just slightly less than $2 \times 10^{-10} h^{-1} M_\odot$, where h is 0.7 for today's Hubble constant $H_0 = 70$ km s^{-1} Mpc^{-1}. For this mass value, the solid lines show that by $z = 10.07$ only ~10^{-3} of the mass had condensed out in haloes as massive as ~$1.5 \times 10^{10} h M_\odot$, whereas by $z = 0$ about half of the mass had condensed into haloes at or above this mass. Courtesy Volker Springel (Sp05).

In arriving at (13–76), we neglected the influence of the cosmological constant Λ. But observations indicate that Λ currently is so small, that it could not have played a major role in structure formation. Some structure is definitely observed even at red shift $z \gtrsim 6$. For constant Λ, the ratio of $\Omega_\Lambda/\Omega_M \sim 2.7(z+1)^{-3}$ at that epoch would have been less than 1%.

13:23 The Internal Structure of Dark Matter haloes

In the Press-Schechter hierarchical depiction of structure formation, density fluctuations lead first to the formation of small aggregates that later seed the formation of larger structures or merge to form more massive aggregates. These aggregates are dominated by dark matter and generally are referred to as *dark matter haloes*. The internal distribution of matter within these haloes has been studied through computer simulations that start out with a large number of randomly arranged masses and follow their subsequent evolution. Navarro, Frenk, and White (Na96, Na97) carried out a series of such simulations and discovered that, independent of the initially assumed halo mass, initial density fluctuation spectrum, or value of the cosmological parameters assumed, the mass density distribution generally tended toward a universal profile:

$$\rho = \frac{\rho_0}{(r/r_s)(1 + r/r_s)^2} \,, \tag{13-78}$$

where ρ_0 is some characteristic density and r_s a scale radius. This so-called *NFW profile* is named after the three authors. Other models have also been suggested (Me05). The observed density profiles in galaxies and clusters of galaxies are in rough, though not entire, agreement with these various models, but an analytic explanation of how the observed profile arises is still lacking.

13:24 Protogalactic Cooling

Until now, we have acted as though gravitational collapse must obviously lead to galaxy formation. However, collapse can just as readily lead to a rebound if there is no way to dissipate kinetic energy. An important question is how the collapsing primordial gas manages to cool itself rapidly enough to form galaxies and eventually stars. In today's Universe, as we saw in Chapter 10, much of the cooling is done by impurities — carbon atoms, CO, H_2O, or dust grains. The early Universe had none of these; carbon and oxygen were synthesized only later in stars. Figure 12.5 indicates that primordial baryonic matter consisted of $\sim$76% hydrogen and $\sim$24% helium by mass, with negligible amounts of deuterium, lithium, and beryllium. As Fig. 13.11 shows, cooling rates due to these primordial atomic or ionic constituents drop to extremely low values once the temperature of the Universe falls below 10^4 K. However, as Fig. 13.12 indicates, hydrogen molecules are formed in trace quantities by the time the temperature has dropped to 1100 K, at $z \sim 400$. Two principal reactions are responsible.

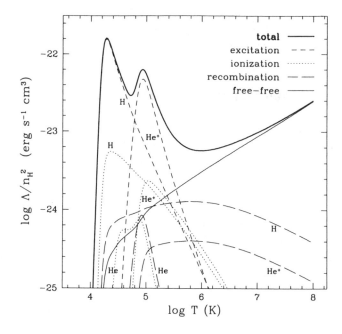

Fig. 13.11. Cooling rates for primordial atomic and ionic constituents. The rates are given as a function of temperature for a primordial composition of gases in collisional equilibrium. The heavy solid line shows the total cooling rate. The cooling is dominated by collisional excitation and re-emission (short-dashed lines) at low temperatures and by free–free emission (thin solid line) at high temperatures. Long-dashed lines and dotted lines, respectively, show the contribution of recombination and collisional ionization. The assumed primordial composition is 76% hydrogen and 24% helium by mass, with no ionizing background (We97a).

$$e^- + H \rightarrow H^- + h\nu \ ,$$
$$H^- + H \rightarrow H_2 + e^- \ . \tag{13-79}$$

in which e^- acts as a catalyst, and

$$H + H^+ \rightarrow H_2^+ + h\nu \ ,$$
$$H_2^+ + H \rightarrow H_2 + H^+ \tag{13-80}$$

in which H^+ is the catalyst. A variety of competing reactions can also destroy H_2, but a fractional abundance of order $n(H_2)/n(H) \sim 3 \times 10^{-5}$ ensues.

13:25 Formation of the First Stars

Current belief is that the first condensations gave rise to stars with several hundred solar masses (Ha96). These are commonly referred to as *Population III stars*. Initially, the abundance of H_2 is too low to effectively cool the condensation, and the temperature rises adiabatically in accord with (4–129). The density also increases and competing reactions among the prime constituents H, e^-, H^+, H^-, H_2, and H_2^+ lead to an increasing fractional abundance of H_2. Eventually as the collapse progresses, a shock ensues as infalling matter piles up. At the shock, the temperature abruptly rises to the virial value and reaches 200 to 500 K, well above the temperature of the microwave background at this epoch, $z \sim 20$ to 30. H_2 rotational states are collisionally excited and radiate away energy in spontaneous transitions to the

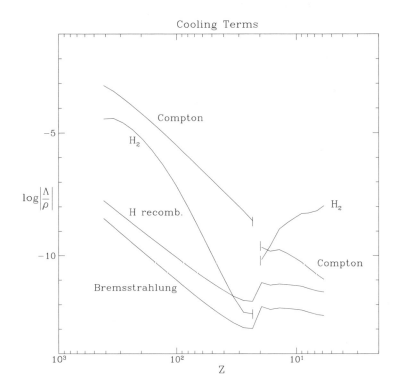

Fig. 13.12. Cooling and heating rates as a function of red shift in a volume enclosing the innermost 12% of the baryonic mass of a spherical aggregate collapsing to form a Population III star. Compton scattering of electrons in close thermal contact with the microwave background radiation can heat the gas as long as the temperature of the microwave background radiation exceeds that of the gas. Radiative excitation of molecular hydrogen by the background radiation also is a strong source of heating at high red shifts. Radiation released in the formation of hydrogen atoms is indicated as recombination. Free–free emission (*Bremsstrahlung*) by ambient plasma is a minor sources of heating. Initially, when the microwave background temperature exceeds the gas temperature in a collapsing cloud, all four mechanisms act to heat the collapsing cloud. But as the gravitational collapse heats the gas, and its temperature rises above that of the background — at red shifts indicated by the vertical tick marks on the plot — H_2 radiative transitions and Compton scattering start to cool the gas. The units of the cooling rate Λ/ρ are ergs s^{-1} g^{-1}. Courtesy of Haiman, Thoul, and Loeb (Ha96).

ground rotational state. The epoch at which this takes place is shown by the tick marks in Fig. 13.12. The cooling by H_2 thereafter lowers the temperature so that further collapse continues until a star forms and can begin to burn its nuclear fuel.

This roughly outlines how the first stars may have formed. Incisive observations will be required to show whether this view is correct.

13:26 Population III Stars

Can the Population III stars be directly observed? To date they have not been found. The oldest, most distant quasars known all have appreciable metallicities. So, they must already represent quite a late stage of evolution. The most metal-poor stars found in the Galaxy, having iron abundances thousands of times lower than the Sun, also exhibit at least some admixture of heavy elements, as Fig. 8.12 illustrates. These stars generally are members of the *Galactic halo* — a tenuous aggregate of stars evenly distributed around the Galaxy's center and ranging out to far greater distances than stars belonging to the disk. In the Sun's neighborhood, within the Galaxy's disk, metal-poor stars are virtually absent. The metal abundance of most stars in our neighborhood is rather similar to that of the Sun. The metallicity of the interstellar gas appears to have changed only slowly, judging by the age of some of the stars that appear to have formed when the Galaxy was young. These observations suggest that metal production was much higher at early cosmic epochs than it is today. The high abundance of heavy elements in the hot gas gravitationally trapped in galaxy clusters is also consistent with this view. These pieces of observational evidence suggest that stars formed at early epochs were far more massive than stars formed today. They copiously produced heavy elements, and explosively ejected them into their environments (La98).

As the earliest stars formed, H_2 was the principal coolant. But the lowest excited state of H_2 is a rotational state that lies at $\sim 500\,K$ above the ground state. To begin radiating away energy the temperature of adiabatically collapsing gas had to reach a temperature of this order. The Jeans mass corresponding to such a high temperature is significantly higher than it is for stars forming today from gas that contains impurities excitable at far lower temperatures. Figure 13.2, equation (13–39), and Problem 13–5 all are consistent with more detailed simulations suggesting that the first generation of stars was very massive, the initial mass function reaching masses of order $\sim 1000 M_\odot$ (Ab02, He02).

In Chapter 12 we saw that primordial chemical evolution stops at mass 7 amu in the early universe, because nuclei with mass 8 are unstable. The triple-α process does not take place because densities are too low. In massive Population III stars, however, this process can go forward.

Studies of nonrotating extremely massive stars show that the proton–proton reaction is unable to produce sufficient heat to stop such stars from contracting before their central temperatures rise to $>10^8$ K, where the triple-α process sets in as shown in Fig. 13.13.

Once carbon is formed, the CNO cycle can start to supply sufficient heat to stem further contraction. Hydrogen keeps burning in the compact hot core until fully exhausted. For stars with masses in the range $\sim 100 - 140\ M_\odot$, further contraction then produces a pulsational instability — an *electron–positron pair instability* — at central temperatures where collisions between photons just begin to form electron–positron pairs.

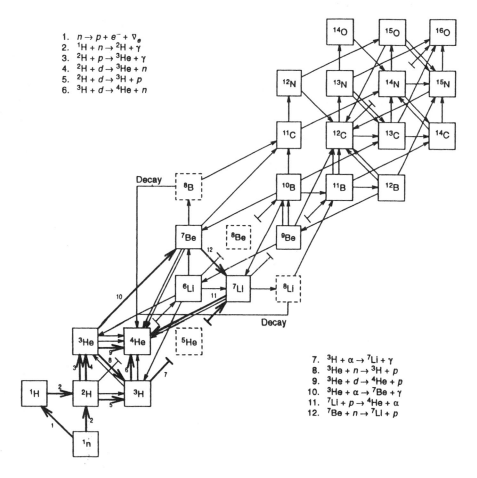

1. $n \to p + e^- + \nabla_e$
2. $^1H + n \to {}^2H + \gamma$
3. $^2H + p \to {}^3He + \gamma$
4. $^2H + d \to {}^3He + n$
5. $^2H + d \to {}^3H + p$
6. $^3H + d \to {}^4He + n$

7. $^3H + \alpha \to {}^7Li + \gamma$
8. $^3He + n \to {}^3H + p$
9. $^3He + d \to {}^4He + p$
10. $^3He + \alpha \to {}^7Be + \gamma$
11. $^7Li + p \to {}^4He + \alpha$
12. $^7Be + n \to {}^7Li + p$

Fig. 13.13. Reactions of importance in nucleosynthesis during early stages of an evolving universe and in Population III stars. The exergonic directions are indicated, although rates are often rapid in both directions. The most important reactions are numbered and have bold arrows. The broken boxes for mass 5 and 8 indicate that all nuclides of this mass are very unstable. Reactions beyond these two mass values would not have occurred in the early Universe, but are believed to have taken place in Population III stars. Sometimes competing reactions lead from one nucleus to another (after (Wa67), (Co95)). Reprinted with permission from *Science* ©1995 American Association for the Advancement of Science.

At threshold, the photons' energy is converted into electron–positron pairs with little kinetic energy. As a result, the pressure suddenly drops, and the ensuing collapse raises the temperature to where even–even nuclei form through the successive addition of alpha particles to oxygen. The heat released as neon, silicon, sulfur, and magnesium are formed produces a rebound and initiates pulsations that eject the outer layers of the star before its interior collapses to form a black hole. In stars with masses $140 - 260M_\odot$ elements as massive as the iron group are formed and a single pair-production pulse may disrupt the star, ejecting all matter and leaving no remnant. Stars of all other masses below $\sim 500M_\odot$, however, appear to remain stable. Figure 13.14 shows the ultimate fate of stars with such high masses (He02).

The ejecta of these very massive stars are predominantly enriched in the even–even nuclei carbon and oxygen, and heavier elements neon, magnesium, silicon, and sulfur reaching up to the iron group formed in the equilibrium process described in Chapter 8. With the extremely energetic explosions expected from such stars, their ejecta may have been able to escape even a massive halo to emerge as constituents of extragalactic space. Indeed, an excess of silicon has been detected in X-ray emission from the hot gases pervading large clusters of galaxies, whose gravitational potential retains any matter ejected from member galaxies (Ba05). Pair instability supernovae, however, do not appear to form the r-process elements evident in lower-mass Galactic stars with extremely low metal abundances. This suggests that stars of lower mass, in the range $8 - 40\ M_\odot$ may also have formed primordially (Tu04). Their explosive ejecta might more readily have been retained in a massive halo and would ultimately have become incorporated in a later generation of low-mass, low-metallicity stars like CS 22892-052 (Fig. 8.12). Population III stars in both of these high-mass ranges would generate sufficient numbers of ionizing photons to reionize intergalactic space.

13:27 Reionization

As the Population III stars light up, their high masses make them highly luminous and hot, generating vast outflows of ultraviolet radiation, which reionize the gas within the larger gaseous halo of which the Population III star may be a part. This may interfere with further formation of nearby Population III stars that were about to form or, alternatively, may generate shocks that promote the collapse and formation of such stars. The complex outcomes of these processes have not yet been satisfactorily determined. Undoubted, however, is the further penetration of ionizing radiation, once it breaks out of a dense halo, and spreads into the ambient Universe. The epoch during which this occurs is referred to as the *reionization era*.

The lower panel of Fig. 13.8 reveals a relatively high optical depth for Thomson scattering, $\tau \sim 0.17$, on large angular scales. This may appear surprising since we saw, in Section 13:5 and Fig. 13.1, that the ionization fraction should have rapidly dropped after decoupling. It suggests that some process must have reionized the universe, at least partially at later times at red shifts around $z \sim 20\pm10$ (Ko03). The estimate for this red-shift range comes in part from the large angular scales — low

ℓ values — over which the correlation between polarization and temperature of the cosmic microwave background is observed. Regions over which sizeable scattering anisotropies can be observed must be associated with late epochs and low red shifts, because they require a large Hubble radius to stretch across substantial regions over the sky.

Two epochs of reionization may have taken place, one around $z \sim 20 \pm 1$ produced by the copious emission of ultraviolet radiation by Population III stars, leading to partial ionization. The other, at $z \sim 6$, coinciding with the formation

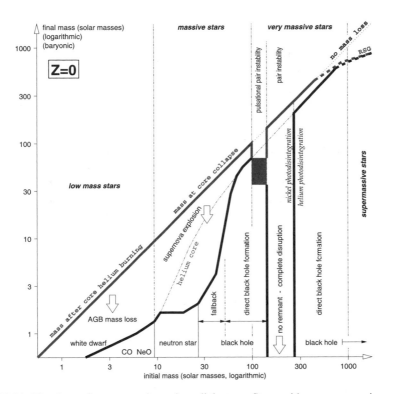

Fig. 13.14. The fate of very massive primordial stars. Stars with masses ranging up to $1000 M_\odot$ may have been common during the epoch of first gravitational collapse, when the prime baryonic constituents were hydrogen and helium, with trace admixtures of deuterium, ^{3}He, lithium, beryllium, and boron. Calculations on nonrotating stars indicate that stars with masses below $10 M_\odot$ would have lost mass during the asymptotic giant branch phase, their remnants ending up as white dwarfs. Stars with initial masses $10 - 100 M_\odot$ would form an iron core and collapse into a neutron star, at the lower end of this mass range, and into a black hole at the higher end. Very massive stars $100 - 1000 M_\odot$ undergo an electron–positron pair instability. In the $100 - 140 M_\odot$ range and possibly also above $500 M_\odot$ these pulsations can shed significant mass from the star's outer layers. In the $140 - 260 M_\odot$ range, the entire star may disrupt leading to the dispersal of elements rich in even–even nuclei. The extremely high ultraviolet luminosities of these stars could have been responsible for the reionization of the intergalactic medium. Courtesy of Alexander Heger (He02).

of quasars, leading to full ionization. The evidence for this second ionization stage comes from observations of quasars at $z \sim 6$. These indicate that the Universe at that epoch contained at least a small but widely distributed component of neutral hydrogen, which is totally lacking in the ambient universe observed today. To see how this conclusion arises we need to refer to the *Gunn–Peterson effect*.

13:28 The Gunn–Peterson Effect

If a distant quasar emits ultraviolet radiation at wavelengths short of the Lyman-α transition, intergalactic neutral hydrogen along the line of sight to the quasar will resonantly scatter radiation redshifted to the wavelength of the Ly-α transition at $1216\,\text{Å}$. The spectrum of the quasar should, therefore, exhibit a sharp cut-off that stretches from $1216\,\text{Å}$ in its own rest–frame to shorter wavelengths. The depth of this absorption can provide information on the amount of intergalactic hydrogen at different distances along the line of sight, as first suggested by Gunn and Peterson (Gu65).

To see this quantitatively, we make use of (7–57) with the added oscillator strength f introduced in (7–59), and take the absorption cross-section for Ly-α to be $\sigma = 2\pi^2 e^2 f/mc$, where e and m are the electron charge and mass and σ has dimensions of area multiplied by frequency. The path length L over which a column of neutral hydrogen with number density n_H provides unit optical depth at red shift z is

$$L = \frac{\Delta\nu}{n_H\sigma} = \frac{mc\Delta\nu}{2\pi^2 e^2 f n_H(z)}\,, \tag{13-81}$$

where $\sigma/\Delta\nu$ is now a cross-section with dimensions of area. As it traverses intergalactic space radiation from the quasar is progressively redshifted with distance ℓ at a rate

$$-\frac{\Delta\nu}{\nu} = \frac{1}{c}\frac{\Delta v}{\Delta\ell}d\ell = \frac{1}{c}H_0(1+z)^{3/2}d\ell\,, \tag{13-82}$$

or

$$d\ell = -\frac{c}{H_0}\frac{\Delta\nu}{\nu}\frac{1}{(1+z)^{3/2}}\,. \tag{13-83}$$

During this transit, the number density changes with z as $n_H(z) = n_H(0)(1+z)^3$, as long as hydrogen atoms are not formed or destroyed through ionization or other processes. The optical depth of the line at red shift z then becomes

$$\tau = \frac{|d\ell|}{L} = \frac{2\pi^2 e^2 f n_H(0)(1+z)^{3/2}}{mH_0\nu}\,. \tag{13-84}$$

The spectra of three distant quasars shown in Fig. 13.15 exhibit broad absorption troughs at red shifts $z \gtrsim 6$. These are not observed in the spectra of quasars at lower red shifts, indicating that the intergalactic medium was largely neutral at early times when the first quasars formed, but then became fully ionized as ultraviolet radiation from these quasars progressively reionized extragalactic space during this second *reionization epoch*.

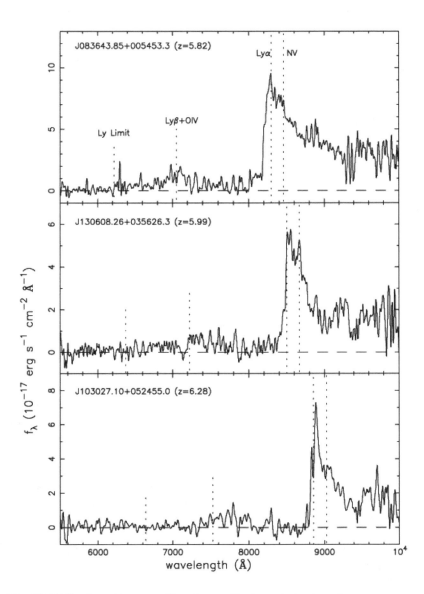

Fig. 13.15. The Gunn–Peterson effect observed in the emission from three quasars at red shift $z \sim 6$. Radiation from the quasars is strongly absorbed at wavelengths short of the red-shifted Ly-α emission and appears to be totally absorbed short of the Lyman limit. Dashed vertical lines indicate the positions of various potential emission lines at the respective red shifts of the quasars. (Courtesy of Fan et al. (Fa01).)

Out to distances of order $z \sim 3$ the Gunn–Peterson effect has also been observed for singly ionized helium whose Ly-α transition lies at 304 Å(Re97). Ionizing neutral helium requires more energetic photons of which fewer exist. Reionization of helium will consequently have taken longer, with traces of neutral helium persisting until $z \sim 3$.

PROBLEM 13–13. For atomic hydrogen, the oscillator strength for the Ly-α transition is $f = 0.416$. What number density $n_H(0)$ would suffice to produce optical depth $\tau = 1$ at $z = 6$, for a Hubble constant $H_0 = 70$? Show that even a very low fraction of neutral hydrogen atoms in a flat universe with $\Omega_B = 0.04$ suffices to produce unit optical depth.

13:29 Quasar Strömgren Spheres

The copious emission of ionizing radiation by quasars leads to the formation of ambient Strömgren spheres.

PROBLEM 13–14. The number density of ambient hydrogen atoms surrounding a quasar at $z = 6$ is $(z + 1)^3$ higher than the mean cosmic hydrogen density today. Using Table 9.2, (a) convince yourself that all but the most energetic ionizing photons are absorbed over distances of order 1 kpc, so that the Strömgren sphere should be almost fully ionized. (b) Show also that the recombination time in the sphere greatly exceeds the age of the Universe at that epoch; once established, the Strömgren sphere persists. (c) Under these conditions, show that the size of the Strömgren sphere can serve as a measure of the total number of ionizing photons the quasar has emitted into the ambient Universe in the course of its existence. (d) Show that Thomson scattering by free electrons is negligible for radiation transiting the fully ionized sphere. (e) As (13–83) shows, photons emitted at wavelengths short of Lyman-α become redshifted by the cosmic expansion, on passing through the Strömgren sphere. Show that the radius of the Strömgren sphere $d\ell$ can be estimated from (13–83), by measuring the frequency range $\Delta\nu$ short of Ly-α in the quasar's rest-frame, in which continuum radiation from the quasar is able to pass through the ionized sphere, emerge into the surrounding neutral gas redshifted to a frequency below Ly-α where the neutral gas cannot absorb it, and hence able to traverse the rest of the Universe to reach us today.

Observations of the continuum emission short of Ly-α thus should provide a measure of the total ionizing radiation emitted by a quasar. Such observations of quasars at different red shifts should provide a history of the growth of their Strömgren spheres over time, and determine the epoch at which the spheres began to touch and the intergalactic medium became fully ionized (Me04).

13:30 Formation of Supermassive Black Holes

Once a Population III star forms, it is highly luminous and generates much of its flux in the ultraviolet. This dissociates molecular hydrogen over vast regions, extending well beyond the radius of the fully ionized circumstellar Strömgren sphere expanding around the star. H_2 is dissociated by photons streaming out of the Strömgren sphere with energies in the range $11.2 - 13.6$ eV. Photons with energies above 13.6 eV ionize hydrogen atoms, and those below 11.2 eV lack the energy to dissociate H_2. The dissociation occurs in two steps. A molecule is first raised into an excited electronic state H_2^*. From this it can decay into the vibrational continuum of the ground electronic state. The vibrational continuum is a state in which the vibrations are so energetic that the molecule is torn apart; it dissociates.

$$H_2 + \nu \rightarrow H_2^* \rightarrow 2H \ . \tag{13-85}$$

Deprived of molecular hydrogen, a region lacks all coolants that could lower the temperature below 10^4 K. The only regions that may still collapse under these conditions are extremely massive domains for which the necessary Jeans mass is in place even at such high temperatures (Br03).

PROBLEM 13–15. The Jeans mass of a halo at temperature T is approximately given by $M_J G/r \gtrsim 3kT/m_H$, where r is the radius of the halo enclosing the mass. Show that at a red shift $z \sim 25$ the density of the halo is high enough for M_J to be below $3 \times 10^9 M_\odot$ even if the temperature is as high as 10^4 K. This can be shown by calculating r and then comparing the density $\sim 3M_J/4\pi r^3$ to the mean ambient density $\rho_0(1 + z)^3$, where the current value ρ_0 is noted in Tables 12.1 and 13.1. Calculate the free-fall time of such a collapse and the approximate age and red shift of the Universe when the halo finally collapses into a black hole.

Although a halo with a mass of $10^9 M_\odot$ could collapse under such conditions, it would not fragment into lower-mass stars in the absence of coolants, because none of the fragments would have masses as high as a Jean mass. For these reasons, it is now generally believed that supermassive black holes formed after the Population III stars had dissociated molecular hydrogen over large regions and the only other available coolant, atomic hydrogen, would not function at temperatures below 10^4 K. In order to rid itself of angular momentum, such a massive collapsing cloud might have to break up into two masses initially spinning about each other, and forming two black holes, rather than one. Eventually, all the angular momentum could be dissipated through interaction with ambient matter or radiated gravitational waves, and the pair would coalesce into a single black hole.

13:31 Accretion Disks Around Supermassive Black Holes

Ambient matter is gravitationally attracted to supermassive black holes. A straight infall, however, is prevented if the material has appreciable orbital angular momentum about the central mass. It may then merely make a close pass by the black hole. However, if the hole already is orbited by an *accretion disk*, infalling matter can be trapped. These disks can grow through accretion of gas, dust, and even stars that are tidally disrupted on close passage by the black hole. Since the inner portions of the accretion disk orbit the black hole more rapidly than the outer portions, the viscous drag due to this velocity gradient transports angular momentum outward. Magnetic fields in the disk can also promote the outward transport of angular momentum. As matter at the inner edge of the disk continues to lose angular momentum, it inexorably spirals into the black hole as described in Section 5:19, adding to the hole's mass and angular momentum. Most supermassive holes, accordingly, appear to be spinning at close to the maximum permitted rate discussed in Section 5:23.

13:32 The Masses of Galaxy Bulges and Central Black Holes

Many galaxies are found to have supermassive black holes at their centers. Their masses range from $\sim 10^6 M_\odot$ for the black hole at the center of the Milky Way to $\gtrsim 10^9 M_\odot$ in galaxies such as M82. The existence of these massive objects is inferred both from the orbital velocities of stars in their immediate environs and from *reverberation mapping*. This second observational technique is particularly useful when the continuum emission from the immediate environs of the black hole is so bright that individual orbiting stars cannot be readily discerned. One then looks for the time delay between a brightness variation of the continuum emitted near the black hole and a corresponding variation in the broad emission lines emanating from gas rapidly swirling about the black hole at greater distances. The time delay tells us the radial distance r of the *broad emission line region, BLR*. The widths of the spectral lines imply an orbital velocity v. The mass of the black hole can then be estimated as $M \sim v^2 r/G$, where G is the gravitational constant. An advantage is that the BLRs in distant galaxies with *active galactic nuclei, AGNs* can be hundreds of times closer to the central black hole than individual stars that could be resolved to determine their Doppler shifts.

Massive black holes appear to have formed very early in the formation of cosmic structures. Quasars and various types of galaxies with active nuclei, some of which are observed at red shifts beyond $z \sim 6$, all appear to harbor massive black holes.

One interesting aspect of these black holes is that they seem to be present in every galaxy that exhibits an elliptically shaped bulge of stars. The *bulge* looks very much like a miniature elliptical galaxy roughly confined to a radius of 0.5 kpc from the central nucleus. The mass of the bulge, determined from the observed dispersion of stellar velocities, is found to be closely related to the mass of the galaxy's central black hole. The velocity dispersion σ in the bulge tends to be of order $\sigma_0 = 200 \, \text{km s}^{-1}$ and the observed mass relation between bulge and black hole is (Tr02)

$$\log_{10}(M_{BH}/M_\odot) = 8 + 4\log_{10}(\sigma/\sigma_0) . \tag{13-86}$$

The source of this relationship is still obscure.

13:33 Growth Through Merging

The earliest formed galaxies observed at high red shifts are more abundant but also smaller than those observed nearby. Today's galaxies appear to have formed through the merger of these smaller younger galaxies. This merging continues to this day. Ultraluminous galaxies exhibiting massive bursts of star formation sometimes have two nuclei and appear to be merging, perhaps along lines indicated in Fig. 13.16. Smaller satellite galaxies are likely to be tidally disrupted, their matter pulled into

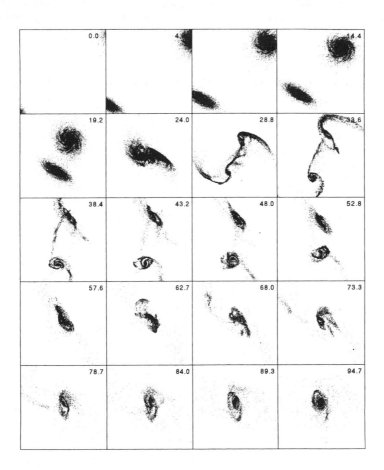

Fig. 13.16. Merger of Galaxies in a computer simulation (Mi96).

larger parent galaxies.

The Galaxy is still interacting with other members of the Local Group. In particular, the Milky Way is gravitationally disrupting the Sagittarius dwarf galaxy, a small satellite system of stars. Four globular clusters, respectively designated M54,

Terzan 7, Terzan 8, and Arp 2, originating in this dwarf galaxy are usually cited as Galactic clusters, indicating that the Galaxy is still accreting metal-poor stars that spent most of their histories in a different, lower-mass system.

The ages and chemical histories of the component galaxies from which the Milky Way formed may have ranged widely. The angular momenta about the Galaxy's center would also vary, depending on how the merger took place. This complicates attempts to establish primordial chemical abundances in the earliest-born stars within the Galaxy. It also becomes difficult to establish reliable genealogies of stars that might have formed early with relatively small abundances of heavy elements — as contrasted to later generations, progressively enriched with material processed by their forebears.

13:34 Chemical Evolution of Galaxies and the Intracluster Medium

The gradual chemical enrichment of galaxies comes about through the ejection of material processed in the interior of stars. Red giants and asymptotic giant branch stars, AGBs, eject considerable amounts of material, seen in dust clouds emanating from these stars. The gas and dust is propelled away by the star's radiation pressure acting on the dust. The dust eventually aggregates in clouds that, on cooling, give birth to new chemically enriched stars. Novae and other cataclysmic variables also enrich the interstellar medium with heavy elements.

Supernova outbursts contribute to the enrichment of heavy chemical elements. However, the blast from these stars can be sufficiently powerful to propel ejecta out of the parent galaxy, thus producing much of the iron that is routinely observed in the X-ray emission spectra from hot gases in clusters of galaxies. A crude estimate suggests that $\sim$90% of the heavy elements found in the Galaxy were ejected from low-mass evolved stars, and only $\sim$10% exploded from supernovae.

The hunt for stars that exhibit abundances of heavy chemical elements many thousands of times lower than the Sun, has permitted the identification, in the Galaxy, of low-mass stars that must have been formed when the Universe was very young and consisted of almost pure hydrogen. The heavy elements found in these stars are magnesium, silicon, calcium, and titanium which, compared to their presence in the Sun, have a higher abundance relative to iron. Although ^{48}Ti, having 22 protons and 26 neutrons must be produced through some other process, Mg, Si, and Ca all are produced by the e-process, discussed in Section 8:12. This process is associated with massive Population III stars that explode as supernovae of type SN II, isolated massive stars that collapse and then explode. The finding of these α-elements in significant quantities in the hot gases that pervade clusters of galaxies indicate that enormous supernova explosions propelled some of their ejecta out into the intracluster medium, where they became trapped in the cluster's gravitational potential (Ba05). Traces of the same elements shown in Fig. 8.12 for low-mass highly depleted Galactic stars, further indicate that these stars constitute a somewhat later generation formed from gases slightly enriched with matter ejected from a precursor Population III (Tr04, Ry96).

13:35 Formation of Our Own Galaxy

In our own galaxy we do have some additional clues about how the initial birth took place. Much of the information lies in the separate components we can identify (Bu00a). An essentially nonrotating *halo* containing around 170 globular clusters may extend out as far as 100 kpc. The clusters orbit the Galactic center along spheroidally distributed Keplerian trajectories. Nested within the central ~25 kpc of the halo, we find stars and open star clusters concentrated in two apparently coplanar flattened disks, respectively called the *thick disk* and *thin disk*. The concentration of stars here is much higher than in the halo. In their common midplane an even thinner disk of gas and dust clouds forms the *extreme disk*. The matter in these three disks has sufficient rotational speed around the Galactic center to keep the disks from plunging into the center. In the innermost parts of the Galaxy there is a bar-shaped aggregate of stars.

Near the Sun, the stars rotate about the Galactic center in nearly circular orbits at speeds of ~200 km s^{-1}. Their velocity components perpendicular to the plane are only ~20 km s^{-1} constraining them gravitationally to a realm no more than ~300 pc above or below the plane. Stars belonging to the thick disk, on average, have lower circular velocities about the galactic center, but maintain their distance from the center through a compensating higher-velocity component perpendicular to the plane that permits them to reach distances of ~1 kpc from the plane.

The metallicities of the stars in these different components are a hint to the history of their origins. In general, we expect stars richer in heavy elements to have been formed more recently, and metal-poor stars to have been formed earlier. This would be particularly true if all the matter we now find in the Galaxy had always been part of it. But it is possible that some Galactic stars were gravitationally captured through tidal disruption of smaller companion galaxies in which the history of heavy element production was very different. A one-to-one correspondence between metallicity and date of birth would then no longer hold. Nevertheless, despite such caveats, we find that globular cluster stars are metal poor.

Judged by their Hertzsprung–Russell diagrams, the ages of most globular clusters, inferred from the masses of stars that by now have turned off the main sequence, are ~12 Gyr — though certain globular clusters appear to be much younger. The thick disk has a metallicity similar to that of globular clusters, while stars in open clusters within the thin disk have appreciably higher concentrations of heavy elements and ages of ~1 Gyr. Even younger stars are found forming in the dusty regions within the extreme disk. The central bulge seems to contain a mixture of stars of different ages and metallicities.

An early picture of the history of the Galaxy was compiled by Eggen, Sandage, and Lynden-Bell, who noted the orbital parameters of the very oldest stars. These stars are deficient in metals, and this gives them an excessively large ultraviolet

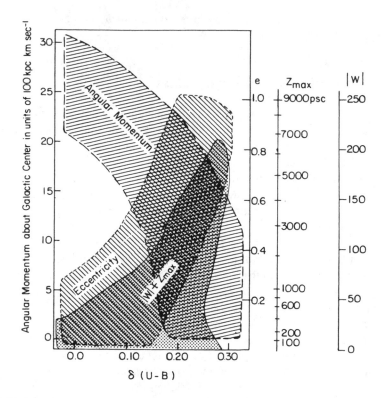

Fig. 13.17. Age of stars and their orbital characteristics. Angular momentum about the Galactic center, eccentricity e of the orbits about the center, the velocity $|W|$ in km s^{-1} perpendicular to the Galactic plane, and height z_{max} risen above the plane. Large values of $\delta(U - B)$ are associated with old stars in the Galaxy (after Eggen, Sandage, and Lynden Bell (Eg62), see text).

magnitude relative to more recently formed metal-rich stars. The difference between U and B magnitudes, $\delta(U - B)$, increases with increasing age (Eg62). As indicated in Fig. 13.17, the three authors found that the oldest stars in the Galaxy have highly eccentric, low angular momentum orbits with high velocities perpendicular to the Galactic plane. This suggests an initial, almost radial collapse toward the center. As pointed out by Oort (Oo65) the angular momentum per unit mass, measured about the Galactic center, is a factor of eight lower for highly metal-deficient RR Lyrae stars than for disk or spiral arm stars in the neighborhood of the Sun. Von Hoerner (vo55) found that globular clusters have orbital characteristics similar to the halo stars. The disk population seems to have had a different origin, apparently due to material that fell into the Galaxy at later epochs. This would be consistent with material from the *Magellanic stream* — gas orbiting the Galaxy along trajectories similar to those of the Magellanic Clouds — and other aggregates of gas falling into the plane of the Galaxy even today. Large-scale mergers observed in other galaxies suggest that either gradual or catastrophic aggregation of matter could have further influenced the structure and chemical composition of the Galaxy.

13:36 Radioactive Dating

An approximate sequence for heavy element production in the Galaxy can be obtained on the assumption that *carbonaceous chondrites* represent material from which the Solar System was formed. The chondrites can be analyzed for their content of thorium and uranium isotopes, ^{232}Th, ^{235}U, and ^{238}U. These elements are formed in the r-process (Section 8:12) in a ratio 1.6 : 1.6 : 1. Their α decay *half-lives* are 1.4×10^{10}, 7.1×10^8, and 4.5×10^9 yr, respectively. The present ^{232}Th : ^{238}U Solar System ratio shown in Table 1.1 is 3.7:1. The ^{238}U : ^{235}U ratio is 1 : 0.007.

PROBLEM 13–16. Current estimates of the age of the Galaxy are $\sim 11 \times 10^9$ yr. If a fraction x of the uranium and thorium was formed in the birth of the Galaxy and $(1 - x)$ was formed continuously between the birth of the Galaxy and the Solar System's birth $\sim 5 \times 10^9$ yr ago, show that $x < 10\%$. If all the metal formation had taken place continuously at an even rate, show that the Galaxy would have to be >25 Gyr old.

Answers to Selected Problems

13–1. Equation (13–22) provides the reciprocal of the damping time when used with values for the radiation density from Table 12.1 and the ionization fraction from Fig. 13.1. Table 12.3 provides ages as a function of red shift. Where the tables do not provide the required information, obtain the extrapolated values using the scaling relations of Section 11:12.

13–2. (a) Photons can only travel a distance $n_e \sigma_e$ before being scattered. Table 12.1 shows that the hydrogen density today is $n_H \sim 1.6 \times 10^{-6}$ cm^{-3}. For a fully ionized gas at $z = 10^4$, the electron density would be z^3 times higher, $n_e \sim 1.6 \times 10^6$ cm^{-3}, and photons could only travel a distance $(n_e \sigma_e)^{-1} \sim 10^{18}$ cm before being scattered.

(b), (c) Because we have a mixture of radiation and matter there is no single value of γ that can be used with the last expression on the right of (9–25). We must use the expression $\partial P / \partial \rho$ instead, noting that $\rho = \rho_r + \rho_B$. For the radiation-dominated era $P_r = aT^4/3$, $\rho_r = aT^4 c^{-2}$, and because of radiation domination, the adiabatic expansion in (4–129) has to be used with $\gamma = 4/3$ to obtain $\rho_B \propto 1/V \propto T^3$. After decoupling, the baryonic pressure $P_B = nkT = \rho_B kT(1+x)/m_H$ dominates the behavior of matter. The main results to be shown can be directly derived from (13–23).

13–5 (a) Using the speed of sound given in (13–23), the Jeans length of (13–39), and recognizing that dark matter dominates the mass density, the baryonic Jeans mass M_{BJ} after decoupling becomes

$$M_{BJ} \sim \frac{\pi}{6} \rho_B(t_0)(z+1)^3 \left[\left(\frac{5kT_B(t_0)(z+1)^2}{9m_H} \right)^{1/2} \left(\frac{\pi}{G\rho_M(t_0)(z+1)^3} \right)^{1/2} \right]^3$$

(b) During decoupling, as seen in Problem 13–2, the speed of sound falls from $c_s(before) \sim 1.4 \times 10^{10}$ cm s^{-1} to $c_s(after) \sim (6kT_B(t)/9m_H)^{1/2} \sim 2.5 \times 10^5$ cm s^{-1} at ~ 3000 K. For $\rho_M(t_0) = (3H_0^2/8\pi G)\Omega_M(t_0) \sim 3 \times 10^{-30}$ g cm^{-31}, this makes $M_{BJ}(after) \sim 2.5 \times 10^3 M_\odot$, while $M_{BJ}(before) \sim M_{BJ}(after)[c_s(before)/c_s(after)]^3 \sim 5 \times 10^{17} M_\odot$.

(c) The Hubble radius at $z \sim 1100$ is $c/(H_0(z+1)^{3/2}) \sim 3.5 \times 10^{23}$ cm. The radius enclosing $M_{BJ}(after)$ is $\lambda_J/2 \sim 1.5 \times 10^{19}$ cm. The mass in a Hubble sphere, therefore is considerably higher than the Jeans mass after decoupling. While the Hubble radius remains almost unchanged, the Jeans length increases by a factor of 5.6×10^4 to ~ 800 kpc, as one goes back in time from the decoupled to the fully coupled regime.

(d) From (c) it is apparent that the volume containing the Jeans mass before decoupling is larger than the Hubble sphere.

(e) The Jeans length during the radiation-dominated epoch diminishes as $(1+z)^{-2}$ with increasing red shifts, so that the volume of the sphere of radius $\lambda_J/2$ declines as $(z+1)^{-6}$, while the enclosed baryon density increases as $(z+1)^3$. This leads to an overall decline in the enclosed baryonic mass in proportion to $(z+1)^{-3}$. During the same epoch, $H \propto (1+z)^2$, the Hubble radius declines as c/H, and $\rho_B \propto (1+z)^3$. So, the mass within the Hubble radius diminishes as $(4\pi/3)\rho_B(c/H)^3 \propto (1+z)^{-3}$ with increasing red shift. The enclosed baryonic mass within a Hubble sphere is always less than the Jeans mass during the radiation-dominated epoch. Figure 13.2 illustrates this sequence of events.

13–6. Substitute $d\ell^2 = r^2 d\Omega^2 + (1 - kr^2/a^2)^{-1} dr^2$ from (11–7) and (11–14) into the expression on the left in (11–18). Then introduce a new variable $r = a(t_e)\psi$, with $\psi = \sigma(\chi)$ and $\sigma(\chi) = \sin\chi$, χ or $\sinh\chi$ depending on whether $k = 1, 0$ or -1.

13–8. (a) This can be done, for example, by substituting the expressions (11–72) and (11–74), respectively, into (13–59) and (13–57). (b) Note that the parameters $x \ll 1, y \ll 1$ in equations (11–72) and (11–74) correspond to early times. The equality of densities as well as expansion rates at early times, when the curvature term k was negligibly small, implies that the quantities $a_{(+o)}$ and $a_{(-o)}$ must assume the same value in equations (11–72) and (11–74). Similarly, expanding $x - \sin x$ and $\sinh y - y$ shows that the parameters x in equation (11–72) and y in (11–74) also must have a one-to-one correspondence because for small values, $x \ll 1, y \ll 1$, the times t then agree.

13–9. The scale factor at the time of maximum expansion in the $k = +1$ domain is obtained by first setting the time for this epoch the same for all three domains, $k = 0, \pm 1$. Then, according to (11–72), $x = \pi$ for the maximum scale factor of a_+, and $t_- = t_+ = t_0 = a_{(+o)}\pi/c = (\pi/2)a_m/c$. Because $a_{(-o)} = a_{(+o)} = a_m/2$, we then obtain from (11–74) $y \sim 2.42$, from the expansion $\sinh y = y + y^3/3! + y^5/5! + \dots$ and $a_- = 2.3a_m$ by virtue of the expansion $\cosh y = 1 + y^2/2! + y^4/4! \dots$. From (13–58) we also have $a_0 = 1.77a_m$. Since we postulated that the collapse occurred at $(z + 1) = 20$, today's scale factor in a flat universe would be $a_0 \sim 35a_m$, and the corresponding time would be $\sim 45\pi a_m/c$.

13–10. (a) From Problem 13–9, we have a scale factor $a = 1.77a_m$ in the $k = 0$ ambient medium, at $(z + 1) = 20$, when the $k = +1$ domain reaches maximum expansion. (13–58) then tells us that, at this epoch, the age of the universe measured in the $k = 0$ ambient universe is $t_{20} = (2/3)(a_m/c)(1.77)^{3/2}$. At decoupling, $(z+1) \sim 1100$, the scale factor and the radius of the vacuole are 55 times smaller in the $k = 0$ space, $a_{1100} = 0.0322a_m$ and $t_{1100} = (2/3)(a_m/c)(1.77/55)^{3/2}$. Setting t_+ equal to this in (11–72), we obtain $x \sim 0.3773$ which yields $a_+ \sim 0.0320a_m$, less than 1% lower than the radius of the vacuole in the $k = 0$ domain. The density contrast then is ~ 1.02. (b) Recalling that the red shift, or ratio of wavelengths, is proportional to the ratio of time intervals required to traverse a wavelength at the speed of light, $\Delta t = \lambda/c$, we can use equation (13–56) to calculate the red shift. We note that the cluster radius today $a\chi_o$ is of order 10^{-3} the distance to the horizon, implying that the term $[(1 - k_e\chi_o^2)/(1 - k_i\chi_o^2)]^{1/2}$ must have a value no greater than $\sim(1+10^{-6})$. In contrast, for $M = 10^{49}$ g, $a_i\chi_o \sim 2 \times 10^{23}$ cm and, $\Delta a/a \sim 10^{-2}$, the term $[1 + (2MG(a_e - a_i)/c^2\chi_o a_e a_i)] \sim (1 + 7.5 \times 10^{-5})$. This conforms to the order of the observed microwave background fluctuations.

13–11. Radiation traversing portions of the vacuole devoid of matter do not participate in the cosmic expansion. However, both on entering and exiting the vacuole, radiation encounters an expanding interface that is comoving with the external medium. The transition across these two surfaces results in a red shift that just equals the red shift that radiation bypassing the vacuole experiences in traversing the same distance. Changes in clock rates at interfaces between the vacuole and the homogeneous $k = 0$ and enclosed $k = 1$ regions can result in either a blue or red shift. For radiation of a given frequency, each increment of time dt can be considered the unit required for one wavelength to pass by. In regions where dt is large, the wavelength will be large in the same proportion. Ratios of time increments dt provided by equations such as (13–52) and (13–46) thus translate into proportions of wavelengths at transition boundaries or across regions with varying gravitational potentials. For radiation bypassing the central k_+ condensation, the wavelength shift due to entry into and exit from the vacuole is thus determined by the respective ratios $dt_v/dt_e]_{ve}$ on entry and on exit given by (13–52), and the gravitational field within the vacuole giving rise to a gravitational wavelength shift $dt_{v,in}/dt_{v,out}$ obtainable from (13–46). We then have

$$\frac{dt_{ve,in}}{dt_{ve,out}} = \left[\frac{dt_e}{dt_v}\bigg|_{in}\right]\left[\frac{dt_{v,in}}{dt_{v,out}}\right]\left[\frac{dt_v}{dt_e}\bigg|_{out}\right] = \frac{(1 - 2MG/r_{e,in})}{(1 - 2MG/r_{e,out})},$$

which gives rise to a red shift. A ray traversing a distance $\sim$2 Mpc through the vacuole, finds that the vacuole has expanded by $\sim 2.5 \times 10^{23}$ cm by the time it exits. For $M \sim 10^{49}$g and $r_{e,in} = 2$ Mpc, the red shift then amounts to $\Delta\lambda/\lambda \sim 10^{-5}$, and the apparent temperature fluctuation is in the same proportion. For radiation passing centrally through the homogeneous interior condensation, a comoving expansion or contraction of the interior $k+$ medium can, respectively, result in a corresponding blue or red shift.

13–12. The expression (9–91) derived previously for the force field generated by the disk of a galaxy can be used here and leads directly to the result if we note that $\ddot{z} = -2\pi\sigma G$. Successively integrating with respect to t leads to $z_0 - z_f = \pi\sigma Gt^2$ where $z_f << z_f$, and the subscripts, respectively, refer to the initial and final disk heights measured from the central plane. But $\sigma = 2z_0\rho_0$, the factor 2 reflecting the disk stretching from $-z_0$ to z_0. Hence, $t = (2\pi\rho_0 G)^{-1/2}$.

13–13. To produce significant optical depth at $z = 6$ the atomic hydrogen density $n_H(0)$, today, would have to be $\sim 1.5 \times 10^{-13}$ cm^{-3}. This has to be compared to the number density in a flat universe with $\Omega_B = 0.04$, which is $n_H(0) \sim 2 \times 10^{-7}$. A neutral fraction as low as 10^{-6} would suffice to produce unit optical depth at $z = 6$. At $z = 0$ an amount roughly 20 times higher would be needed $\sim 2 \times 10^{-5}$.

13–15. Following the procedure suggested, the radius of a Jeans mass of $3 \times 10^9 M_\odot$ is $r \lesssim 1.5 \times 10^{22}$ cm. The mean density of the mass is then $\rho \gtrsim 3 \times 10^{-26}$ g cm^{-3}, which is comparable to the ambient cosmic mass density of dark and baryonic matter at that red shift. The free-fall time at this density given by (13–64) is of order 250 Myr, which is considerably longer than the age of the Universe at $z \sim 25$. The approximate red shift at final collapse at age $\sim$250 Myr is $z \sim 15$.

13–16. For continuous formation of ^{232}Th and ^{238}U at rates dn_T/dt and dn_U/dt, respectively, since a time t in the past, the present abundance ratio R should be

$$R = \left[\int_0^t \frac{dn_T}{dt} 2^{-t/\tau_T} dt\right] \left[\int_0^t \frac{dn_U}{dt} 2^{-t/\tau_U} dt\right]^{-1}$$

where τ_T and τ_U are the half-lives, in Gyr, and $2^{-t/\tau} = e^{-0.693t/\tau}$. At $t \sim 27$ Gyr,

$$R = \frac{14}{4.5} 1.6 \frac{[1 - e^{-0.693t/\tau_T}]}{[1 - e^{-0.693t/\tau_U}]} = 3.7 \,,$$

but this does not agree with the ^{238}U : ^{235}U ratio. If a fraction x of the material had been generated at time $t = 11$ Gyr, when the Galaxy may have formed, and $(1 - x)$ had been generated between 11 and 5 Gyr ago, the ratio would be

$$R = 1.6 \left[x\frac{e^{-0.693t/\tau_T}}{e^{-0.693t/\tau_U}} + (1 - x)\frac{\tau_T}{\tau_U}\left[\frac{e^{-3.45/\tau_T} - e^{-0.693t/\tau_T}}{e^{-3.45/\tau_U} - e^{-0.693t/\tau_U}}\right]\right] = 3.7 \,,$$

which suggests $x \sim 8\%$ of the uranium and thorium having formed at the birth of the Galaxy and 92% formed at a steady rate between that time and the formation of the Solar System $\sim$ 5 Gyr ago.

14 Life in the Universe

14:1 Introduction

Since historic times, we have wondered where we came from and where life originated. As it became apparent that the Earth was just one planet orbiting the Sun, that the Sun was just one star among $\sim 10^{11}$ in our galaxy, and that the Galaxy itself was only one such object among $\sim 10^{11}$ similar systems populating the Universe out to a cosmic horizon, with perhaps countless more lying beyond, it became clear that life on other planets, near some other star, in some other galaxy was possible. The cosmological principle (Section 11:3) also makes this idea philosophically attractive. It would suggest that life is some general state of matter that prevails throughout the Universe. The probability of finding some form of life, however primitive, on other planets either within the Solar System or around nearby stars seems very high from this point of view. Nevertheless, we are unable to predict where life should exist, mainly because we do not yet understand the thermodynamics of living organisms and what different forms life may take.

14:2 Thermodynamics of Biological Systems

Thermodynamics distinguishes between three types of systems. *Isolated systems* exchange neither energy nor matter with their surroundings. *Closed systems* exchange energy but not matter, and *open systems* exchange both matter and energy with their surroundings. Biological systems are always open, but in carrying out some of their functions, they may act as closed systems.

Biological processes also exhibit a well-defined time dependence. As we saw in Section 11:18, some physical processes could take place equally well whether time runs forward or backward. If we viewed a film of a clock's pendulum, we would not be sure whether the film was running forward or back. Only if the film also showed the ratchet mechanism that advances the hands of the clock, would we be able to tell whether it was running in the right direction. The pendulum motion is reversible but the action of a ratchet is an *irreversible process*. Biological processes are invariably irreversible.

In an irreversible process, *entropy*, a measure of disorder, always increases. If a cool interstellar grain absorbs visible starlight and re-emits the radiation thermally, it does so by giving off a large number of low-energy photons. In equilibrium the total energy of emitted photons equals the energy of the absorbed starlight; but the entropy of the emitted radiation is larger. The increased entropy is a measure of the disorder associated with a large number of low-energy photons moving in unpredictable arbitrary directions. The initial state of a single photon carrying a large amount of energy is more orderly and, hence, characterized by a lower entropy.

Biological systems thrive on order. They convert order in their surroundings into disorder. In doing so, however, they also increase their own internal degree of order. The entropy in the surroundings increases, the internal entropy can decrease, but the total entropy of system plus surroundings always increases. The second law of thermodynamics states that the overall entropy change in the entire Universe, in any process, is always positive.

It may seem strange that biological systems can increase their internal order in this way, but actually we encountered a similar process in the alignment of interstellar grains (Section 9:13). We saw that the randomly directed kicks from the formation of hydrogen molecules on a grain's surface tended to orient the grain relative to the ambient magnetic field; it ends up spinning with its major moment of inertia axis lying along the direction of the field. An oriented set of grains shows greater order than randomly oriented dust. The decrease in the grains' entropy is produced through the absorption of low-entropy, high-momentum kicks, and the ultimate dissipation of heat through emission of high-entropy, roughly isotropic, infrared radiation.

These interstellar dust grains are in a state of *stationary nonequilibrium*. Such a state is characterized by transport of energy between a source at high temperature (the heat of formation of molecular hydrogen) and a sink at low temperature (the Universe). There is no systematic change of the system in time, although statistical fluctuations in the orientation, angular momentum, and other properties of the grains do take place.

We may hope that the study of stationary nonequilibrium processes will lead to a better understanding of the behavior of biological systems (Pr61). Most biological processes follow this pattern. When a plant absorbs sunlight — photons whose energy is typically $\sim 2\,eV$ — and thermally re-emits an equal amount of energy in the form of 0.1 eV photons, it is acting as a stationary nonequilibrium system.[1] Pendulum clocks are also stationary systems. Low-entropy energy in the form of a wound-up spring is irreversibly turned into high-entropy heat. As Schrödinger pointed out (Sc44), living organisms and clocks have a thermodynamic resemblance.

Nonequilibrium characterizes virtually all astrophysical processes. Energy flows out of highly compact sources into vast empty spaces. Any biological system stationed near one of these sources could make good use of this energy flow. It would therefore seem that the conditions necessary for the existence of life in one form or another would be commonplace. Maybe life does abound in the Universe; perhaps we only fail to recognize it.

---

[1] The photochemistry of green plants is a complex process that also involves the buildup of large molecules.

In his science fiction novel *The Black Cloud*, Fred Hoyle (Ho57) speculated that interstellar dust clouds might be alive. From a thermodynamic viewpoint the situation would be ideal. We know that dust clouds absorb perhaps half of the starlight emitted in a spiral galaxy like ours. The grain temperatures are so low that a maximum increase in entropy can be produced. What is uncertain, however, is whether the grains are not too cold to make good use of the available energy. At the 10 to 20 K temperatures that might be typical of interstellar grains, the mobility of atoms within the grains is so low that the normal characteristics we associate with life might be ruled out (Pi66).

A thermodynamically similar scheme was suggested by Freeman Dyson (Dy60) who proposed that intelligent civilizations would build thin shells around stars to trap starlight, extract useful energy, and then radiate away heat in the infrared. So far, infrared astronomical observations have not identified such structures, though they probably would have if they were common.

Our experience on Earth is that life will proliferate until stopped by a lack of resources or excessive toxins. It would perhaps be surprising if no form of life had adapted itself sufficiently to make use of the huge outpouring of energy that goes on in the Universe and is apparently just going to waste.

A search for unknown forms of life might concentrate on striking examples of nonequilibrium. A Martian astronomer, for example, would find only two pieces of evidence for life on Earth. The first is a radio wave flux that would correspond to a nonequilibrium temperature of some millions of degrees. This is produced by radio, television, and radar transmitters. The second is an excess of methane, CH_4, which is very short-lived in the presence of atmospheric oxygen. It is converted into CO_2 and H_2O. Its nonequilibrium concentration, which could be spectroscopically detected from Mars, is rapidly replenished by methane bacteria that live in marshes and in the bowels of cows and other ruminants (Sa70b). Some terrestrial CH_4, however, is known to also emanate from mid-ocean volcanic vents, today, and this process could have been more prevalent in earlier times, making CH_4 a somewhat ambiguous marker of life. Consequently, oxygen molecules, O_2, ozone, O_3, and water are also major biomarkers of current interest as a search for molecular indicators of life on planets outside the Solar System, the *exoplanets*, is being planned (De02a).

Water certainly is a necessary ingredient for all life on Earth. The search for life elsewhere accordingly tends to concentrate on planets able to sustain water in liquid form. This requires the planet to be sufficiently far from its parent star so that water on its surface will not evaporate, yet sufficiently near that it will not freeze. This range of distances and temperatures is referred to as the *habitable zone* around the star.

14:3 Organic Molecules in Nature and in the Laboratory

Granted that we do not specifically know how to search for exotic forms of life, could we not find indications of extraterrestrial life in a form familiar on Earth? All terrestrial living matter contains organic molecules of some complexity — proteins and nucleic acids, for example — and we might expect to find either traces of such molecules or at least of their decay products.

We know of two quite distinct locations in which complex molecules are found. There may be many more. First, observations of interstellar molecules by means of their microwave spectra have revealed the existence of such organic molecules as hydrogen cyanide, methyl alcohol, formaldehyde, and formic acid. Larger molecules, such as the sugar glycolaldehyde, CH_2OHCHO, have also been found to be quite prevalent in interstellar space. Infrared observations similarly have shown the existence of the even larger, polycyclic aromatic hydrocarbon molecules. Increasingly complex organic molecules continue to be discovered (Ho01).

Many of these molecules have been synthesized in the laboratory under simulated interstellar conditions, and the mechanisms involved are being studied quantum chemically. Not only the formation, but also the stability of these molecules needs to be understood for the harsh conditions of interstellar space, where prevailing ultraviolet radiation continually threatens to destroy larger molecules. If we are to unravel the origins of life, all this attests to the central importance of understanding the basic chemical processes active in environments quite different from those found on Earth.

Second, an analysis of a meteorite — a *carbonaceous chondrite* — that fell near Murchison in Australia on September 28, 1969, showed the presence of many hydrocarbons and of 17 amino acids, including six that are found in living matter (Kv70). One such amino acid was alanine. It has the form

$$\text{Alanine}: \qquad \underset{\underset{NH_2}{|}}{CH_3—CH—} \overset{\overset{O}{\|}}{C} —OH \qquad (14–1)$$

All *organic acids* are marked by the group of atoms

$$\overset{\overset{O}{\|}}{—C} —OH$$

and *amino acids* contain the additional characterizing *amino group* NH_2.

Contamination by terrestrial amino acids seems to be ruled out by three features of these observations.

(a) Alanine can occur in two different forms: one in which the CH_3, NH_2, COOH, and H surrounding the central carbon atom are arranged in a configuration that causes polarized light to be rotated in a left-handed screw sense; the other in which polarized light would be rotated in the opposite sense. These are, respectively, labeled L- and D-alanine. The symbol L stands for levo — left — and D for dextro — right.

All amino acids can be chemically derived from alanine. If derived from L-alanine such an acid is called an L-amino acid, and if derived from D-alanine, a

D-amino acid. All amino acids found in proteins in terrestrial living matter are L-amino acids. Although not all of them rotate light in a left-handed screw sense, they all can be structurally derived from L-alanine.

The Murchison meteorite showed D- and L-forms in essentially equal abundances. These amino acids are therefore very unlikely to have been biogenic contaminants.

On Earth, amino acids are overwhelmingly in the left-handed form. Why this should be is a mystery. Chemically, the right- and left-handed forms are equally probable. They are mirror images of each other. Perhaps evolutionary considerations have played a role. It might be that primitive life existed in a *racemic* mixture — having both L and D forms — and that the L-form won out in a competition for raw materials essential to life. It might also have been impossible for racemic life to exist in an effective way. The search for nutrients would be inefficient. A bolt in search of a nut is more readily satisfied if all nuts and bolts have a right-handed thread; trying to match nuts and bolts from a racemic mixture would be vexing.

(b) A second distinctive feature of the Murchison material was that the ratio of carbon isotopes ^{13}C to ^{12}C was about twice as high as normally found in terrestrial material. This too indicated that contamination could be ruled out.

(c) Finally some of the amino acids found in the material consisted of nonprotein amino acids. They could not have been contaminants.

The Murchison meteorite more recently has also been shown to contain traces of more than a dozen types of sugar-related organic compounds, molecules with a number of OH groups attached to their carbon structure. These polyhydroxylated compounds, or *polyols*, including sugars, sugar alcohols, and sugar acids are essential to all recognized forms of life. They are components of nucleic acids, discussed in Section 14:5 below, and are also found in cell membranes and can be a source of energy. It is possible that these and other organic compounds delivered to the young Earth billions of years ago through infall of interplanetary matter played a role in the origin of life (Co02).

We still have to ask how these molecules arise. Is their fabrication simply achieved under normal astrophysical conditions? The answer to this seems to be "Yes."

A series of experiments that had their foundations in the work of S. L. Miller (Mi57a), (Mi59) has shown that amino acids and other molecules found in living organisms can be produced artificially if mixtures of gases such as ammonia NH_3, methane CH_4, and water vapor H_2O are irradiated with ultraviolet radiation, subjected to electrical discharges or shocks, or to X-ray, gamma-ray, electron or alpha-particle bombardment. Experiments by two groups have also shown that amino acids form when different mixtures of water, ammonia, methanol, carbon monoxide, carbon dioxide, and hydrogen cyanide, HCN, are frozen under high vacuum conditions onto a surface at a temperature of only 12 to 15 °K and irradiated by ultraviolet radiation (Be02), (Ca02). Thus far, the laboratory-produced molecules have all resulted in racemic mixtures.

Since the laboratory experiments approximate conditions known to prevail on interstellar grains, there is a possibility that the amino acids originated well before the Solar System formed. They could then have become part of the protosolar nebula, and eventually found their way into asteroids and comets and their meteoritic fragments that continue to bombard Earth. The simplest of the amino acids, glycine, NH_2CH_2COOH, has been found in the hot molecular cores of three interstellar regions, where the molecules exhibit rotational temperatures ($\sim$100 K) with column densities of order 10^{14} molecules cm^{-2} (Ku03). However, laboratory simulations suggest that, once formed, the amino acids may also be rapidly destroyed by ultraviolet radiation in space (Eh01). A precarious balance may, therefore, be involved, in which amino acids can only survive if well-shielded inside a dark cloud or a larger body, such as a comet or meteorite.

Because all the gases that are used in the cited experiment are abundant in the Universe, where high-energy irradiation is also common, accounting for the presence of prebiotic molecules no longer seems difficult. This should be true not only in the Solar System, but also in other circumstellar or planetary environments.

Although biogenic molecules are readily formed by energetic bombardment, they are also readily destroyed by it. Such a molecule captured into Earth's atmosphere might therefore be destroyed unless it were rapidly removed to a safer place. On Earth, rain could have washed molecules out of the atmosphere and into the oceans where they would be shielded from destructive irradiation by a protective layer of water.

We note that the conditions for forming life — or highly ordered biogenic molecules — are those that seem thermodynamically favorable (Section 14:2). There is a source of low-entropy energy in solar ultraviolet or cosmic-ray irradiation and a possibility of converting this energy into a higher entropy form through collisions with atmospheric molecules or through radiation at long wavelengths.

14:4 Origins of Life on Earth

Before we can make a rough guess about the origins of life on Earth, we should know something about the Earth's atmosphere and oceans during the æons immediately following the birth of the Solar System.

The origin of water on Earth is still under debate. The deuterium abundance on Earth differs widely from that of the Sun, the planets Jupiter and Saturn, and the comets. The solar D/H ratio is $\sim$2 × 10^{5}. Jupiter and Saturn exhibit similar ratios. On Earth this ratio is $\sim 1.5 \times 10^{-4}$. Spectroscopic studies of water sublimated from comets show considerably higher D/H ratios $\sim 3.8 \times 10^{-4}$. This level is similar to the abundance ratio observed in the Martian mantle, suggesting a cometary origin for water on Earth and Mars. The only other source of deuterium that seems to closely match that found on Earth is in clays associated at submicrometer scales with organic materials in carbonaceous meteorites. The water on Earth may, therefore, have its origins in a few giant impacts either with comets or asteroidal bodies that had D/H ratios comparable to those of the carbonaceous chondrite clays, having originated in the same part of the Solar System, the asteroidal belt. This conclusion, however, is still quite uncertain (Ro01).

When the Earth formed ∼4.5 billion years ago, its initial atmosphere was devoid of molecular oxygen. The most prevalent atmospheric molecules were those strongly reduced by abundant hydrogen — methane, ammonia, water, and ethane. Uncertainty persists on precisely when the earliest forms of life originated on Earth. There is common agreement that life existed when bacterial fossils formed in the 1.9-billion-year-old Gunflint Formation of Ontario, in Canada (Mo05). But it may have flourished appreciably earlier. Studies of ancient rocks, soils, and microfossils show that atmospheric O_2 dramatically increased roughly 2.3 billions years ago. This oxygen is thought to have been produced by *cyanobacteria*, the only organisms existing at the time capable of photosynthetically producing O_2. The overall reaction of interest is

$$CO_2 + 2H_2O \rightarrow CH_4 + 2O_2 \ .$$

The CH_4 is *photolyzed*, destroyed by sunlight, and the hydrogen escapes to space, leaving a higher concentration of O_2. Although the first evidence for cyanobacteria dates back 2.7 billion years and possibly earlier, the rise in oxygen did not take place until 400 million years later. Initially, some of the oxygen may have been quickly depleted in oxidizing minerals on the Earth's surface but, whatever the reason, the production of oxygen eventually appears to have overtaken the depletion, leading to a rapid change from a reducing to an oxygen-rich atmosphere (Ka01, Ca01).

The earliest forms of life coexisted with the earlier reducing atmosphere and were anaerobic. Presumably the very first organism to be formed found itself in a rich environment of large organic molecules (Op61a, b) that had been built up by ultraviolet radiation, lightning in thunderstorms, and other sources of low-entropy energy. Such an organism could feast and procreate at will, until the supply of organic molecules dwindled. Those organisms that obtain energy by breaking down pre-existing molecules are called *heterotrophs*. Clearly they would be at a disadvantage compared to *autotrophs*, organisms which in addition could also make use of energy in other forms; autotrophs that make use of sunlight are called *photoautotrophs*. The autotrophs probably soon took over. Initially they must have been *anaerobes*, but with the escape of hydrogen these were at a disadvantage compared to *aerobes* from which all the higher organisms later evolved. When the oxygen concentration in the atmosphere became roughly one percent of its present abundance, respiration should have become a more efficient process than fermentation and the aerobes may have originated at that time.

Living organisms naturally suffer genetic mutations — alterations in the code that defines the makeup of the *progeny*. The mutation rate can be artificially increased through X-ray and other destructive bombardment. The aerobes probably arose from anaerobes through mutations. Able to thrive on atmospheric oxygen, they soon became the dominant form of life. Anaerobes today proliferate only where atmospheric oxygen is somehow excluded.

The balance between stability and mutability appears particularly important to the success of life. Without mutability a species cannot adapt to changes in its environment, but without some stability, higher forms could not evolve either. In Darwin's theory of survival of the fittest, these fittest are likely to be produced through occasional mutations of rather stable forms.

The death of individuals also appears essential in order that life may evolve. Yet, for life to evolve optimally, each fit individual should attempt to survive — resist death. Presumably there is an optimal eugenic life span that varies from species to species. Some male spiders are devoured immediately after mating. For humans, a longer life span may be desirable because they are needed to help rear the young.

A grouping of primitive cellular life led to symbiotic multicellular forms carrying out complementary, specialized tasks. Ultimately the process led to the higher forms of life encountered today. Even now, however, symbiotic relations are widely encountered between fungi and plants. Symbiotic associations between fungi and animals may also be pervasive (Bℓ00). Interestingly, irreversible thermodynamics should play a role not only in helping us understand the metabolism of life, as in Section 14:2. It is likely that the growth of more highly organized life forms and relationships can also be described using the methods of irreversible thermodynamics.

14:5 The Chemical Basis of Terrestrial Life

All life on Earth contains rather similar organic compounds. The proteins, including the enzymes acting as biological catalysts are formed from a set of 20 ubiquitous amino acids, and, more rarely, from one of at least two additional ones discovered to date (Sr02, Ha02). All organisms also carry the genetic information required for their propagation in the form of *RNA* or *DNA* chains. RNA stands for *ribonucleic acid*, DNA for *deoxyribonucleic acid*. Both RNA and DNA are long-chain molecules, each built from four different basic building blocks, three of which are common to both RNA and DNA. DNA is characterized by sequences of the four nitrogen-containing bases, adenine (A), guanine (G), cytosine (C), and thymine (T). RNA substitutes uracil (U) for thymine. Though genetic information is generally handed down from generation to generation by means of the DNA code, primitive life is believed to have sprung up reliant on RNA. RNA sequences consisting of triplets of these bases or *nucleotides* — AAC, UGC, GGC,... — provide an alphabet that could have spelled out a set of genetic instructions to determine the characteristics of a progeny. A similar triplet code would have later evolved for DNA.

Each biological species propagates a common set of genetic instructions to its progeny. These differ from the genetic makeup of any other species. Though individuals within a species carry a set of slight variants in their genetic makeup — their *genome* — they differ from each other far less than does one species from all others. The genome of different species has been analyzed, and a family tree has been established that shows the relationship between different species — as judged by the similarity of, or differences between, their genetic codes (Fig. 14.1). Automated procedures for tracing the prevalence of different genetic markers among an increasing variety of species, are leading to the construction of progressively more accurate depictions of the tree of life (Ci06).

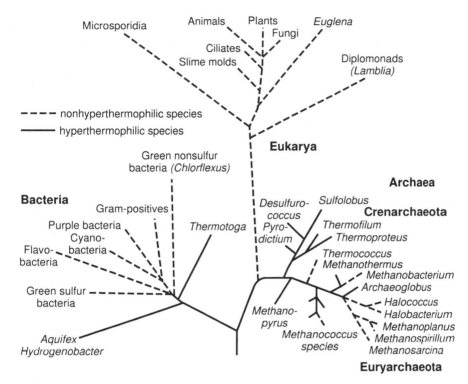

Fig. 14.1. The tree of life showing the three domains of life and the genetic relation among species (after Woese *et al*). Drawing after Otto Kandler (see (Mo97)), reprinted with permission from *Science* ©1997 American Association for the Advancement of Science.

Most interesting about this tree is that it shows the common ancestry of three broad domains of life, the *eukarya* including fungi, plants, and animals; the *bacteria*; and the *archaea* which had not even been recognized as a separate domain until 1977, when biologist Carl Woese announced their discovery. The eukarya split off from the archae at a very primitive stage, just as their combined branch had somewhat earlier let bacteria go their separate genetic way. All the species shown in Fig. 14.1 still exist today. Their common ancestors have yet to be found. The common feature at the base of the tree, however, is that all of the species are *hyperthermophiles*, heat-loving organisms that live at temperatures of 80 to 110 °C or above. Some hyperthermophiles are known to thrive at temperatures as high as 113 °C, which would permit them to live at depths as great as 5 km beneath the surface of Earth. This thermophilic common heritage suggests that primitive life may have originated deep within the Earth, or in hot springs, or deep in the oceans where volcanic vents spew forth material. Archaea found in deep terrestrial hot springs seem to derive their energy from geothermal hydrogen reacting with carbon dioxide to produce methane (Ch02):

$$4H_2 + CO_2 \rightarrow CH_4 + H_2O \ .$$

These organisms are of interest because they suggest the search for subsurface life on Mars that might subsist on the same sources of energy.

Some rocks as old as 3.85 Gyr, are calcium phosphates known to be produced today as metabolic byproducts. These minerals have recently been claimed as offering evidence of early life. Many types of rocks are known to have been produced by micro-organisms over the æons and it is often difficult to decide whether microscopic rock structures are biologically produced or are inclusions that might have a nonbiological origin.

14:6 Laboratory Syntheses

Most theories of the origins of life assume that rather long molecular chains — polymers consisting of 30 to 60 monomers — are needed. The simplest organisms appear to require enclosure in a membrane, or *vesicle*, to confine and protect the chemical chains capable of replicating themselves. Some of the simplest vesicles are made of *amphiphiles* – long molecules, one end of which is oily and is *hydrophobic*, repelling water, whereas the other end is *hydrophilic*, bonding well to water molecules. Such chains can assemble into a closed bubble, the hydrophobic ends pointing inward and the hydrophilic pointing outward. In the laboratory such simple membranes have been synthesized. They are stable as long as the ambient water has no salinity. This has suggested to some workers that life might have originated in fresh water rather than in the oceans.

The required amphiphiles could have a cometary origin. When a comet is heated by sunlight it releases its carbon-rich ices to interplanetary space. Solar ultraviolet irradiation may then convert some of this material into simple hydrocarbons. According to one suggestion, some of these hydrocarbons eventually fall on Earth, come into contact with water, and coalesce into vesicles (Ir02).

Other experiments suggest that complex molecules that actually are responsible for replication, possibly primitive RNA, could have formed on the surfaces of clays (Fe96).

Some preliminary experiments with RNA-catalyzed RNA polymerization have also been successful (Ek96). Another experiment has shown that at least one amino acid chain is able to replicate itself in a soup of two shorter amino acid chains. This autocatalytic process, carried out under plausible prebiotic conditions, is instructive because self-replication of such proteinlike structures may be a significant step in providing insight on primitive reproduction. The molecules formed in this way contained 32 amino acids in a helical configuration formed from a mixture of 15- and 17-amino-acid fragments. Especially significant is the particular segment produced in this fashion; it occurs as a fragment in a yeast enzyme (Le96).

14:7 Panspermia

Until recently, the chances of terrestrial life springing up unassisted seemed so remote that extraterrestrial origins were considered a distinct possibility. An extraterrestrial origin would not explain how or where life started, but would obviate the need for its independent creation in many different places. If microbial life could be formed once, on some planet where it flourished until an asteroidal impact blasted some primitive forms out into space, it might then migrate over large distances, eventually settling down and proliferating elsewhere. Microbes may be no larger than interstellar dust grains, and many billions of them might possibly survive a major asteroidal impact and be blown into space. Many primitive forms of life have shown themselves to be extremely hardy, able to go through long periods of hibernation before reviving when provided adequate conditions. The panspermia hypothesis has rested on the credo that propagation of life could have taken place in this way. Panspermia may well be possible. But as a more productive strategy, we should continue to investigate whether laboratory experiments of the type described in Section 14:6 might not provide insight into how primitive life could have spontaneously originated on Earth.

14:8 Higher Organisms and Intelligence

The mutation rates now deduced from the paleontological record suggest that once primitive life had sprung up on Earth, more complex forms would naturally follow through a process of mutation and survival of the fittest. Differentiation of cells in higher organisms, permitting the formation of individual organs such as eyes, ears, hands, or a brain, all offered genetic advantages that led to the proliferation of successful mutations. Intelligence certainly has its advantages, though the aggressive tendencies of the human race give rise to fear that we might ultimately destroy ourselves through some catastrophic military invention. Barring that, however, are there ways of searching for other intelligent species elsewhere in the Universe? This is a question that has been asked since prehistoric times. We still have no answers but a *Search for Extra-Terrestrial Intelligence, SETI*, has started in recent years (Br97). It searches for radio signals from nearby stars that might reveal complex coded messages, perhaps somewhat like terrestrial radio broadcasts. Once dismissed as a science fiction dream, this search for intelligent life elsewhere in the Universe is now considered a valid investigation to help us understand our place in the Cosmos.

14:9 Communication and Space Travel

If life exists elsewhere in the Universe, perhaps it also shows intelligence. If it is intelligent, perhaps it has organized into a civilization. How should we exchange information with it and how would others be likely to get in touch with us (Sh66), (Dr62)?

This is a problem in communications. How do we most effectively send messages over large distances? How does the enormous time lag between sending and reception of electromagnetic signals affect the problem of exchanging information? These questions are actively being studied, but no single optimum way has yet been discovered. Much depends on what we would best like to do.

If you like to travel, perhaps a rocket journey at relativistic speeds would suit you. But then you must decide how to stay alive during the long trip. Suggestions have been made for deep-freezing spacemen who would undertake the journey. Thus far, nothing much bigger than a frog has been successfully frozen and revived, and it is not clear whether the technique could be developed for large mammals. Unmanned spaceflights, or flights in which several generations would succeed each other before reaching their destination, are also possibilities.

An even more speculative alternative would be to travel across the Universe through *worm holes*. These are tunnels in hyperspace that provide shortcuts. Kip Thorne and his students have spent considerable effort on understanding whether such tunnels could be constructed and whether, once constructed they might be kept from closing up on themselves (Th94). Deep questions of general relativity and quantum mechanics are involved and, until a viable theory of quantum gravity is in hand, we will not know for sure.

If we cannot travel easily to distant parts of the Universe, might we be better off restricting ourselves to communicating through transmission of signals? If so, are radio or visible signals, or perhaps infrared or X-ray messages the best choice? Is there any one electromagnetic frequency that is optimal?

If such a frequency is found, we still must ask ourselves whether its characteristics are optimal only because of our specialized technological resources, or whether there is some more fundamental reason for choosing one particular means of communicating. If we choose the wrong means for transmitting our signals, no one is likely to receive them. We are also likely to miss messages sent by other civilizations if we do not know what means they might employ and what kind of receiver we must build.

What about communication at speeds greater than light? This question has also received serious attention (Cr95). In Chapter 5 we mentioned tachyons, particles that might travel faster than light. If they existed they would have many desirable properties. They might travel at millions of times the speed of light and could make meaningful two-way conversations a possibility. Moreover, high-speed tachyons should require only low-transmission energy, as seen from (5–34) and, might therefore be economical. Finally, tachyons would apparently free us from the limitations imposed by cosmic horizons (Section 11:15). The one disadvantage might be that tachyons seem not to readily interact with normal matter — otherwise they should probably have been synthesized or detected by now. Construction of suitable transmitters and receivers might therefore be difficult.

There are apparently endless sets of questions to be answered before optimal means for communicating with other civilizations can be ascertained. To show what these might be, we may consider two quite different questions in Problems 14–1 and

14–2 below. They are chosen almost at random, but together they show that space engineers of the future will need to have a deep understanding, both of naturally occurring astrophysical phenomena and of fundamental processes, that might permit them to open new communications links across the Universe, much as we have managed to build such links here on Earth.

PROBLEM 14–1. A spaceship slowly accelerates on its voyage from Earth to a distant galaxy. As it accelerates to ever higher speeds, it suffers collisions with interstellar gas and dust, with photons crisscrossing space, with magnetic fields, and with cosmic-ray particles. Estimate the effects of these and other possible particles and fields of interstellar and intergalactic space on the momentum of the spaceship, electric charges deposited on the ship and the effect of these charges, the abrasion and ablation effects on the hull of the ship, heating effects, and so on. What are the most serious limitations? Almost everything discussed in Chapters 6 and 9 bears on this problem.

PROBLEM 14–2. The rate at which messages can be transmitted and received is normally proportional to the area of the transmitter A and to the solid angle Ω subtended by the receiver at the point of transmission. Let us now assume that the transmitted particles or waves have a momentum range Δp, and that the number of message bits that can be transmitted per unit time — the *bit rate* — equals the rate at which distinct phase cells are transmitted (4–65).

(a) For an electromagnetic wave show that the bit rate is

$$\text{photon bit rate} = A\Omega\frac{\nu^2}{c^2}\Delta\nu, \tag{14-2}$$

where ν is the frequency, $\Delta\nu$ is the bandwidth of the transmitted beam, and the antenna only transmits photons of one polarization.

(b) For a tachyon system, show that if (4–65) is applicable,

$$\text{tachyon bit rate} = \left|\frac{A\Omega}{h^3}m^3c^4\frac{\Delta N}{N^3}\right| \quad \text{for } N \gg 1, \tag{14-3}$$

where we have assumed that transmission occurs for tachyon velocities ranging from $V = Nc$ to $V = (N + \Delta N)c$, where N is a large number. If the tachyon mass is of the order of the electron mass and the radiation frequency is that of visible light, show that the tachyon bit rate for $N \lesssim 10^7$, $\Delta N \sim 0.5N$ is several orders of magnitude greater than the electromagnetic bit rate. Show that for $N \sim 10^8$, however, the bit rate and energy expenditure would be comparable to visible light. Equation (5–34) is useful in tackling this problem.

Answer to a Selected Problem

This problem is highly speculative, particularly in view of some of the difficulties cited in Chapter 5.

We assume that phase-space arguments determine the distinguishability of tachyons and that the number of distinguishable tachyons transmitted per unit time determines the bit rate. For a receiver with area A and reception solid angle Ω, the volume from which tachyons are received per unit time is ANc, where N is the tachyon speed measured in units c, the speed of light. The momentum space volume occupied by these tachyons is $\Omega p^2\, dp$, per mode of polarization. The number of distinguishable tachyons incident on the detector in unit time (here referred to as bit rate) therefore would be

$$\left| \frac{ANc\Omega p^2\, dp}{h^3} \right|.$$

We make use of the relativistic expression

$$\mathcal{E}^2 = p^2 c^2 + m^2 c^4 = m^2 c^4 (1 - N^2)^{-1}$$

relating energy $\mathcal{E}$ and rest–mass m to momentum and velocity. This leads to the (imaginary) momentum value

$$p = \frac{N}{\sqrt{1 - N^2}} mc$$

and the bit rate obtained for a velocity range $c\, dN$ reads

$$\left| \frac{A\Omega}{h^3} m^3 c^4 \frac{dN}{N^3} \right| \quad \text{for} \quad N \gg 1.$$

The corresponding expression for electromagnetic radiation is

$$\frac{A\Omega \nu^2\, d\nu}{c^2},$$

where $d\nu$ is the frequency of the radiation. If we take the frequency to be that of visible light, and take m to be an electron mass, the tachyon bit rate is seen to be many magnitudes greater than the electromagnetic bit rate, as long as N remains less than about 10^7 and $dN/N \sim d\nu/\nu$. At that speed, the energy per tachyon would be about $10^{-7} mc^2$ corresponding to about 0.1 eV, while the visual radiation would require a transmission energy about an order of magnitude higher.

If $N \sim 10^8$, the bit rate and energy expenditure per message is comparable to that for visible light, but communication across the Universe can be achieved in times of the order of 100 years. Even if tachyons exist, it is not clear whether they are stable (Be71). If they exist but are unstable they would not be suitable information carriers.

Tachyons are speculative; but they remind us of how much we still need to learn before we understand the contents of our Universe or the prospects of finding intelligent extraterrestrial life.

Epilogue

At crucial points in this book we have been stopped by unsolved problems. Some of the most important questions that remain unanswered involve the delicate balance among the laws of Nature and the contents of the Universe that have made possible the emergence of stars, planets and life.

Had the primordial density fluctuations in the Universe that gave rise to galaxies and stars been of the order of $\delta\rho/\rho \lesssim 10^{-6}$, with the current ratio of baryons to photons $n_B/n_\gamma \sim 10^{-9}$ and the current cosmological constant Λ, gas accumulated in growing condensations would have been so dilute that it could not have cooled radiatively. No stars would have formed, no planets would exist, and life as we know it could not have emerged. Had these density fluctuations been as high as 10^{-3}, under the same circumstances, major cosmic structure would have collapsed to form giant black holes. Stars and planets would then be largely lacking.

If the cosmological constant had been higher by an order of magnitude than observed, other conditions in the Universe remaining unaffected, the cosmic expansion would have been so rapid that no galaxies could have formed. And if the ratio of the gravitational to the electromagnetic forces had been vastly different, the evolution of the Universe would also have taken a quite different course.

If the values of such fundamental physical parameters as the masses of protons and electrons m_p and m_e, the value of the gravitational constant and the electric charge G and e, the proton-to-photon ratio n_B/n_γ, the cosmological constant Λ, and the amplitude of primordial fluctuations $\delta\rho/\rho$, were arbitrary, the emergence of life and astrophysicists capable of studying the Universe would have been quite improbable (Li06)?

Some cosmologists have argued that an *anthropic principle* must have been at work to shape the parameters characterizing our Universe so as to make the existence of life likely. One way of explaining this is that our universe is just one among myriad others, each with its own fundamental constants and corresponding history, each entirely detached from our own Universe. Among these uncountable numbers of universes, only a minuscule fraction would be capable of supporting the existence of life, and the emergence of astrophysicists seeking to comprehend the Cosmos. Our universe, the anthropic principle claims, just happens to be one of these.

The anthropic principle does not provide any indication that the means will ever be found to detect these other universes. Nor is it clear that it has the predictive virtues demanded of all other scientific principles. Time will tell whether it is useful.

In the meantime, we continue to pursue whatever paths may lead us toward a deeper understanding of the Cosmos. Among the most important unanswered questions are:

(1) Do the laws of physics as we know them apply on the largest scales of the Universe and for all time?

(2) Did the Universe have a beginning, and what exactly is time?

(3) What is the topology of the Cosmos, and how could we detect the existence of different branes if they exist?

(4) Is there a connection between the structure of the Universe and the structure of elementary particles?

(5) What is dark energy?

(6) What is dark matter?

(7) Why does the Universe have a basic preference for matter over antimatter?

(8) Why are there so many more photons than baryons ?

(9) How did life originate and do other intelligent civilizations exist?

(10) Are we even asking the right kinds of questions?

A Astronomical Terminology

A:1 Introduction

When we discover a new type of astronomical entity on an optical image of the sky or in a radio-astronomical record, we refer to it as a new *object*. It need not be a star. It might be a galaxy, a planet, or perhaps a cloud of interstellar matter. The word "object" is convenient because it allows us to discuss the entity before its true character is established. Astronomy seeks to provide an accurate description of all natural objects beyond the Earth's atmosphere.

From time to time the brightness of an object may change, or its color might become altered, or else it might go through some other kind of transition. We then talk about the occurrence of an *event*. Astrophysics attempts to explain the sequence of events that mark the evolution of astronomical objects.

A great variety of different objects populate the Universe. Three of these concern us most immediately in everyday life: the Sun that lights our atmosphere during the day and establishes the moderate temperatures needed for the existence of life, the Earth that forms our habitat, and the Moon that occasionally lights the night sky. Fainter, but far more numerous, are the stars that we can only see after the Sun has set.

The objects nearest to us in space comprise the *Solar System*. They form a gravitationally bound group orbiting a common center of mass. The Sun is the one star that we can study in great detail and at close range. Ultimately it may reveal precisely what nuclear processes take place in its center and just how a star derives its energy. Complementing such observations, the study of planets, comets, and meteorites may ultimately reveal the history of the Solar System and the origins of life. Both of these are fascinating problems.

Beyond the Solar System lies the rest of the Universe, the grand structure of which we form a minuscule part.

A:2 The Sun

The Sun is a star. Stars are luminous bodies whose masses range from about 10^{32} to 10^{35} g. Their *luminosity* in the visual part of the spectrum normally lies in the range between 10^{-4} and 10^6 times the Sun's energy outflow. The *surface temperatures* of these stars may range from no more than ~ 1000 K to about 50,000 K. Later in this Appendix, we will see just how we can determine the relative brightness of stars, and the difference between stars and their lower-mass counterparts, the *brown dwarfs*. The determination of temperatures is discussed in Chapter 4.

The Sun, viewed as a star, has the following features.

(a) Its radius is 6.96×10^{10} cm. Although occasional prominences jut out from the solar surface, its basic shape is spherical. The equatorial radius is only a fractional amount larger than the polar radius: $[(r_{eq} - r_{pol})/r] \simeq 6 \times 10^{-6}$ (Di86).

(b) The Sun's radiant luminosity, the rate at which it emits electromagnetic energy, is 3.85×10^{33} erg s^{-1}. Nearly half of this radiation is visible, but an appreciable fraction of the power is emitted in the near ultraviolet and near infrared parts of the spectrum. Solar X-ray and radio emission make only slight contributions to the total luminosity.

(c) The Sun's mass is 1.99×10^{33} g.

(d) Three principal layers make up the Sun's atmosphere. They are the photosphere, chromosphere, and corona.

(i) The *photosphere* is the layer from which the Sun's visible light emanates. It has a temperature of about 6000 K.

(ii) The *chromosphere* is a layer some ten to fifteen thousand kilometers thick. It separates the relatively cool photosphere from the far hotter corona.

(iii) The *corona*, whose temperature is $\sim 1.5 \times 10^6$ K, extends from $1.03 R_\odot$, or about 20,000 km above the photosphere, out to at least several solar radii. Its outer edge merges continuously into the *solar wind* — interplanetary gas, mainly protons and electrons — that streams out from the Sun at speeds of several hundred kilometers per second.

(e) *Sunspots* and sunspot groups, cool regions on the solar surface, move with the Sun as it rotates, and allow us to determine a 27-day rotation period. This period is only an apparent rotation rate as viewed from the Earth which itself orbits the Sun. The actual rotation period with respect to the fixed stars is only about $25\frac{1}{2}$ days at a latitude of $15°$ and varies slightly with latitude; the solar surface does not rotate as a solid shell. The Sun exhibits an 11 year *solar cycle* during which the number of sunspots increases to a maximum and then declines to a minimum. There are special ways of counting sunspots, and a continuous record is kept through the collaborative effort of observatories. At maximum the sunspot number can range to 150. At minimum it can be zero.

The 11 year cycle is actually only half of a longer 22 year cycle that takes into account the polarity and arrangement of magnetic fields in sunspot pairs.

(f) A variety of different events can take place on the Sun. Each type has a name of its own. One of the most interesting is a *flare*, a brief burst of light near a sunspot group. Associated with the visible flare is the emission of solar cosmic-ray particles, X-rays, ultraviolet radiation, and radio waves. Flares are also associated with the emission of clouds of electrons and protons that greatly amplify the solar wind. After a day or two, required for the Sun-to-Earth transit at a speed of $\sim 10^3$ km s^{-1}, these particles can impinge on the Earth's *magnetosphere* (magnetic field and ionosphere), giving rise to *magnetic storms* and *aurorae* that corrugate the ionosphere, disrupting radio communication that depends on smooth ionospheric reflection, sometimes for as long as a day.

A:3 The Solar System

A variety of different objects orbit the Sun. Together they make up the *Solar System*. The Earth is representative of planetary objects. *Planets* are large bodies orbiting the Sun. They are seen primarily by reflected sunlight. The majority emit hardly any radiation themselves. In order of increasing distance from the Sun, the planets are Mercury, Venus, Earth, Mars, Jupiter, Saturn, Uranus, Neptune, and Pluto. All the planets orbit the Sun in one direction; this direction is called *direct*. Bodies moving in the opposite direction are said to have *retrograde orbits*. Table 1.4 gives some of the more important data about planets. It shows that the different planets are characterized by a wide range of size, surface temperature and chemistry, magnetic field strength, and other properties. One of the aims of astrophysics is to understand such differences, perhaps in terms of the history of the Solar System.

Besides the nine planets we have listed, there are many more minor planets, or *asteroids*, orbiting the Sun. Many of these travel along paths lying between the orbits of Mars and Jupiter, a region known as the *asteroidal belt*. The largest asteroid is Ceres. Its radius is 350 km. Its mass is about one ten-thousandth that of Earth.

Comets are objects that, on approaching the Sun from large distances, disintegrate through solar heating: gases that initially were in a frozen state are evaporated off and dust grains originally held in place by these volatile substances are released. The dust and gas, respectively, are seen in reflected and re-emitted sunlight. They make the comet appear diffuse (Fig. A.1). Comet tails are produced when freshly released dust and gas that becomes ionized are repelled from the Sun, respectively, by the pressure of sunlight and by the solar wind. The Solar System may contain as many as 10^{11} comets, most of them in a giant cloud stretching into interstellar space but still gravitationally bound to the Sun. This *Oort cloud* is named after Jan Oort, the Dutch astronomer who originally proposed its existence. Comets are named after their discoverers. Many comets and asteroids have aphelion distances near Jupiter's orbit, and Jupiter has a controlling influence on the shape of the orbits and may have "captured" comets from parabolic orbits into short-period orbits.

A number of objects collectively known as *Centaurs* are intermediate in diameter between typical comets and small icy planets or planetary satellites. They have short-period orbits intermingled with those of the outer planets. Their diameters are estimated at 30 to 200 km and they appear to be drawn from the *Kuiper belt*, a region beyond the outer planets inhabited by perhaps a hundred thousand objects with diameters greater than 100 km and orbits between 50 and 100 AU. Discovered as the first of this group in 1992, is $1992QB_1$, with a diameter of 180 km and a stable, nearly circular orbit about the Sun, some 14 AU beyond Neptune. Pluto and its companion Charon may have originated in the Kuiper Belt. More recently a number of comets have also been discovered at Kuiper belt distances, and estimates suggest that the belt may contain several hundred million to a billion smaller cometary bodies (St96).

Many of the smaller known asteroids, whose orbits lie mainly between Mars and Jupiter, have diameters of the order of a kilometer. They number in the thousands and there must be many more orbiting masses that are too small to have been observed. Among these are bodies that might only be a few meters in diameter or smaller. From time to time, some of these approach the Earth and survive the journey through the atmosphere. Such an object that actually impacts on the Earth's surface is called a *meteorite*. Meteorites are studied with great interest because they are a direct means of learning about the physical and chemical history of at least a small class of extraterrestrial Solar System objects.

Even smaller than the meteorites are grains of dust that also circle the Sun along orbits similar to those of planets. When a dust grain enters the atmosphere, much of it may burn, through heat generated by friction and its high initial velocity. The particle becomes luminous through combustion and can be observed as a *meteor*, historically called a *shooting star*. The dust from a comet tail produces a meteor shower when the Earth passes through the remnants of the tail.

In contrast to meteoritic material, meteoric matter does not generally reach the Earth's surface in recognizable form. However, some fragments do appear to survive and are believed to contribute to a shower of fine dust that continually rains down on the Earth. Most of this dust has a micrometeoritic origin. *Micrometeorites* are micrometer-sized grains of interplanetary origin. They have a large surface-to-mass ratio and are easily slowed down in the upper atmosphere without becoming excessively hot. Once they have lost speed they gradually drift down through the air. Some of these grains may be formed in the burn-up of larger meteors; others may come in unchanged from interplanetary space. Collections of these grains can be made from the arctic snows or deep ocean sediments, far from sources of industrial smoke.

The cloud of dust giving rise to most of these grains permeates the space between the planets. Some of the grains may be trapped, moving along Earth's orbit about the Sun. The dust reflects sunlight and gives rise to a glow known as the *zodiacal light*. The zodiacal light can be seen, on clear nights, as a tongue-shaped glow jutting up over the western horizon after sunset or the eastern horizon before sunrise. The glow is concentrated about the *ecliptic*, the plane in which the Earth orbits the Sun.

Direct measurements of the influx rate of micrometeoritic material have been obtained from grain impact rates on the *Long Duration Exposure Facility* satellite and amount to deposition, over the entire surface of Earth, of $\sim$100 tons of material per day (Lo93). This mass is largely concentrated in particles $\sim$100 μm in radius, and is comparable to the amount of mass hitting Earth in rare catastrophic asteroidal impacts occurring only once every few million years (Ce92). A very small fraction of the micrometeoritic dust in the $\sim 1 - 40$ μm radius range appears to be impinging on Earth from interstellar space. These grains could represent ejecta from supernova explosions transported through interstellar space at velocities of tens of kilometers per second (Me02).

We recognize that these planetary and interplanetary objects continually interact. Planets and their satellites have often collided with asteroids. The surfaces of Mercury, the Moon, and asteroids, are pockmarked by impact craters. Earth too shows vestiges of such bombardment; but our atmosphere erodes away and destroys crater outlines in a time of the order of several million years, whereas on the Moon erosion times are of the order of billions of years. A giant asteroid is believed to have collided with the Earth about 4.5 billion years ago, tearing out a big chunk that became the Moon. The impacts of other large asteroids may have led to great climatic changes and the extinction of wide-ranging forms of life.

We should note that in talking about planets, meteorites, meteors, and micrometeoritic dust grains we are enumerating different-sized members of an otherwise homogeneous group. The major difference between these objects is their size. Other differences can be directly related to size. For example, it is clear that planets may have atmospheres whereas micrometeorites do not. But this difference arises because only massive objects can retain a surrounding blanket of gas. The gravitational attraction of small grains just is not strong enough to retain gases at temperatures encountered in interplanetary space. The different names given to these different-sized objects have arisen because they were initially discovered by a variety of different techniques; and although we have known the planets, meteorites, meteors, and other interplanetary objects for a long time, we have just recently come to understand their origin and interrelation.

A set of objects similar to the planets are the *satellites* or *moons*. A satellite orbits its parent planet and these two objects together orbit the Sun. In physical makeup and size, satellites are not markedly different from planets. The planet Mercury is only four times as massive as our Moon. Ganymede, one of Jupiter's satellites, Titan, one of Saturn's satellites, and Triton, one of Neptune's satellites, all are nearly twice as massive as the Moon. Titan even has an atmosphere. Many other satellites are less massive; they look very much like asteroids. An extreme of the moon phenomenon is provided by the rings of Saturn, Jupiter, and Uranus, consisting of clouds of fine dust — micrometeoritic grains, all orbiting the parent planet like minute interacting moons.

Evidently there are great physical similarities between satellites and planetary objects of comparable size. The main difference lies in the orbital motion of the two classes of objects. Some asteroids may have been gravitationally captured by Jupiter and become Jovian satellites.

A:4 Extrasolar Planetary Systems

Our Solar system is not unique. More than a hundred planets are known, by now, to orbit nearby stars. Many stars are also known to be orbited by disks that can have dust densities thousands of times greater than our zodiacal cloud. It is from these disks that planets are believed to form. The *extrasolar planets* and planetary systems are of great interest for determining how common planetary systems are, how they form, how they evolve, and how varied they may be.

A:5 Stars and Brown Dwarfs

The somewhat vague distinction between planets and interplanetary objects is not unique. Differences between stars and planets are also somewhat vague. We talk about *binaries* in which two stars orbit about a common center of gravity. Often one of these is much less massive than the other, sometimes no more than one thousandth the mass of the dominant partner. This is similar to the ratio of Jupiter's mass to that of the Sun.

Stars and planets, however, do differ from each other. Stars are bodies sufficiently massive to generate high temperatures and pressures in their interior where nuclear reactions can convert hydrogen into helium. Intermediate between giant planets, such as Jupiter, and stars somewhat less massive than the Sun are *brown dwarfs* which, though not sufficiently massive to convert hydrogen into helium can, for a short period, release energy through the thermonuclear burning of deuterium, ^7Li, and ^3He, before settling down and radiating slowly through gravitational contraction. The dividing line below which conversion of hydrogen into helium is not possible lies at masses of $0.08\ M_\odot$. This distinction in mass separates brown dwarfs from stars. The dividing line between planets and brown dwarfs lies at roughly $0.0075\ M_\odot \sim 75\ M_J$, where M_J is the mass of Jupiter. Below this mass, the hydrostatic pressure at a body's center is insufficient to overcome the Coulomb repulsion that normally prevent solids from being compressed. Planet-sized bodies are not sufficiently massive to overcome Coulomb forces. Brown dwarfs do overcome these forces but are kept from indefinite collapse by electron degeneracy pressures discussed in Chapter 8 (Ku97a).

A:6 Stellar Systems and Galaxies

Before we turn to a description of individual stars, we should first consider the groupings in which stars occur.

Stars are often assembled in a number of characteristic configurations, and we classify these systems primarily according to their size and appearance. Many stars are single. Others have no more than one companion; such pairs are called *binaries*. There exist many ternaries consisting of three stars; and higher multiple systems are not uncommon. About 30% of all stars are multiple systems. For stars more massive than the Sun, these fractions are considerably higher.

Depending on their separation and orientation, binary stars can be classified as *visual*, *spectroscopic*, or *eclipsing*. The limit of visual resolution of a binary is given by available optical techniques. Refinements are continually being made, and interferometric techniques now allow us to resolve stars only milliarcseconds apart. For smaller separations, we cannot use interferometric techniques. The two stars in

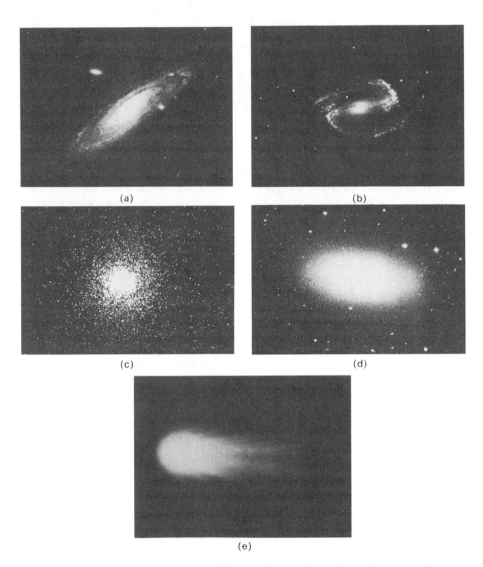

Fig. A.1. (a) The Andromeda Nebula, NGC 224, Messier 31, a spiral with two smaller companion galaxies, one of which, the elliptical galaxy NGC 205, is shown enlarged (d). The barred spiral galaxy (b) is NGC 1300. Its spiral classification is SBb. These three pictures were photographed at the Mount Wilson Observatory. The globular cluster (c) is Messier 3 (M3), also known as NGC 5272. The comet (e) is comet Brooks: the photograph was taken on October 21, 1911. Only the region of the comet around the head is shown. Such heads, called comas, typically have diameters of $\sim 10^4 - 10^5$ km. The portion of M31 apparent in (a) stretches ~ 15 kpc along its major axis. Photographs (c) and (e) were taken at the Lick Observatory.

such a close pair constitute a spectroscopic binary and have to be resolved indirectly by means of their differing spectra. We sometimes encounter a special but very important type of spectroscopic binary in which the stars orbit about each other roughly in a plane that contains the observer's line of sight. One star may then be seen eclipsing the other and a change in brightness is observed. An eclipse of this kind becomes probable only when the two companions are very close together, no more than a few radii apart. We call such systems *eclipsing binaries*. Binaries are important because they provide the only means for determining accurate mass values for stars (other than the Sun). How these masses are determined is shown in the discussion of orbital motions (Section 3:5).

Close binaries are also important because if one of the two stars begins to expand as it moves onto the red-giant branch of the Hertzsprung–Russell diagram (see Section A:7(g) and Figs. 1.4 and 1.7), its surface material may become more strongly attracted and flow over to the companion star. Portions of the giant star previously in its interior are thus revealed. This allows us to check for systematic production of the heavy elements in the star and also to test the theory of chemical evolution and energy production in stars (Section 8:12). If the companion to the giant is compact, the infalling material can radiate X-rays on impact (see Section 5:19).

Sometimes stars form an aggregate of half a dozen or a dozen members. This is called a stellar *group*. Stellar *associations* are larger groupings of some 30 stars mutually receding from one another. Associations appear to have had a common origin and to have become separated shortly after formation. By observing the size of an association and the rate at which it is expanding, we can determine how long ago the expansion started and how old the stars must be.

Two principal groupings are called clusters: *galactic clusters* and *globular clusters*. Galactic clusters usually comprise 50 to several hundred stars loosely and amorphously distributed but moving with a common velocity through the surrounding field of stars. Globular clusters (Fig. A.1(c)) are much larger, contain several hundred thousand stars, and have a striking spherical (globular) appearance. Stars in a cluster appear to have had a common origin and common history. Binaries and higher multiples and groups of stars often form small subsystems in clusters.

Normally stars and clusters are members of *galaxies*. These are more or less well-defined, characteristically shaped systems containing between 10^8 and 10^{12} stars (Fig. A.1(a,b,d)). Some galaxies appear elongated and are called elliptical or E galaxies. Highly elongated ellipticals are designated E7. If no elongation can be detected and the galaxy has a circular appearance, it is called a globular galaxy and is classified as E0. Other numerals, between 0 and 7, indicate increasing apparent elongation. The observed elongation need not correspond directly to the actual elongation of the galaxy because the observer on Earth can only see the galaxy in projection.

Elliptical galaxies show no particular structure except that they are brightest in the center and appear fainter at the periphery. *Spiral galaxies* (S) and *barred spiral galaxies* (SB) exhibit a spiral structure denoted by a symbol O, a, b, or c following the spiral designation to indicate increasing openness of the spiral arms. In this notation, a compact spiral is designated SO and a barred spiral with far-flung spiral arms and quite open structure is designated SBc (see Figs. A.1(a,b,d))

Not all galaxies can be described by designations, E, S, or SB. Some are classified as *irregular* and designated by the symbol Ir. Peculiar galaxies of one kind or another are denoted by a letter p following the type designation, for example, E5p.

Galaxies do not contain stars alone. In some spiral galaxies the total mass of interstellar gas and dust is comparable to the total stellar mass observed. Gas may be detected in absorption or through emission of radiation. Through spectroscopic studies in the radio, infrared, visible, ultraviolet, and X-ray domains, the spectra of many ions, atoms, and molecules can be identified, and their temperature, density, and radial velocity determined. Dust clouds can be detected through their extinction, which obscures the view of more distant stars. Dust also absorbs optical and ultraviolet radiation and re-emits at long infrared wavelengths. This process is so effective that some galaxies radiate far more strongly in the infrared than in all other spectral ranges combined.

The major fraction of a galaxy's mass is normally concentrated in *dark matter*, a mysterious form of matter that makes itself known solely through the gravitational attraction it exerts (see Section 1:12).

Galaxies are not the largest aggregates in the Universe. Galaxies often occur in pairs and groups. Figures A.1(a) and 1.11 show such groupings. The Sun is one of billions of stars in the *Milky Way*, often referred to as *the Galaxy*, spelled with a capital "G". The Galaxy is a member of the *Local Group* that contains more than two dozen galaxies of which the Andromeda Nebula and the Galaxy are the dominant members accounting for most of the mass (Table 1.5).

Larger *clusters of galaxies* containing up to several thousand galaxies also exist. Groupings on a larger scale include filamentary structures composed of tenuous chains of galaxies, enormous voids surrounded by denser concentrations — walls of galaxies and possibly superclusters — entire groupings of clusters of galaxies. Beyond that scale, no further clustering is apparent. On the largest scales, the Universe can best be described as consisting of randomly grouped aggregates and voids (see Figs. 1.11, 1.12, and 13.6).

The scheme of classification of galaxies leaves a number of borderline cases in doubt. Small E0 galaxies are not appreciably different from the largest globular clusters. Merging galaxies sometimes cannot be distinguished from irregular ones; and the distinction between a group or a cluster of galaxies may also be a matter of taste. The classification is useful nevertheless; it gives handy names to frequently found objects without making any attempt to provide rigorous distinctions.

Crossing the vast spaces between the galaxies are quanta of electromagnetic radiation and highly energetic cosmic-ray particles that travel at almost the speed of light. These are the carriers of information that permit us to detect the existence of the distant objects.

On a photographic plate or *charge-coupled device, CCD*, we expect images of nearby galaxies to appear larger than more distant objects. On this assumption, the angular diameter of a galaxy can be taken to be a rough indicator of its distance.

When the spectra of such galaxies are correlated with their assumed relative distances, we find that a few nearby galaxies have blue-shifted spectra, but all distant galaxies have spectra that are systematically shifted toward the red part of the spectrum. Galaxies at progressively larger apparent distances exhibit increasing red shifts. This correlation is so well established that we now take an observation of a remote galaxy's red shift as a standard indicator of its distance, and attribute the *red shift* to a high recession velocity. The galaxies appear to be flying apart. The Universe expands!

A:7 Brightness of Stars

(a) The Magnitude Scale

A casual look at the sky reveals that some stars appear brighter than others. The eye can clearly distinguish the brightness of two objects only if one of them is approximately 2.5 times as bright as the other. The factor of 2.5 can therefore serve as a rough indicator of apparent brightness, or *apparent visual magnitude m_v* of stars. This has established the *magnitude scale*.

Stars of first magnitude, $m_v = 1$, are brighter by a factor of $\sim$2.5 than stars of second magnitude, $m_v = 2$, and so on. The visual magnitude scale extends into the region of negative values; but the Sun, Moon, Mercury, Venus, Mars, Jupiter, occasional bright comets, and the three stars, Sirius, Canopus, and α Centauri are the only objects bright enough to have apparent visual magnitudes less than zero.

Normally it would be cumbersome to use a factor of 2.5 in computing the relative brightness of stars of different magnitudes. Since this factor has arisen not because of some feature peculiar to the stars that we study, but is quite arbitrarily dependent on a property of the eye, we are tempted to discard it altogether in favor of a purely decimal system; but a brightness ratio of 10 is not useful for visual purposes. As a result, a compromise that accommodates some of the advantages of each of these systems is in use. We define a magnitude in such a way that stars whose brightness differs by precisely five magnitudes have a brightness ratio of exactly 100. Because $100^{1/5} = 2.512$, we still have reasonable agreement with what the eye sees, and for computational work we can use standard logarithms to the base 10.

(b) Color

The observed brightness of a star depends on whether it is seen by eye, recorded on a photographic plate, or detected by means of a radio telescope. For different astronomical objects the *spectral energy distribution, SED*, the ratio of energy emitted, e.g., in the optical domain, the infrared, or radio regime, varies widely. The *color* or SED of an object can be roughly described by observing it through a variety of filters or with different detectors in several different spectral regions. The apparent magnitudes obtained in these measurements can then be compared. Several standard filters and instruments have been developed for this purpose so that we may compare and contrast data from observatories all over the world. The resulting brightness indicators are listed below:

m_v denotes *visual magnitude*. m_{pg} denotes *photographic magnitude*. Although photographic plates have now been all but displaced by detector arrays, the need for standardization to follow long-term trends has required the maintenance of traditional wavelength bands in modern photometry. A photographic plate is more sensitive to blue light than the eye; photographic brightness is usually labeled B, for blue. An older designation is m_{pg}.

V or m_{pv} denotes photovisual magnitude obtained with a photographic plate and a special filter used to pass yellow light and reject some of the blue light. Modern usage generally refers to V as visual magnitude.

U denotes the *ultraviolet magnitude* obtained with a particular ultraviolet transmitting filter.

I denotes *infrared magnitude* obtained in the photographic part of the infrared. At longer wavelengths photographic plates are no longer sensitive, but a number of infrared spectral magnitudes have been defined so that results obtained with indium antimonide, mercury cadmium telluride, and other infrared detectors might be compared by different observers. These magnitudes are labeled J, K, L, M, N, and Q.

Table A.1 lists the wavelengths at which these magnitudes are determined.

Table A.1. Effective Wavelength for Standard Brightness Measurements.

Symbol	Effective Wavelength	Symbol	Effective Wavelength
U	$0.36\,\mu$m	K	$2.2\,\mu$m
B	0.44	L	3.4
V	0.55	M	5.0
R	0.70	N	10.2
I	0.90	Q	21
J	1.25		

$1\,\mu$m (pronounced *micron* or *micrometer*) $= 10^{-6}$ m $= 10^{-4}$ cm $= 10^4$ Å (Ångstrøms).

m_{bol} denotes the total apparent magnitude of an object integrated over all wavelengths. This *bolometric magnitude* is the brightness that would be measured by a bolometer — a detector equally sensitive to energy radiated at any wavelength.

(c) Color Index

The difference in brightness as measured with different filters gives an indication of a star's color. The ratio of blue to yellow light received from a star is given by the difference in magnitude — $\log_{2.5}$ of the brightness ratio — of the star measured with blue and visual filters. This quantity is known as the *color index*:

$$C = B - V.$$

Differences such as $U-B$ are also referred to as color indexes.

The comparison of colors involved in producing a reliable color index can only be achieved if we can standardize detectors and filters used in the measurements. And even then errors can creep into the comparison. For this reason some standard stars have been selected to define a point where the color index is zero. These stars are denoted by the spectral-type symbol A0 (see Section A:8).

(d) Bolometric Correction

Normally the bolometric brightness of a star can only be obtained by means of observations spanning the entire spectrum. The *bolometric correction*, BC, is defined as the difference between the bolometric and visual magnitudes of a star. The bolometric correction is always positive

$$\mathrm{BC} = m_v - m_{\mathrm{bol}} .$$

(e) Absolute Magnitude

For many purposes we need to know the absolute magnitude rather than the apparent brightness of a star. We define the *absolute magnitude* of a star as the apparent magnitude we would measure if the star were placed a distance of 10 pc from an observer. (1 pc = 3×10^{18} cm. See Section 2:2.)

Suppose the distance of a star is r pc. Its brightness diminishes as the square of the distance between star and observer. The apparent magnitude of the star will therefore be greater, by an additive term $\log_{2.5} r^2/r_0^2$, than its absolute magnitude.

$$m = M + \log_{2.5} \frac{r^2}{r_0^2} = M + 5\log\frac{r}{r_0} ,$$

where the logarithm is taken to the base 10 when no subscript appears. Because $r_0 = 10$ pc, we have the further relation for the *distance modulus*, μ_0,

$$\mu_0 \equiv m - M = 5\log r - 5. \tag{A-1}$$

Thus far no attention has been paid to the extinction of light by interstellar dust. The apparent magnitude is increased through extinction — the star appears fainter — and a positive factor A has to be subtracted from the right side of equation (A–1) to restore M to its proper value

$$M = m + 5 - 5\log r - A. \tag{A-2}$$

Obtaining the star's distance, r, is often less difficult than assessing the interstellar extinction A. We discuss this difficulty in Section A:8(a) below.

The detector and filter used in obtaining the apparent magnitude m in equation (A–2) determines the value of the absolute magnitude M. We can therefore use subscripts, v, pg, pv, and bol for absolute magnitudes in exactly the same way as for apparent magnitudes.

(f) Luminosity

Once we have obtained the bolometric absolute magnitude of a star, we can obtain the rate at which it radiates energy, i.e., its *luminosity*, L, directly in terms of the solar luminosity, $L_\odot$:

$$\log\left(\frac{L}{L_\odot}\right) = \frac{1}{2.5}\left[M_{\mathrm{bol}_\odot} - M_{\mathrm{bol}}\right] . \tag{A–3}$$

The luminosity of the Sun, $L_\odot$, is 3.85×10^{33} erg sec^{-1} and the solar bolometric magnitude, $M_{\mathrm{bol}\odot}$, is 4.6. The luminosity of stars varies widely. For a brief interval of a few days, a supernova explosion can be as luminous as all the stars in a galaxy. The brightest stable stars are a million times more luminous than the Sun. At the other extreme, a *white dwarf* may be a factor of a thousand times fainter than the Sun; and brown dwarfs, stars with masses ranging down to $\sim M_\odot/60$, may have luminosities $10^{-7}L_\odot \lesssim L \lesssim 10^{-4}L_\odot$ as they settle down to contract and slowly radiate away gravitational potential energy, over billions of years (Ku97a).

(g) Hertzsprung–Russell and Color-Magnitude Diagrams

One of the most useful diagrams in all astronomy is the *Hertzsprung–Russell, H–R diagram* (Figs, 1.4 to 1.6). It presents a plot of luminosity and temperature for detected stars. A related set of diagrams the *color-magnitude diagrams* (Figs. 1.3 and 1.7) plots the magnitudes of stars against color index. The ordinate on such a plot can show either M_v, or M_{bol}, or luminosity. When only a comparison of stars all of which are known to be equally distant is needed, it suffices to plot the apparent magnitude. Figure A.2 shows a color-magnitude diagram for the Pleiades star cluster. The Pleiades are among the most recently born Galactic stars. Figure 1.7 plots the characteristics of M3, an old Galactic globular cluster. The age difference is reflected in the appearance of the two diagrams.

These figures all show that stars appear only in select areas of the H–R and color-magnitude diagrams. The largest number of stars cluster about a fairly straight band called the *main sequence*. This is particularly clear for the Pleiades cluster. The main sequence runs from the upper left to the lower right end of the diagram, or from bright blue down to faint red stars. To the right and above the main sequence (Fig. 1.4) lie bright red stars along a track called the *red-giant branch*. There is also a *horizontal branch* that joins the far end of the red-giant branch to the main sequence. These two branches show up particularly in Fig. 1.7. In the horizontal

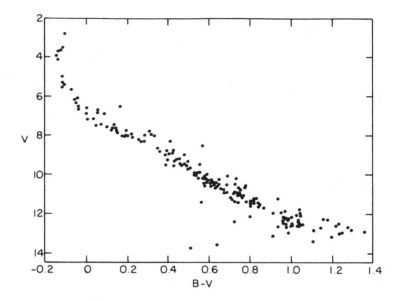

Fig. A.2. Color-magnitude diagram of the Pleiades cluster stars, after correction for interstellar extinction effects. The Pleiades cluster contains some of the most recently formed stars in the Galaxy (after Mitchell and Johnson (Mi57b)).

branch, we find stars whose brightness varies periodically. Finally, some faint *white dwarf stars* lie below and to the left of the main sequence. The rest of the diagram is usually empty.

A:8 Classification of Stars

(a) Classification System

The classification of stars is a complex task, primarily because we find many special cases hard to fit into a clean pattern. Currently a "two-dimensional" scheme is widely accepted. One of these "dimensions" is a star's spectrum; the other is its luminosity. Each star is assigned a two-parameter classification code. Although the object of this section is to describe this code, we should note that the ultimate basis of the classification scheme is an extensive collection of spectra such as those shown in Fig. A.3. Each spectrum is representative of a particular type of star.

Stars are classified primarily according to their spectra, which are related to their color. Although the primary recognition marks are spectral, the sequence of the classification is largely in terms of decreasing stellar surface temperature — that is, a shift in the star's radiation to longer wavelengths. The bluest common stars are labeled O, and increasingly red stars are classed according to the sequence (Table A.2)

Table A.2. Spectral Classification of Stars.[a]

Type	Main Characteristics	Subtypes	Spectral Criteria	Typical Stars
Q	Nova: sudden brightness increase by 10 to 12 magnitudes			T Pyx Q Cyg
P	Planetary nebula: hot star with intensely ionized gas envelope			NGC 6720 NGC 6853
W	Wolf–Rayet stars: hot stars		Broad emission of OIII to OVI, NIII to NV, CII to CIV, and HeI and HeII.	HD 191765
O	Hot stars, continuum strong in UV (O5 to O9)		OII λ 4650 dominates HeII λ 4686 dominates $\Big\}$ emission Lines narrower $\quad$ lines Absorption lines dominate; only HeII, CII in emission SiIV λ4089 at maximum OII λ 4649, HeII λ 4686 strong	BD +35°4013 BD +35°4001 BD +36°3987 ζ Pup, λ Cep 29 CMa τ CMa
B	Neutral helium dominates	B0 B1 B2 B3 B5 B8 B9	CIII/4650 at maximum HeI λ 4472 > OII λ 4649 HeI lines are maximum HeII lines are disappearing Si λ 4128 > He λ 4121 λ 4472 = Mg λ 4481 HeI λ 4026 just visible	ε Ori β CMa, β Cen δ Ori, α Lup π^4 Ori, α Pav 19 Tau, ϕ Vel β Per, δ Gru λ Aql, λ Cen
A	Hydrogen lines decreasing from maximum at A0	A0 A2 A3 A5	Balmer lines at maximum CaII K = 0.4 Hδ K = 0.8 Hδ K > Hδ	α CMa S CMa, ι Cen α PsA, τ^3 Eri β Tri, α Pic
F	Metallic lines becoming noticeable	F0 F2 F5 F8	K = H + Hδ G band becoming noticeable G band becoming continuous Balmer lines slightly stronger than in Sun	δ Gem, α Car π Sgr α CMi, ρ Pup β Vir, α For
G	Solar-type spectra	G0 G5	Ca λ 4227 = Hδ Fe λ4325 > Hγ on small-scale plates	α Aur, β Hya κ Gem, α Ret
K	Metallic lines dominate	K0 K2 K5	H and K at maximum strength Continuum becoming weak in blue G band no longer continuous	α Boo, α Phe β Cnc, ν Lib α Tau
M	TiO bands		TiO bands noticeable Bands conspicuous Spectrum fluted by the strong bands Mira variables, Hγ, Hδ	α Ori, α Hya ρ Per, γ Cru W Cyg, RX Aqr χ Cyg, o Cet
R, N	CN, CO, C_2 bands		CN, CO, C_2 bands appear instead of TiO. R stars show pronounced H and K lines.	
S	ZrO bands		ZrO bands	R Gem

[a] Compiled mainly from Keenan in *Stars and Stellar Systems*, K.A. Strand (ed.), with permission from the University of Chicago Press (Ke63) (based on Cannon and Pickering (Ca24)) and also from Allen (Aℓ55). This table, which is based on the Henry Draper classification scheme, is a rough guide to the spectral features of stars. The classification of stars, however, remains an ongoing process and changes occur. (With the permission of Athlone Press of the University of London, 2nd ed. © C. W. Allen, 1955 and 1963, and with the permission of the University of Chicago Press.)

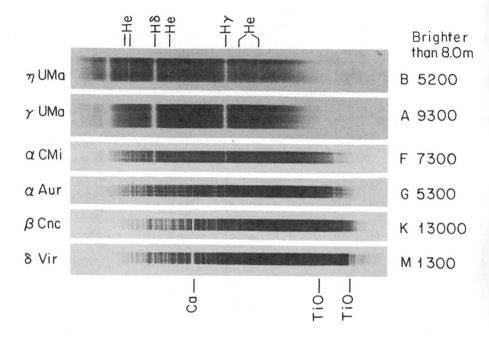

Fig. A.3. Schematic diagram of spectra of typical stars representing different spectral types. The number of stars brighter than the eighth magnitude in each class is listed on the right, next to the star's spectral type. (With the permission of the Yerkes Observatory, University of Chicago.)

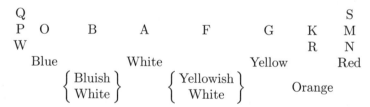

Over 99% of all stars belong to the basic series O, B, A, F, G, K, and M. Stars with designation R, N, and S are comparatively rare. The classes R and N denote stars containing unusually strong molecular bands of diatomic carbon, C_2, and cyanogen, CN. S stars are characterized by bands of titanium oxide, TiO, and zirconium oxide, ZrO. Spectral type Q denotes *novae* — stars that suddenly brighten by many orders of magnitude becoming far brighter than any nonvariable star. P denotes *planetary nebulae*, hot stars with surrounding envelopes of intensely ionized gas. W refers to Wolf–Rayet stars, intensely hot stars that exhibit broad emission bands of ionized carbon, nitrogen, and helium. These stars appear to consist of a nuclear-processed interior exposed by extreme surface mass loss.

Stars classed as W, O, B are sometimes said to be *early types*, whereas stars of class G, K, M, R, N, S are designated *late types*. Globular cluster stars, and stars that make up the Galaxy's spherical *halo*, are primarily late type stars often referred to as *Population II* stars. Early type stars are designated *Population I* and are largely found in the Galaxy's *disk* — the Milky Way plane.

The transition from one spectral class to another proceeds in ten smaller steps. Each spectral class is subdivided into ten subclasses denoted by Arabic numerals after the letter. A5 lies intermediate between spectral types A0 and A9; and F0 is just slightly redder than A9. A Roman numeral following the spectral type designation indicates a star's luminosity class. Each of these has a name:

I — Supergiant
II — Bright Giant
III — Normal Giant
IV — Subgiant
V — Main Sequence

The Sun has spectral type G2 V indicating that it is a yellow main sequence star.

Table A.3. Effective Stellar Temperatures.[a]

| Types | Main-Sequence Subgiants | | Giants | | | Supergiants |
	V	IV	III	II	Ib	Ia
			T_e (° K)			
O4	48670		48180			47690
O8	38450		37090			35730
B0	33340		31540		25700	
B5		15400		14800	13100	
A0		10000		9700	10200	
A3			8500			
F0			7200			
F5	6700	6600	6500	6350	6200	....
G0	6000	5720	5500	5350	5050	....
G5	5520	5150	4800	4650	4500	....
K0	5120	4750	4400	4350	4100	....
K5	4350	....	3700	3600	3500	....
M0	3750	....	3500	3400	3300	....
M2	3350	....	3100	2050	....	....

[a] Adapted from Keenan (Ke63), Böhm-Vitense (Bö81), and Vacca et al. (Va96). See also text.

Sometimes we find classes I, II, and III collected under the heading "giant" while stars of group V are called "dwarfs." Letters "g" or "d" are placed in front of the spectral class symbol to denote these types. Similarly placed letters "sd" and "w", denote subdwarfs and white dwarf stars. Another classification feature concerns supergiants, which are often separated into two luminosity classes Ia and Ib depending on whether they are bright or faint.

A letter "e" following a spectral classification symbol denotes the presence of emission lines in the star's spectrum. There is one exception to this. The combination Oe5 denotes O stars in the range O5 to O9; it has no further connection to emission.

A letter "p" following the spectral symbol denotes that the star has some form of peculiarity.

The color designation (stellar spectral type) given here is nearly linear in the color index $B-V$. It is not however linear in $U-V$ nor do the $U-V$ values decrease monotonically with increasingly late spectral type. Small differences in color indexes exist for giants and main sequence stars of the same spectral type. This unfortunate difficulty has arisen for historical reasons.

We might still see how well stellar colors approach those of a blackbody. The closeness of fit is shown in Fig. A.4, called a color-color diagram. Four factors are responsible for the rather large deviations from a blackbody. (i) For stars around spectral type A, where the fit to the blackbody spectrum is poorest, absorption by hydrogen atoms in their first excited states produces a deviation. We talk about the *Balmer jump* in connection with the sharp rise in absorption at wavelengths corresponding to the Balmer continuum produced by these excited atoms in the outer atmosphere of a star. (ii) Cool stars have H$^-$ ions in the outer atmospheres. These ions absorb radiation selectively, making the star appear bluer. (iii) The relatively high abundance of metals in Population I stars produces a number of absorption lines that change the color of a star, moving it toward the lower right of Fig. A.4. (iv) Finally, no star looks completely black, because its outer layers are not equally opaque at all wavelengths. Light at different wavelengths therefore reaches us from different depths within the star, and these levels are at different temperatures. The resulting spectrum of starlight therefore corresponds to a mixture of temperatures, rather than to blackbody radiation at one well-defined temperature.

Determination of the spectral type of a star by means of its color index alone would be very difficult, because proper account would have to be taken of the changes in color produced by interstellar dust. Small dust grains tend to absorb and scatter blue light more strongly than red. Light from a distant star therefore appears much redder than when emitted. To discover the true color index of the star a correction has to be introduced for interstellar reddening. However, in order to make this correction, we have to know how much interstellar dust lies along the line of sight to the star, and to what extent a given quantity of dust changes the color balance. None of this information is normally available. Instead, we have to make use of a circular line of reasoning. We know that nearby stars of any given spectral type exhibit characteristic absorption or emission lines in their spectra. Since these stars are near, there is little intervening interstellar dust, and their spectra can be taken to be unreddened. We can therefore draw up tables listing the spectral lines featuring each color class. A distant star can then be classed in terms

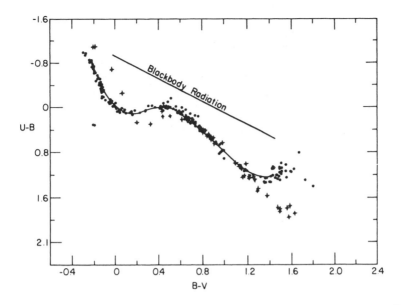

Fig. A.4. The relation between the color systems $U-B$ and $B-V$ for unreddened main-sequence stars (dots) and little-reddened supergiants and yellow giants (crossed dots). The line along which blackbody radiators would fall is also shown (after Johnson and Morgan (Jo53)).

of its spectral lines rather than its color index and the color index can be used to verify the class assignment. If the color is redder than expected, we have an indication of reddening by interstellar dust. Whether dust is actually present can then be checked — in many instances — by seeing whether other stars in the immediate neighborhood of the given object all are reddened by about the same amount. If they are, we have completed the analysis. The results give the correct spectral identification of stars in the chosen region and, in addition, we are given the extent to which interstellar dust changes the color index. A similar analysis can also be applied to determine the extent to which the overall brightness of the star is diminished through extinction by interstellar dust. This analysis allows us to determine the amount of obscuration in all the spectral ranges for which observations exist.

As already stated, the color and spectrum of a star depend on its surface temperature. Table A.3 gives the effective temperature for representative stars and Figure A.5 relates stellar temperatures to color. As discussed in Chapter 4, the effective temperature is measured in terms of the radiant power emitted by the star over unit surface area. By analyzing the spectra of stars we can obtain their speed of rotation from the broadening of stellar spectral lines. If the axis of rotation of a star is inclined at an angle i relative to the line of sight we obtain a measure of $v_e \sin i$, where v_e is the equatorial velocity of the star. Only those stars whose spin axes are perpen-

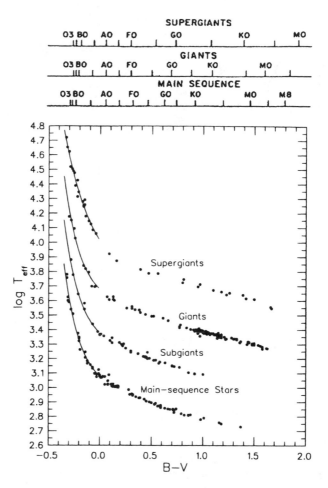

Fig. A.5. Effective temperatures and colors for all stars separated by luminosity class. For clarity, temperatures of giants, subgiants, and main-sequence stars are lowered by 0.3 in log T_{eff} with respect to the next more luminous class. The lines are shown to help guide the eye along the steep portions of the curve (after Flower (Fℓ96)).

dicular to the line of sight exhibit the full Doppler broadening due to the rotation of the star; but by analyzing the distribution of line widths, we can statistically determine both the rotational velocity and the distribution function of the angle i (Hu65). As far as we can tell, rotation axes of stars are randomly oriented with respect to the Galaxy's rotation axis. Table A.4 gives some typical values of v_e for different types of stars.

Table A.4. Stellar Rotation for Stars of Luminosity Classes V, III, and I (after Cox, Co00).

Spectral Type	Mean v_e (km s^{-1}) V	III	I
O8	200		125
B0	170	100	102
B5	240	130	40
A0	180	100	40
A5	170		38
F0	100		30
F5	30		< 25
G0	10	30	< 25
K, M	< 10	< 20	< 25

(b) Variable Stars

Two main types of variable stars can be listed. *Extrinsic variables* can be: (i) close binary stars whose combined brightness varies because one star eclipses the other; or (ii) stars that are eclipsed by, or periodically illuminate, ambient ejecta or remnants. *T Tauri variables*, named after the star in which this second type of behavior was first noted, are young stars orbited by dust clouds from which they were formed.

Intrinsic variables are stars whose luminosity actually changes with time. The brightness variations may be repetitive as for periodic variables, erratic as for irregular variables, or semiregular. The distinction is not always clear-cut. A brief summary of some characteristics of periodic or pulsating variables is given in Table A.5. The brighter of these stars are important in the construction of a reliable cosmic distance scale.

Other types of intrinsic variables include exploding stars such as novae, recurrent novae, supernovae, dwarf novae, and shell stars.

The brightness of a nova rises 10 to 12 magnitudes in a few hours. The return to the star's previous low brightness may take no more than a few months, or it may take a century. Both extremes have been observed. The absolute photographic brightness at maximum is about -7.

Recurrent novae brighten by about 7.5 magnitudes at periods of several decades. Their peak brightness is about the same as that of ordinary novae. The brightness decline usually takes 10 to 100 days but sometimes lies outside this range.

Supernovae are about ten magnitudes brighter than novae. Their luminosities may rival that of their parent galaxies. Two major types have been recognized. Supernovae of type II exhibit spectral lines of hydrogen in their optical spectra, whereas supernovae of type I do not. SNe I occur in all galaxies, where they have the spatial distribution of older stars; typically their absolute magnitudes are $M_v = -16$ at maximum. SNe II occur only in the arms of spiral galaxies, are associated with

Table A.5. Properties of Pulsating Variables.

Type	Range of Period, P (days)	Spectral Type	Mean Brightness M_v and Variation ΔM_v	Remarks
RR Lyrae (Cluster Variables)	<1	A4 to F4	$M_v = 0.6$ $\Delta M_v \sim 1.0$	Found in the halo of the Galaxy
Classical Cepheids	1–50	F to K	$M_v = -2.6$ to -5.3 $M_v, \Delta M_v$ depend on P $\Delta M_V \sim 0.4$ to 1.4	Found in the disk of the Galaxy
W Virginis Stars (Type II Cepheids)	>10	F, G	$M_v =$ one or two mag. less luminous than Class. Ceph. of similar period. $\Delta M_v = 1.2$	Halo population
Mira Stars (Long Period Variables)	100–1000	Red giant	$M_v \sim$ from -2.2 to 0, $\Delta M_v =$ from 3 to 5 for increasing period	Intermediate between disk and halo
Semiregular Variables	40–150	Red giant	$M_v = 0$ to -1 $\Delta M_v \sim 1.6$	Disk population

populations of young stars, and have $M_v = -14$ at maximum. The two types of supernovae can be subdivided into several subtypes, but roughly 80% of SNe I are of a type designated as SN Ia, whose light curves are all remarkably similar. This makes them useful distance indicators.

On exploding, a supernova can thrust many solar masses of matter into interstellar space at initial speeds of tens of thousands of kilometers per second. Often these gaseous shells persist as *supernova remnants* for several thousand years. On photographic plates they appear as filamentary arcs surrounding the point of initial explosion.

Dwarf novae brighten by about four magnitudes to a maximum absolute brightness of $M_v +4$ to $+6$. Their spectral type normally is A. Their outbursts are repeated every few weeks.

Shell stars are B stars having bright spectral lines. The stars seem to shed shells. A rise in brightness of one magnitude can occur.

Flare stars sporadically brighten by ~ 1 magnitude over intervals measured in tens of minutes. They then relapse. These stars are yellow or red dwarfs of low luminosity. The flares may well be similar to those seen on the Sun, except that they occur on a larger scale. In extreme cases the star brightens a hundredfold.

R Coronae Borealis stars are stars that suddenly dim by as much as eight magnitudes and then within weeks return to their initial brightness. At maximum the spectrum is of class R, rich in carbon.

The variable stars are not very common, but they are interesting for two reasons. First, some of the variable stars have a well-established brightness pattern that allows us to use them as distance indicators (see Chapter 2). Second, the intrinsic variables show symptoms of unstable conditions inside a star or on its surface. In that sense the variable stars provide important clues to the structure of stars and to the energy balance or imbalance at different stages of stellar evolution.

Novae, T Tauris, and some stars at the extreme end of the giant branch, the *Asymptotic giant branch stars, AGB,* are found to be strong emitters of infrared radiation. The novae and AGB stars eject material that forms dust on receding from the parent star, while *T Tauris* are largely embedded in the dust clouds from which they formed. Some of the evolved, dust-shrouded giant stars also emit extremely narrow, luminous, and highly polarized spectral lines in water vapor, OH, and SiO radio transitions, making them recognizable as cosmic *masers* (See Section 7:11).

Table A.6. Stellar Velocities Relative to the Sun, and Mean Height Above Galactic Plane.[a]

Objects	Velocity[b], v $\mathrm{km\,s^{-1}}$	Density, ρ $10^{-3}\,M_\odot\,\mathrm{pc}^{-3}$	Height, h pc
Interstellar clouds			
Large clouds	8		
Small clouds	25		
Early main sequence stars:			
O5–B5	10 ⎫		50
B8–B9	12 ⎭ 0.9		60
A0–A9	15	1	115
F0–F5	20	3	190
Late main sequence stars:			
F5–G0	23		
G0–K6	25	12 ⎫	
K8–M5	32	30 ⎭	350
Red-giant stars:			
K0–K9	21	0.1	270
M0–M9	23	0.01	
High-velocity stars:			
RR Lyrae variables	120	10^{-5}	
Subdwarfs	150	1.5	
Globular clusters	120–180	10^{-3}	

[a] Stellar velocities collected by Spitzer and Schwarzschild from other sources (Sp51a). Densities ρ, and heights h, after (Aℓ64). (With the permission of the Athlone Press of the University of London, 2nd ed.© C.W. Allen 1955 and 1964.)

[b] Root mean square value for component of velocity projected onto the Galactic plane.

A:9 The Distribution of Stars in Space and Velocity

We judge the radial velocities of stars by their spectral line shifts. Transverse velocities can be obtained for nearby stars from their *proper motion* — their angular velocity across the sky — and from their distance, if known. We find that stars of different spectral type have quite different motions. Stars in the Galactic plane have low relative velocities, while stars that comprise the *Galactic halo* have large velocities relative to the Sun. In practice there is no clear-cut discontinuity between these populations. This is rather well illustrated by the continuous variation in velocities given in Table A.6. A star's velocity is correlated with its mean height above the Galactic plane. By noting the distribution of stars in the solar neighborhood, we at least obtain some idea about how many stars of a given kind have been formed in the Galaxy. If we can compute the life span of a star, as outlined in Chapter 8, then we can also judge the rate at which stars are born. For short-lived stars such birth

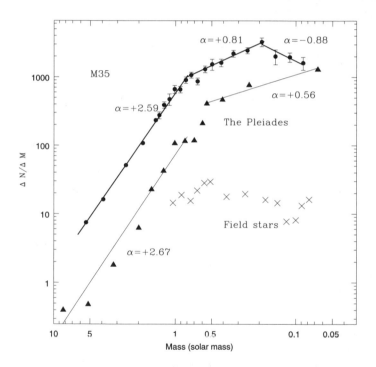

Fig. A.6. The mass functions of the young open clusters M35 and the Pleiades. This plot indicates the rates at which stars of different masses are born in the Galaxy today. From such a plot we can obtain the *Salpeter birth rate function* ψ giving the rate of star formation in unit volume of the Galaxy (see Section 8:2). The values of α shown are for a birth rate $\psi \propto M^{-\alpha}$, and show that a single value of $\alpha \sim +2.6$ fits stars with masses $\gtrsim 0.8 M_\odot$, but that inclusion of the birth rates of low-mass stars requires a more complex relation (Ba01).

rates represent current formation rates; and we can look for observational evidence to corroborate estimates of longevity once the spatial number density of a given type of star has been established (Fig. A.6).

Such studies are still in relatively preliminary stages, because we are not quite sure what the appearance of a star should be at birth, particularly if it is still surrounded by some of the dust from which it has been formed (Section 1:4).

As we look to increasing distances across the Universe and are able to detect galaxies at large red shifts, their colors and luminosities begin to tell us the numbers of stars that are shining there and the lengths of their life spans. From such surveys we are beginning to trace the star formation rates in the Universe from early times to the present. Figure A.7 provides an estimate for these rates. It shows that current

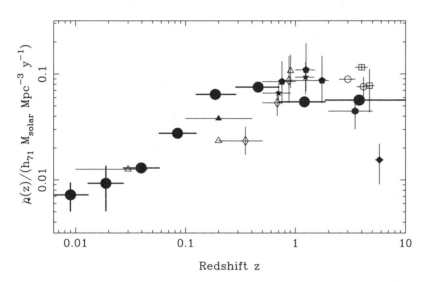

Fig. A.7. The cosmic star formation rate at different red shifts as determined by means of a number of complementing observations. The rate of star formation per year is plotted per unit red-shift interval $\Delta z = 1$ in terms of solar masses formed per comoving cubic megaparsec per year. The notation on the ordinate, $h = 0.71$, indicates that the formation rates assume a Hubble constant of 71 km s^{-1} Mpc^{-1} (He04a). Courtesy of Alan Heavens.

star formation rates may be roughly a factor of ten lower than at their peak when the Universe had attained only one-third its present age.

A:10 Pulsars and X-Ray Sources

(a) Pulsars

Isolated pulsars are radio sources that emit pulsed radiation with clocklike regularity. Pulsars are identified primarily in the radio wavelength region. The central star in the Crab Nebula or the powerful gamma-ray emitter Geminga, both of which emit visible light as well as gamma rays, are notable exceptions. Where pulsars are companions to giant stars, whose atmosphere they are tidally stripping away, they can also be strong intermittent X-ray sources. For isolated pulsars the regularity of the pulses is constant to about one part in 10^8 per year. Pulses are typically spaced anywhere between a few milliseconds and a few seconds apart.

The coherence and pulse rates tell us that these sources are small compared to normal stars. Pulsars are *neutron stars*, whose cores consist of closely packed, degenerate neutrons. In such a star, more than a solar mass is packed into a volume about 10 km in diameter. A pulsar's rotation period is given by the interval between the main pulses. The radiation is emitted in a direction tangential to the charged particles moving with the rotating star and, hence, there is a loss of angular momentum and a corresponding slowdown of the star's rotation and of the pulse rate (Go68). Pulsars also are sources of highly relativistic particles and thus contributors to the Galactic cosmic-ray component.

A small number of pulsars are associated with known *supernova remnants*. One such remnant is in the constellation Vela. Another is the Crab Nebula, remnant of a supernova seen in 1054 AD. It was identified as the stellar remnant of the supernova, more than 25 years before the pulsar's discovery. The Crab pulsar now pulses every 33 ms. Using its present slowdown rate, we can make a rough linear extrapolation of the pulsar's period backward in time and see that this is indeed a remnant of the object that exploded in 1054 AD. Slowly pulsating pulsars are thought to be old.

Within the past three decades a class of binary pulsars has been discovered — two compact sources orbiting each other. Many are neutron-star/neutron-star binaries. The constancy of their periods can be better than one part in 10^{10} per year. The most rapidly spinning pulsar in any kind of binary rotates with a period of 1.4 ms (He06).

Many hundreds of pulsars are by now known. They are concentrated toward the plane of the Galaxy (see Fig. 6.6).

(b) X-Ray Stars

The most readily observed X-ray sources are Galactic. Figure A.8 shows a clustering of the sources about the Galactic plane. These sources are associated with stars and fall into several groups.

(i) The Crab pulsar emits extremely regular X-ray pulses at its 33 millisecond radio pulsation rate. Other pulsars also emit a regular stream of pulses, but this is somewhat of an exception among stellar X-ray sources.

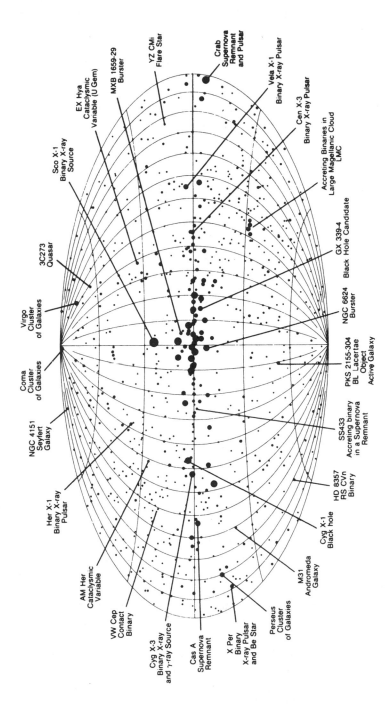

Fig.A.8. The X-ray sky. Map of the X-ray sky known in 1987 plotted in galactic coordinates. Note the concentration toward the galactic center and galactic plane. Bright sources are shown as larger circles. A number of individual sources are identified both by their respective catalogue designations and by source type. By 1997, the roughly 10⁶ registered extragalactic sources would have totally saturated this plot. (With the permission of Kent S. Wood and the Naval Research Laboratory, as well as of Jay M. Pasachoff and W.B. Saunders, Co.)

(ii) Many other X-ray sources are associated with neutron stars (Pa96). Matter tidally stripped from a companion star crashes onto the neutron star surface releasing energy re-emitted as X-rays. X-rays are similarly released when tidally captured matter impinges onto an *accretion disk* orbiting a *black hole*, a star in an ultimate state of collapse (Section 8:16). Matter composing the accretion disk is prevented by an excess of angular momentum from falling into the black hole. As it sheds the angular momentum, it gradually spirals inward and disappears in the black hole.

(iii) As the sensitivity of X-ray instrumentation has improved many classes of ordinary stars have also been detected.

Millions of extragalactic sources also emit X-rays.

A:11 Quasars and Active Galactic Nuclei, AGNs

The central elliptical galaxy in massive clusters is often embedded in a halo of hot, X-ray emitting gas — possibly ejecta propelled out of the galaxy by powerful supernova explosions. Many quasars, radio galaxies, and galaxies with *active galactic nuclei, AGNs* also are powerful X-ray sources. The term "quasar" and "QSO" — for "quasi-stellar object" — are often used interchangeably.

Quasars have many features in common with some types of radio galaxies; in particular, their visible spectra bear a strong resemblance to spectra of the nuclei of *Seyfert galaxies*, spiral galaxies with compact nuclei that emit strongly in the infrared and exhibit highly broadened emission lines from ionized gases. In both the quasars and Seyfert nuclei, we find highly ionized gases with spectra indicating temperatures of the order of 10^5 to 10^6 K and number densities $\sim 10^6 \, \mathrm{cm}^{-3}$. The conditions resemble those found in the solar corona. In the quasars and Seyfert nuclei the spectra of these gases show velocity differences of the order of 1000 or $2000 \, \mathrm{km \, s}^{-1}$, indicating either: (a) that gases are being shot out of these objects at high velocity; (b) that they are falling in at high speed; (c) that there is fast rotation; or (d) that there is a great deal of turbulent motion present. Most likely, a combination of two or three factors is involved.

The quasars and active nuclei of Seyfert galaxies sometimes show brightness variations on a time scale of hours. These highly luminous nuclei are, therefore, believed to either radiate into narrowly collimated beams emanating from rotating sources, or to be less than a few light-hours or days $\sim 10^{14}$ to 10^{16} cm in diameter. This argument assumes that the brightness changes are coherent. It would be invalidated if the variations were due to independent outbursts in different portions of a rather larger source.

Quasars have spectra that are highly red-shifted, indicating that they are at extreme distances and hence must be extremely luminous to appear as bright as they do. Only extreme infrared galaxies, whose peak emission occurs at wavelengths of $\sim 100 \, \mu$m, are comparably luminous. Some quasars have luminosities exceeding $10^{46} \, \mathrm{erg \, s}^{-1}$ — a hundred times higher than the Galaxy. Since these objects are compact, their surface brightness must be some ten orders of magnitude greater than that of normal galaxies. Extremely high X-ray luminosity also characterizes many quasars and active galactic nuclei. Many quasars, AGNs, and *blazars* are also gamma-ray sources. Blazars resemble quasars in most respects, except that their spectra are largely featureless.

A:12 Gamma-Ray Bursts

Gamma-ray bursts are short outbursts of gamma rays, in the energy range from 50 keV to 1 MeV, generally lasting from a fraction of a second to one hundred seconds. In other energy ranges the bursts have sometimes been observed to last longer. An outburst that occurred on February 17, 1994, was observed to emit gamma radiation at an energy of 30 GeV, for about an hour and a half. An outburst on May 8, 1997, was observed to brighten at optical wavelengths over the following two days, before fading over the following three or four days (Dj97). It lies far out in the Universe, beyond a red shift $z = 0.835$ (Me97a).

The more than one thousand bursts observed to date, appear to arrive from random directions in the sky (Fig. A.9). The most distant gamma-ray burst discovered

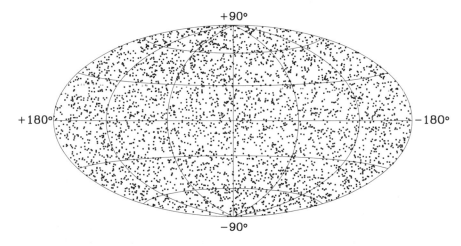

Fig. A.9. Distribution on the sky of the 3096 gamma-ray bursts (GRBs) registered by the Burst and Transient Source Experiment (BATSE) on board the Compton Gamma-Ray Observatory, CGRO (St01).

to date, GRB 050904, was observed on September 4, 2005 to erupt in a galaxy at red shift $z = 6.295$ (Ka06). Powerful gamma-ray bursts thus appear to be observable out to the most distant galaxy observed so far or the most distant quasar, respectively, at $z = 6.578$ and $z = 6.42$ (Ra06). The energy of a burst appears to be beamed into a relatively narrow solid angle. Bursts lasting longer than 2 s are now known to arrive from remote galaxies and appear to originate in extremely powerful supernova explosions. Short bursts sometimes lasting only a tenth of a second are also observed in distant galaxies, and believed to be emitted in the final merger of two compact stars, a neutron star binary or a neutron-star/black-hole binary, as the stars coalesce as a single black hole (see Section 8:18).

A small handful of GRBs are repeating bursters identified with neutron stars associated with supernova remnants. The first to be discovered lies in the Large Magellanic Cloud, only 50 kpc distant. The others have been identified with Galactic supernova remnants. Their luminosity is far lower than that of ordinary GRBs.

Table A.7. Energy and Number Densities of Photons and Cosmic Rays.

	Cosmic-Ray Particles	Visible Light	Microwave Background
Energy density in Galaxy (ergs cm^{-3})	10^{-12}	$\sim 2 \times 10^{-13}$	$\sim 5 \times 10^{-13}$
Extragalactic energy density (ergs cm^{-3})	?	$\sim 2 \times 10^{-14}$	$\sim 5 \times 10^{-13}$
Number density in Galaxy (cm^{-3})	$\sim 10^{-9}$	$\sim 10^{-1}$	$\sim 10^{3}$
Extragalactic number density (cm^{-3})	?	$\sim 10^{-2}$	$\sim 10^{3}$

A:13 Photons and Cosmic-Ray Particles

The Earth, the Solar System, and the Galaxy are all bathed in streams of photons and highly relativistic particles. Within the Galaxy photon densities are higher than outside since starlight and infrared emission make a strong local contribution. Outside the Galaxy, there is a ubiquitous microwave component that fills the Universe with the spectrum of a blackbody at 2.73 K (Fi96).

Cosmic-ray particles, highly energetic electrons and nucleons, constitute a denser energy bath in the Earth's vicinity than starlight and microwave photons combined. We do not know how the particles are distributed in extragalactic space, but believe that lower-energy cosmic rays are trapped in the Galaxy's magnetic field, and are locally generated in supernova explosions. The highest-energy cosmic rays with energies of $\sim 3 \times 10^{20}$ eV cannot be constrained by the Galaxy's magnetic field, and most probably are generated in violent explosions in distant quasars or active galaxies.

Table A.7 shows the energy densities of some of these components. X-rays and gamma rays, highly energetic photons, have far smaller energy densities than visible and microwave radiation (see also Fig. A.10).

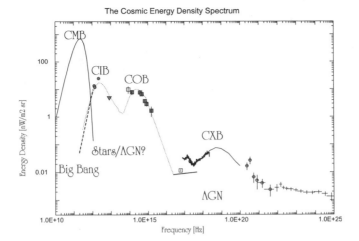

Fig. A.10. The spectrum of the diffuse cosmic background. The microwave background (CMB), infrared background (CIB), optical background (COB), and X-ray background (CXB) are indicated. At highest frequencies, there is also a diffuse gamma-ray background. At intervening frequencies the background has not yet been measured, largely because foreground sources have hindered detection. The CMB is a relic of the primordial hot universe. The infrared background is largely due to absorption and re-emission by dust grains heated by visible light from stars or higher-frequency radiation and cosmic rays. The X-ray background is mainly contributed by active galactic nuclei (AGN). Spectral frequencies are expressed in powers of 10. Courtesy of Günther Hasinger (Ha00).

A:14 Background Radiation

The view of the distant Cosmos is marred by several diffuse sources of foreground radiation. The Earth's atmosphere emits faintly even at visual wavelengths. In the far-infrared it emits a bright thermal glow and is totally opaque at many wavelengths. X-rays and gamma rays are also largely absorbed. Launching telescopes above the atmosphere helps, but zodiacal dust grains scatter and absorb sunlight, re-emitting energy at infrared wavelengths, providing a foreground glow through which more distant sources must be viewed. Even if we were to launch spacecraft entirely out of the Solar System we would still have a near-infrared foreground glow from the Galaxy's billions of red stars; heated dust clouds emit strongly in the mid- and far-infrared; and the Galactic plane is aglow with diffuse radio, X-ray, and γ-radiation.

To reveal the faint isotropic glow from distant portions of the Universe, we must first come to understand all these nearby sources of diffuse radiation, so we can compensate for them. This has been an arduous task. We have not yet succeeded at all frequencies, but what we do know is gathered in Figure A.10.

B Astrophysical Constants

B:1 Physical Constants

Speed of light	$c = 2.998 \times 10^{10}$ cm s^{-1}
Planck constant	$h = 6.626 \times 10^{-27}$ erg s
Gravitational constant	$G = 6.674 \times 10^{-8}$ cm^3 g^{-1} s^{-2}
Electron charge	$e = 4.803 \times 10^{-10}$ esu
Mass of electron	$m_e = 9.1094 \times 10^{-28}$ g
Mass of proton	$m_P = 1.6726 \times 10^{-24}$ g
Mass of hydrogen atom	$m_H = 1.6735 \times 10^{-24}$ g
Mass of neutron	$m_N = 1.6749 \times 10^{-24}$ g
Atomic mass unit	amu $= (1/12)\, m_{12\text{C}} = 1.6605 \times 10^{-24}$ g
Avogadro's number	6.0221×10^{23}
Boltzmann constant	$k = 1.3807 \times 10^{-16}$ erg K^{-1}
Thomson scattering cross-section	$\sigma_e = 6.652 \times 10^{-25}$ cm^2
Radiation density constant	$a = 7.566 \times 10^{-15}$ erg cm^{-3} K^{-4}
Stefan–Boltzmann constant	$\sigma = 5.670 \times 10^{-5}$ erg cm^{-2} K^{-4} s^{-1}
Rydberg constant	$R_\infty = 2.1799 \times 10^{-11}$ erg
Fine structure constant	$\alpha = 7.29735 \times 10^{-3}$

B:2 Astronomical Constants

Year	3.156×10^7 s
Astronomical unit, AU	1.49598×10^{13} cm
Parsec, pc	3.086×10^{18} cm
	3.262 light years
Solar mass, $M_\odot$	1.989×10^{33} g
Solar radius, $R_\odot$	6.957×10^{10} cm
Solar luminosity, $L_\odot$	3.845×10^{33} erg s^{-1}
Luminosity of star with $M_{\text{bol}} = 0$	2.97×10^{35} erg s^{-1}
Cosmic microwave background temperature	2.725 K

B:3 Units

Æon	10^9 yr $\equiv$ 1 Gyr
Ångstrøm unit, Å	10^{-8} cm
Atmosphere, atm	1.013×10^6 dyn cm^{-2} = 760 torr
Calorie	4.184×10^7 erg
Electron Volt, eV	1.602×10^{-12} erg
Hertz, Hz	1 s^{-1}
Jansky, Jy	10^{-26} W m^{-2} Hz^{-1}
Megahertz, MHz	10^6 Hz
Micron, μm	10^{-6} m = 10^{-4} cm

List of References

Articles and books that are cited in the text are referred to by the first two letters of the author's name and the last two digits of the year in which the publication appears. An article by Johnson published in 1928 is designated (Jo28). Where more than four individuals have co-authored a work, only the first author's name is used. A publication that appeared prior to 1908 carries a numeral designation –, for example Newton's *Principia* carries the designation (Ne–). Publications that appeared after 1999 are listed sequentially, as in the progression (Aℓ96), (Aℓ99), (Aℓ01), (Aℓ04). An asterisk following a given designation, for example (He67)*, implies that the reference is important for the entire section in which it is cited.

The following abbreviations have been used for frequently cited journals:

A + A: Astronomy and Astrophysics.

Ann. New York Acad. Sci.: Annals of the New York Academy of Sciences.

Ann. Rev. Astron. and Astrophys.: Annual Reviews of Astronomy and Astrophysics.

AJ: Astronomical Journal (USA).

ApJ: Astrophysical Journal (USA). If the letter L precedes the page number, the article appeared in the affiliated journal Astrophysical Journal Letters. The Astrophysical Journal also publishes a supplement, denoted by "ApJS"

Astrophys. Lett.: Astrophysical Letters.

B.A.N.: Bulletin of the Astronomical Society of the Netherlands.

I.A.U.: International Astronomical Union. The organization issues symposium proceedings and a variety of other publications.

J. Roy. Astron. Soc. Canad.: Journal of the Royal Astronomical Society of Canada.

J. Phys. USSR: Journal of Physics of the Soviet Union.

MNRAS: Monthly Notices of the Royal Astronomical Society (London).

PASP: Publications of the Astronomical Society of the Pacific.

Phil. Mag.: Philosophical Magazine (London).

Phys. Rev.: Physical Reviews (USA).

Phys. Rev. Lett.: Physical Review Letters (USA).

Proc. Nat. Acad. Sci.: Proceedings of the National Academy of Sciences (USA).

Proc. Roy. Soc.: Proceedings of the Royal Society (London).

PASJ: Publications of the Astronomical Society of Japan.

Quart. J. Roy. Astron. Soc.: Quarterly Journal of the Royal Astronomical Society (London).

Rev. Mod. Phys.: Reviews of Modern Physics (USA).

Soviet AJ: Astronomical Journal (Soviet Union). This journal also appears in English translation.

Z. Astrophys.: Zeitschrift für Astrophysik (Germany).

(Ab84) L. F. Abbott & M.B. Wise, "Large-Scale Anisotropy of the Microwave Background and the Amplitude of Energy Density Fluctuations in the Early Universe," *ApJ, 282*, L47 (1982).

(Ab02) T. Abel, G. L. Bryant & M. L. Norman, "The Formation of the First Star in the Universe," *Science, 295*, 93 (2002).

(Aℓ48) R. A. Alpher & R. Herman, "Evolution of the Universe," *Nature, 162*, 774 (1948).

(Aℓ53) R. A. Alpher, J. W. Follin, Jr. & R. Herman, "Physical Conditions in the Initial Stages of the Expanding Universe," *Phys. Rev., 92*, 1347 (1953).

(Aℓ55) C. W. Allen, *Astrophysical Quantities*, Athlone Press, London (1955).

(Aℓ63) L. H. Aller, "Astrophysics" in *The Atmospheres of the Sun and Stars*, 2nd ed., Ronald Press, New York (1963).

(Aℓ64) C. W. Allen, *Astrophysical Quantities*, 2nd ed., Athlone Press, London (1964).

(Aℓ73) C. W. Allen, *Astrophysical Quantities*, 3rd ed., Athlone Press, London (1973).

(Aℓ96) C. Alcock, *et al.*, "The Macho Project: Limits on Planetary Mass Dark Matter in the Galactic Halo from Gravitational Microlensing," *ApJ, 471*, 774 (1996).

(Aℓ97) C. Alcock, *et al.*, "The Macho Project: 45 Candidate Microlensing Events from the First-Year Galactic Bulge Data," *ApJ, 479*, 119 (1997).

(Aℓ99) L. Allen, M. J. Padgett & M. Babiker, "The Orbital Angular Momentum of Light," *Progress in Optics, 39* 291 (1999).

(Aℓ01) J. F. Alves, C. J. Lada & E. A. Lada, "Internal structure of a cold dark molecular cloud inferred from the extinction of background light," *Nature, 409*, 159 (2001).

(Aℓ01a) R. A. Alpher and R. Herman, *Genesis of the Big Bang*, Oxford University Press (2001).

(Aℓ04) E. Alécian & S. M. Morsink, "The Effect of Neutron Star Gravitational Binding Energy on Gravitational Radiation-Driven Mass-Transfer Binaries."*ApJ, 614*, 914 (2004).

(An72) E. Anders, "Physico-Chemical Processes in the Solar Nebula, as inferred from Meteorites" in *l'Origine du Système Solaire*, H. Reeves (Ed.), Centre National de la Recherche Scientifique, Paris (1972).

(An89) E. Anders & N. Grevesse, "Abundances of the elements: Meteoritic and solar," *Geochimica et Cosmochimica Acta, 53*, 197 (1989).

(Ao06) W. Aoki, *et al.*, "HE 1327–2326, An Unevolved Star with [Fe/H] < -5.0. I. A Comprehensive Abundance Analysis," *ApJ, 639*, 897 (2006).

(Ar65) H. Arp, "A Very Small, Condensed Galaxy," *ApJ, 142*, 402 (1965).

(Ar70) W. D. Arnett & D. D. Clayton, "Explosive Nucleosynthesis in Stars," *Nature, 227*, 780 (1970).

(Ar00) S. Arzumanov, *et al.*, "Neutron life time value measured by storing ultracold neutrons with detection of inelastically scattered neutrons," *Physics Letters B, 483*, 15 (2000).

(Ar02) N. Arkani-Hamed, S. Dimopoulos & G. Dvali, "Large Extra Dimensions: A New Arena for Particle Physics," *Physics Today* February issue 35 (2002).

(As71) M. E. Ash, I. I. Shapiro & W. B. Smith, "The System of Planetary Masses," *Science, 174*, 551 (1971).

(Ba72) J. N. Bahcall & R. L. Sears, "Solar Neutrinos," *Ann. Rev. Astron. and Astrophys.* (1972).

(Ba73) J. M. Bardeen, B. Carter & S. W. Hawking, "The Four Laws of Black Hole Mechanics," *Commun. math. Phys., 31*, 161, (1973).

(Ba84) J. N. Bahcall, "K Giants and the Total Amount of Matter Near the Sun," *ApJ, 287*, 926 (1984).

(Ba96) J. N. Bahcall, F. Calaprice, A. B. McDonald & Y. Totsuka, "Solar Neutrino Experiments: the Next Generation, *Physics Today, 49/7*, 30 (1996).

(Ba97) C. H. Bagley & G. G. Luther, "Preliminary Results of a Determination of the Newtonian Constant of Gravitation; A Test of the Kuroda Hypothesis," *Phys. Rev. Lett., 78*, 3047 (1997).

(Ba01) D. Barrado y Navascués, *et al.*, "From the Top to the Bottom of the Main Sequence: A Complete Mass Function of the Young Open Cluster M35," *ApJ, 546*, 1006 (2001).

(Ba03) J. N. Bahcall & C. Peña-Garay, "A road map to solar neutrino fluxes, neutrino oscillation parameters, and tests for new physics," *Journal of High Energy Physics, 0311*, 004 (2003).

(Ba05) W. H. Baumgartner, M. Lowenstein, D. J. Horner & R. F. Mushotzky, "Intermediate-Element Abundances in Galaxy Clusters," *ApJ, 620*, 680 (2005).

(Ba05a) R. Barvainis & R. Antonucci, "Extremely luminous Water Vapor Emission from a Type-2 Quasar at Redshift $z = 0.66$," *ApJ, 628*, L89 (2005).

(Be39) H. A. Bethe, "Energy Production in Stars," *Phys. Rev., 55*, 434 (1939).

(Be71) A. Bers, R. Fox, C.G. Kuper & S.G. Lipson, "The Impossibility of Free Tachyons" in *Relativity and Gravitation*, C.G. Kuper & A. Peres (Eds.), Gordon and Breach, New York (1971).

(Be73) J. D. Bekenstein, "Black Holes and Entropy," *Phys. Rev. D, 7*, 2333 (1973).

(Be97) G. Bernstein, P. Fischer, J. A. Tyson & G. Rhee, "Improved Parameters and New Lensed Features for Q0957+561 From WFPC2 Imaging," *ApJ, 483*, L79 (1997).

(Be02) M. P. Bernstein, *et al.*, "Racemic amino acids from the ultraviolet photolysis of interstellar ice analogues," *Nature, 416*, 401 (2002).

(Be03) C. L. Bennett, *et al.* "First-Year Wilkinson Microwave Anisotropy Probe (WMAP) Observations: Preliminary Maps and Basic Results," *ApJS, 148*, 1 (2003).

(Bh95) A. Bhattacharjee & Y.Yuan, "Self-Consistent Constraints on the Dynamo Mechanism," *ApJ, 449*, 739 (1995).

(Bi23) G. D. Birkhoff, *Relativity and Modern Physics*, §§ 4, 5, 11, Harvard University Press, Cambridge, MA (1923).

(Bi87) R. M. Bionta, *et al.*, "Observation of a Neutrino Burst in Coincidence with Supernova 1987A in the Large Magellanic Cloud," *Phys. Rev. Lett., 58*, 1494 (1987).

(Bi95) D. J. Bird, *et al.*, "Detection of a Cosmic Ray with Measured Energy Well Beyond the Excited Spectral Cutoff Due to Cosmic Microwave Radiation," *ApJ, 441*, 144 (1995).

(Bi97) P. L. Biermann, "The origin of the highest energy cosmic rays," *J. Physics G: Nucl. Part. Phys., 23*, 1 (1997).

(Bℓ61) A. Blaauw, "On the Origin of the O- and B- type Stars with High Velocities (The 'Run-away' Stars), and Some Related Problems," *B.A.N., 15*, 265 (1961).

(Bℓ99) S. A. Bludman & D. C. Kennedy, "Analytic Models for the Mechanical Structure of the Solar Core," *ApJ, 525*, 1024 (1999).

(Bℓ00) M. Blackwell, "Terrestrial Life — Fungal from the Start?" *Science, 289*, 1884 (2000).

(Bℓ03) M. R. Blanton, *et al.*, "The Galaxy Luminosity Function and Luminosity Density at Redshift $z = 0.1$," *ApJ, 592*, 819 (2003).

(Bo48) H. Bondi & T. Gold, "The Steady State Theory of the Expanding Universe," *MNRAS, 108*, 252 (1948).

(Bo52) H. Bondi, *Cosmology*, Cambridge University Press, Cambridge, England (1952).

(Bo56) W. B. Bonnor, "Boyle's Law and Gravitational Instability" *MNRAS, 116*, 351 (1956).

(Bo64a) S. Bowyer, E. Byram, T. Chubb & H. Friedman, "X-ray Sources in the Galaxy," *Nature, 201*, 1307 (1964).

(Bo64b) S. Bowyer, E. Byram, T. Chubb & H. Friedman, "Lunar Occultation of X-ray Emission from the Crab Nebula," *Science, 146*, 912 (1964).

(Bo71) B. J. Bok, "Observational Evidence for Galactic Spiral Structure" in *Highlights in Astronomy*, DeJager (Ed.), I.A.U. (1971).

(Bö81) E. Böhm-Vitense, "The Effective Temperature Scale" *Ann. Rev. Astron. and Astrophys., 19*, 295 (1981).

(Bo83) C. F. Bohren & D. R. Huffman, *Absorption and Scattering of Light by Small Particles*, John Wiley & Sons, New York (1983).

(Bo85) A. M. Boesgaard & G. Steigman, "Big Bang Nucleosynthesis: Theories and Observations," *Ann. Rev. Astron. and Astrophys., 23*, 319 (1985).

(Bo86) J. Bouvier, C. Bertout, W. Benz & M. Mayor, "Rotation in T Tauri stars I. Observations and immediate analysis," *A + A, 165*, 110 (1986).

(Bo95) S. P. Boughn & J. M. Uson, "Constraints on the Dwarf Star Content of Dark Matter," *Phys. Rev. Lett., 74*, 216 (1995).

(Bo96) A. P. Boss, "Extrasolar Planets," *Physics Today, 49/9*, 32 (1996); "Giants and dwarfs meet in the middle," *Nature, 379*, 397 (1996).

(Br97) W. J. Broad, "Scanning the Cosmos For Signs of Alien Life," *The New York Times* B7, August 26, 1997.

(Br03) V. Bromm & A. Loeb, "Formation of the First Supermassive Black Holes," *ApJ, 596*, 34 (2003).

(Bu57) E. M. Burbidge, G.R. Burbidge, W.A. Fowler & F. Hoyle, "Synthesis of the Elements in Stars," *Rev. Mod. Phys., 29*, 547 (1957).

(Bu00) A. Burrows, "Supernova Explosions in the Universe," *Nature, 403*, 727 (2000).

(Bu00a) R. Buser, "The Formation and Early Evolution of the Milky Way Galaxy," *Science, 287*, 69 (2000).

(By67) E. T. Byram, T. A. Chubb & H. Friedman, "Cosmic X-ray Sources, Galactic and Extragalactic," *Science, 152*, 66 (1967).

(Ca24) A. J. Cannon & E. C. Pickering, "The Henry Draper Catalogue," *Ann. Astron. Obs. Harvard College, 91* to *99* (1924).

(Ca48) H. B. G. Casimir, "On the attraction between two perfectly conducting plates," *Proc. Koninklijke Akademie van Wetenschappen, 51*, 793 (1948).

(Ca66) D. Cattani & C. Sacchi, "A Theory on the Creation of Stellar Magnetic Fields," *Nuovo Cimento, 46B*, 8046 (1966).

(Ca68) A. G. W. Cameron, "A New Table of Abundances of the Elements in the Solar System," in *Origin and Distribution of the Elements*, L.H. Ahrens, (Ed.), Pergamon Press, New York (1968), p. 125.

(Ca96) P. J. Callanan, *et al.*, "On the Mass of the Black Hole in GS 2000+25," *ApJ, 470*, L57 (1996).

(Ca96a) J. E. Carlstrom, M. Joy & L. Grego, "Interferometric Imaging of the Sunyaev-Zeldovich effect at 30 GHz," *ApJ, 456*, L75 (1996).

(Ca01) D. C. Catling, K. J. Zahnle & C. P. McKay, "Biogenic Methane, Hydrogen Escape, and the Irreversible Oxidation of Early Earth," *Science, 293*, 839 (2001).

(Ca02) G. M. Muñoz Caro, *et al.*, "Amino Acids from ultraviolet irradiation of interstellar ice analogues," *Nature, 416*, 403 (2002).

(Ce92) Z. Ceplecha, "Influx of interplanetary bodies onto Earth," *A + A, 263*, 361 (1992)

(Ce95) C. Cervetto, B. Barsella & A. Ferrara, "On the Charge Distribution Function of Dust Grains in a Hot Gas," *ApJ, 443*, 648 (1995)

(Ch39) S. Chandrasekhar, *An Introduction to the Study of Stellar Structure*, University of Chicago Press, Chicago (1939), Dover, New York (1957).

(Ch43) S. Chandrasekhar, *Principles of Stellar Dynamics*, University of Chicago Press, Chicago (1943), Dover, New York (1960).

(Ch50) S. Chandrasekhar, *Radiative Transfer*, University of Chicago Press, Chicago (1950), Dover, New York (1960).

(Ch58) S. Chandrasekhar, "On the Continuous Absorption Coefficient of the Negative Hydrogen Ion," *ApJ, 128*, 114 (1958).

(Ch64) J. H. Christenson, J. W. Cronin, V. L. Fitch & R. Turlay, "Evidence for the 2π Decay of the K_2^0 Meson," *Phys. Rev. Lett., 13*, 138 (1964).

(Ch70) W. Y. Chau, "Gravitational Radiation and the Oblique Rotator Model," *Nature, 228*, 655 (1970).

(Ch83) S. Chandrasekhar, *The Mathematical Theory of Black Holes*, Clarendon Press, Oxford (1983).

(Ch96) R. Y. Chiao, A. E. Kozhekin & G. Kurizki, "Tachyonlike Excitations in Inverted Two-Level Media," *Phys. Rev. Lett., 77*, 1254 (1996).

(Ch97) B. D. G. Chandran, "The Effects of Velocity Correlation Times on the Turbulent Amplification of Magnetic Energy," *ApJ, 482*, 156 (1997).

(Ch02) F. H. Chapelle, *et al.*, "A hydrogen-based subsurface microbial community dominated by methanogens," *Nature, 415*, 312 (2002).

(Ch03) J. Chiaverini, *et al.* "New Experimental Constraints on Non-Newtonian Forces below $100\mu m$," *Phys. Rev. Lett., 151101*, (2003).

(Ci06) F. D. Ciccarelli, *et al.*, "Toward Automatic Reconstruction of a Highly Resolved Tree of Life," *Science, 311*, 1283 (2006).

(Cℓ68) D. D. Clayton, *Principles of Stellar Evolution and Nucleosynthesis*, McGraw-Hill, New York (1968).

(Co57) T. G. Cowling, *Magnetohydrodynamics*, Interscience, New York (1957).

(Co68) C. C. Counselman III & I. I. Shapiro, "Scientific Uses of Pulsars," *Science, 162*, 352 (1968).

(Co70) R. Cowsik, Y. Pal & T. N. Rengarajan, "A Search for a Consistent Model for the Electromagnetic Spectrum of the Crab Nebula," *Astrophys. and Space Sci., 6*, 390 (1970).

(Co71a) D. P. Cox & E. Daltabuit, "Radiative Cooling of a Low Density Plasma," *ApJ, 167*, 113 (1971).

(Co71b) R. Cowsik & P. B. Price, "Origins of Cosmic Rays," *Physics Today, 24/10*, 30 (1971).

(Co72) J. J. Condon, "Decimetric Spectra of Extragalactic Radio Sources," Doctoral Dissertation in Astronomy, Cornell University (1972).

(Co95) C. J. Copi, D. N. Schramm & M. S. Turner, "Big-Bang Nucleosynthesis and the Baryon Density of the Universe," *Science, 267*, 192 (1995).

(Co95a) L. L. Cowie & A. Songaila, "Astrophysical Limits on the Evolution of Dimensionless Physical Constants over Cosmological Time," *ApJ, 453*, 596 (1995).

(Co96) W. N. Colley, J.A. Tyson & E.L. Turner, "Unlensing Multiple Arcs in 0024+1654: Reconstruction of the Source Image," *ApJ, 461*, L83 (1996).

(Co97) P. S. Coppi & F. A. Aharonian, "Constraints on the Very High Energy Emissivity of the Universe From the Diffuse GeV Gamma-Ray Background," *ApJ, 487*, L9 (1997).

(Co00) A. N. Cox, (Ed.) "Allen's Astrophysical Quantities," fourth edition, Springer (2000).

(Co02) G. Cooper, *et al.*, "Carbonaceous meteorites as a source of sugar-related organic compounds for the early Earth," *Nature, 414*, 879 (2002).

(Co04) N. J. Cornish, D. N. Spergel, G. D. Starkman & E. Komatsu, "Constraining the Topology of the Universe," Submitted to *Phys. Rev. Lett., 92*, 201302, (2004).

(Co04a) A. Coc, *et al.*, "Updated Big Bang Nucleosynthesis Compared with *Wilkinson Microwave Anisotropy Probe* Observations and the Abundance of Light Elements," *ApJ, 600*, 544 (2004).

(Co04b) J. J. Cowan & F.-K. Thielemann, "R-Process Nucleosynthesis in Supernovae," *Physics Today, 57/10* 47, October 2004.

(Cr95) I. A. Crawford, "Some Thoughts on the Implications of Faster-Than-Light Interstellar Space Travel," *Quart. J. Roy. Astron. Soc., 36*, 205 (1995).

(Cr05) S. M. Croom *et al.*, "The 2dF QSO Redshift Survey - XIV. Structure an evolution from the two-point correlation function," *MNRAS, 356*, 438 (2005).

(Cu00) C. L. Curry & C. F. McKee, "Composite Polytrope Models of Molecular Clouds. I. Theory," *ApJ, 528*, 734 (2000).

(Da51) L. Davis & J. L. Greenstein, "The Polarization of Starlight by Aligned Dust Grains," *ApJ, 114*, 206 (1951).

(Da68) R. Davis, Jr., D. S. Harmer & K.C. Hoffman, "Search for Neutrinos from the Sun," *Phys. Rev. Lett., 20*, 1205 (1968).

(Da69) K. Davidson & Y. Terzian, "Dispersion Measures of Pulsars," *AJ, 74*, 849 (1969).

(Da70) K. Davidson, "The Development of a Cocoon Star," *Astrophys. and Space Sci., 6*, 422 (1970).

(Da96) A. F. Davidsen, G. A. Kriss & W. Zheng, "Measurement of the opacity of ionized helium in the intergalactic medium," *Nature, 380*, 47 (1996).

(da00) L. N. da Costa, *et al.*, "Redshift-Distance Survey of Early-Type Galaxies: Dipole of the Velocity Field," *ApJ, 537*, L81 (2000).

(Da04) T. M. Davis & C. H. Lineweaver, "Expanding Confusion: Common Misconceptions of Cosmological Horizons and the Superluminal Expansion of the Universe," *Publications of the Astronomical Society of Australia 21*, 97 (2004).

(Da05) S. Dado & A. Dar, "The Double-Peak Spectral Energy Density of Gamma-Ray Bursts and the True Identity of GRB 031203," *ApJ, 627*, L109 (2005).

(de17) W. deSitter, "On Einstein's Theory of Gravitation and its Astronomical Consequences," *MNRAS, 78*, 3 (1917).

(De68) S. F. Dermott, "On the Origin of Commensurabilities in the Solar System–II. The Orbital Period Relation," *MNRAS, 141*, 363 (1968).

(De73) S. F. Dermott "Bode's Law and the Resonant Structure of the Solar System" *Nature Phys. Sci., 244*, 18 (1973).

(De94) D. J. DePaolo, "Strange bedfellows," *Nature, 372*, 131 (1994).

(De95) C. D. Dermer & N. Gehrels, "Two Classes of Gamma-Ray Emitting Active Galactic Nuclei," *ApJ, 447*, 103 (1995).

(De97) S. J. Desch & W. G. Roberge, "Ambipolar Diffusion and Far-Infrared Polarization from the Galactic Circumnuclear Disk," *ApJ, 475*, L115 (1997).

(De02) C. P. Deliyannis, A. Steinhauer, & R.D. Jeffries, "Discovery of the Most Lithium-Rich Dwarf: Diffusion in Action," *ApJ, 577*, L39 (2002).

(De02a) D. J. Des Marais, *et al.*, "Remote Sensing of Planetary Properties and Biosignatures on Extrasolar Terrestrial Planets" *Astrobiology, 2*, 153 (2002).

(De04) S. F. Dermott, "How Mercury got its spin," *Nature, 429*, 814 (2004).

(Di31) P. A. M. Dirac, "Quantized Singularities in the Electro-magnetic Field," *Proc. Roy. Soc., A, 133*, 60 (1931).

(Di38) P. A. M. Dirac, "A New Basis for Cosmology," *Proc. Roy. Soc., A, 165*, 199 (1938).

(Di67) R. H. Dicke, "Gravitation and Cosmic Physics," *Am. J. Phys., 35*, 559 (1967).

(Di86) R. H. Dicke, J. R. Kuhn & K. J. Libbrecht, "The variable oblateness of the sun: measurements of 1984," *ApJ, 311*, 1025 (1986).

(Di97) L. K. Ding, *et al.*, "Reexamination of Cosmic-Ray Composition around 10^{18} eV from Fly's Eye Data,"*ApJ, 474*, 490 (1997).

(Dj97) S. G. Djorgovski, *et al.*, "The optical counterpart to the γ-ray burst GRB970508," *Nature, 387*, 876 (1997).

(Do97) M. A. Dopita, *et al.*, "*Hubble Space Telescope* Observations of Planetary Nebulae in the Magellanic Clouds V. Mass Dependence of Dredge-up and the Chemical History of the Large Magellanic Cloud," *ApJ, 474*, 188 (1997).

(Dr62) F. D. Drake, "Intelligent Life in Space," Macmillan, New York (1962).

(Dr96) L. O'C. Drury,"Free-fall Timescale in an Arbitrary Number of Spatial Dimensions," *Quart. J. Roy. Astron. Soc., 37*, 255 (1996).

(Dr98) B. T. Draine & A. Lazarian, "Electric Dipole Radiation from Spinning Dust Grains," *ApJ, 508*, 157 (1998).

(Dr99) B. T. Draine & A. Lazarian, "Magnetic Dipole Microwave Emission from Dust Grains," *ApJ, 512*, 740 (1999).

(Du62) S. Dushman & J. M. Lafferty, "Scientific Foundations of Vacuum Technique," Wiley, New York (1962).

(Dw97) E. Dwek, *et al.*,"Detection and Characterization of Cold Interstellar Dust and Polycyclic Aromatic Hydrocarbon Emission, from *COBE* Observations," *ApJ, 475*, 565 (1997).

(Dy60) F. J. Dyson, "Search for Artificial Stellar Sources of Infrared Radiation," *Science, 131*, 1667 (1960).

(Eb55) R. Ebert, "Über die Verdichtung von HI Gebieten," *Z. Astrophys., 37*, 217 (1955).

(Ed30) A. S. Eddington, "On the Instability of Einstein's Spherical World," *MNRAS, 90*, 668 (1930).

(Eg62) O. Eggen, D. Lynden-Bell & A. Sandage, "Evidence From the Motions of Old Stars that the Galaxy Collapsed," *ApJ, 136*, 748 (1962).

(Eg03) K. Eguchi, *et al*, "First Results from KamLAND: Evidence for Reactor Antineutrino Disappearance," *Phys. Rev. Lett. 90*, 021802 (2003).

(Eh01) P. Ehrenfreund, *et al.*, "The photostability of Amino Acids in Space," *ApJ, 550*, L95 (2001).

(Ei–a) A. Einstein, "On the Electrodynamics of Moving Bodies," *Ann. d. Phys., 17*, 891 (1905), translated and reprinted in *The Principle of Relativity*, A. Sommerfeld, (Ed.), Dover, New York.

(Ei–b) A. Einstein, "Does the Inertia of a Body Depend Upon Its Energy Content?" *Ann. d. Phys., 18*, 639 (1905), translated and reprinted in *The Principle of Relativity*, A. Sommerfeld, (Ed.), Dover, New York.

(Ei11) A. Einstein, "On the Influence of Gravitation on the Propagation of Light," *Annalen der Physik, 35*, 898 (1911) (translated in *The Principle of Relativity*, A. Sommerfeld, (Ed.), Dover, New York.

(Ei16) A. Einstein, "The Principle of General Relativity" *Ann. d. Phys., 49*, 769 (1916), translated and reprinted in *The Principle of Relativity*, A. Sommerfeld, Ed., Dover, New York.

(Ei17) A. Einstein, "Cosmological Considerations on the General Theory of Relativity," *S. B. Preuss. Akad. Wiss.*, p. 142. (1917), translated in *The Principle of Relativity*, A. Sommerfeld, (Ed.), Dover, New York.

(Ei45) A. Einstein & E. G. Straus,"The Influence of the Expansion of Space on the Gravitation Fields Surrounding the Individual Stars," *Rev. Mod. Phys., 17*, 120 (1945); *Rev. Mod. Phys., 18*, 148 (1946).

(Ei03) F. Eisenhauer, *et al.*, "A Geometric Determination of the Distance to the Galactic Center," *ApJ 597*, L121 (2003).

(Ek96) E. H. Ekland & D. P. Bartel, "RNA-catalysed RNA polymerization using nucleoside triphosphates," *Nature, 382*, 373 (1996).

(El04) M. F. El Eid, B. S. Meyer & L.-S. The, "Evolution of Massive Stars up to the End of Central Oxygen Burning," *ApJ, 611*, 452 (2004).

(Ev96) D. S. Evans, "Standards of Angular Diameter," *Observatory, 116,* 230 (1996).

(Ez65) D. Ezer & A. G. W. Cameron, "The Early Evolution of the Sun," *Icarus, 1,* 422 (1963).

(Fa76) S. M. Faber & R. E. Jackson, "Velocity Dispersions and Mass-to-Light Ratios for Elliptical Galaxies," *ApJ, 204,* 668 (1976).

(Fa85) A. C. Fabian, "Gas in elliptical galaxies," *Nature, 314,* 130 (1985).

(Fa96) D. Fadda *et al.,* "The Observational Distribution of Internal Velocity Dispersions in Nearby Galaxy Clusters," *ApJ, 473,* 670 (1996).

(Fa01) X. Fan, *et al.,* " A Survey of $z > 5.8$ Quasars in the Sloan Digital Sky Survey. I. Discovery of Tree New Quasars and the Spatial Density of Luminous Quasars at z~ 6," AJ, 122, 2833 (2001).

(Fe96) J. P. Ferris, A. R. Hill Jr., R. Liu & L. E. Orgel, "Synthesis of long prebiotic oligomers on mineral surfaces," *Nature, 381,* 59 (1996).

(Fi96) D.J. Fixsen *et al.,* "The Cosmic Microwave Background Spectrum from the Full *COBE* FIRAS Data Set," *ApJ, 473,* 576 (1996).

(Fi02) D. J. Fixsen & J. C. Mather "The Spectral Results of the Far-Infrared Absolute Spectrophotometer Instrument on COBE," *ApJ, 581,* 817 (2002).

(Fi02a) D. P. Finkbeiner, D. J. Schlegel, C. Frank & C. Heiles, "Tentative Detection of Electric Dipole Emission from Rapidly Rotating Dust Grains," *ApJ, 655,* 898 (2002).

(Fi04) M. Fischer, *et al.,* "New Limits on the Drift of Fundamental Constants from Laboratory Measurements," *Phys. Rev. Lett., 92,* 230802 (2004).

(Fi05) L. A. Fisk "Journey into the Unknown Beyond," *Science, 309,* 2016 (2005).

(Fi05a) D. F. Figer, "An upper limit to the masses of stars," *Nature, 434,* 192 (2005).

(Fℓ96) P. J. Flower, "Transformations from Theoretical Hertzsprung–Russell Diagrams to Color-Magnitude Diagrams: Effective Temperatures, $B–V$ Colors, and Bolometric Corrections," *ApJ, 469,* 355 (1996).

(Fo01) E. B. Fomalont, B. J. Geldzahler & C. F. Bradshaw, "Scorpius X-1: The Evolution and Nature of the Twin Compact Radio Lobes," ApJ, 558, 283 (2001).

(Fr22) A. Friedmann, "Über die Krümmung des Raumes," *Z. Phys., 10,* 377 (1922).

(Fr96) H. T. Freudenreich, "The Shape and Color of the Galactic Disk," *ApJ, 468,* 663 (1996).

(Fr03) C. L. Fryer & P. Mészáros, "Neutrino-Driven Explosions in Gamma-Ray Bursts and Hypernovae," *ApJ, 588,* L25 (2003).

(Fu04) M. Fukugita & P. J. E. Peebles, "The Cosmic Energy Inventory," *ApJ, 616,* 643 (2004).

(Ga–) Galileo Galilei, *Dialogues Concerning Two New Sciences*, translated by H. Crew & A. de Salvio, Northwestern University Press (1946).

(Ge05) N. Gehrels, *et al.* "A short γ-ray burst apparently associated with an elliptical galaxy at redshift $z = 0.225$, *Nature, 437*, 851 (2005).

(Gh03) A. M. Ghez, *et al.*, "The First Measurements of Spectral Lines in a Short-Period Star Bound to the Galaxy's Central Black Hole: A Paradox of Youth," *ApJ, 586*, L127 (2003).

(Gi62) R. Giacconi, H. Gursky, F. Paolini & B. Rossi, "Evidence for X-rays From Sources Outside the Solar System," *Phys. Rev. Lett., 9*, 439 (1962).

(Gi64) V. L. Ginzburg & S. I. Syrovatskii, *The Origin of Cosmic Rays*, Pergamon Press, New York (1964).

(Gi69) V. L. Ginzburg, '*Elementary Processes for Cosmic Ray Astrophysics*, Gordon & Breach, New York (1969).

(Gi97) R. Giovanelli, *et al.*, "The Tully–Fisher Relation and H_0," *ApJ, 477*, L1 (1997).

(Gℓ97) N. K. Glendenning, *Compact Stars – Nuclear Physics, Particle Physics, and General Relativity*, Springer–Verlag, New York (1997).

(Gn97) O. Y. Gnedin & J. P. Ostriker, "Destruction of the Galactic Globular Cluster System," *ApJ, 474*, 223 (1997).

(Go62) T. Gold, "The Arrow of Time," *Am. J. of Phys., 30*, 403 (1962).

(Go68) T. Gold, "Rotating Neutron Stars as the Origin of the Pulsating Radio Sources," *Nature, 218*, 731 (1968).

(Go73) P. Goldreich & W. R. Ward, "The Formation of Planetesimals," *ApJ, 183*, 1051 (1973).

(Go95) A. A. Goodman & D. C. B. Whittet, "A point in Favor of the Superparamagnetic Grain Hypothesis," *ApJ, 455*, L181 (1995).

(Gr66) K. Greisen, "End to the Cosmic Ray Spectrum," *Phys. Rev. Lett., 16*, 748 (1966).

(Gr68) J. M. Greenberg, "Interstellar Grains" in *Nebulae and Interstellar Matter*, B. M. Middlehurst & L. H. Aller, (Eds.), University of Chicago Press, Chicago (1968).

(Gr80) I. S. Gradshteyn & I. M. Ryzhik, "Tables of Integrals, Series, and Products," Corrected and Enlarged Edition, Academic Press, Inc., New York Section 3.411 (1980).

(Gr97) N. Grosso, *et al.*, "An X-ray superflare from an infrared protostar," *Nature, 387*, 56 (1997).

(Gu54) S. N. Gupta, "Gravitation and Electromagnetism," *Phys. Rev., 96*, 1683 (1954).

(Gu65) J. E. Gunn & B. A. Peterson, "On the Density of Neutral Hydrogen in Intergalactic Space," *ApJ, 142*, 1633 (1965).

(Gu66) H. Gursky, *et al.*, "A Measurement of the Location of the X-ray Source Sco X-1," *ApJ, 146*, 310 (1966).

(Gu81) A. H. Guth, "Inflationary Universe: A Possible Solution to the Horizon and Flatness Problem," *Phys. Rev. D, 23*, 347 (1981).

(Ha54) R. Hanbury Brown & R. Q. Twiss, "A New Type of Interferometer for Use in Radio Astronomy," *Phil. Mag., 45*, 663 (1954).

(Ha62) C. Hayashi, R. Hoshi & D. Sugimoto, "Evolution of the Stars," *Progress of Theor. Physics*, Suppl. 22 (1962).

(Ha65) E. R. Harrison, "Olbers' Paradox and the Background Radiation Density in an Isotropic Homogeneous Universe," *MNRAS, 131*, 1 (1965).

(Ha66) C. Hayashi, "Evolution of Protostars," *Ann. Rev. Astron. and Astrophys., 4*, 171 (1966).

(Ha70) M. Harwit, "Is Magnetic Alignment of Interstellar Dust Really Necessary?," *Nature, 226*, 61 (1970).

(Ha70a) E. R. Harrison, "Fluctuations at the Threshold of Classical Cosmology," *Phys. Rev. D., 1*, 2726 (1970).

(Ha75) S. W. Hawking, "Particle Creation by Black Holes," *Commun. Math. Phys., 43*, 199 (1975).

(Ha81) M. Harwit, *Cosmic Discovery — The Search, Scope and Heritage of Astronomy*, Basic Books, New York, (1981)

(Ha92) M. Harwit, "Cosmic Curvature and Condensation," *ApJ, 392*, 394 (1992).

(Ha95) M. Harwit, "Time and its Evolution in an Inhomogeneous Universe," *ApJ, 447*, 482 (1995).

(Ha96) Z. Haiman, A. A. Thoul & A. Loeb, "Cosmological Formation of Low-Mass Objects," *ApJ, 464*, 523 (1996).

(Ha99) H. J. Habing, *et al.*, "Disappearance of stellar debris disks around main-sequence stars after 400 million years," *Nature, 401*, 456 (1999).

(Ha99a) I. Hachisu, M. Kato & K. Nomoto, "A Wide Symbiotic Channel to Type Ia Supernovae," *ApJ, 522*, 487 (1999).

(Ha00) G. Hasinger "X-ray Surveys of the Obscured Universe," *Lecture Notes in Physics, 548*, 423 (2000).

(Ha02) B. Hao, *et al.*, "A New UAG-Encoded Residue in the Structure of a Methanogen Methyltransferase," *Science, 296*, 1462 (2002).

(Ha04) J. R. Hargis, E. L. Sandquist & M. Bolte, "The Luminosity Function and Color-Magnitude Diagram of the Globular Cluster M12," *ApJ, 608*, 243 (2004).

(He44) G. Herzberg, *Atomic Spectra and Atomic Structure*, Dover, New York (1944).

(He50) G. Herzberg, *Molecular Spectra and Molecular Structure I. Spectra of Diatomic Molecules*, 2nd ed., Van Nostrand, Princeton, NJ, (1950).

(He62) O. Heckmann & E. Schücking, "Relativistic Cosmology," in *Gravitation*, L. Witten, (Ed.), John Wiley, New York. (1962).

(He67) G. Herzberg, "The Spectra of Hydrogen and Their Role in the Development of Our Understanding of the Structure of Matter and of the Universe," *Trans. Roy. Soc.*, Canada V, Ser IV, 3 (1967).

(He68) C. E. Heiles, "Normal OH Emission and Interstellar Dust Clouds," *ApJ, 151,* 919 (1968).

(He68a) A. Hewish, S. J. Bell, J. D. H. Pilkington, P. F. Scott, & R. A. Collins, "Observation of a Rapidly Pulsating Radio Source," *Nature, 217,* 709 (1968).

(He69) C. E. Heiles, "Temperatures and OH Optical Depths in Dust Clouds," *ApJ, 157,* 123 (1969).

(He02) A. Heger & S. E. Woosley, "The Nucleosynthetic Signature of Population III," *ApJ, 567,* 532 (2002).

(He04) F. Herwig "Dredge-up and Envelope Burning in Intermediate-Mass Giants of Very Low Metallicity," *ApJ, 605,* 425 (2004).

(He04a) A. Heavens, B. Panter, R. Jimenez & J. Dunlop, "The star-formation history of the Universe from stellar populations of nearby galaxies," *Nature, 428,* 625 (2004).

(He06) J. W. T. Hessels, *et al.,* "A Radio Pulsar Spinning at 716 Hz," *Science. 311,* 1901 (2006).

(Hi71) J. M. Hill, "A Measurement of the Gravitational Deflection of Radio Waves by the Sun," *MNRAS, 153,* 7p (1971).

(Hi87) K. Hirata, *et al.,* "Observation of a Neutrino Burst from the Supernova SN 1987A," *Phys. Rev. Lett., 58,* 1490 (1987).

(Hi96) R. H. Hildebrand, "Problems in Far-Infrared Polarimetry," in *Polarimetry of the Interstellar Medium,.* W. G. Roberge & D. C. B. Whittet (Eds.), *ASP Conference Series, 97,* 254 (1996).

(Hi98) W. R. Hix, A. M. Khokhlov, J. C. Wheeler & F.-K. Thielemann, "The Quasi-Equilibrium–Reduced α-Network," *ApJ, 503,* 332 (1998).

(Ho48) F. Hoyle, "A New Model for the Expanding Universe," *MNRAS, 108,* 372 (1948).

(Ho57) F. Hoyle, "The Black Cloud," Signet, New York (1957).

(Hö65) B. Höglund, & P. G. Mezger, "Hydrogen Emission Line $n_{110} \rightarrow n_{109}$: Detection at 5009 Megahertz in Galactic HII Regions," *Science, 150,* 339 (1965).

(Ho69) L. M. Hobbs, "Regional Studies of Interstellar Sodium Lines," *ApJ, 158,* 461 (1969).

(Ho71) D. Hollenbach & E.E. Salpeter, "Surface Recombination of Hydrogen Molecules," *ApJ, 163,* 155 (1971).

(Ho74) G. t'Hooft, "Magnetic Monopoles in Unified Gauge Theories," *Nuclear Physics B, 79,* 276 (1974).

(Ho91) D. J. Hollenbach, T. Takahashi & A.G.G.M. Tielens, "Low-Density Photodissociation Regions," *ApJ, 377,* 192 (1991).

(Ho96) M. Hof, *et al.,* "Measurement of Cosmic-Ray Antiprotons from 3.7 to 19 GeV," *ApJ, 467,* L33 (1996).

(Ho01) J. M. Hollis, *et al.,* "The Spatial Scale of Glycolaldehyde in the Galactic Center," *ApJ, 554,* L81 (2001).

(Ho02) J. B. Holberg, T. D. Oswalt & E. M. Simon, *ApJ, 571*, 512 (2002).

(Hu29) E. Hubble, "Distance and Radial Velocity Among Extragalactic Nebulae," *Proc. Nat. Acad. Sci., 15*, 168 (1929).

(Hu65) S. Huang, "Rotational Behavior of the Main-Sequence Stars and its Plausible Consequences Concerning Formation of Planetary Systems I and II," *ApJ, 141*, 985 (1965) and *ApJ, 150*, 229 (1967).

(Hu75) R. A. Hulse & J. H. Taylor, "Discovery of a Pulsar in a Binary System," *ApJ, 195*, L51 (1975).

(Hu97) W. Hu, N. Sugiyama, & J. Silk, "The physics of microwave background anisotropies," *Nature, 386*, 37 (1997).

(Hu02) W. Hu & S. Dodelson, "Cosmic Microwave Background Anisotropies," *Ann. Rev. Astron. & Astrophys., 40*, 171 (2002).

(Ib65) I. Iben, Jr., "Stellar Evolution–I. The Approach to the Main Sequence," *ApJ, 141*, 993 (1965).

(Ib70) I. Iben, "Globular-Cluster Stars," *Scientific American*, July 1970, p. 27.

(Ib03) A. I. Ibrahim, J. H. Swank & W. Parke, "New Evidence for Proton-Cyclotron Resonance in a Magnetar Strength Field from SGR 1806-20," *ApJ 584*, L17 (2003).

(Ir02) R. Irion, "Astrobiologists Try to 'Follow the Water to Life'," *Science, 296*, 647 (2002).

(Já50) L. Jánossy, *Cosmic Rays*, 2nd ed., The Clarendon Press, Oxford, England (1950).

(Ja70) D. A. Jauncey, A. E. Niell & J. J. Condon, "Improved Spectra of Some Ohio Radio Sources with Unusual Spectra," *ApJ, 162*, L31 (1970).

(Je69) E. B. Jenkins, D. C. Morton & T. A. Matilsky, "Interstellar Lα Absorption in β^1, δ, and π Scorpii," *ApJ, 158*, 473 (1969).

(Je70) E. B. Jenkins, "Observations of Interstellar Lyman-α Absorption," *IAU Symposium # 36: Ultraviolet Stellar Spectra and Related Ground Based Observations*, L. Houziaux & H.E. Butler (Eds.), Reidel Publ. Co., Holland (1970).

(Je04) E. L. N. Jensen, R. D. Mathieu, A. X. Donar & A. Dullighan, "Testing Protoplanetary Disk Alignment in Young Binaries," *ApJ, 600*, 789 (2004).

(Jo28) J. B. Johnson, "Thermal Agitation of Electricity in Conductors," *Phys. Rev., 32*, 97 (1928).

(Jo53) H. L. Johnson & W. W. Morgan, "Fundamental Stellar Photometry for Standards of Spectral Type on the Revised System of the Yerkes Spectral Atlas," *ApJ, 117*, 313 (1953).

(Jo67) R. V. Jones & L. Spitzer, Jr., "Magnetic Alignment of Interstellar Grains," *ApJ, 147*, 943 (1967).

(Jo04) D. H. Jones, *et al.*, "The 6dF Galaxy Survey: samples, observational techniques and the first data release," *MNRAS, 355*, 747 (2004).

(Ka54) F. D. Kahn, "The Acceleration of Interstellar Clouds," *B.A.N., 12*, 187 (1954).

(Ka59) N. S. Kardashev, "On the Possibility of Detection of Allowed Lines of Atomic Hydrogen in the Radio-Frequency Spectrum," *Astronomicheskii Zhurnal, 36*, 838 (1959), *Soviet AJ, 3*, 813 (1959).

(Ka68) F. D. Kahn, "Problems of Gas Dynamics in Planetary Nebulae," *I.A.U., Symposium on Planetary Nebulae*, D. Osterbrock & C.R. O'Dell (Eds.), Springer-Verlag, New York (1968).

(Ka87) R. Kawabe, *et al.*, "High-Resolution Observations of CO from the Bipolar Nebula CRL 2688," *ApJ, 314*, 322 (1987).

(Ka97) P. Kaaret & E. C. Ford, "Using Neutron Stars and Black Holes in X-ray Binaries to Probe Strong Gravitational Fields," *Science, 276*, 1386 (1997).

(Ka01) J. F. Kasting, "The Rise of Atmospheric Oxygen," *Science, 293*, 819 (2001).

(Ka03) V. M. Kaspi, *et al.*, "A Major Soft Gamma Repeater-Like Outburst and Rotation Glitch in the No-Longer-So-Anomalous X-Ray Pulsar IE 2259+586," *ApJ, 588*, L93 (2003).

(Ka06) N. Kawai, *et al.*, "An optical spectrum of the afterglow of a γ-ray burst at a redshift of $= 6.295$," *Nature, 440*, 184 (2006).

(Ke63) P. C. Keenan, "Classification of Stellar Spectra" in *Stars and Stellar Systems*, Vol. 3, K. A. Strand (Ed.) (1963).

(Ke65) F. J. Kerr & G. Westerhout, "Distribution of Interstellar Hydrogen" in *Stars and Stellar Systems V: Galactic Structure*, A. Blaauw & M. Schmidt (Eds.), University of Chicago Press, Chicago (1965).

(Ke69) K. I. Kellermann, I. I. K. Pauliny-Toth, & P. J. S. Williams, "The Spectra of Radio Sources in The Revised 3C Catalogue," *ApJ, 157*, 1 (1969).

(Ke71) K. I. Kellermann, *et al.*,"High Resolution Observations of Compact Radio Sources at 6 and 18 Centimeters," *ApJ, 169*, 1 (1971).

(Ki05) B. Kim, *et al.*, "The Velocity Field of Baryonic Gas in the Universe,"ApJ, 625, 599 (2005).

(Kℓ73) R. W. Klebesadel, I. B. Strong & R. A. Olson, "Observations of Gamma-Ray Bursts of Cosmic Origin," ApJ, 182, L85 (1973).

(Ko90) E. W. Kolb & M. S. Turner, *The Early Universe*, Addison-Wesley, Reading, MA (1990).

(Ko96) E. W. Kolb, A. Linde & A. Riotto, "Grand-Unified-Theory Baryogenesis after Preheating," *Phys. Rev. Lett., 77*, 4290 (1996).

(Ko03) A. Kogut, *et al.*, "First-Year Wilkinson Microwave Anisotropy Probe (WMAP) Observations: Temperature-Polarization Correlation," *ApJS, 148*, 161, 2003.

(Ko04) M. Konacki & B. F. Lane, "The Visual Orbit of the Spectroscopic Binaries HD 6118 and HD 27483 from the Palomar Testbed Interferometer," *ApJ, 610*, 443 (2004).

(Kr68) K. S. Krishna Swamy & C. R. O'Dell, "Thermal Emission by Particles in NGC 7027," *ApJ, 151*, L61 (1968).

(Kr94) P. P. Kronberg, "Extragalactic Magnetic Fields," *Reports on Progress in Physics, 57*, 325 (1994).

(Kr96) R. C. Kraan-Korteweg, *et al.*, "A nearby massive galaxy cluster behind the Milky Way," *Nature, 379*, 519 (1996).

(Kr01) F. Krennrich, *et al.*, "Cutoff in the TeV Energy Spectrum of Markarian 421 During strong Flares in 2001," *ApJ, 560*, L45 (2001).

(Ku97) R. M. Kulsrud, R. Cen, J. P. Ostriker, & D. Ryu, "The Protogalactic Origin for Cosmic Magnetic Fields," *ApJ, 480*, 481 (1997).

(Ku97a) S. R. Kulkarni, "Brown Dwarfs: A Possible Missing Link Between Stars and Planets," *Science, 276*, 1350 (1997).

(Ku97b) T. Kundić, *et al.*, "A Robust Determination of the Time Delay in 0957+561A, B and a Measurement of the Global value of Hubble's Constant," *ApJ, 482*, 75 (1997).

(Ku03). Y.-J. Kuan, *et al.*, "Interstellar Glycine," *ApJ, 593*, 848 (2003).

(Kv70) K. Kvenvolden, *et al.*,"Evidence for Extraterrestrial Amino-acids and Hydrocarbons in the Murchison Meteorite," *Nature, 228*, 923, (1970).

(La51) L. Landau & E. Lifshitz, *The Classical Theory of Fields*, Addison-Wesley, New York (1951).

(La67) J. W. Larimer, "Chemical fractionations in meteorites-I. Condensation of the elements," *Geochimica et Cosmochimica Acta, 31*, 1215 (1967).

(La67a) J. W. Larimer & E. Anders "Chemical fractionations in meteorites-II. Abundance patterns and their interpretation," *Geochimica et Cosmochimica Acta, 31*, 1239 (1967).

(La72) R. B. Larson, "Infall of Matter in Galaxies," *Nature, 236*, 21 (1972).

(La74) K. Lang, *Astrophysical Formulae*, Springer-Verlag, New York (1974), p47.

(La94) T. R. Lauer & M. Postman, "The Motion of the Local Group with Respect to the 15,000 Kilometer per Second Abell Cluster Inertial Frame," *ApJ, 425*, 418 (1994).

(La96) R. J. Laureijs, *et al.*, "Very cold dust associated with molecular gas," *A + A, 315*, L317 (1996).

(La97) S. K. Lamoreaux, "Demonstration of the Casimir Force in the 0.6 to 6 μm Range," *Phys. Rev. Lett., 78*, 5 (1997).

(La98) R. B. Larson, "Early star formation and the evolution of the stellar initial mass function in galaxies," *MNRAS, 301*, 569 (1998).

(La99) A. Lazarian & B. T. Draine, "Nuclear Spin Relaxation within Interstellar Grains," *ApJ, 520*, L67 (1999).

(Le31) G. Lemaître, "A Homogeneous Universe of Constant Mass and Increasing Radius accounting for the Radial Velocity of Extra-Galactic Nebulae," *MNRAS, 91*, 483 (1931).

(Le50) G. Lemaître, *The Primeval Atom*, Van Nostrand, Princeton, NJ (1950).

(Le56) T. D. Lee & C. N. Yang, "Question of Parity Conservation in Weak Interactions," *Phys. Rev., 104*, 254 (1956).

(Le72) T. J. Lee, "Astrophysics and Vacuum Technology," *Journal of Vacuum Science and Technology*, Jan. – Feb. (1972).

(Le96) D. H. Lee, *et al.*, "A self-replicating peptide," *Nature, 382*, 525 (1996).

(Li46) E. M. Lifshitz, "On the Gravitational Stability of the Expanding Universe," *J. Phys. USSR, 10*, 116 (1946).

(Li67) C. C. Lin, "The Dynamics of Disk-Shaped Galaxies," *Ann. Rev. Astron. & Astrophys., 5*, 453 (1967).

(Li06) M. Livio & M. J. Rees, "Anthropic Reasoning," *Science, 309*, 1022 (2006).

(Lo–) H. A. Lorentz, "Electromagnetic Phenomena in a System Moving with Any Velocity Less Than Light," *Proceedings of the Acad. Sci.*, Amsterdam 6, 1904, translated and reprinted in *The Principle of Relativity*, A. Sommerfeld (Ed.), Dover, New York.

(Lo78) M. Longair, "Cosmological Aspects of Infrared and Millimetre Astronomy," in *Infrared Astronomy*,G. Setti & G.G. Fazio(Eds.), Reidel, Dordrecht, Holland (1978).

(Lo93) S. G. Love & D. E. Brownlee, " A Direct Measurement of the Terrestrial Mass Accretion Rate of Cosmic Dust," *Science, 262*, 550 (1993).

(Lo03) K. Lodders, "Solar System Abundances and Condensation Temperatures of the Elements," *ApJ, 591*, 1220 (2003).

(Lu03) J.-P. Luminet, J. R. Weeks, A. Riazuelo, R. Lehoucq & J.-P. Uzan, "Dodec ahedral space topology as an explanation for weak wide-angle temperature correlations in the cosmic microwave background." *Nature, 425*, 593 (2003).

(Lu03a) T. Ludlam & L. McLerran, "What Have We Learned From the Relativistic Heavy Ion Collider?" *Physics Today, 56/10*, 48 October (2003).

(Ma70) D. S. Mathewson & V. L. Ford, "Polarization Observations of 1800 Stars," *Memoirs of the Royal Astronomical Society, 74*, 139 (1970).

(Ma72a) P. M. Mathews & M. Lakshmanan, "On the apparent Visual Forms of Relativistically Moving Objects," *Nuovo Cimento, 12B*, 168 (1972).

(Ma72b) R. N. Manchester, "Pulsar Rotation and Dispersion Measures and the Galactic Magnetic Field," *ApJ, 172*, 43 (1972).

(Ma97) P. D. Mannheim, "Are Galactic Rotation Curves Really Flat?" *ApJ, 479*, 659 (1997).

(Mc01) J. E. McClintock, *et al.*, "A Black Hole Greater than 6 $M_\odot$ in the X-Ray Nova XTE J1118 +480," *ApJ, 551*, L147 (2001).

(Me68) P. G. Mezger, "A New Class of Compact, High Density HII Regions" in *Interstellar Ionized Hydrogen*, Y. Terzian (Ed.), Benjamin, Menlo Park, CA (1968).

(Me69) P. Meyer, "Cosmic Rays in the Galaxy," *Ann. Rev. Astron. Astrophys., 7*, 1 (1969).

(Me96) M. Metcalfe, *et al.*, "Galaxy formation at high redshifts," *Nature, 383*, 236 (1996).

(Me97) D. M. Mehringer & K M. Menten, "44 GHz Methanol Maser and Quasi-Thermal Emission in Sagittarius B2," *ApJ, 474*, 346 (1997).

(Me97a) M. R. Metzger, *et al.*, "Spectral constraints on the redshift of the optical counterpart to the γ-ray burst of 8 May 1997," *Nature, 387*, 878 (1997).

(Me02) D. D. Meisel, D. Janches & J. D. Mathews, "Extrasolar Micrometeors Radiating from the Vicinity of the Local Interstellar Bubble," *ApJ, 567*, 323 (2002).

(Me02a) B. S. Meyer, "r-Process Nucleosynthesis without Excess Neutrons," *Phys. Rev. Lett., 89*, 231101 (2002).

(Me04) A. Mesinger, Z. Haiman & R. Cen, "Probing the Reionization History Using the Spectra of High-Redshift Sources," *ApJ, 613*, 23 (2004).

(Me05) D. Merritt, J. F. Navarro, A. Ludlow & A. Jenkins, "A Universal Density Profile for Dark and Luminous Matter," *ApJ, 624*, L85 (2005).

(Mi08) H. Minkowski, "Space and Time" an address delivered to the German Natural Scientists and Physicians (1908), translated and reprinted in *The Principle of Relativity*, A. Sommerfeld (Ed.), Dover, New York.

(Mi57a) S. L. Miller, "The Formation of Organic Compounds on the Primitive Earth," *Ann. New York Acad. Sci., 69*, 260 (1957).

(Mi57b) R. I. Mitchell & H. L. Johnson, "The Color-Magnitude Diagram of the Pleiades Cluster," *ApJ, 125*, 418 (1957).

(Mi59) S. L. Miller & H. C. Urey, "Organic Compound Synthesis on the Primitive Earth," *Science, 130*, 245 (1959).

(Mi73) C. W. Misner, K. S. Thorne, & J. A. Wheeler, *Gravitation*, Freeman, San Francisco (1973).

(Mi95) M. Miyoshi, *et al.*, "Evidence for a black hole from high rotation velocities in a sub-parsec region of NGC4258," *Nature, 373*, 127 (1995).

(Mi95a) M. Milgrom, "MOND and the Seven Dwarfs," *ApJ, 455*, 439 (1995).

(Mi96) J. C. Mihos & L. Hernquist, "Gasdynamics and Starbursts in Major Mergers," *ApJ, 464*, 641 (1996).

(Mi98) I. F. Mirabel & L. F. Rodriguez, "Microquasars in our galaxy," *Nature, 392*, 673 (1998).

(Mo67) D. C. Morton, "Mass Loss from Three OB Supergiants in Orion," *ApJ, 150*, 535 (1967).

(Mo97) V. Morell, "Microbiology's Scarred Revolutionary," *Science, 276*, 699 (1997).

(Mo97a) A. Moiseev, *et al.*, "Cosmic-Ray Antiproton Flux in the Energy Range from 200 to 600 MeV," *ApJ, 474*, 479 (1997).

(Mo05) S. Moorbath, "Dating earliest life," *Nature, 434*, 155 (2005).

(Mu97) R. Mushotzky, "How one galaxy can be a cluster," *Nature, 388*, 126 (1997).

(Mü05) P. Müschlegel, *et al.*, "Resonant Optical Antennas," *Science, 308*, 1607 (2005).

(My98) P. C. Myers, F. C. Adams, H. Chen & E. Schaff, "Evolution of the Bolometric Temperature and Luminosity of Young Stellar Object," *ApJ, 492, 703*, (1998).

(Na96) J. F. Navarro, C. S. Frenk & S. D. M. White, "The Structure of Cold Dark Matter Halos," *ApJ, 462*, 563 (1996).

(Na97) J. F. Navarro, C. S. Frenk & S. D. M. White, "A Universal Density Profile from Hierarchical Clustering," *ApJ, 490*, 493 (1997).

(Na97a) R. Narayan, M. R. Garcia & J. E. McClintock, "Advection-Dominated Accretion and Black Hole Event Horizons," *ApJ, 478*, L79 (1997).

(Ne–) Isaac Newton, *Mathematical Principles of Natural Philosophy & Systems of the World*, revised translation by Florian Cajori, University of California Press, Berkeley (1962).

(Ne95) D. A. Neufeld, S. Lepp & G. Melnick, "Thermal Balance in Dense Molecular Clouds: Radiative Cooling Rates and Emission-Line Luminosities," *ApJ Suppl., 100*, 132 (1995).

(Ne97) R. Neuhäuser, "Low-Mass Pre-Main Sequence Stars & Their X-ray Emission," *Science, 276*, 1363 (1997).

(Ne02) V. V. Nesvizhevsky, *et al.*, "Quantum states of neutrons in the Earth's gravitational field," *Nature, 415*, 297 (2002).

(Ni05) D. J. Nice, *et al.* "A $2.1 M_\odot$ Pulsar Measured by Relativistic Orbital Decay," *ApJ 634*, 1242 (2005).

(Ni95) M. M. Nieto, *et al.*, "Theoretical Motivation for Gravitation Experiments on Ultra-Low Energy Antiprotons & Antihydrogen," *Third Biennial Conference on Low-Energy Antiproton Physics, LEAP'94* G. Kernel, P. Krizan & M. Mikuz (Eds.), World Scientific, Singapore (1995), p. 606.

(Ni96) P. D. Nicholson, *et al.*, "Observations of Saturn's Ring-Plane Crossings in August and November 1995," *Science, 272*, 509 (1996).

(Ni05) F. Nicastro, *et al.*, "The mass of the missing baryons in the X-ray forest of the warm-hot intergalactic medium," *Nature, 433*, 495 (2005).

(No96) K. Nordtvedt, "From Newton's Moon to Einstein's Moon," *Physics Today, 49/5*, 26 (1996).

(No97) D. Normile, "New Experiments Step up Hunt for Neutrino Mass," *Science, 276*, 1795 (1997).

(Nu84) H. Nussbaumer & W. Schmutz, "The hydrogenic 2s $\rightarrow$ 1s two photon emission," *A + A, 138*, 495 (1984).

(Ny28) H. Nyquist, "Thermal Agitation of Electric Charge in Conductors," *Phys. Rev., 32*, 110 (1928).

(Od65) M. Oda, *et al.*, "Angular Sizes of the X-ray Sources in Scorpio and Sagittarius," *Nature, 205*, 554 (1965).

(O'H98) T. O'Halloran, P. Sokolsky & S. Yoshida, "The Highest -Energy Cosmic Rays," *Physics Today, 51/1* 31 (1998).

(Oi97) M. Oishi, "Molecules in Astrophysics: Probes and Processes", *IAU Symposium 178* E. F. Van Dieshoeck (Ed.), Kluwer, Dordrecht (1997), p.61.

(Oℓ03) C. M. Oliveira, *et al.*, "Interstellar Deuterium, Nitrogen, and Oxygen Abundances toward GD 246, WD 2331-475, HZ 21, and Lanning 23: Results from the *FUSE* Mission," *ApJ, 587*, 235 (2003).

(Oo27a) J. H. Oort, "Observational Evidence Confirming Lindblad's Hypothesis of a Rotation of the Galactic System," *B.A.N., 3*, 275 (1927).

(Oo27b) J. H. Oort, "Investigations Concerning the Rotational Motion of the Galactic System, Together with New Determinations of Secular Parallaxes, Precession and Motion of the Equinox," *B.A.N., 4*, 79 (1927).

(Oo65) J. H. Oort, "Stellar Dynamics" in *Galactic Structure*, A. Blaauw & M. Schmidt (Eds.), University of Chicago Press, Chicago (1965), p. 455.

(Op39a) J. R. Oppenheimer & G.M. Volkoff, "On Massive Neutron Cores," *Phys. Rev., 55*, 374 (1939).

(Op39b) J. R. Openheimer & H. Snyder, "On Continued Gravitational Contraction," *Phys. Rev., 56*, 455 (1939).

(Op61a) A. I. Oparin, *Life, Its Nature, Origin and Development*, Academic Press, New York (1961).

(Op61b) A. I. Oparin & V. G. Fessenkov, *Life in the Universe*, Foreign Languages Publishing House, Moscow; also Twayne and Co., New York (1961).

(Pa57) E. N. Parker, "Sweet's Mechanism for Merging Magnetic Fields in Conducting Fluids," *Journal of Geophysical Research, 62*, 509 (1957).

(Pa68) F. Pacini, "Rotating Neutron Stars, Pulsars and Supernova Remnants," *Nature, 219*, 145 (1968).

(Pa96) D. Page & A. Sarmiento, "Surface Temperature of a Magnetized Neutron Star and Interpretation of the *ROSAT* Data. II.," *ApJ, 473*, 1067 (1996).

(Pa01) V. Pankonin, E. Churchwell, C. Watson, & J.H. Bieging, "A Methyl Cyanide Search for the Earliest Stages of Massive Protostars," *ApJ, 558*, 194 (2001).

(Pa03) L. Page, *et al.*, "First-Year Wilkinson Microwave Anisotropy Probe (WMAP) Observations: Interpretation of the TT and TE Angular Power Spectrum Peaks," *ApJS 148*, 233 (2003).

(Pe64) P. C. Peters, "Gravitational Radiation and the Motion of Two Point Masses," *Phys. Rev., 136*, B1224 (1964).

(Pe65) A. A. Penzias & R.W. Wilson, "A Measurement of Excess Antenna Temperature at 4080 Mc/s," *ApJ, 142*, 420 (1965).

(Pe85) R. C. Peterson, "Radial Velocities of Remote Globular Clusters: Stalking the Missing Mass," *ApJ, 297*, 309 (1985).

(Pe93) P. J. E. Peebles, *Principles of Physical Cosmology*, Princeton University Press, Princeton, NJ (1993).

(Pe95) M. A. C. Perryman, *et al.*, " Parallaxes and the Hertzsprung-Russell diagram from the preliminary Hipparcos solution H30," *A + A, 304*, 69 (1995).

(Pe98) K. Peach, "Time's Broken Arrow," *Nature, 396*, 407.

(Pe98a) M. A. C. Perryman, *et al.*, "The Hyades: distance, structure, dynamics and age," *A + A, 331*, 81 (1998).

(Pe98b) S. Perlmutter, *et al.*, "Discovery of a supernova explosion at half the age of the Universe," *Nature, 391*, 51 (1998).

(Pe01) M. Pettini & D. V. Bowen, "A New Measurement of the Primordial Abundance of Deuterium toward Convergence with the Baryon Density from the Cosmic Microwave Background," *ApJ, 560*, 41 (2001).

(Pe05) S. Perlmutter, private communication, August 27, 2005.

(Pi54) F. A. E. Pirani, "On the Influence of the Expansion of Space on the Gravitational Field Surrounding an Isolated Body,"*Proceedings of the Cambridge Philosophical Society, 50*, 637 (1954).

(Pi66) G. C. Pimentel, *et al.*,"Exotic Biochemistries in Exobiology" in *Biology and the Exploration of Mars*, C. S. Pittendrigh, W. Vishniac, & J. P. Pearman (Eds.), Nat. Acad. Sci., NRC, Washington (1966).

(Po74) A. M. Polyakov, "Particle Spectrum in Quantum Field Theory," *JETP Lett., 20*, 194 (1974).

(Pr61) I. Prigogine, "Thermodynamics of Irreversible Processes," John Wiley, New York (1961).

(Pr74) W. H. Press & P. Schechter, "Formation of Galaxies and Clusters of Galaxies by Self-Similar Gravitational Condensation," *ApJ, 187*, 425 (1974).

(Pr95) J. D. Prestage, R. L. Tjoelker & L. Maleki, "Atomic Clocks and Variations of the Fine Structure Constant," *Phys. Rev. Lett., 74*, 3511 (1995).

(Pr04) A. Prestwich, personal communication September 2004.

(Pu79) E. M. Purcell, "Suprathermal Rotation of Interstellar Grains," *ApJ, 231), 404* (1979).

(Qu03) H. R. Quinn, "The Asymmetry Between Matter and Antimatter," *Physics Today, 56/2*, 30 (2003).

(Ra71) D. M. Rank, C. H. Townes & W. J. Welch, "Interstellar Molecules and Dense Clouds," *Science, 174*, 1083 (1971).

(Ra95) R. S. Raghavan, "Solar Neutrinos — From Puzzle to Paradox," *Science, 267*, 45 (1995).

(Ra99) L. Randall & R. Sundrum, " An Alternative to Compactification," *Phys. Rev. Lett., 83*, 4690 (1999).

(Ra05) S. Rainville, *et al.*, "A direct test of $E = mc^2$," *Science, 438*, 1096 (2005).

(Ra06) E. Ramirez-Ruiz, "Ancient blast comes to light," *Nature, 440*, 154 (2006).

(Re64) S. Refsdal, "The Gravitational Lens Effect," *MNRAS, 128*, 295 (1964).

(Re64a) S. Refsdal, "On the Possibility of Determining Hubble's Parameter and the Masses of Galaxies from the Gravitational Lens Effect," *MNRAS, 128*, 307 (1964).

(Re67) M. J. Rees, "Studies in Radio Source Structure I. A Relativistically Expanding Model for Variable Quasi-Stellar Radio Sources," *MNRAS, 135*, 345 (1967).

(Re68a) M. J. Rees & D.W. Sciama, "Large-Scale Density Inhomogeneities in the Universe," *Nature, 217*, 511 (1968).

(Re68b) M. J. Rees, "Proton Synchrotron Emission from Compact Radio Sources," *Astrophys. Lett., 2*, 1 (1968).

(Re68c) M. J. Rees & W. L. W. Sargent, "Composition and Origin of Cosmic Rays," *Nature, 219*, 1005 (1968).

(Re71) M. J. Rees, "New Interpretation of Extragalactic Radio Sources," *Nature, 229*, 312 (1971).

(Re87) N. Reid, "The stellar mass function at low luminosities," *MNRAS, 225*, 873 (1987).

(Re88) A. Renzini & F. F. Pecci, "Tests of Evolutionary Sequences using Color-Magnitude Diagrams of Globular Clusters, *Annual Reviews of Astronomy and Astrophysics 26*, 199 (1988).

(Re97) D. Reimers, *et al.*, "Patchy intergalactic He II absorption in HE 2347-4342," *A & A, 327*, 890 (1997).

(Re04) Reid, M. J. & Brunthaler, A. " The Proper Motion of Sagittarius A*. II. The mass of Sagittarius A*," *ApJ, 616*, 872 (2004).

(Ri56) W. Rindler, "Visual Horizons in World Models," *MNRAS, 116*, 662 (1956).

(Ri95) A. G. Riess, W. H. Press, & R. P. Kirshner, "Determining the Motion of the Local Group Using Type Ia Supernovae Light Curve Shapes," *ApJ, 445*, L91 (1995).

(Ri99) C. Ritossa, E. García-Berro & I. Iben, Jr., "On the Evolution of Stars that Form Electron-Degenerate Cores Processed by Carbon-Burning. V. Shell Convection Sustained by Helium Burning, Transient Neon Burning, Dredge-out, URCA cooling, and other properties of an $11 M_\odot$ Population I Model Star," *ApJ, 515*, 381 (1999).

(Ri01) A. G. Riess, *et al.*, "The Farthest Known Supernova: Support for an Accelerating Universe and a Glimpse of the Epoch of Deceleration," *ApJ, 560*, 49 (2001).

(Ri04) Riess, A. G., *et al.*, "Type Ia Supernova Discoveries at $z > 1$ From the *Hubble Space Telescope*: Evidence for Past Deceleration and Constraints on Dark Energy Evolution," *ApJ, 607*, 665 (2004).

(Ro33) H. P. Robertson, "Relativistic Cosmology," *Rev. Mod. Phys., 5*, 62 (1933).

(Ro55) H. P. Robertson, "The Theoretical Aspects of the Nebular Red Shift," *PASP, 67*, 82 (1955).

(Ro64) P. G. Roll, R. Krotkov, & R. H. Dicke, "The Equivalence of Inertial and Passive Gravitational Mass," *Annals of Physics* (USA), *26*, 442 (1964).

(Ro65) F. Rosebury, "Handbook of Electron Tube and Vacuum Techniques," Addison-Wesley, Reading, MA (1965).

(Ro68) H. P. Robertson & T. W. Noonan, *Relativity and Cosmology*, Sanders, (1968).

(Ro01) F. Robert, "The Origin of Water on Earth," *Science, 293*, 1056 (2001).

(Ru71) M. Ruderman, "Solid Stars," *Scientific American*, February (1971), p. 29.

(Ry71) G. R. Rybicki & A. P. Lightman, "Radiative Processes in Astrophysics," John Wiley & Sons, New York (1971).

(Ry96) S. G. Ryan, J. E. Norris, & T. C. Beers, "Extremely Metal-Poor Stars. II. Elemental Abundances and the Early Chemical Enrichment of the Galaxy," *ApJ, 471* 254 (1996).

(Sa52) E. E. Salpeter, "Nuclear Reactions in Stars Without Hydrogen," *ApJ, 115*, 326 (1952).

(Sa55) E. E. Salpeter, "The Luminosity Function and Stellar Evolution," *ApJ, 121*, 161 (1955).

(Sa55a) E. E. Salpeter, "Nuclear Reactions in Stars II. Protons on Light Nuclei," *Phys. Rev., 97*, 1237 (1955).

(Sa57) A. Sandage, "Observational Approach to Evolution–II.A Computed Luminosity Function for K0-K2 Stars from $M_v = +5$ to $M_v = -4.5$," *ApJ, 125*, 435 (1957).

(Sa58) A. Sandage, "Current Problems in the Extragalactic Distance Scale," *ApJ, 127*, 513 (1958).

(Sa66) A. R. Sandage, *et al.*, "On the Optical Identification of Sco X-1," *ApJ, 146*, 316 (1966).

(Sa67) E. E. Salpeter, "Stellar Structure Leading up to White Dwarfs and Neutron Stars" in "Relativity Theory and Stellar Structure," Chapter 3 in *Lectures in Applied Mathematics*, Vol. 10., American Mathematical Society (1967).

(Sa67a) R. K. Sachs & A. M. Wolfe, "Perturbations of a Cosmological Model and Angular Variations of the Microwave Background," *ApJ, 147*, 73 (1967).

(Sa68) D. H. Sadler, "Astronomical Measures of Time," *Quarterly Journal, Roy. Astron. Soc., 9*, 281 (1968).

(Sa70a) E. E. Salpeter, "Solid State Astrophysics," in *Methods and Problems of Theoretical Physics* J.E. Bowcock, Ed., North-Holland, Amsterdam (1970).

(Sa70b) C. Sagan, "Life," in *Encyclopedia Britannica* (1970).

(Sa97) R. W. Sayer, D. J. Nice, & J. H. Taylor, "The Greenbank Northern Sky Survey for Fast Pulsars," *ApJ, 474*, 426 (1997).

(Sa97a) M. Samland, G. Hensler, & Ch. Theis, "Modeling the Evolution of Disk Galaxies. I. The Chemodynamical Method and the Galaxy Model," *ApJ, 476*, 544 (1997).

(Sc16) K. Schwarzschild, "Über das Gravitationsfeld eines Massenpunktes nach der Einsteinschen Theorie," *Sitzungsberichte der Königlich- Preussischen Akademie der Wissenschaften, VII*, 189 (1916).

(Sc44) E. Schrödinger, *What is Life*, Cambridge University Press, Cambridge, England (1944).

(Sc54) E. Schücking, "Das Schwarzschildsche Linienelement und die Expansion des Weltalls," *Zeitschrift für Physik, 137*, 595 (1954).

(Sc58a) L. I. Schiff, "Sign of the Gravitational Mass of a Positron," *Phys. Rev. Lett., 1*, 254 (1958).

(Sc58b) M. Schwarzschild, "Structure and Evolution of the Stars," Princeton University Press, Princeton, NJ (1958).

(Sc92) G. Schaller, D. Schaerer, G. Meynet, & A. Maeder, "New grids of stellar models from 0.8 to 120 $M_\odot$ at $Z = 0.020$ and $Z = 0.001$" *A + A Suppl, 96*, 269 (1992).

(Sc96) B. Schwarzschild, "New Measurements of Ancient Deuterium Boost the Baryon Density of the Universe," *Physics Today, 49/8*, 17 (1996).

(Sc97) M. Schmidt, *et al.*, "Experimental Determination of the Melting Point and Heat Capacity for a Free Cluster of 139 Sodium Atoms," *Phys. Rev. Lett., 79*, 99 (1997).

(Sc99) B. Schwarzschild, "Two Experiments Observe Explicit Violation of Time-Reversal Symmetry," *Physics Today, 52/2*, 19 (1999).

(Sc02) B. Schwarzschild, "Direct Measurement of the Sun's Total Neutrino Output Confirms Flavor Metamorphosis," *Physics Today, 55/7*, 13 (2002).

(Sc03) R. Schödel, *et al.*, "Stellar Dynamics in the Central Arcsecond of our Galaxy," *ApJ, 596*, 1015 (2003).

(Sc05) S. J. Schwartz, *et al.*, "The Gamma-Ray Giant Flare from SGR 1806-20: Evidence of Crustal Cracking via Initial Timescales," *ApJ, 627*, L129 (2005).

(Se65) P. A. Seeger, W. A. Fowler, & D. D. Clayton, "Nucleosynthesis of Heavy Elements by Neutron Capture," *ApJS, 11*, 121 (1965).

(Se97) J. M. M. Senovilla & R. Vera, "Impossibility of the Cylindrically Symmetric Einstein–Straus Model," *Phys. Rev. Lett., 78*, 2284 (1997).

(Se99) S. Seager, D. D. Sasselov & D. Scott, "A new Calculation of the Recombination Epoch," *ApJ, 523* L1 (1999).

(Se04) K. R. Sembach, *et al.*, "The Deuterium-to-Hydrogen Ratio in a Low-Metallicity Cloud Falling onto the Milky Way," *ApJS, 150*, 387 (2004).

(Sh60) I. S. Shklovskii, "Cosmic Radiowaves," Harvard University Press, Cambridge, MA (1960).

(Sh66) I. S. Shklovskii & C. Sagan, "Intelligent Life in the Universe," Delta, New York (1966).

(Sh97) F. H. Shu, H. Shang, A. E. Glassgold & T. Lee, " X-rays and Fluctuating X-Winds from Protostars," *Science, 277*, 1475 (1997).

(Si68) J. Silk, "Cosmic Black-Body Radiation and Galaxy Formation," *ApJ, 151* 459 (1968).

(So95) A. Songaila, E. M. Hu & L. L. Cowie, "A population of very diffuse Lyman-α clouds as the origin of the He$^+$ absorption signal in the intergalactic medium," *Nature, 375*, 124 (1995).

(Sp51a) L. Spitzer, Jr. & M. Schwarzschild, "The Possible Influence of Interstellar Clouds on Stellar Velocities," *ApJ, 114*, 394 (1951).

(Sp51b) L. Spitzer, Jr. & J. L. Greenstein, "Continuous Emission from Planetary Nebulae," *ApJ, 114*, 407 (1951).

(Sp62) L. Spitzer, Jr., *The Physics of Fully Ionized Gases*, Interscience, New York (1962).

(Sp97) D. Sprayberry, C. D. Impey, M. J. Irwin & G. D. Bothun, "Low Surface Brightness Galaxies in the Local Universe III. Implications for the Field Galaxy Luminosity Function," *ApJ, 481*, 104 (1997).

(Sp03) D. N. Spergel, *et al.*, "First Year Wilkinson Microwave Anisotropy Probe (WMAP) Observations: Determination of Cosmological Parameters", *ApJS, 148*, 175 (2003).

(Sp05) V. Springel, *et al.*, "Simulations of the formation, evolution and clustering of galaxies and quasars," *Nature, 435*, 629 (2005).

(Sr98) P. Sreekumar, *et al.*, "Egret Observations of the Extragalactic Gamma-Ray Emission," *ApJ, 494*, 523 (1998).

(Sr02) G. Srinivasan, C. M. James & J. A. Krzycki, "Pyrrolysine Encoded by UAG in Archaea: Charging of a UAG-Decoding Specialized tRNA," *Science, 296*, 1459 (2002).

(St39) B. Strömgren, "The Physical State of Interstellar Hydrogen," *ApJ, 89*, 526 (1939).

(St69) T. P. Stecher, "Interstellar Extinction in the Ultraviolet II," *ApJ, 157*, L125 (1969).

(St72) L. J. Stief, *et al.*, "Photochemistry and Lifetimes of Interstellar Molecules," *ApJ, 171*, 21 (1972).

(St96) A. Stern & H. Campins, "Chiron and the Centaurs: escapees from the Kuiper belt," *Nature, 382*, 507 (1996).

(St97) R. B. Stothers & C.-W. Chin, "The Mixing Length in Convective Stellar Envelopes is Proportional to Distance from the Convective Boundary," *ApJ, 478*, L 103 (1997)

(St98) C. C. Steidel, *et al.*, "A Large Structure of Galaxies at Redshift $z \sim 3$ and its Cosmological Implications, "*ApJ, 492*, 428 (1998).

(St01) B. E. Stern, *et al.*, "An Off-Line Scan of the BATSE Daily Records and a Large Uniform Sample of Gamma-Ray Bursts," *ApJ, 563*, 80 (2001).

(St02) D. Steeghs & J. Casares, "The Mass Donor of Scorpius X-1 Revealed," *ApJ, 568*, 273 (2002).

St04) C. H. Storry, *et al.*, "First Laser-Controlled Antihydrogen Production," *Phys. Rev. Lett., 93*, 263401 (2004).

(Su72) R. A. Sunyaev & Ya. B. Zel'dovich, "The Observation of Relic Radiation as a Test of the Nature of X-ray Radiation from the Clusters of Galaxies," *Comments on Astrophysics & Space Science, 4*, 173 (1972).

(Su04) T. Suda, *et al.*, "Is HE 0107-5240 A Primordial Star? The Characteristics of Extremely Metal-Poor Carbon-Rich Stars," *ApJ, 611*, 476 (2004).

(Sw58) P. A. Sweet, "The Neutral Point Theory of Solar Flares," *I.A.U. Symposium 6, Electromagnetic Phenomena in Cosmical Plasma*, B. Lehnert, ed., New York: Cambridge University Press) P 123 (1958).

(Sw93) S. P. Swordy, J. L'Heureux, P. Meyer, & D. Müller, "Elemental Abundances in the Local Cosmic Rays at High Energies," *ApJ, 403*, 658 (1993).

(Ta82) J. H. Taylor & J. M. Weisberg, "A New Test of General Relativity: Gravitational Radiation and the Binary Pulsar PSR 1913 + 16," *ApJ, 253*, 908 (1982).

(Ta95) Y. Tanaka, *et al.*, "Gravitationally redshifted emission implying an accretion disk and massive black hole in the active galaxy MCG-6-30-15," *Nature, 375*, 659 (1995).

(Ta98) T. Tanimori, *et al.*, "Detection of Gamma Rays of up to 50 TeV from the Crab Nebula," *ApJ, 492*, L33 (1998).

(Ta98a) M. Takeda, *et al.*, "Extension of the Cosmic-Ray Energy Spectrum beyond the Predicted Greisen-Zatsepin-Kuz'min Cutoff," *Phys. Rev. Lett., 81*, 1163 (1998)

(Te59) J. Terrell, "Invisibility of the Lorentz Contraction," *Phys. Rev., 116*, 1041 (1959).

(Te72) Y. Terzian, *A Tabulation of Pulsar Observations*, Earth and Terrestrial Sciences, Gordon & Breach, New York (1972).

(Te04) M. Tegmark, *et al.*, "The Three-Dimensional Power Spectrum of Galaxies from the Sloan Digital Sky Survey," *ApJ, 606*, 702 (2004).

(Th75) K. Thorne & A. N. Żytkow, "Red Giants and Supergiants with Degenerate Neutron Cores," *ApJ, 199*, L19 (1975).

(Th93) S. E. Thorsett, Z. Arzoumanian, M. M. McKinnon & J.H. Taylor, "The Masses of Two Binary Neutron Star Systems," *ApJ, 405*, L29 (1993).

(Th94) K. S. Thorne, *Black Holes and Time Warps — Einstein's Outrageous Legacy*, W. W. Norton, New York (1994).

(Th96) S. E. Thorsett, "The Gravitational Constant, the Chandrasekhar Limit, and Neutron Star Masses," *Phys. Rev. Lett., 77*, 1432 (1996).

(To47) C. H. Townes, "Interpretation of Radio Radiation from the Milky Way," *ApJ, 105*, 235 (1947).

(To97) G. Torres, R. P. Stefanik & D. W. Latham, "The Hyades Binary 51 Tauri: Spectroscopic Detection of the Primary, the Distance to the Cluster, and the Mass–Luminosity Relation," *ApJ, 474*, 256 (1997).

(To02) T. Totani & A. Panaitescu, "Orphan Afterglow of Collimated Gamma-Ray Bursts: Rate Predictions and Prospects for Detection," *ApJ, 576*, 120 (2002).

(Tr02) S. Tremaine, *et al.*, "The Slope of the Black Hole Mass versus Velocity Dispersion Correlation," *ApJ, 574*, 740 (2002).

(Tr04) C. Travaglio, *et al.*, "Galactic Evolution of Sr, Y and Zr: A Multiplicity of Nucleosynthetic Processes," *ApJ, 601*, 864 (2004).

(Tu97) M. S. Turner, "Inflationary Cosmology," in *Relativistic Astrophysics*, B.J.T. Jones & D. Marković (Eds.), Cambridge University Press, Cambridge, England (1997).

(Tu04) J. Tumlinson, A. Venkatesan & J. M. Shull, "Nucleosynthesis, Reionization, and the Mass Function of the First Stars," *ApJ, 612*, 602 (2004).

(Ty98). J. A. Tyson, G. P. Kochanski & I. P. Dell'Antonio, " Detailed Mass Map of CL 0024+1654 from Strong Lensing," *ApJ, 498*, L107 (1998).

(Un69) A. Unsöld, "Stellar Abundances and the Origin of Elements," *Science, 163*, 1015 (1969).

(Va70) R. S. Van Dyck, Jr., C. E. Johnson & H. A. Shugant, "Radiative Lifetime of Metastable 2^1S_0 State of Helium," *Phys. Rev. Lett., 25*, 1403 (1970).

(Va96) W. D. Vacca, C. D. Garmany & J. M. Shull, "The Lyman-Continuum Fluxes and Stellar Parameters of O and Early B-Type Stars," *ApJ, 460*, 914 (1996).

(vB02) K. von Braun, *et al.*, "Photometry results for the globular clusters M10 and M12: Extinction maps, color-magnitude diagrams, and variable star candidates," *AJ 124*, 2067 (2002).

(vd68) S. van den Bergh, "Galaxies of the Local Group," *J. Roy. Astron. Soc. Canad., 62*, 145, 219 (1968).

(vd72) S. van den Bergh, "Search for Faint Companions to M31," *ApJ, 171*, L31 (1972).

(vd94) S. Van den Bergh, "The Local Group," *Proceedings of the Third CTIO/ESO Workshop on The Local Group: Comparative and Global Properties*, A. Layden, R. C. Smith, & J. Storm (Eds.), ESO Conference and Workshop Proceedings, No. 51 (1994), page 3.

(vd02) F. F. S. van der Tak, *et al.*, "Triply deuterated ammonia in NGC 1333," *A + A, 388*, L53 (2002).

(Vi99) E. T. Vishniac & A. Lazarian, "Reconnection in the Interstellar Medium," *ApJ, 511*, 193 (1999).

(Vi05) J. S. Villasenor *et al.* "Discovery of the short γ-ray burst GRB 050509," *Nature, 437*, 855 (2005).

(vo55) S. von Hoerner, "Über die Bahnform der Kugelförmigen Sternhaufen," *Z. Astrophys., 35*, 255 (1955).

(Wa34) A. G. Walker, "Distance in an Expanding Universe," *MNRAS, 94*, 159 (1934).

(Wa67) R. V. Wagoner, "Cosmological Element Production," *Science, 155*, 1369 (1967).

(Wa97) A. Watson, "To Catch a WIMP," *Science, 275*, 1736 (1997).

(Wa03) R. V. Wagoner, "Heartbeats of a neutron star," *Nature, 424*, 27 (2003).

(We68) V. Weidemann, "The HR Diagram of White Dwarfs," *Ann. Rev. Astron. & Astrophys., 6*, 351 (1968).

(We71) A. S. Webster & M. S. Longair, "The Diffusion of Relativistic Electrons from Infrared Sources and Their X-ray Emission," *MNRAS, 151*, 261 (1971).

(We87) P. S. Wesson, K. Valle, & R. Stabell, "The Extragalactic Background Light and a Definitive Resolution of Olbers's Paradox," *ApJ, 317*, 601 (1987).

(We97) J. K. Webb, *et al.*, "A high deuterium abundance at redshift $z = 0.7$," *Nature, 388*, 250 (1997).

(We97a) D. H. Weinberg, L. Hernquist, & N. Katz, "Photoionization, Numerical Resolution and Galaxy Formation," *ApJ, 477*, 8 (1997).

(Wh64) F. L. Whipple, "The History of the Solar System," *Proc. Nat. Acad. Sci., 52*, 565 (1964).

(Wh89) J. C. Wheeler, C. Sneden, & J. W. Truran, Jr., "Abundance Ratios as a Function of Metallicity," *Ann. Rev. Astron. & Astrophys., 27*, 279 (1989).

(Wh96) D. C. B. Whittet, *et al.*, "An ISO SWS view of interstellar ices — First results." *A + A, 315*, L357 (1996).

(Wh04) B. A. Whitney, *et al.*, "A GLIMPSE of Star Formation in the Giant H II Region RCW 49," *ApJS, 154*, 315 (2004).

(Wi58) D. H. Wilkinson, "Do the 'Constants of Nature' Change with Time?" *Phil. Mag.*, Ser 8:*3*, 582 (1958).

(Wi96) Jennifer J. Wiseman & P. T. P. Ho, "Heated Gaseous Streamers and Star Formation in the Orion Molecular Cloud," *Nature, 382*, 139 (1996).

(Wi98) F. Wilczek, "Liberating Quarks and Gluons," *Nature, 391*, 330 (1998).

(Wi99) F. Wilczek, "Getting its from bits," *Nature, 397*, 303 (1999).

(Wi04) J. G. Williams, S. G. Turyshev & D. H. Boggs, "Progress in Lunar Laser Ranging Tests of Relativistic Gravity," *Phys. Rev. Lett., 93*, 261101 (2004).

(Wo57) L. Woltjer, "The Crab Nebula," *B.A.N., 14*, 39 (1957).

(Yo04) J. Yoo, J. Chanamé & A. Gould, "The end of the Macho Era: Limits on Halo Dark Matter from Stellar Halo Wide Binaries," *ApJ 601*, 311 (2004).

(Za54) H. Zanstra, "A simple Approximate Formula for the Recombination Cocffi- cient of Hydrogen," *Observatory, 74*, 66 (1954).

(Ze72) Ya. B. Zel'dovich, "A Hypothesis, Unifying the Structure and the Entropy of the Universe," *MNRAS, 160*, 1P (1972).

(Zu04) V. Zubko, *et al.*, "Observations of Water Vapor Outflow from NML Cygnus," *ApJ, 610*, 427 (2004); *ApJ, 617*, 1371 (2005).

(Zw97) E. G. Zweibel & C. Heiles, "Magnetic fields in galaxies and beyond," *Nature, 385*, 131 (1997).

Index

ASTRONOMY AND
ASTROPHYSICS LIBRARY

ASTRONOMY AND
ASTROPHYSICS LIBRARY